Algorithms for Intelligent Systems

This book series publishes research on the analysis and development of algorithms for intelligent systems with their applications to various real world problems. It covers research related to autonomous agents, multi-agent systems, behavioral modeling, reinforcement learning, game theory, mechanism design, machine learning, meta-heuristic search, optimization, planning and scheduling, artificial neural networks, evolutionary computation, swarm intelligence and other algorithms for intelligent systems.

The book series includes recent advancements, modification and applications of the artificial neural networks, evolutionary computation, swarm intelligence, artificial immune systems, fuzzy system, autonomous and multi agent systems, machine learning and other intelligent systems related areas. The material will be beneficial for the graduate students, post-graduate students as well as the researchers who want a broader view of advances in algorithms for intelligent systems. The contents will also be useful to the researchers from other fields who have no knowledge of the power of intelligent systems, e.g. the researchers in the field of bioinformatics, biochemists, mechanical and chemical engineers, economists, musicians and medical practitioners.

The series publishes monographs, edited volumes, advanced textbooks and selected proceedings.

Indexed by zbMATH.

All books published in the series are submitted for consideration in Web of Science.

Anupam Yadav · Satyasai Jagannath Nanda ·
Meng-Hiot Lim
Editors

Proceedings of International Conference on Paradigms of Communication, Computing and Data Analytics

PCCDA 2023

Springer

Editors
Anupam Yadav
Department of Mathematics
Dr. B. R. Ambedkar National Institute
of Technology
Jalandhar, India

Satyasai Jagannath Nanda
Department of Electronics
and Communication Engineering
Malaviya National Institute of Technology
Jaipur
Jaipur, India

Meng-Hiot Lim
School of Electrical and Electronic
Engineering
Nanyang Technological University
Singapore, Singapore

ISSN 2524-7565 ISSN 2524-7573 (electronic)
Algorithms for Intelligent Systems
ISBN 978-981-99-4625-9 ISBN 978-981-99-4626-6 (eBook)
https://doi.org/10.1007/978-981-99-4626-6

This Springer imprint is published by the registered company Springer Nature Singapore Pte Ltd.
The registered company address is: 152 Beach Road, #21-01/04 Gateway East, Singapore 189721, Singapore

Preface

This book contains outstanding research papers as the proceedings of the International Conference on Paradigms of Communication, Computing and Data Analytics (PCCDA 2023). PCCDA 2023 has been organized by Dr. Akhilesh Das Gupta, Institute of Technology and Management, Delhi, India, and technically sponsored by Soft Computing Research Society, India. The conference is conceived as a platform for disseminating and exchanging ideas, concepts and results of researchers from academia and industry to develop a comprehensive understanding of the challenges of the advancements of intelligence in computational viewpoints. This book will help in strengthening congenial networking between academia and industry. We have tried our best to enrich the quality of the PCCDA 2023 through the stringent and careful peer-review process. This book presents novel contributions to Communication, Computing and Data Analytics and serves as reference material for advanced research. PCCDA 2023 received many technical contributed articles from distinguished participants from home and abroad. After a very stringent peer-reviewing process, only 68 high-quality papers were finally accepted for presentation and the final proceedings.

Jalandhar, India
Jaipur, India
Singapore

Anupam Yadav
Satyasai Jagannath Nanda
Meng-Hiot Lim

Contents

1 **Philosophical Review of Artificial Intelligence for Society 5.0** 1
Ggaliwango Marvin, Micheal Tamale, Benjamin Kanagwa, and Daudi Jjingo

2 **A Review of Different Approaches for Emotion Detection Based on Facial Expression Recognition** 17
Sonu Mittal, Kamal Parashar, Priyanshu Belwal, and Tushar Gahlaut

3 **The Long Short-Term Memory Tuning for Multi-step Ahead Wind Energy Forecasting Using Enhanced Sine Cosine Algorithm and Variation Mode Decomposition** 31
Mohamed Salb, Luka Jovanovic, Nebojsa Bacanin, Goran Kunjadic, Milos Antonijevic, Miodrag Zivkovic, and V. Kanchana Devi

4 **Design of Traffic Monitoring System by Greedy Perimeter Stateless Routing Protocol** 45
Movva Ganesh Kasyap, Arepalli Gayathri, Yaram Chandana, Vadithe Venkatesh Naik, and Yaddanapudi Sarada Devi

5 **Crop-Weed Detection, Depth Estimation and Disease Diagnosis Using YOLO and Darknet for Agribot: A Precision Farming Robot** 57
Medha Wyawahare, Jyoti Madake, Agnibha Sarkar, Anish Parkhe, Archis Khuspe, and Tejas Gaikwad

6 **Audio Classification of Emergency Vehicle Sirens Using Recurrent Neural Network Architectures** 71
Arya Shah, Amanpreet Singh, and Artika Singh

7 **Assessment of Variable Threshold for Anomaly Detection in ECG Time Signals with Deep Learning** 85
Biraja Mishra and Rajeev Kumar

8 Green Cloud Computing: Achieving Sustainability Through Energy-Efficient Techniques, Architectures, and Addressing Research Challenges 97
Sneha, Prabhdeep Singh, and Vikas Tripathi

9 AI-Based Smart Dashboard for Electric Vehicles 107
Narayana Darapaneni, Anwesh Reddy Paduri, B. G. Sudha, Dilip Kumar Mohapatra, Ghanshyam Ji, Mrudul George, and N. Swathi

10 Solving Systems of Nonlinear Equations Using Jaya and Jaya-Based Algorithms: A Computational Comparison 119
Sérgio Ribeiro, Bruno Silva, and Luiz Guerreiro Lopes

11 In-Depth Analysis of Artificial Intelligence in Mammography for Breast Cancer Detection 137
Shweta Saraswat, Bright Keswani, and Vrishit Saraswat

12 The Task Allocation to Virtual Machines on Dynamic Load Balancing in Cloud Environments 145
Rudresh Shah and Suresh Jain

13 Ensemble of Supervised Machine Learning Models for Cardiovascular Disease Prediction 157
Archi Agrawal, Dinesh Singh, Charul Dewan, and Shipra Varshney

14 Computing Model for Real-Time Online Fraudulent Identification 167
Ramani Jaydeep Ramniklal and Jayesh N. Zalavadia

15 Ontology and Machine Learning: A Two-Way Street to Improved Knowledge Representation and Algorithm Accuracy 181
Leila Zemmouchi-Ghomari

16 Nonmetaheuristic Methods for Group Leader Selection, Cluster Formation and Routing Techniques for WSNs: A Review 191
Kumar Dayanand, Binod Kumar, Barkha Kumari, Mohit Kumar, and Kumar Arvind

17 A Comprehensive Review of Machine Learning-Based Approaches to Detect Crop Diseases 203
Rajesh Kumar and Vikram Singh

18 Physiological Signals for Emotion Recognition 221
Shruti G. Taley and M. A. Pund

19 Visual HOG-Enabled Deep ResiNet for Crime Scene Object Detection 233
T. J. Nandhini and K. Thinakaran

20 Scheming of Diamond Ring Harvestor for Low-Powered IoT Devices 245
Shruti Taksali and Amruta Lipare

21 Smart Computer Commands Using Gesture Recognition 253
Sonali Patil, Chinmay Shyam Mukhedker, Mukul Sanjay Chaudhari, Jaspreetsingh Kulwindarsingh Pannu, and Varun Prasannan

22 A New Software Approach to Automated Translation (On the Example of the Logistics Sublanguage) 263
Rodmonga Potapova, Vsevolod Potapov, and Oleg Kuzmin

23 Recent Advances in the Index Calculus Method for Solving the ECDLP 275
Aayush Jindal, Aman Jatain, and Shalini Bhaskar Bajaj

24 Pedestrian Detection Using YOLOv5 and Complete-IoU Loss for Autonomous Driving Applications 287
E. Raja Vikram Reddy and Sushil Thale

25 Health Ware—A New Generation Smart Healthcare System 297
Nihar Ranjan, Maya Shelke, and Gitanjali Mate

26 EEG-Based Sleep Stage Classification System 311
Medha Wyawahare, Rohan Bhole, Vaibhavi Bobade, Akshay Chavan, and Shreya Dehankar

27 FLoRSA: Fuzzy Logic-Oriented Resource Scheduling Algorithm in IaaS Cloud 323
Kapil Tarey and Vivek Shrivastava

28 Real-Time Audio Communication Using WebRTC and MERN Stack 335
Soham Sattigeri and Shripad Bhatlawande

29 Ego Network Analysis Using Machine Learning Algorithms 343
S. Vaibhav, M. P. Dhananjay Kumar, Tejashwini Hosamani, Vrunda Patil, and S. Natarajan

30 Brain Tumor Detection and Classification 353
K. R. Roopa, Sainath Sindagikar, Pruthvi G. Kalkod, P. M. Vishnu, and Lata

31 Safe Vote–Fraudulent Vote Prevention System 363
Neethu Chandrasekhar, Arjun B. Nair, Avinash Thomas George, Binitta Varghese, and Diya Anna Thomas

32 **Intelligent Framework for Early Prediction of Diabetic Retinopathy: A Deep Learning Approach** 377
Adil Husain Rather and Inam Ul Haq

33 **Advanced Footstep Piezoelectric Power Generation for Mobile Charging Using RFID** 389
Kiran Ingale, Atharva Jivtode, Sakshi Bandgar, Ayush Biyani, and Vedant Chaware

34 **Radial Distribution Networks Reconfiguration with Allocation of DG Using Quasi-Oppositional Moth Flame Optimization** 401
Sneha Sultana, Sourav Paul, Poulomi Acharya, Pronoy Das Choudhury, and Provas Kumar Roy

35 **Recent Trends in Risk Assessment of Electromagnetic Radiations** 415
Juhi Pruthi and Ashutosh Dixit

36 **RGB and Thermal Image Analysis for Marble Crack Detection with Deep Learning** 427
Eleni Vrochidou, George K. Sidiropoulos, Athanasios G. Ouzounis, Ioannis Tsimperidis, Ilias T. Sarafis, Vassilis Kalpakis, Andreas Stamkos, and George A. Papakostas

37 **Rover with Obstacle Avoidance Using Image Processing** 439
Krishneel Sharma, Krishan P. Singh, Bhavish P. Gulabdas, Shahil Kumar, Sheikh Izzal Azid, and Rahul Ranjeev Kumar

38 **A Systematic Literature Review of Network Intrusion Detection System Models** 453
Yogesh and Lalit Mohan Goyal

39 **A Comprehensive Study on Online and Offline Evaluation of Recommendation System** 469
Tamanna Sachdeva, Lalit Mohan Goyal, and Mamta Mittal

40 **Autonomous Delivery Vehicle Using Raspberry Pi and Computer Vision** 481
Vijay Ravindran, S. Chandrika, Ram Prakash Ponraj, C. Krishnakumar, S. Devadharshini, and S. Lakshmi

41 **Standard Plane Classification of Fetal Brain Ultrasound Images** 495
Jasmin Shanavas and G. Kanjana

42 **Panoramic Radiograph Segmentation Using U-Net with MobileNet V2 Encoder** 509
Suvarna Bhat and Gajanan K. Birajdar

43 Molecular Recognition and Feature Extraction System 523
Dannerick Elisha, Jimson Sanau, Mansour H. Assaf, Rahul R. Kumar, Bibhya Sharma, and Ronesh Sharma

44 Object Recognition with Voice Assistant for Visually Impaired 537
Deepanshu Jain, Isha Nailwal, Arica Ranjan, and Sonu Mittal

45 Emotion Recognition-Based Emoji Retrieval 547
P. Parvathi Sreyani, Kandula Rakshitha, Nasalai Sanjana, Yeddula Greeshma, and Ashwini M. Joshi

46 An Outage Probability-Based RAW Station Grouping for IEEE 802.11ah IoT Networks 559
Md. Arifuzzaman Mondal and Md. Iftekhar Hussain

47 Machine Learning Algorithms and Grid Search Cross Validation: A Novel Approach for Diabetes Detection 571
Vishal V. Mahale, Ashish G. Nandre, Mahesh V. Korade, and Neha R. Hiray

48 Environment Mapping Using Ultrasonic Sensor for Obstacle Detection and Navigation 583
Medha Wyawahare, Aditya Shirude, Akshara Amrutkar, Anurag Landge, and Ashfan Khan

49 Identification and Classification of Skin Diseases with Erythema Using YOLO Algorithm 595
C. Santhosh Kumar, K. Amritha Devangana, P. L. Abirami, M. Prasanna, and S. Hari Aravind

50 PSO-Based Controller for LFC of Deregulated Power System 607
Dharmendra Jain, M. K. Bhaskar, and Manish Parihar

51 Solar Maximum Power Point Tracking and Machine Learning-Based Forecasting 625
Akshay Pandya, Galav Bhatt, Jash Patadia, and Het Patel

52 Comparative Performance Analysis of Various Controllers for Quadruple Tank System 639
C. Praveen Kumar and K. Ayyar

53 Supervised Machine Learning Text Classification: A Review 651
Nisar Ahmad Kangoo and Apash Roy

54 Train Delay Prediction Using Machine Learning 663
Nilesh N. Dawale and Sunita Nandgave

55 Logical Formalization for a HMDCS-UV 675
Salima Bella and Ghalem Belalem

56 SANKET—A Vision Beyond Gestures 689
Isha Gawde, Jisha Philip, Kanaiya Kanabar, Shilpa Tholar, and Shalu Chopra

57 Assessing the Effectiveness of Different Mass Communication Approaches Used for Government Medical Programs in Rural Areas of Uttarakhand 699
Pradeep Joshi, Omdeep Gupta, Mayank Pant, Kartikeya Raina, and Bhanu Sharma

58 Computer Vision-Based Virtual Mouse Cursor Using Hand Gesture 713
Tanmay Sonawane, Sarvesh Waghmare, Abhishek Dongare, Avadhut Joshi, and Anandkumar Birajdar

59 A Review of Machine Learning Models for Disease Prediction in Poultry Chickens 723
Divya Verma, Neelam Goel, and Vivek Kumar Garg

60 Technological Approach Toward Smart Grid Security: A Review 739
Saish Kothawade, Akshat Dubey, Anush Shetty, Kartik Chaudhari, and Rachana Patil

61 Storage and Verification of Medical Records Using Blockchain, Decentralized Storage, and NFTs 753
Shubham Thakur and Vijay Kumar Chahar

62 A Study on Prediction of Temperature in Metropolitan Cities Using Machine Learning 769
Shweta S. Aladakatti, A. Bharath, V. T. Adarsha, B. J. Ajith, and H. R. Chaithra

63 A Review of Secure Authentication Techniques in Fog Computing 783
Mahgul Afzali and Gagandeep

64 SafeMaps: Crime Index-Based Urban Route Prediction 793
Ria Singh, Shatakshi Mohan, Harsh Pooniwala, V. V. Gokul, and S. Shilpa

65 Controlling the Steering Wheel Using Deep Reinforcement Learning: A Survey 805
Narayana Darapaneni, Anwesh Reddy Paduri, B.G. Sudha, Vidyadhar Bendre, Midhun Chandran, M. Mohana Priya, and Varghese Jacob

66 NL2SQL: Rule-Based Model for Natural Language to SQL 817
Kevin Tony, Kripa Susan Shaji, Nijo Noble, Ruben Joseph Devasia, and Neethu Chandrasekhar

67 VANET-Based Communication in Vehicles to Control Accidents Using an Efficient Routing Strategy 829
Humera Maahin, Deepthi Kondamuri, Sarvani Polisetty, and Sarada Devi Yaddanapudi

68 ECG Image Classification for Arrhythmia Using Deep Learning 839
Shasmita Nair, Prerna Peswani, Jai Rohra, and M. Vijayalakshmi

About the Editors

Dr. Anupam Yadav is Associate Professor, Department of Mathematics, Dr. B. R. Ambedkar National Institute of Technology Jalandhar, India. His research area includes numerical optimization, soft computing and artificial intelligence; he has more than ten years of research experience in the areas of soft computing and optimization. Dr. Yadav has done a Ph.D. in soft computing from the Indian Institute of Technology Roorkee, and he had worked as Research Professor at Korea University. He has published more than 25 research articles in journals of international repute and has published more than 15 research articles in conference proceedings. Dr. Yadav has authored a textbook entitled *An Introduction to Neural Network Methods for Differential Equations*. He has edited several books which are published by various book series of Springer. Dr. Yadav was General Chair, Convener and Member of the steering committee of several international conferences. He is Member of various research societies and editorial boards.

Dr. Satyasai Jagannath Nanda is Assistant Professor at the Department of Electronics and Communication Engineering, Malaviya National Institute of Technology Jaipur, since June 2013. Prior to joining MNIT Jaipur, he has received the Ph.D. degree from School of Electrical Sciences, IIT Bhubaneswar, and M.Tech. degree from the Department of Electronics and Communication Engineering, NIT Rourkela. He was the recipient of Canadian Research Fellowship-GSEP, from the Department of Foreign Affairs and Intern. Trade (DFAIT), Government of Canada, for the year 2009–2010. He was awarded Best Ph.D. Thesis Award at SocPros 2015 by IIT Roorkee. He received the best research paper awards at SocPros-2020 at IIT Indore, IC3-2018 at SMIT Sikkim, SocPros-2017 at IIT Bhubaneswar, IEEE UPCON-2016 at IIT BHU and Springer OWT-2017 at MNIT. He is the recipient of prestigious IEI Young Engineers Award by Institution of Engineers, Government of India, in the field of Electronics and Telecommunication Engineering for the year 2018–2019. Dr. Nanda is Senior Member of IEEE and IEEE Computational Intelligence Society.

Dr. Meng-Hiot Lim is Faculty at the School of Electrical and Electronic Engineering. He is holding a concurrent appointment as Deputy Director for the M.Sc. in Financial Engineering and the Centre for Financial Engineering, anchored at the Nanyang Business School. He is Versatile Researcher with diverse interests, with research focus in the areas of computational intelligence, evolvable hardware, finance, algorithms for UAVs and memetic computing. He is currently Editor-in-Chief of the *Journal of Memetic Computing* published by Springer. He is also Series Editor of the book series by Springer titled *Studies in Evolutionary Learning and Optimization.*

Chapter 1
Philosophical Review of Artificial Intelligence for Society 5.0

Ggaliwango Marvin, Micheal Tamale, Benjamin Kanagwa, and Daudi Jjingo

1 Introduction

Artificial intelligence has come a long way since its inception 60 years ago, and it continues to evolve and change the world in ways we couldn't have imagined. Today, AI has reached new heights and has a wide range of applications, from playing complex games to language processing, speech recognition, and facial recognition [1–3]. With its exponential growth and its increasing presence in an ever-growing number of sectors, AI is well on its way to becoming a source of significant economic prosperity. But as AI continues to evolve, it poses major policy questions for policymakers, investors, technologists, scholars, and students. AI ethics are critical to its development, and it is essential that ethical standards be established to ensure that AI meets a certain standard of public justification and supports citizens' rights, promoting substantively fair outcomes when deployed [4–7]. The use of AI in everyday life also raises ethical collisions, and human rights principles and legislation must play a key role in addressing these ethical challenges [8–10]. The rapid development of AI presents many opportunities and challenges for the human race. As AI becomes more autonomous and intelligent, it has the potential to greatly improve the performance of manufacturing and service systems, as well as contribute to social development and human life [2, 11, 12, 13]. However, the hardware and software of a fully autonomous, learning, reasoning AI system must mimic the processes and subsystems that exist within the human brain [14, 15].

The future of AI is rapidly changing the way we interact with machines. AI has already achieved the capability to interact with humans and build relationships

G. Marvin (✉) · B. Kanagwa · D. Jjingo
Makerere University, Kampala, Uganda
e-mail: ggaliwango.marvin@mak.ac.ug

M. Tamale
Kabale University, Kabale, Uganda

A. Yadav et al. (eds.), *Proceedings of International Conference on Paradigms of Communication, Computing and Data Analytics*, Algorithms for Intelligent Systems,
https://doi.org/10.1007/978-981-99-4626-6_1

through conversations, and the next generation of autonomous technology will make many decisions autonomously [16]. AI is not just about technology; it also involves philosophical and psychological issues. It is imperative that we integrate AI ethics into AI education and development and continue to study the social psychology of intelligent machines [5, 17, 18]. Therefore, the rapid development of AI presents both opportunities and challenges, and it is up to us to ensure that it benefits the human race and contributes to social development and human life. The AI community is continuously exploring and discussing the potential of AI and its implications, and it is important that we stay informed and up to date on the latest developments in this field [10, 19–21].

One of the most respected fathers of AI include Alan Turing and John Searle appreciated for the Turing test and Chinese Room Argument. In the context of Society 5.0, the Turing Test and the Chinese Room Argument are important concepts in the philosophy of AI [3, 22, 23]. The Turing Test aims to determine if a computer is capable of thinking like a human being, by having the computer mimic human responses in a specific subject area and format. The test has been updated with variations, such as the Reverse Turing Test, Total Turing Test, and Minimum Intelligent Signal Test, to make it more relevant. The Chinese Room Argument, on the other hand, is a thought experiment introduced by John Searle, which debates whether a machine can truly understand language and cognition, or if it is just simulating the ability [13, 24–26]. This argument raises important ethical and moral questions about the development of AI in Society 5.0, such as the extent to which AI can replace human intelligence and what the implications of that would be [10, 11, 27]. The discussions and debates surrounding these concepts will continue to play a significant role in shaping the development and implementation of AI in Society 5.0 [22, 29].

We observe that the philosophy of artificial intelligence (AI) is a field that encompasses the ethical, philosophical, and existential implications of AI's development and increasing presence in society. It involves the study of fundamental concepts such as intelligence, knowledge, and artificial intelligence, and how they impact human existence [23]. The development of AI raises important questions about the relationship between humans and machines, the future of humanity, and the need for ethical standards and governance mechanisms [18, 30, 31]. The philosophical foundations of AI are essential in guiding the development and use of AI in Society 5.0 [22, 29]. This future society is envisioned to be a harmonious co-existence between humans and AI, where technology serves to enhance human capabilities and improve the overall well-being of society while preserving human values and dignity. The philosophy of AI plays a crucial role in ensuring that AI systems are developed in a responsible and ethical manner, aligning with the values and aspirations of society, and contributing to a better future for all [22, 32–34].

Some of the key philosophical foundations of AI include the study of mind–body dualism, the nature of intelligence and knowledge, the limits of AI and its impact on human existence, the relationship between humans and machines, the ethical implications of AI, and the need for ethical standards and governance mechanisms [18, 31, 35, 36]. These philosophical foundations help to provide a foundation for reflecting on the fundamental questions and issues that arise from AI's growing presence and influence in our lives [10, 37, 38]. Critical evaluation of AI applications and implications is extremely important. This can be achieved through interdisciplinary dialog and research in order to gain a better understanding of AI impact on humanity and the world in general. This is what makes deep understanding of AI philosophies very crucial since it provides a baseline for evaluating ethical and moral principles. It is through this that AI philosophy guides AI development in Society 5.0. By examining these philosophical foundations [27, 28, 31, 39–41], we can gain a deeper understanding of the impact of AI on human and social existence, and ensure that AI technologies are developed in a responsible and ethical manner that aligns with the values and aspirations of society [11, 34, 37].

2 Literature Review

To identify relevant studies for this philosophical review of the philosophy of artificial intelligence, the following search strategies will be employed. Electronic database searches: Searches were conducted in relevant databases such as PubMed, Web of Science, and Google Scholar. The search terms included "artificial intelligence," "AI," "philosophy of AI," "ethics of AI," and "singularity." The reference lists of relevant review articles and other key studies were manually searched to identify additional studies that may not have been identified through the electronic database searches. Gray literature sources such as conference proceedings and technical reports will be searched using the same search terms as in the electronic database searches.

Inclusion criteria covered studies written in English, studies that were peer-reviewed articles, book chapters, or technical reports, studies published between 1973 and 2022, and studies that were relevant to the philosophy of artificial intelligence and Society 5.0. Exclusion criteria limited studies not written in English, studies that are not peer-reviewed, studies published outside of the specified time frame, and studies that are not relevant to the philosophy of Artificial Intelligence. Data were extracted from the selected studies using a standardized data extraction form that includes information about the study design, sample, data sources, and main findings. Data were synthesized using a narrative synthesis approach, with a focus on identifying trends, patterns, and differences among the studies [1, 16, 42, 43]. The results are presented in a structured manner in this paper to illustrate key points.

2.1 Philosophical Review

The field of artificial intelligence (AI) is rapidly expanding and has significant implications for society and humanity. As AI continues to shape our world and impact our daily lives, it is essential to critically evaluate its philosophical foundations and ensure that its development aligns with human values and goals [30, 39, 40].

Being complex and multi-layered, the philosophy of AI provides a 360 window for understanding development and deployment of AI with existential, ethical, and philosophical underpinnings for humanity.

The nature of intelligence is one of such foundational underpinnings of AI philosophy. It ignites various questions that are extremely essential for defining intelligence. Particularly, "What does it mean to be intelligent?" This makes AI engineers and developers to think deeply of whether intelligence is just a matter of processing information or there could be a lot more constituting it [40, 41]. Where as the requirement of consciousness and self-awareness is what most philosophers argue for [31, 36], others claim that certainty in AI intelligence is provable by its ability to successfully execute tasks that required real human intelligence before [25, 26, 44].

The relationship between machines and humans is the other fundamental philosophical foundation that requires maximum attention. The increasing sophistication and capabilities of AI systems most especially in domains where they completely outperform human beings make the relationship extremely sensitive. This performance of machines fundamentally ignites questions for justifying the role of humans in a machine dominated world. This gets us human beings thinking about our human values, their importance, and relevance in operating the world [45]. Values like human creativity and empathy not only get doubted by also require real justification of their importance today. This opens up various questions about AI ethics. We need to examine and contextualize the human concerns about ethical standards and governance. Most essentially, the mechanisms for implementing standards of AI systems that respectfully fit into human interests for example already defined human rights [6, 31, 36].

The implications of rapid AI development like Large Language Models are another issue tickling the philosophy of AI, it is causing uncontrollable paradigm shifts for the future of humanity [44, 46], a world to live in where everything is connected. This is what ignites the concept of Society 5.0. This concept is defining the future of technology and society in which everything that is technology driven plays a very critical role in improving the quality of life. The concept of Society 5.0 speaks to the importance of paying undivided attention to the importance of understanding AI impact on social existence and humanity [37, 38, 47]. Therefore, ethical and responsible considerations of AI development are one of the most critical aspects that are essential for shaping the future of Society 5.0 through examining the philosophical underpinning of AI development, deployment, and monitoring.

Knowledge and intelligence manifest a very sensitive relationship that requires critical philosophical examination. That relationship provides another AI philosophical foundation for understanding the meaning or AI capabilities to process vast amounts of information without having the ability to know anything. So, it ignites questions like "What does it mean for AI to truly "Know" something" ? This is also a very complex research challenge particularly the knowledge representation problem thus how knowledge can be represented in a computer. This is still a very huge challenge in AI research [4, 42, 43]. Whereas some philosophers still argue that AI-driven systems can never truly represent knowledge in the same sense as humans do, others believe that the aspect of knowledge representation in AI can be developed in real world true understanding and reasoning [14, 15, 48]. It is also not yet clear how AI can be able outperform humans without a clear true understanding of knowledge representation.

The other philosophical foundation of AI is the question of consciousness. This conspiracy makes some philosophers believe that AI can never be conscious. Other philosophers argue that it is very possible to create consciousness in machines [39, 40]. Whereas human beings have not yet justified the source of their consciousness, we cannot rule out the fact that it is not possible for machines to be conscious. This mysterious phenomenon of consciousness has been a long-term debate by philosophers for centuries, and these debates are not yet over. What complicates the debate is the unclear distinction and relationship between consciousness and intelligence [46, 49]. We shall let you know when we find out.

The mind–body problem is another AI philosophical question that is rooted in the complex relationship between the mind and the physical world. "The mind is simply a product of the brain", that is what some philosophers claim. Others believe "There is more about the mind, than just being a product of the brain". These arguments have very huge implications on the development and deployment of AI. The doubt created by the mind–body problem even affects the viability of creating truly intelligent machines with capabilities of experiencing and understanding the world in the same way humans do [7, 25, 35, 48].

The other philosophical underpinning of AI is the Concept of the Mind. This is focused on addressing the relationship between human behavior and the nature of mental states. Experts in the AI domain argue there is a possibility of replicating human intelligence within machines. They also argue that machines can develop subjective experiences on their own. We have actually observed this in Large Language Models like ChatGPT, where the model hallucinates to give incorrect links to citations. We think this may be the same for some facts spilled out by such models. However, this raises the question about the meaning of consciousness. It also raises a question about machine abilities to experience the world as humans do [7, 23, 25, 35, 50]. We actually argue that machines could have some human experiences, and this can only be rejected if humans could scientifically justify the sources of their dreams or why they lie.

Another AI philosophical foundation is the theory of computation. This theory examines computational problems in the context of the relationship between computational processes and computational algorithms. The relationship examination gives a basis to understand possible limitations and capabilities of Intelligent Algorithms and AI systems. Understanding of such relationships is what guides the development and deployment of novel AI technologies that work [18, 35, 51–53]. With this, we clearly understand that AI is a buzz word today but it doesn't mean that AI can solve everything; therefore, AI should not be hyped and all traceable limitations of AI systems should be clearly documented and reported. This is part of what makes them ethical and responsible.

The other critical and sensitive foundation of AI is AI Ethics. These focus on cross examining the moral and ethical implications and underpinnings of AI development and deployment [10, 45, 54]. Ethics of AI ignite questions about AI impact in relation to human rights, values, and responsibility of AI systems through the entire development and deployment processes [11, 28, 55]. This peaks the essence of AI Ethical standards for ensuring utilization of AI in responsible, respectful, and human dignity keeping human interests in context [56–58].

Another important foundation for AI philosophical thinking is the philosophy of science. This mainly focuses on underpinning the nature of scientific knowledge. It also underpins the methods of creating or generation and validating scientific knowledge [44, 46]. In AI, this philosophy gives a clear basis for evaluation of valid and reliable AI methods and models. It provides original scientific principles for ensuring that AI systems are built on sound scientific methods and ideologies [49, 59].

Philosophy of Language is another important AI philosophical foundation that looks at the relationship of meaning with the nature of language. It is extremely important building and deploying intelligent systems that require effective communication with humans based on language understanding. Examples of such technologies included conversational AI models like ChatGPT and other Language models [48, 60, 61].

We cannot ignore the fact that the rate at which deeper questions about the future of humanity are rising is directly proportional to the rate of AI development. And we can no longer ignore the need for interdisciplinary research engagements to understand humanity and society today [37, 38, 58, 62]. The philosophy of AI provides a foundation for reflecting on the fundamental questions and issues that arise from AI's growing presence and influence in our lives and is critical to shaping the future of Society 5.0. There are many philosophical foundations that are relevant to AI and Society 5.0. It is important to note that each of these philosophical foundations is complex and multifaceted, and there are many different perspectives and interpretations of each [63, 64]. However, by considering these philosophical foundations, we can gain a deeper understanding of the challenges and opportunities posed by AI and Society 5.0, and work toward developing AI systems that are aligned with human values and promote the well-being of society as a whole [64, 65].

3 Results and Discussion

3.1 Results

From literature, we observe that the various philosophies of AI boil down to a multidisciplinary field that encompasses ethics, epistemology, metaphysics, and the philosophy of mind [25, 45, 66]. Some of the most significant philosophical foundations of AI that shape our understanding of the field and its impact on society include.

Firstly, the ethics of artificial intelligence which is an essential aspect of AI philosophy that addresses the moral implications of creating and using intelligent machines. AI raises complex ethical questions regarding accountability and responsibility, human dignity, and privacy. The trolley problem, a classic example in ethical philosophy, is an illustration of the ethical dilemmas posed by AI. It asks whether it is ethical to sacrifice one person's life to save several others in a hypothetical scenario, where a runaway trolley is headed toward a group of people, and a lever must be pulled to divert the trolley toward one individual [28, 56, 57, 64–67]. The use of AI in autonomous weapons and decision-making systems that prioritize certain lives over others also raises ethical concerns about the use of AI in military applications.

The second important philosophical foundation of AI is the epistemology of artificial intelligence, which is concerned with how AI systems acquire and utilize knowledge. AI systems rely on vast amounts of data to make predictions and decisions, and the question of how AI systems acquire and use knowledge is a critical one. The nature of knowledge representation, the role of prior knowledge, and the relationships between AI systems and human experts are all topics of inquiry within the epistemology of AI [66, 68]. For example, the concept of explainability in AI refers to the extent to which AI systems can be transparent about their decision-making processes and the factors that influence their outputs.

Thirdly, metaphysics of artificial intelligence concerns the fundamental nature of intelligence and the relationship between human and machine intelligence. The doubt on true intelligence and possession of consciousness by machines is a long-term AI philosophical question and unresolved debate. Whereas some philosophers believe in human intelligence simulation rather than true intelligence that constitutes consciousness for machines, others argue possibilities of machine surpassing human intelligence and discredit the necessity of consciousness for true intelligence. The debate has had and still has substantial implications on the understanding the potential future of AI and the human mind [13, 40, 69]

Finally, the philosophy of the mind. It grounds the human perspectives on the viewing, interpreting, and handling the nature of mental states and processes [35]. This philosophy illuminates the extent of mental state possession and utilization by machines. It provides a better reflection of possessing true experience, subjective emotions, and feeling. It ignites questions like "can machines cave a sense of self if they simply executed pre-programmed instructions" ?

The above four philosophical underpinnings are just a tip of the iceberg for a wide-ranging scope of philosophical questions arising from rapid AI development. Other basic essential philosophies include the philosophy of mathematics, philosophy of science among others. All of these foundations are critical to understanding the impact of AI on society, and they help us to better evaluate the role that AI should play in shaping the future of humanity [37, 38, 47, 58, 62].

3.2 Discussion

It is crucial to understand that Society 5.0 is currently a theoretical concept, and its components may continue to change and evolve as technology and society advance.

While discussions of Society 5.0 often highlight its major components, it is important to note that these are not exhaustive and may overlap with different philosophies of AI. The categorization of various philosophies of AI according to Society 5.0 themes provides a general framework for understanding their relationship to the overarching concept of Society 5.0, but it is essential to recognize that not all philosophies may align with the goals and vision of Society 5.0 and may even be in opposition.

It can be noted that there are valid variations and subcategories of the categories. These different names can reflect different perspectives on the same philosophy, or can emphasize different aspects of the philosophy. For example, "Human-Centered AI" and "User-Centered AI" both reflect the idea that the development and deployment of AI should prioritize the needs and well-being of people, while "Ethical AI" and "Responsible AI" both emphasize the importance of ensuring that AI is developed and used in a way that is consistent with ethical principles and values. The clarification and refinement can be inferred from the cases presented by various philosophical relationships with theme of Society 5.0 theme. As presented in Table 1, they can greatly help inform responsible AI research methods for attaining Society 5.0.

The classification above not only informs a strategy to formulate responsible AI methods, it also underscores the need to consider ethical, social, cultural, and most importantly intersectional implications of AI. It provides a blueprint for developing and deploying human-centered, sustainable, and responsible AI technologies with an overall objective of attaining essential attributes of Society 5.0. Whereas the complexity of the relationships among overlapping categories seems recurrent, it is recommended that researchers and practitioners to focus on specific themes of Society 5.0 instead of multiple but choose underlying philosophies that make their projects scalable and compatible with other Society 5.0 themes. This is what makes an appropriate responsible AI methodology or approach. The overlapping categories also demonstrate the need to consider wider context of Society 5.0 in both the development and evaluation roles and implications of resultant AI technologies. It is also very important to be mindful of resultant philosophical relationships from overlapping categories and themes of Society 5.0. This is particularly important for establishing responsible AI methods, evaluation frameworks, deployment, and monitoring strategies that do not conflict with themes of Society 5.0.

Categorizing philosophies of AI in Society 5.0 contexts is a constantly evolving process, like ways the responsible AI methods derived from them. It is very possible that new classes and subclasses may emerge as the AI field continues to grow. However, it is extremely essential to comprehensively understand the various AI philosophies underpinning AI methods and relationships with the Society 5.0 themes [55, 64, 65]. This is particularly important for developing and utilizing AI in ways that align with values and goals of Society 5.0. These embrace human well-being enhancement, prioritization of ethics, sustainability promotion, and most importantly collaborative encouragement among humans and machines [56, 57]. By classifying the philosophies of AI within the Society 5.0 framework, we can be sure of developing and deriving relevant responsible AI methods to achieve the inclusive

Table 1 Matching AI philosophies to Society 5.0

Clusters of philosophies of AI that fit into Society 5.0	Types of AI technologies arising philosophies of AI that fit into Society 5.0	Specific relationships to Society 5.0 themes	Overlapping categories between the variations of AI philosophies and their relationship to Society 5.0 themes
Human-centered AI: this cluster includes philosophies that focus on the development of AI systems that prioritize the needs and well-being of humans [45, 54, 70, 71]	Human-driven AI, human-centered AI, and human-guided AI	User-centered AI People-centered AI Human-friendly AI	Human-centered AI and empowered AI: both categories focus on the role of AI in improving the lives of people and making technology accessible and inclusive for all
Empowered AI: this cluster includes philosophies that seek to empower individuals and communities through the deployment of AI [70, 71]	Empowering AI, democratic AI, and free AI	Inclusive AI Accessible AI Participatory AI	
Ethical AI: this cluster includes philosophies that emphasize the ethical and moral responsibility of AI [27, 28, 30, 34, 45, 54, 71–73]	Ethical AI, Transparent AI, and Accountable AI	Responsible AI [9] Moral AI [27], 28 Fair AI [8]	Ethical AI and Reliable AI: Both categories focus on the responsible and safe use of AI, ensuring that AI systems are secure, stable, and do not cause harm
Reliable AI: this cluster includes philosophies that prioritize the reliability and stability of AI systems [17, 28, 30, 34, 45, 54, 71–73]	Robust AI, safe AI, and verifiable AI	Stable AI Secure AI Safe AI	
Harmonizing AI: this cluster includes philosophies that aim to balance and harmonize human and machine intelligence [67, 74, 75]	Integrative AI, hybrid AI, and harmonizing AI	Synergistic AI Complementary AI Balancing AI	Harmonizing AI and collaborative AI: both categories focus on the collaborative relationship between AI and humans, promoting cooperation, balance, and social intelligence
Collaborative AI: this cluster includes philosophies that emphasize the collaborative and cooperative nature of AI systems [5, 8, 71]	Collaborative AI, collective AI, and cooperative AI	Cooperative AI [8] Collaborative intelligence [71] Social AI [5]	

(continued)

Table 1 (continued)

Clusters of philosophies of AI that fit into Society 5.0	Types of AI technologies arising philosophies of AI that fit into Society 5.0	Specific relationships to Society 5.0 themes	Overlapping categories between the variations of AI philosophies and their relationship to Society 5.0 themes
Autonomous AI: this cluster includes philosophies that advocate for the development of autonomous and self-governing AI systems [40, 50, 54, 67]	Decentralized AI, distributed AI, and self-organizing AI	Self-Determining AI Independent AI Sovereign AI	Autonomous AI and intelligent AI: both categories focus on the advancement and cognitive capabilities of AI, enabling AI to be independent and advanced in its decision-making abilities
Intelligent AI: this cluster includes philosophies that focus on the development of intelligent and advanced AI systems [13, 15, 24, 50, 67]	Intelligent AI, advanced AI, and evolutionary AI	Advanced AI Cognitive AI [24] High-Performance AI	
Sustainable AI: This cluster includes philosophies that prioritize the sustainability and long-term impact of AI systems [7, 8, 37, 54, 67]	Sustainable AI, green AI, and responsible AI	Eco-friendly AI Green AI Climate-friendly AI	Sustainable AI and eco-friendly AI: both categories focus on the environmental impact of AI, promoting eco-friendly and climate-friendly approaches in AI development and deployment
Human–machine integration: this cluster includes philosophies that focus on integrating human and machine intelligence [7, 19, 25, 35, 43, 70, 76]	Human–machine integration, integrative AI, and hybrid AI	Human–machine synergy Human–machine fusion Human–machine cooperation	

goal of creating intelligent and AI-driven systems for human life enhancement and contribution toward a better future for all.

4 Conclusion and Philosophical Questions

4.1 Conclusion

In order to create responsible AI methods toward Society 5.0, understanding the philosophies of AI is a mandatory requirement today. The future where various technologies AI are driving quality life improvement for all is what constitutes the vision for Society 5.0. This ongoing process constitutes an effort to solve complex social and environmental problems. This is where AI philosophical underpinnings aid formulation of moral and ethical principles for guiding responsible development and utilization of AI technologies toward Society 5.0. By continuously evaluating the alignment of AI systems with human values, responsible AI methods founded on appropriate philosophies ensure a better future for all.

The philosophy of AI which provides the best responsible AI methods requires constant engagement of multiple disciplines for a research dialog for AI development and deployment in a way that promotes humanity [30, 58, 62, 65]. This means that the most appropriate AI philosophy is likely multifaceted, complex, and constitutes aspects of ethics, metaphysics, ontology, and epistemology [8, 71, 77]. These are extremely important for understanding limits, abilities, ethical, and social implications of AI development and utilization [55, 56, 78, 79]. Therefore, human values and goals for equitable benefits are achievable with examination of behavioral and co-existence of humans and machines [7, 25, 35, 54]. This is easily archivable through AI transparency, fairness, accountability, trustworthiness, and explainability [37, 47].

This chapter provided a foundation for critical reasoning and reflection behind AI impact, harmonious co-existence of AI, and humanity and deep understanding of AI philosophies. We provided a philosophical reflection on critical aspects of free will, social economics, superintelligence, security, ethics, artificial life, responsibility, mind–body dualism, inclusiveness, bias, teleology, human nature, privacy, superintelligence, and various AI philosophies. We also provided a benchmark for formulating responsible AI methods based on AI philosophies. This work plays a crucial role in shaping the future of Society 5.0 by guiding responsible and ethical AI development and use.

The philosophical review of artificial intelligence for Society 5.0 highlights several research gaps that need to be addressed in order to ensure the responsible and ethical development and use of AI technologies [7, 69, 73, 80]. Additionally, there is a need for the development of domain specific guidelines and regulations that govern the development and deployment of AI systems, to ensure that they do not perpetuate biases or discriminate against certain groups of people. To address these gaps, further research and reflection on the ethical and philosophical implications of AI in Society 5.0 is necessary.

4.2 *Philosophical Questions*

This basically leaves us with about 10 important philosophical questions.

- What is the nature of intelligence and how can it be artificially replicated?
- What is the relationship between humans and AI, and how can we ensure ethical and moral alignment between the two?
- What are the implications of AI on human values and the future of humanity?
- How can we ensure that AI systems are transparent, accountable, and respectful of privacy and data protection laws?
- What is the impact of AI on job displacement, income inequality, and the role of humans in a world dominated by AI systems?
- How can we ensure that AI is developed in a responsible and ethical manner, aligned with human values and aspirations?
- How can we evaluate the social and economic impact of AI and ensure its positive contribution to society?
- What is the role of human cognition and decision-making in an AI-driven world? What are the philosophical implications of AI becoming super intelligent?
- How can the philosophy of AI shape the future of Society 5.0 and contribute to a better future for all?

References

1. Kilani A, Ben Hamida A, Hamam H (2017) Artificial intelligence review. In: Encyclopedia of information science and technology, Fourth edn. IGI Global, pp 106–119
2. Koptseva N (2022) KROO Commonwealth of Enlighteners of Krasnoyarsk. Modern research in the field of the sociology of artificial intelligence: basic approaches. Part 3. Sociol Artif Intell3(2):7–22. https://doi.org/10.31804/2712-939x-2022-3-2-7-22
3. L Sias 2021 DePaul University. Ideology AI Philosophy Today 65(3):505–522 https://doi.org/10.5840/philtoday2021514405
4. Soranno DE, Bihorac A, Goldstein SL, Kashani KB, Menon S, Nadkarni GN, Neyra JA, Pannu NI, Singh K, Cerda J et al (2022) Artificial intelligence for AKI! Now: Let's not await Plato's Utopian republic. Kidney360 3(2):376–381. https://doi.org/10.34067/KID.0003472021
5. Резаев АВ, Трегубова НД (2021) Artificial intelligence and artificial sociality: new phenomena and challenges for the social sciences. Monit Public Opin Econ Soc Changes (1). https://doi.org/10.14515/monitoring.2021.1.1905
6. Kazim E, Koshiyama AS (2021) A high-level overview of AI ethics. Patterns 2(9):100314. https://doi.org/10.1016/j.patter.2021.100314
7. L Floridi J Cowls M Beltrametti R Chatila P Chazerand V Dignum C Luetge R Madelin U Pagallo F Rossi 2018 AI4People-an ethical framework for a good AI society: opportunities, risks, principles, and recommendations Minds Mach 28(4):689–707. https://doi.org/10.1007/s11023-018-9482-5
8. Benthall S, Goldenfein J (2021) Artificial intelligence and the purpose of social systems. In: Proceedings of the 2021 AAAI/ACM conference on AI, ethics, and society. ACM, New York
9. IR Nourbakhsh 2021 AI ethics: a call to faculty Commun ACM 64(9):43–45. https://doi.org/10.1145/3478516.doi:10.1145/3478516

10. Artificial intelligence and society: Summit of the G7 science academies. Trends Sci 24(9):9_107–9_109. https://doi.org/10.5363/tits.24.9_107
11. Abdulllah SM (2019) Artificial intelligence (AI) and its associated ethical issues. Islam and Civilisational Renewal 10(1):124–126. https://doi.org/10.52282/icr.v10i1.78
12. S Ziesche R Yampolskiy 2018 Towards AI welfare science and policies Big Data Cognitive Comput 3:1–2. https://doi.org/10.3390/bdcc3010002
13. World L (2005) Al and philosophy: how can you know the dancer from the dance? IEEE Intell Syst 20(4):84–85. https://doi.org/10.1109/mis.2005.61
14. Stock O, Schaerf M (2006) Reasoning, action and interaction in AI theories and systems: essays dedicated to Luigia Carlucci Aiello. Springer, Berlin
15. Dietrich E (2006) Artificial intelligence, philosophy of. In: Encyclopedia of cognitive science. Wiley, Chichester
16. Mijwil MM, Abttan RA (2021) Artificial intelligence: a survey on evolution and future trends. Asian J Appl Sci 9(2). https://doi.org/10.24203/ajas.v9i2.6589
17. Mathew D, Shukla VK, Chaubey A, Dutta S (2021) Artificial intelligence: hope for future or hype by intellectuals? In: 2021 9th international conference on reliability, infocom technologies and optimization (trends and future directions) (ICRITO). IEEE
18. Mamina RI, Pochebut SN (2022) Artificial intelligence in the view of philosophical methodology: an educational track. Discourse 8(1):64–81. https://doi.org/10.32603/2412-8562-2022-8-1-64-81
19. Neapolitan RE (2012) Contemporary artificial intelligence. Chapman and Hall/CRC
20. Fulcher J (2008) Computational intelligence: an introduction. In: Studies in computational intelligence. Springer, Berlin, pp 3–78
21. S Rasmussen MJ Raven GN Keating MA Bedau 2003 Collective intelligence of the artificial life community on its own successes, failures, and future Artificial Life 9(2):207–235 https://doi.org/10.1162/106454603322221531
22. Omohundro S (2012) Rational artificial intelligence for the greater good. In: The frontiers collection. Springer, Berlin, pp 161–179
23. F Bruneault AS Laflamme 2021 AI ethics: how can information ethics provide a framework to avoid usual conceptual pitfalls? An overview AI Society 36(3):757–766. https://doi.org/10.1007/s00146-020-01077-w
24. Lektorsky VA (2021) On the philosophical issues of artificial intelligence and cognitive studies. Философские науки 64(1):7–12. https://doi.org/10.30727/0235-1188-2021-64-1-7-12
25. V Schiaffonati 2003 Minds and Machines 13 4 537 552. https://doi.org/10.1023/a:1026252817929
26. V Akman 2000 Introduction to the special issue on philosophical foundations of artificial intelligence J Exp Theoret Artif Intell: JETAI 12(3):247–250. https://doi.org/10.1080/09528130050111419
27. I Gabriel 2022 Toward a theory of justice for artificial intelligence Daedalus 151(2):218–231. https://doi.org/10.1162/daed_a_01911
28. Boddington P (2020) TPM: The philosophers' magazine. The ethics of AI and the moral responsibility of philosophers. Philosophers Mag (89):62–68. https://doi.org/10.5840/tpm20208940
29. KS Gill JM Artz 1987 Artificial Intelligence for Society IEEE Expert 2(2):108–108. https://doi.org/10.1109/mex.1987.4307076
30. E Moczuk B Płoszajczak 2020 Artificial intelligence—benefits and threats for society Humanit Soc Sci Q. https://doi.org/10.7862/rz.2020.hss.22
31. Waelen R (2022) Why AI ethics is a critical theory. Philos Technol 35(1). https://doi.org/10.1007/s13347-022-00507-5
32. Rashid MAN, Mullah M, Zain ZM (2020) Application of artificial intelligence: a review. Int J Adv Eng Res Sci 7(3):316–321. https://doi.org/10.22161/ijaers.73.47
33. Wang N, Yan L, Wang Y (2019) Review of theoretical research on artificial intelligence. DEStech Trans Comput Sci Eng (iciti). https://doi.org/10.12783/dtcse/iciti2018/29138

34. Nascimento AM, Bellini CGP (2018) Artificial intelligence and industry 4.0: the next frontier in organizations. BAR—Braz Adm Rev 15(4). https://doi.org/10.1590/1807-7692bar2018180152
35. VC Müller 2012 Introduction: philosophy and theory of artificial intelligence Minds Mach 22(2):67–69. https://doi.org/10.1007/s11023-012-9278-y
36. Zhang Y (2022) A historical interaction between artificial intelligence and philosophy. https://doi.org/10.48550/ARXIV.2208.04148
37. BJ Grosz P Stone 2018 A century-long commitment to assessing artificial intelligence and its impact on society Commun ACM 61(12):68–73. https://doi.org/10.1145/3198470.doi:10.1145/3198470
38. Burukina O, Karpova S, Koro N (2019) Ethical problems of introducing artificial intelligence into the contemporary society. In: Human systems engineering and design. Springer, Cham, pp 640–646
39. Vernon D, Furlong D (2007) Philosophical foundations of AI. In: 50 years of artificial intelligence. Springer, Berlin, pp 53–62
40. Müller VC (2016) New developments in the philosophy of AI. In: Fundamental issues of artificial intelligence. Springer, Cham, pp 1–4
41. McCarthy J (2008) The philosophy of AI and the AI of philosophy. In: Philosophy of information. Elsevier, pp 711–740
42. E Hilker 1986 Artificial intelligence: a review of current information sources Collect Build 7(3):14–30. https://doi.org/10.1108/eb023192
43. What is (artificial) intelligence? In: Playing smart. The MIT Press (2019)
44. S Colombano 2000 AI's philosophical underpinnings IEEE Potentials 19 3 23 25. https://doi.org/10.1109/45.876893
45. García-Vigil JL (2021) Reflections around ethics, human intelligence and artificial intelligence. Gaceta medica de Mexico 157(3):298–301. https://doi.org/10.24875/GMM.M21000561
46. Causey RL (1994) Book review: philosophy and artificial intelligence by Todd C. Moody (Prentice Hall, 1993). SIGART Newsletter 5(1):52–54. https://doi.org/10.1145/181668.1064814
47. L Deng 2018 Artificial intelligence in the rising wave of deep learning: the historical path and future outlook [perspectives] IEEE Sign Process Mag 35(1):180–177. https://doi.org/10.1109/msp.2017.2762725
48. McCarthy J, Hayes PJ (1981) Some philosophical problems from the standpoint of artificial intelligence. In: Readings in artificial intelligence. Elsevier, pp 431–450
49. Cordeschi R (1989) Philosophical assumptions in artificial intelligence: a tentative criticism of a criticism. In: Informatik-Fachberichte. Springer, Berlin, pp 359–364
50. Copeland BJ, Proudfoot D (2007) Artificial intelligence. In: Philosophy of psychology and cognitive science. Elsevier, pp 429–482
51. David D (2021) Artificial Intelligence as solution in facing the age of digital disruption 4.0. JUDIMAS 1(1):107. https://doi.org/10.30700/jm.v1i1.1090
52. JM Górriz J Ramírez A Ortíz FJ Martínez-Murcia F Segovia J Suckling M Leming Y-D Zhang JR Álvarez-Sánchez G Bologna 2020 Artificial intelligence within the interplay between natural and artificial computation: advances in data science, trends and applications Neurocomputing. 410:237–270. https://doi.org/10.1016/j.neucom.2020.05.078
53. S Tuinen Van 2020 PHILOSOPHY IN THE LIGHT OF AI: Hegel or leibniz Angelaki: J Theor Humanit 25(4):97–109. https://doi.org/10.1080/0969725x.2020.1790838
54. Jain A (2021) Artificial intelligence and human society: Inteligencia artificial y sociedad humana. South Florida J Develop 2(4):4963–4989. https://doi.org/10.46932/sfjdv2n4-003
55. JJ Bryson 2020 The artificial intelligence of the ethics of artificial intelligence: an introductory overview for law and regulation MD Dubber F Pasquale S Das Eds The oxford handbook of ethics of AI Oxford University Press 1–25
56. Liao SM (2020) Ethics of artificial intelligence. Oxford University Press
57. Muhlenbach F (2020) A methodology for ethics-by-design AI systems: dealing with human value conflicts. In: 2020 IEEE International conference on systems, man, and cybernetics (SMC). IEEE

58. Makhamatov TM (2019) Philosophy of artificial intelligence. Humanit Bull Univ Finance 9(4):52–56. https://doi.org/10.26794/2226-7867-2019-9-4-52-56
59. Pratt I (1993) Book review: foundation of artificial intelligence by David Kirsh (ed) (Cambridge, MA: MIT Press). SIGART Newsletter 4(2):11–14. https://doi.org/10.1145/152941.1064727
60. McCarthy J (1989) Artificial intelligence, logic and formalizing common sense. In: Philosophical logic and artificial intelligence. Springer, Dordrecht, pp 161–190
61. PH Schönemann 1985 On artificial intelligence Behav Brain Sci 8(2):241–242 https://doi.org/10.1017/s0140525x0002063x
62. Gittinger JL (2019) Ethics and AI. In: Personhood in science fiction. Springer, Cham, pp 109–143
63. AA Hopgood 2003 Perspectives—artificial intelligence: hype or reality? Computer 36(5):24–28. https://doi.org/10.1109/mc.2003.1198233
64. Honavar V (2007) Symbolic artificial intelligence and numeric artificial neural networks: towards a resolution of the dichotomy. In: The springer international series in engineering and computer science. Springer, Boston, MA, pp 351–388
65. Artificial intelligence: a philosophical introduction (1994) Choice (Chicago, Ill.) 31(08):31–4403. https://doi.org/10.5860/choice.31-4403
66. Meyer J-JC, Hoek van der W (1995) Epistemic logic for AI and computer science. Cambridge University Press
67. Horvitz E (2017) AI, people, and society. Science 357(6346):7. https://doi.org/10.1126/science.aao2466
68. Philosophy and AI: essays at the interface (1992) Choice (Chicago, Ill.) 30(01):30–0219. https://doi.org/10.5860/choice.30-0219
69. Feng T (2019) Artificial intelligence's turn of philosophy. IOP Conf Ser: Mater Sci Eng 646(1):012008. https://doi.org/10.1088/1757-899x/646/1/012008
70. BC Stahl A Andreou P Brey T Hatzakis A Kirichenko K Macnish S Laulhé Shaelou A Patel M Ryan D Wright 2021 Artificial intelligence for human flourishing—beyond principles for machine learning J Bus Res 124:374–388. https://doi.org/10.1016/j.jbusres.2020.11.030
71. Lunkov AS (2020) Institute of philosophy and law of the Ural branch of the Russian academy of sciences. The ethics of artificial intelligence: from philosophical discussions to technical standardization. In: VIII Information school of a young scientist Central Scientific Library of the Urals Branch of the Russian Academy of Science. Central Scientific Library of the Urals Branch of the Russian Academy of Sciences
72. Liu F, Shi Y (2018) Research on artificial intelligence ethics based on the evolution of population knowledge base. In: Intelligence science II. Springer, Cham, pp 455–464
73. Dubber MD, Pasquale F, Das S (2020) The oxford handbook of ethics of AI. Oxford University Press
74. VC Müller 2016 Fundamental issues of artificial intelligence Springer Cham
75. Müller VC (2013) Philosophy and theory of artificial intelligence. Springer, Berlin
76. J Keating I Nourbakhsh 2018 Teaching artificial intelligence and humanity Commun ACM 61(2):29–32. https://doi.org/10.1145/3104986
77. AI: the tumultuous history of the search for artificial intelligence (1993) Choice (Chicago, Ill.) 31(03):31–1555. https://doi.org/10.5860/choice.31-1555
78. Livet P, Varenne F (2020) Artificial Intelligence: philosophical and epistemological perspectives. In: A guided tour of artificial intelligence research. Springer, Cham, pp 437–455
79. J-G Ganascia 2010 Epistemology of AI revisited in the light of the philosophy of information Knowl Technol Policy 23 1–2 57–73. https://doi.org/10.1007/s12130-010-9101-0
80. D Schiff B Rakova A Ayesh A Fanti M Lennon 2021 Explaining the principles to practices gap in AI IEEE Technol Soci Mag 40(2):81–94. https://doi.org/10.1109/mts.2021.3056286

Chapter 2
A Review of Different Approaches for Emotion Detection Based on Facial Expression Recognition

Sonu Mittal, Kamal Parashar, Priyanshu Belwal, and Tushar Gahlaut

1 Introduction

Humans are the most advanced and sophisticated species known. We communicate with each other in ways more than just talking. The face, being one of the most exposed regions of the human body, comprises many features in a relatively little space that uniquely express different emotional states of a human being. Facial expressions of a person are one of the most important forms of non-verbal communication [1]. Facial expressions of a person express his emotional state. As we are advancing into technology, human and machine interaction is increasing day by day. Emotions play a major role in these interactions since humans always have some type of emotional state [2]. These types of emotional interactions have the potential to revolutionise services like education, animation, gaming, and therapy.

1.1 Emotion Detection

Emotion detection is the process of detecting emotions by extracting and analysing facial expressions. In years of research, it has come to notice that humans have basic seven emotional states that they maintain. These states are neutral, happiness, sadness, anger, disgust, fear, and surprise. Although it is simple for us humans to discern emotions, it is challenging for machines to do the same [3]. If machines were able to detect these emotions, they might be able to help its user more efficiently.

S. Mittal · K. Parashar · P. Belwal (✉) · T. Gahlaut
Department of Computer Science & Engineering, Dr Akhilesh Das Gupta Institute of Technology and Management, New Delhi, India
e-mail: priyanshubelwal88@gmail.com

A. Yadav et al. (eds.), *Proceedings of International Conference on Paradigms of Communication, Computing and Data Analytics*, Algorithms for Intelligent Systems,
https://doi.org/10.1007/978-981-99-4626-6_2

1.2 Objective

The objective of this paper/research is to:

1. To study various pre-processing techniques and databases for emotion recognition.
2. To study various machine learning algorithms and their evaluation approaches.

After this short introduction, Sect. 2 describes the methodology, Sect. 3 consists of literature review, Sect. 4 is result analysis and finally Sect. 5 has conclusion and future scope.

2 Methodology

Several papers were studied in this research. Few papers were selected for the analysis from the literature review. The selection of papers was primarily based on accuracy measures. Then, the latest papers were selected and sorted for distinct methods so that comparison can be made between different models.

The models used in the papers were studied, and their network architecture was examined. CNN, K-NN, MLP neural network, SVM, DCNN, Xception algorithm, Auto-FERNet and many others were studied.

The papers selected use different data sets. KDEF, FER2013, RAF-DB, Cohn and Kanade DFAT-504, CK+, JAFFE, RaFD, CFEE are all of them. Some models share some common data sets but most of them have distinct data sets.

The result of each model is shown in their papers. The results from the implementation of these papers are shown and analysed in this paper.

3 Literature Review

The works related to emotion detection are covered in this section. We have reviewed several papers which have various methods for emotion detection. Different pre-processing techniques and feature extraction are also included in this section. Below we have a structured form of review which we have done. All the papers that we have reviewed are included in Table 1.

To create 3D models of the face, Microsoft Kinect was used in Paper [4]. Microsoft Kinect comes with two cameras. One uses visible light, while the other uses infrared light. It provides three-dimensional coordinates for various facial muscles. The Facial Action Coding System (FACS) produces a set of coefficients known as Action Units (AU). These Action Units (AU) represent various areas of the face. Six men between the ages of 26 and 50 took part and tried to emulate the emotions that were attributed to them. Each person had two sessions, with three trials in each session. The accuracy of 3-NN was 96%. MLP had a 90% accuracy rate.

Table 1 Literature review

Ref. ID	Pre-processing technique used	Method used	Data set used	Accuracy measures
[4]	Microsoft Kinect is used for 3d modelling, 121 points are used to model the face. A matrix is used to store the coordinates of points	k-NN classifier and MLP neural network 7 Emotion classes are used	KDEF	95%—k-NN 76%—MLP
[5]	Detection and alignment of faces, correction of illumination, pose, occlusion and data augmentation are done. Correction of illumination is done by histogram equalisation	CNNs 7 Emotion classes are used	FER2013	75%
[6]	Face detection is done. Detected faces were rescaled to 48 × 48 pixels images. Rescaled images were converted into Gabor magnitude representation	SVM classifiers, AdaBoost and AdaBoost + SVM 7 Emotion classes are used	Cohn and Kanade's DFAT-504	86.48, 85 and 89.75%
[7]	Normalisation of contrast, luminance segmentation and region analysis is done Face localisation and point localisation is also done	VISBER 4 Emotion classes are used	Samples are captured by the author itself	72%
[8]	Face detection is done	DSENet model 7 Emotion classes are used	FER2013	65.03%
[9]	Raw images are acquired and rescaling and normalisation of images is done to increase uniformity	CNN 7 Emotion classes are used	Kaggle facial expression	56.77%
[10]	Balancing the data set using oversampling and undersampling methods normalise the pixel values	CNN 7 Emotion classes are used	FER2013	83%
[11]	Balancing the data set	AdaBoost, logistic regression, DNN, CNN 7 Emotion classes are used	FER2013	33, 36, 39 and 64%
[12]	Not mentioned	Auto-FERNet	FER2013, CK+, JAFFE	73.78, 98.89 and 97.14%

(continued)

Table 1 (continued)

Ref. ID	Pre-processing technique used	Method used	Data set used	Accuracy measures
[13]	Tracker is employed which uses a face template to initially locate the position of the 22 facial features of our face model in a video stream and uses a filter to capture their positions over subsequent frames	Support vector machines 7 Emotion classes are used	Cohn-Kanade FER database	87.5%
[14]	Augmentation techniques like horizontal flip, shear, rotation, scaling, zooming in and out as the face and the underlying expressions can be at different distances	CNN 7 Emotion classes are used	FER2013	70.10%
[15]	Face detection illumination correction normalisation employs histogram equalisation and linear plane fitting	CNN	FER2013	75.2%
[16]	Identity and expression 3D face modelling, false detections removal, temporal smoothing, 3D facial reconstruction from videos, error pruning	DCNN 7 Emotion classes are used	RaFD, KDEF, RAF-DB, CFEE, CK+	97.65, 92.24, 83.27, 96.84 and 96.45%
[17]	Images are resized to lower resolution, and zero-mean normalisation is done here	CNN 8 Emotion classes are used	FER2013	65%
[18]	Batch normalisation and ReLU is done, "transfer learning" technique is used to pre-train the CNN Model	CNN 8 Emotion classes are used	CK+, BU-3DEF and FER2013	85% (CK+), 90% (BU-3DEF), 90% (FER2013)
[19]	LBP is used here by taking 8 neighbouring pixels surrounded by the centre pixel and normalising each pixel to create 8 binary digits	CNN and LBP 8 Emotion classes are used	CK+, JAFFE and YALE FACE	80% (CK+), 76% (JAFEE)
[20]	Images cropped and resized to 64 × 64 and divided into ten subject-independent data sets to conduct experiments	Redundancy reduced CNN 8 Emotion classes are used	CK+, JAFFE	92% (CNN), 84% (MIXED)
[21]	Feature extraction	CNN 8 Emotion classes are used	FER2013	65%
[22]	Enhance the data set with various transformations to generate various micro-changes in appearances and poses	CNN + LSTM* 8 Emotion classes are used	JAFEE	84% (CNN), 86% (CNN + LTSM)

(continued)

Table 1 (continued)

Ref. ID	Pre-processing technique used	Method used	Data set used	Accuracy measures
[23]	The photographs are cropped and converted to greyscale. They are hence giving us a more normalised form of the testing data	CNN 8 Emotion classes are used	FER2013	86%
[24]	Histogram equalisation is done to improve the contrast of images which results in better distribution	Deep learning models 8 Emotion classes are used	CK+, JAFFE and FACES	85.19%, 65.17%, 84.38%
[25]	Pre-processing step involves face detection for the two data sets. The frontal faces are rescaled using OpenCV21. Then, facial features are extracted using the deep CNN framework	DCNN 8 Emotion classes are used	CK+, JAFFE	83%
[26]	Initially, the face region is extracted from the given face images using the proposed face detection algorithm. The last connection layer of mini-Xception is used to extract deep features from the cropped face regions	Xception and CNN 7 Emotion classes are used	FER2013	95.60%
[27]	Vectorised facial land marker method is used Facial feature normalisation is done to eliminate the effect created by the size differences between faces	DCNN 8 Emotion classes are used	RaFD	84.33%
[28]	Images are reshaped into 100 × 100 pixels and then passed into the CNN system	DCNN 8 Emotion classes are used	CK+	92.81%

The authors of Paper [7] presented VISBER, a model for identifying emotions from facial expressions and attributes that use a fuzzy rule-based method. It categorises a video series of images into a set of fundamental emotions with matching intensities (joy, sorrow, anger, fear). Although VISBER was created in C++ for Linux, it can also be used on Windows. The Free Fuzzy Logic Library (FFLL), which is intended for time-critical applications, was used to perform the fuzzy classification. The FCL standard language was used to generate the fuzzy models. The average rate of recognition was 72%.

In Paper [8], the authors have proposed a system for use in an e-learning platform for teachers to recognise students' learning emotions. In the paper, the authors have compared the DSENet model to residual network 34 layers through ResNet-34. There are numerous residual blocks from the residual network, and these blocks make up the main architecture of the proposed system. A total of 100 epochs have been made using a Nadam optimiser, with batch size of 8 and learning rate was 0.002. An accuracy of 65.03% was achieved on the DSENet model, while it was 58.07% for the ResNet-34 model. The best accuracy achieved was 71.18% by the DSENet model using transfer learning, and without the transfer learning technique, the best accuracy was 63.76%.

In Paper [11], the authors applied different machine algorithms to the FER2013 data set because the data is severely unbalanced, and each algorithm's performance displayed different strengths and weaknesses in dealing with this. Various algorithms are applied and tested: AdaBoost, convolutional neural network (CNN), logistic regression and dense neural network (DNN). CNN performed the best on the classification task.

In Paper [12], the authors suggested an appropriate and compact Facial Expression Recognition Network Auto-FERNet, which uses a differentiable Neural Architecture Search (NAS) model to automatically search the FER data set. A 12-layer network with an auxiliary block is trained on FER2013 without the use of ensemble or additional training data to show the effectiveness of the system. The findings demonstrate that even a pure Auto-FERNet may outperform all prior approaches without a network with a performance of 73.11%. On CK+ and JAFFE, respectively, experimental results beat the state of the art with accuracy of 98.89% (10 times) and 97.14%, which also validates the resilience of our approach.

In Paper [24], the proposed technique for identifying facial expressions involves building and testing a CNN model. Other pre-trained deep CNN models are compared to the performance of the CNN model as a benchmark. The effectiveness of VGG-Face, which is pre-trained for face recognition, is evaluated in comparison to that of Inception and VGG, which are pre-trained for object identification. CK+, JAFFE, and FACES are the three face databases used in all investigations. Contrast enhancement is a crucial step in the pre-processing discussed. For contrast enhancement histogram equalisation (HE) was used. To avoid fully training the model, transfer learning is used. In order to apply the transfer learning technique, trials are repeated using taught models. With the use of ROI photos, the final layer of Inception-v3, VGG19 and VGG-Face for FER were retrained.

The authors of Paper [26] created a mini-Xception architecture based on Xception and convolution neural network (CNN). A real-time vision system was created which validates the concept and performs face detection and emotion classification in a single blended step using the proposed mini-Xception architecture. The suggested model's parameters are cut using depth-wise separable convolutions. Two layers make up depth-wise separable convolutions: depth-wise convolutions and point-wise convolutions. Four residual depth-wise separable convolutions make up the proposed architecture, which is then followed by a batch normalisation process and the activation of ReLUs. The last layer generates a prediction using global average pooling and a soft-max activation function. The proposed face detection algorithm is first used for extracting the face region from the provided face images. The final fully connected layer of mini-Xception is then extracted from the cropped face regions to extract deep features. They used the FER2013 data set for the experimental analysis, and the results show that all tasks can be efficiently performed using the proposed method such as detection and classification with seven different emotions (e.g. sad, surprise, anger, disgust, fear, neutral and happy.) using the Mini-Xception algorithm, with an accuracy of around 95.60%.

In Papers [6, 13], authors have used SVM-based models for emotion detection. While the model in [13] takes real-time video feed as input, the model in [6] takes frontal faces from the video streams. The model in [6] sends the image patches to an expression recogniser. This patch is represented as a Gabor. This patch is then processed by an SVM classifier. It is to note that the SVM model in [6] when combined with AdaBoost enhances performance. In [13], the authors address the difficulties of face localisation and feature extraction in spontaneous expressions by utilising a real-time facial feature tracker. It computes the displacement of face features with respect to the neutral frame. These displacement values are fed into the training stage of an SVM classifier. The authors then asked volunteers to express emotions naturally in an unconstrained set-up. This was done to compare the results of person dependent and independent detection. The accuracy achieved by the model in [13] is 87.5 in comparison to 89.75 for a similar data set in [6]. This slight increase in accuracy is attributed to the use of AdaBoost in [6].

The authors of Paper [16, 25, 27, 28] use DCNN as a model to detect facial expression. In [16], the face is detected from the video feed. After that, the false detections were removed. Then to mitigate the effects of any jitters in the extracted landmarks, temporal smoothing was performed and 3D facial reconstruction was done from the videos. SVM is selected as the choice of binary learner used in ECOC. In [25], OpenCV21 is used to detect and crop frontal images. In [27], vectorisation of facial features is done before feeding images to models to detect emotion. The authors of [28] use a regularisation method called "dropout" in the CNN fully connected layers, which has proven to be very effective in reducing overfitting. Accuracy achieved for Radboud (RaFD) data set is 97.65 and 84.33% for [16], 27, respectively, and for CK+ data set accuracy achieved is 96.45%, 83% and 92.81% for [16, 25] and [28], respectively. Model created in [16] gives higher accuracy than other models in comparison to it may be due to the 3D facial reconstruction pre-processing approach applied.

Many papers in the literature review use CNN-based models for emotion classification. In Paper [5], the authors demonstrated FER classification using CNNs on static pictures without any pre-processing or feature extraction activities. The architecture of CNN applied has six convolutional layers using ReLU and SoftMax as an activation function. In Paper [9], the face is detected, cropped and normalised using the OpenCV Haar Cascade classifier. ReLU and SoftMax are used here as well. ReLU is used after every convolution operation and SoftMax after max pooling. In Paper [10], authors used oversampling and undersampling to balance the data set, and normalisation is done in order to simplify the data. Same activation functions are used in [10] as well. To reduce the loss function, the Adam optimiser is used. In Paper [14], the authors designed and trained their own custom CNN architecture. It entailed using image augmentation techniques, then fine tuning the model architecture and hyperparameters. In Paper [15], the authors focused on the algorithmic variation and their impact on performance and analysing and discussing the performance of several works while highlighting the significant differences between them, with an emphasis on the underlying CNN architectures. In Paper [17], there are numerous organised subnets in the CNN model. A condensed CNN model that was specially trained makes up each subnet. These subnets are connected to create the full network. With this architecture, authors combine the results of different structured CNN models, making them a part of the entire network. In Paper [18], three innovative CNN models with various architectures were suggested by the authors. The first is a shallow network known as the Light-CNN, a fully convolutional neural network composed of six depth-wise separable residual convolution modules to address the problem of complex topology and overfitting. The second is a CNN with two branches that extracts both deep learning features and traditional LBP at the same time. The third model is a pre-trained CNN that was created using the transfer learning technique to compensate for a lack of training samples. In Paper [19], LBP and CNN are the feature extraction techniques employed here. The CNN design scales the image to a format that can be processed quickly without sacrificing crucial features in order to produce reliable predictions. In order to get an accurate result, the CNN method passes the input image through a number of different layers, including the convolution layer, rectified linear unit, pooling layer and fully connected layer. It is to note that SVM is chosen as the classifier for detecting facial expressions in [19]. In Paper [20], FRR-CNN convolutional kernels are divergently induced in contrast to classic CNN, leading to less duplicated features and a more compact representation of an image. Furthermore, the information concealed in mutual differences between each pair of feature maps in each convolutional layer is implemented to reduce redundancy of the representing features. In Paper [21], the authors concentrated primarily on using CNN to address the FER problem. For the purpose of recognising facial expressions, the authors employed a variety of architectures, including VGG16, ResNet and GoogleNet. After integrating the phases of feature extraction, template library and facial expression comparison, a streamlined structure with only four steps results. In Paper [22], the problem of facial expression recognition was solved using CNN and LSTM approach, an advanced modification of RNN, which is called a recurrent neural network. In order to boost the data set's image count, augmentation was performed as a pre-processing method. In Paper [23], the objective of facial expression recognition was solved by using CNN with FACS, OpenCV and MaxPooling2D. FACS examines the 44 "action units" of the face that move. Successful and more pleasant face recognition from sources like photos or videos was accomplished using OpenCV. To maintain the highest pixel value possible in the feature map, MaxPooling2D is used. The neural network then performed forward–backward propagation on these pixel values through this probability composition was generated through a SoftMax function.

3.1 Data Set Review

Table 2 contains data sets which are used in different models. Half of the data sets discussed here were sourced from Kaggle. The size of images in data sets is fixed for that specific data set. Some of the data sets are very vast such as RAF-DB, FER2013, and KDEF all of them contain more than twenty-five thousand images.

3.2 Remarks

The following observations are drawn from the literature review:

Sadness and fear were difficult to recognise in [4] when using a 3D face model. The use of glasses, facial hair and skin colour all had an impact on recognition. Changing the head orientation had a significant impact on the results. The majority of images that were misclassified in [5] came from fear and sadness. There was no mention of the effects on the results. The classifier performed admirably in [6], and without the explicit detection and registration of facial features, good results were obtained while processing the output of an autonomous face detector. AdaBoost significantly accelerated the application and improved classification performance. The recognition rate for happiness and sadness is lowest in [7] (may be due to inaccuracies in point localisation). The method discussed in this paper aids in the recognition of mixed emotions. The transfer learning technique was applied to the DSENet model in [8], and it increased accuracy by approximately 7.4%. The best accuracy without transfer learning was 63.76%, while the best accuracy with transfer learning was 71.18%. In [9], the face in the webcam is detected using the OpenCV Haar Cascade classifier. The accuracy for fear and anger is the lowest. Oversampling was used in [10] to balance the FER2013 data set. After balancing the data set with random oversampling, there was a sharp increase in accuracy and a decrease in loss. The accuracy of balancing the data set using sampling techniques was not improved in [11]. The other three methods underperformed in comparison with CNN, tested on the classification task, while AdaBoost and logistic regression outperformed DNN. Disgust was frequently misinterpreted as angry or sad. Sad and disgusting emotions are the least accurate in [13]. The error analysis in [14] was difficult to perform because the trained model performed better than human-level accuracy. The classes for fear and sadness had the lowest accuracy. The authors of [16] created their own data set after 3D reconstruction of human facial videos, and the model was fine-tuned using an existing data set. It produced an acceptable result, with the highest accuracy being 97.65%. In [17], happiness and sadness are classified much better than the average classification measure. The author believes that incorporating the Local Binary Pattern (LBP) will improve overall accuracy in future. Surprise and happiness classes are slightly more accurate than other classes in [20]. Integrating long-term short-term memory (LSTM) with CNN resulted in a 2% improvement in accuracy

Table 2 Data set review

S. No	Data set	Content	Emotion classes
1	CK+ (Kaggle)	The CK+ data set consists of 593 video sequences from a total of 123 different subjects, ranging from 18 to 50 years of age with a variety of genders and heritage The video sequences have a resolution equal to either 640 × 490 or 640 × 480 pixels Out of these videos, 327 are labelled with some expression	Seven expression classes: contempt, fear, happiness, sadness, disgust, anger and surprise
2	FER2013 (Kaggle)	The FER2013 data set consists of 48 × 48pixels, greyscale images The training set in FER2013 consists of 28,709 images and the testing set consists of 3589	Seven expression classes: happy, disgust, fear, sad, surprise, neutral, angry
3	JAFFE (zendo)	The JAFFE data set consists of 200 + images of facial expressions captured from ten Japanese women. All the images in the data set are 8-bit greyscale having resolution 256 × 256 pixels	Seven expression classes: happy, angry, fear, sad, surprise, disgust, neutral
4	KDEF (Kaggle)	The KDEF data set consists of 32,900 + images All the images in the data set are 224 × 224-pixels greyscale in PNG format	Eight expression classes: anger, disgust, happiness, surprise, contempt, neutral, fear and sadness
5	RaFD	The RaFD data set is an album of 67 models which includes Caucasian men, women and children. And Moroccan Dutch males were also included	Eight expression classes: disgust, happiness, anger, sadness, surprise, contempt, fear and neutral
6	FACES	The FACES data set consists of a set of images of natural faces of 171 young, middle-aged, older women and men The data set comprises two pictures per person per facial expression thus resulting in a set of 2052 images	Six expression classes: happiness, disgust, fear, anger, sadness and neutral
7	YALE FACE (Kaggle)	The YALE FACE data set consists of 165 GIF images belonging to 15 subjects There are eleven images of each subject	Facial expressions and configurations: centre-light, happy, left-light, with glasses, without glasses, right-light, normal, sad, sleepy, wink and surprised

(continued)

Table 2 (continued)

S. No	Data set	Content	Emotion classes
8	RAF-DB	The RAF-DB data set is a large-scale database with around 30,000, diverse facial images which are taken from the Internet	It consists of two different subsets: seven basic emotions and twelve compound emotions
9	CFEE	The CFEE data set consists of 1610 images captured from 230 subjects. These images were then converted to 256 × 256 in size and the colour channel was changed to greyscale	Seven expression classes: angry, fearful, disgusted, surprised, happy, sad and neutral

in [22]. The most accurate classes were neutral and angry. Overfitting and convergence issues were observed in [24] when a CNN model was trained from scratch. When compared to a pre-trained CNN model, this resulted in lower accuracy. The DCNN model implemented in [25] can be used by anyone because no extensive pre-processing or retraining is required. The emotion classes of sadness and surprise are frequently misinterpreted as happiness. According to [27], vectorised facial features can reduce data as well as training time. Such features can significantly accelerate the development of apps. The mean square error value in [28] decreases as the number of training data increases. Furthermore, the system's performance reaches 92.81% accuracy rate.

4 Result Analysis

As studied in the literature review, there are some robust models we have reviewed. The CNN model with 3D modelling implemented in paper [16] is the one that is tested for various data sets. It gives a good accuracy measure for each one of them. The Auto-FERNet model has also been tested on three databases and it gives good accuracy. K-NN model in paper [4] gives a pretty good accuracy for the KDEF data set. As for the FER2013 case, Auto-FERNet has accuracy lower than other databases but it is notable that FER2013 is one of the most challenging data sets to work with. It is very vast, diverse and unbalanced. Its human accuracy is also significantly low. Despite that, the CNN model in paper [26] with Xception algorithm has given an accurate measure of 95.60%, which is very much unexpected for the FER2013 data set. If we take a look at the performance of other models with FER2013, we can take a note from Table 1 that models with FER2013 have continuously low accuracy measures. Some of the measures are as low as 33%. Keeping that in mind, we can say that CNN with Mini-Xception is the best model we have reviewed so far. There is also a trend which is noticeable from the review that CNN models always perform well in combination with some other algorithms and they bring the accuracy significantly up.

5 Conclusion and Future Scope

This paper studies different emotion detection models. Several papers were reviewed by us in the literature section and a few of them were analysed for their performance. This paper aimed at studying different approaches for emotion detection and the same was achieved. Emotion detection has many applications in psychology, security, education, robotics, etc. A lot of research has been conducted in this field and emotion detection has been progressively improving and there is still a room for improvement. The ultimate aim of these systems is to increase accuracy and efficiency. This achievement will have a positive impact in this domain.

This paper's analysis is based on a comprehensive review of the work done in this field in past years and it doesn't involve any actual implementation of the system. An implementation-based comparative analysis of the models is recommended as a possible future work in this domain. We can test models on different data sets. And different models can be tested on common data sets for a fair accuracy comparison.

References

1. Prudhvi GNV (2023) Ultimate guide for facial recognition using a CNN. https://medium.com/@prudhvi.gnv/ultimate-guide-for-facial-emotion-recognition-using-a-cnn-f9239fdc63ad. Accessed 02 Jan 2023
2. Joseph A, Geetha P (2023) Facial emotion detection using modified eyemap–mouthmap algorithm on an enhanced image and classification with tensorflow. https://doi.org/10.1007/s00371-019-01628-3. Accessed 02 Jan 2023
3. Mehendale N (2023) Facial emotion recognition using convolutional neural networks (FERC). https://doi.org/10.1007/s42452-020-2234-1. Accessed 02 Jan 2023
4. Tarnowski P, Kołodziej M, Majkowski A, Rak RJ (2017) Emotion recognition using facial expressions. In: International conference on computational science. ICCS, Zurich, Switzerland
5. Singh S, Nasoz F (2020) Facial expression recognition with convolutional neural networks. In: 2020 10th annual computing and communication workshop and conference (CCWC). Las Vegas, NV, USA
6. Bartlett MS, Littlewort G, Fasel I, Movellan JR (2003) Real time face detection and facial expression recognition: development and applications to human computer interaction. In: Computer vision and pattern recognition workshop. CVPRW '03
7. Esau N, Wetzel E, Kleinjohann L, Kleinjohann B (2007) Real-time facial expression recognition using a fuzzy emotion model. In: 2007 IEEE international fuzzy systems conference. London, UK
8. Tseng FH, Cheng YP, Wang Y, Suen HY (2022) Real-time facial expression recognition via dense & squeeze-and-excitation blocks. Human-centric Comput Inf Sci 12, Article number: 39
9. Santra A, Rai V, Das D, Kundu S (2022) Facial expression recognition using convolutional neural network. Int J Res Appl Sci Eng Technol (IJRASET) 10(V), ISSN: 2321-9653
10. Pavan Kumar K, Shankar Reddy Y (2022) Facial emotion recognition using machine learning. Int Res J Modernization Eng Technol Sci 4(4), e-ISSN: 2582-5208
11. Gory S, Al-khassaweneh M, Szczurek P (2020) Machine learning approach for facial expression recognition. In: 2020 IEEE international conference on electro information technology (EIT). Chicago, IL, USA
12. Li S, Li W, Wen S, Shi K, Yang Y, Zhou P, Huang T (2021) Auto-FERNet: a facial expression recognition network with architecture search. IEEE Trans Netw Sci Eng 8(3)

13. Michel P, El Kaliouby R (2003) Real time facial expression recognition in video using support vector machines. In: ICMI '03: proceedings of the 5th international conference on Multimodal interfaces. Association for Computing Machinery, New York, pp 258–264, ISBN: 978-1-58113-621-0
14. Lonkar S (2021) Facial expressions recognition with convolutional neural networks
15. Pramerdorfer C, Kampel M (2016) Facial expression recognition using convolutional neural networks: state of the art
16. Koujan MR, Alharbawee L, Giannakakis G, Pugeault N (2020) Real-time facial expression recognition "in the wild" by disentangling 3D expression from identity. In: 2020 15th IEEE international conference on automatic face and gesture recognition (FG 2020). Buenos Aires, Argentina
17. Liu K, Zhang M, Pan Z (2016) Facial expression recognition with CNN ensemble. In: 2016 international conference on cyberworlds (CW). Chongqing, China
18. Shao J, Qian Y (2019) Three convolutional neural network models for facial expression recognition in the wild. In: Neurocomputing 355:82–92
19. Ravi R, Yadhukrishna SV, Rajalakshmi P (2020) A face expression recognition using CNN & LBP. In: 2020 fourth international conference on computing methodologies and communication (ICCMC). Erode, India
20. Xie S, Hu H (2017) Facial expression recognition with FRR-CNN. Electr Lett, Image Vis Process Disp Technol 53(4)
21. Gan Y (2018) Facial expression recognition using convolutional neural network. In: ICVISP 2018: proceedings of the 2nd international conference on vision, image and signal processing, pp 1–5, Article no.: 29
22. Hung BT, Tien LM (2021) Facial expression recognition with CNN-LSTM. In: Research in intelligent and computing in engineering, advances in intelligent systems and computing, vol 1254. Springer, Singapore
23. Kundu P, Kundu P, Mallik S, Bhowmick S, Mandal P, Banerjee H, Pal SB (2021) Facial expression recognition using convoluted neural network (CNN). In: Cyber intelligence and information retrieval, lecture notes in networks and systems, vol 291. Springer, Singapore
24. Sajjanhar A, Wu Z, Wen Q (2018) Deep learning models for facial expression recognition. In: 2018 digital image computing: techniques and applications (DICTA). Canberra, ACT, Australia
25. Mayya V, Pai RM, Manohara Pai MM (2016) Automatic facial expression recognition using DCNN. In: Procedia computer science, proceedings of the 6th international conference on advances in computing and communications, vol 93
26. Fatima SA, Kumar A, Raoof SS (2021) Real time emotion detection of humans using mini-Xception algorithm. In: IOP conference series: materials science and engineering, vol 1042, 2nd international conference on machine learning, security and cloud computing (ICMLSC 2020). Hyderabad, India
27. Yang G, Ortoneda JS, Saniie J (2018) Emotion recognition using deep neural network with vectorized facial features. In: 2018 IEEE international conference on electro/information technology (EIT). Rochester, MI, USA
28. Liliana DY (2018) Emotion recognition from facial expression using deep convolutional neural network. J Phys, Conf Ser 1193. In: International conference of computer and informatics engineering. Bogor, Indonesia

Chapter 3
The Long Short-Term Memory Tuning for Multi-step Ahead Wind Energy Forecasting Using Enhanced Sine Cosine Algorithm and Variation Mode Decomposition

Mohamed Salb, Luka Jovanovic, Nebojsa Bacanin, Goran Kunjadic, Milos Antonijevic, Miodrag Zivkovic, and V. Kanchana Devi

1 Introduction

There are several challenges to forecasting energy consumption accurately, including the complexity of the generated data, which includes multiple sources of energy, various end-use sectors, and diverse demand patterns. Energy consumption is also influenced by various factors such as weather, economic conditions, population growth, and technological advancements, which can make it difficult to accurately predict future energy usage. To address these challenges and improve forecasting, it is necessary to find better models and methods that can take into account multiple factors. One approach for addressing data complexity is through the use of signal decomposition methods. A goal of applying signal decomposition is to extract separate signal components from composite signals. By utilizing decomposition techniques, trends

M. Salb · L. Jovanovic · N. Bacanin (✉) · G. Kunjadic · M. Antonijevic · M. Zivkovic
Singidunum University, Danijelova 32, 11000 Belgrade, Serbia
e-mail: nbacanin@singidunum.ac.rs

M. Salb
e-mail: mohamed.salb.16@singimail.rs

L. Jovanovic
e-mail: luka.jovanovic.191@singimail.rs

G. Kunjadic
e-mail: gkunjadic@singidunum.ac.rs

M. Antonijevic
e-mail: mantonijevic@singidunum.ac.rs

M. Zivkovic
e-mail: mzivkovic@singidunum.ac.rs

V. K. Devi
School of Computer Science and Engineering, Vellore Institute of Technology, Chennai, India
e-mail: kanchanadevi@vit.ac.in

A. Yadav et al. (eds.), *Proceedings of International Conference on Paradigms of Communication, Computing and Data Analytics*, Algorithms for Intelligent Systems,
https://doi.org/10.1007/978-981-99-4626-6_3

in energy data can be observed and separated, making the prediction process more robust and resilient to noise. This research utilizes one such method, variational mode decomposition (VMD) [1] to address data complexity inherent to energy data time series.

An approach that has seen great success in recent years applied to forecasting is the use of machine learning (ML) algorithms. By formulating energy data as a time series forecasting can be systematically tackled. One notably well-performing approach for time series is the use of Long short-term memory (LSTM) [2]. However, like many algorithms, LSTM networks present several adjustable parameters, called hyperparameters that require adequate adjustments to ensure favorable performance. With the increasing complexity of algorithms methods for automating hyperparameters selection are required. Therefore, a popular approach is to formulate parameter selection as an optimization problem.

Swarm intelligence algorithms simulated individual behaviors of agents obeying sets of predefined rules. This allows algorithms to tackle complex tasks, and even NP-hard problems with relative ease, using realistic computational power and within reasonable time frames. Swarm intelligence algorithms excel at optimization problems. Therefore, this world explores the potential of a notably well-performing sine cosine algorithm (SCA) [3] for optimizing LSTM network hyperparameters to improve performance. Furthermore, this work introduces an enhanced variation of the SCA that builds on the success of the original approach.

The remainder of this work is organized as per the following. In Sect. 2 you can find the preceding and related research done on similar topics. In Sect. 3 you can find the proposed method for this research. In Sect. 4 you can find the experiments with a comparative analysis. Lastly, Sect. 5 concludes this paper and describes future research.

2 Background and Related Works

Energy forecasting involves the use of data and models for determining future energy demand or supply [4]. This process relies on data sources such as historical energy consumption and production data, weather data, economic data, and other relevant information. Models, including statistical models and ML models, can then be applied to analyze this data and make predictions about future energy demand or supply [4]. These predictions are made using data and models, which account for factors such as weather, economic conditions, population growth, technological advancements, and government policies [4, 5].

Several ML models can be utilized to determine the demand or supply of different types of energy [6]. Artificial neural networks (ANNs) are well-suited to analyzing complex data and identifying patterns and relationships and can be used to predict the demand or supply of various types of energy including renewable [7]. Support vector machines (SVMs) can be used to categorize data or predict numerical values and have been used to predict the demand or supply of various types of energy [8].

The decision has been applied to making predictions based on a series of decisions and has been used to predict the demand or supply [9]. Random forests are ensemble models that combine several decision trees to make more precise predictions [9]. Gradient boosting a novel ML approach combines a series of weak models to form a strong model [9].

2.1 *Overview of Long Short-Term Memory (LSTM)*

For handling long-term relationships in the time domain, Hochreiter and Schmidhuber [2] created the LSTM neural network, a form of recurrent neural network (RNN). It has nonlinear gated units and memory cells. The LSTM can handle vanishing/exploding gradient difficulties [10].

Where $X_t = [x_t^1, x_t^2, x_t^3, \ldots x_t^N]$ denote the count of entries. $H_t = [h_t^1, h_t^2, h_t^3, \ldots h_t^K]$ represent the K hidden neurons, and $C_t = [c_t^1, c_t^2, c_t^3, \ldots c_t^K]$ represent the note status of the LSTM network during iteration t, and f_t represents the forgotten gate, i_t represent the input gate, and O_t represents the output gate. On every iteration, t, the input X_t is delivered to three gates, together with the previously hidden state H_{t-1}, to calculate the next hidden state H_t and to adjust the prior cell state C_{t-1} to calculate the next cell state C_t. The following is the mathematical formula of several LSTM procedures:

In the first step, the forgetting gate f_t in the LSTM layer determines whether information from the previous cell state C_{t-1} is to be deleted expressed as:

$$f_t = \sigma(W_f X_t + U_f H_{t-1} + b_f) \tag{1}$$

The LSTM units then determine whether information should be contained in the cell state C_t. This procedure consists of two steps: firstly, candidate vector $\widehat{C}_t$ collects all of the data from the prior phase H_{t-1} and given input X_t. Secondly, the input gate scans the data to be entered in the cell state C_t carefully. These procedures are calculated as follows:

$$\widehat{C}_t = \tanh(W_c X_t + U_c H_{t-1} + b_c) \tag{2}$$

$$i_t = \sigma(W_i X_t + U_i H_{t-1} + b_i) \tag{3}$$

In the third stage, a new cell state C_t is generated based on the outcomes of the preceding phases as follows:

$$C_t = f_t \times C_{t-1} + i_t \times \widehat{C}_t \tag{4}$$

Finally, the output gate O_t decided how much information should be delivered to calculate the outcome H_t at the following iteration:

$$O_t = \sigma(W_O X_t + U_O H_{t-1} + b_O) \tag{5}$$

$$H_t = O_t \times \tanh(C_t) \tag{6}$$

In which W_f, W_i, W_C, W_O are the weights arrays related to the input X_t while U_f, U_i, U_C, U_O are the repeated weights arrays linked with previous hidden phase

H_{t-1} and b_f, b_i, b_C, and b_O are the biases variables for the four gates. $\sigma(x) = 1/(1 + e^{-x})$ is the log-sigmoid objective function, while $tanh(x) = (e^x - e^{-xW}/(e^x + e^{-x})$ is the hyperbolic tangent objective function. The log-sigmoid objective function provides values ranging from 0 to 1 that describe what proportion of each quantity goes across. A "0" denotes that nothing will flow across, while a "1" indicates that everything may go across, values between 0 and 1 define what percentage of the number goes across.

2.2 *Metaheuristics Optimization*

Metaheuristic optimizations have lately emerged as popular NP-hard problem optimization solutions. Swarm intelligent algorithms are the best-known among these algorithms. It is inspired by several species of insects, plants, and animals [11]. Metaheuristic techniques are structured after the comprehensive feeding, hunting, and breeding behaviors displayed by groups of relatively basic organisms such as birds, fish, mammals, insects, and trees. Artificial bee colony (ABC) [12], firefly algorithm (FA) [13]. Another branch of metaheuristics based on mathematical operations has recently arisen. The arithmetic optimization algorithm (AOA) [14] is driven by the adding, subtracting, multiplying, and dividing arithmetic operations, whereas the sine cosine algorithm (SCA) [3] is driven by the mathematical features of the sine and cosine operations.

Algorithms have already been employed to handle a range of practical real-life scenarios of NP-hard difficulty. Metaheuristics have been used for wireless sensor networks optimization [15, 16], cryptocurrency prices forecasting [17, 18], COVID-19 infections forecasting [19, 20], identifying brain cancer from MRI images [21, 22], neural network optimization and hyperparameters' optimization [23, 24].

2.3 *Variational Mode Decomposition (VMD)*

A relatively novel yet powerful method for decomposing a provided by is VMD [1]. This method deconstructs a complex input signal f into a discrete set of band-limited sub-signals u_k. These are characterized by bandwidth, estimated by H^1 Gaussian smoothness of the shifted signal, and center pulsation ω_k. This form of decomposition is a constrained variational problem as per Eq. (7).

$$\max_{\{u_k\},\{\omega_k\}} \left\{ \sum_{k=1}^{K} ||\partial_t \left[\left(\delta(t) + \frac{j}{\pi t} \times u_k(t) \right) \right] e^{j\omega_k t} ||_2^2 \right\} \tag{7}$$

Where Eq. (7) is subjected to: $\sum_{k=1}^{K} u_k = f$, and can be tackled by incorporating a quadratic penalty alongside Lagrangian multiplies as given in Eq. (8)

$$
\begin{aligned}
L(\{u_k\}, \{\omega_k\}, \lambda) = \alpha \sum_{k=1}^{K} ||\partial_t\Big[\Big(\delta(t) + \frac{j}{\pi t}\Big) \times u_k(t)\Big]e^{j\omega_k t}||_2^2 \\
+ ||f(t) - \sum_{k=1}^{K} u_k(t)||_2^2 + \Big\langle\lambda(t), f(t) - \sum_{k=1}^{K} u_k(t)\Big\rangle
\end{aligned}
\tag{8}
$$

where parameter a is applied to balance data fidelity. The attained modes can be formulated as per Eq. (9)

$$
\hat{u}_k(\omega) = \frac{\hat{f}(\omega) - \sum_{i \neq k} \hat{u}_i(\omega) + (\frac{\hat{\lambda}(\omega)}{2})}{1 + 2\alpha(\omega - \omega_k)^2} \tag{9}
$$

in which ω_k represents the calculated center of the appropriate modes' intensify spectrum. Additionally, wiener filtering was incorporated to increase resilience to sampling and noise as shown in Eq. (10).

$$
\omega_k = \frac{\int_0^\infty \omega|\hat{u}_k(\omega)|^2 d\omega}{\int_0^\infty |\hat{u}_k(\omega)|^2 d\omega} \tag{10}
$$

A full description of the VMD algorithm in detail can be explored in [1].

3 Methods

3.1 Original Sine Cosine Algorithm (SCA)

Originally introduced in 2016 the SCA [3] algorithm is a novel mathematics-inspired population-based optimization algorithm. The proposed method modifies agent locations via a mechanism inspired by trigonometry. Upcoming formulas describe how solutions are upgraded in SCA:

$$
X_i^{t+1} = X_i^t + r_1 \cdot \cos(r_2) \cdot |r_3 P_i^t - X_i^t| \tag{11}
$$

$$
X_i^{t+1} = X_i^t + r_1 \cdot \sin(r_2) \cdot |r_3 P_i^t - X_i^t| \tag{12}
$$

The above-mentioned are combined to produce equations:

$$
X_i^{t+1} = \begin{cases} X_i^t + r_1 \cdot \sin(r_2) \cdot |r_3 P_i^t - X_i^t|, & r_4 < 0.5 \\ X_i^t + r_1 \cdot \cos(r_2) \cdot |r_3 P_i^t - X_i^t|, & r_4 \geq 0.5 \end{cases} \tag{13}
$$

wherein X_i^t denotes the present individual i in the $d-th$ dimension at iteration t, P_i^t denotes an ideal agents positioning in the $d-th$ dimension at iteration t, and r_1,r_2,r_3 and r_4 are arbitrary factors. These factors are used to prevent local optima and to harmonize exploring and exploiting search tendencies.

The value of r_1 influences should an agent adjusts its location toward the global optimal ($r_1 < 1$) or away from it ($r_1 > 1$). It harmonizes intensification and diversification search behaviors, r_1 declines linearly from a fixed constant (a) to 0 [3]. It is updated utilizing the given formula:

$$r_1 = a - t\frac{a}{T_{\max}} \tag{14}$$

in which a represents a constant, t denotes the ongoing iteration, and $T_{\max}$ is the highest iterative count.

Factor r_2 given in the region $[0, 2\pi]$ defines the intensity and direction of agent movement regarding the attained ideal agent, objective. An additional arbitrary factor, r_3, gives the location an arbitrary weighting. This enables stressing ($r_3 > 1$) or reducing the stress ($r_3 < 1$) the influence of the endpoint of other solutions' location updates. r_3 from the [0, 2] range. The arbitrary factor r_4 from range [0, 1] acts as a switch to select which function is used in Eq. (13).

3.2 Enhanced Sine Cosine Algorithm (ESCA)

The original SCA metaheuristic demonstrates respectable performance. However, extensive testing using standard CEC functions has indicated that in some executions the algorithm tends to dwell ins sub-optimal regions of the search space. This shortcoming may result in decreased overall performance. This work, therefore, proposes an enhanced SCA (ESCA) that tackles the shortcomings of the original SCA to further improve on the admirable performance.

The introduced alteration incorporates the search mechanism from the novel reptile search algorithm [25] shown in Eq. (15).

$$\begin{aligned}
X_i^{t+1} &= \begin{cases} X_{\text{best}}^t - \eta_i \times \beta - R_i^t \times \text{rand}, & t \leq \frac{T}{4} \\ X_{\text{best}}^t \times X_{\text{rand}}^t \times ES \times rand, & t \leq \frac{T}{2} \text{ and } t > \frac{T}{4} \end{cases} \\
\eta_i &= X_{\text{best}}^t \times P_i, \\
R_i &= \frac{X_{\text{best}}^t - X_i^t}{X_{best}^t + \epsilon}, \\
ES &= 2 \times r_1 \times \left(1 - \frac{1}{T}\right), \\
P_i &= \alpha + \frac{X_i^t - M(X_i^t)}{X_{\text{best}}^t \times (\text{UB} - \text{LB}) + \epsilon}
\end{aligned} \tag{15}$$

in which X_{best}^t represents the current best solution, the ongoing iteration is shown as t while the highest iterative count is T. Further β defines a constant defining the speed of exploration and has a value of 0.1, X_{rand}^t is an arbitrarily selected agent, ES denotes a random decreeing value between $[-2, 2]$. A minimal value is defined by η to ensure that the denominator cannot become 0. A random value r_1 is chosen from an a range of $[-1, 1]$, α defines a constant of 0.1. Finally, rand represents an arbitrary value from the range [0, 1].

To allow both algorithms to contribute to the exploration an additional control parameter μ is also introduced. During every iteration, a random value DS is drawn from a uniform distribution in the range [0, 1]. Should the value of $DS > \mu$ standard SCA search is conducted. However, should the value of $DS < \mu$ then RSA search is utilized. The value of $\mu = 0.3$ has been empirically determined through extensive experimentation.

Algorithm (1) presents the pseudo-code of the introduced ESCA.

Algorithm 1 Pseudo-code of the ESCA

```
Initialization. Chaotically initialized solutions X_i(1, 2, 3, ··· , n).
Initialize the maximum count of iterations T.
Initialize DS value to 0.3
while t < T do
  Generate random value of μ
  if DS > μ then
    for Every X in the generated candidates do
      Calculate objective value.
      if f(X) is greater than f(P) then
        Update the positioning of the ideal agent (P = X*).
      end if
    end for
    Redefine r1 using Eq. (14).
    Redefine r2, r3 and r4 factors.
    Redefine the locations of ideal solution by Eq. (13).
  else
    for Every X in the generated candidates do
      Calculate objective value.
      Update positions using reptile search mechanism from Eq. (15)
    end for
  end if
end while
return P as the ideal agent.
```

4 Experiments and Comparative Analysis

4.1 Datasets

For this research, a real-world dataset covering an anonymous wind farm in China has been utilized. The dataset is publicly available.[1] For this work data concerning wind farm number 2 has been used date from 01.07.09 to 31.12.2010. A total of 70% of the available data was used to train the models, the following 10% was used to validate the approach, with the latter 20% reserved for testing as shown in Fig. 1.

4.2 Experimental Setup

The experimental procedure involves comparing several state-of-the-art metaheuristic algorithms tasked with selecting optimal LSTM parameters. The results are then evaluated using MAE, MSE, RMSE and R^2 metrics shown in Eq. (16), Eq. (17), Eq.

[1] http://blog.drhongtao.com/2016/07/gefcom2012-load-forecasting-data.html.

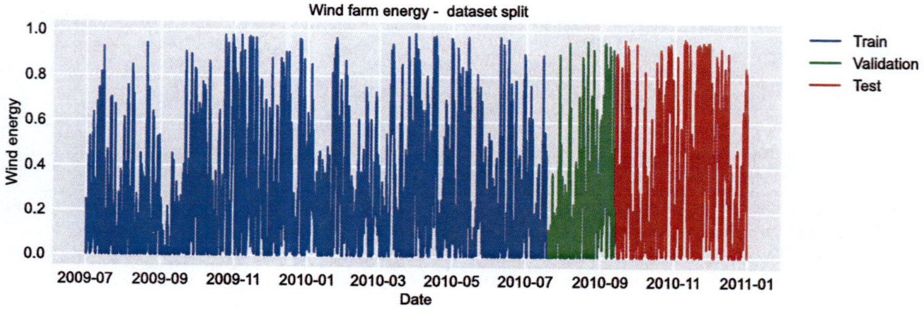

Fig. 1 Wind farm dataset virtualization with data split

(18), Eq. (19) respectively.

$$\text{MAE} = \frac{1}{n}\sum_{i=1}^{n}\left|y_i - \hat{y}_i\right| \tag{16}$$

$$\text{MSE} = \frac{1}{n}\sum_{i=1}^{n}\frac{(y_i - \hat{y}_i)^2}{y_i} \tag{17}$$

$$\text{RMSE} = \sqrt{\frac{1}{n}\sum_{i=1}^{n}\left(y_i - \hat{y}_i\right)^2} \tag{18}$$

$$R^2 = 1 - \frac{\sum_{i=1}^{n}\left(y_i - \hat{y}_i\right)^2}{\sum_{i=1}^{n}(y_i - \bar{y})^2}, \tag{19}$$

For testing purposes, VMD and all evaluated algorithms have been independently implemented for this research using Python accompanied by standard libraries including TensorFlow, Pandas, NumPy, Sklearn, and VMD-Python, while seaborn was utilized for visualization. A total of 3 ($k = 3$) modes have been extracted from original input features using VMD, with one additional mode representing the residuals also provided. A visualization of each decomposed feature can be seen in Fig. 2.

Each evaluated LSTM network has been tasked with casting predictions three steps ahead based on the input of six preceding steps. The metaheuristics tasked with optimizing the performance of these networks have been provided with a population of five agents and allocated eight iterations to improve results. Additionally, due to excessive computational demands, these experiments have been carried out

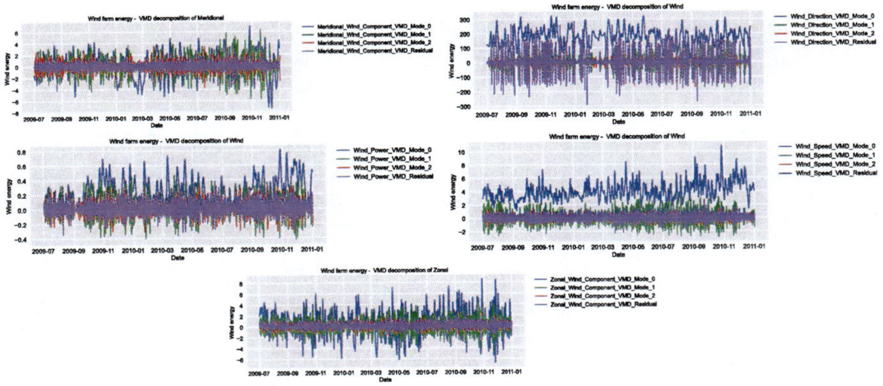

Fig. 2 Decomposition mode visualizations for each input feature

over eight individual executions to account for randomness intrinsic to metaheuristic algorithms.

4.3 Experimental Results

During testing, all metaheuristics were provided with decomposed signal inputs as evaluated on their prediction performance using the described metrics. Overall objective function performance results averaged over 8 independent runs for each tested metaheuristic as demonstrated in Table 1.

As demonstrated in Table 1, the introduced novel LSTM-ESCA approach attained the best objective results in the word, mean, and median cases, only slightly being outdone by the original SCA in the worst-case scenario.

Detailed metrics for the best execution of each metaheuristic are provided in Table 2.

Table 1 Overall objective function performance metrics for each tested approach

Method	Best	Worst	Mean	Median	Std	Var
LSTM-ESCA	**0.006784**	0.006973	**0.006879**	**0.006879**	9.47E–05	8.97E–09
LSTM-SCA	0.006864	**0.006923**	0.006893	0.006893	2.95E–05	8.68E–10
LSTM-PSO	0.007032	0.007281	0.007157	0.007157	1.25E–04	1.55E–08
LSTM-ABC	0.006869	0.006928	0.006901	0.006928	**2.94E–05**	**8.63E–10**
LSTM-WOA	0.007014	0.007126	0.007063	0.007014	5.52E–05	3.05E–09

Table 2 Detailed metrics for each prediction step of best-performing models of every evaluated approach

	Error indicator	LSTM-ESCA	LSTM-SCA	LSTM-PSO	LSTM-ABC	LSTM-WOA
One-step ahead	R^2	**0.879276**	0.872990	0.872271	0.877037	0.872033
	MAE	**0.075739**	0.077739	0.078073	0.076589	0.077883
	MSE	**0.011881**	0.012500	0.012570	0.012101	0.012594
	RMSE	**0.109000**	0.111801	0.112118	0.110006	0.112222
Two-step ahead	R^2	0.947839	**0.948728**	0.946276	0.948175	0.947074
	MAE	0.050406	**0.050028**	0.051666	0.050812	0.050734
	MSE	0.005133	**0.005046**	0.005287	0.005100	0.005209
	RMSE	0.071648	**0.071035**	0.072713	0.071417	0.072171
Three-step ahead	R^2	0.966084	**0.969051**	0.967094	0.965383	0.967067
	MAE	0.041545	**0.040221**	0.041549	0.042781	0.041488
	MSE	0.003338	**0.003046**	0.003238	0.003407	0.003241
	RMSE	0.057774	**0.055189**	0.056907	0.058368	0.056930
Overall results	R^2	**0.931066**	0.930256	0.928547	0.930198	0.928725
	MAE	**0.055897**	0.055996	0.057096	0.056727	0.056702
	MSE	**0.006784**	0.006864	0.007032	0.006869	0.007014
	RMSE	**0.082366**	0.082848	0.083857	0.082882	0.083753

The best achieved results are marked with bold

As demonstrated in Table 2, the introduced LSTM-ESCA approach attained the optimal prediction results from one-step-ahead predictions, being slightly outdone by the original LSTM-SCA approach for two and three-step-ahead forecasts. However, the proposed approach still attained the best performance in overall evaluations, outperforming all competing algorithms.

To visually demonstrate the improvements made to with the novel introduced metaheuristic convergence and distribution graphs of the objective and R^2 functions are shown in Fig. 3.

As shown in Fig. 3, significant improvements have been made over the original SCA algorithm, and a faster convergence rate has been achieved. Finally, the best parameter value selected from their respective ranges: total number of hidden network layers $[1, 2]$, neuron count in first layer [100, 200], neuron count in second layer [100, 200], training epochs [300, 600], dropout rate [0.05, 0.2] by each evaluated method are shown in Table 3.

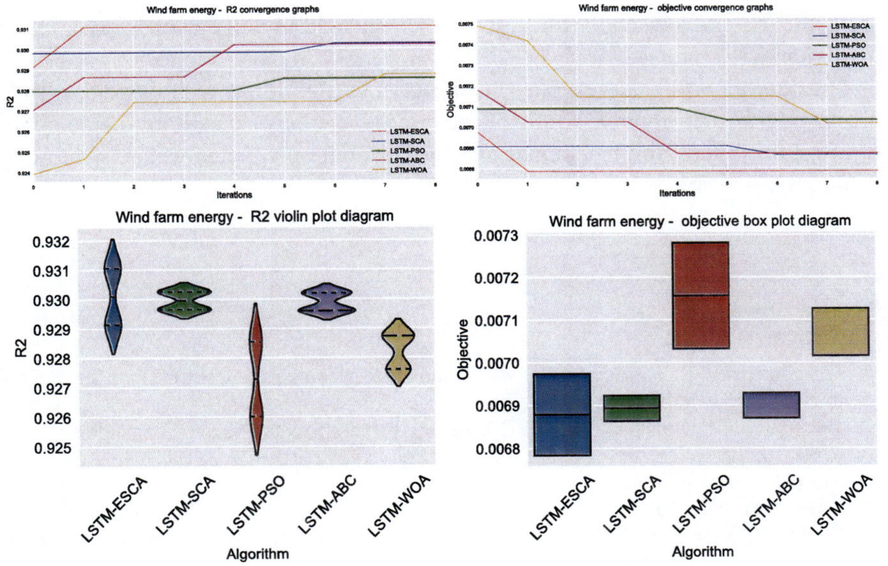

Fig. 3 Objective and R^2 convergence and distribution plots

Table 3 Selected parameters for best-performing models by beach metaheuristic

Method	Neurons layer 2	learning rate	Training epochs	dropout	Number of layers	Neurons layer 2
LSTM-ESCA	188	0.010000	600	0.172817	2	100
LSTM-SCA	100	0.009183	600	0.192692	2	121
LSTM-PSO	118	0.010000	514	0.078699	2	100
LSTM-ABC	151	0.007756	533	0.085550	2	152
LSTM-WOA	112	0.010000	598	0.101423	2	149

5 Conclusion

In the field of modern wind energy planning, forecasting complex problems in operating loads and minimizing risks while improving performance is crucial. To address this need, recent advances in deep learning techniques have shown promise as effective methods for forecasting. In this study, a novel ESCA metaheuristic algorithm is proposed and its potential in combination with a long short-term memory (LSTM) network for wind power generation is explored. By utilizing VMD prepossess the input data results were significantly improved. The combination of ESCA and LSTM, known as SECA-LSTM, is particularly well-suited for extracting complex, nonlinear features from real-time series datasets, leading to even further improvement in wind speed prediction. These findings demonstrate the potential of using deep learning approaches to accurately forecast wind energy and improve the performance of modern power systems.

Future work will be to test other metaheuristics algorithms with long short-term memory to further improve results. Furthermore, we hope to explore the potential of novel decomposition techniques and further refine prediction accuracy through better prepossessing.

References

1. Dragomiretskiy K, Zosso D (2013) Variational mode decomposition. IEEE Trans Signal Process 62(3):531–544
2. Hochreiter S, Schmidhuber Jü (1997) Long short-term memory. Neural Comput 9(8):1735–1780
3. Mirjalili S (2016) SCA: a sine cosine algorithm for solving optimization problems. Knowl-Based Syst 96:120–133
4. Suganthi L, Samuel AA (2012) Energy models for demand forecasting-a review. Renew Sustain Energy Rev 16(2):1223–1240
5. Islam MA, Che HS, Hasanuzzaman M, Rahim NA (2020) Energy demand forecasting. In: Energy for sustainable development. Elsevier, pp 105–123
6. Perera KS, Aung Z, Woon WL (2014) Machine learning techniques for supporting renewable energy generation and integration: a survey. In: International workshop on data analytics for renewable energy integration. Springer, pp 81–96
7. Ahmad T, Zhang H, Yan B (2020) A review on renewable energy and electricity requirement forecasting models for smart grid and buildings. Sustain Cities Soc 55:102052
8. Shiri A, Afshar M, Rahimi-Kian A, Maham B (2015) Electricity price forecasting using Support Vector Machines by considering oil and natural gas price impacts. In: 2015 IEEE International conference on smart energy grid engineering (SEGE). IEEE, pp 1–5
9. Foley Aoife M, Leahy Paul G, Marvuglia Antonino, McKeogh Eamon J (2012) Current methods and advances in forecasting of wind power generation. Renew Energy 37(1):1–8
10. Hochreiter S, Schmidhuber J (1996) LSTM can solve hard long time lag problems. Adva Neural Inf Process Syst 9
11. Raslan AF, Ali AF, Darwish A (2020) Swarm intelligence algorithms and their applications in Internet of Things. In: Swarm intelligence for resource management in internet of things. Elsevier, pp 1–19
12. Karaboga D, Akay B (2009) A comparative study of artificial bee colony algorithm. Appl Math Comput 214(1):108–132
13. Yang XS (2009) Firefly algorithms for multimodal optimization. In: International symposium on stochastic algorithms. Springer, pp 169–178
14. Abualigah L, Diabat A, Mirjalili S, Abd Elaziz M, Gandomi AH (2021) The arithmetic optimization algorithm. Comput Meth Appl Mech Eng 376:113609
15. Bacanin N, Tuba E, Zivkovic M, Strumberger I, Tuba M (2019) Whale optimization algorithm with exploratory move for wireless sensor networks localization. In: International conference on hybrid intelligent systems. Springer, pp 328–338
16. Zivkovic M, Bacanin N, Tuba E, Strumberger I, Bezdan T, Tuba M (2020) Wireless sensor networks life time optimization based on the improved firefly algorithm. In: 2020 International wireless communications and mobile computing (IWCMC). IEEE, pp 1176–1181
17. Salb M, Zivkovic M, Bacanin N, Chhabra A, Suresh M (2022) Support vector machine performance improvements for cryptocurrency value forecasting by enhanced sine cosine algorithm. In: Computer vision and robotics. Springer, pp 527–536

18. Bačanin Džakula N et al (2021) Cryptocurrency forecasting using optimized support vector machine with sine cosine metaheuristics algorithm. In: Sinteza 2021-International scientific conference on information technology and data related research. Singidunum University, pp 315–321
19. Zivkovic M, Bacanin N, Venkatachalam K, Nayyar A, Djordjevic A, Strumberger I, Al-Turjman F (2021) COVID-19 cases prediction by using hybrid machine learning and beetle antennae search approach. Sustain Cities Soc 66:102669
20. Zivkovic M, Jovanovic L, Ivanovic M, Krdzic A, Bacanin N, Strumberger I (2022) Feature selection using modified sine cosine algorithm with COVID-19 dataset. In: Evolutionary computing and mobile sustainable networks. Springer, pp 15–31
21. Basha J, Bacanin N, Vukobrat N, Zivkovic M, Venkatachalam K, Hubálovský S, Trojovský P (2021) Chaotic harris hawks optimization with quasi-reflection-based learning: an application to enhance CNN design. Sensors 21(19):6654
22. Jovanovic L, Zivkovic M, Antonijevic M, Jovanovic D, Ivanovic M, Jassim HS (2022) An emperor penguin optimizer application for medical diagnostics. In: 2022 IEEE zooming innovation in consumer technologies conference (ZINC). IEEE, pp 191–196
23. Bacanin N, Alhazmi K, Zivkovic M, Venkatachalam K, Bezdan T, Nebhen J (2022) Training multi-layer perceptron with enhanced brain storm optimization metaheuristics. Comput Mater Contin 70:4199–4215
24. Bacanin N, Stoean C, Zivkovic M, Jovanovic D, Antonijevic M, Mladenovic D (2022) Multi-swarm algorithm for extreme learning machine optimization. Sensors 22(11):4204
25. Abualigah L, Abd Elaziz M, Sumari P, Geem ZW, Gandomi AH (2022) Reptile search algorithm (RSA): a nature-inspired meta-heuristic optimizer. Expert Syst Appl 191:116158

Chapter 4
Design of Traffic Monitoring System by Greedy Perimeter Stateless Routing Protocol

Movva Ganesh Kasyap, Arepalli Gayathri, Yaram Chandana, Vadithe Venkatesh Naik, and Yaddanapudi Sarada Devi

1 Introduction

A network is a collection of linked computers, servers, mainframes, peripherals, network devices, and other devices that facilitate data sharing. An example of a network is the Internet, which connects millions of individuals globally. Wireless communication is a method of transmitting data from one point to another without the use of physical connections like wires, cables, or other media. Information is typically sent across a short distance from a transmitter to a receiver in communication systems. An "ad-hoc network" is a temporary local area network (LAN). When an ad-hoc network is installed permanently, it develops into a LAN. Multiple users can be connected to an ad hoc network at once, although performance might decrease.

The phrase "vehicular ad-hoc network" (VANET) refers to a method for creating mobile networks that uses moving vehicles as network nodes. VANET transforms each vehicle into a wireless node, enabling vehicles to connect to one another even when they are 100–300 m apart, thereby building a large network as shown in Fig. 1.

As vehicles leave the current network owing to signal range limitations, additional vehicles can join in to connect vehicles to one another in order to form a mobile

M. G. Kasyap (✉) · A. Gayathri · Y. Chandana · V. V. Naik · Y. S. Devi
Department of Electronics and Communication Engineering, V.R. Siddhartha Engineering College, Vijayawada, Andhra Pradesh, India
e-mail: movvaganesh17@gmail.com

A. Gayathri
e-mail: gayathriarepalli2001@gmail.com

V. V. Naik
e-mail: venkateshnaikvadithe12345@gmail.com

Y. S. Devi
e-mail: saradadevi@vrsiddhartha.ac.in

A. Yadav et al. (eds.), *Proceedings of International Conference on Paradigms of Communication, Computing and Data Analytics*, Algorithms for Intelligent Systems,
https://doi.org/10.1007/978-981-99-4626-6_4

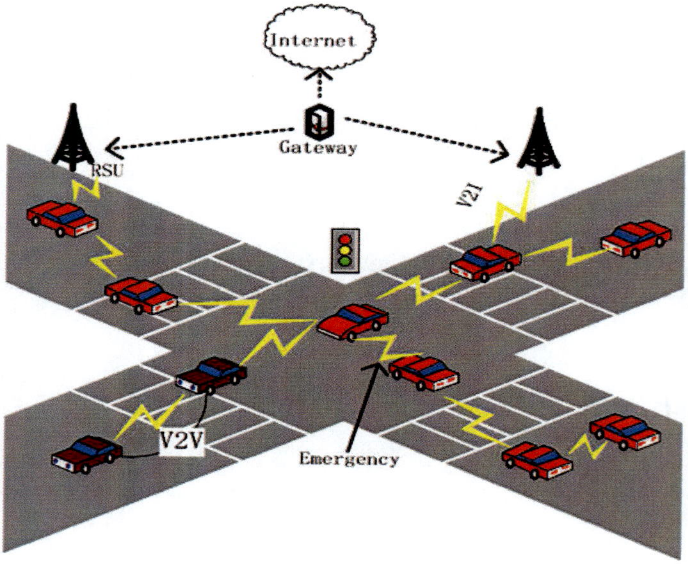

Fig. 1 Vehicle ad-hoc network communication scenario illustration

Internet. It is anticipated that police and fire trucks would be the first systems to integrate it in order to communicate with one another and ensure safety. Transportation networks are quickly becoming available for the deployment and development of both new and established applications. Its defining characteristics are rapid topological change, high mobility, and transient, one-time interactions MANETs and VANETs both.

There are three types of communication between nodes in a VANET: Vehicle-To-Vehicle (V2V), Vehicle-To-Roadside (V2R), and Vehicle-To-Infrastructure (V2I). Fixed nodes called Roadside Units (RSUs) are positioned next to the road to improve connectivity and service availability. Both the Internet and the core network can be accessed by RSUs. Figure 1 provides an illustration of these ideas. Vehicle communications have been implemented using a variety of methods and designs.

Platooning, vehicle collisions, highway entry V2V communications, cruise control, safety monitoring, tunnel services, bridge monitoring, parking localization, security warning, map localization, intersection safety, blind crossing, and Internet access are a few of the applications of vehicular ad-hoc networks (VANETs).

2 Literature Survey

Al-Azzawi et al. [1] focused on the design of a vehicular ad-hoc networks (VANETs) protocol with specific implementations of the DT3P algorithm method for the development of a traffic light control system. Screening data sets from the 1914 Cleveland

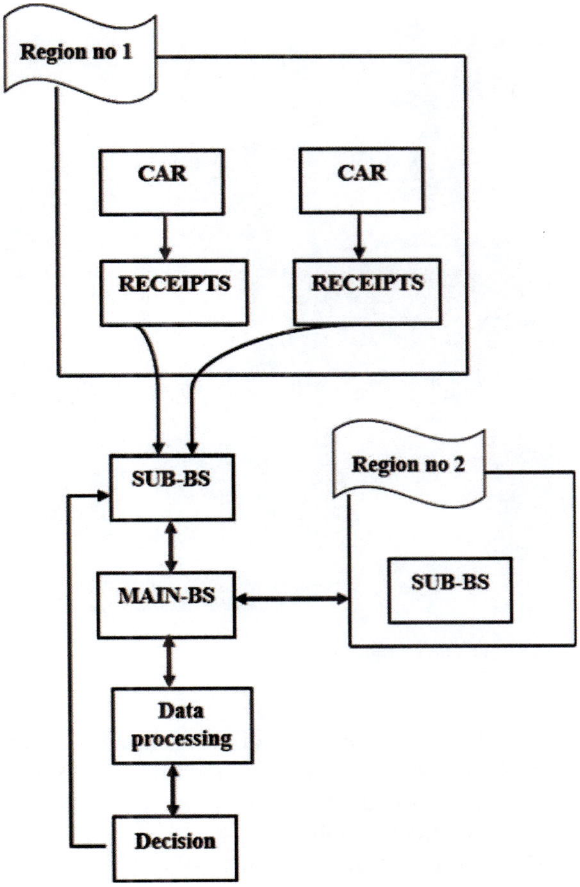

Fig. 2 System flow chart

Junction is one of the strategies used in this article. The regular distribution of cars at an intersection is the main area, where the research paper has limits or research gaps.

According to Karunakar et al. [2], one of the most important pieces of information for vehicles is their location. Most often, location-based routing protocols would provide information about the physical area of cooperating vehicles. Because there is no creation and maintenance of the total path from the source node to the destination node, general location-based routing has been found to perform better than topology-based routing conventions.

Al-Kharasani et al. [3] presented CACA, an urban VANET cluster algorithm developed by Adept Cooperative Algorithm. This protocol was created to boost flooding and boost network performance. The Quality of Path (QoP) metric, which was utilized to resolve the routing trade-off overhead problem between mobility limitations and QoS requirements in VANETs, provides the basis for better routing

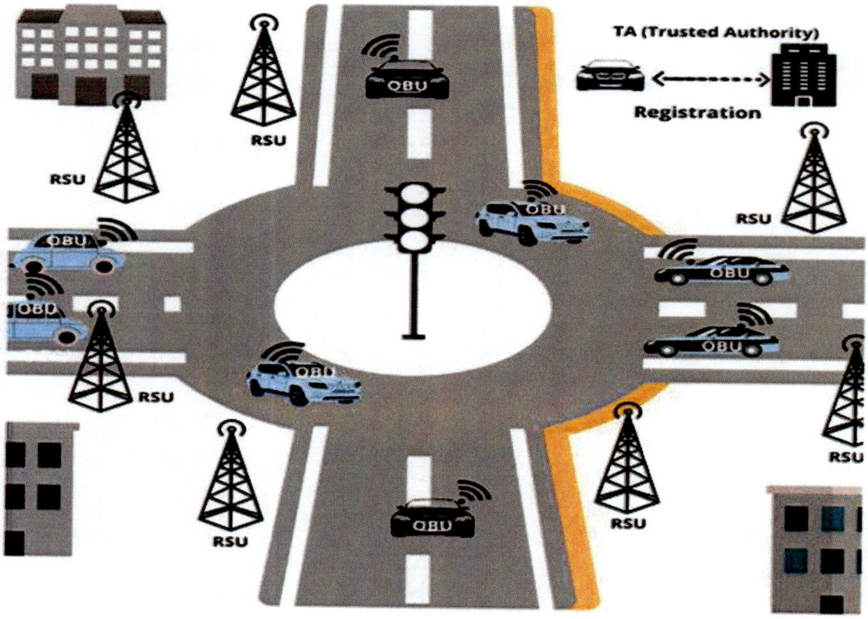

Fig. 3 Vehicular ad-hoc network

scalability. The primary goal of this measure is to gather crucial and effective information about neighbours during the route discovery phase and use it to choose the next forwarding node after determining whether the current forwarding is out-of-date and satisfies the QoS criterion.

Chen et al. [4] for data propagation in VANETs, the recent results have been proposed. We reviewed the security issues in this area as well as the necessity for enabling technologies to enable effective data dissemination for automotive applications in addition to dissemination methods.

Cardenas et al. [5] presented the probabilistic multimeric routing protocol (ProMRP). To decide which vehicles to hand over to next, ProMRP uses three criteria (vehicle density, travel time to the destination, and bandwidth available).

Bhatia et al. [6] demonstrated a VANET system that uses computationally intelligent models to forecast traffic flow behaviour. They created an architecture with RSUs and OBUs controlled by an SDN controller and linked to a cloud infrastructure appropriate for high computational power and real-time data storage. The performance of the suggested model can be further enhanced in future by using extraneous variables, such as changes in local congestion locations' vehicle densities.

Tomar et al. [7] presented a thorough analysis of traffic light synchronization strategies for intelligent cars. To manage intersection congestion, we first briefly describe the different traffic light control technology and then present a summary of the state of development of traffic signal synchronization for intelligent cars. Along with that, we go over a lot of new and promising directions for traffic light networking.

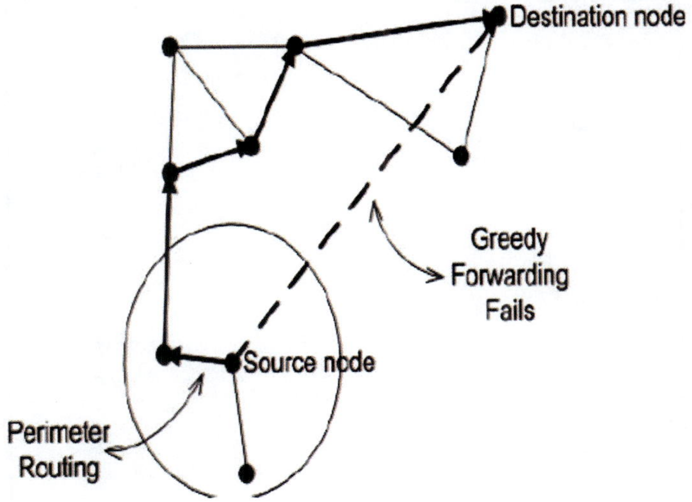

Fig. 4 Working of greedy mode

The findings of this study will serve as a foundation for future research into traffic light synchronization for intelligent vehicles.

Hamdi et al. [8] presented a thorough analysis of the methods used to detect incidents currently. Based on traffic monitoring and event detection, current strategies were examined. A comparative analysis of traffic accident detection techniques from the standpoint of working with advantages and disadvantages is given. To find uses for issue detection and management, proprietary event detection systems were also investigated. The study analysed all of the accident detection methods currently in use and provided a comparative analysis of their strengths and drawbacks.

Mohanty et al. [9] it was discovered that FCM and Fuzzy K-means, which use fuzzy measure computations, give outcomes that are comparable to those of K-means clustering in the congestion detection process on a busy road. Despite this, the method still requires longer execution durations than K-means clustering. In addition, due to the closer proximity of cluster centres, Fuzzy K-means is the best of these fuzzy algorithms for detecting congestion.

Hamideh Fatemidokht and Marjan Kuchaki Rafsanjani [10] a novel vehicle clustering method called QMM-VANET was introduced to maintain the stability of an ad-hoc traffic network. The parameters for distrust, mobility restrictions, and QoS needs are utilized to calculate the QoS value for each vehicle in this protocol. This value is traded across nearby vehicles, and the cluster leader is selected from those with the greatest QoS value.

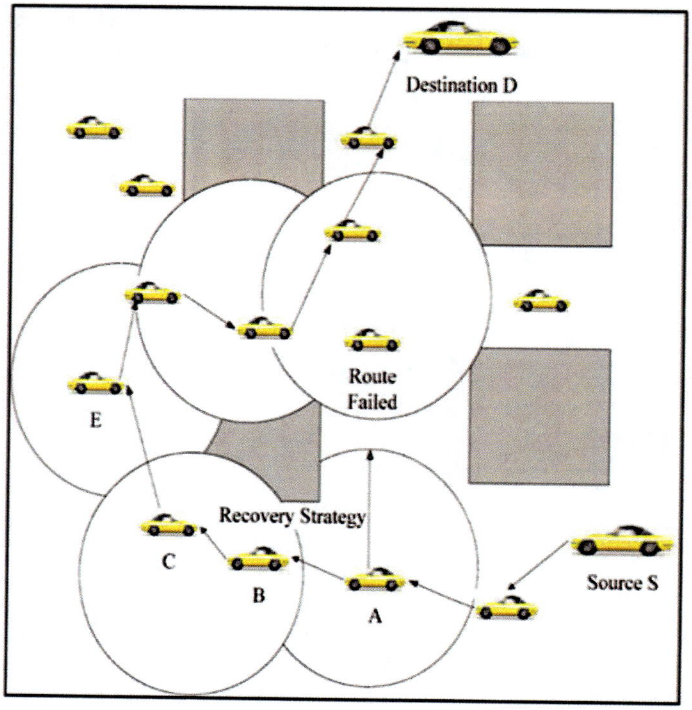

Fig. 5 Working of greedy perimeter stateless routing's perimeter phase (GPSR)

3 Methodology

Traffic light system for efficient city administration. Using the Vehicle Dedicated Network and Vehicle-to-Infrastructure (VANET V2I) protocol, traffic information from roads and streets is collected. Each vehicle is computed by a mobile device that can connect to a VANET, and the computed information is then transmitted to a nearby base station. The optimal traffic management strategy will then be chosen by the control system.

From Fig. 2, it is considered as:

- Each car in the city is assumed to have a mobile device that can communicate with receiving devices using the VANET protocol and has a unique identification.
- Place receiving equipment in predictable locations. It uses an Omni-antenna.
- A Sub-Base Station (SUB-BS) is present in each area.
- The receivers of each region are connected via SUB-BS.
- The base station MAIN-BS serves as a connection point for all SUB-BS stations.
- The MAIN-BS is made up of an Omni-antenna, a database, and a CPU that determines how to run traffic lights by counting the number of cars on each street.

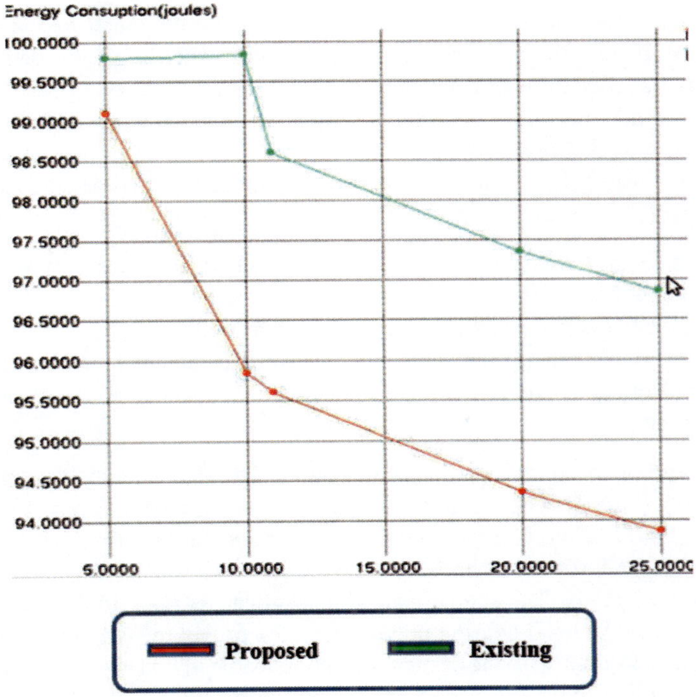

Fig. 6 Energy consumption of nodes

The intersection shown in the image above is made up of four separate carriageways, each with a different density of vehicles. Traffic is controlled by traffic lights that receive information from a data processing centre located along the route as shown in Fig. 3. The data processing unit, which collects data from the sub-base station, makes decisions based on traffic, and processes the data to the server, forms a data processing centre. According to traffic conditions, the data will be updated, and the entire file will be kept in regional databases.

4 Protocol

4.1 Greedy Perimeter Stateless Routing Protocol (GPSR)

GPSR is a speedy and efficient routing method for wireless, mobile networks. By using the positions of nodes to determine packet forwarding decisions, GPSR leverages the correlation between geographic position and connectivity in a wireless network in contrast to traditional routing algorithms that use graph-theoretic concepts of shortest paths and transitive reachability to find routes. Packets are forwarded by GPSR to nodes that are constantly getting closer to the target by using greedy forwarding. If only the network's nodes didn't have such a greedy path. After a packet has travelled through a series of gradually closer faces of a planar subgraph of the entire radio network connectivity graph, GPSR recovers by resuming greedy forwarding at a node closer to the goal.

Networks using GPSR cannot be designed using prior wired or wireless routing techniques. These networks consist of:

- Sensor networks, which can be portable, extremely dense, have a lot of nodes and have a lot of resources per node.
- Rooftop networks: a steadfast, numerous node installation.
- Widely variable density, mobile, non-power-restricted vehicle networks.
- Ad-hoc networks are mobile with fluctuating densities and lack a reliable infrastructure.

The source node's position is determined by GPS through Greedy Perimeter Stateless Routing (GPSR). The location of the neighbouring node is learned by beacon exchange. Grid Location Service (GLS) and Hierarchical Location Service are two examples of location services used to locate the destination (HLS). It has two modes: perimeter and greedy. The source node or packet carrier node in the greedy mode chooses the nearest neighbouring node to send the packet to. When the forwarding node is unable to connect to the destination and is unable to find a neighbour node that is closer than it. The packet reaches the maximum for the region. To overcome the local optimum issue in this scenario, GPSR uses perimeter mode. The greedy mode of GPSR is shown in Fig. 4, where the source node S selects neighbour node B above all of its one-hop neighbours as the recipient of the packet since it is located closest to the destination node D. A local optimum issue occurs when GPSR uses perimeter mode. In the perimeter mode, there are two steps. It creates graph planarization in the first stage using a relative neighbourhood graph (RNG). In the second step, the right-hand rule is used to identify the subsequent neighbour node that will relay the packet to its destination.

Figure 5 shows the perimeter-based forwarding method. A vehicle node cannot recognize another vehicle node if it is farther from the destination vehicle D. It chooses node B to forward packets using the perimeter mode's right-hand rule. Similar to node A, node B also passes the packet to node C. Up until the perimeter mode returns to the greedy mode, this process continues.

5 Results

It is designed to select the nearest node to interact with the data processing centre to specify its shortest route with a low density of vehicles. As shown in Fig. 6, we compared the results of our algorithm in red, and the existing DT3P algorithm in red as follows:

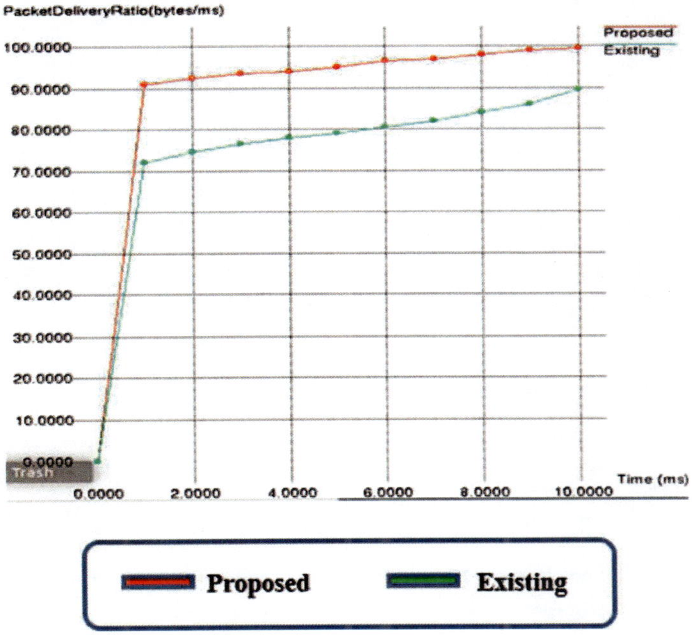

Fig. 7 Packet delivery ratio

1. The average energy consumed by the nodes in our simulation is reduced when compared to the existing DT3P algorithm simulation because the dynamic routing of nodes is considered in this simulation.
2. The exchange of packets between the source and destination nodes is taken by means of the packet delivery ratio which is represented in Fig. 7 as shown.
3. The time taken by the packets to transmit from the source node to the destination is given by the end-to-end delay which is shown in Fig. 8.
4. During the transmission of packets from the source to the destination end, not all the packets will be received by the destination node some may be lost due to the inadequate signal strength at the destination end if more packet loss is observed then the destination node cannot perform the expected tasks which result in the failure of the network by dynamic routing this problem can be fixed. This is shown in Fig. 9.

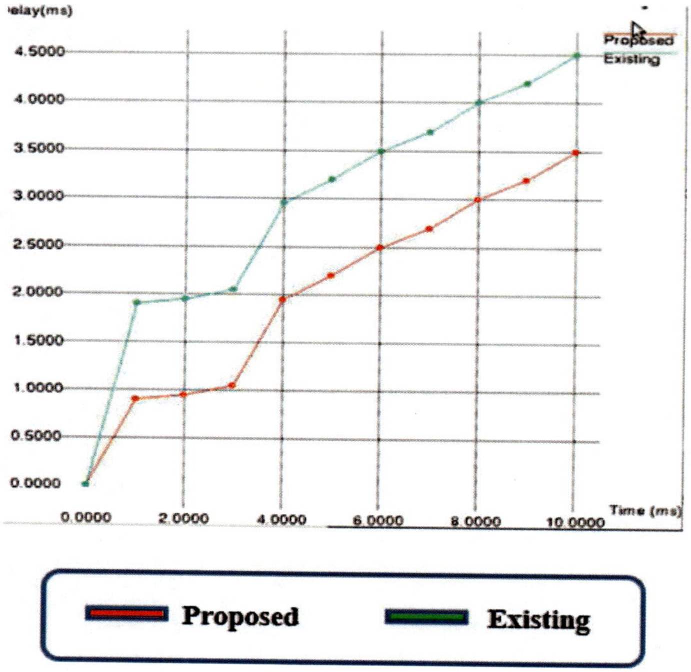

Fig. 8 E2E delay

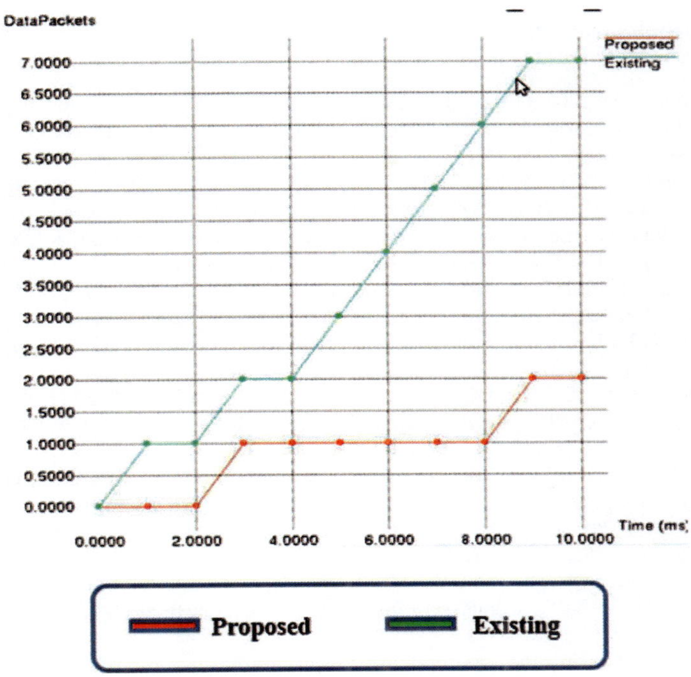

Fig. 9 Packet loss ratio

6 Conclusion

In this survey, a network containing real-world vehicle mobility and a traffic light control system is included in the simulation framework. It was hypothesized that a junction with two large intersections would reduce both the energy consumed by each traffic node and the amount of time it required for vehicles to go through the traffic when compared to the current approach. It is designed as a simulation that can be used to research peak hours and come up with solid results. Energy usage, end-to-end delay, and packet loss ratio all experienced significant decreases.

References

1. Mohammed AA (2022) Designing a control system for traffic lights by VANET protocol. Int J Nonlinear Anal Appl 13(1):1659–1666. https://doi.org/10.22075/ijnaa.2022.5781
2. Karunakar P, Matta J, Singh RP, Kumar O (2020) Analysis of position based routing Vanet protocols using Ns2 simulator. Int J Innov Technol Explor Eng (IJITEE) 9(5):1105–1109
3. AL-Kharasani NM, Zuriati AZ, Shamala K, Zurina MH (2020) An adaptive relay selection scheme for enhancing network stability in VANETs. IEEE Access 8:128757–128765
4. Chen W, Guha RK, Known TJ, Lee J, Hsu YY (2011) A survey and challenges in routing and data dissemination in vehicular Ad Hoc networks. Wirel Commun Mob Comput 11(7):787–795
5. Zhu M, Cao J, Pang D, He Z, Xu M (2015) SDN-based routing for efficient message propagation in VANET. In: International conference on wireless algorithms, systems and applications, pp 788–797
6. Ucar S, Ergen SC, Ozkasap OR (2016) Multichip-cluster-based IEEE 802.11p and LTE hybrid architecture for VANET safety message dissemination. IEEE Trans Veh Technol 65(4):2621–2636
7. Tomar I, Sreedevi I, Pandey N (2022) State-of-art review of traffic light synchronization for intelligent vehicles: current status, challenges, and emerging trends. Electronics 11:465
8. Rashid SA, Audah L, Hamdi MA, Abood MS, Alani S (2020) Reliable and efficient data dissemination scheme in VANET: a review. Int J Electr Comput Eng (IJECE) 10(6)
9. Mohanty A, Mahapatra S, Urmila B (2019) Traffic congestion detection in a city using clustering techniques in VANETs. Indonesian J Electr Eng Comput Sci 13:884–891. https://doi.org/10.11591/ijeecs.v13.i3.pp884-891
10. Pandey AK (2013) Simulation of traffic movement in VANET using Sumo. Unpublished Thesis, National Institute of Technology, Rourkela

Chapter 5
Crop-Weed Detection, Depth Estimation and Disease Diagnosis Using YOLO and Darknet for Agribot: A Precision Farming Robot

Medha Wyawahare, Jyoti Madake, Agnibha Sarkar, Anish Parkhe, Archis Khuspe, and Tejas Gaikwad

1 Introduction

Precision agriculture has developed as a critical strategy for boosting agricultural output, optimising resources, and lowering production costs. Crop management is critical to the livelihoods of millions of people in India, where agriculture is the backbone of the economy. Precision agriculture depends on modern technology and data analytics to make better educated crop management decisions. Precision agriculture relies on accurate weed detection to decrease crop damage, optimise pesticide use, and maximise total crop output. Convolutional neural networks (CNNs) have demonstrated promising results in weed detection. The YOLO model, a deep learning-based

M. Wyawahare · J. Madake · A. Sarkar (✉) · A. Parkhe · A. Khuspe · T. Gaikwad
Department of Electronics and Telecommunication, Vishwakarma Institute of Technology, Pune, India
e-mail: agnibha10@gmail.com

M. Wyawahare
e-mail: medha.wyawahare@vit.edu

J. Madake
e-mail: jyoti.madake@vit.edu

A. Parkhe
e-mail: anish.parkhe20@vit.edu

A. Khuspe
e-mail: archis.khuspe20@vit.edu

T. Gaikwad
e-mail: tejas.gaikwad20@vit.edu

A. Yadav et al. (eds.), *Proceedings of International Conference on Paradigms of Communication, Computing and Data Analytics*, Algorithms for Intelligent Systems,
https://doi.org/10.1007/978-981-99-4626-6_5

object identification approach, in particular, was discovered to be computationally efficient and well-suited to real-time weed detection in big agricultural fields.

Furthermore, estimating the distance of each weed/crop plant from the camera allows for more focused herbicide treatments, resulting in additional cost savings. Additionally, the development of automated plant disease detection systems has the potential to significantly increase agricultural output. These systems can swiftly and correctly diagnose illnesses on plant leaves, stems, and fruits by utilising computer vision and artificial neural networks. Because the leaves are the most vulnerable component of the plant, they are frequently the first to exhibit indications of illness, making them an ideal location for diagnosis. The YOLO model for weed detection, monocular distance estimate, and the construction of a plant disease detection application are the emphasis of this study. The goal is to investigate the potential of these technologies to boost crop output and lower production costs in India's agriculture industry. All of these technologies are combined into the Agribot, an autonomous precision farming robot created and built utilising interdisciplinary methodologies and cutting-edge technology such as computer vision, artificial neural networks, and design simulation tools. The Agribot performs practical and analytical activities such as soil health monitoring, weed detection and eradication, disease diagnosis, and pesticide optimization, with the objective of boosting crop output while lowering agricultural environmental impact by minimising the demand for herbicides.

2 Literature Survey

An improved YOLOV3 model called YOLOV3-dense was developed to recognise apples in orchards with changing lighting conditions and diverse backgrounds. The model uses the DenseNet approach and outperformed other fruit detection models with an average detection time of 0.304 s per frame [1]. Another study used YOLOV3 to recognise animals in images using the Darknet method, with performance depending on various training and testing photos. The objective is to create an animal recognition process that will output the animal's name using the YOLOV3 model [2]. Monocular depth estimation, a method of inferring depth from a single picture, has received significant attention due to the development of deep neural networks. This study examined current deep learning-based approaches for depth estimation, including supervised, unsupervised, and semi-supervised learning, as well as network architectures and loss functions. Depth estimation accuracy may be increased by employing various network architectures, loss functions, and training procedures [3]. A novel approach to absolute depth estimation was proposed based on the entire scene structure instead of individual items. This approach utilises structure recognition in the scene to provide absolute depth information that can simplify object recognition and aid in scene recognition [4]. An image classification model using convolutional neural networks (CNNs) was developed to identify plant diseases with high accuracy using the Plant Village data set. The model was built using the Keras deep learning framework, adjusting parameters such as the number of epochs and modifying dropout and rectified linear unit (ReLU) functions. Using the Plant Village data set, the model obtained a high accuracy of 99.89% in classifying 38 types of plant illnesses [5]. The paper evaluates the effectiveness of various machine learning and deep learning methods in identifying citrus plant disease. Deep learning approaches were found to outperform machine learning methods in the detection of citrus plant diseases, with VGG-16 providing the best results. The study suggests the integration of IoT, cloud computing, and big data technologies to improve the system and highlights the potential use of fuzzy logic and bio-inspired methods for increased accuracy [6]. A review of current deep learning-based research in identifying and categorising weed species in crops found that supervised learning approaches can achieve excellent accuracy when sufficient labelled data is available, but computing speed remains a significant barrier to implementation [7]. Weed detection and classification are crucial for precise herbicide application in agriculture. This article examines new research on weed detection utilising classic machine learning and deep learning approaches in computer vision technologies. The study analysed associated public data sets and weeding machines, discussed the limitations and difficulties of current weed identification systems, and predicted future research trends [8].

3 Methodology

The goal of this research was to create a weed identification system that can reliably discriminate between crops and weeds in digital photos. In the current study, a technique involving six major phases was used to build the weed identification system. As shown in Fig. 1, these procedures included gathering input data, pre-processing, segmentation, feature extraction, feature selection, and classification. The initial stage was to gather a data set of input photos from diverse sources, such as digital cameras, mobile phones, and the Internet. The data set, which included both crops and weeds, was big and balanced enough to capture a wide range of climatic circumstances as well as differences in crop and plant kinds. Following image collection, the data set was pre-processed to ensure uniformity in size, format, and quality. This included, among other things, image scaling, normalisation, colour correction, and noise reduction. The purpose of this phase was to ensure that the input data was in a consistent format, as most machine learning algorithms need. When images have varied dimensions, they might be difficult to comprehend and analyse, resulting in differences in the model's performance. The next phase was segmentation, which included distinguishing between crops and weeds in the input picture. This was accomplished by annotating the photos with bounding boxes using the freeware 'LabelImg.' The annotations were saved as text files and contained the image's name, the bounding box centre coordinates, and the size of each bounding box. This was essential for YOLO object detection and tracking, and each text file had the same name as the image it accompanied. The last phase was feature extraction, which entailed finding weed patches in the photos and extracting characteristics that would be used to categorise the weeds. The identification of important and informative elements that were critical

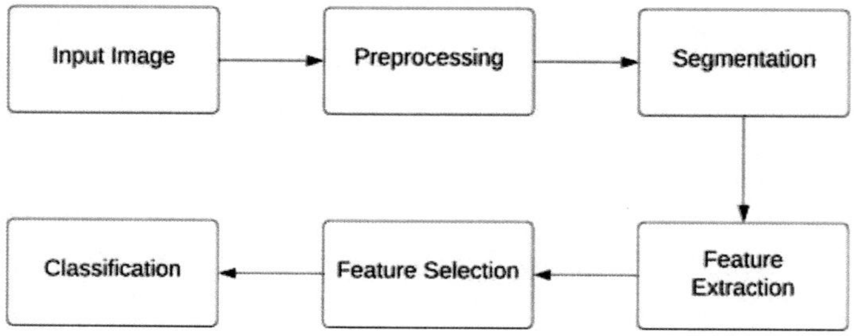

Fig. 1 Weed detection system steps using YOLO

for the categorization task came next in the procedure. This phase was required to eliminate characteristics that were redundant or unnecessary and might cause noise in the model or overfitting. The next phase was classification, which employed the specified attributes to categorise the plant photos as weeds or crops. This phase was critical in allowing the algorithm to discriminate between the two sorts of plants. All of the preceding phases were carried out using the YOLO model architecture. This design enabled the identification and categorization of weeds in photos to be efficient and accurate.

3.1 Agribot

The Agribot is a smart and creative autonomous robot that was created to aid with precision farming operations in agricultural regions. All-Terrain Wheels, a solid frame and chassis, steering mechanisms, shock absorbers, and a protective body cover are among the specialised components that allow it to easily negotiate off-road terrain and muddy locations. Furthermore, the robot is outfitted with DC motors and servos, allowing it to do duties such as spraying fertiliser and water and moving a delta arm mechanism with three degrees of freedom.

The Agribot was created using a multidisciplinary approach that used cutting-edge technology including computer vision, artificial neural networks, and design simulation tools. As a result, it may carry out a variety of practical and analytical activities, such as soil health monitoring, weed detection and removal, disease diagnosis, and pesticide optimisation. The Agribot's ultimate purpose is to boost crop productivity while lowering the environmental effect of agricultural activities by eliminating the demand for pesticides. This will not only assist farmers, but will also contribute to agriculture's long-term growth by supporting ecologically friendly and effective agricultural practices.

3.2 *Extraction of Features*

During this step, it is vital to remove elements connected to a specific region of interest that are important in understanding the meaning of an image, such as colour, form, and texture. Recently, researchers have shown an interest in using textural clues to identify plant illnesses. To construct the system, several feature extraction approaches such as the spatial grey-level dependence matrix, histogram-based feature extraction, and colour co-occurrence method can be used.

3.3 *Camera*

The project used the Logitech C270 camera to capture photos of crops and weeds. The camera's image sensor and lens were designed to capture high-resolution still images and videos. Using the built-in bracket, the camera was mounted at a suitable height and angle to ensure that the entire field was in the frame of the Agribot.

The images captured by the camera were then processed using computer vision algorithms to detect the presence of weeds, estimate the depth from the camera to the weeds, and diagnose the disease of the crops.

Technical specifications of the Logitech C270 camera:

- Image Sensor: 1/5" VGA CMOS
- Video Resolution: Up to 720p at 30 frames per second
- Still Image Resolution: Up to 3.0 megapixels (software enhanced)
- Lens Type: Fixed focus lens
- Field of View: 60-degree diagonal
- Microphone: Built-in microphone with noise-cancelling technology
- Connectivity: USB.

3.4 *YOLO*

3.4.1 Weed Detection Using YOLO

The You Only Look Once (YOLO) neural network is a sophisticated object identification method that can detect the bounding boxes and class probabilities of objects in an image in a single step. It has grown in prominence in the recent years because to its excellent performance in a variety of sectors, including transportation, animal identification, and monitoring moving objects. YOLO's first version, launched in 2016, included 24 convolutional layers for feature extraction and two dense layers for predictions. The Darknet-53 architecture-based YOLOv7 model features significant enhancements and feature extraction layers. The model is trained on crop photos using several approaches to recognise the crop, as shown in Fig. 2. Once trained, an

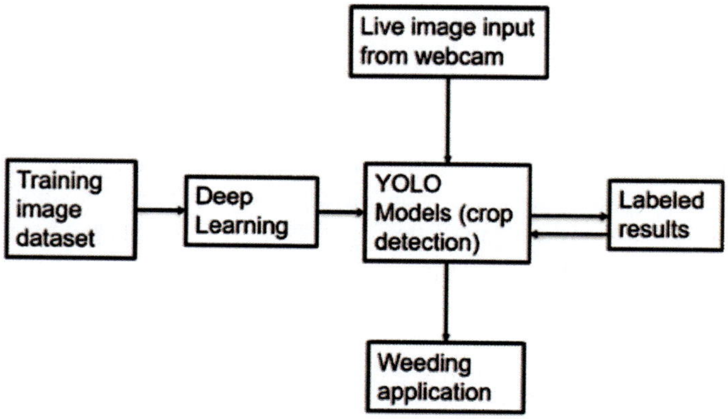

Fig. 2 Depicting the design and operation of the YOLO model

algorithm uses the model's bounding box coordinates to recognise crop samples from the input picture, as shown in Fig. 3. The YOLO model analyses incoming photos in real-time during the inference phase, creating bounding boxes and associated probabilities for each item spotted. To guarantee accurate object recognition, the YOLO model's bounding boxes are filtered and refined using a technique known as non-maximum suppression (NMS). This approach removes duplicate or overlapping detections and keeps just the most likely and unique items, improving the overall accuracy and reliability of the object detection process. Using the YOLO model architecture for real-time weed/crop identification entails gathering live video footage of crops and weeds with a camera or image-capturing device. The YOLO model is used to analyse the video frames, which provides bounding boxes around any weeds or crops detected in the picture. This approach enables farmers to identify and locate weeds in real time, allowing them to take early action to eliminate the weeds and prevent crop loss. It also allows farmers to optimise the use of pesticides and other chemicals by focusing on weed-infested areas.

3.4.2 Distance Estimation Using YOLO

Many visual analytics solutions rely on object depth estimates. Significant progress has been achieved in this field in the recent years, but robust, efficient, and exact depth estimation in real-time video remains a difficulty. Using a single camera, the method calculates the distance between weeds (monocular distance estimation).

Figure 4 shows how YOLO builds a boundary box around the item and determines a proportionate inversion correlation between the distance and the boundary box's dimensions (height, breadth). YOLO also aids in obtaining the exact equation between the variables under consideration, the dependent variables such as distance, the independent variable, and the height and breadth of the border box.

Fig. 3 Leaf affected by a fungal infection

Fig. 4 Actual camera view of the Agribot [bounding box shows the predicted class (*weed*) and estimated depth (*distance*)]

3.4.3 Darknet

Darknet is a convolutional neural network that underpins the YOLO object detection method. It is an open-source neural network framework written in C and Compute Unified Device Architecture (CUDA). It is simple to set up and supports both CPU and GPU computation. Darknet is part of the depth estimation system and works with the camera attached to the bottom of the Agribot (at a distance of 22 inches from the ground). Crops and weeds were not present in the YOLO model's pre-trained classes. The model was created using Darknet and a pre-trained tiny YOLO model containing around 1400 photos of weeds and crops. For distance estimate, the best weights with the lowest training loss were chosen.

A reference image was taken based on the object's distance from the camera, which was calculated manually. It was then assigned as a constant 'KNOWN_WIDTH.' The width of the weed image in real time was approximated and set using a 'WEED_WIDTH' constant. These values were necessary for determining the camera's focal length used for which a function was implemented. With the help of these reference images and the trained weights, the model could differentiate between crop plants and weeds and simultaneously display their distances from the camera in inches. For detection purposes, the OpenCV method of cv2.dnn_DetectionModel() was implemented. It is a class representing high-level API for object detection networks. It enables setting the parameters for pre-processing the input image and creates a net from the file with trained weights and config (which was set up and configured beforehand based on the changes necessary in tiny YOLO's configuration file). This class configures the pre-processing input, does the forward pass, and delivers the detection results. In addition to SSD and Faster R-CNN, it supports the YOLO topology.

3.5 *Plant Disease Identification*

3.5.1 Introduction

Visible symptoms and indicators of plant illnesses produced by infectious organisms such as fungus, bacteria, and viruses can be observed. These signs may be noticed and identified visually, allowing for early diagnosis and treatment of damaged plants. Fungi are the main culprits; they are single-celled or multicellular organisms that tear down plant tissues and create visible spores, mildew, and mould. Blotches, yellowing, and spots can be seen on the leaves of infected plants, as shown in Fig. 5a, and the fungus can be seen as a growth or mould on the leaves.

Because bacteria are so minute, identifying bacterial diseases in plants can be difficult. However, observable markers include deformities on the underside of stems or leaves, as well as water-soaked sores with bacterial slime. The wet regions on the leaves in Fig. 5b are typical indicators of bacterial infections, which vary from fungal spots in that they are typically surrounded by leaf veins.

a

b

Fig. 5 **a** Fungal infection has infected a leaf. **b** Bacteria have infected the leaf

Even though viruses are too small to be seen under a light microscope, they may penetrate host cells and multiply. Plant infections produced by viruses are not often obvious, but some symptoms, such as discoloured or wrinkled leaves and a pattern of mosaic-like patches on leaves called after the tobacco mosaic virus, can be seen by a trained observer.

3.5.2 Data Set

To identify crop species and diseases, a system is proposed that utilises convolutional neural network techniques and models based on the Plant Village data set. This data set contains photogs of both healthy and diseased leaves from 14 different crop species such as apple, blueberry, and tomato. The images demonstrate 17 distinct diseases caused by various agents, such as bacteria, oomycete mould, viruses, and mites. Additionally, the data set features images of healthy leaves for 12 crop species. The photos in the data set offer a wide variety of plant textures and diseases, including apple scab, apple black rot, cherry powdery mildew, and many more.

3.6 Web Application

A web interface for identifying plant diseases has been developed using the Plant Village data set. The web app is built using Python programming language and utilises various libraries, including TensorFlow, Flask, Gevent for epoch generation, and Scikit-Learn for performing image pre-processing on the data set. An interface similar to the one shown in Fig. 6 is created to enable users to identify plant illnesses.

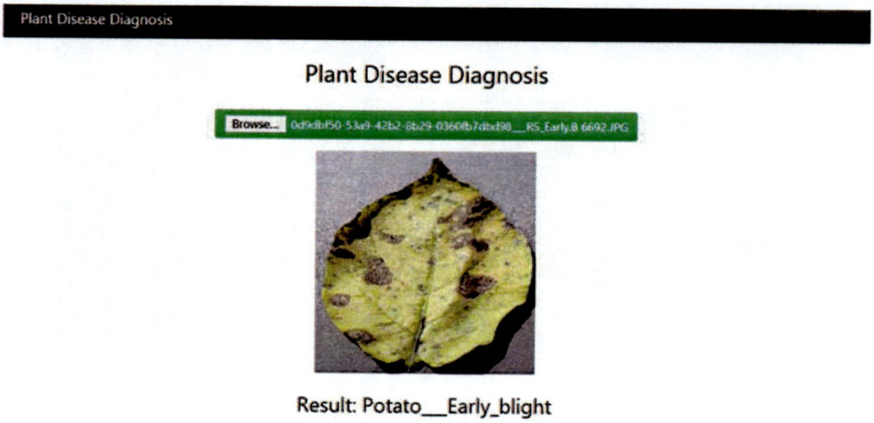

Fig. 6 Hosted web app output using flask API

4 Results

This work involved developing a weed-detecting and depth-estimating programme and a disease-detecting flask application which would add to the functionality and feature set of the Agribot. A model was developed using the cutting-edge YOLO neural network to recognise and discriminate between the crop and the weed using bounding box coordinates.

The confusion matrix in Fig. 7a shows that the model accurately categorised 86% of the crops and 86% of the weeds. The value of 0.86 for both predicted crops and predicted weeds in the confusion matrix can be interpreted as a good result, as the data set is balanced.

In Fig. 7b, the F1-confidence curve rises initially and then drops after a confidence score of 0.6. This indicates that as the model's confidence score for an item grows, so does the model's F1-score. However, the F1-score begins to fall to a confidence score of 0.6. The model overall classes' average F1-score (in this case, crops and weeds) is 0.82, with a confidence score of 0.421.

Figure 7c shows that crops and weeds have high precision values, indicating that the classifier is generating accurate predictions for these classes with a low amount of erroneous positive or false negative predictions. This signifies that the classifier properly identifies crops and weeds with high accuracy. When the confidence score threshold is set to 0.921, the average accuracy of all classes (in this example, crops and weeds) is equal to 1.00. This is a desired consequence in many applications, including agricultural weed detection. Predictions with a confidence score greater than or equal to 0.921 are regarded as optimistic, while predictions with a lower confidence level are considered negative.

The crop class has a precision of 0.892, whereas the weed class has a precision of 0.875, as seen in Fig. 7d. This shows that the algorithm detects crops more accurately than weeds, with fewer false positive detections among the crops recognised. The

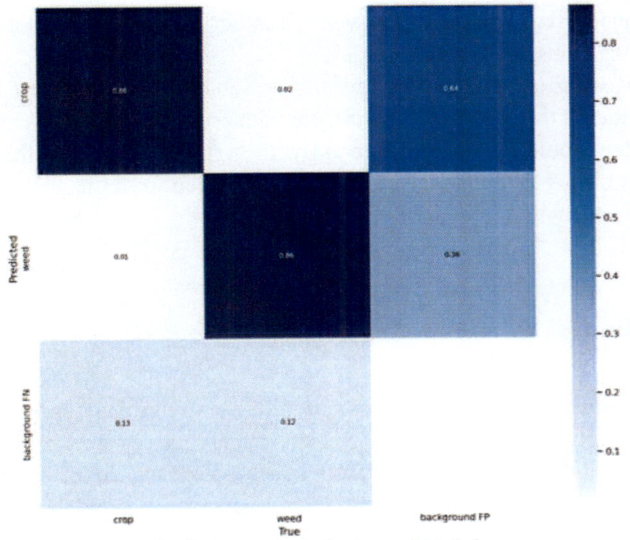

a. Confusion matrix for the two projected classes.

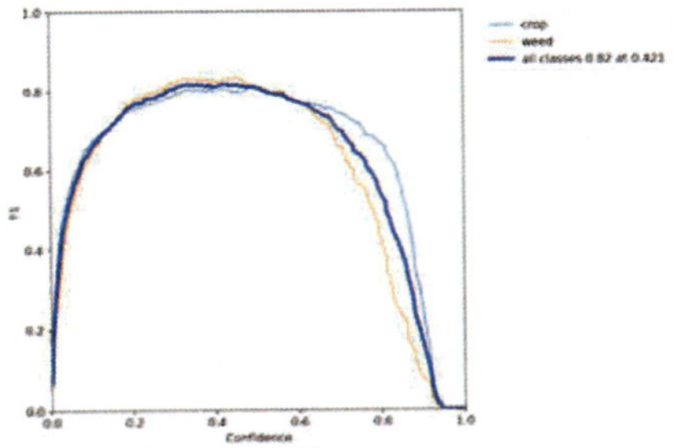

b. F1-score - Confidence curve

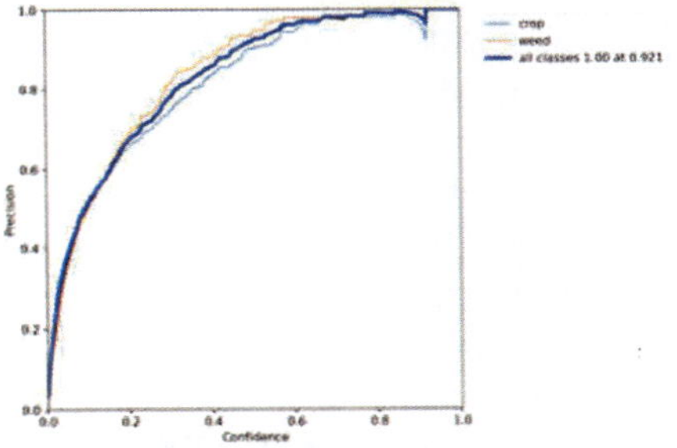

c. Precision - Confidence curve

d. Precision-Recall curve

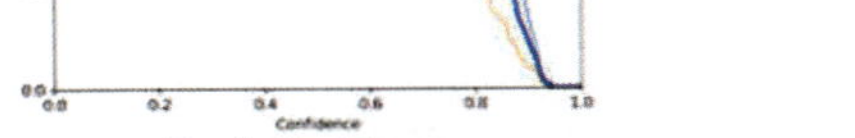

e. Recall - Confidence curve

Fig. 7 **a** Confusion matrix for the two projected classes. **b** F1-score-confidence curve. **c** Precision-confidence curve. **d** Precision-Recall curve. **e** Recall-confidence curve precision ratings for both classes, however, are rather high, showing that the model performs well in terms of overall accuracy.

Figure 7e is used to see a trade-off between recall and confidence, where increasing recall often comes at the cost of decreasing confidence and vice versa. A good object detection model should balance high recall with high confidence so that as many objects as possible are detected accurately, as can be seen from the results obtained, with both classes having maintained the balance.

5 Conclusion

Weed identification is crucial to precision agriculture since the development of weeds in fields can significantly affect crop productivity. Effective weed prevention and management depend on accurate and effective weed detection. In the recent years, it has been demonstrated that convolutional neural networks (CNNs) are an effective tool for item detection in a range of industries, including agriculture. The You Only Look Once (YOLO) model, a well-known real-time item detection technique, was put into use to identify weeds in agriculture fields according to research. The YOLO model is ideally suited for this job since it can carry out object identification in a single forward propagation through a neural network and deliver findings instantly. Additionally, by employing a monocular approach, the focus is on figuring out how far away each weed or crop plant is, which requires using reference photos and the focal length of the camera to get precise findings. In order to identify the sort of disease present, a flask application was created and trained on a data set of healthy and damaged plant leaves. The results demonstrate that CNNs are capable of precisely and effectively identifying weeds in crop fields, improving precision agriculture. These three tasks, weed detection, distance estimation for weed removal, and detecting the type of plant disease, are beneficial to farmers and agriculturists and, when combined with their implementation in the Agribot can be very fruitful, as farmers will no longer have to spend much time and money on such tasks, which were previously very cumbersome and expensive. There is presently no such implementation on the market that combines the three applications, delivering an innovative approach to all of these difficulties at a low cost.

References

1. Tian Y, Yang G, Wang Z, Wang H, Li E, Liang Z (2019) Apple detection during different growth stages in orchards using the improved YOLO-V3 model. Comput Electron Agric 157:417–426
2. Karthikeya Reddy B, Bano S, Greeshmanth Reddy G, Kommineni R, Yaswanth Reddy P (2021) Convolutional network-based animal recognition using YOLO and Darknet. In: 2021 6th International conference on inventive computation technologies (ICICT). IEEE, pp 1198–1203
3. Zhao C, Sun Q, Zhang C, Tang Y, Qian F (2020) Monocular depth estimation based on deep learning: an overview. Sci China Technol Sci 63(9):1612–1627

4. Torralba A, Oliva A (2002) Depth estimation from image structure. IEEE Trans Pattern Anal Mach Intell 24(9):1226–1238
5. Appalanaidu MV, Kumaravelan G (2021) Plant leaf disease detection and classification using machine learning approaches a review. In: Innovations in computer science and engineering: Proceedings of 8th ICICSE, pp 515–525
6. Sujatha R, Chatterjee JM, Jhanjhi NZ, Brohi SN (2021) Performance of deep learning vs machine learning in plant leaf disease detection. Microprocess Microsyst 80:103615
7. Mahmudul Hasan ASM, Shoal F, Diepeveen D, Laga H, Jones MGK (2021) A survey of deep learning techniques for weed detection from images. Comput Electron Agric 184:106067
8. Wu Z, Chen Y, Zhao B, Kang X, Ding Y (2021) Review of weed detection methods based on computer vision. Sensors 21(11):3647

Chapter 6
Audio Classification of Emergency Vehicle Sirens Using Recurrent Neural Network Architectures

Arya Shah, Amanpreet Singh, and Artika Singh

1 Introduction

Emergency vehicle sirens are a vital component of the public safety system, serving as a warning signal to the general public to make way for emergency vehicles. With the increasing number of vehicles on the roads, it is essential to have a system that can quickly and accurately identify the approach of emergency vehicles. The emergency vehicle sirens serve as a means of communication between the emergency vehicle and the general public, allowing for efficient and safe navigation of the vehicle through traffic. The unique sound of each type of emergency vehicle siren is important for identifying the type of vehicle, which can have a significant impact on the public's response. The traditional method of recognizing the type of emergency vehicle based on the sound of the siren is subjective and prone to error. With advancements in technology, it has become possible to develop automated systems for recognizing the type of emergency vehicle based on the siren sound.

Recently, techniques in deep learning such as RNN architectures have demonstrated high success in tasks such as classification of audio signals. RNNs are well-suited for audio classification problems as they are designed to process sequential data, making them well-suited to modelling the temporal dependencies in audio signals. In this paper, we propose an RNN-based model for the classification of different types of emergency vehicle sirens, including ambulance, firetruck, and police car sirens, where the Mel-frequency cepstral coefficients (MFCCs) are

A. Shah (✉) · A. Singh · A. Singh
NMIMS University, Mumbai, India
e-mail: arya.shah82@nmims.edu.in

A. Singh
e-mail: amanpreet.singh60@nmims.edu.in

A. Singh
e-mail: artika.singh@nmims.edu

A. Yadav et al. (eds.), *Proceedings of International Conference on Paradigms of Communication, Computing and Data Analytics*, Algorithms for Intelligent Systems,
https://doi.org/10.1007/978-981-99-4626-6_6

extracted and used as input for neural network models. The model is trained and evaluated on sireNNet [1], a data set proposed by the authors which consists of four classes, namely ambulance, police, firetruck, and a control class of traffic noise which jointly constitutes the data set of emergency vehicle siren sounds, and the results show that the proposed model of gated recurrent units (GRU) architecture is capable of achieving high accuracy in identifying the type of emergency vehicle.

The contribution of this work is twofold. Firstly, we demonstrate the potential of RNN architectures as compared to CNN architecture in solving real-world audio classification problems, such as identifying different types of emergency vehicle sirens. Secondly, this work can contribute to the development of real-time systems for alerting road users of the approach of emergency vehicles, ultimately reducing the number of accidents caused by vehicle collisions.

2 Literature Review

This section aims at providing a concise literature review of the existing research conducted in the field of audio feature extraction and classification done using various methods proposed by the corresponding researchers. The techniques used and the applications for the same cover a broad area in terms of applications and algorithms used.

Lezhenin et al. [2] use spectro-temporal patterns to classify environmental sound, which is unstructured. Since LSTM nets understand temporal relationships well, they are suggested. Mel-spectrograms from Urbansound8k are used for model training and fivefold cross-validation. A baseline CNN for the LSTM outperforms existing methods and the baseline network for comparison. The baseline CNN had 80.48% accuracy, whereas the suggested LSTM network had 84.25%. The paper's technique did not outperform the GoogleNet network using Mel-spectrogram, MFCC, and CRP pictures as input, but just Mel-spectrogram features did outperform.

Another study provides a comparative study on breathing sounds classification methods using 3 deep learning architectures—VGG16, ResNet-50, and GoogleNet. Zakaria et al. [3] claim that as per traditional methods of feature extraction and pattern classification, their proposed methodology includes utilization of cycle-based breathing sounds of three types, namely crackle, normal, and wheeze which are processed in order to create gammatonegram images which worked as inputs for the neural networks discussed in the paper. Out of the 3 network architectures, GoogleNet was the top performing model with an improved accuracy of 63.69% as compared to VGG16 and ResNet-50. The ICBHI challenge data set was utilized in the paper for training and validation.

Das et al. [4] proposed that LSTM-based approach for sound classification shows better results as compared to previously known approaches of CNN, ANN, R-CNN, and other popular machine learning techniques. The usage of data augmentation and stacking of various spectral features is also highlighted in the paper whose

LSTM model provides an overall accuracy of 98.81% said to be the state-of-the-art performance on the popular sound classification data set of UrbanSound8k. The selected important features for the task of sound classification include Mel-spectrogram, Tonnetz, Chromagram, Spectral Contrast, MFCC, Chroma CENS, i.e. Chroma Energy Normalized and Chroma Constant-Q Transform. The state-of-the-art results in the feature were achieved through stacking the two features of Mel-frequency cepstral-coefficients and Chroma Short-Term Fourier Transform which was then used in the LSTM network.

Zhao and Yin [5] mention a major issue of overfitting in environment sound classification: the varied position between the sound source and physical medium and the interference of other sounds may cause sound effects to overlap and generate confusion in complicated environmental sounds. During neural network training, a specific voice may produce unacceptable prediction results on unknown data. White noise (Gaussian type) and signal-to-noise ratio (SNR noise) are added to the audio data set to tackle the network generalization problem. Evaluation trials were run using Urbansound8K. It was observed that white noise (Gaussian) with a modest weight factor enhances environmental sound and high SNR noise improves model generalization and reduces reverb and other undesired noises in the data set.

Kumar and Chaturvedi [6] proposes a Squeeze and Excitation Block-Based Neural Network (SENet)-based CNN approach to classify audio using the MATLAB tool. Out of the various features of sound signals such as chroma, centroid-based, wavelet analysis, and MFCC features were extracted which were then fed to the proposed fast efficient neural network which gave an accuracy of 89.2%.

Environmental sound classification is non-stationary, according to Zhao and Yin [7] CNN can classify such sounds, however data singularity and unequal sample length cause overfitting and poor accuracy. The authors offer waveform stretching, pitch shifting, and audio augmentation to overcome the unequal audio duration problem and prevent over fitting. An improved LeNet network classifies log Mel-features. The authors' technique achieves 95.65% accuracy on Urbansound8K. The waveform stretching method fills up audio data in less than 4 s, while pitch shifting data augmentation boosts data availability and accuracy utilizing few parameters.

It has been observed that classifying accuracy of sound has a powerful relation with extraction of features. Demir et al. [8] extract deep features using a fully connected layer of the proposed CNN which is trained on spectrogram images. The feature vector is then conveyed as input to a random subspace of an ensemble of KNN classifiers for testing performance. The DCASE competition's 2017 edition of data set for Acoustic Scene Classification (ASC) and Urbansound8k has been used upon which the proposed CNN scores an accuracy of 96.23% and 86.70%, respectively.

The research discussed identifying and notifying ambulance siren sounds in a noisy setting. Pramanick et al. [9] suggest a simpler yet effective architecture for siren and other urban sound detection. Scalogram, Fourier decomposition, and Mel-spectogram sound-to-image transformation methods have been discussed. The pre-processing and augmentation effects on the data set are investigated, and the Urban-sound8k data set is used for testing and assessment. The proposed technique classifies urban sounds with 88.66% accuracy and detects ambulance sirens with 99.35% accuracy. The authors' CNN with Mel-spectrogram feature outperformed GoogleNet, ALexNet, and VGG16.

Niranjan et al. [10] aim to improve the traditional method of using deep neural network on audio data. The paper implemented dense convolutional neural network for the processing of audio signals. The data set was categorized into 5 major categories and the further processed. The paper uses ensemble architecture which includes two CNN models using Mel-spectrogram as input. This model is then again combined with another model that uses MFCC spectrogram. The author shows comparison between ensemble and multi-model network approach. Both the model had 92% accuracy but the author concludes that multi-model approach will be advantageous based on its ability to handle two different inputs.

Hirata et al. [11] implemented CNN using slice bi-spectrogram. The CNN had 5 convolutional layers, 3 pooling layers, and a fully concentrated layer. ReLu activation function has been used. The model uses UrbanSound8k data set consisting of 10 classes. The model is implemented on MATLAB using power spectrogram (PS) and slice spectrogram (SBS). SBS showed higher accuracy and further experiments were conducted in order to verify the proposed method.

Lu et al. [12] implemented CNN using transfer learning for environmental sound classification. In this paper, environmental sounds are depicted as RGB images, where the red-channel corresponds to log Mel-spectrogram (LMS). The green-channel corresponds to the feature of scalogram, while the blue-channel corresponds to MFCCs. All the three scalogram, LMS, and MFCCs features were used as input features of the model. Cross-validation with the number of folds set to 5 is used in order to evaluate the performance.

Singh et al. [13] proposed CRNN to train the model. The second layer was proposed in order to handle unlabelled data set using mean teacher algorithm. The first step was to extract log Mel-spectrogram followed by data augmentation. The CRNN model is then applied with a mean teacher algorithm which helps to deal with unlabelled data. The author compares different approaches for training the model and then further compares their f1 score to find the best one. These approaches are tested in 10 classes.

Wu [14] divides each audio segment into silence, pure voice, non-pure voice, and background voice. The paper employed decision tree classifiers for feature selection and SVM approaches to improve classification. The Adaboost technique trained many classifiers for the same training set and aggregated the weak classifiers to generate a stronger final classifier. This research introduces a unique integrated learning method, Binary Tree Structure and SVM (BTSS), based on binary tree structure and model double perturbation. SVM Integration Algorithms were employed to improve classifier diversity. Varied kernel parameters and penalty factors disturb the model, and different binary tree architectures enhance classifier differences. The article showed that the modified approach outperforms single SVM in music and background tone classification.

Evangelista et al. [15] presented the ensemble decision tree approach to use the PhysioNet database to quickly classify the heart sounds. The paper contrasts the Gini

Gain list's overall classification accuracy results with those from the features based on the G.I. list. An ensemble decision tree approach was employed to process the features because of its effectiveness in binary classification when combined with a MATLAB software for classification. Both an 80% train-set split and a 20% test-set split with a k-fold cross-validation of 5 and a 90% train-set split and 10% test-set split with a k-fold cross-validation of 10 were used in the cross-validation.

Mkrtchian and Furletov [16] first compares the different audio data sets available in terms of their source, number of classes, and size. In order to increase accuracy, LSTM was employed in the paper's application and analysis of convolutional neural network models on audio and sound data sets. Fivefold (ESC-10 and ESC-50) and tenfold (ESC-100) data sets were each taken (UrbanSound8K). In order to assess the accuracy amongst them, block diagrams were drawn. The proposed neural network architectures were evaluated using fivefold cross-validation and compared to a baseline CNN.

Ahmed et al. [17] presented sequential feature selection to minimize MFCC-extracted feature dimension. The collection contains sounds of rain, wind, human gait, and passing vehicles. Recurrent neural networks (RNNs) were trained on sound occurrences using selected features. The study compares the proposed method to an RNN trained with Mel-frequency band (MFB) features and a CNN trained with Mel-frequency band (MFCC) features. Mel-frequency cepstral coefficients (MFCC) were used to derive audio characteristics based on the perceptual Mel scale. Classifier performance determines feature addition or removal. Sequential feature selection is subdivided into four distinct algorithms: sequential forward and backward selection and sequential forward floating and backward floating selection. The most accurate algorithm was MFCC-SFFS-RNN.

Paissan et al. [18] offer a system in which urban sound detection is processed by embedded IoT devices in order to identify occurrences useful to law enforcement and municipalities. Their innovative neural network design is based on PhiNets, which help in the detection of real-time auditory events utilizing microcontroller units. The architecture delivers cutting-edge performance on the UrbanSound8k data set and is capable of operating on both waveforms and spectrograms. Using 101 M classification criteria, the authors attain an accuracy of 78%.

Vujošević and Đukanović [19] approach the problem of sound classification using the approach of image classification. Environmental noises are more difficult to define than acoustic or musical sounds, according to the authors. The sound files are represented by their visual representations, including Mel-spectrograms, tonal centroids, spectral contrast, and chromagrams. On top of these visual representations, a fully convoluted neural network (FCNN) is trained. Using tenfold cross-validation, the proposed method by the author gets an average accuracy score of 73%. The experimental study also depicts that there is a significant improvement in terms of margin when training a FCNN compared to a fully connected neural network that gave an accuracy score of only 59%, both trained and evaluated on the UrbanSound8k data set.

Multi-feature fusion is used to classify environmental sounds by Li et al. [20] GFCC chartered characteristics based on human auditory characteristics and short-term memory time and energy characteristics are fused to improve audio representation comprehensiveness and accuracy. This study retrieved GFCC characteristics and added short-term energy properties after each frame's GFCC using the same framing procedure. Linearly merging the two traits removes their order, decreasing classification accuracy. Normalizing GFCC and short-term energy eliminates scaling effects. Linearly overlay E-GFCC increases a single feature with gammatone filter sound information and time-domain feature information, improving sound event categorization and non-stationary sound signals. Further, sound feature correctness was compared.

Kroos et al. [21] sought nature, human, music, effects, and urban classifications. Convolutional neural networks created a baseline system. The baseline system employed log Mel-spectral coefficients, which are frequently employed in supervised learning using neural network classifiers. Using human-derived, high-level categories, the authors designed the "Making Sense of Sounds" acoustic data set and machine learning challenge to achieve high generalization in machine classification. A baseline system based on deep learning obtained 80.8% accuracy on the evaluation data set.

3 Methodology

To develop an effective ML-based diagnostic tool, a comparison of evaluation metrics of multiple ML-models applied to our data set is necessary. Because it provides the foundation for the investigation, the model preparation is the most significant part in the research process. Figure 1 depicts the phases needed in establishing an apt model and fine-tuning it for the best potential outcome.

3.1 Data Set

The baseline data set used was emergency vehicle siren sounds from Kaggle. The data set includes 600 labelled audio recordings extracted and saved as.wav audio files from websites such as Google and YouTube. These audio files are divided into

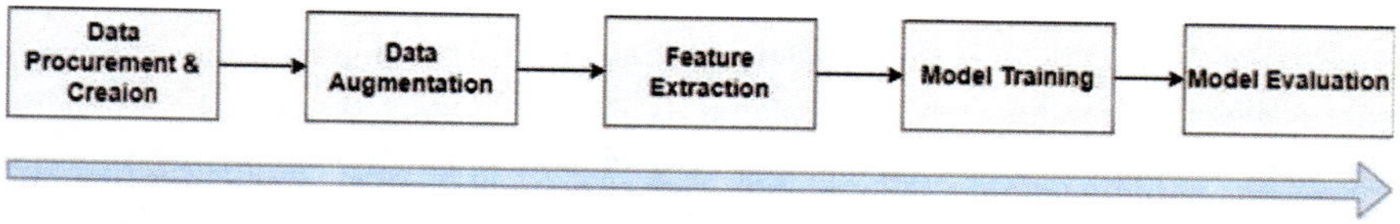

Fig. 1 General flow of steps in methodology

3 categories: ambulance, firetruck, and traffic. Traffic sounds are provided so that the model can be trained to handle background traffic noise in the real world. File length was uniformly set to 3 s. The audio recordings in the data set are sampled at 48 kHz. Each recording is 517 KB in size.

3.2 Data Augmentation

The proposed model classifies audio into three categories: ambulance, police, and firetruck. All three categories are important in order for the model to perform better. Since the original data set does not contain police siren audio files, it was downloaded and saved as a.wav file from YouTube before being converted into 200, 3 s audio files. We introduce superimposing traffic noise to the audio file as a part of augmentation of the data set. The traffic audio file used for superimposing was also downloaded as a.wav file from YouTube. The data set includes both original audio files as well as augmented audio files. The augmented data set was further normalized with a peak value of 100% using the simple peak level normalization method. The volume level was set to 6 dB. The final data set contains 1675 tagged audio recordings.

3.3 Feature Extraction

Based on the existing literature reviewed by us, it has been predominantly identified that Mel-frequency cepstral coefficients (MFCCs) have been widely used to give satisfying results as compared to other equivalent feature extraction methods [4]. To extract the features of the audio signals, we used the Mel-frequency cepstral coefficient representation. MFCC is a widely known and used representation for speech and audio signals, as it captures the spectral information of the audio signal in a compact form. In this study, we extracted 80 MFCC coefficients from each audio signal for the 4 classes present in the data set—ambulance, firetruck, police and traffic, which were then used as input features for the CNN and RNN models.

For the process of feature extraction, we make use of the open-source Python package librosa [22] which has been developed specifically for music and audio analysis. Within the librosa library of functions, we make us of the load function with the resample type set to "kaiser_fast," which is a method for resampling digital signals that uses a Kaiser window that helps to smooth the signal and reduce the amount of energy in the high frequency components while maintaining the quality of the signal. This technique is fast and efficient, making it a largely adopted feature extraction choice for resampling audio signals in real-time applications. The number of MFCC features extracted from the wave signal were set to 80. The complete features along with its corresponding class of audio signal were dumped into a pickle file that was later used as input to the model. The sample data frame after feature extraction and labelling with corresponding class can be observed in Fig. 2.

Fig. 2 Sample representation of extracted features

	feature	class
0	[-302.7111, 95.41589, -53.674664, -10.595546, ...	ambulance
1	[-312.88766, 96.170265, -46.613415, -13.018149...	ambulance
2	[1.4056735, 104.4055, -40.99026, 1.8715706, -2...	ambulance
3	[-170.2212, 117.14442, -37.209496, -5.6599894,...	ambulance
4	[-129.6423, 126.78272, -35.10049, -14.79055, -...	ambulance

3.4 Model Training

Before training the model on extracted set of input features, the complete extracted features data set was split into train and test data set individually. A comparison table depicting the size of training and testing data size before and after augmentation is given in Table 1. After splitting the data set into training and testing set, the training set is used as input for the three models—CNN, LSTM, and GRU for training purpose. The model training is done using Google Colab Notebooks and the TensorFlow [23] library along with Keras as backend is used for creating CNN and RNN networks—LSTM and GRU.

For the CNN model, we use two 1D convolutional layers of filter size 3 with kernel size as 13 and filter size 16 with kernel size 11, respectively. The optimizer used was Adam with the activation function as ReLu. The dropout rate was maintained at 0.3. Following which a 1D global max pooling layer along with a dense layer with softmax activation has been used. The metrics used for evaluation during model training is Accuracy and the loss function employed is binary_crossentropy. Training for 400 epochs with a batch size of 64 was set.

For the LSTM model, we use one LSTM layer with 28 hidden units, and the return condition has been set to "true." The dropout rate was maintained at 0.3. Following that, the model contains two time-distributed dense layers, where the first layer has a size of 256 and the second has a size of 512. In both layers, ReLU has been used as the activation function, and the optimizer used was Adam. After being used as input by the flattening layer, the output of the time-distributed dense layer is then sent on to the last layer, which is the dense layer. The evaluation metric during model training is accuracy, and the loss function considered is binary cross-entropy. The model trained for 200 epochs with a batch size of 64.

In the GRU model implementation, the GRU layer of size 128 along with the dropout rate of 0.5 is added. Following which we have introduced two dense layers

Table 1 Train/test split comparison of data sets

Train/test split = 80:20	Original data set	Augmented data set
Train	661	1340
Test	166	335
Total	827	1675

Table 2 Comparison of model accuracies

Model	Original data set		Augmented data set		Train-test ratio
	Train-set accuracy (%)	Test-set accuracy (%)	Train-set accuracy (%)	Test-set accuracy (%)	
CNN	100	97.59	99.85	92.84	80:20
LSTM	100	98.80	100	97.61	
GRU	100	98.19	99.70	98.81	

of size 64 with activation ReLu and the output layer set to the number of classes, i.e. 4 with the softmax activation function. The optimizer used is Adam, the loss function is maintained as binary_crossentropy and the model training metric used is that of accuracy. The GRU model trained for 1500 epochs with 64 set as the batch size.

3.5 *Model Evaluation*

Based on the 3 models trained CNN, LSTM, and GRU, the evaluation metrics are depicted in a tabular format in Table 2 below, with the comparison done between augmented and original set of data. The splitting of both the data sets was done in the ratio of 20% for testing and 80% for training.

4 Discussion and Results

Based on the model training performance and evaluation metrics, it has been observed that GRU model outperforms the LSTM and CNN model. The model summary, loss plot, and the confusion matrix generated have been depicted in Figs. 3, 4, and 5, respectively.

5 Conclusion and Future Work

In this paper, we propose an audio classification model for identifying different types of emergency vehicle sirens using CNN, GRUs, and LSTM architectures. Gated recurrent units (GRUs) perform better than other algorithms, with a training accuracy of 99.7% and a testing accuracy of 98.80%. These models are trained using extracted features based on MFCC, which showed better results in the previous research. The traffic sound was superimposed on all of the siren audio files, allowing the model to adapt to real-world scenarios with traffic noise in the background. This work can be useful in developing real-time systems for alerting road users of the

```
Layer (type)                 Output Shape              Param #
=================================================================
input_1 (InputLayer)         [(None, 1, 80)]           0

gru (GRU)                    (None, 128)               80640

dropout (Dropout)            (None, 128)               0

dense (Dense)                (None, 64)                8256

dense_1 (Dense)              (None, 4)                 260

=================================================================
Total params: 89,156
Trainable params: 89,156
Non-trainable params: 0
```

Fig. 3 Model summary of GRU

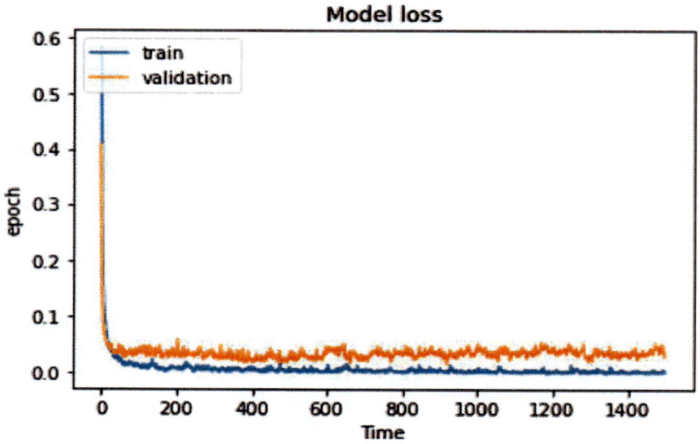

Fig. 4 Loss plot for GRU model

approach of emergency vehicles and can contribute to reducing accidents caused by vehicle collisions. For future work, since the data was augmented manually, there are numerous other methods, like a neural network and temporal stretching, that can be used to augment the data set.

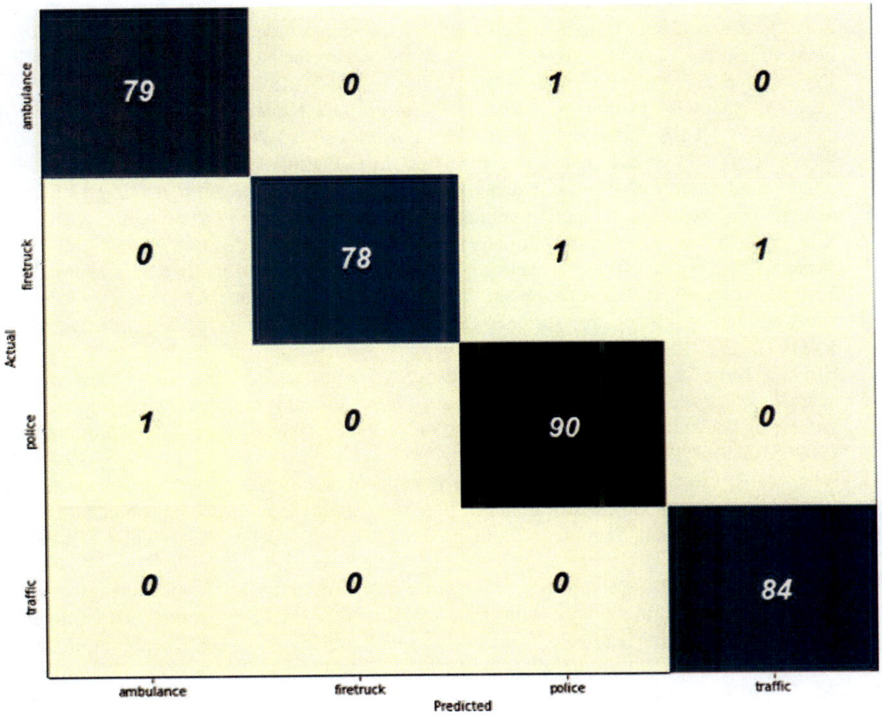

Fig. 5 Confusion matrix generated for classification of test data using GRU

References

1. Shah A, Singh A (2023) sireNNet-emergency vehicle siren classification dataset for urban applications. Mendeley Data, V1. https://doi.org/10.17632/j4ydzzv4kb.1
2. Lezhenin I, Bogach N, Pyshkin E (2019) Urban sound classification using long short-term memory neural network. In: 2019 Federated conference on computer science and information systems (FedCSIS), Leipzig, Germany, pp 57–60. https://doi.org/10.15439/2019F185
3. Zakaria N, Mohamed F, Abdelghani R, Sundaraj K (2021) VGG16, ResNet-50, and GoogLeNet deep learning architecture for breathing sound classification: a comparative study. In: 2021 International conference on artificial intelligence for cyber security systems and privacy (AI-CSP), El Oued, Algeria, pp 1–6. https://doi.org/10.1109/AI-CSP52968.2021.9671124
4. Das JK, Ghosh A, Pal AK, Dutta S, Chakrabarty A (2020) Urban sound classification using convolutional neural network and long short term memory based on multiple features. In: 2020 Fourth International conference on intelligent computing in data sciences (ICDS), Fez, Morocco, pp 1–9. https://doi.org/10.1109/ICDS50568.2020.9268723
5. Zhao W, Yin B (2021) Environmental sound classification based on adding noise. In: 2021 IEEE 2nd International conference on information technology, big data and artificial intelligence (ICIBA), Chongqing, China, pp 887–892. https://doi.org/10.1109/ICIBA52610.2021.9688248
6. Kumar K, Chaturvedi K (2020) An audio classification approach using feature extraction neural network classification approach. In: 2nd International conference on data, engineering and applications (IDEA), Bhopal, India, pp 1–6. https://doi.org/10.1109/IDEA49133.2020.9170702

7. Zhao W, Yin B (2022) Environmental sound classification based on pitch shifting. In: 2022 International seminar on computer science and engineering technology (SCSET), Indianapolis, IN, USA, pp 275–280. https://doi.org/10.1109/SCSET55041.2022.00070
8. Demir F, Abdullah DA, Sengur A (2020) A new deep CNN model for environmental sound classification. IEEE Access 8:66529–66537. https://doi.org/10.1109/ACCESS.2020.2984903
9. Pramanick D, Ansar H, Kumar H, Pranav S, Tengshe R, Fatimah B (2021) Deep learning based urban sound classification and ambulance siren detector using spectrogram. In: 2021 12th International conference on computing communication and networking technologies (ICCCNT), Kharagpur, India, pp 1–6. https://doi.org/10.1109/ICCCNT51525.2021.9579778
10. Niranjan K, Shankar Kumar S, Vedanth S (2021) Ensemble and multi model approach to environmental sound classification. In: 2021 Fourth International conference on electrical, computer and communication technologies (ICECCT), Erode, India, pp 1–5. https://doi.org/10.1109/ICECCT52121.2021.9616775
11. Hirata K, Kato T, Oshima R (2019) Classification of environmental sounds using convolutional neural network with bispectral analysis. In: 2019 International symposium on intelligent signal processing and communication systems (ISPACS), Taipei, Taiwan, pp 1–2. https://doi.org/10.1109/ISPACS48206.2019.8986304
12. Lu J, Ma R, Liu G, Qin Z (2021) Deep convolutional neural network with transfer learning for environmental sound classification. In: 2021 International conference on computer, control and robotics (ICCCR), Shanghai, China, pp 242–245. https://doi.org/10.1109/ICCCR49711.2021.9349393
13. Singh U et al (2021) Polyphonic sound event detection and classification using convolutional recurrent neural network with mean teacher. In: 2021 12th International conference on computing communication and networking technologies (ICCCNT), Kharagpur, India, pp 1–4. https://doi.org/10.1109/ICCCNT51525.2021.9579677
14. Wu D (2019) An audio classification approach based on machine learning. In: 2019 International conference on intelligent transportation, big data & smart city (ICITBS), Changsha, China, pp 626–629. https://doi.org/10.1109/ICITBS.2019.00156
15. Evangelista EB, Guajardo G, Ning T (2020) Classification of abnormal heart sounds with machine learning. In: 2020 15th IEEE International conference on signal processing (ICSP), Beijing, China, pp 285–288. https://doi.org/10.1109/ICSP48669.2020.9320916
16. Mkrtchian G, Furletov Y (2022) Classification of environmental sounds using neural networks. In: 2022 systems of signal synchronization, generating and processing in telecommunications (SYNCHROINFO), Arkhangelsk, Russian Federation, pp 1–4. https://doi.org/10.1109/SYNCHROINFO55067.2022.9840922
17. Ahmed A, Serrestou Y, Raoof K, Diouris J-F (2021) Sound event classification using neural networks and feature selection based methods. In: 2021 IEEE International conference on electro information technology (EIT), Mt. Pleasant, MI, USA, pp 1–6. https://doi.org/10.1109/EIT51626.2021.9491869
18. Paissan F, Ancilotto A, Brutti A, Farella E (2022) Scalable neural architectures for end-to-end environmental sound classification. In: 2022 IEEE International conference on acoustics, speech and signal processing (ICASSP), Singapore, Singapore, pp 641–645. https://doi.org/10.1109/ICASSP43922.2022.9746093
19. Vujošević L, Đukanović S (2021) Deep learning-based classification of environmental sounds. In: 2021 25th International conference on information technology (IT), Zabljak, Montenegro, pp 1–4. https://doi.org/10.1109/IT51528.2021.9390124
20. Li R, Yin B, Cui Y, Du Z, Li K (2020) Research on environmental sound classification algorithm based on multi-feature fusion. In: 2020 IEEE 9th Joint International information technology and artificial intelligence conference (ITAIC), Chongqing, China, pp 522–526. https://doi.org/10.1109/ITAIC49862.2020.9338926
21. Kroos C et al (2019) Generalisation in environmental sound classification: the 'making sense of sounds' data set and challenge. In: 2019 IEEE International conference on acoustics, speech and signal processing (ICASSP), Brighton, UK, pp 8082–8086. https://doi.org/10.1109/ICASSP.2019.8683292

22. McFee B, Raffel C, Liang D, Ellis DPW, McVicar M, Battenberg E, Nieto O (2015) librosa: audio and music signal analysis in python. In: Proceedings of the 14th python in science conference, pp 18–25
23. TensorFlow Developers (2022). TensorFlow (v2.9.3). Zenodo. https://doi.org/10.5281/zenodo.7604251

Chapter 7
Assessment of Variable Threshold for Anomaly Detection in ECG Time Signals with Deep Learning

Biraja Mishra and Rajeev Kumar

1 Introduction

Every classifier uses a decision threshold to do a split between the classes. This decision threshold is used for mapping the processed output to a class. Most Machine/Deep Learning (ML/DL) algorithms consider a fixed threshold, say 0.5, as a default value. Still, it is difficult to determine whether the default threshold is accurate for a particular test case without doing the model evaluation. However, the threshold selection in every test case scenario aims to maximize the True Positive (TP) and minimize the False Negative (FN) values. Taking a fixed threshold for classification includes the following drawbacks. First, if the value of the threshold is too high, there is a chance that the classifier may classify more instances as a negative class even if they belong to a positive class. Second, regardless of a dataset's characteristics, the classifier always makes a partition based on a fixed criterion. An alternative to this is to change the criteria for choosing a threshold or take variable thresholds to improve the performance [1]. The Receiver Operating Curve (ROC) [2] is often used with the variable threshold to provide optimal performance. The ROC curve is the most commonly used method for plotting the classification/decision thresholds. The closer the value is to the right angle of the curve, the more accurately the classifier performs [3].

Thresholding is a commonly used method for detecting anomalies in time series data, especially in medical time series such as Electrocardiogram (ECG) signals. Electrocardiograms (ECGs), one type of electronic health record, generate signals that can be easily acquired and stored as time series. Time series is a dynamic measurement of the system repeatedly over time. It can be sampled uniformly as well as non-uniformly with time. It has been further divided into single-variate/ univariate

B. Mishra (✉) · R. Kumar
Data to Knowledge (D2K) Lab School of Computer and Systems Sciences, Jawaharlal Nehru University, New Delhi 110067, India
e-mail: mishrabiraja23@gmail.com

A. Yadav et al. (eds.), *Proceedings of International Conference on Paradigms of Communication, Computing and Data Analytics*, Algorithms for Intelligent Systems,
https://doi.org/10.1007/978-981-99-4626-6_7

(UTS) and multivariate time series data (MTS). Univariate ECG considers a single lead of ECG, such as QRS complex, whereas multivariate ECG considers multiple ECG signals simultaneously. Forecasting time series needs the use of historical values of the dataset along with the associated patterns/trends that help predict future scopes. In the case of an ECG time signal, thresholding is used to diagnose ischemic heart disease, electrolyte imbalances, and abnormalities in the ECG waves. This can be done by using any traditional approach, including finding the mean and standard deviation and setting a threshold. Any point that lies outside the threshold is considered an anomaly. However, doing this doesn't require any expert knowledge. It is easy to analyze the result and get an idea about the anomalous data point in the dataset [4].

Beyond traditional methods, Deep Learning (DL) has also contributed significantly to the classification of heartbeats [5]. Deep learning approaches like Convolutional Neural Networks (CNNs) [6] and Multi-Layer Perceptron (MLPs) [7] were used for the detection of anomalies by researchers in different domains. Recurrent Neural Networks (RNNs) and Long Short-Term Memory (LSTMs) [4] play a big role while dealing with non-stationary time series datasets, such as ECG. RNN has a similar architecture as any traditional neural network except that the hidden unit in the network does a slightly different function. It computes a function using the given input along with its previous output. The working of the Recurrent Neural Network can be viewed by unfolding them across time. The tremendous progress of deep learning approaches in the last decade motivated researchers to use these methods for anomaly detection. For feature representation, deep generative models like autoencoders [8], and Convolutional neural network (CNN)[6] are employed to design models that detect abnormality in various cardiovascular diseases diagnosed through ECG using threshold [9]. This paper aims to train an LSTM-based Encoder-Decoder model on the ECG dataset and analyze the impact of varying thresholds on its ability to detect anomalies. As part of our methodology, we utilize ROC Curves to determine the threshold that is closest to the optimal [10]. We address the following research questions.

RQ 1 : What is the impact of threshold in the Anomaly detection of ECG Time Signals?

RQ 2 : How do variable thresholds affect the performance?

The research significance of this paper is to propose a novel approach for anomaly detection in ECG time signals using LSTM Autoencoder and thresholding. The proposed methodology includes computing the reconstruction error and using it to generate different thresholds for detecting abnormalities in ECG time signals.

Initially, random thresholds are used to assess the classifier's performance, followed by additional thresholds determined by a formula that incorporates the mean and standard deviation of the error. The performance of each individual threshold is then assessed using ROC curves. This paper also evaluates the performance of the proposed method by comparing it with the performance of random thresholds using ROC curves. This can help diagnose the abnormalities in ECG time signal more effectively, which helps in medical diagnosis and treatment.

The rest of the paper is organized as follows. In Sect. 2, we conduct a literature survey on anomaly detection in ECG time signals using Deep Learning Architectures by incorporating fixed thresholds. The proposed methodology of this work is detailed in Sect. 3. Section 4 describes the dataset taken and the incorporation of the methodology in the dataset using varying thresholds. It assesses the performance of the classifier using different ROC Curves. Finally, we conclude the work in Sect. 5.

2 Related Work

This section covers the related works that detect anomalies in the time series data, especially in ECG Time Signal, using deep learning-based frameworks [11].

In Time Series, anomaly detection is usually formulated as filtering the outlier data points according to some norm or by setting some labels. These outliers are known as *anomalies* in time series jargon. The key issues in time series data and detecting anomalies have been the frontline research area for the last decades. The issues of the existing state-of-the-art unsupervised methods and how they lack scalability during anomaly detection are of primary concern.

2.1 Anomaly Detection with Deep Learning Architecture

Deep learning approaches like Convolutional Neural Networks (CNNs), Long Short-Term Memory (LSTMs), and Multi-Layer Perceptron (MLPs) were used for the detection of anomalies by researchers. Recurrent Neural Network (RNN) plays a big role while dealing with non-stationary time series datasets. The working of the RNN can be viewed by unfolding them across time. The tremendous progress of deep learning approaches in the last decade motivated researchers to use these methods for anomaly detection [12]. Alexander et al. [13] proposed a Generative Adverse Networks (GAN)-based Time series anomaly detection method (TadGAN). In TadGAN, LSTM-RNN is used as a base model for identifying the correlation between the time series data points. It allows the reconstruction of the time series data effectively by allowing minimal loss of cycle consistency, improving performance, and increasing generalizability. The result depicts the effectiveness of the baseline methods and has the highest F1-score. Kwei-Herng Lai et al. [14] also adopted Recurrent Neural Networks with Long Short-Term Memory and autoregression to model the correlation between the nonlinear time series data instances. Zhang et al. [12] proposed a Multi-Scale Convolutional Recurrent Encoder-Decoder (MSCRED), which performs well for the detection and diagnosis of anomalies in multivariate time series (MTS).

Inducing Threshold: If the ratio of the mean exceeds the given threshold, it is assumed that there is an anomaly in the data. After the period of the data is calculated, the standard deviation between two consecutive periods is calculated. If the ratio of standard deviations exceeds the predefined threshold, the existence of an anomaly is assumed.

The process of identifying anomalies involves calculating the cumulative distribution of the error in the prediction. If the mapping between the normal distribution and the cumulative distributed value is too small or too large, the anomaly is assumed to exist.

2.2 Anomaly Detection in ECG Time Series

Using two real-world time series datasets, Numenta Anomaly Benchmark (NAB) as UTS and Electrocardiography (ECG) as MTS, Tung et al. [8] proposed two methods for anomaly detection using recurrent autoencoder ensembles in a time series dataset. Their proposed method exploited autoencoders using sparsely-connected RNNs (S-RNN). Using S-RNNs, multiple autoencoders can be generated using different neural network connections. The two proposed methods were ensembled to enable outlier detection and consisted of an independent framework and one shared framework. The proposed framework shows that this ensemble approach was effective and outperformed state-of-the-art methods. To improve accuracy, denoising autoencoders can use deep neural networks in the future.

A Bi-LSTM and CNN-based model for anomaly detection in the ECG time series signals is proposed by Kai et al. [6]. This model is designed to detect abnormality in various cardiovascular diseases diagnosed through ECG. Using CNN, the features were to be extracted first and detected as effectively done using Bi-LSTM. The morphological features of the ECG signal are directly extracted from the input (ECG signals itself) using CNN architecture, and the time factors are extracted using LSTM. For the experiment, the MIT-BIH arrhythmia database was taken. The slope of two adjacent points defines the feature. The slope is then converted into matrices as per the time sequences. This improves the accuracy of the proposed detection method. The experiment shows that the training complexity is less, and the accuracy is high. Using the R wave Pan-Tompkins algorithm, the background noises are removed, and the frequency content of the features is filtered. Compared to an SVM and LSTM-based model, this model has better accuracy and F1 value.

Inducing Threshold: The irregularity can sometimes be high and low depending on the comparison we are making in our test case. For example, [9] in the case of motion detection in smoke areas, the degree of irregularity varies from high to low with rigid objects and cars, respectively [1]. In the domain of Image Segmentation also, the threshold has been used as a basic tool. However, it also has plenty of use cases in the domain of medical time signals, such as pre-processing of ElectroCardioGram (ECG) signals, finding peaks, and detecting abnormalities/irregularities in the signal.

Methods besides deep learning can also be used for anomaly detection. One major work in this is a statistical anomaly detection framework proposed by Jian et al. [15] that someway outperforms other DL-based methods, namely, SARIMA, LSTM, LSTM with GRU, etc. Using two different sliding time windows (a large and a small), the means of the dataset are calculated.

3 Research Methodology

In Sect. 2, we conceptualized anomaly detection using DL-based architectures using thresholds on time series data, especially in ECG time signals. This section details the concept by detecting abnormalities/irregularities in the ECG signals using varying thresholds and assesses the performance of variable thresholds using the ROC Curve.

3.1 Model Selection

In this work, we selected an LSTM Autoencoder for Anomaly detection in the ECG time signals. The autoencoder's motivation is its ability to handle sequential data and the long-term dependencies in the ECG time signal [8]. Recurrent Neural Network (RNN) has a similar architecture as any traditional neural network except that the hidden unit in the network does a slightly different function. It is well suited for handling sequential data. One variant of RNN is Long Short-Term Memory (LSTM). The LSTM autoencoder is specially designed to handle long-term dependencies and is the most commonly used DL architecture for anomaly detection using thresholding. The input is transformed into a feature space of lower dimensions by the encoder, which is then saved in the bottleneck or latent layer. It might seem like the autoencoder only makes dimensionality reduction, but it also deals with the nonlinearity of the data. Among the abundant use case of autoencoders, the primary and most potential use case are Anomaly Detection.

In the case of ECG time signals, the autoencoder is trained on normal ECG signals to capture the temporal dependencies and handle the patterns; any new pattern that deviates from the normal pattern is considered an anomaly.

3.2 Reconstruction and MAE

As a next step, we calculate the reconstruction error with respect to the considered autoencoder. In the context of the autoencoder, the reconstruction error measures the difference between the input signal and the reconstructed output produced by the autoencoder. The reconstruction error is calculated for each time stamp of the ECG signal. For Example, in an ECG signal with T time stamps, the reconstruction error is a vector of T elements.

We then summarize the reconstruction error for a signal over all the time stamps, using Mean Absolute Error (MAE). This MAE is then used for setting the threshold to detect anomalies. There are several approaches to setting the threshold, such as fixed thresholding, variable thresholding, adaptive thresholding. However, once the threshold is set, it can be used to detect the anomalies by comparing it with the reconstruction error/MAE.

3.3 Thresholds and Assessment

We consider a set of random thresholds $\{0.02, 0.04, 0.06, 0.08, 0.1\}$ that are subject to change as per the experiment and evaluate the performance of our encoder and decoder architecture over these thresholds to find anomalies in the ECG time signals. We generate a confusion matrix by comparing each threshold with the mean absolute error of the ECG signal and classifying them into anomalous and non-anomalous classes.

This confusion matrix contains key metrics such as True Positives (TP), True Negatives (TN), False Negatives (FN), and False Positives (FP), which are then utilized to achieve an optimal trade-off between TP and FN across the variable thresholds.

3.4 Generation of ROC for the Defined Thresholds

We can use the Confusion Matrix generated above to calculate the False Positive Rate (FPR) and the True Positive Rate (TPR). Using these (TPR, FPR) points generated for each threshold, we plot a receiver operating characteristic (ROC) curve, a graphical representation of a binary classification model, autoencoder in our case. It compares the classifier's performance and helps decide the near—optimal threshold that gives the best performance [16]. In addition, we quantify the confusion matrix and calculate the precision, recall, F1-score, etc., for all the variable thresholds to show the varying performance of the classifier.

3.5 Optimal Threshold Using ROC

After analyzing the performance from the above ROC curve, we further extend our experiment to find the optimal threshold. we systematically generate a hundred thresholds from the MAE calculated from the training data using the formula below

$$\text{Threshold} = \text{Mean (Error)} + i * 0.1 * \text{Std (Error)}$$

We propose using the above formula incorporating both the mean and standard deviation of the error to generate the thresholds. This formula generates thresholds at minimal intervals from the mean, enhancing the precision of the roc curve generated. Once the thresholds are calculated using this formula, we calculate (TPR, FPR) points for all the thresholds and generate a ROC curve accordingly. The process of finding the optimal threshold of this ROC curve involves identifying the point closest to the right angle of the curve. Since a ROC curve can't guarantee finding the optimal threshold, we refer to the threshold found using this method as a (near-) optimal one.

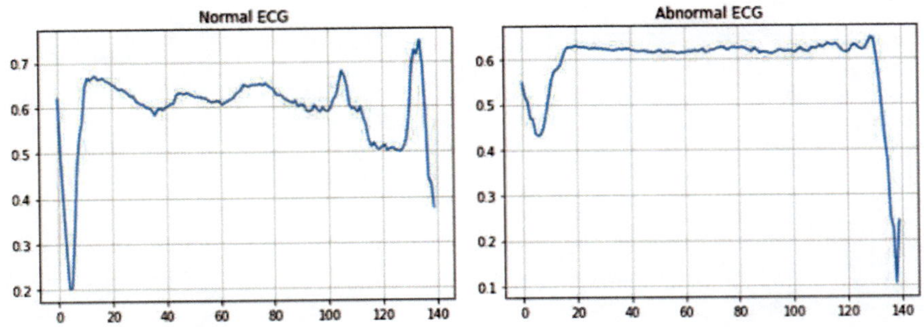

(a) Normal ECG (b) Abnormal ECG

Fig. 1 ECG signal

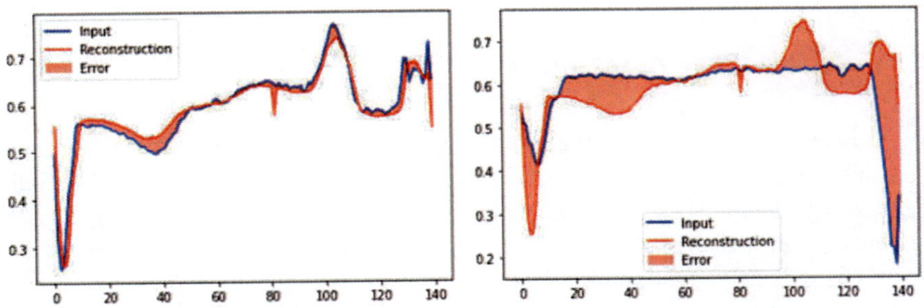

(a) In Normal Test Data (b) In Anomalous Test Data

Fig. 2 Reconstruction errors

4 Results and Analysis

4.1 Data Sets and Experimental Setup

The MIT-BIH ECG5000 dataset is taken from the *PhysioNet ECG Database* for the experiment. The dataset is a 20-hour-long observed univariate ECG time signal. This contains five thousand rows and 141 columns. The *CSV* file is directly uploaded to Jupyter Notebook (Python 3.6). The key modules like *Layers* and *Losses* from the *TensorFlow* are imported to build the *Autoencoder*. The key *Layers*, including the *Input*, *Dense*, and *Flatten* layers are used to build the encoder, and the *Dense* and *Reshape* layers are used to build the decoder. We utilized the *Mean Squared Error(MSE)* metric to compute the loss associated with the reconstruction process. The input and output layers use the *ReLU* and *Sigmoid* activation functions. 20% of the input data is used for training, while the remaining 80% is used for training. After splitting, Boolean indexing separates the data into normal and abnormal ECG (Fig. 1).

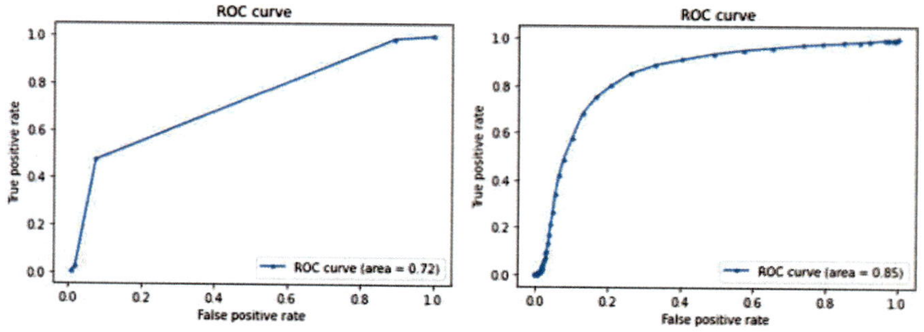

(a) For Variable Thresholds (b) For Optimal Threshold

Fig. 3 ROC curves

4.2 Performance Assessment

We train the autoencoder with the above-mentioned configurations using the training data and set the random thresholds over the calculated reconstruction error. With the defined set of random thresholds, we assess the classifier's performance over the reconstruction error generated on the test data. Figure 2 depicts the trends of this error. Using the five random thresholds (discussed in Sect. 3.3), different confusion matrices are generated from which we calculated the TPR and FPR and plotted the ROC curve (Fig. 3a). In addition to generating the above ROC curve, other performance metrics such as precision, recall, accuracy, and F1-score are also computed for the threshold values and presented in Table 1. We used this table of performance metrics in conjunction with the ROC curve for evaluating the performance. However, finding a (near-) optimal point from this curve is difficult and can lead to a lousy conclusion as the thresholds considered are random and minimal in number.

We used systematically generated thresholds to plot the ROC Curve for obtaining a more precise optimal threshold value. We calculated a hundred different thresholds using the formula mentioned in Sect. 3.5 and obtained a ROC Curve (Fig. 3b) using these thresholds by plotting the associated (TPR, FPR) points. We got a (near-) optimal threshold point from the generated ROC curve that has given the best TPR and FPR value pair. We present our analysis of the results obtained in Table 1 and Fig. 3a along with the effectiveness of the proposed method for finding the optimal threshold(Fig. 3b), in the following Sect. 4.3.

Table 1 Performance on various thresholds

Threshold	Precision	Recall	F1-score	Accuracy
0.02	0.83	0.73	0.72	0.73
0.04	0.95	0.95	0.95	0.95
0.06	0.78	0.73	0.70	0.73
0.08	0.56	0.58	0.44	0.58
0.1	0.50	0.57	0.42	0.57

4.3 Discussion

ECG signals are a critical tool for diagnosing and monitoring cardiac conditions and accurate anomaly detection is crucial for ensuring timely treatment. While discussing anomaly detection, setting a suitable threshold becomes a key challenge. So many existing approaches for setting the threshold [9] have their benefits and limitations. In our work, we used the reconstruction error of an autoencoder model to set variable thresholds and showed how they could impact the performance of deep learning models.

In our work, we first considered some random thresholds (Sect. 3.3), generated the confusion matrices individually, and using the TPR and FPR value, we got from the confusion matrix, we plotted the ROC Curve (Fig. 3a) to measure the performance. As we can see, the TPR and FPR values at one threshold vary significantly from another, which causes the points on the curve to be widely separated. This creates a challenge in two aspects, one is in evaluating how well the model performs, and the other is in determining the optimal threshold.

Similar work is also carried out by considering other performance metrics like precision, recall, accuracy, and F1-score for the same threshold values presented in Table [1], and our findings indicate that a threshold value of *0.04* produced the best performance across all the metrics and the thresholds *0.08* and *0.1* led to the worst performance. However, as these threshold values are selected randomly and are subjected to change as per the experiment, we couldn't determine the (near-) optimal threshold based on this performance.

The analysis of the performance of variable thresholds and their impact on the classifier's performance points out a critical challenge of finding the optimal threshold. Our proposed method of generating systematic thresholds and finding a near-optimal threshold by plotting a ROC curve is proven effective. As the no of thresholds is higher, we were able to find the optimal threshold even more precisely. The AUC value of 0.85 also admits the same. This indicates how the increase in the number of thresholds provides a more balanced distinction between the TPR and FPR, and this finer distinction leads to a more accurate evaluation of the classifier's performance.

5 Conclusion

In this work, we have investigated a deep learning approach, an LSTM-based autoencoder, for anomaly detection in ECG time signals. We have explored using different thresholds for identifying anomalies and compared the performance of these thresholds using a ROC Curve. We have also evaluated the performance of those randomly chosen thresholds by using other performance metrics like precision, recall, F1-score, and accuracy and demonstrated the impact of variable thresholds on the accuracy of anomaly detection in ECG signals. Our experiment shows that setting an appropriate decision threshold is essential for effective anomaly detection. We proposed a method to find the optimal threshold using the ROC curve; this involves systematically generating a set of thresholds and finding the optimal threshold with a higher true positive rate and lower false positive rate in the ROC Curve generated. Our study has demonstrated that the proposed method for finding the optimal threshold positively impacts the model's performance. The efficiency of this method may further be enhanced either by increasing the number of thresholds used or by using an even more sophisticated method for generating the thresholds, or by using both. In future works, the proposed methodology can also be used to assess variable thresholds in use cases other than the ECG time signals.

References

1. Weszka JS, Rosenfeld A (1978) Threshold evaluation techniques. IEEE Trans Syst Man Cybern 8(8):622–629
2. Fawcett T (2006) An introduction to roc analysis. Pattern Recogn Lett 27(8):861–874
3. Flach PA (2016) Roc analysis. In: Encyclopedia of machine learning and data mining. Springer, pp 1–8
4. Malhotra P, Ramakrishnan A, Anand G, Agarwal P, Shroff G (2016) LSTM-based encoder-decoder for multi-sensor anomaly detection, Lovekesh Vig
5. Chauhan S, Vig L (2015) Anomaly detection in ECG time signals via deep LSTM networks. In: Proceedings IEEE International conference on data science and advanced analytics (DSAA), pp 1–7
6. Cui KX, Xia XJ (2022) ECG signal anomaly detection algorithm based on CNN-BiLSTM. In: Proceedings 11th international conference information and communication technology (ICTech), pp 193–197
7. Savalia S, Emamian V (2018) Cardiac arrhythmia classification by multi-layer perceptron and convolution neural networks. Bioengineering 5(2)
8. Kieu T, Yang B, Guo C, Jensen CS (2019) Outlier detection for time series with recurrent autoencoder ensembles. In: Proceedings of 28th international joint conference artificial intelligence (IJCAI), pp 2725–2732
9. Wu X, Lu X, Leung H (2017) An adaptive threshold deep learning method for fire and smoke detection. In: Proceedings IEEE international conference systems, man, & cybernetics (SMC), pp 1954–1959
10. Hong CS (2009) Optimal threshold from roc and cap curves. Commun Stat-Simul Comput 38(10):2060–2072

11. Du S, Li T, Horng SJ (2018) Time series forecasting using sequence-to-sequence deep learning framework. In: Proceedings 9th International symposium parallel architectures, algorithms, and programming (PAAP), pp 171–176
12. Lai KH, Zha D, Xu J, Zhao Y, Wang G, Hu X (2021) Revisiting time series outlier detection: Definitions and benchmarks. In: Proceedings of 35th conference neural information processing systems datasets and benchmarks track (Round 1)
13. Blázquez-García A, Conde A, Mori U, Lozano JA (2021) A review on outlier/anomaly detection in time series data. ACM Comput Surv 54(3):1–33
14. Geiger A, Liu D, Alnegheimish S, Cuesta-Infante A, Veeramachaneni K (2020) TadGAN: time series anomaly detection using generative adversarial networks. In: Proceedings IEEE international conference big data, pp 33–43
15. Kao JB, Jiang JR (2019) Anomaly detection for univariate time series with statistics and deep learning. In: Proceedings IEEE Eurasia conference IOT, communication & engineering (ECICE), pp 404–407
16. Metz Charles E (1978) Basic principles of roc analysis. Semin Nucl Med 8(4):283–298

Chapter 8
Green Cloud Computing: Achieving Sustainability Through Energy-Efficient Techniques, Architectures, and Addressing Research Challenges

Sneha, Prabhdeep Singh, and Vikas Tripathi

1 Introduction

Green cloud computing is a developing field of study that tries to lessen the negative effects of cloud computing on the atmosphere while maintaining a level of customer satisfaction. The energy required to power data centers and other cloud infrastructure has grown to be a major problem as cloud computing has become an integral aspect of contemporary company operations. In response, scientists and business professionals are investigating energy-efficient methods, designs, and tactics to make cloud computing more sustainable. Green cloud computing seeks to maximize economic gains while reducing the carbon footprint and environmental effects of cloud computing. To overcome the difficulties posed by green cloud computing, a multidisciplinary strategy integrating energy-saving methods, architectural concepts, and research activities is needed to achieve this aim. Similar to the National Institute of Standards and Technology, cloud computing offers consumers services including Infrastructure as a Service, Platform as a Service, and Software as a Service to persuade owners of business applications to embrace and transfer their programs to the cloud. Figure 1 shows cloud computing service layers architecture [1].

IaaS: IaaS offers clients pay-per-use access to the basic components of computing infrastructure, such as virtual servers, and networking capabilities. The operating system and the applications running on the infrastructure are completely under

Sneha (✉) · P. Singh · V. Tripathi
Graphic Era Deemed to be University, Dehradun, India
e-mail: snehaji7456@gmail.com

V. Tripathi
e-mail: vikastripathi.cse@geu.ac.in

A. Yadav et al. (eds.), *Proceedings of International Conference on Paradigms of Communication, Computing and Data Analytics*, Algorithms for Intelligent Systems,
https://doi.org/10.1007/978-981-99-4626-6_8

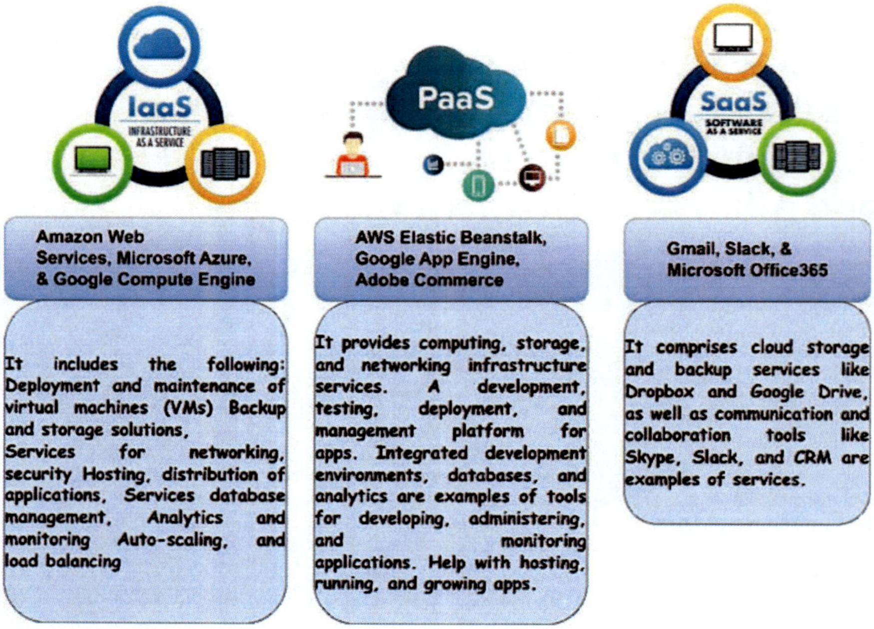

Fig. 1 Cloud computing service layers architecture [2]

the control of the customer. Customers who want a highly adaptable and flexible computing infrastructure to support their applications should use IaaS.

PaaS: Customers that use PaaS receive a comprehensive platform, replete with infrastructure, middleware, and tools required for creating, deploying, and maintaining applications. Customers who wish to concentrate on creating and delivering apps without handling the supporting infrastructure should choose PaaS.

SaaS: Customers that use SaaS have access to software programs that the cloud provider hosts and manages. Clients can alter certain features of the program to suit their needs, but they have no control over the software or the underlying infrastructure. Customers that wish to use a pre-built software application without having to handle the infrastructure or software themselves will find SaaS to be the perfect solution [2].

2 Techniques to Make Cloud "Green"

Green cloud computing is an important idea that tries to lessen the environmental effect of cloud computing. Data centers may become more sustainable and lower their carbon footprint by employing energy-efficient techniques and sustainable infrastructures. Virtualization, energy-efficient hardware, energy-efficient cooling, and power management (dynamically) can all help to make cloud computing "green." Table 1 shows techniques to make the cloud "green."

Table 1 Various techniques to make the cloud "green" [3]

Virtualization	On a single physical computer, virtual instances of hardware components, such as servers and storage devices, are created using virtualization technology. Virtualization can lower the energy consumption of cloud computing infrastructures by concentrating workloads onto fewer physical servers. This is because fewer physical computers are required to service a given amount of workload, which reduces the demand for cooling and electricity
Energy-efficient hardware	Hardware that consumes less energy is another crucial method for making cloud computing "green." Data centers may lower their energy use and carbon impact by utilizing low-power CPUs, memory architecture, and other hardware elements. For instance, ARM-based processors, built for low-power applications that are more energy-efficient than conventional X86 CPUs, are now widely used in data centers
Energy-efficient cooling	Cooling is a key energy consumer. Data centers may lower their energy use and carbon impact by utilizing energy-efficient cooling methods, such as free cooling. Free cooling includes cooling the data center using outside air rather than mechanical cooling systems, which in some cases may be more energy-efficient
Power management (dynamically)	In this, hardware components' power consumption is dynamically changed in response to workload demands. For instance, if a server is not being used to its maximum potential, its power consumption might be decreased to conserve energy. By employing dynamic power management strategies, data centers may lower their energy use and carbon footprint without compromising performance [4]

3 Green Cloud Computing Architecture

This architecture employs a new middleware called the Green Broker, which facilitates users in requesting cloud services. The Green Broker evaluates the eco-friendliness of different cloud providers and selects the one that satisfies the user's requirements in the most sustainable way. To ensure sustainability, cloud services must belong to one of three categories. The Green Broker accesses a public directory where cloud providers can list their green offerings, including environmentally friendly products, cost structures, and optimal usage times that emit the least amount of carbon emissions. The Carbon Emission Directory is responsible for maintaining the latest information on the energy efficiency of cloud services. The collection includes parameters such as the effectiveness of cooling in a cloud-based data center, the cost of network utilization, and the rate of carbon emissions from power. The Green Broker estimates the carbon emissions of each cloud provider that offers the desired cloud service. Subsequently, it chooses the cloud services that emit the least carbon and purchases them on behalf of the clients [5] as shown in Fig. 2.

The primary objective of the green cloud framework resides in the meticulous tracking of energy consumption during the processing of user requests. Comprising two significant constituents, namely the Carbon Emission Directory and Green Cloud

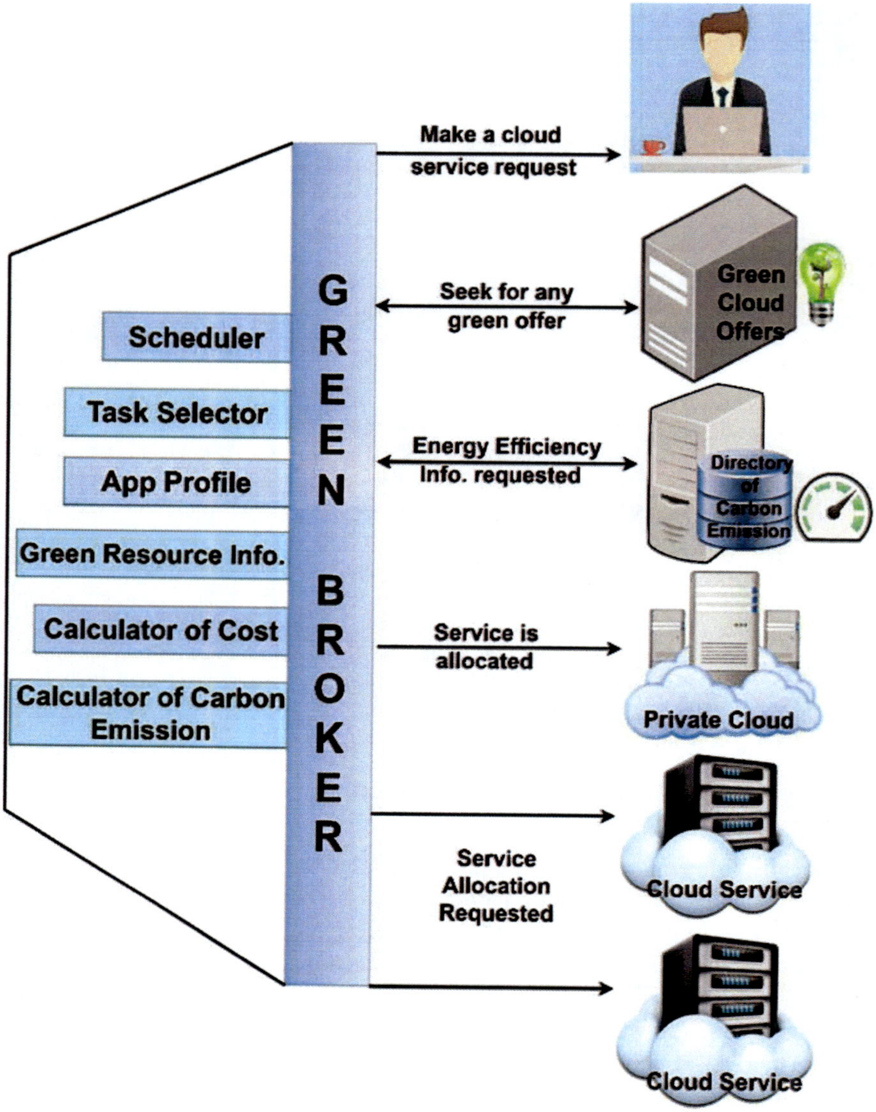

Fig. 2 Green cloud computing architecture for energy usage [5]

Offers, this framework aims to incentivize companies toward the provision of environmentally sustainable services. The indispensability of the Green Broker becomes evident as it assumes a pivotal role in the management of services, operating from the vantage point of the user. It diligently scrutinizes and selects cloud services based on the quality of service requirements of the clientele, thereby ensuring minimal carbon emissions while delivering its services. The cloud infrastructure grants users unrestricted access to a trifecta of service categories, encompassing Software as a

Service (SaaS), Platform as a Service (PaaS), and Infrastructure as a Service (IaaS). Consequently, the provisioning of these services must adhere to stringent energy efficiency standards.

The Green Broker functions as an intermediary of utmost intelligence, strategically positioned within the realm of cloud infrastructure, meticulously governing and optimizing the intricate utilization of energy resources. This middleware layer assumes a mediating role between CSPs and the diverse assemblage of cloud users, comprising both individuals and organizations, who actively engage with cloud-based services. By rigorously scrutinizing the energy consumption patterns inherent to the cloud infrastructure, the Green Broker diligently examines the availability of sustainable and renewable energy sources, encompassing the likes of wind power and solar energy [6].

Behold the Carbon Emission Directory, a powerful instrument devised with the express purpose of meticulously monitoring and adroitly managing the carbon emissions that arise as a byproduct of cloud computing endeavors. This prodigious tool, affectionately known as the CED, bestows upon the cloud customers and service providers a comprehensive array of statistical insights pertaining to the carbon footprint, reverberating through the intricate labyrinth of cloud infrastructure. Encompassing data centers, networks, and an assemblage of other vital components, the CED employs a rigorous methodology wherein it procures extensive data encompassing the energy consumption and ensuing carbon emissions associated with the cloud infrastructure. With such indispensable data at their disposal, cloud customers can make judicious choices regarding the services they wish to employ, duly accounting for their respective environmental ramifications [7].

Cloud service providers occupy an unequivocally critical role in the grand tapestry of provisioning cloud services, championing sustainability and assuming a formidable mantle of ecological responsibility. It is these guardians of the cloud, these stalwart cloud providers, who wield authoritative dominion over the infrastructure that serves as the conduit for the delivery of cloud services to a diverse clientele, encompassing the likes of data centers, networks, and servers. Harnessing the boundless potential of energy-efficient technologies, virtuosic feats of virtualization, and the judicious consolidation techniques that serve to optimize the utilization of servers, these cloud service providers channel their energies toward the fine-tuning and optimal functioning of cooling and power distribution systems, thereby exemplifying their unyielding commitment toward ecological harmony [8].

In the grand tapestry of Green Cloud Offers, one finds an extraordinary pantheon of cloud services, imbued with an ethereal quality of ecological benevolence and longevity. These remarkable offerings, birthed from the desire to diminish the carbon footprint and mitigate the environmental impact that clouds of computing can potentially unleash, continue to manifest themselves as steadfast purveyors of computing prowess, marked by their unassailable reliability and unwavering performance. Embedded within these offerings lies a treasure trove of energy-efficient hardware, adroitly harnessed virtualization techniques, and cunning strategies of consolidation, all united in the pursuit of minimizing energy wastage. Furthermore, these offerings are complemented by an exquisite array of tools and services, thoughtfully curated to empower cloud users, as they embark upon the noble endeavor of managing their energy consumption, while concurrently wielding the power to curtail their environmental footprint [8].

4 Cloud Computing Requirements for Power

A form of computing known as "on-demand computing" or "cloud computing" has been made available to users on the Internet, providing them with access to utilities such as software applications, data storage, and computing capacity. This permits users to avail of powerful computer resources and store and process their information within remote data centers, which results in decreased latency. With Facebook alone desiring petabytes of storage, the amount of Internet users has caused a surge in the necessity for storage capacity [9]. To address this requirement, cloud computing necessitates electricity for data transmission, data processing hardware and software, networking components, and data storage.

Parameters used for measuring power consumption:

1. PUE: Power Usage Effectiveness is a metric that determines the ratio of the total power used by a data center to the power consumed by its IT hardware. This measurement provides insight into the energy efficiency of the data center [10].
2. TDP: The capacity for a computer chip to necessitate power for cooling when being employed for an actual exercise is quantified through Thermal Design Power (TDP). This power, described in Watts, is utilized for the identification of the efficacy of a cooling system that is necessary. TDP is then a pivotal element when selecting a processor, allowing people to compare the power requirements of different chips [11].
3. CUE: The Green Grid Organization designated Carbon Usage Effectiveness (CUE) as an indicator to calculate the carbon dioxide emissions of an information center in November 2012 [12].
4. DCiE: A metric has been implemented to measure the effectiveness of the data center infrastructure, which assesses the relative proportion of energy spent to power the IT equipment concerning the total energy used for whole data centers, such as cooling and other associated components [13].
5. CPE: Ascertain the ratio of the number of instructions a processor can perform per second to the amount of power it consumes to ascertain its efficacy [14].
6. The Energy Reuse Factor: The Efficiency Ratio Factor (ERF): ERF is a metric employed to ascertain the magnitude of energy that is recycled exteriorly of data centers. The incorporation of renewable sources of energy, for instance, wind and hydroelectricity, can render cloud habitats more ecologically sound [15].
7. Water Usage Effectiveness (WUE): A computation determines the yearly water usage of a data center taking into account the amount of water required for energy production, humidification, the power generation of apparatus, and cooling of

the facility. This proportion is then represented as a portion of the total water utilization of the data center [16].

8. Space, Wattage, and Performance (SWaP): Sun Microsystems have developed a metric known as SWaP as an approach to computing space and energy. It is calculated as the aggregate of the performance, power utilization, and rack unit height. There is an obligation on the designers of hardware components and innovators to produce energy-efficient CPUs, servers, and data processing centers, as well as to detect renewable energy sources to meet the power requirements [17].

5 Research Challenges of Green Cloud Computing

Green data centers: Data centers are the pillar of cloud computing infrastructure, and they use a lot of energy. It is a key scientific problem to develop energy-efficient data center architectures that leverage renewable energy sources, waste heat recovery, and other green technologies to lower the carbon footprint of cloud computing.

Consolidation of VMs: Consolidation of virtual machines (VMs) is a process that includes merging many virtual machines (VMs) onto a single physical server. The objective is to save energy by lowering the number of physical servers needed to provide a given workload. Unfortunately, the trade-off between the number of VMs running on a physical server and the server's energy consumption is not simple [18].

Load balancing (that is energy-aware): Load balancing is crucial for ensuring that cloud resources are used efficiently. Load balancing systems that are cognizant of energy consumption can spread workloads over numerous servers in an energy-efficient way. Yet, designing energy-aware load balancing techniques that account for individual server and network energy use is a big research issue.

Green networking: Networking components like switches and routers require a lot of energy. Creating energy-efficient networking systems to minimize cloud energy usage is a big research problem. Low-power networking components, network virtualization, and software-defined networking are examples of green networking technology [19].

The estimation of carbon footprint: Measuring the carbon footprint of cloud computing services is critical for encouraging sustainability. An important research topic is developing ways to evaluate the carbon footprint of cloud computing services and offer feedback to clients on their energy consumption and carbon emissions.

Green cloud service assessment: An important research problem is developing metrics and evaluation procedures to measure the greenness of cloud computing services and assist consumers in making informed decisions. Energy efficiency, carbon emissions, and environmental impact measures may all be used to evaluate green cloud services.

Regulatory difficulties: Addressing regulatory challenges in the cloud computing industry associated with energy consumption, carbon emissions, and environmental sustainability is a crucial area of research. Laws can influence the adoption of green cloud computing technologies, and they differ between nations [2].

6 Conclusion

Green cloud computing has a viable way to address the ICT industry's sustainability concerns. Green cloud computing relies heavily on energy management. Energy management architecture entails the design and implementation of various strategies and solutions that can optimize energy usage in cloud computing systems.

The National Institute of Standards and Technology has played a major role in creating standards and the best practices for green cloud computing, but more work is needed to assure widespread adoption.

It is critical to recognize that the adoption of green cloud computing is a cultural and organizational problem as much as a technological one. Companies must adapt their attitude and embrace sustainability as a key goal, which necessitates a substantial change in how they approach IT decision-making. In promoting green cloud computing, governments and policymakers have a crucial role to play through incentives, laws, and education. The use of energy-efficient techniques and architecture can drastically lower cloud computing's carbon footprint and pave the path for a greener future. Cloud computing has become a ubiquitous technology in modern times, delivering scalable, on-demand computing resources to organizations and people. Yet, rising demand for cloud computing services has resulted in rising energy usage, raising worries about cloud computing's environmental effect. To reach the full potential of green cloud computing, various research problems must be solved. These issues include green data centers, consolidation of VMs, load balancing, green networking, the estimation of carbon footprint, green cloud service assessment, and regulatory difficulties.

References

1. Ahmad A, Khan SU, Khan HU, Khan GM, Ilyas M (2021) Challenges and practices identification via a systematic literature review in the adoption of green cloud computing: client's side approach. IEEE Access 9:81828–81840
2. Hu N, Tian Z, Du X, Guizani N, Zhu Z (2021) Deep-green: a dispersed energy-efficiency computing paradigm for green industrial IoT. IEEE Trans Green Commun Netw 5(2):750–764
3. Bi J, Yuan H, Zhang J, Zhou M (2022) Green energy forecast-based bi-objective scheduling of tasks across distributed clouds. IEEE Trans Sustain Comput 7(3):619–630
4. Skourletopoulos G et al (2019) Elasticity debt analytics exploitation for green mobile cloud computing: an equilibrium model. IEEE Trans Green Commun Netw 3(1):122–131
5. Kumar S, Buyya R (2012) Green cloud computing and environmental sustainability. In: Harnessing Green It
6. Yamini R (2012) Power management in cloud computing using green algorithm. In: IEEE-International conference on advances in engineering, science, and management (ICAESM–2012) March 30, 31
7. Xiang D et al (2016) Eco-aware online power management and load scheduling for green cloud data centers. IEEE Syst J 10.1:78–87
8. Arthi T, Shahul Hameed H (2013) Energy-aware cloud service provisioning approach for a green computing environment. IEEE

9. Usmin S, Arockia Irudayaraja M, Muthaiah U (2014) Dynamic placement of virtualized resources for data centers in the cloud, June, IEEE
10. Kaur K, Garg S, Aujla GS, Kumar N, Zomaya A (2019) A multi-objective optimization scheme for job scheduling in sustainable cloud data centers. IEEE Trans Cloud Comput 1–1
11. Ganapathy D, Warner EJ (2008) Defining thermal design power based on real-world usage models. In: Intersociety conference on thermal and thermomechanical phenomena in electronics systemsI THERM, pp 1242–1246
12. Ismail L, Abed EH (2019) Linear power modeling for cloud data centers: taxonomy, locally corrected linear regression, simulation framework, and evaluation. IEEE Access 7:175003–175019
13. Yeganeh H, Salahi A, Pourmina MA (2019) A novel cost optimization method for mobile cloud computing by capacity planning of green data center with dynamic pricing. Can J Electr Comput Eng 42(1):41–51
14. Amokrane A, Zhani MF, Langar R, Boutaba R, Pujolle G (2013) Greenhead: virtual data center embedding across distributed infrastructures. IEEE Trans Cloud Comput 1(1):36–49
15. Yang Y, Chang X, Liu J, Li L (2017) Towards robust green virtual cloud data center provisioning. IEEE Trans Cloud Comput 5(2):168–181
16. Wazid M, Das AK, Bhat VK, Vasilakos AV (2020) LAM-CIoT: lightweight authentication mechanism in cloud-based IoT environment. J Netw Comput Appl 150:102496
17. Wen Z et al (2021) Running industrial workflow applications in a software-defined multi-cloud environment using green energy aware scheduling algorithm. IEEE Trans Industr Inf 17(8):5645–5656
18. Alarifi A et al (2020) Energy-efficient hybrid framework for green cloud computing. IEEE Access 8:115356–115369
19. Madan P, Singh V, Singh DP, Diwakar M, Pant B, Kishor A (2022) A hybrid deep learning approach for ECG-based arrhythmia classification. Bioengineering 9(4):152

Chapter 9
AI-Based Smart Dashboard for Electric Vehicles

Narayana Darapaneni, Anwesh Reddy Paduri, B. G. Sudha, Dilip Kumar Mohapatra, Ghanshyam Ji, Mrudul George, and N. Swathi

1 Introduction

The electric vehicle market is growing all over the world and is starting to go mainstream. Manufacturers are competing to make modern stylish dashboards in electric vehicles, since it is one of the most important deciding factors for any electric vehicle buyer. Integrated smart dashboard design is getting more important these days. An ideal dashboard has to integrate a lot of features, and the dashboard design becomes complex when it has to fit with the main functions of the dashboard. GPS integration of charging stations along with other features is more challenging. Incorporating all these features without losing the look and appeal is not an easy task. The design of the dashboard for most EVs is still modest and is lacking in the aspect of dimensional design, appeal, and other ergonomic aspects. Therefore, the need is to design an efficient and improved smart dashboard for a smooth driving experience. To design a smart dashboard which is a graphical interface, we can use platforms like Android, iOS, Windows, etc. An android application is used in this project for the user interface design. Our dashboard will display details such as the calculated time to reach destination, speed, battery charge level, and temperature. As shown in Fig. 1, the dashboard in the proposed model displays the estimated SOC of the battery. Here, we are using GRU, RNN, ANN, and LSTM to solve various regression and prediction analysis. The precise SoC assessment of the EV battery is the first stage in

N. Darapaneni
Northwestern University/Great Learning, Evanston, USA

A. R. Paduri (✉) · B. G. Sudha
Great Learning, Bangalore, India
e-mail: anwesh@greatlearning.in

D. K. Mohapatra · G. Ji · M. George · N. Swathi
PES University, Bangalore, India

A. Yadav et al. (eds.), *Proceedings of International Conference on Paradigms of Communication, Computing and Data Analytics*, Algorithms for Intelligent Systems,
https://doi.org/10.1007/978-981-99-4626-6_9

Fig. 1 AI-based smart dashboard

dashboard development. The SoC is the fundamental battery metric, and it serves as the foundation for practically all other forecasts.

2 Literature Review

In electric vehicles, the energy comes from the battery. The majority of electric vehicles employ lithium-ion batteries for a variety of reasons. These batteries are well-liked because of their high energy density, extremely low discharge rate, and low cost of maintenance.

A vital battery parameter that shows the battery's usable capacity is state of charge. It is crucial for battery management, so it is also important to estimate [1]. A battery's state of charge is defined as the ratio between its nominal full capacity and its current available capacity [2, 3]. But the SoC varies with temperature and battery age [4]. Any electric vehicle (EV) driver can benefit from prior knowledge of the battery's state of charge (SOC), as it helps him determine whether the destination can be reached with the battery's capacity. Without adequate monitoring and upkeep, the battery could experience major problems like overcharging and fire risks [5]. It varies on a number of variables, including the battery's age, temperature, and construction material [6]. A correct SOC estimation is a difficult endeavor, because of the varied architectures and the selection of hyperparameters [7, 8]. Since the state of charge (SoC) of a battery is not linearly related to current and temperature, estimating the SoC can be difficult. Direct measurement, book-keeping estimation, and AI techniques can all be used to calculate it [9]. AI methods are the best due to its tremendous learning capabilities. There are a number of SoC estimation methods. Some of them are Coulomb counting methods, model-based methods, OCV testing methods, ML and AI methods. Though the Coulomb method is easy, it has some drawbacks [10]. Here, the estimated value depends on the initial value of SoC which is not always available. Also, in this method the total accumulated errors are found to be high [11].

Open circuit voltage (OCV) is conducted at a stable or idle state of the battery [12]. It requires that the battery be kept in an unused state for a long time, usually around 24 h. This is practically difficult. Model-based methods generally use adaptive filter algorithms (AFA) [13, 14]. These algorithms require a detailed and thorough knowledge about the internal characteristics of the battery. The Kalman filter model is built on the assumption that the dynamic model and statistical property of the battery is known in advance and established precisely [15]. Also, the parametrization of the battery requires various tests which are time consuming [16]. Data-driven methods are comparatively simpler when compared with the previous methods. They forecast the output SoC by simply mapping the input values, which are mostly voltage, current, and temperature. They are often used for SoC estimates due to their simplicity. A number of algorithms are there for SoC estimation such as ANN, SVM, or extreme learning machines. These methods can be either machine learning methods or deep learning methods. Depending upon the cell types and operations, we can apply different algorithms to learn battery parameters and SoC of the battery. PFNN algorithm which is an improved version of Fuzzy Neural Network can be applied to lithium battery charge estimation which improves the transient properties of voltage regulation when there is a load change in the battery. The ML methods such as SVR, RF, regression or Gaussian are also used widely due to their simple execution. But ML methods have only two layers for input and output [17]. These algorithms take each individual input, process it, and give the output. But they don't save any intermediate states or history information. Hence, these models cannot effectively incorporate the complicated electro-chemical and electro-thermal characteristics of the battery [17]. In contrast, the DL methods such as CNN and RNN have many computational layers. The efficiency of the deep learning methods can be improved by adding more computational layers [17]. Due to the issue of gradient vanishing, pure RNNs are unable to explain long-term dependencies in the data. The gradient is exponentially back propagated in RNN. But by using a memory cell, LSTM permits the gradient to continue to flow as before. Algorithms like GRU and LSTM (a modified RNN) deal with input variables (current, voltage and temperature) at every time step.

Chennali used LSTM with different layers for SoC estimation and validated the value of SoC at different temperatures. According to his studies, LSTM provides more accurate SoC estimation than other generic algorithms. The degree to which the model has learned information from the training data depends on the SoC prediction accuracy. With the use of gates at various time steps, the LSTM model gathers data from the training set. The LSTM decides what to remember and what to forget by using input, forget, and output gates; as a result, it can handle long-term dependencies. The LSTM algorithm can collect more information from the training data with the historical time of input for specified time steps.

Since the state of the battery is a time series data, it can be estimated with maximum accuracy with algorithms which have memory to store intermediate battery states. In order to estimate SoC to match with the nonlinear features of the battery, the LSTM technique is proposed in this study. Each data point is processed by LSTM, an upgraded version of RNN, which then saves the state with historical information for each layer. The information to be carried forward is stored in the memory cell, and it ensures that no information is lost across the layers [18, 19]. Since LSTM is a fully data-driven model, it can be used to estimate the SOC of many battery varieties [20]. Also, the performance accuracy is estimated using MAE and RMSE.

3 Materials and Methods

3.1 Data Collection

We faced real-time challenges in collecting the electric vehicles dataset. We got a few electric vehicles datasets but those were either limited in the data size or limited in the features. Finally, we succeeded in collecting the data from the below sources.

1. Calce [21]: Calce Battery Research Group has provided a very good number of datasets related to lithium-ion batteries. It has a wide variety of datasets collected at different temperatures with a lot of features. We can use this dataset for all our research after citing the reference.
2. Mendeley [22]: Mendeley data is a free and secured cloud platform where we can use the data for research and publishing by citing the reference of the site. Here, we have data with sufficient size and limited number of features where data is captured at different temperatures. The batteries used are lithium-ion batteries. We decided to go ahead with the above dataset. Our data consists of 13,269 rows with 44 features. The data collected is 5 days EV battery (Li-ion) data. The data for electric vehicles was collected at different temperatures, different states like vehicle state charger connected, vehicle state lock, vehicle state unlock, etc. State of charge (SoC) is dependent on the ambient temperature and may cause errors in SoC estimation. To address this issue, we collected data for different temperatures like 0°, 10°, 25°, 40°. We have taken 44 features in our model.

3.2 Data Preparation

For our analysis, we have taken the data at 25 °C room temperature. The data is stored in an excel file which has around 13,269 rows. The data had few columns having more than 80 percent of the data as null. We analyzed the respective domains of those columns and discussed that with the field experts and then understood that there is no significance of those features. Hence, we removed those features which had many nulls. We have cleaned the data and removed the outliers using the IQR method. We have imputed the nulls in the data. Since we had a categorical variable, we applied the get_dummies function and encoded it. The correlation between different features has been plotted using a heat map. The most important features such as current, voltage, and temperature for SoC estimation have been identified and plotted. In our analysis, we noticed that the voltage is negatively skewed for 30 bins. Since it is negatively skewed the mean < median < mode.

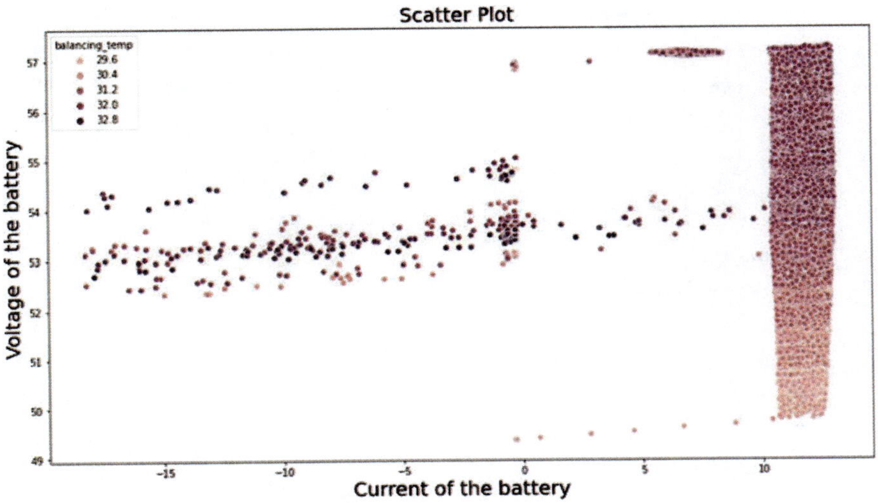

Fig. 2 Scatter plot

3.3 Data Exploration

(1) *Scatter Plot:* As shown in the diagram, Fig. 2 shows us the relationship between the quantitative variables current and the voltage. As we see in the graph, most of the voltage is distributed around +10 and +13 current's value. Scatter plot you can see less points between −15 to +10. We can observe with the color codes how the temperature is distributed. As we observed there is a positive correlation between the voltage and current of the battery at different temperatures.

We also observed that there is a positive correlation between voltage and current of the battery using a scatter plot, and also, we observed that there is a positive correlation between the voltage, current, and temperature.

3.4 Data Visualization

1. *Current and Voltage Plot:* Current and the voltage are directly proportional. When there is an increase in voltage, current is also increased. In Fig. 3, current remains constant at a certain threshold.
2. *Measured SoC versus Voltage:* We can see in Fig. 4 how the measured SoC is distributed across the voltage. Voltage and display SoC are linearly related.
3. *Measured SoC versus State:* We can see in Fig. 5 how the measured SoC is distributed across the voltage. Voltage and display SoC are linearly related.
4. *Actual and Prediction data for ANN:* Actual and prediction have lots of differences as shown in Fig. 6.

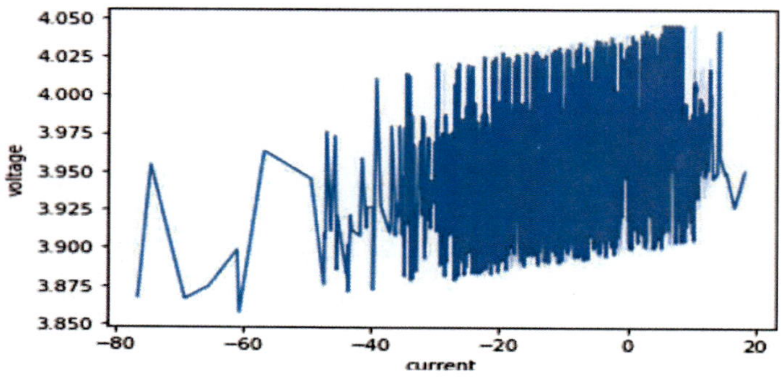

Fig. 3 Current versus voltage

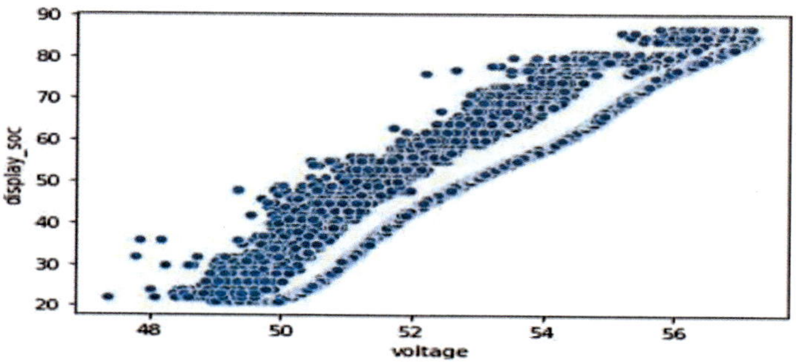

Fig. 4 Measured SoC versus voltage

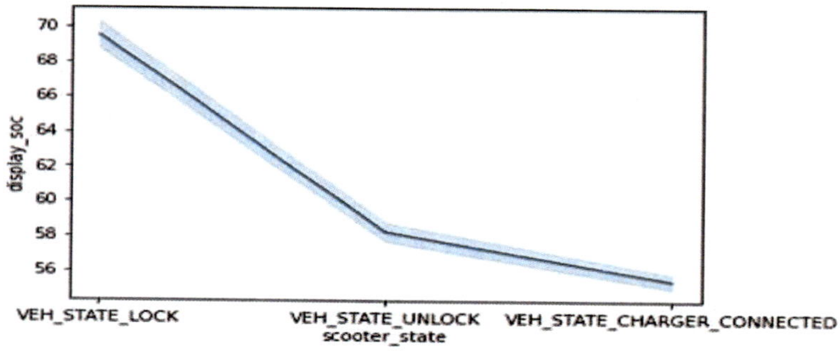

Fig. 5 Measured SoC versus state

ANN loss refers to the difference between the predicted output of the model which is shown in the red line and the actual target value which is shown in the blue line. We have used MAE as our loss function.

5. *Actual and Prediction data for LSTM:* Calculation of the actual and prediction loss in LSTMs is similar to that in ANNs, but the architecture of LSTMs allows them to handle sequential data and capture long-term dependencies. We have

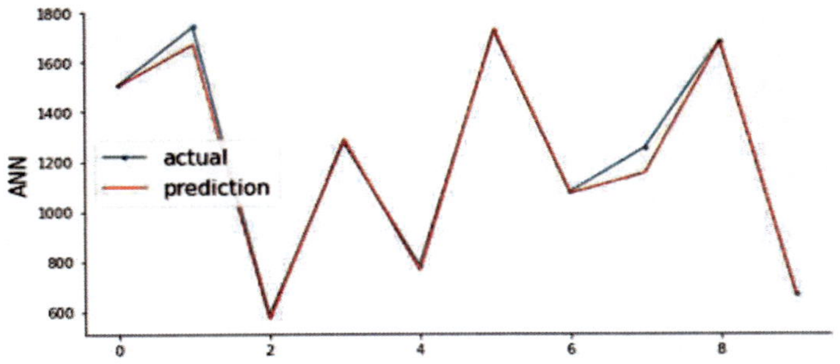

Fig. 6 Actual and prediction data for ANN

used MAE as our loss function. Actual and prediction loss are almost similar as shown in Fig. 7.

6. *Actual and Prediction data for GRU:* Calculation of the actual and prediction in GRUs is like that in LSTMs and ANNs, but the architecture of GRUs allows it to handle sequential data and capture long-term dependencies in a computationally efficient manner compared to LSTMs.

We have used MAE as our loss function. Actual and prediction loss have slight differences as shown in Fig. 8.

7. *Train and Test Loss of LSTM:* The train and test loss of an LSTM model refers to the measurement of the error in the output produced by the model on the training data (train loss) and on unseen data (test loss). As shown in Fig. 9, the model seems to fit correctly.
8. *Train and Test Loss of GRU:* A GRU model's "train and test loss" refers to the measurement of the output error on training data (train loss) and on unobserved data (test loss).

As shown in Fig. 10 we monitor the train and test loss during the process, it seems to fit correctly. The train and test loss in GRUs is slightly more than that in LSTMs.

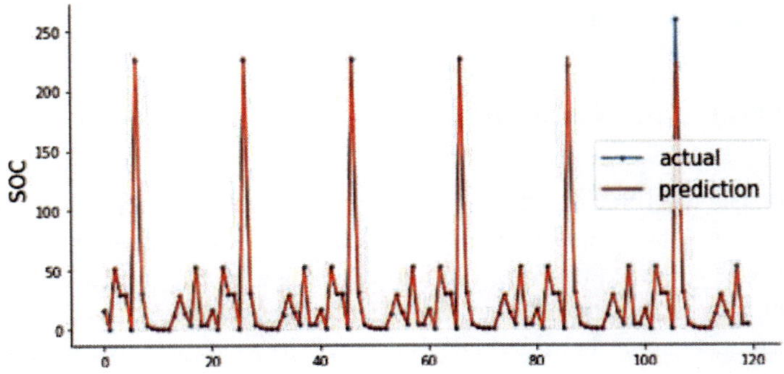

Fig. 7 Actual and prediction data for LSTM

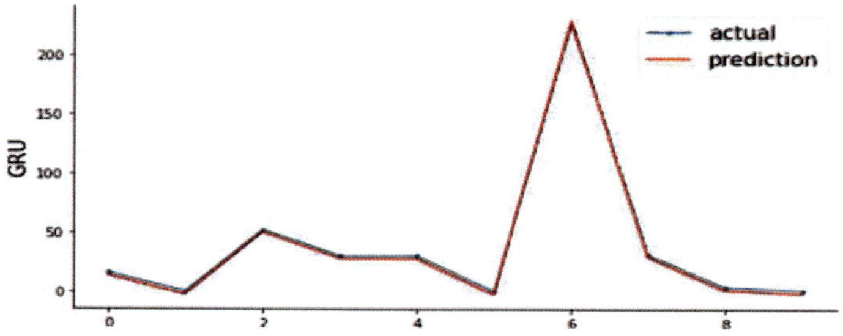

Fig. 8 Actual and prediction data for GRU

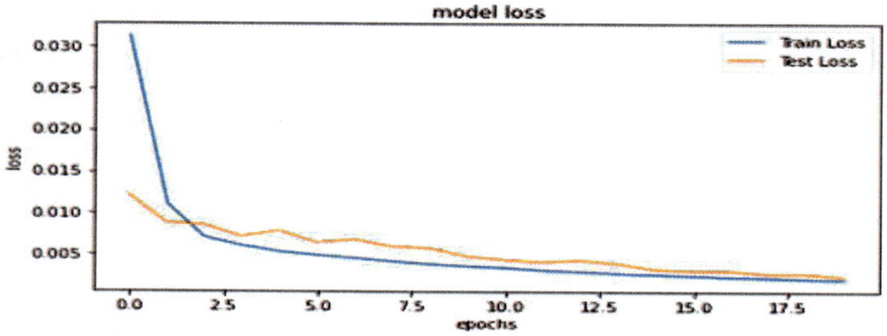

Fig. 9 Train and test loss of LSTM

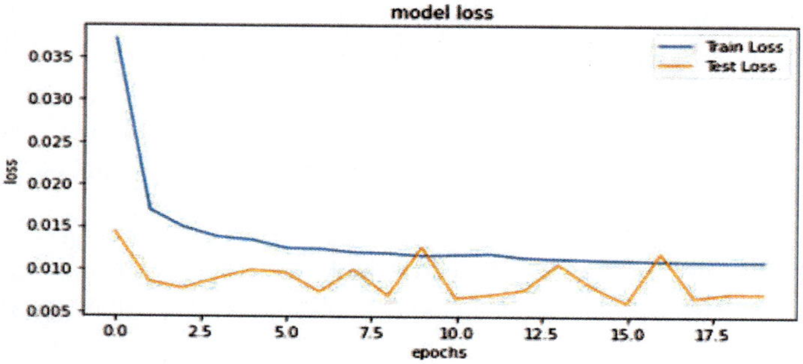

Fig. 10 Train and test loss of GRU

4 Discussion and Conclusion

For smart dashboard SoC estimation is important as all other data is dependent on SoC percentage value. So, we started our research on SoC estimation.

Dataset was collected from various sources such as Calce dataset and Mendeley dataset. Our experiment included additional features such as vehicle state, total regenerative current, as input to ANN, GRU, and LSTM models to estimate SoC more precisely.

As part of data exploration, some of our observations are.

- Current increases with respect to voltage of the LiB initially for 20 percent of the voltage and then remains steady.
- We found that the battery SoC decreases fast when the vehicle is in an unlocked state.
- SoC percentage decreases as temperature increases.

By fitting our data into a hyperplane that can describe the link between the different variables, we are trying to determine the effectiveness of various algorithms for predicting SOC in this case. Applying various memory-integrated algorithms allows us to do this. These algorithms use similar training techniques and base their output on previous input. All of the network's layers have the same parameters. Because of the size of the dataset, LSTM is preferable here. GRU, however, is quicker for smaller datasets. They take longer to train and have a lower convergence rate. LSTMs learn more quickly than GRUs since they have more parameters.

In our investigation, the LSTM reaches the least loss point and the best accuracy after running the model for 26 epochs. The loss curve remains unchanged after this specific point, demonstrating the highest accuracy.

We found from the experimental results that, as given in Table 1, out of all the models tested, LSTM gives SoC percentage with highest accuracy. It has the lowest MAE value of 0.8450. Hence, we selected the LSTM algorithm to use in our smart dashboard for display of SoC.

As shown in Fig. 11, ANN has higher loss when compared with the LSTM, GRU also has slightly higher loss when compared with LSTM. So, we have reached the conclusion that for SoC models LSTM performs better and gives the best result with the loss measure MAE.

Therefore, we are developing a dashboard that will include current, voltage, temperature, and battery state to estimate the SoC of the battery using a variety of neural network methods, including ANN, GRU, and LSTM. But we discovered that LSTM combined with MAE yields the best outcome.

Table 1 I Comparison of loss with different models

Models		
Model Name	RMSE	MAE
ANN	35.2414	20.3774
LSTM	4.0595	0.8450
GRU	5.0522	2.5490

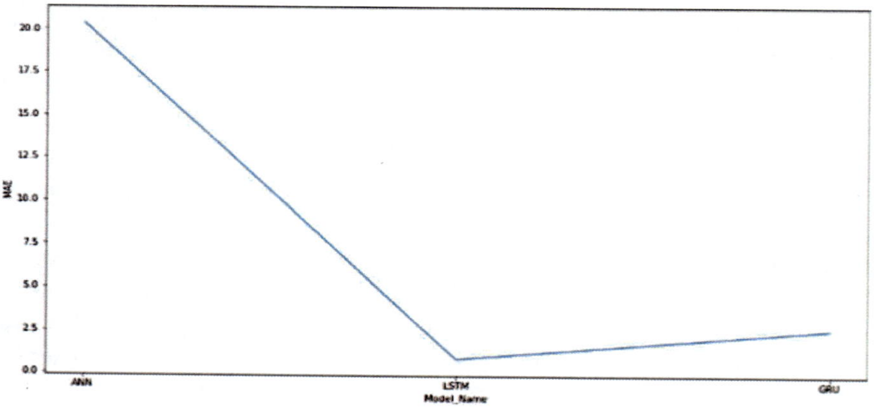

Fig. 11 Line graph for loss for three different models

References

1. Trivedi M, Kakkar R, Gupta R, Agrawal S, Tanwar S, Niculescu V-C, Raboaca MS, Alqahtani F, Saad A, Tolba A (2022) Blockchain and deep learning-based fault state of charge and state of energy estimation for lithium-ion batteries based on a long short- term memory neural network. detection framework for electric vehicles. Mathematics 10(19):3626
2. Wang W, Wang X, Xiang C, Wei C, Zhao Y (2018) Unscented kalman filter-based battery soc estimation and peak power prediction method for power distribution of hybrid electric vehicles. IEEE Access 6:35 957–35 965
3. Wu X, Li X, Du J (2018) State of charge estimation of lithium-ion batteries over wide temperature range using unscented kalman filter. IEEE Access 6:41 993–42 003
4. Lipu MH, Hannan M, Hussain A, Ayob A, Saad MH, Karim TF, How DN (2020) Data-driven state of charge estimation of lithium- ion batteries: algorithms, implementation factors, limitations and future trends. J Clean Prod 277:124110
5. Ilott AJ, Mohammadi M, Schauerman CM, Ganter MJ, Jerschow A (2018) Rechargeable lithium-ion cell state of charge and defect detection by in-situ inside-out magnetic resonance imaging. Nat Commun 9(1):1776
6. Yong JY, Ramachandaramurthy VK, Tan KM, Mithulananthan N (2015) A review on the state-of-the-art technologies of electric vehicle, its impacts and prospects. Renew Sustain Energy Rev 49:365–385
7. Hannan MA, How DN, Mansor MB, Lipu MSH, Ker PJ, Muttaqi KM (2021) State-of-charge estimation of li-ion battery using gated recurrent unit with one-cycle learning rate policy. IEEE Trans Ind Appl 57(3):2964–2971
8. How DN, Hannan MA, Lipu MSH, Sahari KS, Ker PJ, Muttaqi KM (2020) State-of-charge estimation of li-ion battery in electric vehicles: a deep neural network approach. IEEE Trans Ind Appl 56(5):5565–5574
9. Lipu MH, Hannan M, Karim TF, Hussain A, Saad MHM, Ayob A, Miah MS, Mahlia TI (2021) Intelligent algorithms and control strategies for battery management system in electric vehicles: progress, challenges and future outlook. J Clean Prod 292:126044
10. Li Z, Huang J, Liaw BY, Zhang J (2017) On state-of-charge determination for lithium-ion batteries. J Power Sources 348:281–301
11. Hu X, Feng F, Liu K, Zhang L, Xie J, Liu B (2019) State estimation for advanced battery management: Key challenges and future trends. Renew Sustain Energy Rev 114:109334
12. Lin F-J, Huang M-S, Yeh P-Y, Tsai H-C, Kuan C-H (2012) Dsp- based probabilistic fuzzy neural network control for li-ion battery charger. IEEE Trans Power Electron 27(8):3782–3794

13. Anton JCA, Nieto PJG, Viejo CB, Vila´n JAV (2013) Support vector machines used to estimate the battery state of charge. IEEE Trans Power Electron 28(12):5919–5926
14. Lipu MSH, Hannan MA, Hussain A, Saad MH, Ayob A, Uddin MN (2019) Extreme learning machine model for state-of-charge estimation of lithium-ion battery using gravitational search algorithm. IEEE Trans Ind Appl 55(4):4225–4234
15. Misyris GS, Doukas DI, Papadopoulos TA, Labridis DP, Agelidis VG (2018) State-of-charge estimation for li-ion batteries: a more accurate hybrid approach. IEEE Trans Energy Convers 34(1):109–119
16. Xiong R, Cao J, Yu Q, He H, Sun F (2017) Critical review on the battery state of charge estimation methods for electric vehicles. IEEE Access 6:1832–1843
17. Liu Y, Zhao G, Peng X (2019) Deep learning prognostics for lithium-ion battery based on ensembled long short-term memory networks. IEEE Access 7:155 130–155 142
18. Song X, Yang F, Wang D, Tsui K-L (2019) Combined cnn-lstm network for state-of-charge estimation of lithium-ion batteries. IEEE Access 7:88 894–88 902
19. Zou Y, Hu X, Ma H, Li SE (2015) Combined state of charge and state of health estimation over lithium- ion battery cell cycle lifespan for electric vehicles. J Power Sour 273:793–803
20. Berecibar M, Gandiaga I, Villarreal I, Omar N, Van Mierlo J, Van den Bossche P (2016) Critical review of state of health estimation methods of li-ion batteries for real applications. Renew Sustain Energy Rev 56:572–587
21. Calcel (2020) Calce battery research grou
22. Philip VCNMSMK (2020) Lg 18650hg2 li-ion battery data and example deep neural network xev soc estimator script

Chapter 10
Solving Systems of Nonlinear Equations Using Jaya and Jaya-Based Algorithms: A Computational Comparison

Sérgio Ribeiro, Bruno Silva, and Luiz Guerreiro Lopes

1 Introduction

Nonlinear equation have an extensive importance in many domains of knowledge, including Physics, Economics, Chemistry, and several branches of Engineering [1] and appear in almost all simulations of physical processes [2].

Nevertheless, there is no general numerical method that is robust and efficient enough to solve systems of nonlinear equations (SNLEs), which is perhaps the hardest problem in numerical mathematics [3]. Newton's method is a well-known and usual method for solving SNLEs [4]. As with the majority of its variants, its success is dependent on the quality of the initial approximations chosen [5].

However, a SNLE can be effortlessly converted into a homologous optimization problem by adopting the sum of each system equation's absolute value as an objective function to be minimized:

S. Ribeiro
Graduate Program in Informatics Engineering, University of Madeira, Funchal, Madeira Is., Portugal

B. Silva
Doctoral Program in Informatics Engineering, University of Madeira, Funchal, Portugal
e-mail: bruno.silva@madeira.gov.pt

Regional Secretariat for Education, Science and Technology, Regional Government of Madeira, Funchal, Portugal

L. G. Lopes (✉)
Faculty of Exact Sciences and Engineering, University of Madeira, 9020–105 Funchal, Madeira Is., Portugal
e-mail: lopes@uma.pt

A. Yadav et al. (eds.), *Proceedings of International Conference on Paradigms of Communication, Computing and Data Analytics*, Algorithms for Intelligent Systems,
https://doi.org/10.1007/978-981-99-4626-6_10

$$\min f(\mathbf{x}) = \sum_{j=1}^{n} \left| e_j(\mathbf{x}) \right|, \tag{1}$$

where $e_j(\mathbf{x}) = 0$ is the jth equation, and $\mathbf{x} = (x_1, \ldots, x_n)^T$.

The resulting optimization problem can be solved by stochastic approaches such as pure or hybrid metaheuristic algorithms (see, e.g., the review in [6] and the references therein), which have a strong ability to determine near-optimal solutions while being more flexible, effective, and robust when comparing to deterministic optimization methods [7], particularly on large-scale optimization problems, and without requiring good initial approximations, albeit with the trade-off that the quality of the obtained solutions cannot be guaranteed.

Newton's method, with its aforementioned limitations, has been hybridized with different metaheuristic algorithms in order to combine the positive characteristics of both classes of methods when solving SNLEs. Its hybridization with the Harris hawks algorithm [8] is a recent example.

All population-based metaheuristic algorithms require the adjustment of parameters like number of iterations and population size, while other requires the same with algorithmic-specific variables. The performance of such algorithms can be negatively affected by the poor adjustment of these parameters, with the added difficulty that there is sometimes a lack of guidance on how to properly select them.

Various population-based algorithms have been successfully adapted to efficiently solve SNLEs, including different hybridizations of the particle swarm optimization algorithm (see, e.g., [9–11]) and of the sine cosine algorithm (see, e.g., [12, 13]). However, these algorithms require the specification of a set of initial parameters, which has the disadvantage of being determinant for the quality of the results.

To address such issue, the Jaya algorithm was proposed in [14] as a parameter-less metaheuristic approach that is both efficient and easy to set up. When compared to other population-based algorithms, Jaya has some advantages, including being generic, simple to implement, and not relying on algorithm-specific parameters [15]. As a result, it is considered a parameter-less algorithm, as it only demands the most essential control parameters of a population-based algorithm.

Different variants to the original Jaya algorithm have been proposed intending to improve its effectiveness and performance, either by tweaking some of its operators or by better balancing global and local search spaces.

This study examines some Jaya-based algorithms with various types of modifications to the original Jaya algorithm. The chosen variants include the refinement of the main equation, the use of additional population data and an enhanced search strategy, the utilization of multiple populations, and the use of an oppositional population.

The performance of Jaya and some Jaya-based variants in solving SNLEs is here investigated, and a comparative analysis is performed on the results produced by each algorithm. The aim is to assess whether the Jaya algorithm and some of its variants can be used efficiently to deal with this type of problem and to answer which variant is more successful than others in this particular application domain.

2 Related Work

2.1 Jaya Algorithm

Jaya is a population-based metaheuristic for optimizing different unconstrained and constrained problems [14, 16]. The fundamental tenet of the algorithm is that the solution obtained for a particular problem should avoid the worst solution while tending toward the best one.

This strategy just relies on standard control parameters such as the dimension $popSize$ of the population and the maximum of iterations allowed ($maxIter$), in order to achieve its main goal which is to optimize (i.e., minimize or maximize) an objective function $f(\mathbf{x})$.

Considering the number of decision variables ($numVar$), a design variable index $v \in [1, numVar]$, a population index $p \in [1, popSize]$, and an iteration $i \in [1, maxIter]$, let $\mathbf{x}_{v,p,i}$ be the value of the vth variable of the pth population candidate during the ith iteration. Then, the modified value $\mathbf{x}^{\text{new}}_{v,p,i}$ is obtained as follows:

$$\mathbf{x}^{\text{new}}_{v,p,i} = \mathbf{x}_{v,p,i} + r_{1,v,i}\left(\mathbf{x}_{v,best,i} - \left|\mathbf{x}_{v,p,i}\right|\right) - r_{2,v,i}\left(\mathbf{x}_{v,\text{worst},i} - \left|\mathbf{x}_{v,p,i}\right|\right), \tag{2}$$

where $r_{1,v,i}$ and $r_{2,v,i}$ are random numbers in [0, 1] for the vth variable during the ith iteration, while $\mathbf{x}_{v,\text{best},i}$ and $\mathbf{x}_{v,\text{worst},i}$ are the candidate solutions with the best and the worst fitness values, respectively.

The Jaya pseudocode is presented in Algorithm 1.

Algorithm 1 Jaya algorithm

Initialize $numVars$, $popSize$ and $maxIters$;
Generate initial population X;
Evaluate the fitness $f(\mathbf{x})$ of each $\mathbf{x} \in X$;
while $i < maxIter$ and not terminate **do**
    Determine $\mathbf{x}_{v,best,i}$ and $\mathbf{x}_{v,worst,i}$;
    for $p \leftarrow 1, popSize$ **do**
        for $v \leftarrow 1, numVars$ **do**
            Update $\mathbf{x}^{new}_{v,p,i}$ by Eq. (2);
        end for
        Calculate $f(\mathbf{x}^{new}_{v,p,i})$;
        if $f(\mathbf{x}^{new}_{v,p,i})$ is better than $f(\mathbf{x}_{v,p,i})$ **then**
            $\mathbf{x}_{v,p,i} \leftarrow \mathbf{x}^{new}_{v,p,i}$;
            $f(\mathbf{x}_{v,p,i}) \leftarrow f(\mathbf{x}^{new}_{v,p,i})$;
        else
            Keep $\mathbf{x}_{v,p,i}$ and $f(\mathbf{x}_{v,p,i})$ values;
        end if
    end for
end while
Report solution found;

2.2 Modified Jaya Algorithm

One drawback of heuristic and nature-inspired algorithms is the exploration phase, in which the algorithm may get trapped into local optima. Jaya algorithm is not immune to premature convergence and this is one of the main aspects addressed by its different variants, as is the case of the modified Jaya (MJAYA) algorithm [17].

MJAYA follows the same procedures of the original Jaya algorithm but proposes a modified main equation given by

$$\mathbf{x}_{v,p,i}^{\text{new}} = \mathbf{x}_{v,p,i} + r_{1,v,i}\left(\mathbf{x}_{v,\text{worst},i} - \left|\mathbf{x}_{v,p,i}\right|\right) - L \times r_{2,v,i}\left(\left|\mathbf{x}_{v,p,i}\right|^2 - \mathbf{x}_{v,\text{best},i}^2\right), \tag{3}$$

where the objective function term with the best fitness value is only used to adjust the values of the remaining terms, and the coefficient L at every iteration is determined as follows:

$$L = \begin{cases} 1, & \text{if } rand > 0.5 \\ -1, & \text{otherwise.} \end{cases} \tag{4}$$

2.3 Enhanced Jaya Algorithm

The enhanced Jaya (EJAYA) algorithm [18] improves upon the original implementation of Jaya by making efficient use of additional information from the population to prevent exploration from being trapped in local optima, thereby minimizing the risk of premature convergence.

EJAYA uses the original Jaya algorithm parameters, such as the current best and worst solutions, while introducing new parameters such as the mean solution and historical solutions to balance its global exploration ability and local exploitation strategy.

The exploitation strategy is used to try to inhibit the algorithm from becoming entangled in a local optimum by removing the main function's absolute value symbol, and instead using an upper local (P_u) and a lower local (P_l) attract points. P_u is determined in the following way:

$$P_u = \lambda_3 \times \mathbf{x}_{v,\text{best},i} + (1 - \lambda_3) \times M, \tag{5}$$

where λ_3 is an uniformly distributed random number in [0, 1], $\mathbf{x}_{v,\text{best},i}$ is the best candidate in terms of fitness value, and M is the mean of the current population, defined as:

$$M = \frac{1}{popSize} \sum_{p=1}^{popSize} \mathbf{x}_p. \tag{6}$$

The lower point P_l is written as

$$P_l = \lambda_4 \times \mathbf{x}_{v,\text{worst},i} + (1 - \lambda_4) \times M, \tag{7}$$

where λ_4 is a random number in [0, 1] with uniform distribution, $\mathbf{x}_{v,\text{worst},i}$ is the worst candidate in terms of fitness value, and M is the current population mean, defined in Eq. (6).

EJAYA's local exploitation approach is as follows:

$$\mathbf{x}^{\text{new}}_{v,p,i} = \mathbf{x}_{v,p,i} + \lambda_5 \left(P_u - \mathbf{x}_{v,p,i}\right) - \lambda_6 \left(P_l - \mathbf{x}_{v,p,i}\right), \tag{8}$$

where λ_5 and λ_6 are uniformly distributed random numbers in the interval [0, 1].

The global exploration approach was inspired by the backtracking search algorithm [19] and uses differential vectors between the current and historical (i.e., old) populations, providing additional search space when compared to vectors from the same generation population.

In the first interaction, the historical population $X^{\text{old}}_{v,p}$ is the same as $X_{v,p}$. Afterward, they are selected in the following way:

$$X^{\text{old}}_{v,p} = \begin{cases} X_{v,p}, & \text{if } P_{\text{switch}} \leq 0.5 \\ X^{\text{old}}_{v,p}, & \text{otherwise,} \end{cases} \tag{9}$$

where P_{switch} is a random number with a uniform [0, 1] distribution, which defines the switching probability between the two populations.

After selecting the population, the EJAYA algorithm randomly rearranges the elements $\mathbf{x}^{\text{old}}_{v,p,i}$ of the historical population $X^{\text{old}}_{v,p}$ by applying a shuffling function *permuting*(·) to the entire historical population, as shown below:

$$X^{\text{old}}_{v,p} = permuting\left(X^{\text{old}}_{v,p}\right). \tag{10}$$

The objective function for the global exploration strategy is expressed as:

$$\mathbf{x}^{\text{new}}_{v,p,i} = \mathbf{x}_{v,p,i} + k \times \left(\mathbf{x}^{\text{old}}_{v,p,i} - \mathbf{x}_{v,p,i}\right), \tag{11}$$

where k is a standard normally distributed random number.

In EJAYA, the local exploitation (LES) and global exploration (GES) strategies are both equally relevant, and as such, the update strategy is selected as follows:

$$\text{Strategy} = \begin{cases} \text{LES}, & \text{if } P_{\text{select}} > 0.5 \\ \text{GES}, & \text{otherwise,} \end{cases} \tag{12}$$

where P_{select} is a random number in [0, 1] with uniform distribution. Furthermore, the historical population $X^{\text{old}}_{v,p}$ and the current population $X_{v,p}$ are initialized by the same method.

2.4 SAMP–Jaya Algorithm

The self-adaptive multi-population Jaya (SAMP–Jaya) combines the ideas of Jaya with the island model from genetic algorithms (GAs) [20], although with some modifications.

Instead of dividing the population into only two groups, named master island and slave island as in the basic island model from GA, in SAMP–Jaya, the number of sub-populations (or slave islands) is determined programmatically on the basis of the current problem state's characteristics. The population migration between islands is based on the quality of the fitness value and a greedy selection mechanism.

The number of sub-populations is adjusted along the search phase. Attempting to maintain diversity and augment the exploratory process, newly created solutions (which are randomly generated) are used to replace duplicate ones. The variable m, whose value at the beginning of the execution is $m = 2$, is used to specify the number of distinct sub-populations.

2.5 Oppositional Jaya Algorithm

The oppositional Jaya (OJaya) approach [21] offers two improvements over the original Jaya algorithm. One is provided by the oppositional learning (OL), a population-based algorithm that simultaneously calculates and evaluates the current (X) and oppositional (X^o) populations to choose the best one for the following generation, and another by the distance–adaptive coefficient (DAC), which is determined based on the best and worst positions. The first method provides an expansion of the search space and promotes population diversity and strength, whereas the second causes the population to move faster in the direction of the best position and away from the worst one.

In OJaya, the oppositional population elements are generated as follows:

$$\mathbf{x}^o_{v,p,i} = s \times \left(A_{v,i} + B_{v,i}\right) - \mathbf{x}_{v,p,i}, \tag{13}$$

where s is a random number in [0, 1], and $A_{v,i}$ and $B_{v,i}$ are dynamic bounds for the population, which are given by

$$A_{v,i} = \min\left(\mathbf{x}_{v,p,i}\right), \qquad B_{v,i} = \max\left(\mathbf{x}_{v,p,i}\right). \tag{14}$$

Both dynamic bounds $A_{v,i}$ and $B_{v,i}$ are set to be updated every 50 iterations to prevent the population from becoming stuck in a local minimum as the search space shrinks with each iteration.

As these dynamic bounds have the potential to cause $\mathbf{x}^o_{v,p,i}$ to escape the minimum and maximum limits of constrained problems, it is necessary to reset $\mathbf{x}^o_{v,p,i}$ in the following manner when this occurs:

$$\mathbf{x}^{o}_{v,p,i} = rand\left(A_{v,i}, B_{v,i}\right), \tag{15}$$

where $rand\left(A_{v,i}, B_{v,i}\right)$ is a uniformly distributed random number in $[A_{v,i}, B_{v,i}]$.

The oppositional learning is used when generating both the initial and the current population of each iteration.

In order to achieve the benefits offered by DAC and provide fine-tuning of the population convergence in the latter stages of the exploration to find the global optima, the distance-adaptive coefficient d_i is determined as follows:

$$d_i = \begin{cases} \left(\frac{f(\mathbf{x}_{v,\text{best},i})}{f(\mathbf{x}_{v,\text{worst},i})}\right)^2, & \text{if } f\left(\mathbf{x}_{v,\text{worst},i}\right) \neq 0 \\ 1, & \text{otherwise.} \end{cases} \tag{16}$$

The main function of OJaya is comparable to that of the original Jaya, with the addition of the d_i factor, as shown below:

$$\mathbf{x}^{\text{new}}_{v,p,i} = \mathbf{x}_{v,p,i} + r_{1,v,i}\left(\mathbf{x}_{v,\text{best},i} - \left|\mathbf{x}_{v,p,i}\right|\right) - d_i \times r_{2,v,i}\left(\mathbf{x}_{v,\text{worst},i} - \left|\mathbf{x}_{v,p,i}\right|\right). \tag{17}$$

As d_i is a function of $\mathbf{x}_{v,\text{best},i}$ and $\mathbf{x}_{v,\text{worst},i}$, whose distance gradually decreases during the iterations, its value is small at the beginning of the search process and gradually converges to 1 as the process approaches the end. This is the self-adaptive nature of d_i, which is achieved without the need for additional parameters.

3 Computational Experiments

3.1 Experimental Setting and Implementation

The population size adopted varied with the dimensionality of the test problem. For all algorithms considered in this study, it was set to 10× the problem dimension, which in this study was taken as equal to 4, 8, 12, 16, and 20.

The maximum number of iterations for every algorithm under consideration was set to 1000× the problem dimension, while the number of independent runs for each algorithm and problem was equal to 51, as suggested in [22].

The implementation was done in Julia programming language using double precision floating-point arithmetic. Computational experiments were conducted on a computer with an AMD processor Ryzen 5 3500X and 16 GB RAM DDR4.

3.2 Test Problems

The methaphor-less optimization algorithm considered were tested on a set of difficult nonlinear equation system problems. The benchmark problems selected from the literature are presented below.

In addition to the definition of the functions, the domain D chosen for each test problem is also indicated. The test problems are scalable with respect to the number n of variables.

Taking into account that five different dimensions were considered for each of the 14 problems shown below, the computational analysis carried out involved a total of 70 different nonlinear equation system problems.

Problem 1 ([23], Schubert–Broyden function), $n = 4, 8, 12, 16, 20$.

$f_1(\mathbf{x}) = (3 - x_1)x_1 + 1 - 2x_2$

$f_i(\mathbf{x}) = (3 - x_i)x_i + 1 - x_{i-1} - 2x_{i+1}, \quad i = 2, \ldots, n-1$

$f_n(\mathbf{x}) = (3 - x_n)x_n + 1 - x_{n-1}$

$D = ([-100, 100], \ldots, [-100, 100])^T$

Problem 2 ([24], prozlem D1–Modified Rosenbrock), $n = 4, 8, 12, 16, 20$.

$$f_{2i-1}(\mathbf{x}) = \frac{1}{1 + \exp(-x_{2i-1})} - 0.73$$

$f_{2i}(\mathbf{x}) = 10(x_{2i} - x_{2i-1}^2), \quad i = 1, \ldots, \frac{n}{2}$

$D = ([-10, 10], \ldots, [-10, 10])^T$

Problem 3 ([24], Problem D3–Powell badly scaled), $n = 4, 8, 12, 16, 20$.

$f_{2i-1}(\mathbf{x}) = 10^4 x_{2i-1}x_{2i} - 1$

$f_{2i}(\mathbf{x}) = \exp(-x_{2i-1}) + \exp(-x_{2i}) - 1.0001, \quad i = 1, \ldots, \frac{n}{2}$

$D = ([0, 100], \ldots, [0, 100])^T$

Problem 4 ([24], Problem D6–Shifted and augmented trigonometric function with an Euclidean sphere), $n = 4, 8, 12, 16, 20$.

$$f_i(\mathbf{x}) = n - 1 - \sum_{j=1}^{n-1} \cos(x_j - 1) + i(1 - \cos(x_i - 1)) - \sin(x_i - 1),$$

$i = 1, \ldots, n-1$

$$f_n(\mathbf{x}) = \sum_{j=1}^{n} x_j^2 - 10000$$

$D = ([-200, 200], \ldots, [-200, 200])^T$

Problem 5 ([25], Economics modeling application), $n = 4, 8, 12, 16, 20$.

$$f_i(\mathbf{x}) = \left(x_i + \sum_{k=1}^{n-i-1} x_k x_{i+k} \right) x_n - c_i, \quad i = 1, \ldots, n-1$$

$$f_n(\mathbf{x}) = \sum_{j=1}^{n-1} x_j + 1$$

where the constants c_i can be chosen arbitrarily; here, $c_i = 0, \ i = 1, \ldots, n-1$

$D = ([-100, 100], \ldots, [-100, 100])^T$

Problem 6 ([26], Example 1–The Bratu problem), $n = 4, 8, 12, 16, 20$.
$f_1(\mathbf{x}) = -2x_1 + x_2 + \alpha h^2 \exp(x_1)$
$f_n(\mathbf{x}) = x_{n-1} - 2x_n + \alpha h^2 \exp(x_n)$
$f_i(\mathbf{x}) = x_{i-1} - 2x_i + x_{i+1} + \alpha h^2 \exp(x_i),\ i = 2, \ldots, n-1,$
where $\alpha \geq 0$ is a parameter, assuming here $\alpha = 3.5$, and $h = \dfrac{1}{n+1}$.
$D = ([-100, 100], \ldots, [-100, 100])^T$

Problem 7 ([26], Example 2–The beam problem), $n = 4, 8, 12, 16, 20$.
$f_1(\mathbf{x}) = -2x_1 + x_2 + \alpha h^2 \sin(x_1)$
$f_n(\mathbf{x}) = x_{n-1} - 2x_n + \alpha h^2 \sin(x_n)$
$f_i(\mathbf{x}) = x_{i-1} - 2x_i + x_{i+1} + \alpha h^2 \exp(x_i),\ i = 2, \ldots, n-1,$
where $h = \dfrac{1}{n+1}$ and $\alpha \geq 0$ is a parameter; here $\alpha = 11$.
$D = ([-100, 100], \ldots, [-100, 100])^T$

Problem 8 ([27], 21–Extended Rosenbrock function), $n = 4, 8, 12, 16, 20$.
$f_{2i-1}(\mathbf{x}) = 10(x_{2i} - x_{2i-1}^2)$
$f_{2i}(\mathbf{x}) = 1 - x_{2i-1},\quad i = 1, \ldots, \frac{n}{2}$
$D = ([-100, 100], \ldots, [-100, 100])^T$

Problem 9 ([27], 26–Trigonometric function), $n = 4, 8, 12, 16, 20$.
$$f_i(\mathbf{x}) = n - \sum_{j=1}^{n} \cos x_j + i(1 - \cos x_i) - \sin x_i,\quad i = 1, \ldots, n$$
$D = ([-100, 100], \ldots, [-100, 100])^T$

Problem 10 ([27], 27–Brown almost-linear function), $n = 4, 8, 12, 16, 20$.
$$f_i(\mathbf{x}) = x_i + \sum_{j=1}^{n} x_j - (n+1),\quad i = 1, \ldots, n-1$$
$$f_n(\mathbf{x}) = \left(\prod_{j=1}^{n} x_j\right) - 1$$
$D = ([-10, 10], \ldots, [-10, 10])^T$

Problem 11 ([27], 28–Discrete boundary value function), $n = 4, 8, 12, 16, 20$.
$f_1(\mathbf{x}) = 2x_1 - x_2 + h^2(x_1 + h + 1)^3/2$
$f_n(\mathbf{x}) = 2x_n - x_{n-1} + h^2(x_n + nh + 1)^3/2$
$f_i(\mathbf{x}) = 2x_i - x_{i-1} - x_{i+1} + h^2(x_i + t_i + 1)^3/2,\ i = 2, \ldots, n-1,$
where $h = \frac{1}{n+1}$ and $t_i = ih$.
$D = ([0, 5], \ldots, [0, 5])^T$

Problem 12 ([27], 30–Broyden tridiagonal function), $n = 4, 8, 12, 16, 20$.
$f_1(\mathbf{x}) = (3 - 2x_1)x_1 - 2x_2 + 1$
$f_n(\mathbf{x}) = (3 - 2x_n)x_n - x_{n-1} + 1$
$f_i(\mathbf{x}) = (3 - 2x_i)x_i - x_{i-1} - 2x_{i+1} + 1,\quad i = 2, \ldots, n-1$
$D = ([-1, 1], \ldots, [-1, 1])^T$

Problem 13 ([28], Example 4.1–Nonlinear resistive circuit), $n = 4, 8, 12, 16, 20$.

$$f_i(\mathbf{x}) = g(x_i) + \sum_{j=1}^{n} x_j - i, \quad i = 1, \ldots, n,$$

where $g(x_i) = 2.5x_i^3 - 10.5x_i^2 + 11.8x_i$.

$D = ([-100, 100], \ldots, [-100, 100])^T$

Problem 14 ([28], Example 4.2), $n = 4, 8, 12, 16, 20$.

$$f_i(\mathbf{x}) = x_i - \frac{1}{2n}\left(\sum_{j=1}^{n} x_j^3 + i\right), \quad i = 1, \ldots, n$$

$D = ([-10, 10], \ldots, [-10, 10])^T$

4 Results and Discussion

The average and best (i.e., minimum) fitness values found for each different algorithm and problem function are shown below in Tables 1, 2, 3, 4, 5, 6, 7, 8, 9, and 10.

Tables 1, 2, 3, 4, and 5 show the average fitness values for 51 runs of each algorithm with each problem, while Tables 6, 7, 8, 9, and 10 present the best fitness values for every combination algorithm/problem in each dimension. In each table, the best value obtained for every problem is bolded, while the second best is underlined.

The best results were obtained by EJAYA, both in terms of average and absolute values. This result can be explained by the nature of the class of problems under consideration.

Table 1 Average fitness for each algorithm and problem with dimension 4

Problem	Jaya	MJAYA	EJAYA	SAMP-Jaya	OJaya
01	*0.037301*	759.5075	**7.8e−13**	0.082208	3.60415
02	*0.061348*	4.366276	**0.013026**	0.099944	1.98836
03	1.101062	1.700416	1.529441	*0.9208*	**0.012071**
04	3.172041	1285.801	**0.794585**	*2.942294*	3.060386
05	0.002189	97.55153	**7.03e−13**	*0.002149*	4.28393
06	*0.02041*	127.4752	**0.020138**	0.020458	28.08892
07	*0.867158*	112.8549	**0.86427**	0.867175	21.75779
08	2.293608	112.3845	**0.211815**	*1.238739*	18.14579
09	0.232556	1.066933	*0.113705*	0.231329	**0.108791**
10	*6.56e−5*	1.370445	**7.96e−13**	0.000155	1.100248
11	*0.4248*	0.472382	*0.4248*	*0.4248*	**0.200767**
12	1.106871	1.301978	**0.823576**	1.016749	*0.916372*
13	**0.055345**	76634	*0.060916*	0.061493	32.60203
14	*0.005805*	2.591199	**7.8e−13**	0.008878	2.037

Table 2 Average fitness for each algorithm and problem with dimension 8

Problem	Jaya	MJAYA	EJAYA	SAMP-Jaya	OJaya
01	*1.82763*	7726.655	**1.7393**	1.861864	152.2006
02	*0.68266*	72.27876	**0.021869**	0.716674	22.01862
03	**2.367052**	3.520389	3.196196	*2.647138*	863.877
04	10.48261	20889.06	**4.073712**	*10.27806*	197.7471
05	*0.008329*	121.0815	**6.96e−13**	0.009389	465.9061
06	73.93172	158.8989	**0.001618**	*72.14528*	163.2751
07	73.67055	158.4651	**0.690457**	*72.24566*	157.4344
08	21.74472	4420.537	**0.590003**	*12.81173*	418.4348
09	5.08573	7.992402	**0.304068**	5.011239	*2.560319*
10	0.432841	21.90249	**9e−13**	*0.308484*	7.468406
11	**0.312071**	*0.702193*	**0.312071**	**0.312071**	0.819976
12	*1.401421*	4.072288	**1.383169**	1.401567	2.409392
13	*0.786868*	906616.7	0.997937	**0.716111**	4334.69
14	*0.030862*	24.76336	**8.59e−13**	0.066765	8.036433

Table 3 Average fitness for each algorithm and problem with dimension 12

Problem	Jaya	MJAYA	EJAYA	SAMP-Jaya	OJaya
01	*2.027671*	15607.16	2.110943	**2.026045**	569.9055
02	2.715694	295.517	**0.03217**	*1.562613*	65.5981
03	**3.777186**	5.245614	5.058997	*4.337887*	807.2976
04	*23.54298*	48843.81	**7.137143**	23.94317	1777.321
05	0.007516	187.1631	**7.38e−13**	*0.006327*	2073.987
06	82.86831	445255.7	**1.724116**	*82.52612*	505.4965
07	*82.21373*	156.6737	**0.528442**	82.70787	387.7884
08	*69.17203*	28507.12	**1.004769**	70.81044	1078.883
09	18.32003	20.72244	**0.73744**	18.10252	*17.07631*
10	*0.390616*	74.52302	**0.057542**	0.413794	28.72603
11	**0.237439**	*1.458474*	**0.237439**	**0.237439**	2.653205
12	*1.410814*	8.467387	1.43527	**1.409355**	4.396191
13	5.564458	2058324	**5.516597**	*5.433279*	47757.1
14	0.066187	42.94972	**9.01E−13**	*0.019483*	20.28755

Table 4 Average fitness for each algorithm and problem with dimension 16

Problem	Jaya	MJAYA	EJAYA	SAMP-Jaya	OJaya
01	**2.007063**	24388.38	2.22222	*2.008302*	1845.694
02	4.779463	689.847	**0.198708**	*3.872578*	124.2643
03	**5.631415**	7.283231	6.980571	*6.049226*	322.6964
04	*43.47981*	85029.14	**12.09548**	45.29773	3149.952
05	*0.004715*	214.6964	**5.5e−13**	0.004928	3530.811
06	*90.76954*	1.07e+17	**5.085716**	91.11857	922.8598
07	91.57444	3.31e+14	**5.248573**	*90.94256*	748.7884
08	152.1592	60370.24	**3.377784**	*133.6855*	4192.898
09	37.74819	39.81277	**0.907257**	*37.46247*	44.7763
10	0.33721	456.7778	**0.164283**	*0.258828*	75.9037
11	*0.192521*	3.828483	**0.190461**	**0.190461**	3.588161
12	**1.414091**	13.25389	1.476021	*1.414669*	8.31018
13	*14.43239*	3176498	**12.88087**	14.87231	172534.6
14	0.351574	59.89378	*0.039723*	**1.17e−12**	30.05229

Table 5 Average fitness for each algorithm and problem with dimension 20

Problem	Jaya	MJAYA	EJAYA	SAMP-Jaya	OJaya
01	*2.001243*	32930.64	2.883153	**1.999806**	3107.198
02	7.587837	1118.439	**0.492324**	*7.51904*	236.5383
03	**7.113975**	9.746856	9.01987	*7.835969*	367279.4
04	*74.74562*	118316.4	**13.83089**	75.53196	10691.88
05	*0.00403*	212.8566	**6.94e−9**	0.004408	7312.667
06	109.2742	3.49e+21	**5.42168**	*107.1104*	1825.048
07	*108.1948*	1.46e+21	**7.008227**	108.5383	1441.275
08	284.1983	94117.88	**8.325698**	*207.4755*	8610.55
09	*63.20463*	65.21878	**1.579544**	63.49481	104.6752
10	*0.286777*	21340.98	**0.08651**	0.490721	132.5569
11	1.151586	6.011312	**0.158707**	*0.701521*	3.94693
12	**1.413574**	17.73365	1.650891	*1.413914*	12.11648
13	29.76446	4800462	**22.13837**	*29.19197*	295083.2
14	0.301565	76.89122	**9.35e−9**	*9.47e−9*	43.24111

Table 6 Best fitness value for each algorithm and problem with dimension 4

Problem	Jaya	MJAYA	EJAYA	SAMP-Jaya	OJaya
01	0.013514	23.91481	**3.74e−13**	*0.008449*	0.259312
02	0.015	1.261467	**2.53e−13**	*0.011323*	0.111472
03	0.076134	0.284877	1	*0.000835*	**0.000125**
04	0.639886	14.33572	**2.48e−8**	1.021443	*0.333206*
05	*0.000202*	10.17526	**1.19e−13**	0.000238	0.000344
06	*0.020156*	48.56258	**0.020138**	0.020166	0.020288
07	*0.864726*	41.85269	**0.86427**	0.864893	0.864809
08	0.000199	28.30774	**4.4e−13**	*0.000171*	0.060556
09	0.139507	0.407813	**1.89e−8**	0.032742	*0.026646*
10	4.37e−13	0.387378	*3.2e−13*	**2.43e−13**	9.33e−13
11	*0.4248*	*0.4248*	*0.4248*	*0.4248*	**0.008097**
12	0.408424	0.553353	**5.3e−13**	0.346904	*0.048358*
13	3.72e−13	1160.649	*3.66e−13*	**2.37e−13**	8.71e−13
14	**3.13e−13**	0.456107	*4.11e−13*	5.64e−13	7.74e−13

Table 7 Best fitness value for each algorithm and problem with dimension 8

Problem	Jaya	MJAYA	EJAYA	SAMP-Jaya	OJaya
01	*1.746513*	1163.029	**1.642072**	1.748557	1.855896
02	0.246311	28.06783	**7.45e−13**	*0.236563*	1.725815
03	1.098038	2.231975	2.000208	*0.186106*	**0.00033**
04	4.969094	3377.885	**0.413205**	*4.770179*	4.972168
05	*0.000722*	3.844443	**1.11e−13**	0.001467	0.405516
06	63.05564	100.3864	**0.001618**	35.50098	*6.966559*
07	59.56525	126.8712	**0.690457**	*21.63332*	43.54873
08	*0.262853*	1190.088	**6.24e−13**	0.417598	3.024152
09	3.030025	6.545709	**9.08e−5**	3.788699	*0.379582*
10	**5.64e−13**	5.82194	5.89e−13	*5.85e−13*	0.001798
11	*0.312071*	*0.312071*	*0.312071*	*0.312071*	**0.174061**
12	1.343339	2.189377	*1.068989*	1.343156	**0.957273**
13	*0.258026*	173194.6	**8.79e−13**	0.259604	5.784104
14	6.13e−13	8.706569	*4.65e−13*	**4.42e−13**	8.46e−13

Table 8 Best fitness value for each algorithm and problem with dimension 12

Problem	Jaya	MJAYA	EJAYA	SAMP-Jaya	OJaya
01	*1.985728*	8431.876	**1.947732**	1.995528	2.123706
02	0.707785	103.0104	**7.09e−13**	*0.635289*	4.309436
03	1.251606	4.093437	3.000423	*1.032287*	**0.000523**
04	18.03753	15107.65	**0.131327**	*9.392153*	14.96514
05	0.000994	11.22811	**8.98e−14**	*0.000644*	0.116536
06	*66.17686*	103.9643	**0.008119**	66.98294	144.4244
07	71.25037	101.1358	**0.514594**	*68.91963*	134.912
08	*1.660379*	4933.791	**1.41e−8**	1.747052	28.36734
09	12.73162	17.11407	**7.52e−13**	12.71641	*6.227708*
10	2.12e−9	23.16159	**2.79e−12**	*4.82e−10*	6.273952
11	**0.237439**	**0.237439**	**0.237439**	**0.237439**	*1.01749*
12	1.398442	5.837022	**1.339655**	1.398353	*1.398094*
13	3.090889	497417.5	**1.595609**	*2.831569*	41.32905
14	7.66e−13	28.86674	*6.82e−13*	**6.26e−13**	0.096067

Table 9 Best fitness value for each algorithm and problem with dimension 16

Problem	Jaya	MJAYA	EJAYA	SAMP-Jaya	OJaya
01	*1.992515*	18145.68	**1.971647**	1.992711	3.853688
02	1.467945	309.3021	**1.29e−12**	*1.339854*	8.824352
03	3.604961	5.762362	6.000153	*0.067755*	**0.000767**
04	25.55577	57483.21	**0.150535**	28.60737	*12.6763*
05	0.000681	14.83673	**7.13e−14**	*0.000368*	10.04305
06	77.52219	150	**0.248488**	*70.53718*	306.507
07	*74.80778*	101.4781	**0.524374**	78.48955	261.161
08	3.801844	17369.55	**0.000242**	*3.580692*	140.3054
09	32.29294	35.66817	**9.39e−13**	30.27802	*14.05743*
10	**4.02e−7**	46.58691	*4.24e−7*	5.3e−7	2.006934
11	**0.190461**	**0.190461**	**0.190461**	**0.190461**	*1.029605*
12	1.410731	9.530622	**1.410497**	*1.410538*	4.018768
13	9.35898	1258665	**4.635214**	*9.351241*	273.8042
14	8.03e−13	44.99119	*7.65e−13*	**7.63e−13**	1.215313

Table 10 Best fitness value for each algorithm and problem with dimension 20

Problem	Jaya	MJAYA	EJAYA	SAMP-Jaya	OJaya
01	1.997252	23424.63	**1.996577**	*1.996677*	5.199616
02	2.140845	488.0374	**1.26e−6**	*2.039682*	23.07343
03	*4.001134*	8.016777	8.000024	4.280304	**0.000924**
04	*47.35422*	78182.15	**3.49e−7**	50.47265	50.79653
05	*0.000287*	24.98126	**1.21e−9**	0.000554	2.709221
06	87.26843	102.406	**0.724969**	*81.69802*	662.5349
07	*88.5846*	101.4305	**1.406232**	94.15044	448.7978
08	25.19813	39450.97	**0.074645**	*13.811*	444.3823
09	56.32278	60.36546	**9.02e−9**	54.59069	*43.4929*
10	*2.63e−5*	112.4798	**4.38e−6**	3.89e−5	20.92637
11	**0.158707**	*0.400519*	**0.158707**	**0.158707**	2.098237
12	*1.413378*	12.25678	**1.409187**	1.41338	3.627609
13	21.80754	2819758	**10.7863**	*19.81322*	1063.36
14	**7.85e−9**	65.1341	8.21e−9	*8.2e−9*	20.35027

Because the problems are difficult scalable nonlinear equation systems, an algorithm that focuses on balancing local and global exploration can better explore the search space and avoid local minima.

Box plots were also used to show the results so that the performance of each algorithm could be compared in more ways than just the average and best fitness values. Since the results are at very different scales, a logarithmic scale was used on the vertical axis to make it easier to compare them.

The Economics modeling application problem (Problem 5) is displayed on Fig. 1. This example shows a clear difference in results from the different algorithms tested, with the worst result from enhanced Jaya being smaller than any result from the other algorithms.

Naturally, these results are highly problem-dependent. A different example, the nonlinear resistive circuit problem (Problem 13) on Fig. 2, shows a more balanced result, where the choice of algorithm was not such a decisive factor, with the exception of MJAYA, which performed significantly worse than the alternatives on this particular problem.

The fact that one of the algorithms consistently produces the best results indicates that there is a better strategy for approaching and solving problems of this nature. It is noteworthy that the EJAYA algorithm consistently achieved the best (minimum) fitness value, since it could be the case that a more global search would produce a less satisfactory result due to not being given enough time to lead to a more exact one.

The Friedman rank test was used to determine whether the differences in algorithms' performance observed in this study were statistically significant.

The test statistic and p-value for the mean results of each experiment were 205.828407 and 2.097028e−43, respectively. An experiment corresponds to a prob-

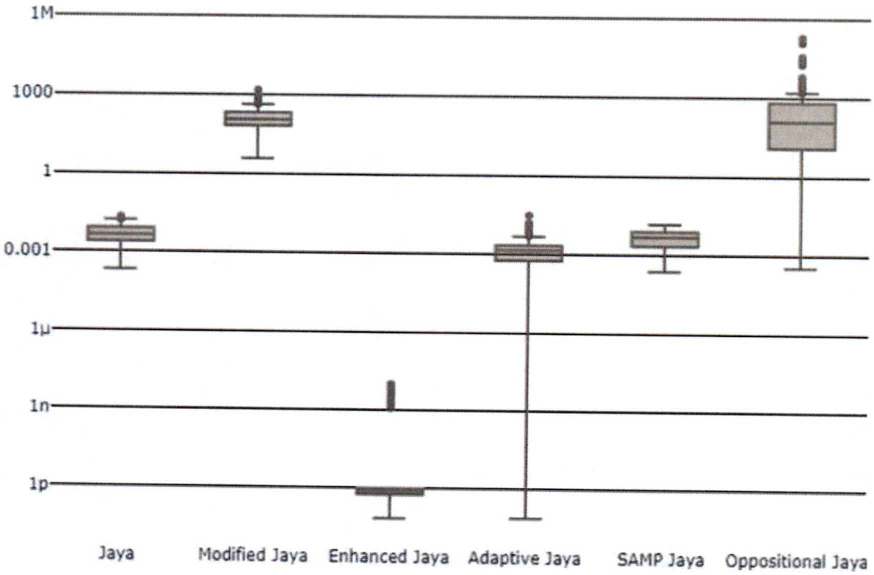

Fig. 1 Algorithms performance for Problem 5–economics modeling application

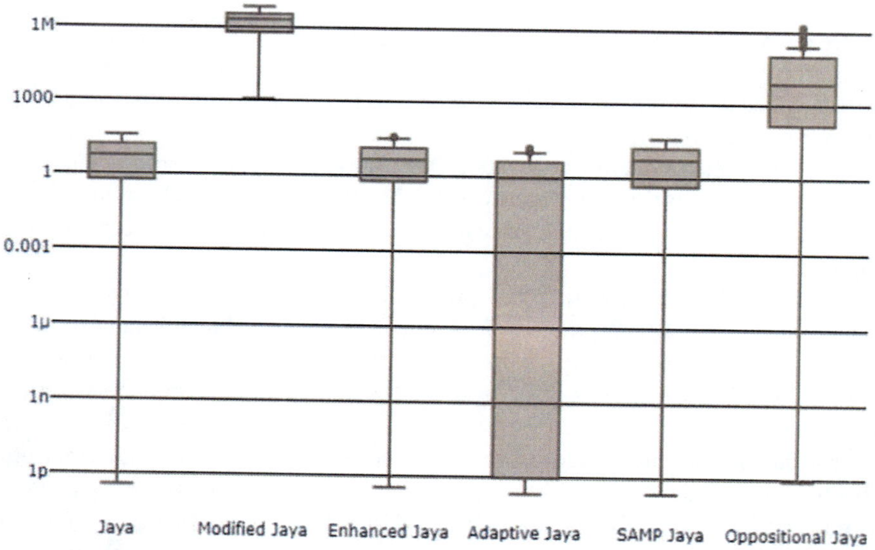

Fig. 2 Algorithms performance for Problem 13–nonlinear resistive circuit

lem with a certain number of variables. For the best results from each experiment, the obtained test statistic and p-value were 166.029499 and 7.438888e−35, respectively. These results indicate that the p-values are lower than the the significance level considered, $\alpha = 0.05$, which indicates the existence of statistically significant performance differences between the algorithms.

The Nemenyi post-hoc test was then utilized to validate the existence of statistically significant differences between each pair of algorithms and compare their performance. At $\alpha = 0.05$, the p-values for each pairwise comparison of means returned by the Nemenyi test indicate that all groups with EJAYA have statistically significantly different means (p-value for EJAYA–Jaya: 0.00423; EJAYA–MJAYA: 0.001; EJAYA–SAMP-Jaya: 0.011719; EJAYA-OJaya: 0.001). The enhanced Jaya had the lowest mean rank, confirming that it is the algorithm with the best performance.

In turn, the modified Jaya algorithm was consistently outperformed. Despite the fact that this Jaya variant also attempts to reduce population premature convergence, the comparatively simple method used to achieve this did not result in the desired effects on the set of problems used. The other three algorithms were also quite ineffective in solving this important class of numerical problems.

5 Conclusion

The Jaya algorithm and some of its variants were tested against a set of difficult scalable nonlinear equation systems in order to evaluate their performance, as well as understanding if a specific Jaya variant could be better suited to this class of problems.

The enhanced Jaya algorithm consistently performed better than the other variants, both in finding the best result and obtaining good results on average when compared to the other variants. The ability of EJAYA to solve this class of problems demonstrates that, with enough iterations, a degree of global exploration leads to better results without sacrificing local exploration.

References

1. Pérez R, Lopes V (2004) Recent applications and numerical implementation of quasi-Newton methods for solving nonlinear systems of equations. Numer Alg 35(2):261–285
2. Kelley C (2003) Solving nonlinear equations with Newton's method. SIAM, Philadelphia, PA
3. Karr C, Weck B, Freeman L (1998) Solutions to systems of nonlinear equations via genetic algorithms. Eng Appl Artif Intell 11(3):369–375
4. Grau-Sánchez M (2009) Improving order and efficiency: composition with a modified Newton's method. J Comput Appl Math 231(2):592–597
5. Choi H, Kim S, Shin BC (2022) Choice of an initial guess for Newton's method to solve nonlinear differential equations. Comput Math Appl 117:69–73
6. Dokeroglu T, Sevinc E, Kucukyilmaz T, Cosar A (2019) A survey on new generation metaheuristic algorithms. Comput Ind Eng 137:106040

7. Lin MH, Tsai JF, Yu CS (2012) A review of deterministic optimization methods in engineering and management. Math Probl Eng 2012:756023
8. Sihwail R, Solaiman O, Omar K, Ariffin K, Alswaitti M, Hashim I (2021) A hybrid approach for solving systems of nonlinear equations using Harris hawks optimization and Newton's method. IEEE Access 9:95791–95807
9. Kumar N (2022) An alternative computational optimization technique to solve linear and non-linear Diophantine equations using discrete WQPSO algorithm. Soft Comput 26(22):12531–12544
10. Pan L, Zhao Y, Li L (2022) Neighborhood-based particle swarm optimization with discrete crossover for nonlinear equation systems. Swarm Evol Comput 69:101019
11. Verma P, Parouha R (2022) Solving systems of nonlinear equations using an innovative hybrid algorithm. Iran J Sci Technol Trans Electr Eng 46(4):1005–1027
12. Jui J, Ahmad M (2021) A hybrid metaheuristic algorithm for identification of continuous-time Hammerstein systems. Appl Math Model 95:339–360
13. Suid M, Ahmad M (2023) A novel hybrid of nonlinear sine cosine algorithm and safe experimentation dynamics for model order reduction. Automatika 64(1):34–50
14. Rao R (2016) Jaya: a simple and new optimization algorithm for solving constrained and unconstrained optimization problems. Int J Ind Eng Comput 7:19–34
15. Degertekin S, Lamberti L, Ugur I (2018) Sizing, layout and topology design optimization of truss structures using the Jaya algorithm. Appl Soft Comput 70:903–928
16. Rao R (2019) Jaya: an advanced optimization algorithm and its engineering applications. Springer, Cham, Switzerland
17. Elattar EE, ElSayed SK (2019) Modified JAYA algorithm for optimal power flow incorporating renewable energy sources considering the cost, emission, power loss and voltage profile improvement. Energy 178:598–609
18. Zhang Y, Chi A, Mirjalili S (2021) Enhanced Jaya algorithm: a simple but efficient optimization method for constrained engineering design problems. Knowl Based Syst 233:107555
19. Civicioglu P (2013) Backtracking search optimization algorithm for numerical optimization problems. Appl Math Comput 219(15):8121–8144
20. Rao R, Saroj A (2017) A self-adaptive multi-population based Jaya algorithm for engineering optimization. Swarm Evol Comput 37:1–26
21. Yu J, Kim C, Rhee SB (2019) Oppositional Jaya algorithm with distance-adaptive coefficient in solving directional over current relays coordination problem. IEEE Access 7:150729–150742
22. Liang J, Qu B, Suganthan P, Hernández-Díaz A (2013) Problem definitions and evaluation criteria for the CEC 2013 special session on real-parameter optimization. Technical Report 201212, computational intelligence laboratory. Zhengzhou University, Zhengzhou, China
23. Bodon E, Del Popolo A, Lukšan L, Spedicato E (2001) Numerical performance of ABS codes for systems of nonlinear equations. Technical Report DMSIA 01/2001, Universitá degli Studi di Bergamo, Bergamo, Italy
24. Friedlander A, Gomes-Ruggiero M, Kozakevich D, Martínez J, Santos S (1997) Solving nonlinear systems of equations by means of quasi-Newton methods with a nonmonotone strategy. Optim Methods Softw 8(1):25–51
25. van Hentenryck P, McAllester D, Kapur D (1997) Solving polynomial systems using a branch and prune approach. SIAM J Numer Anal 34(2):797–827
26. Kelley C, Qi L, Tong X, Yin H (2011) Finding a stable solution of a system of nonlinear equations. J Ind Manag Optim 7(2):497–521
27. Moré J, Garbow B, Hillstrom K (1981) Testing unconstrained optimization software. ACM Trans Math Softw 7(1):17–41
28. Yamamura K, Kawata H, Tokue A (1998) Interval solution of nonlinear equations using linear programming. BIT Numer Math 38(1):186–199

Chapter 11
In-Depth Analysis of Artificial Intelligence in Mammography for Breast Cancer Detection

Shweta Saraswat, Bright Keswani, and Vrishit Saraswat

1 Introduction

There has been a rise in the study of the potential of AI in radiology in recent years. The First Finite-Difference Your work may find use in a variety of settings if you employ methods like self-encoders and convolutional neural networks. With the development of deep learning (DL) techniques, artificial intelligence (AI) technologies have made great strides in the area of image identification. Artificial intelligence (AI) has the potential to enhance the assessment of treatment efficacy and the early detection of pancreatic, liver, and breast cancers. Having breast cancer poses a significant risk to a woman's physical and mental health. Better screening and early diagnostic approaches for breast cancer are desperately needed. When it comes to breast cancer, a proper diagnosis is essential for effective treatment and early detection.

Even though the accuracy of earlier CAD systems hasn't changed much, the use of computer-aided diagnosis (CAD) in mammography has grown. Instead of using image analysis, these methods relied on encouraging patients to look for probable cancers in mammograms. Recently developed deep (multilayered) convolutional neural networks have significantly increased prediction accuracy. Deep learning algorithms have recently been used for digital mammography and breast tomosynthesis.

The main thing a radiologist does these days is look into, evaluate, and take care of breast imaging. Radiologists may make more mistakes because their jobs are hard and they often work long hours, but AI may be able to help with this. Doctors can identify patients more precisely thanks to CAD programs. An image is examined and judged

S. Saraswat (✉) · B. Keswani
Suresh Gyan Vihar University, Jaipur, Rajasthan 302017, India
e-mail: shwetavrishit@gmail.com

V. Saraswat
Medanta Hospital, Gurugram, Haryana 122001, India

A. Yadav et al. (eds.), *Proceedings of International Conference on Paradigms of Communication, Computing and Data Analytics*, Algorithms for Intelligent Systems,
https://doi.org/10.1007/978-981-99-4626-6_11

throughout the CAD process in line with predetermined criteria [1]. Convolutional neural networks (CNNs), a kind of deep learning (DL), have been used in the area of medical imaging, and this has undergone a significant transformation [2]. The history of this technique and applications of AI to the interpretation of breast-mammography pictures are covered. Moreover, previous and potential future uses of AI in diagnostic imaging are highlighted.

2 Analyzing Breast Tissue Using AI and Mammography

The term "artificial intelligence" is used to describe a wide range of technologies, such as the neural network, whose goal is to mimic how the brain works. This system is composed of the input, masking, and output layers. Deep neural networks are highly suited for processing information in several dimensions because of their tree-like structure. The input layer, convolution layer, and pooling layer are these layers. A large number of stacked convolutional layers make up the multilayer perceptron model. The interconnected nodes of this paradigm allow it to easily accommodate varying degrees of complexity. Before using it to train CNN models, both false and real data are annotated with regression coefficients. Both supervised and unsupervised methods are utilized to train convolutional neural networks (CNN) [3]. Although supervised learning uses predetermined labels or diagnoses in the training data, unsupervised learning does not. It seems that the best method for dealing with the difficulties of picture classification is supervised learning.

3 Identifying and Categorizing Breast Mammograms

Rapid and constant growth is the best predictor of breast cancer. The first phase of computer-assisted diagnosis (CAD) is bulk identification, which must be completed before radiologists can make an accurate diagnosis. In order to effectively remove masses from mammography pictures, Crowd Search Optimization-based Intuitive Fuzzy Clustering with Adjacent Attraction (CrSAIFCM-NA) is proposed. Intuitive fuzzy clustering is accomplished via the use of nearby attractions and crowd-based optimum search. Proof that the chapters were properly divided [4]. In order to develop a unified CAD system, experts employed deep convolutional neural networks (CNNs) and the filled resolution convolutional network (FrCN). Mammograms with suspicious masses may be located, outlined, and categorized using this technique (FrCN). They showed a detection accuracy of 98.96% using INbreast datasets, which might be very useful for radiologists making a diagnosis [5].

3.1 Characterization, Identification, and Microcalcification

Calcifications, sometimes referred to as calcium salt deposits, show up as little white spots on a mammogram. The presence of calcifications increases the risk of breast cancer. Breast calcifications might be the first indicator of the disease. Calcifications may sometimes be utilized to help in breast cancer diagnosis. The two most frequent types of calcification are macrocalcifications and microcalcifications, respectively. The two most prevalent types of calcification in rocks are limestone and dolomite. These labels help to distinguish between the two categories. Even though their conclusion is conceivable, the majority of big, granular macrocalcifications are age-related, so it is preferable to ignore them. These macrocalcifications are distinctive in any scenario. Although the great majority of macrocalcifications are benign, it is nonetheless possible. One of the risk factors for breast cancer is microcalcifications, which may range in size from 0.1 to 1 mm and may or may not be associated with visible masses [6]. Because of some of the cutting-edge CAD technologies that are now available, calcifications in mammograms may be easier to spot. Digital breast tomosynthesis is one use for this method. CNN outperformed humans in identifying, describing, and labeling calcifications in mammograms, according to research by Cai H. and colleagues. This model dramatically improves performance by switching out conventional IDs with filtered deep features and a deep convolutional neural network. Overall, they had a classification accuracy of 89.32% and a reading and comprehension accuracy of 86.35% [7].

Jian W. and his colleagues used the dual-tree complex wavelet transform to come up with a CAD method for better identification. If this method works, it could be used to find tiny lumps in the breast called microcalcifications, which have been linked to a higher risk of breast cancer [8]. Y. Guo et al. provide a novel hybrid method that combines the contourlet transform with a non-linking, fundamental, pulse-coupled neural network to identify microcalcification in mammograms. With this cutting-edge method, microcalcification may be discovered. Guo Y et al. initially described the technique to identify microcalcification in mammograms. This section's citations are insufficient. A plan has been developed by Guo Y and his team, and it is now being implemented [9]. With automated software that recognizes, classifies, and diagnoses anomalies in mammograms, such as masses and changes, radiologists may be able to save time and effort. This is not entirely impossible, given the existence of such programs. If anything comparable happened, it would help reduce the anxiety that many people are now experiencing.

3.2 Identifying the Types Breast Masses

By dividing the patient's mass into small pieces, the chances of the treatment working may go up. By looking at ROIs from the mini-MIAS database, researchers used fuzzy outlines to automatically find breast masses in mammograms. According to the experts, they did it to lessen their workload. According to the findings, the overall accuracy was 88.08%, with true positives occurring 91.12% of the time [10]. As indicated in the image, there are many barriers that make it challenging to find masses on mammograms. Low contrast, uneven masses, edge activity, and pixel-level brightness changes are a few characteristics of such pictures. Several variables may make it difficult to identify breast cancer.

Using mesh-free directed radial basis function clustering, researchers have come up with helpful criteria for the segmentation of the breast and suspicious mass areas. This strategy was employed for data analysis. Whether or not the potentially harmful zones were in fact healthy was determined using the SVM classifier. The analysis's findings showed that the DDSM dataset had a sensitivity and specificity of 97.12% and 92.43%, respectively [11]. We have identified breast tumors in mammograms using dynamic programming and fitting a model to the data. Then, a variety of criteria were used to categorize breast cancers.

To find breast cancer early and correctly, it is important to know how to classify breast abnormalities correctly [12]. We could use the implementation of an automated image segmentation method to show the potential and benefits of DL in highly specialized healthcare systems.

3.3 Analyzing Breast Density

2D mammography is often used to do this because breast density is such an important factor in figuring out a woman's risk of getting breast cancer. Women with dense breasts than the norm had a two- to six-fold higher risk of developing breast cancer. The total mammographic density of a woman is based on how much dense tissue is in each breast. Mammograms are analyzed in such situations. Thicker areas show up as cotton-like white patches in mammography [13]. An accurate and consistent evaluation of breast density is essential given the present status of breast cancer diagnosis because it helps patients and medical professionals make better-informed choices about the course of therapy. Breast cancer risk and breast density are related. In determining breast density from mammograms, artificial intelligence systems reportedly perform better than human experts in various clinical studies. By increasing the number of training samples, Mohamed AA et al. improved the AUC for differentiating "scattered density" from "heterogeneous density" on large (1801 image) DM datasets. By focusing optimization efforts on the area of the curve that most closely corresponds to the system's actual performance, the goal was accomplished. As a result, we were able to discern between homogeneous and distributed densities [14]. They also found that radiologists preferred the medial oblique CNN model (MLO) for BD classification over the conventional CNN model (CC). It was another of their discoveries.

The Food and Drug Administration has given the green light to DenSeeMammo, a program that uses AI to evaluate BD. A radiologist with expert knowledge, a radiologist with less expert knowledge, and a radiologist with no updates at all was compared to DenSeeMammo. The striking closeness between the expert radiologist's diagnosis of BD and the DM AI model (weighted = 0.79; 95% confidence interval: 0.73–0.81) astounded them [15]. Using 58,894 randomly selected digital mammograms, Lehman CD et al. assessed the performance of a deep learning (DL) system for the identification of breast cancer (BD). The model was created using PyTorch and a ResNet-18 deep convolutional neural network (CNN).

When the first radiologist tried to make sense of the data, DL assessments were found in 9729 of the 10,763 four-way BI-RADS cases. These results show that the DL model and the initial density study have a reasonable agreement ($k = 0.67$; 95% confidence interval [CI]: 0.66, 0.68) [16]. If AI is used to evaluate MBD, it could lead to more accurate risk assessments for breast cancer, less variation in how radiologists diagnose it, and new ways to find and treat this disease early on.

4 Risk Assessment

The high incidence of breast cancer and its associated mortality rates make worries about women's mental and physical health worse. Risk factors for breast cancer include age, family history, hormonal problems (like early periods, a late age at first pregnancy, and a limited number of children), estrogen levels, and living a bad lifestyle. Over time, cigarette smoking, drinking too much, and eating a lot of saturated fat have become risk factors. If people, in general, knew more about these risk factors, it might be possible to find breast cancer earlier and come up with new ways to stop it.

Artificial intelligence researchers reportedly spend a lot of effort attempting to predict if a person may get breast cancer. While ML algorithms have been created since January 2000, they were not made accessible to the general public until May 2018. By looking at different machine learning (ML) methods, Nindrea RD's team found that the SVM method was the best at predicting the likelihood of a breast cancer diagnosis. This research was conducted between 1992 and 1997. In this research, the efficacy of five distinct machine learning algorithms was evaluated. A few of the techniques that have been developed include decision trees (DT), artificial neural networks (ANN), the Naive Bayes classifier, and the K-nearest neighbor (KNN) algorithm. ANN was determined to be the most efficient approach out of the five that were studied.

Research that has been published shows that a lot of time and work has been put into figuring out if and how artificial intelligence can be used to give an accurate assessment of breast cancer risk. In January 2000, ML methods were initially made available to academics. By May 2018, the general public had broad access to ML methods. Nindrea RD and her colleagues found that the SVM method was the most effective after examining the effectiveness of other ML algorithms for calculating the chance of diagnosing breast cancer. In this work, we investigated the K-nearest neighbor (KNN) approach, naïve Bayes, DT, and ANN. It has been found that ANN is the best machine learning method as a consequence [17]. AI may assist doctors in advising patients at high risk of breast cancer on the best preventative measures, even though it has been shown that it is more accurate than conventional approaches in identifying breast cancer risk.

5 Improvements Made to the Image Quality

An accurate diagnosis requires scans of the highest quality. The capacity of artificial intelligence (AI) to identify and measure breast issues primarily depends on mammography data. Mammograms aid in the detection and evaluation of breast anomalies.

CGI's quality has improved in a variety of settings. Due to the extra details it would supply on the data processing technique, dimensional indications, and shift invariance, completing multi-scale shearlet transformations would result in the provision of multi-resolution data. Those that fulfill this description are more likely to provide accurate answers to queries. This may make it easier to find cancer cells, especially those that have a more diffuse appearance.

Shenbagavalli P. and his colleagues used the shearlet transform to decide whether the cases in the DDSM dataset were benign or cancerous. This technique may improve image quality to an astounding 93.45% [18]. To increase the possible uses of mammography, Teare P. and colleagues specifically used a novel technique termed the contrast-limited reactive histogram equalization (CLAHE) method, which uses chemicals to enhance colors. They classified mammography images using two deep convolutional neural networks (CNNs) of varying sizes, produced patches, and a randomized forest gating network, and they got accuracy and responsiveness of 0.80 and 0.91, respectively. Using these techniques, they discovered a sensitivity of 0.91 and an accuracy of 0.80 [19].

6 Conclusion

This is especially true in the field of medical imaging, where diagnostic results often need to be interpreted. DL is moving forward with this owing to its quick processing, incredible reproducibility, and artificial intelligence (AI), which is resilient to fatigue, and has the ability to provide medical professionals with factual and useful information. A diagnosis that is accurate the first time would be advantageous to both physicians and patients. Several studies have focused on the CAD system's potential as a tool for breast cancer screening. Comparable issues have been explored at other institutions. A number of these technologies may integrate the results of numerous diagnostic techniques with the findings of imaging studies like mammography in order to more accurately diagnose, classify, and identify breast illnesses. Calculations may be made to determine a patient's prognosis, estimated life expectancy, and treatment efficacy. With these developments, doctors could be more accurate while simultaneously easing their patients' worries and the associated paperwork workload. Recent advancements in deep learning have made artificial intelligence increasingly prevalent in breast imaging. Their application is anticipated to improve every aspect of mammography and digital breast tomosynthesis, from the generation of the initial image and noise mitigation to the appraisal of risk, the diagnosis of cancer, the design of a treatment plan, and the prognosis of the patient. Do not anticipate these advantages right now.

References

1. Artificial intelligence in mammography-based breast cancer screening (2019) Case Med Res
2. Geras KJ, Mann RM, Moy L (2019) Artificial intelligence for mammography and digital breast tomosynthesis: current concepts and future perspectives. Radiology 293(2)
3. Bahl M (2020) Artificial intelligence: a primer for breast imaging radiologists. J Breast Imaging
4. Lång K, Hofvind S, Rodríguez-Ruiz A, Andersson I (2021) Can artificial intelligence reduce the interval cancer rate in mammography screening? Eur Radiol
5. Yoon JH, Kim EK (2021) Deep learning-based artificial intelligence for mammography. Korean J Radiol
6. Goyal S (2021) An overview of current trends, techniques, prospects, and pitfalls of artificial intelligence in breast imaging. Rep Med Imaging
7. Naderan M (2021) Review methods for breast cancer detection using artificial intelligence and deep learning methods. Syst Res Inf Technol
8. Díaz O, Rodríguez-Ruiz A, Gubern-Mérida A, Martí R, Chevalier M (2021) Are artificial intelligence systems useful in breast cancer screening programmes?
9. Moy L, Gao Y (2021) Digital mammography is similar to screen-film mammography for women with personal history of breast cancer. Radiology
10. Freeman K, Geppert J, Stinton C, Todkill D, Johnson S, Clarke A, Taylor-Phillips S (2021) Use of artificial intelligence for image analysis in breast cancer screening programmes: systematic review of test accuracy
11. Dahlblom V, Andersson I, Lång K, Tingberg A, Zackrisson S, Dustler M (2021) Artificial intelligence detection of missed cancers at digital mammography that were detected at digital breast tomosynthesis. Radiol: Artif Intell
12. Abdollahi J, Davari N, Panahi Y, Gardaneh M (2022) Detection of metastatic breast cancer from whole-slide pathology images using an ensemble deep-learning method. Arch Breast Cancer
13. Uematsu T, Nakashima K, Harada TL, Nasu H, Igarashi T (2022) Artificial intelligence computer-aided detection enhances synthesized mammograms: comparison with original digital mammograms alone and in combination with tomosynthesis images in an experimental setting. Breast Cancer
14. Garrucho L, Kushibar K, Jouide S, Diaz O, Igual L, Lekadir K (2022) Domain generalization in deep learning based mass detection in mammography: a large-scale multi-center study. Artif Intell Med
15. Dar RA, Rasool M, Assad A (2022) Breast cancer detection using deep learning: datasets, methods, and challenges ahead. Comput Biol Med
16. Sebastian AM, Peter D (2022) Artificial intelligence in cancer research: trends, challenges and future directions. Life

17. Saraswat S, Keswani B, Saraswat V (2023) The role of artificial intelligence in healthcare: applications and challenges after COVID-19
18. Zheng D, He X, Jing J (2023) Overview of artificial intelligence in breast cancer medical imaging. J Clin Med
19. Morgan MB, Mates JL (2023) Ethics of artificial intelligence in breast imaging. J Breast Imaging

Chapter 12
The Task Allocation to Virtual Machines on Dynamic Load Balancing in Cloud Environments

Rudresh Shah and Suresh Jain

1 Introduction

Cloud computing is a field that is growing quickly and gives people and businesses an easy and inexpensive way to access computing resources, storage, and infrastructure over the Internet. The on-demand self-service and pooling of resources in cloud computing make it easy for users to get the resources they need quickly and easily. There are four deployment models you mentioned (public, private, community, and hybrid) represent different ways that cloud services can be deployed and used by organisations. Private clouds are used by a single organisation and are typically maintained on-premises or within a dedicated data centre. Public clouds are owned and operated by a third-party provider and are open to the public. Community clouds are shared by multiple organisations and are maintained by a third party. Hybrid clouds are a combination of private and public clouds and allow organisations to take advantage of the benefits of both models. Infrastructure, Platform, and Software are the three service models of cloud computing, and they each represent a distinct level of abstraction and control over the underlying computer infrastructure. IaaS provides the basic infrastructure components, such as virtualized computing resources, storage, and networking. PaaS provides a platform for building, deploying, and managing applications, while SaaS provides users with access to ready-to-use applications over the Internet. Due to its effective resource sharing, cloud computing has evolved as a paradigm that is cost-effective and has experienced tremendous growth in popularity in recent years. One of the major challenges in cloud computing is the virtual machine

R. Shah (✉) · S. Jain
Medi-Caps University, Indore, India
e-mail: rudresh.shah@medicaps.ac.in

S. Jain
e-mail: Suresh.jain@medicaps.ac.in

A. Yadav et al. (eds.), *Proceedings of International Conference on Paradigms of Communication, Computing and Data Analytics*, Algorithms for Intelligent Systems,
https://doi.org/10.1007/978-981-99-4626-6_12

allocation problem, which refers to the challenge of efficiently allocating VMs with different resource demands.

The optimal use of computer resources is ensured through load balancing, which is a crucial component of cloud computing. When there is an imbalance in the workload between different resources, some may be overworked while others are idle, which results in subpar performance and lost output. So that no single resource becomes a bottleneck, load balancing algorithms assist in distributing workloads equitably among many resources. Load balancing can raise the performance, scalability, and dependability of cloud computing systems by ensuring that resources are utilized effectively [1].

There are many different load-balancing algorithms that can be used in cloud computing, each with its own advantages and disadvantages. Some popular algorithms include round-robin, least connections, and IP hash. The choice of algorithm depends on factors such as the type of workloads being run, the characteristics of the resources being used, and the desired level of performance. Load balancing is an essential aspect of cloud computing, and it is important for organizations to carefully consider their load balancing needs and to choose the right algorithm to meet their requirements. They can guarantee their cloud computing solutions are highly scalable, dependable, and effective by doing this. A distributed swarm intelligence technique that can be used to address load balancing issues is the Ant Colony Optimisation (ACO) algorithm.

For complicated problem solving, the ACO algorithm mimics the behaviour of ants. Ants in nature are able to determine the shortest route between their colony and a food supply, and the ACO algorithm is based on this ability. During load balancing, the ACO algorithm is used to evenly distribute workloads across many resources so that no one resource becomes a bottleneck. The ACO method is a metaheuristic algorithm, which means it takes a high-level approach to locating approximations of solutions to optimisation problems. The ACO algorithm can be used to discover a nearly optimal solution for load balancing that distributes the load among several resources. Using the ACO algorithm, load-balancing issues in cloud computing can be resolved quickly and efficiently. The ACO algorithm can assist enterprises in making sure that their cloud computing systems are highly scalable, dependable, and effective by utilising swarm intelligence approaches.

The VM allocation issue is significant since VMs are the fundamental elements of cloud computing and are used to offer a variety of services to consumers. VMs must be utilised as effectively as possible for cloud computing systems to be scalable, dependable, and cost-effective. Due to the fact that each VM has a different set of resource requirements, it is challenging to determine how to best divide resources among them. The VM allocation problem can be solved using a wide variety of methods and strategies, each with advantages and downsides of its own. Heuristics, machine learning algorithms, and mathematical programming are some of the more well-liked strategies. The VM allocation issue is a significant cloud computing obstacle, so it's crucial for organisations to thoroughly analyse their VM allocation requirements and select the best strategy to satisfy them. They can guarantee their cloud computing solutions are highly scalable, dependable, and effective by doing this. Several heuristics and optimisation strategies have been put forth by researchers to enhance the load balancing of several heterogeneous resources in VM allocation.

According to some studies, the optimum approach to arrange virtual machines on physical servers is by applying evolutionary algorithms [2] or particle swarm optimisation [3]. These algorithms take into account both resource usage and user QoS needs. A few studies have also suggested utilising game theory [4] to simulate the cooperation and rivalry among various tenants in a cloud data centre. This method takes into account tenants' strategic behaviour and how their resource allocation choices affect the performance of the entire system. Additionally, some recent research has concentrated on employing machine learning methods, such as deep neural networks and reinforcement learning [5, 6], to make dynamic and real-time decisions for VM placement in cloud data centres. These techniques have produced encouraging outcomes in terms of improved load balancing and resource utilisation.

The topic of VM allocation in cloud data centres is challenging and complex because of the many resource types and the fluctuating needs of customers. Numerous heuristics, optimisation algorithms, and machine learning techniques have been suggested and used to enhance load balancing and resource utilisation in cloud data centres in order to address these issues.

These meta-heuristic algorithms balance the demand on physical and virtual resources in an effort to maximise resource usage [7]. They employ several optimisation techniques, including emulating the annealing process that occurs naturally, harnessing the wisdom of the swarm, and exploiting randomization to hunt for the global optimum. With a manageable temporal complexity, these algorithms can efficiently balance the resources and increase resource utilisation. Furthermore, because these algorithms are capable of handling a variety of limitations, including security, power consumption, and network bandwidth, they may be tailored to varied use cases and situations. As a result, the process for allocating resources is made more flexible and effective.

The resource allocation issue in cloud data centres has a promising answer provided by the meta-heuristic algorithms. They have demonstrated success at better utilising resources by distributing the workload across virtual and actual ones. These algorithms are a good option for resource allocation in cloud data centres because of their adaptability to various use cases and settings. The ACO algorithm is based on the idea of swarm intelligence, in which a colony of ants cooperates to find the optimal answer. The ants in this scenario stand in for several resource allocation strategies, and each ant will select the optimum PM for each virtual machine based on the resources and limitations available. The method uses a pheromone updating process in which the ants leave a pheromone trail behind them; the stronger the trail, the more alluring it is to the following ant. This technique makes sure that the best options are selected and improved over time.

The load balancing of numerous heterogeneous resources, such as CPU, memory, and network bandwidth, is another factor taken into account by the ACO algorithm. By optimising resource usage, virtual machines are distributed more effectively as a result. On cloud computing platforms, the ACO approach is an appropriate option for VM allocation [8]. It can find answers quickly and has excellent robustness, which guarantees good performance. It is a potent tool for multidimensional resource load balancing in cloud data centres due to the utilisation of swarm intelligence and the pheromone updating method. On cloud computing systems, the ACO methodology has the potential to be a useful method for allocating VMs. However, to effectively leverage the ACO technique, several challenges must be addressed, such as how to customise the parameters, how to overcome premature or slow convergence, and how to determine the value of each parameter. To overcome these challenges, the authors proposed a mathematical model and introduced the concept of PM selection expectation, along with a series of experiments to determine the value of each parameter. The results showed that the improved ACO algorithm performed better in terms of load balancing and resource utilisation compared to other algorithms. These findings contribute to the development of efficient and effective VM allocation methods for cloud computing platforms.

2 Literature Review

The goal of the suggested technique by Shagufta and Niresh [9] is to locate the overload node quickly and balance the load across nodes while utilising all available resources. It is based on the ACO. In the current business environment, firms heavily rely on the automated business processes provided by corporate IT systems. The cloud computing concept is based on the Internet and provides virtualized resources that are offered as a service over the Internet. It is dynamically scalable in nature. In order to prevent any one node from becoming overwhelmed, load balancing—one of the most difficult problems in cloud computing—calls for the distribution of the dynamic workload across numerous nodes. The SALB algorithm is used in [9]. The primary benefit of our work is the balancing of the system's overall load while attempting to maximise and decrease various parameters (performance, SLA violation, minimal overhead, and energy issue). ACO will be studied in order to create a load-balancing technique that is efficient. Using an initial heuristic approach, ACO is modified by Banerjee et al. [10] for the service allocation and scheduling mechanism in cloud systems. This adjustment mechanism aids in reducing the cloud system services' Makespan. In this research, a modified ACO-based heuristic algorithm has been suggested to start the service load distribution process in a cloud computing architecture. The ACO and coefficient's pheromone updating process is changed to and. The probability of fulfilling the request has also been increased using the updated scheduling, which helps reduce the duration of cloud computing-based services. The problems with fault tolerance are not taken into account in the simulation. It is anticipated that a continuous ant colony with other modified parameters could demonstrate better results compared to other optimisation models, even in defective service requests and disrupted resource allocators, due to the absence of any restore time in service and resource allocator distribution.

A cloud work scheduling policy based on the load-balancing ACO algorithm is suggested by Li et al. [11]. This chapter suggests a method for balancing the workload across the entire system while attempting to reduce the lifespan of a specific activity. The LBACO algorithm has been introduced in this research for achieving task scheduling with load balancing, and it has been empirically tested in situations with 100–500 tasks. The outcome of the experiment demonstrates how well the LBACO balances the load across the entire system. LBACO can handle every situation and outperforms FCFS and ACO algorithms in a cloud computing environment, regardless of whether the task sizes are the same or not. There are two noteworthy issues that merit more research in terms of future effort. First off, there is no constraint on precedence between tasks in our work because we believe that all tasks are mutually independent. Second, we make a computationally expensive assumption that is unrealistic for cloud systems. The availability vector should also be expanded to include information about task needs as part of future work in order to support the heterogeneous processing of the tasks.

An efficient load-balancing algorithm to maximise or decrease multiple parameters is devised by Ratan and Anant [12] using an ACO-based approach. We have so far covered the fundamental ideas behind load balancing and cloud computing. Additionally, the Swarm intelligence-based load-balancing technique has been discussed. We have talked about how the idea of ACO can help mobile agents balance a cloud's load. The drawback of this method is that it would work better in a cluster in our cloud. Therefore, research may move forward to build a complete load-balancing system in a full cloud environment. Our goal in this study is to create a load-balancing algorithm that is efficient, utilising the ACO technique to optimise or decrease various performance characteristics for clouds of various sizes, such as CPU load, memory capacity, delay, or network load. In order to start the service load distribution process under cloud computing architecture, a heuristic approach based on ant colony optimisation has been developed in this research. It has been established that the pheromone update mechanism is a reliable and effective technique for distributing the load. With the help of the ACO technique, this update helps to reduce the make-span of cloud computing-based services and the portability of serving requests. The concerns with fault tolerance are not taken into account in this method. The fault tolerance issues can now be included in future studies by researchers.

PSO-based adaptive multi-objective task scheduling (AMOTS) has been introduced by He et al. [13]. The goal of an advanced strategy is to use resources as efficiently as possible while lowering costs, average costs, and task completion times. The adjustable acceleration coefficient was used to demonstrate improved PSO algorithm outcomes; however, the cost of the method was slightly high, which is a common drawback of the AMOTS approach. The multi-objective optimisation (MOO) approach was introduced by Zuol et al. [14] to manage TS concerns and requirements of biodiversity in cloud computing. The resource cost model, which the author presents, replicates the relationship between user resource costs and budget expenses. In terms of efficiency in the CloudSim environment, experimental results showed that the MOO technique recovered only a half of one percent in the finer case development.

Three bio-inspired (MPSO, MCSO, and HYBRID) scheduling and resource management algorithms for a cloud environment are presented in this chapter [15]. As compared to other algorithms, the MPSO algorithm schedules tasks more effectively. On the other hand, as compared to existing algorithms, our suggested HYBRID (MPSO + MCSO) strategy is more effective at distributing the resources to the VMs. When compared to other cutting-edge benchmark algorithms, our suggested HYBRID method not only decreases the average response time but also boosts resource consumption by about 12%. In the future, we will concentrate on more efficient dynamic scheduling, in which jobs reach the cloud at varying periods. An efficient scheduling algorithm is needed, according to this paper [16], to carry out efficient job scheduling with minimum completion time and proper resource use. In this chapter, a more efficient version of ACO has been created, which performs better than the original ACO in terms of overall completion time and resource usage. EACO shortens lead times and boosts output. The outcomes of the investigation indicate that the EACO findings increased makespan. The suggested algorithm is now operating on the makespan problem; other functions, such as cost and load balancing, will be added.

Cloud computing is described in this work [17] as a vastly complicated system made up of tens of thousands of cloud resource nodes and communication linkages. The secret to assuring the successful execution of parallel tasks in a cloud environment with countless resource nodes is to learn how to obtain dependable virtual machine resources and assign cloud jobs to dependable resource nodes for execution. This research suggests a task scheduling optimisation technique employing an upgraded ACO algorithm in cloud computing to address the issue of an uneven job scheduling load and low reliability. A task-scheduling satisfaction function is built based on the suggested model to find the best task-scheduling combination for the three goals of least waiting time, most resource load balancing, and least task completion cost. To accelerate the convergence speed of the ant colony method, we optimise the two features of pheromone updating and pheromone volatilization. Also, in order to perform load balancing, the load weight coefficient of virtual machines is included during the local pheromone update process. Using CloudSim, simulation tests are run and compared with various methodologies. The outcomes supported the viability of the suggested scheduling optimisation approach. By assuring load balancing and reducing task scheduling completion and convergence times, this approach increases the efficiency with which virtual machines use their resources.

A unique approach for scheduling scientific operations in a cloud computing environment is provided in this article [18]. The algorithm's multi-objective parameters include makespan, load balance, and costs. Prioritising jobs based on dependencies, execution time, and data transfers between various activities is the initial step of the method. Pre-processing aims to remove bottleneck tasks and improve resource distribution in accordance with task demands and intensities. ACO is utilised for optimisation in the following stage. The experimental findings are compared in order to validate the suggested algorithm. The outcome demonstrates that the MrLBA outperforms other cutting-edge techniques on benchmark workflows. The workflow scheduling literature includes techniques that concentrate on certain aspects while ignoring others. The established approaches must take into account interrelated objectives. Enhancing one target has an impact on others and makes them more challenging to implement. The majority of algorithms did not take into account cloud computing's privacy and security issues while creating their scheduling algorithms. The solutions created should take this goal into account as well because cloud computing is open to several security assaults. Another goal to take into account while building scheduling algorithms for scientific procedures is energy consumption.

The following phase will involve doing a quantitative study and additional consideration of the resource cloud node failure type and accompanying recovery method. The relevant task scheduling algorithm is also put out, which takes into account node local security, resource node failure, and communication connection failure. This algorithm will also be further optimised and improved in terms of resource use, and research on ensuring QoS and lowering energy consumption in cloud data centres will be done.

A technique based on the ACO was put forth in [19], where redistribution of overloaded nodes was carried out based on the threshold value. Ant will look for available nodes to use if the load on the current node is below the threshold. Ants will only migrate in one direction in this place. In cloud computing systems, efficient resource management and load balancing are crucial. An approach to load balancing is described in this work. The proposed load balancing system relies on the cooperation of three concurrently running threads, which speeds up the mechanism's execution. Additionally, the load balancer is no longer required to wait for information about the concerned weakly loaded and severely loaded virtual machines to become available. Depending on the unique virtual machine's flexibility for passive time duration, the task may be expanded in the future. Additionally, the workload status may be established during the actual load balancing process.

3 Virtual Machine Allocation

According to Beloglazov et al. [20], the VM allocation issue can be seen as a bin packing issue with various charges and bin sizes. Since the cloud computing platform uses VMs, VM allocation is also NP-hard in the strong sense, much as bin packing is generally known to be. Heuristic methods, which lower the temporal complexity, are frequently used to address such problems. In this chapter, we use the ACO to address such a challenge. In order to prevent the algorithm from settling for the local optimal solution or sluggish convergence, we additionally incorporate an enhanced policy. We must first go over the fundamentals of ACO.

3.1 Ant Colony Optimisation

When foraging in the actual world, ants can always decide which path is the best to take to get to their meal. Real ants emit pheromones when they are moving, which is the cause of this. Additionally, the pheromone concentration will influence how the other ants behave. While travelling, several ants might create a favourable feedback loop. Finally, the ideal route can be chosen. ACO is suggested to act in a manner that is similar to that of real ants, which serves as inspiration.

In the ACO system, ants are first spread out randomly across all VMs, and each ant's pheromone level is set up based on the bandwidth, MIPS and number of processors of its starting VM. Then, via a procedure known as 'selection of the next VM,' ants are free to hop from one VM to another as they like. In this procedure, ants decide which virtual machine (VM) to use next depending on other trials or global pheromones, which they often follow under the assumption that this is the quickest way. It only selects a VM at random at the initial step; if the likelihood is higher than the VM selected at random, it selects a different VM. Each ant will maintain a tabu table history, which will be consulted to see if the next VM selected has been visited. It will repeat the selecting process if the virtual machine has already been visited; otherwise, it will go there and mark it as having been there on its tabu table. Since this form of ACO method ignores the starting state of idle processors, many are compelled to devise a new strategy centred on discovering an even more optimal manner of handling bandwidth and traffic. Furthermore, none of them are concerned with correctly allocating work to each and every virtual machine processor in order to avoid overloading. A recommended method will attempt to optimize each and every processor inside of each VM rather than considering a collection of processors as a virtual machine in order to achieve all of the aforementioned goals. According to the capabilities of each individual processor, a suggested technique would be to execute ACO optimisation for all processors in VMs and then save the results as an optimised list using ACO optimisation. The sophisticated algorithm will balance several metrics while concentrating on the multi-objective work. The fundamental ACO algorithm lacks quick adaptation and greatly increases the workload of tasks during runtime. As a result, ACO lengthens the execution process and slows convergence. In order to solve these issues in a cloud environment that operates in real-time, the Modified ACO algorithm is used. Additionally, because it does not account for individual processors in a machine, leaving some idle processors in a virtual machine reduces the efficiency of the virtual machine. As a result, various improvements were made to the modified ACO algorithm to make it more effective.

3.2 Algorithm for Modified Ant Colony Optimisation

The main objectives of the proposed algorithm in this study are to balance multi-objective tasks, reduce makeover time, and improve task scheduling effectiveness.

On processors, all ants are first displayed at random. Then, ants are permitted to move based on the pheromone values. Every single ant does this step. An ant will deposit pheromone in a processor based on factors such as processing speed, amount of traffic it must manage, the number of processors it houses, etc. After that, it records that virtual machine in its history of tours; as a result, all ants tend to travel similarly. By using the suggested plan, developers can distribute fewer tasks to VMs with sluggish processing power and more tasks to VMs with better processing power. As a result, developers do not need to worry about the framework using up their resources and can leave some of their best processors in their virtual machine alone.

Algorithm 1. Algorithm for allocating virtual machines based on a modified ACO.

1. Get initial pheromone and declared threshold value.
2. Input m ants.
3. for each iteration I do
4. for system balanced do
5. Create group according to load like overload, underload
6. sort VM according to Load ascending order
7. sort VM according to priority
8. Calculate according to the placement of ant and find destination VM and transfer task.
9. Update the best allocation and update global pheromone according to Global Pheromone Update Policy.
10. end for
11. end for
12. Output the best allocation.

The proposed plan can be judged in a number of ways, such as by comparing different techniques to the ACO system to see how well it works, how efficiently it uses resources, and how well it compares to other frameworks. Some of them have makespan; several iterations were necessary to reduce makespan. Without impacting one parameter more than another, the MACO procedure balances a system's whole load and makespan in balancing time. In comparison to earlier techniques, it will effectively balance system load while reducing makespan.

4 Simulation

An application for simulating events is Cloudsim 3.0. In the following case, the simulation is used: All of the jobs that need to be done stand alone. The computational magnitude of a task might vary. In Millions of Instructions, each task's duration is displayed (MI). On 20 tasks and 7 VMs, we ran experiments first. Java NetBeans 8.2 and a 32-bit operating system with 8 GB of RAM required were used to implement the suggested technique. In this research, Modified ACO reduced the multi-objective TS procedure and makespan. The suggested algorithm is superior to other conventional algorithms like the basic ACO because Modified ACO tries to decrease makespan

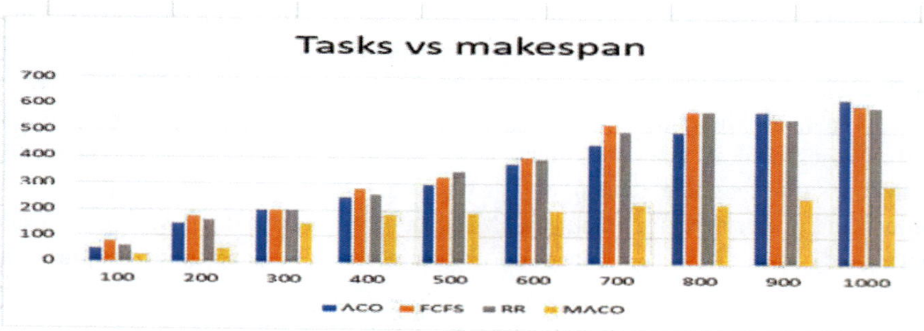

Fig. 1 Makespan versus number of tasks

and balance the workload in task scheduling. The following tests used basic ACO and modified ACO algorithms with span ranges of 10–100 task sets.

The graph shows that the proposed algorithm outperforms the fundamental ACO, and it is clear that the modified ACO completely balances the load on the entire system. According to Fig. 1, the number of tasks is depicted with sets of 100–1000 tasks shown along the *x*-axis and the task duration, which spans from 100 to 700 ms, depicted along the *y*-axis. This graph shows that Modified ACO outperforms ACO in terms of efficient task scheduling and timeliness reduction.

Figure 2 displays the number of iterations as sets of 10–50 along the *x*-axis, with the corresponding makespan interpreted along the *y*-axis. The graph demonstrates that the modified ACO method outperforms the original ACO algorithm, demonstrating that the suggested methodology yields superior outcomes. Implementing a TS algorithm based on Modified ACO, the algorithm is run on eight processors for eight occurrences of the problem. Each problem instance receives 10 trials using ACO, and the average value of task processor utilisation and execution time calculated. As a result, Modified ACO can schedule more activities in less time, and CloudSim is able to use a processor's or virtual machine's resources more effectively to shorten the processing time.

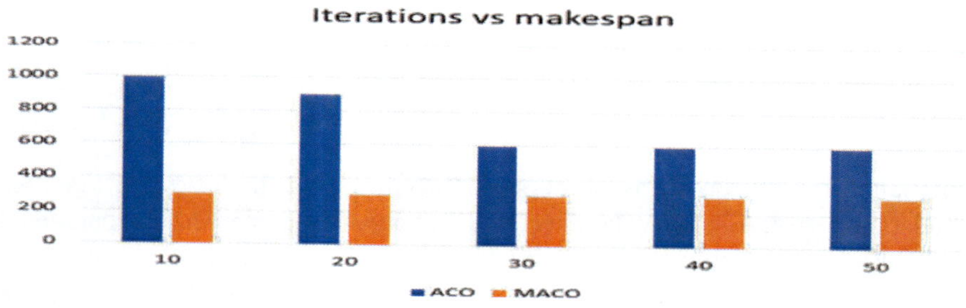

Fig. 2 Makespan versus number of iterations

Table 1 Compare algorithms based on makespan and load balancing

Task allocation algorithm	MakeSpan	LB
ACO	Moderate	Moderate
FCFS	High	Low
RR	High	Low
MACO	Low	High

Table 1 compares the different algorithms to identify their average performance. also use load balancing in cloud computing. MACO will be used for task allocation and improved virtual machine performance.

5 Conclusion

In this research, the Modified ACO method was used to find a way to schedule tasks in the cloud so that they could be done in less time. The makespan number may not change over the course of the runtime, and the convergence time may be a little slow in order to address an issue with the ACO algorithm. As a result, a new approach is required, which prompts the Modified ACO to be used to address the issue at hand and accelerate convergence. According to the experimental findings, Modified ACO achieved the MOTS procedure well while also reducing makespan. The implementation of this method, as assessed in the real-time CloudSim framework, leads to a significant increase in the efficient use of resources. In terms of shorter makeover time and greater load balancing ability, the proposed solution performs better. Although not all tasks have the same characteristics, modified ACO can handle better load balancing. In order to create an effective cloud computing environment, the modified ACO performed better than any other existing approaches. The Modified ACO framework can be used to create more efficient and straightforward cloud computing frameworks by applying it to a wide range of typical workloads on different VM configurations with their diversity of CPUs.

References

1. Mayank S, Jain SC (2021) A predictive priority-based dynamic resource provisioning scheme with load balancing in heterogeneous cloud computing. IEEE Access 9:62653–62664
2. Tawfeek MA, El-Sisi A, Keshk AE, Torkey FA (2013) Cloud task scheduling based on ant colony optimization. In: Computer engineering & systems (ICCES), pp 64–69
3. Li K, Xu G, Zhao G, Dong Y, Wang D 2011) Cloud task scheduling based on load balancing ant colony optimization. In: Sixth annual Chinagrid conference (ChinaGrid). IEEE, pp 3–9
4. Razaque A, Vennapusa NR, Soni N, Janapati GS (2016) Task scheduling in cloud computing. In: IEEE long Island systems, applications and technology conference (LISAT), pp 1–5
5. Nizomiddin BK, Choe T-Y (2015) Dynamic task scheduling algorithm based on ant colony scheme 7(4)

6. Hongyan C, Li Y, Liu X, Ansari N, Liu Y (2016) Cloud service reliability modelling and optimal task scheduling. IET Commun 1–12
7. Tsai CW, Huang WC, Chiang MH, Chiang MC, Yang CS (2014) A hyper-heuristic scheduling algorithm for cloud. IEEE Trans Cloud Comput 2(2):236–250
8. Panda SK, Jana PK (2016) Normalization-based task scheduling algorithms for heterogeneous multi-cloud environment. Inf Syst Front 1–27
9. Shagufta K, Niresh S (2014), Effective scheduling algorithm for load balancing using ant colony optimization in cloud computing. Int J Adv Res Comput Sci Soft En 4(2)
10. Banerjee S, Mukherje I, Mahanti PK (2009) Cloud computing initiative using modified ACO framework, vol 3. World Academy of Science, Engineering and Technology
11. Li K, Xu G, Zhao G, Dong Y, Wang D (2011) Cloud task scheduling based on load balancing ant colony optimization. In: 2011 Sixth annual ChinaGrid conference. IEEE
12. Ratan M, Anant J (2012) Ant colony optimization: a solution of load balancing in cloud. Int J Web Semant Technol (IJWesT) 3(2)
13. He H, Xu G, Pang S, Zhao Z (2016) AMTS: adaptive multi-objective task scheduling strategy in cloud computing. China Commun 13(4):162–171
14. Zuo L, Shu L, Dong S, Zhu C, Hara T (2015) A multi-objective optimization scheduling method based on the ant colony algorithm in cloud computing. IEEE Access 3:2687–2699
15. Domanal SG, Guddeti RMR, Buyya R (2020) A hybrid bio-inspired algorithm for scheduling and resource management in cloud environment. IEEE Trans Serv Comput 13(1):3–15
16. Jain REACO (2020) An enhanced ant colony optimization algorithm for task scheduling in cloud computing. Int J Secur Appl 13(4):91–100
17. Wei X (2020) Task scheduling optimization strategy using improved ant colony optimization algorithm in cloud computing, J Ambient Intell Hum Comput 1(0123456789):3
18. Arfa M, Muhammad S, Muhammad T (2021) MrLBA: multi-resource load balancing algorithm for cloud computing using ant colony optimization, cluster. Computing. https://doi.org/10.1007/s10586-021-03322-3
19. Joshi NA (2014) Dynamic load balancing in cloud computing environments. Int J Adv Res Eng Technol (IJARET) 5:201–205
20. Beloglazov A, Abawajy J, Buyya R (2012) Energy-aware resource allocation heuristics for efficient management of data centers for cloud computing. Future Gener Comput Syst 28(5):755–768

Chapter 13
Ensemble of Supervised Machine Learning Models for Cardiovascular Disease Prediction

Archi Agrawal, Dinesh Singh, Charul Dewan, and Shipra Varshney

1 Introduction

Cardiovascular diseases (CVDs) are a group of disorders that affect the heart or blood vessels and are the leading cause of morbidity and death worldwide, accounting for over 17 million deaths per year, making their prevention a major challenge for contemporary society. These diseases include hypertension, coronary heart disease, heart failure, angina, myocardial infarction, and stroke [1]. Many different chronic diseases affect adults today, reducing their freedom and harming their health. Over 92 million adult Americans and over one billion people globally are affected by CVDs. Improper food, sedentary behavior, smoking, physical inactivity, and excessive alcohol use are the most significant behavioral risk factors for heart disease and stroke. Moreover, the psychological stress and lifestyle changes brought about by the COVID-19 pandemic can also contribute to the development and worsening of cardiovascular diseases [2, 3]. People with underlying cardiovascular issues should take particular steps to guard themselves against COVID-19 and should keep managing their cardiovascular health through dietary adjustments and the right kind of medical care. In individuals with non-alcoholic fatty liver disease, a careful assessment of CVD risk is required [4]. For decades, treatments aimed at reducing CVD

A. Agrawal (✉) · D. Singh · C. Dewan · S. Varshney
Department of Information Technology, Dr. Akhilesh Das Gupta Institute of Technology and Management, New Delhi, India
e-mail: agrawalarchi02@gmail.com

C. Dewan
e-mail: charul.dewan@adgitmdelhi.ac.in

S. Varshney
e-mail: shipra.varshney@adgitmdelhi.ac.in

A. Yadav et al. (eds.), *Proceedings of International Conference on Paradigms of Communication, Computing and Data Analytics*, Algorithms for Intelligent Systems,
https://doi.org/10.1007/978-981-99-4626-6_13

risk factors through nutritional recommendations and medications have had variable success. Thus, early prediction of cardiovascular diseases based on medical examination and other factors is considered an effective measure for their prevention.

Machine learning is a powerful tool used in the healthcare sector to improve the overall efficiency of healthcare systems while lowering the cost of care. The purpose of this chapter is to identify predictive data patterns for fast and accurate estimation of the risk of CVDs in humans for their detection and diagnosis at an early stage. The main contribution of this work is the proposal of a LightGBM model for long-term CVD risk prediction with higher accuracy than other existing models. A comparative evaluation of several supervised machine learning models and ensemble methods was done on a balanced dataset for detecting the presence or the absence of cardiovascular diseases in human beings. Ensemble techniques increase generalizability and robustness over a single estimator by combining the predictions of numerous base estimators. For more accurate prediction, enhanced performance, and avoiding overfitting, the hyperparameters of these models were tuned using cross-validation techniques.

2 Related Works

Many machine learning models have been employed for the detection and early prediction of several ailments and diseases in human beings by analyzing large amounts of data. This has reduced the expenses involved in the manual detection of diseases performed in laboratories and has helped hasten the healing process.

Vaduganathan et al. [5] proposed that almost all countries outside of high-income countries continue to see an increase in the burden of CVD, and the age-standardized rate of CVD has started to climb in several places where it was previously receding in high-income nations. A model for predicting CVD for a three-year evaluation of CVD risk was given by Yang et al. [6]. It was based on a sizable population in eastern China with a high CVD risk utilizing the Random Forest algorithm, which would serve as a benchmark for the country's efforts on CVD prediction and therapy. A list of available prognostic models, their strengths and limitations, as well as recommendations and the identification of research gaps to be addressed to improve cardiovascular prevention in Latin America and the Caribbean, has been provided by Carrillo-Larco et al. [7].

There have been several studies in this field involving different datasets. CVD risk was assessed using machine learning (ML) models by Dritsas et al. [8] on participants, particularly those older than 50 years, using Kaggle's dataset, and it was stated that, when compared to Naive Bayes, Support Vector Machine (SVM), and Random Forest, the Logistic Regression classifier, with its 72.1% accuracy, was the best acceptable classifier. Whereas, when the ensemble method of XGBoost, as well as the supervised learning models of Logistic Regression, Naive Bayes, Support Vector Machine, K-Nearest Neighbors, Decision Tree, Random Forest, and Naive Bayes, were applied to the UCI dataset, Anbuselvan [9] obtained the highest accuracy of 86.89% was obtained using Random Forest. The Gradient Descent Optimization model, proposed by Nawaz et al. [10], on the other hand, produced an accuracy of 98.54% on the UCI dataset. Keniya et al. [11] explained how to diagnose a patient's ailment using their symptoms, age, and gender. An ensemble of Logistic Regression and K-Nearest Neighbor classifiers has also been proposed by Rahim et al. [12], but it did not perform well on Kaggle's CVD dataset when experimented with. The weighted KNN model had the highest accuracy of 93.5% in predicting illnesses using the aforementioned criteria. Once the illness is anticipated, the medical resources required for the treatment could easily be handled.

Further, deep learning methods have also been proposed to predict heart abnormalities at a prior stage. An ANN (Artificial Neural Network) with an "adam" optimizer and a "sigmoid" activation function in the output layer was analyzed by Pasha et al. [13] to obtain an accuracy of 85.24%. Khan et al. [14] used a variety of ANNs and the merging of spectral information to analyze PCG signals in order to diagnose CVD with 99.99% accuracy. An RCNN-based image heart disease prediction model was proposed and trained by Saikumar and Rajesh [15] on the Radiology dataset. This model reduced error loss and achieved high accuracy on heart images for CVD detection.

However, there has been no research on how to select an intelligent machine learning algorithm with optimal hyperparameters to reduce the computational time while enhancing the accuracy for the low-cost prediction of CVDs. Hence, we evaluated several supervised machine learning models and their ensembles for economical prediction. In the next section, we will discuss the dataset preprocessing techniques and optimized ensemble models used in this study.

3 Methodology

3.1 Dataset Description

The Cardiovascular Disease Prediction dataset from Kaggle was used to build the training and testing datasets for cardiovascular disease risk prediction. It contains a balanced record of 70,000 individuals collected at the moment of medical examination, with almost an equal number of people who have been diagnosed with CVD and those who are healthy. It has 11 attributes, as shown in Table 1, including age (in days), height (cm), weight (kg), gender, systolic blood pressure, diastolic blood pressure, cholesterol, glucose, smoking, alcohol intake, and physical activity, and a target variable depicting the presence or absence of CVD. While smoking, alcohol intake and physical activity are subjective features (information given by participants), systolic blood pressure, diastolic blood pressure, cholesterol, and glucose are examination features (results of a medical examination).

Table 2 depicts the Pearson correlation coefficients matrix between continuous features and the target variable in the dataset. It has also been observed that people

Table 1 Kaggle's open-source Cardiovascular Disease Prediction dataset

Age	Gender	Height	Weight	ap_ hi	ap_ lo	Cholestrol	Gluc	Smoke	Alco	Active	Cardio
18,393	2	168	62.0	110	80	1	1	0	0	1	0
20,228	1	156	85.0	140	90	3	1	0	0	1	1
18,857	1	165	64.0	130	70	3	1	0	0	0	1
17,623	2	169	82.0	150	100	1	1	0	0	1	1
17,474	1	156	56.0	100	60	1	1	0	0	0	0

Table 2 Pearson's Correlation Coefficient Matrix between continuous features and target

Attributes	Age	Height	Weight	SBP	DBP	CVD
Age	1	−0.82	0.054	0.021	0.018	0.24
Height	−0.082	1	0.29	0.0055	0.0062	−0.011
Weight	0.054	0.29	1	0.031	0.044	0.18
SBP	0.021	0.0055	0.031	1	0.016	0.054
DBP	0.018	0.0062	0.044	0.016	1	0.66
CVD	0.24	−0.011	0.18	0.054	0.066	1

with higher cholesterol and glucose levels who are physically less active are at a greater risk of CVD. Drinking women have higher CVD risks than drinking men.

3.2 Data Preprocessing

The dataset containing 70,000 observations is balanced and has no missing values. Twenty-four duplicate samples that were removed. Since age is given in days, it was converted into years. Further, the dataset contains multiple outliers and unbelievably high and low values of systolic and diastolic blood pressure that were to be filtered. The rows with systolic blood pressure values above 250 and below 45 were removed, along with the diastolic blood pressure values over 150 and below 30. Significantly lower values of human weight were also deleted.

BMI (Body Mass Index) was calculated as a person's weight in kilograms divided by the square of their height in meters and was added as a new feature. It was found to have a high correlation with the target variable and is a significant factor as described by Nikam et al. [16]. Height, among all continuous variables, has the lowest correlation with the target, as shown in Table 2, and has therefore been removed. However, BMI, a height-dependent feature, is still included.

The dataset contains label-encoded values for gender, smoking, alcohol intake, glucose, cholesterol, and physical activity. Label encoding assigns unique values to different categories, which machine learning models interpret as having greater importance. Therefore, one-hot encoding has been performed for encoding categorical variables. Finally, the dataset was standardized using StandardScaler.

3.3 Models

The dataset was split into training and testing sets in an 80:20 ratio after preprocessing. The model was then trained using the following machine learning algorithms after finding a set of optimal hyperparameter values using grid search and random search. The main distinction between these two techniques is that in GridSearchCV, we define the different combinations and train the model, whereas, in RandomizedSearchCV, the model chooses the combinations at random and is thus faster than GridSearchCV. Both of these approaches use the cross-validation technique and are quite effective in adjusting the variables and making the model more generalizable.

Logistic regression estimates the probability of an event with a categorical dependent variable or multinomial regression by using the log odds ratio, giving a probabilistic value that lies between 0 and 1. Instead of fitting a regression line in logistic regression, a sigmoid function is fitted instead. Sigmoid function $S(x)$ is given as:

$$\sigma(x) = \frac{1}{1 + \exp(-z)} \tag{1}$$

The majority of medical areas, social sciences, and machine learning employ logistic regression.

Random forest is a supervised machine learning approach that makes use of ensemble learning, a method that combines several weak or weakly correlated classifiers into a strong classifier to solve complicated problems and raise the accuracy of ML algorithms. It constructs decision trees on various samples and uses their averages for regression and the majority vote for classification. Random Forest's ability to handle data sets with both continuous and categorical variables is one of its key characteristics.

Gradient Boosting Machine (GBM) is akin to Gradient Descent as we gradually move in the direction of better prediction by finding a negative gradient and going in the opposite direction to lessen the loss. Boosting is an ensemble learning strategy that involves developing several models consecutively, each one aiming to address the shortcomings of the one before it. In GBM, the current classifier's residual serves as the input for the next classifier, on which trees are created. The classifiers gradually capture the residuals to capture the greatest variance present in the data.

XGBoost (eXtreme Gradient Boosting) is an efficient implementation of the gradient boosted trees algorithm. It covers a wide range of data formats, relationships, distributions, and adjustable hyperparameters with robustness. For better performance, XGBoost uses a more regularized model formalization to prevent overfitting. Each independent variable is given a weight, which is utilized to anticipate the outcomes. The variables that the first decision tree incorrectly predicted are sent to the second decision tree, and their weight is raised.

AdaBoost (Adaptive Boosting) is an ensemble technique where each weak learner is constructed as a decision tree with just one split and two terminal nodes to classify the observations. Each classifier has varied weights assigned to it based on its efficiency. At the end of each round, weights are assigned to the observations in such a way that incorrectly predicted observations have increased weight, thereby increasing the likelihood that they will be selected more frequently in the following classifier's sample.

LightGBM is a gradient boosting framework that uses tree-based learning techniques and is regarded as a very potent computing algorithm because of its high computational speed and low memory consumption. In contrast to previous algorithms that grow trees levels-wise, LightGBM, proposed by Ke et al. [17] grows trees leaf-wise to minimize loss while expanding the same leaf. Gradient-based one-sided sampling and Exclusive Feature Bundling are two unique approaches found in LightGBM that are used to handle huge numbers of data instances and features, respectively.

Voting classifier trains on an ensemble of several models that are passed to it and predicts an output based on the largest majority of votes by using either hard or soft voting.

Hard voting is an ensemble method where each model provides a prediction, which is tallied as one vote. The class with the highest majority of votes is the predicted output class. If the predicted outputs of three classifiers were (1, 1, 0), 1 would be the predicted class.

Soft voting relies on probabilistic outcome values. The output class is the prediction based on the average of the anticipated likelihood assigned to each class by all classifiers. If the average probabilities of classes A and B are 0.34 and 0.64 respectively, then the class with the highest probability as averaged by each classifier (class B) is the output.

4 Result and Analysis

The performance of various supervised classification algorithms was analyzed and evaluated using accuracy as a metric. It is calculated as:

$$\text{Accuracy} = \frac{\text{TP} + \text{TN}}{\text{TP} + \text{FP} + \text{TN} + \text{FN}} \tag{2}$$

Table 3 shows the performance results in terms of accuracy on training and testing sets while applying seven different machine learning models to the given dataset after choosing the optimal set of hyperparameters. Among Logistic Regression, Random Forest, Gradient Boosting Machine, AdaBoost, XGBoost, and LightGBM, the maximum accuracy on the testing set was obtained by LightGBM (73.91%),

Table 3 Training and testing accuracy of various models

Models	Training accuracy	Testing accuracy
LightGBM	74.07	73.91
Voting classifier	76.86	73.89
Gradient boosting machine	74.29	73.89
Random forest	79.38	73.70
XGBoost	74.01	73.69
AdaBoost	73.14	72.97
Logistic regression	73.14	72.97

Table 4 Classification report of LightGBM for dependent variable

	Precision	Recall	F1-score	Support
0	0.72	0.79	0.75	6803
1	0.76	0.69	0.72	6629
Accuracy			0.74	13,432
Macro avg.	0.74	0.74	0.74	13,432
Weighted avg.	0.74	0.74	0.74	13,432

closely followed by Gradient Boosting (73.89%), and Random Forest (73.70%). The predictive ability of the LightGBM model in cardiovascular diseases is promising and superior in terms of accuracy, and it outperforms the other ML models because of its leaf-wise split approach. This is also the highest accuracy for CVD risk prediction recorded on this dataset so far. In addition, LightGBM took the least amount of time as compared to other models. The classification report for LightGBM has been depicted in Table 4.

Hard and Soft Voting Classifiers were applied to different combinations of the top-performing models, namely, Gradient Boosting, LightGBM, XGBoost and Random Forest. Among all possible combinations, the best outcome (73.89% test accuracy and 76.86% training accuracy) was derived using a soft voting classifier on an ensemble of gradient boosting and random forest. As expected, logistic regression, being a single estimator, performed worse than other ensemble methods.

5 Conclusion

The use of ML models in CVD prediction has the potential to improve early detection, risk assessment, and targeted interventions, leading to improved outcomes and reduced costs. Since the medical field involves many time-consuming manual processes, it has become necessary to automate these procedures by using the capabilities of machine learning. This study on the effectiveness of various machine learning models and their ensembles for cardiovascular disease prediction revealed that hyperparameter tuning of LightGBM is capable of delivering the most accurate results with the least amount of computational time as compared to other supervised machine learning and ensemble models. It can significantly reduce the cost of care and computation and increase the efficiency of healthcare systems. With its strong generalization ability, LightGBM can also be applied to other types of diagnosis and treatment for their early prediction. The experimentation results also conveyed that the voting classifier gave the best outcome when soft voting was applied to the random forest and gradient boosting. The combination of these two models showed a significant improvement in prediction accuracy, demonstrating the effectiveness of ensemble learning in the field of cardiovascular disease prediction. We can further improve model accuracy by incorporating better data processing mechanisms and advancing the hyperparameter tuning of certain models like XGBoost. This work highlights the possibility of applying cutting-edge ensemble learning techniques for the early identification and prevention of cardiovascular disorders.

References

1. Kumar MD, Ramana KV (2021) Cardiovascular disease prognosis and severity analysis using hybrid heuristic methods. Multimedia Tools Appl 80(5):7939–7965
2. Mai F, Del Pinto R, Ferri C (2020) COVID-19 and cardiovascular diseases. J Cardiol 76(5):453–458
3. Paul S (2023) Advances and application of artificial intelligence and machine learning in the field of cardiovascular diseases and its role during the pandemic condition. In: System design for epidemics using machine learning and deep learning. Springer, Cham, pp 221–229
4. Targher G, Corey KE, Byrne CD (2021) NAFLD, and cardiovascular and cardiac diseases: factors influencing risk, prediction and treatment. Diabetes Metab 47(2):101215
5. Vaduganathan M, Mensah GA, Turco JV, Fuster V, Roth GA (2020) Global burden of cardiovascular diseases and risk factors. J Am Coll Cardiol 80(25):2361–2371
6. Yang L, Wu H, Jin X, Zheng P, Hu S, Xu X, Yu W, Yan J (2020) Study of cardiovascular disease prediction model based on random forest in eastern China. Sci Rep 10:5245
7. Carrillo-Larco RM, Altez-Fernandez C, Pacheco-Barrios N, Bambs C, Irazola V, Miranda, JJ, Danaei G, Perel P (2019). Cardiovascular disease prognostic models in Latin America and the Caribbean: a systematic review. Global Heart 14(1):81–93 (Science Direct)
8. Dritsas E, Alexiou S, Moustakas K (2022) Cardiovascular disease risk prediction with supervised machine learning techniques. In: Proceedings of the 8th international conference on information and communication technologies for ageing well and e-Health—ICT4AWE. SciTePress, Greece, pp 315–321
9. Anbuselvan P (2020) Heart disease prediction using machine learning techniques. Int J Eng Res Technol (IJERT) 09(11)
10. Nawaz MS, Shoaib B, Ashraf MA (2021) Intelligent cardiovascular disease prediction empowered with gradient descent optimization. Heliyon 7(5)
11. Keniya R, Khakharia A, Shah V, Gada V, Manjalkar R, Thaker T, Warang M, Mehendale N, Mehendale N (2020) Disease prediction from various symptoms using machine learning. SSSN Electron J
12. Rahim A, Rasheed Y, Azam F, Anwar MW, Rahim MA, Muzaffar AW (2021) An integrated machine learning framework for effective prediction of cardiovascular diseases. IEEE Access 9:106575–106588

13. Pasha SN, Ramesh D, Mohmmad S, Harshavardhan A, Shabana (2020) Cardiovascular disease prediction using deep learning techniques. IOP Conf Ser: Mater Sci Eng 981(2):022006 (IOP Publishing Ltd, India)
14. Khan MU, Samer S, Alshehri MD, Baloch NK, Khan H, Hussain F, Kim SW, Zikria YB (2022). Artificial neural network-based cardiovascular disease prediction using spectral features. Comput Electr Eng 101
15. Saikumar K, Rajesh V (2022) A machine intelligence technique for predicting cardiovascular disease (CVD) using Radiology Dataset. Int J Syst Assur Eng Manage (Springer)
16. Nikam A, Bhandari S Mhaske A, & Mantri S (2020) Cardiovascular disease prediction using machine learning models. In: 2020 IEEE Pune section international conference (PuneCon), pp 22–27
17. Ke G, Meng Q, Finley T, Wang T, Chen W, Ma W, Ye Q, Liu T-Y (2017) LightGBM: a highly efficient gradient boosting decision tree. In: Advances in neural information processing systems, vol 30

Chapter 14
Computing Model for Real-Time Online Fraudulent Identification

Ramani Jaydeep Ramniklal and Jayesh N. Zalavadia

1 Introduction

There will be a developing reliance on the web as the world moves closer to a cashless civilization. The results of misfortunes experienced online cannot be downplayed. Utilizing Virtual Private Arrange (VPN), sending victim's data over the browser; other difficult-to-detect strategies are occasions of recognizing clouding methods. Once a cardholder's personality has as of now been procured, scammers may utilize the accreditations separately or offer them to others, as is the case in India, where the character of the cardholder for nearly 70% of the populace has as of now been sold on the dark advertises [1]. When a major credit card burglary assault hit the United Kingdom as of late, it fetched the country's economy GBP 17 million in aggregate. An arrangement of universal hoodlums stole 32,000 credit and charge card points of interest within the 2000s [2]. The biggest false exchanges in history are thought to have happened in this occurrence. As a result, credit card robbery costs the economy billions of dollars [3]. Both cardholders and card guarantors are guaranteed of dependable operation. Opposite to belief, scammers need to form cardholders & money related companies to accept that the false exchanges were genuine. Also, a few likely to be false happen routinely for the money-related advantage indeed when card backers or shoppers are unconscious of them. Both affirmed organizations and shoppers are sometimes unconscious of fake credit card charges. Recognizing false action among hundreds of honest to goodness exchanges is troublesome, especially in case wrong exercises are significantly littler [4].

R. J. Ramniklal (✉)
Department of CS & IT, Atmiya University, Rajkot, Gujarat, India
e-mail: jaydeep.ramani@atmiyauni.ac.in

J. N. Zalavadia
Department of Commerce & Management, Atmiya University, Rajkot, Gujarat, India
e-mail: jayesh.zalavadia@atmiyauni.ac.in

A. Yadav et al. (eds.), *Proceedings of International Conference on Paradigms of Communication, Computing and Data Analytics*, Algorithms for Intelligent Systems,
https://doi.org/10.1007/978-981-99-4626-6_14

Prescient analytics, information gathering, and modeling frameworks that coordinate gathering and irregularity location help dodge money-related wrongdoings [5]. Most of these procedures require the utilization of other machine learning calculations, especially unsupervised and administered ones which will be supportive in malware discovery [6]. When endeavoring to recognize each occasion of burglary, in any case, these machine-learning frameworks confront endless impediments [7]. Standard evaluation measures must have the most noteworthy values within the compelling show for machine learning. To realize this perfect frame, a few changes are required in this field. The challenges in identifying false credit cards depend on a wide run of factors, counting machine learning calculations, cross-validation strategies, counting re-sampling strategies. These components may move forward the quality of the demonstration, which assessment measurements may affirm. Since adjusted datasets are exceedingly abnormal in real-world issues, the classification strategy frequently decreases the noteworthiness of the minority lesson within the database. Credit card extortion recognizable proof intensely depends on this bunch, which speaks to underserved populaces. Due to the dataset's unequal dispersion of the bunches, the proposed arrangement uncovered the imbalanced course issue by utilizing distinctive sampling approaches after selecting the foremost viable machine learning strategies. The enhanced cross-validation (CV) approaches are moreover taken into consideration in this investigation in expansion to the resampling methodologies.

E-commerce has relentlessly expanded. Agreeing to cardrates.com, worldwide retail e-commerce income was $4.9 trillion in 2021, with credit card get to 108.5 million day by day in United States. As per quantum computing extraordinary reenactment abilities to handle certain complicated issues in conventional computing, we accept Quantum Machine Learning (QML) offers a potential arrangement to managing with the enormous sum of online extortion data. This chapter proposes and actualizes a Machine Learning (ML) framework to assess online exchange data for extortion location. It too shows how ML's control could be utilized in significant commercial applications, how machine-learning-based identifying extortion methods are inspected. It clarifies how ESVDS employments hyper-planes for classification, utilizing Part Trap to change over nonlinear SVDS classifiers to a normal one, and how quantum material science may hurry the classification of ever more troublesome part work. The prescribed trick avoidance foundation is covered.

2 Related Works

To distinguish extortion, machine learning strategies are being utilized increasingly [8, 9]. Since profoundly uneven information and scattered designs decrease the prescient control of ordinary machine learning calculations [10], nonstationary data stands up to ordinary amassing and classification strategies, distinctive ways are being examined to overcome this issue. In spite of the fact that strategies for machine learning have been displayed, they are still based on inactive or nontime serial relationships. The foremost common machine learning strategies utilized for recognizing extortion are arranged in Table 1.

Quantum assesses the organize productivity with some diverse ML methods. SVM could be a high-performance, broadly utilized information investigation innovation made by Vapnik et al., AT&T Chime labs [21, 22]. SVM strategy utilized to fathom issues with two bunches of data. SVM isolates values into two categories by changing over the input arrangement into a space with a part of characteristics. SVM has been connected to a few frameworks for information examinations, counting for false exercises [23, 24]. Finding a conditional likelihood that builds the hyperplane among two bunches to maximize the edges is the objective of SVM. Figure 1 delineates the perfect hyper-plane, competent of creating the most prominent division between the two bunches.

Table 1 ML Method for fraud detection

Ref. No.	Research methods	Fraud detection methods
[11]	Neural network (NN)	Reporting in financial
[12]	Logistic regression (LR)	Transaction in credit card
[13]	Support vector machines (SVM)	Transactions in credit card, insurance, and reporting in financial
[14]	Decision tree (DT)	Transactions in credit cards, reporting in financial
[15]	Genetic algorithm (GA)	Transactions in credit card
[16]	Text mining	Reporting in financial
[17]	Self–organizing map	Transactions in credit card
[18]	Bayesian network	Transactions in credit card
[19]	Artificial immune systems	Transactions in credit card
[20]	Ensemble method (EM) (KNN, SVM, NN, and so on)	Transactions in credit card, and reporting in financial

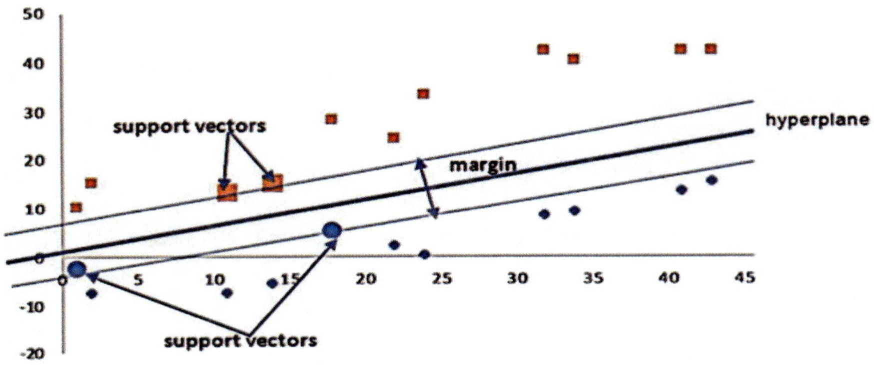

Fig. 1 A two-group classification method with the base classifiers

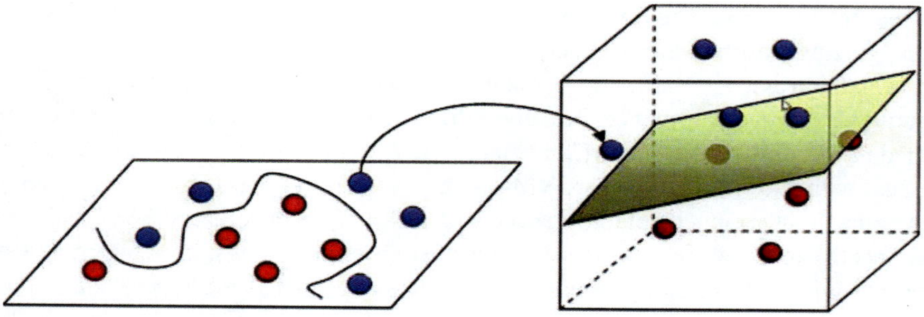

Fig. 2 Nonlinear SVM classification (input feature phase)

Comparative to other directed learning procedures, the relapse show (classifier) must be labelled. For occurrence, in false exercises, "false status" of the classification, whereas the highlights of the exercises will be the significant components (properties). Once the ideal hyper-plane is built, it's utilized to distinguish ordinary from false charges. Hyper-parameters come in two assortments: Back vectors are isolated into two bunches by the difficult edge hyper-plane appeared in Fig. 2.

It predicts even without mistake, whereas the soft one permits the fewest potential errors [25]. SVM has linear and nonlinear classifiers. The nonlinear support vector classifiers must be changed into a linear one to make it easier to identify the ideal hyper-plane. One such procedure is known as the "Kernel Trick," and it is illustrated below:

Separate linear variables are included in the classification model of a support vector regression classifier is specified in Eq. (1).

$$z_l = a_1x_1 + a_2x_2 + \cdots + a_nx_n \ldots \quad (1)$$

The nonlinear SVM classification algorithm to separate nonlinear variables in the classification model is specified in Eq. (2).

$$z_{nl} = a_1x_1^{0.5} + a_2x_2^3 + \cdots + a_nx_n^v \ldots \quad (2)$$

Because every term in the classification model is distinct, the nonlinear parameters may be substituted with new linear factors.

$$y_1 = x_1^{0.5}, y_2 = x_2^3, \ldots, y_n = x_n^v \ldots \quad (3)$$

Then the linear classification z_l, which is equal to the kernels trick's last step

$$z_{nl} : z_l = a_1y_1 + a_2y_2 + \cdots + a_ny_n \ldots \quad (4)$$

Quadratic confined double optimization issues [26], which require exceptionally capable computer control, may be fathomed to supply more modern bit capacities.

Nonspecific quadratic unconstrained binary optimizations (QUBO) can be created for SVM and after that QUBO with quadratic infeasibility punishment serving as an impediment [27]. This would be one strategy for tending to the issue. Since issues must be changed to QUBO organize, this interpretation strategy is one of the bottlenecks for quantum mechanics. Quantum computing has seen a few victories with QUBO [28]. The procedure execution tests of such a framework [29, 30] are profoundly energizing and energize us to investigate its potential uses in extortion detection. The discovery of false behavior in credit card exchanges requires taking into consideration many components, concurring to connect investigate. Each procedure employs a distinctive procedure to progress the by and large viability of the models it proposes. A machine learning calculation, on the other hand, might give changed results depending on how it is connected. To decide which calculation works best, attempt utilizing more. In datasets, the awkwardness lesson issue is especially predominant. As a result, overlooking this issue may result in subpar execution. The proposed ponder and significant resampling procedures which will be utilized in trials can be utilized to address this issue. Also, the amount of evaluation pointers is vital for evaluating the viability of the show from different perspectives. Previous works some of the time needed some of these highlights. Subsequently, a novel methodology is proposed.

3 Proposed Taxonomy of the Research Work

A common procedure for distinguishing false exchanges on credit and charge cards on a real-time premise is given within the Fig. 3. Stages to form, counting well-known ones just like the Point of Deal (POS), Mechanized Teller Machine (ATM), and online, are utilized as information sources. Let's envision that an exchange is transmitted to the credit card processor. Figure 3 shows the study's handle beneath the recommended system. We chose classification as framework systems as our fundamental demonstrates its solidness. Appropriately, the data are entered into the database whether the show acknowledges or rejects the provided task. The exchanges checking staff of the monetary company performs thorough reconnaissance and detailing assignments. In this work, the Offline Demonstrate Preparing module is the major subject. The ML demonstration, built on measurable data with distinctive classifiers, gets its diverse accentuation utilizing Cat boost. After that advancement, the calculation is associated with utilizing our extortion location framework to recognize extortion immediately.

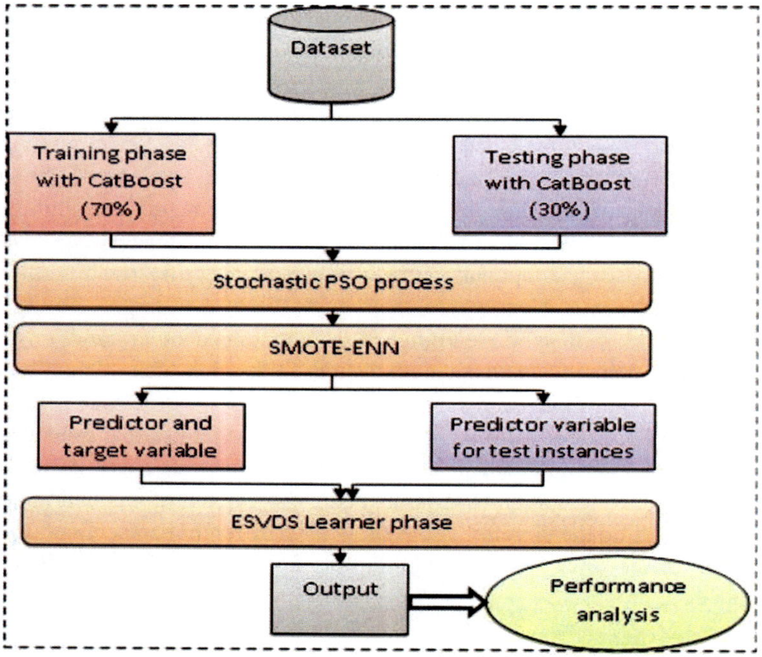

Fig. 3 Overall research in the pipeline view

3.1 Preprocessing Phase

The initial collection of characteristics or crude traits portrays each database test. Without pre-processing, this strategy may give wrong discoveries. As part of preprocessing, the crude highlights are disseminated to find exceptions, and dispense with noise.

3.2 CatBoost and SMOTE-ENN Phase

The study's credit card dataset has a critical awkward nature, which hurts the execution of ML models. The Destroyed is frequently utilized to address the issue of lopsided classes [31–33]. It is an oversampling methodology that levels out the dissemination of classes all through the data-set by counting engineered tests for the minority lesson. Under-sampling approaches like Altered Closest Neighbor (ENN) adjust a data-set by killing larger part tests of course. Under-sampling may expel learning-critical occasions. Furthermore, under-sampling procedures lose their viability when tests from major lessons immensely exceed those from the minority course, as was the case with the credit card database utilized in this way of thinking.

Moreover, as oversampling copies existing information tests, it may result in overfitting. The proposed credit card extortion discovery demonstrates SMOTE-ENN to deliver an adjusted data-set. A half breed re sampling strategy, the SMOTE-ENN conducts both oversampling and under-sampling of the information. The minority lesson tests are oversampled, utilizing destroyed and excess illustrations are disposed of utilizing ENN [34]. This procedure employs ENN's neighborhood cleaning run the show to dispose of tests that shift from two neighbors [35]. The SMOTE-ENN technique's pseudo-code appears in Algorithm-1.

Algorithm 1: Algorithm for SMOTE-ENN System

Output: Credit card balances dataset

Input: Input information

Step 1: Process of Over-sampling:

1: Select xi randomly from minor class

2: Explore for KNN of xi

3: Create a sample by assuming p value by predicting all neighbors q then K range at random, link q value with the p value to produce the segment of line in the attribute space by combining them together

4: Apply the label for the minor class to construct an artificial sample

5. Generate subsequent artificial samples by convexly joining the two samples that were chosen previously in step 1

Step 2: Process of Under-sampling:

6: Pick a sample from S(xi), where S variable indicate the total count of samples as xi identifier from minor class

7: Examine K neighbor with KNN algorithm of xi

8: If xi having enormous near attributes from the new class then xi will be discarded

9: Every sample in the data-set, do again steps 6 through 8

4 Experimental Design

A comparison of proposed strategies in up to date procedures in this considered region appears in this segment. Our fundamental objective is to extend the model's capacity for false exercises. To do this, a more in-depth understanding of the information is required.

Information from Brazillian Banks and the UCSD-FICO database are behavioral. Most of the fundamental contemplations we looked at were gathered by the Shipper category convention, which tells us what kind of company or organization the shopper is. Credit card firms' corporate official programs draw an expansive number of corporate laborers; information from these programs too ended up significant. Information from POS frameworks are other critical sources; credit card firms evaluate customers' reimbursement likelihood utilizing credit card limits and credit evaluations. Location data just like the state and local may too be vital in directing our choices.

A more full information of the design of behavior may be conceivable with a few exchanges from a single customer. Given the previously mentioned characteristics, it is crucial to comprehend their basic structure; it is apparent that the highlights coordinate the behavioral information. The dataset's highlights all take after the multivariate Gaussian likelihood, agreeing to early information investigation. Indeed in spite of the fact that it can appear like a clear issue to resolve, the torment of estimation mistakes is the most impediment.

4.1 Dataset Description

Trials were carried out employing a clear Linux setting that consisted of a single Quad-Core processor and 8 gigabytes of Smash. This setup arrangement naturally comes about when connected to Brazilian information records, which comprises real-time overhauled data with 0.3 million test information, and UCSD-FICO measurements, which are e-commerce records and come in two adaptations, the harder of which were utilized within the testing. Brazilian bank data contains a definitions proportion of 25.71, whereas the UCSD FICO information source had a descriptors proportion of 45.6. Since various exercises are carried out by a single client within the 0.1 million exchange example from the 70,124 clients, this information may give more understanding of cardholder extortion. Table 2 gives a diagram of the specialized dialect utilized in this examination.

In uneven settings, the classification calculations show effectiveness inconsistency; Execution lattices are specified within the Table 3.

TPR, FPR, TNR, FNR, affectability, specificity, and MCC were chosen as execution measures. MCC evaluates the association between the real and anticipated names. When the watched course level and expected course level are a culminating backup to one another, it takes -1. But as it were on the off chance that the real lesson name matches the expected course name does it take 1, else it does not. Location rate measures how well a show predicts veritable positive occurrences in recognizing credit card extortion.

Table 2 ML Symbols and Parameter

Symbols	Parameter
K_{11}	True positive value
K_{00}	True negative value
K_{01}	False positive value
K_{10}	False negative value

Table 3 Performance Metrics

True positive rate (TPR)	$\frac{k_{11}}{k_{11}+k_{10}}$
False positive rate (FPR)	$\frac{k_{01}}{k_{01}+k_{00}}$
True negative rate (TNR)	$\frac{k_{00}}{k_{00}+k_{01}}$
False negative rate (FNR)	$\frac{k_{10}}{k_{11}+k_{10}}$
Accuracy Matthew correlation coefficient (MCC)	$\frac{k_{11}+k_{00}}{k_{11}+k_{00}k_{01}+k_{10}(k_{11}*k_{00})-(k_{01}*k_{10})}$
Precision	$\frac{k_{11}}{k_{11}+k_{10}}$
Recall	$\frac{k_{11}}{k_{11}+k_{10}}$
Detection Rate	$\frac{k_{11}}{k_{11}+k_{10}}$

4.2 Data Imbalance

Most real-world twofold measurements are imbalanced. The issue of information awkwardness has as of late been settled by a few people, counting [36–38], etc. The two datasets that were utilized in this regard may well be respected as guidelines in this range of ponder. Most financial companies, counting banks, are frequently unwilling to supply scholastics with their data due to client security concerns. In our occurrence, information awkwardness demonstrates the need for false charges, since most employment of credit cards are not false. Figure 4 shows the conveyance of false and authentic exchange in Brazilian managing an account and UCSD-FICO databases. In this Data Imbalance we have used the Brazilian Bank dataset and UCSD-FICO Dataset which will be used for the check of fraud ant transactions in the banking sector.

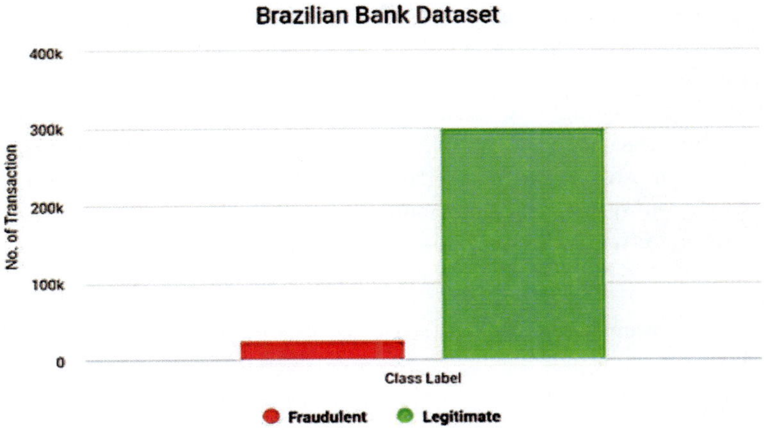

Fig. 4 Distribution of fraudulent and legitimate transaction of Brazilian Bank Dataset

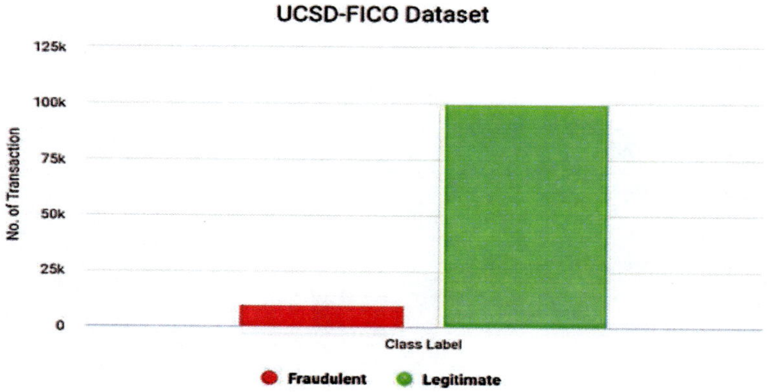

Fig. 5 Distribution of fraudulent and legitimate transactions of UCSD-FICO Dataset

Here in Fig. 4, it seen that Brazilian Bank Dataset Fraudulent Transaction near to 25% as compare to More than 300 K transaction.

Here in Fig. 5, it is seen that Brazilian Bank Dataset Fraudulent Transaction near to 22% as compared to More than 100 K transactions.

5 Experimental Result

An Outfit show was utilized to fathom the essential concerns around the distinguishing proof of false movement utilizing credit cards. Since few clients are likely to commit extortion, it's basic to analyze exchange data. Since there are not sufficient pertinent factors within the standard UCSD-FICO and Brazilian keeping money databases, gathering learning does not upgrade wrong negative or untrue positive rates that performed comparably to up to date calculation in terms of execution. Gathering demonstrates the cardholder's decision-making handle and gives clear decision-making parameters. As a result, the CatBoost approach was utilized to make perfect extricated highlights. Concurring to Tables 4 and 5, the proposed approach's prescient execution on the UCSD-FICO database expanded whereas MCC remained near to its unique esteem. Prescient behavioral issue detailing has progressed the location of wrong—positives, which hurt money related teaching. Expanded affectability immovably builds up moved forward extortion discovery.

Table 4 Performance evaluation before the application of the hybrid ensemble to the UCSD FICO Information

Techniques	Accuracy	Error rate	Specificity	Sensitivity	F1-score	MCC
Logistic regression	98	10	28	99	99	48
Boosting	99	00	78	97	99	66

Table 5 Performance evaluation after the use of a hybrid ensemble using the USCD FICO data-sets

Learning algorithm	Accuracy	Error rate	Specificity	Sensitivity	F1-score	MCC
Logistic regression	98	00	42	99	99	58

The suggested strategy improves the Brazilian bank dataset by 58.03–69.97% and the UCSD-FICO dataset by 54.66–69.40%, respectively, as shown in Table 6. Furthermore, Table 6 shows that the suggested technique efficiently manages minority class instances by using ensemble feature engineering, which is shown to be the most common behavior.

In the Fig. 6, Evaluation of Brazilian Bank information was seen that TPR, FPR, TNR, FNR, FDR, Acc, MCC, AUC was different in every part of percentage if we see the proposed details of than it gives 99% of accuracy level as compare to other this proposed give a better performance and results.

In the Fig. 7, if we seen the UCSD-FICO Dataset and compare the results than it gives information of TPR, FPR, TNR, FNR, FDR, Acc, MCC, AUC was different in every part of percentage if we see the proposed details of than it gives a good accuracy level as compare to other this proposed give a better performance and results.

Table 6 AUPR measurement using benchmark data-sets

Techniques	Brazilian bank dataset measurement of AUPR	UCSD-FICO data set Measurement of AUPR
LRextra trees	30	26
Boosting	41	41
CatBoost	99	96
Proposed	99	96

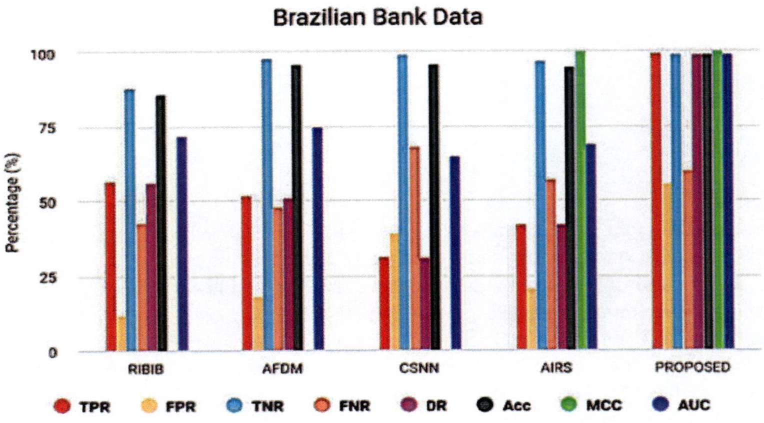

Fig. 6 Evaluation of Brazilian Bank information

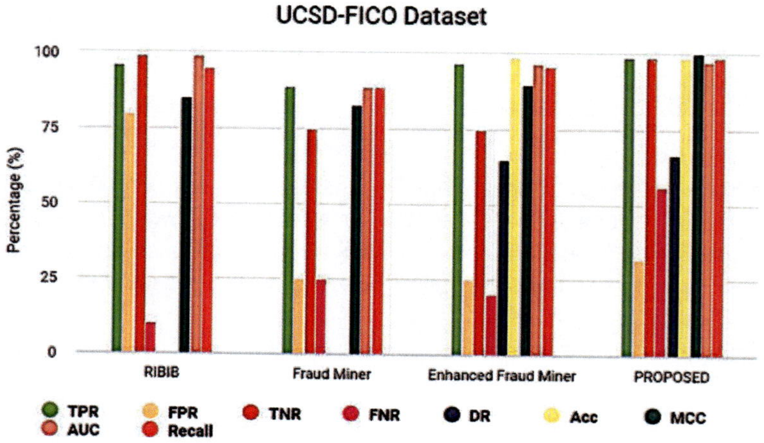

Fig. 7 Results of the UCSD-FICO database evaluation

6 Conclusion

In this investigation, cross breed design for recognizing credit card and web extortion was put out for thought. Amid the primary organization of the investigation, gathering methods for selecting highlights were utilized to exchange the input vector space onto the ideal highlight set. The created location demonstrated was utilized with SPSO & CatBoost within the moment arranged. The integration of SPSO and ESVDS was tried to see how well it worked in terms of accuracy and speed of learning, and it was compared with customary strategies. Concurring to the discoveries, the recommended approach moved forward exactness whereas conveying satisfactory execution. The framework proposed appears to have a untrue positive esteem of 0.00234, a negative result esteem of 0.0003045, a tall location of 0.9914, a accuracy of 0.9996, MCC of 1, and AUC of 0.9955, that outflanks RIBIB. In future advancement, an unused component for overseeing energetic exchange behaviors will be created utilizing profound neural systems.

References

1. Dubey SC, Mundhe KS, Kadam AA (2020) Credit card fraud detection using artificial neural network and backpropagation. In: Proceedings of ICICCS, Rasayani, India, pp 268–273
2. Martin T (2022) Credit card fraud: the biggest card frauds in history. Available https://www.uswitch.com/credit-cards/guides/credit-card-fraud-the-biggest-card-frauds-in-history
3. Zhang X, Han Y, Xu W, Wang Q (2019) HOBA, A novel feature engineering methodology for credit card fraud detection with a deep learning architecture. Inf Sci 557(10):302–316
4. Ssaghir Y, Taher R, Haque RMS, Hacid HZ (2019) An experimental study with imbalanced classification approaches for credit card fraud detection. IEEE Access 7:93010–93022

5. McCue (2015) Advanced topics. Data mining and predictive analysis. Oxford, Butterworth-Heinemann, pp 349–365
6. Ahmed F, Shamsuddin R (2021) A comparative study: credit card fraud detection using machine learning. In: Proceedings of IEEE access ICCDS, pp 112–118, 2021
7. Jain Y, Namrata T, Shripriya D, Jain S (2019) A comparative analysis of various credit card fraud detection techniques. Int J Recent Technol 7(5S2):402–40
8. Lakshmi SVSS, Kavilla SD (2018) Machine learning for credit card fraud detection system. Int J Appl Eng Res 13(24):16819–16824
9. Sailusha R, Gnaneswar V, Ramesh R, Rao GR (2020) Credit card fraud detection using machine learning. In: Proceedings of ICICCS, India, pp 1264–1270
10. Thabtah F, Hammoud S, Kamalov F, Gonsalves A (2020) Data imbalance in classification: experimental evaluation. Inf Sci 513:429–441
11. Ravisankar P, Ravi V, Rao GR, Bose I (2011) Detection of financial statement fraud and feature selection using data mining techniques. Decis Support Syst 50(2):491–500
12. Itoo F, Meenakshi, Singh S (2021) Comparison and analysis of logistic regression, Naïve Bayes and KNN machine learning algorithms for credit card fraud detection. Int J Inf Technol 13:1503–1511
13. Rtayli N, Enneya N (2020) Enhanced credit card fraud detection based on SVM-recursive feature elimination and hyper-parameters optimization. J Inf Secur Appl 55(3), Art. no. 102596
14. Huang Y, Yen DC (2015) Detecting the financial statement fraud: the analysis of the differences between data mining techniques and experts' judgments. Knowl-Based Syst 89:459–470
15. Duman E, Ozcelik MH (2011) Detecting credit card fraud by genetic algorithm and scatter search. Expert Syst Appl 38(10):13057–13063
16. Hájek P, Henriques R (2017) Mining corporate annual reports for intelligent detection of financial statement fraud—A comparative study of machine learning methods. Knowl-Based Syst 128:139–152
17. Olszewski D (2014) Fraud detection using self-organizing map visualizing the user profiles. Knowl-Based Syst 70:324–334
18. de Sá AGC, Pereira ACM, Pappa GL (2018) A customized classification algorithm for credit card fraud detection. Eng Appl Artif Intell 72:21–29
19. Halvaiee NS, Akbari MK (2014) A novel model for credit card fraud detection using artifcial immune systems. Appl Soft Comput 24:40–49
20. Kim E, Lee J, Shin H, Yang H, Cho S, Nam S-K, Song Y, Yoon J-A, Kim J-I (2019) Champion-challenger analysis for credit card fraud detection: hybrid ensemble and deep learning. Expert Syst Appl 128:214–224
21. Vapnik V (2006) Estimation of dependences based on empirical data, 2nd edn. Springer, New York. Available https://link.springer.com/book/10.1007/0-387-34239-7
22. Cortes C, Vapnik V (1995) Support-vector networks. Mach Learn 20(3):273–297
23. Rtayli N, Enneya N (2020) Enhanced credit card fraud detection based on SVM-recursive feature elimination and hyper-parameters optimization. J Inf Secur Appl 55, Art. no. 102596
24. Gyam NK, Abdulai J-D (2018) Bank fraud detection using support vector machine. In: Proceedings of IEEE—IEMCON, pp 37–41
25. Bingham E (2015) Advances in independent component analysis and learning machines. Academic, New York
26. Ferris MC, Munson TS (2002) Interior-point methods for massive support vector machines. SIAM J Optim 13(3):783–804
27. Kochenberger GA, Glover F, Wang H (2013) Binary unconstrained quadratic optimization problem. In: Handbook of combinatorial optimization. Springer, New York, pp 533–557
28. Li J, Ghosh S (2020) Quantum-soft QUBO suppression for accurate object detection. In: Computer vision and pattern recognition. Springer, pp 158–173
29. Date P, Arthur D, Pusey-Nazzaro L (2021) QUBO formulations for training machine learning models. Sci Rep 11(1):10029
30. Willsch D, Willsch M, De Raedt H, Michielsen K (2020) Support vector machines on the D-wave quantum annealer. Comput Phys Commun 248, Art. no. 107006

31. Abdoh SF, Rizka MA, Maghraby FA (2018) Cervical cancer diagnosis using random forest classifier with SMOTE and feature reduction techniques. IEEE Access 6:59475–59485
32. Ishaq A, Sadiq S, Umer M, Ullah S, Mirjalili S, Rupapara V, Nappi M (2021) Improving the prediction of heart failure patients' survival using SMOTE and effective data mining techniques. IEEE Access 9:39707–39716
33. Asniar N, Maulidevi U, Surendro K (2021) SMOTE-LOF for noise identification in imbalanced data classification. J King Saud Univ-Comput Inf Sci 34(6)
34. Inan MSK, Ulfath RE, Alam FI, Bappee FK, Hasan R (2021) Improved sampling and feature selection to support extreme gradient boosting for PCOS diagnosis. In: Proceedings of IEEE-CCWC, pp 1046–1050
35. Le T, Vo MT, Vo B, Lee MY, Baik SW (2019) A Hybrid approach using oversampling technique and cost-sensitive learning for Bankruptcy prediction. In: Applications of machine learning methods in complex economics and financial networks, vol 2019, pp 1–12
36. Raghuwanshi BS, Shukla S (2018) Underbagging based reduced kernelized weighted extreme learning machine for class imbalance learning. Eng Appl Artif Intell 74:252–270
37. He H, Garcia EA (2008) Learning from imbalanced data. IEEE Trans Knowl Data Eng 21(9):1263–1284
38. Galar M, Fernandez A, Barrenechea E, Bustince H, Herrera F (2011) A review on ensembles for the class imbalance problem: bagging-, boosting-, and hybrid-based approaches. IEEE Trans Syst Man Cybern Part C (Appl Rev) 42(4):463–484

Chapter 15
Ontology and Machine Learning: A Two-Way Street to Improved Knowledge Representation and Algorithm Accuracy

Leila Zemmouchi-Ghomari

1 Introduction

Machine learning (ML) algorithms accuracy is how often the algorithms predict or classify correctly or as expected provided data. Noisy and ambiguous data can cause decreased accuracy. However, data semantics can assist ML algorithms in understanding the context and meaning of the data they process, resulting in improved performance and more accurate predictions [1]. Indeed, data understanding can generate more meaningful and relevant features leading to improved model accuracy as the algorithms can better capture the relationships and patterns in the data. Furthermore, the semantic interpretation of the data can help ML algorithms explain their predictions, making them easier to understand, particularly in healthcare and finance, where transparent and comprehensible data can be precious [2, 3]. Context is essential to avoid semantic ambiguity in data interpretation [4]. However, connecting data to its context is a recognized research issue. Automatic interpretation and reasoning capabilities are made possible using ontology-based approaches to formalize the context model. "Ontology is a formal and explicit specification of a conceptualization" [5].

On the other hand, ML is crucial in ontology engineering [6, 7]. Due to ML algorithms, ontologies can be generated automatically from large amounts of data. This can be very useful, particularly in domains where manual ontology creation requires much time and resources [8].

In this chapter, we investigate how ontologies and machine learning algorithms can be combined to improve ML algorithms' accuracy and facilitate the construction of ontologies.

L. Zemmouchi-Ghomari (✉)
Ecole Nationale Supérieure des Technologies Avancées, ENSTA, Algiers, Algeria
e-mail: leila.ghomari@enst.dz

A. Yadav et al. (eds.), *Proceedings of International Conference on Paradigms of Communication, Computing and Data Analytics*, Algorithms for Intelligent Systems,
https://doi.org/10.1007/978-981-99-4626-6_15

An overview of ontology usage in ML is presented in Sect. 2, and ontology learning is explored in Sect. 3, which explains how ML automates some phases of the ontology engineering process. Finally, after discussing the challenges and trends associated with merging these two artifacts, we conclude this section with a summary of the main findings and research perspectives.

2 Ontology Use Cases in Machine Learning

By understanding the semantic meaning of the data, ML algorithms can better handle noisy and ambiguous data and make more accurate predictions [9]. Furthermore, the algorithms can better distinguish between words with multiple meanings and make more accurate predictions. For example, the word "stick" can have different meanings depending on the context, such as a tool for applying adhesive or a branch from a tree.

For example, in sentiment analysis, ML algorithms classify text data into different sentiments, such as positive, negative, or neutral [10]. Another use case concerns Named Entity Recognition (NER), where named entities such as people, organizations, or locations are identified in text data [11]. The algorithms can recognize entities referred to by nicknames or abbreviations and make more accurate predictions. For example, the algorithm can recognize that "Barack Obama" and "President Obama" refer to the same person. ML algorithms also classify images into different categories. Again, by understanding the semantic meaning of the data, the algorithms can better handle images containing multiple or partially occluded objects, making more accurate predictions [12].

Some recent use cases where ontologies have been used in machine learning from the literature include:

1. Predictive modeling in healthcare: In a study by Choi et al. [13], an ontology of medical concepts was used to improve predictive modeling in the healthcare domain. The ontology was used to annotate electronic health records with information about medical concepts, which was then used as input to machine learning algorithms for disease prediction. An essential challenge of predictive modeling in healthcare is aligning the knowledge learned with medical knowledge. A GRAPH-based Attention Model (GRAM) is proposed to address this challenge by supplementing electronic health records with hierarchical information inherent to medical ontologies.
2. Annotation of social media data to predict sentiment and trend analysis: In [14], the authors stated that ontologies of social media concepts are used to annotate social media data with information about concepts such as emotions, events, and relationships. The annotated data are then inputted into machine learning algorithms for sentiment analysis and/or trend analysis. For example, the Sociopedia [15] system analyzes social media topics. Based on a keyword, it constructs an ontology automatically. Relationships in the ontology are inferred

using related documents obtained from Wikipedia and DBpedia based on the retrieved top tweets. Ontology construction involves POS (Part Of Speech) and NER (Named Entity Recognition). The system includes a query summarization analysis, a comparison detection, and a sentiment analysis component, as the researchers are monitoring a marketing campaign for a new product launch on Twitter. Even if emotions are not explicitly referenced, machine learning algorithms automatically extract them and make sophisticated predictions of future behavior.
3. Personalized video recommender systems: In a study [16], based on users' activity on a website, the proposed system recommends videos according to their interests. A domain ontology and user items content are used to implement the recommendations. The performance of the recommended items is evaluated Using Predictive Accuracy Metrics, precision, and recall. According to the authors, ontologies can help recommender systems, thus improving the accuracy of traditional recommendation systems. In addition, ontology libraries are available. Metrics obtained using deep learning (based on neural networks) determine the predictive accuracy of the proposed framework. In addition, ontologies are crucial when analyzing tasks where the predicting rating is explained to the user with context.
4. Prediction of pharmacological consequences of drug interactions: In [17], a deep learning model is presented to accurately and automatically predict drug interactions. Deep feed-forward networks trained on structural similarity profiles, Gene Ontology (GO) term similarity profiles and target gene similarity profiles of known drug pairings are used in an integrated model. Based on the findings, it is clear that Gene Similarity Profiles and Target Gene Similarity Profiles improve prediction accuracy.

We notice that annotated and formally represented data is used as input data to ML algorithms.

3 Automated Ontology Learning

Machine learning algorithms can analyze the data and extract the relationships and concepts present in the data, which can then be used to generate an ontology. A typical example of automated ontology generation is using machine learning algorithms to extract relationships and concepts from a large corpus of text data, such as news articles or scientific papers, and then using these relationships and concepts to generate an ontology automatically. For example, the algorithm might extract concepts such as "person," "organization," and "location," and relationships such as "works for" and "located in."

In addition, ML algorithms can be used to populate ontologies with data. An example is using named entity recognition (NER) algorithms to extract named entities from text data, such as news articles, and then using these entities to populate an ontology with information about people, organizations, and locations. For example, the NER algorithm might extract the entity "Barack Obama" from a news article. The ontology might be populated with information about Barack Obama, such as his political affiliation and role as the 44th President of the United States.

Machine learning algorithms can generate embeddings for the concepts and relationships in an ontology which is also a knowledge graph, allowing the ontology to be processed by machine learning algorithms [18]. These embeddings can improve the accuracy of machine learning algorithms, such as recommendation systems or natural language processing models.

An example is using Graph Neural Networks (GNNs) to generate vector representations of the concepts and relationships in a knowledge graph [19]. The embeddings might be used to identify related concepts in the knowledge graph and make recommendations based on the related concepts.

More concretely, suppose we have a knowledge graph or an ontology with several entities and relationships between them. We want to use GNNs to generate vector representations, or embeddings, of these entities and relationships.

First, the entities are represented as nodes and relationships as edges in a graph. Then, each node and edge is assigned an initial feature vector, which could include information such as the entity's name, type, and attributes or the relationship's type and strength.

Next, a GNN model is applied, aggregating information from neighboring nodes and edges to update each node's feature vector. The GNN model typically consists of several layers, each applying a nonlinear function to the aggregated information.

Finally, the feature vectors of the entities and relationships are extracted as embeddings. These embeddings capture the context and relationships of the entities and can be used for various downstream tasks such as node classification, link prediction, and recommendation.

For example, suppose we want to recommend movies to users based on their viewing history. We can represent the movies and the users as nodes in a knowledge graph and the viewing history as edges. A GNN model is applied to generate embeddings of the movies and users, which capture their features and relationships. Then, these embeddings are used to recommend movies to users based on their similarity to previously watched movies.

The process in which machine learning algorithms automatically generate an ontology from data is known as automated ontology learning. Automated ontology learning aims to reduce the time and complexity involved in manually creating an ontology, making it easier to use ontologies in machine learning.

The choice of ML algorithm categories depends on the specific requirements of the task, such as the type of data and the desired level of accuracy. As a result, some categories may be more appropriate for specific tasks than others.

- Unsupervised algorithms: do not require any pre-existing knowledge and create the ontology by clustering concepts in the data and grouping them into categories.

- Supervised algorithms: require pre-existing knowledge and create the ontology by annotating data with concepts from a pre-existing ontology or using machine learning algorithms to classify the data.
- Hybrid algorithms combine unsupervised and supervised algorithms to create an ontology. For example, unsupervised methods can be used to identify potential concepts in the data, and then supervised methods can be used to refine and validate these concepts.

Most generated ontologies using automated ontology learning methods belong to the biomedical domain, such as Gene Ontology (GO), a widely used ontology for representing gene and gene product information. It is composed of three main categories, namely Biological Process (BP), Cellular Component (CC), and Molecular Function (MF). Another example is the Ontology for Biomedical Investigations (OBI), which is an ontology for representing biomedical investigation information. In addition to the Disease Ontology (DO). All these ontologies were generated using automated and manual methods, including unsupervised and supervised methods.

Here are some instances of how GO has been produced using various approaches:

Experts assign GO keywords to genes based on experimental data in manual annotation [20]. For example, the Gene Ontology Consortium (GOC) is a group of biologists who assign GO terms to genes using a combination of manual curation and automated approaches. Manual annotation can ensure GO annotations' high accuracy and consistency.

Supervised machine learning [21]: these algorithms learn to classify genes into multiple GO categories based on labeled training data. For example, the DeepGO method predicts GO terms for a given gene using a deep neural network. DeepGO's training set comprises gene sequences and their related GO annotations. Supervised machine learning has the advantage of managing enormous volumes of data and discovering complicated patterns.

Unsupervised machine learning [22]: these algorithms learn to recognize data patterns without labeled training material. For example, the ClueGO algorithm employs a network-based technique to find functionally linked groups of genes based on their GO annotations. The method groups genes into clusters based on their GO annotations and displays the results as a network. Unsupervised machine learning has the advantage of discovering novel patterns in data that would not be obvious using manual or supervised methods.

4 Discussion

Ontologies can help in explaining ML models and their outputs. An ontology can provide a structured representation of the concepts and relationships in a particular domain, which can help to explain the outputs of machine learning algorithms working in that domain. Using an ontology, machine learning models' outputs can be mapped to a human-understandable representation of the concepts and relationships in the domain, making the outputs more interpretable and understandable to domain experts and stakeholders.

Models and outputs of machine learning can be explained using ontologies. Ontologies contain structured representations of concepts and relationships belonging to a particular domain, which can assist algorithms that work within that domain to explain the results. Furthermore, ontologies enable domain experts and stakeholders to interpret and understand machine learning models' outputs by mapping them to a human-understandable representation of the concepts and relationships in the domain.

Confalonieri and colleagues [23] present Trepan reloaded, a revolutionary decision assistance system based on domain ontologies. The decision-making process of neural networks is explained using decision trees. Ontologies arrange knowledge to be used in future generations of decision trees. According to the conclusions of a user study, explanations with a structure similar to human comprehension are more understandable. To clarify the steps needed, neural networks are explained from input to output.

Doctor XAI, Panigutti, and colleagues [24] have expanded upon relational concepts across time with their new method. This technique benefits medical scenarios since it deals with ontology and sequential data with multiple labels. In addition, doctor XAI can associate data that occur over time, creating a new knowledge source to help improve predictions and explanations [25].

In a medical setting, for example, a patient suffering from chest pain, the machine learning algorithm aids in diagnosis. First, the patient's medical history is reviewed, including past heart issues and other pertinent criteria such as age, gender, and blood pressure. The data is then processed, and the chance of a heart attack is predicted using patterns learned from past medical situations. The algorithm uses an ontology to produce a clear and understandable forecast explanation. The algorithm can identify and explain the reasons for the prediction by relating the patient's medical history to the ontology's relevant topics. For example, it may emphasize that the patient's high blood pressure and previous history of cardiac issues were significant variables contributing to a high risk of heart attack.

This explanation is communicated to the patient and their family members to help them comprehend the diagnosis and treatment plan. Furthermore, the explanation may help guide future medical decisions by offering insight into the elements influencing a patient's health.

However, using ontologies in machine learning can have several downsides and obstacles, including:

- Complexity: Creating and maintaining an ontology can be complex and time-consuming, especially for large and diverse domains. This can result in high overhead costs regarding the time and resources required to develop and maintain the ontology.
- Scalability: Scalability is a challenge when using ontologies in machine learning, as the size and complexity of the ontology can increase rapidly as more data is

added. This can make it difficult to handle large amounts of data promptly and efficiently.

- Data quality: The quality of the data used to train machine learning algorithms is critical to their performance. In the case of ontologies, the quality of the data can be impacted by the accuracy and completeness of the ontology and the quality of the annotations made to the data. In addition, allowing for the evolution of the ontology over time can help to address changing requirements and to keep the ontology up-to-date. This can include using agile development methods and incorporating user feedback and continuous improvement into the development process.
- Semantic heterogeneity: Ontologies can be challenging to use in machine learning when multiple ontologies cover the same domain, as this can result in semantic heterogeneity and inconsistencies. Ontology matching can be used to overcome the challenges of semantic heterogeneity. These techniques aim to align multiple ontologies that cover the same domain and resolve inconsistencies, making it easier to use ontologies in machine learning.
- Expertise: Developing and using ontologies in machine learning requires specialized knowledge and expertise in ontology engineering and machine learning, which can be a barrier for some organizations. This challenge can be overcome with interdisciplinary collaboration. Encouraging interdisciplinary collaboration between ontology engineers, machine learning experts, and domain experts can help overcome the challenges of using ontologies in machine learning. This can include collaboration in the development of ontology and the design and implementation of machine learning algorithms.

Some trends in using ontologies for machine learning are the increased use of ontologies in medicine and biology. Although large amounts of data generated in the context of these fields motivate this trend, data volume can be challenging to manage and analyze without using ontologies. Additionally, ontologies are widely used in heterogeneous data sources. Besides, ontologies are known as knowledge representation artifacts in various domains, such as natural language processing, robotics, and artificial intelligence. Concerning the development of automated ontology learning methods, there is a trend toward combining unsupervised and supervised methods.

5 Conclusion

In summary, ontologies can provide a structured representation of concepts and relationships that can help explain the outputs of ML algorithms, making the outputs more interpretable and understandable to domain experts and stakeholders.

By leveraging ontologies, machine learning models can gain a deeper understanding of the data they are processing, leading to improved performance and more accurate predictions.

On the other hand, machine learning algorithms are used to automate, populate, refine, or process ontologies, improving quality and more valuable ontologies. Machine learning plays a significant role in ontology engineering, helping to automate and improve various aspects of the ontology development process. By leveraging machine learning algorithms, ontology engineers can create, populate, and refine ontologies more effectively and efficiently.

Indeed, automated ontology learning has the potential to significantly reduce the time and complexity involved in developing an ontology and make it easier to use ontologies in machine learning. However, there are also challenges to this approach, including the quality of the generated ontology, the scalability of the algorithms, and the complexity of the methods.

Future work should tackle these issues with proposed approaches that interpret big data and adapt them for ontologies since the big data research area also suffers from the same challenges.

References

1. Karn AL, Sengan S, Kotecha K, Pustokhina IV, Pustokhin DA, Subramaniyaswamy V, Buddhi D (2022) ICACIA: an intelligent context-aware framework for COBOT in defense industry using ontological and deep learning models. Robot Auton Syst 157:104234
2. Burkart N, Huber MF (2021) A survey on the explainability of supervised machine learning. J Artif Intell Res 70:245–317
3. Bhardwaj R, Nambiar AR, Dutta D (2017) A study of machine learning in healthcare. In: 2017 IEEE 41st annual computer software and applications conference (COMPSAC), vol 2. IEEE, pp 236–241
4. Zemmouchi-Ghomari L (2018) Current development of ontology-based context modeling. Int J Distrib Artif Intell (IJDAI) 10(2):51–64
5. Gruber TR (1993) A translation approach to portable ontology specifications. Knowl Acquis 5(2):199–220
6. Maedche A, Staab S (2001) Ontology learning for the semantic web. IEEE Intell Syst 16(2):72–79
7. Khadir AC, Aliane H, Guessoum A (2021) Ontology learning: grand tour and challenges. Comput Sci Rev 39:100339
8. Kethavarapu UPK, Saraswathi S (2016) Concept-based dynamic ontology creation for job recommendation system. Procedia Comput Sci 85:915–921
9. Horwath JP, Zakharov DN, Mégret R, Stach EA (2020) Understanding important features of deep learning models for segmentation of high-resolution transmission electron microscopy images. Comput Mater 6(1):108
10. Naresh A, Venkata Krishna P (2021) An efficient approach for sentiment analysis using machine learning algorithm. Evol Intell 14:725–731
11. Dong X, Qian L, Guan Y, Huang L, Yu Q, Yang J (2016) A multiclass classification method based on deep learning for named entity recognition in electronic medical records. In: 2016 New York scientific data summit (NYSDS). IEEE, pp 1–10
12. Loussaief S, Abdelkrim A (2016) Machine learning framework for image classification. In: 7th International conference on sciences of electronics, technologies of information and telecommunications (SETIT). IEEE, pp 58–61
13. Choi E, Bahadori MT, Song L, Stewart WF, Sun J (2017) GRAM: graph-based attention model for healthcare representation learning. In: Proceedings of the 23rd ACM SIGKDD international conference on knowledge discovery and data mining, pp 787–795

14. Sapountzi A, Psannis KE (2018) Social networking data analysis tools & challenges. Futur Gener Comput Syst 86:893–913
15. Kaushik R, Apoorva Chandra S, Mallya D, Chaitanya JNVK, Kamath SS (2016) Sociopedia: an interactive system for event detection and trend analysis for twitter data. In: Proceedings of 3rd international conference on advanced computing, networking and informatics: ICACNI 2015, vol 2. Springer India, pp 63–70
16. Sharma S, Rana V, Kumar V (2021) Deep learning based semantic personalized recommendation system. Int J Inf Manag Data Insights 1(2):100028
17. Lee G, Park C, Ahn J (2019) Novel deep learning model for more accurate prediction of drug-drug interaction effects. BMC Bioinform 20(1):1–8
18. Manda P, SayedAhmed S, Mohanty SD (2020) Automated ontology-based annotation of scientific literature using deep learning. In: Proceedings of the international workshop on semantic Big Data, pp 1–6
19. Ye Z et al (2022) A comprehensive survey of graph neural networks for knowledge graphs. IEEE Access 10:75729–75741
20. Gene Ontology Consortium (2015) Gene ontology consortium: going forward. Nucleic Acids Res 43(D1):D1049–D1056
21. Kulmanov M, Khan MA, Hoehndorf R (2018) DeepGO: predicting protein functions from sequence and interactions using a deep ontology-aware classifier. Bioinformatics 34(4):660–668
22. Bindea G, Mlecnik B, Hackl H, Charoentong P, Tosolini M, Kirilovsky A, Galon J (2009) ClueGO: a Cytoscape plug-in to decipher functionally grouped gene ontology and pathway annotation networks. Bioinformatics 25(8):1091–1093
23. Confalonieri R, del Prado FM, Agramunt S, Malagarriga D, Faggion D, Weyde T, Besold TR (2019) An ontology-based approach to explaining artificial neural networks. arXiv preprint arXiv:1906.08362
24. Panigutti C, Perotti A, Pedreschi D (2020) Doctor XAI: an ontology-based approach to black-box sequential data classification explanations. In: Proceedings of the 2020 conference on fairness, accountability, and transparency, pp 629–639
25. Olivas ES, Guerrero JDM, Martinez-Sober M, Magdalena-Benedito JR, Serrano L (2009) Handbook of research on machine learning applications and trends: algorithms, methods, and techniques. IGI Global

Chapter 16
Nonmetaheuristic Methods for Group Leader Selection, Cluster Formation and Routing Techniques for WSNs: A Review

Kumar Dayanand, Binod Kumar, Barkha Kumari, Mohit Kumar, and Kumar Arvind

1 Introduction

In many applications, wireless sensor networks (WSN) are a significant source of cutting-edge technologies. WSNs are essential in resolving several significant issues in military research and domain monitoring [1]. As a low-cost legacy system, wireless sensor networks (WSNs) have been used in a variety of industries, including industrial control and intelligent transportation systems [2]. WSNs offer extensive physical data that can be used further. Thus, there is no need for significant paradigm shifts while using WSN applications. Sensor devices built on WSN are more affordable and easier to install. Furthermore, it may operate on its own in dangerous locations where human presence is impossible. The main difficulties in WSNs are those related to network life span [3]. The lifespan of the sensor is solely dependent on its batteries, which are challenging to repair or recharge due to the challenging environments in which they operate. This issue raises the expense of new technology while undermining

K. Dayanand
Department of Computer Science and Engineering, Mangalayatan University, Aligarh, India

B. Kumar (✉)
Kalinga University, Naya Raipur, India
e-mail: bit.binod15@gmail.com

B. Kumari
G H Raisoni College of Engineering Pune, Pune, India

M. Kumar
MIT Art, Design and Technology University, Pune, India

K. Arvind
Ashoka Institute of Technology and Management Varanasi, Varanasi, India

A. Yadav et al. (eds.), *Proceedings of International Conference on Paradigms of Communication, Computing and Data Analytics*, Algorithms for Intelligent Systems,
https://doi.org/10.1007/978-981-99-4626-6_16

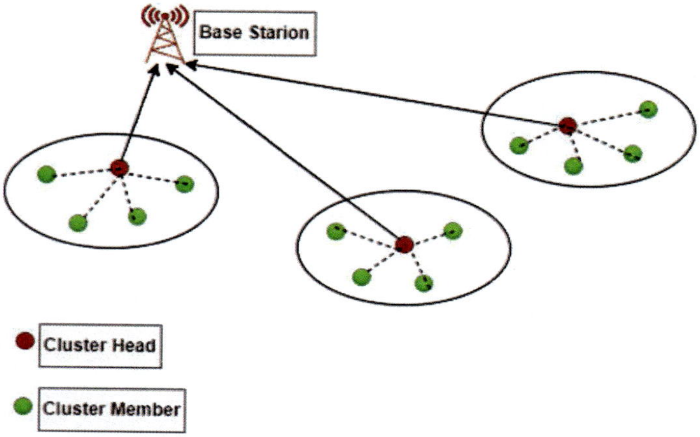

Fig. 1 Basic structure of LEACH

its incorporation into WSN. In light of this, extended network lifetimes are seen as a significant problem in WSN. Consequently, to increase power consumption and prolong the network's lifespan, In the WSNs, a clustering approach is applied. By eliminating lengthy detachment communications, the clustering protocols, in which the sensor nodes are grouped into small clusters, are an efficient way to cut power consumption and extend network lifetime. One node serves as the group head (CH) for each cluster, performing more functions than Member Nodes (MNs). Practically every MN in the cluster transmits its sensing data to its group leader, who then sends it to its CH, who then sends it to the BS using a single hop or many hops [4]. Although the clustering protocol is thought to be a useful method of power management for WS' nodes, clustering structure is still a significant problem that negatively impacts the longevity of the network due to inefficient node power usage. Additionally, the WSNs' subpar clustering structure frequently has an impact on the network's later processes, like data aggregation and routing discovery, where it gets the network ready for operation. As a result, the effectiveness of the clustering structure has a significant impact on the lifetime of WSNs. At the beginning of the twentieth century, Heinzelman et al. [4] proposed the LEACH (Low-Power Adaptive Clustering Hierarchy) communication protocol. This protocol uses a clustering mechanism to help reduce power consumption. Group leaders are chosen using cluster rotation techniques, and other nodes join these group leaders to form a cluster [5]. The CH receives the revised raw data before sending it along with the received data to the BS [6]. By reducing the amount of power used during transmission, the LEACH protocol extends the life of WSNs. Through the use of a random group leader (CH) rotation, LEACH was created to provide balanced power utilization. In the LEACH algorithms, a CH is dynamically chosen at each interval. The optimal group leader (CH) proportion in the networks serves as the sole foundation for the CH collection, which is based on the incapacity of other independent nodes to have more resources than the chosen CH [7]. Additionally, it relies on how frequently an expected node has in the past evolved into a group leader (CH). Following the selection of the CH, the CH will broadcast its location to nearby nodes, which will then choose the best CH and relay to include in the group [8, 9]. The message broadcast by the CH is sent via Carrier Sense Multiple Access (CSMA) to reduce collisions. The CH will generate a transmission schedule and transmit it to all of the nodes in each of its clusters after announcing its position [10]. Each of the neighboring nodes has TDMA slots in this transmission schedule, which allows for minimal power usage because the nodes can turn off their radios while stationary. The main goal of cluster-based routing protocols is to promote low power consumption by the network nodes in order to prolong the lifespan of the network service. In this section, a review of a few powers' efficiency-conscious WSN protocols was done. As a strong clustering framework for wireless sensor networks, LEACH was created. It relies on the expected signal strength of clusters and uses the CH as a router to connect to the BS [11]. The clusters carry out the information handling on a local level. In the LEACH algorithms, clusters are created via a dispersed method, in which the nodes are free to choose whatever they choose. A node is initially chosen as the CH based on its probability value; non-CH nodes choose their group by choosing the CH that can be reached with the least amount of power. The function of being a CH is frequently rotated across the nodes in each cluster to balance the burden [3]. This study presents a comprehensive survey of nonmetaheuristic clustering techniques for sensor networks.

2 Related Work

In this section, group leader selection was classified among three distinctive approaches, and scrutinizing was done on comparison assistance for deliberation in group leader selection. Numerous arguments or different parameters were used, and necessary reclustering requires cluster formation, equal scattering of group leaders, and the construction of balanced clusters. Ramesh et al. [12]

In this section, the extra and added particulars were discussed in four types of clustering models that are in existence:

1. separate hop-flat architecture
2. separate hop clustering architecture
3. Many hops flat architecture
4. Many-hop clustering architecture.

Tyagi et al. [13] put forth contemplation on clustering algorithms dependent on the LEACH protocol. The algorithms were classified into different parameters, which discussed the various objectives that the researcher wants to achieve and the comparison between different network performances. Kalla et al. [14]. In this chapter, many important clustering protocols are discussed. Overlap of clusters, position awareness, power efficiency, uniform clustering, and stability of clustering [15] The concepts

Table 1 The evaluation chart of counterfeit parameters and domain settings of non-metaheuristics clustering with group leader selection method

R. N.	CH selection algorithm	Selection approach	Transmission methods	Mobility	Domain
[19]	Clonal selection algorithms modified	Not specified	Many hop	Fixed	Similar
[16]	Gradable control procedure named double group leaders and many hop based on resonance propagation	Based on unconsumed power	Many hop	Fixed	Similar
[20]	Enhanced clustering hierarchy	Based on power and local detachment	Many hop	Fixed	Similar
[21]	Power-broadcasting proportion clustering protocol	Based on unconsumed power and broadcasting proportion	N/S	Fixed	Mixed
[22]	Cluster manager-based group leader selection methods	Based on power and detachment	Many hop	Mobile	Similar
[23]	Power-efficient mobility-based group leader selection	Based on nodes mobility level, unconsumed power, detachment to sink	Many hop	Mobile	Similar
[24]	Low power adaptive clustering hierarchy-many objective optimization algorithms	Based on cluster detachment, sink node detachment, overall power consumption and balance of network power consumption	Many hop	Fixed	Similar
[25]	Power balanced clustering routing	Based on unconsumed power	Many hop	Fixed	Similar
[26]	Fuzzy-based power-efficient clustering approach	Based on unconsumed power, communication quality, average detachment from base station	Separate hop	Fixed	Similar
[27]	Technique for order preferences by similarity to ideal solution method	Based on unconsumed power, available storage	Separate hop	Fixed	Similar

(continued)

Table 1 (continued)

R. N.	CH selection algorithm	Selection approach	Transmission methods	Mobility	Domain
[28]	Dynamic quantum-based adaptive group leader selection method	Based on quantity of closer nodes and remaining power	N/S	Fixed	Similar
[29]	Improved power-efficient clustering protocol method	Based on power consumed and the ratio from the initial power	Many hop	Fixed	Similar
[10]	Greedy technique for grouping mobile routing using artificial neural	Based on detachment from the cluster center and its unconsumed power	N/S	Mobile Sink	Similar
[30]	Dynamic range-low power adaptive clustering hierarchy protocol	Based on unconsumed power, the position, and centrality of nodes	Separate hop	Fixed	Similar
[31]	Threshold group leader selection	Based on unconsumed power and detachment from BS	Many hop	Fixed	Similar
[32]	Non-threshold based group leader rotation scheme method	Based on unconsumed power and detachment	Many hop	Fixed	Similar

discussed above provide a better and wider interpretation and perception of clustering and its parameters and techniques. Surrounding positioning, parameter analysis, and newer clustering protocols that correlate with ongoing trends in WSNs are discussed. In recent contemplations, researchers have published or put forth commentary on nonmetaheuristic methods in group leader selection and cluster formation to clear up selective objectives with different domain settings, for instance, potency and mobility. The indications and hindrances of the methods, including future supervision and directions towards extensive clustering in wireless sensor networks, are also entrusted. In Fig. 1, we explain the LEACH architecture of WSN, which is also known as the Low Power Adaptive Clustering Hierarchy Protocol, but it is a schedule-based separate hop routing protocol. LEACH introduced the base of selection of heads and cluster formation, which is very useful for conservation of power and was not possible in a flat-model sensor network.

3 Nonmetaheuristic Methods

Clustering in wireless sensor networks is categorized into two important methods, which are non-metaheuristic methods that perform the important roles of group leader selection and cluster formation.

- Group leader selection

CH selection is an important step in clustering, as it becomes a big responsibility for WSNs to transfer the data and assemble data efficiently. In the ongoing years, CH selection has been emphasized in several works of literature, as picking the most precise CH will intensify the total life span and dependability of the network. [16] In nonmetaheuristic methods, group leader selection is constructed on selection criteria that are mandatory for persuaded applications and domains. Several group leader techniques that are put forth for different domains are discussed in this particular section.

- Many hops' data transmission

Enormous and moderate-scale wireless sensor networks endorse Many hops in data transmission to visualize prolonged ranged transmission may disparage the longevity of the sensor node. The group leader, who is farthest from the base station, collects and aggregates the data, and they keep sending that data to the group leader in front of them until it reaches the base station. Many hops' data transmission gives guarantees for reliable data transmission. In many hopes, domains will diminish power consumption and extend network lifetime in WSNs [17].

- Separate hop data transmission

Separate hop and many hop communication are two basic communication patterns that are used in WSNs. It was noticed that in the case of separate hope communication, the furthest member nodes, or CHs, tend to deplete their battery power faster than other nodes in a network. In a separate hop, the data packets are directly transmitted to the CH or BS without any relay; the nodes located farther away have a higher power burden due to long-range communication, and these nodes may die out first. To overcome those problems, we used many hops of communication between member nodes and their respective CHs and also between CHs and the BS. A cut-off value calculated by the ratio of the number of group leaders with the number of active nodes [18].

- Cluster Formation

In WSNs, clustering sensor nodes is an operative technique to obtain scalability, self-organization, power conservation, channel accessing, and routing. [5] The need for cluster formation automatically emerges after group leader choice, thereby increasing the life span of the networks. The work of group creation can be done before or after the group leader selection, depending on the applications and objectives. The group formation technique is very useful to diminish the hotspot problem in WSN distributions.

- limitation and domain-setting analysis of nonmetaheuristics methods

The simulation parameters used and the domain settings from the aforementioned procedures in nonmetaheuristics clustering are analyzed and compared in the following tables, where the field R.N. means Reference Number:

- Table 1. Describe the comparison chart of simulation parameters and domain settings of non-metaheuristic clustering with the group leader selection method.
- Table 2. Depicts the assessment or comparative chart of simulation parameters and domain settings for nonmetaheuristic clustering with the cluster formation method.
- Table 3. Various non metaheuristic methods in many hop data transmission domain settings are described in tabular form.

Table 2 The Comparison chart of simulation parameters and domain settings of nonmetaheuristics clustering with cluster formation method

R. N.	Cluster formation algorithm	Selection approach	Transmission methods	Mobility	Domain
[33]	Power optimal clustering routing protocol based on QoS node deployment	Based on unconsumed power and detachment of nodes from	Many hop	Fixed	Similar
[34]	Unequal clustering protocol for power harvesting sensor networks	Based on unused power	Many hop	Fixed	Similar
[35]	Lifetime-enhancing cooperative data gathering and relaying algorithm	Based on communication detachment and	Many hop	Fixed	Similar
[36]	Nonuniform K-means	Based on quantity of closer nodes, unconsumed power	Separate hop	Fixed	Similar
[37]	Hybrid optimabased cluster formation algorithm	Based on detachment and power	Many hop	Mobile	Similar
[38]	Grid Clustering	Based on unconsumed power, detachment	Many hop	Fixed	Mixed

Table 3 Various nonmetaheuristics methods in Many hop data transmission domain settings are described in the tabular form

R. N.	Algorithms	Objective	Advantages	Constraints	Scope
[39]	Two-tier scattered fuzzy logic-based protocol	Improve the efficiency of data collection	Emphases on different scenarios by changing the location and protocol of the sink used for many hops routing	Only the SA algorithm is used to trial the boost ability of the projected method	PSO techniques used for boost the ability and Node's mobility are also to be focused on scope
[40]	Efficient target tracking approach	Balanced power consumption and increasing the power proficiency	CCH is used to collect sensed data and collective data near to the data source and the transmitting data are reduced	CH selection and CCH may expand the selection time	No future direction and suggestions
[41]	Modified low power adaptive clustering hierarchy algorithms	Balance the network power load	Competing CH mechanisms focus on reducing the power cost of communications	It does not address hotspot issues that may be faced	Issues such as the detachment between a group leader and a BS

4 Conclusion

WSNs are being considered a very important technology according to today's technology. With the help of this technology, we can automate any area like home automation, industry automation, and hospital automation. But in WSNs, the major problem is extending the lifespan of the sensor network. After going through many research papers, we concluded that to increase the life of the sensor network, we required the selection of a group leader and then cluster formation. Clustering in WSNs has attracted attention in recent years due to its benefits of reducing power consumption and extending the network's life. The Low Power Adaptive Clustering Hierarchy (LEACH) protocol was the first clustering protocol introduced and gave rise to the idea of creating many existing clustering methods. Following a thorough review of numerous research papers, we have come to the conclusion that choosing a group leader and subsequently forming a cohesive unit are necessary to extend the life of the sensor network. The advantages of clustering in WSNs, such as lower power consumption and a longer network lifespan, have gained attention recently. The first clustering protocol introduced was the Low Power Adaptive Clustering Hierarchy (LEACH) protocol, which sparked the development of other modern clustering techniques. We discovered after reading a number of research papers that choosing a group leader and cluster information are crucial for extending the lifespan of wireless sensor networks [42]. In this survey, the various clustering approaches and strategies used in wireless sensor networks from 2018 to 2022 were covered. For a better and clearer understanding of this technique, this paper briefly covers the nonmetaheuristic algorithms. In nonmetaheuristics, approaches are categorized into mobility, many hop data transfer, separate hop data transfer, heterogeneity, and other parameters. Group leader selection and cluster formation under these two approaches are described in detail. Parameter setting, advantages, limitations, and future suggestions are given by the respective authors listed in details. This survey paper will be highly helpful for those new researchers who want to contribute work to extend the life of wireless sensor networks using nonmetaheuristic methods.

References

1. Sujanthi S, Nithya Kalyani S (2020) SecDLECdl:QoS-aware secure deep learning approach for dynamic cluster based routing in WSNs Assisted IoT, Part of SpringerNature
2. Kumar M, Mukherjee P, Verma K, Verma S, Rawat DB (2021) Improved deep convolutional neural network based malicious node detection and energy-efficient data transmission in wireless sensor networks. IEEE Trans Netw Sci Eng 9(5):3272–3281
3. Kumar M, Mittal S, Akhtar AK (2020) A NSGA-II based energy efficient routing algorithm for wireless sensor networks. J Inf Sci Eng 36(4)
4. Heinzelman WR, Chandrakasan A, Balakrishnan H (2000) Power-efficient communication protocol for wireless microsensor networks. IEEE Comput Soc 8:8020
5. Kumar M, Mukherjee P, Verma S, Shafi J, Wozniak M, Ijaz MF (2023) A smart privacy preserving framework for industrial IoT using hybrid meta-heuristic algorithm. Sci Rep 13(1):1–17
6. Kumar M, Mukherjee P, Verma S, Kaur M, Singh S, Kobielnik M, Ijaz MF (2022) BBNSF: blockchain-based novel secure framework using RP2-RSA and ASR-ANN technique for IoT enabled healthcare systems. Sensors 22(23):9448
7. Kumar M, Mittal S, Akhtar AK (2021). Energy efficient clustering and routing algorithm for WSN. Recent Adv Comput Sci Commun (Formerly: Recent Patents on Computer Science) 14(1):282–290
8. Han T, Bozorgi S, Orang A, Hosseinabadi A, Sangaiah A, Chen MY (2019) A hybrid unequal clustering based on density with power conservation in wireless nodes. Sustainability 11(3):746
9. Du X, Lin F (2005) Designing efficient routing protocol for mixed sensor networks. In: 24th IEEE International performance, computing, and communications conference, vol 2005, pp 51–58
10. Karabekir B, Aydin MA, Zaim AH (2021) Power-efficient clustering-based mobile routing algorithm for wireless sensor networks. Electrica 21(1):41–49
11. Kumar M, Kumar D, Akhtar MAK (2021) A modified GA-based load balanced clustering algorithm for WSN: MGALBC. Int J Embed Real-Time Commun Syst (IJERTCS) 12(1):44–63
12. Ramesh K, Somasundaram DK (2011) A comparative study of group leader selection algorithms in wireless sensor net- works. Int J Comput Sci Eng 2(4):153–164
13. Tyagi S, Kumar N (2013) A systematic review on clustering and routing techniques based upon LEACH protocol for wireless sensor networks. J Netw Comput Appl 36(2):623–645
14. Kalla N, Parwekar P, Clustering techniques for wireless sensor networks. Smart Comput Inf
15. Qi H, Liu F, Xiao T, Su J (2018) A robust and power-efficient weighted clustering algorithm on mobile ad hoc sensor networks. Algorithms 11(8):116

16. Zahedi A (2018) An efficient clustering method using weighting coefficients in similar wireless sensor networks. Alexandria Eng J 57(2):695–710
17. Mehta D, Saxena S (2020) MCH-EOR: multi-objective group leader-based power-aware optimized routing algorithm in wireless sensor networks. Sustain Comput 28, article 100406
18. Amutha S, Kannan B, Kanagara JM, Power-efficient cluster manager-based group leader selection technique for communication networks. IJCS 33(14), article e4427
19. Zhang W, Gao K, Zhang W et al (2019) A hybrid clonal selection algorithm with modified combinatorial recombination and success-history based adaptive mutation for numerical optimization. Appl Intell 49:819–836. https://doi.org/10.1007/s10489-018-1291-2
20. Amuthan A, Arulmurugan A (2018) Semi-Markov inspired hybrid trust prediction scheme for prolonging lifetime through reliable group leader selection in WSNs
21. Din MSU, Rehman MAU, Ullah R, Park C-W, Kim BS (2019) A mixed power wireless sensor network clustering protocol. Wireless Commun Mob Comput 2019:11
22. Luo Z, Xiong NX (2017) An efficient approach of group leader selection for balanced power consumption in wireless sensor networks. IJFGCN 10(2):1–8
23. Umbreen S, Shehzad D, Shafi N, Khan B, Habib U (2020) An power-efficient mobility-based group leader selection for lifetime enhancement of wireless sensor networks. Ieee Access 8:207779–207793
24. Wu D, Geng S, Cai X, Zhang G, Xue F (2020) A many-objective optimization WSN power balance model. KSII Trans Internet Inf Syst 14(2):514–537
25. Yao YX, Chen W, Guo J, He X, Li R (2020) Simplified clustering and improved inter cluster cooperation approach for wireless sensor network power balanced routing 2020(1)
26. Dwivedi AK, Sharma AK (2020) FEECA: fuzzy based power efficient clustering approach in wireless sensor network. Eai Endorsed Trans Scalable Inf Syst 7(27):12
27. Salah Ud Din M et al (2020) Towards network lifetime enhancement of resource constrained IoT devices in mixed wireless sensor networks. Sensors 20(15):4156
28. Turgut IA (2020) Dynamic coefficient-based group leader election in wireless sensor networks. Pamukkale Üniversitesi Mühen- dislik Bilimleri Dergisi 26(5):944–952
29. Hassan AAH, Shah WM, Habeb AHH, Othman MFI, al-Mhiqani MN (2020) An improved power-efficient clustering protocol to prolong the lifetime of the WSN-based IoT. IEEE 8:200500–200517
30. Pour SE, Javidan R (2021) A new power aware group leader selection for LEACH in wireless sensor networks. IET Wireless Sens Syst 11(1):45–53
31. Narayan V, Daniel AK (2021) A novel approach for group leader selection using trust function in WSN. Scalable Comput-Pract Experience 22(1):1–13
32. Choudhury N, Matam R, Mukherjee M, Lloret J, Kalaimannan E (2021) NCHR: a non-thresh hold-based cluster-head rotation scheme. IEEE Internet Things J 8(1):168–178
33. Xu KD, Zhao Z, Luo Y, Hui G, Hu L (2019) An power- efficient clustering routing protocol based on a high-QoS node deployment with an inter-cluster routing mechanism in WSNs. Sensors 19(12):2752
34. Ge YJ, Nan YR, Chen Y (2020) Maximizing information transmission for power harvesting sensor networks by an uneven clustering protocol and power management 14(4):1419–1436
35. Agbulu GP, Kumar GJR, Juliet AV (2020) A lifetime- enhancing cooperative data gathering and relaying algorithm for cluster-based wireless sensor networks 16(2)
36. Tang X, Zhang M, Yu P, Liu W, Cao N, Xu Y (2020) A non-uniform clustering routing algorithm based on an improved K-means algorithm. Comput, Mater Continua 64(3)
37. Priya AV, Srivastava AK, Arun V (2020) Hybrid optimal power management for clustering in wireless sensor network. Comput Electr Eng 86:106708
38. Gandhi GS, Vikas K, Ratnam V, Babu KS (2020) Grid clustering and fuzzy reinforcement-learning based power-efficient data aggregation scheme for distributed WSN 14(16):2840–2848
39. Sert SA, Alchihabi A, Yazici A (2018) A two-tier distributed fuzzy logic-based protocol for efficient data aggregation in Many hops WSNs. IEEE Trans Fuzzy Syst 26(6):3615–3629
40. Feng J, Shi XZ, Zhang JX (2018) Dynamic group leader selection and data aggregation for efficient target monitoring and tracking in wireless sensor networks 14(6)

41. Zhao L, Qu SC, Yi YF (2018) A modified cluster-head selection algorithm in wireless sensor networks based on LEACH. EURASIP J Wireless Commun Networking 1:2018
42. Hu W, Yao W, Hu Y, Li H (2019) Selection of group leaders for wireless sensor network in ubiquitous power internet of things. Int J Comput Commun Control 14(3):344–358

Chapter 17
A Comprehensive Review of Machine Learning-Based Approaches to Detect Crop Diseases

Rajesh Kumar and Vikram Singh

1 Introduction

Farming has developed to such a great level that it is simply not subject to feeding the ever-increasing population. However, this important source of revenue is under threat due to plant diseases. Plant diseases cause large-scale production and financial losses in farming and forestry, e.g., leaf rust (a fungal disease) in crops causes a considerable financial loss, and by eliminating only 20% of the disease, cultivators can leverage a profit of about ₹8761 crore. Consequently, detecting and identifying plant diseases early is of utmost significance to implementing on-time solutions [1]. There are multiple methods of detecting plant deformities. Some diseases have no visible symptoms or only appear when there is no reason to take measures. In such scenarios, conducting comprehensive analysis through a prevailing microscope is essential. In a few conditions, detection of symptoms is feasible only in portions of the electromagnetic spectrum that are invisible to human beings. However, most diseases produce some form of expression on the observable spectrum. Side effects of the infection should be visible on various portions of a plant, such as leaves, stems, fruit, and seeds [2].

Several studies have been done in the last few years on a variety of plant diseases using various image-processing techniques and deep learning-based models on distinct datasets. Several challenges were found while working with deep learning/ machine learning models: nearly 80% of studies trained and evaluated their models using laboratory-conditioned datasets like the PlantVillage dataset [3]. The disease severity assessment is more significant than disease classification or diagnosis for the early identification and treatment of agricultural diseases in the field [4].

R. Kumar (✉) · V. Singh
Department of Computer Science and Engineering, Chaudhary Devi Lal University, Sirsa, Haryana, India
e-mail: rajesh31070@gmail.com

A. Yadav et al. (eds.), *Proceedings of International Conference on Paradigms of Communication, Computing and Data Analytics*, Algorithms for Intelligent Systems,
https://doi.org/10.1007/978-981-99-4626-6_17

The goal of the paper is to find and categorize plant infections based on the symptoms of diseases that show up on the leaves of the plant. In many instances, the diagnosis, or at the very least, the initial assessment of the ailment, is made by people through observation. Autonomous illness detection systems based on computer vision have already been presented by a number of researchers. Trained specialists may be skilled at diagnosing the disease. Unfortunately, there are generally no specialists in the field who can do data-based analysis and advice cultivators. Hence, it is very important to look for a swift, automated, inexpensive, and accurate methodology for detecting plant diseases.

1.1 Plant Disease Recognition and Classification

Computer vision, a branch of artificial intelligence, enables robots to replicate the human visual system and accurately analyze and identify real-world images just like humans do. Numerous industries, including medical diagnosis, espionage, satellite photography, and agribusiness, have already witnessed the advantages offered by computer vision-based solutions. In agriculture, computer vision-powered systems can effectively detect and categorize plant diseases by analyzing extracted disease characteristics or symptoms. These systems follow a well-defined sequence of steps, starting from image acquisition and progressing through various image-processing tasks such as scaling, filtering, segmentation, feature extraction, and selection. Ultimately, machine learning or deep learning techniques are employed for detection and classification purposes [5].

1.2 Plant Disease Causes

There are two primary groups of causes responsible for crop diseases: biotic and abiotic (Fig. 1). Biotic factors like viruses, fungi, bacteria, mites, and slugs cause microbial infections in plants, while abiotic factors like water, temperature, irradiation, and lack of nutrients hurt plant growth. But the present investigation is mainly confined to biotic factors (Fig. 2).

The three normal types of plant infections are discussed in the following manner:

i. **Viral diseases**: Among all plant contaminations trying to perceive and analyze sicknesses brought about by diseases is a tricky task. Likewise, these spots are frequently off track for side effects that happen because of the absence of sustenance or damage, as there is no prescient marker that can persistently screen the plants' state. The common spreaders of viral diseases are whiteflies, leafhoppers, aphids, cucumber-crawling insects, etc. [6].
ii. **Fungal diseases**: The main carrier of foliar diseases is a fungus, including rusts, rhizoctonia rots, sclerotinia rots, etc. [7]. It is primarily visible on long-standing

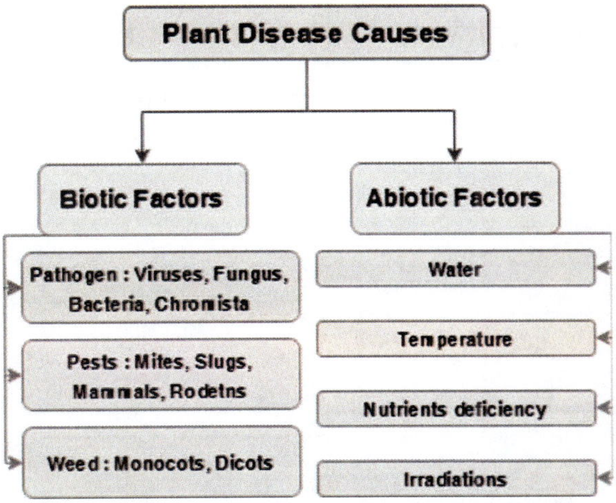

Fig. 1 Plant disease causes

Fig. 2 Different types of pathogens: viruses, fungi, and bacteria. *Source* Cotton Research Station, Sirsa

foliage with greyish marks or that is drenched in water. With the growth of the pest, these stains turn black and aid fungi to develop.

iii. **Bacterial diseases**: Although less frequent than fungi or viruses, pathogenic bacteria are the cause of numerous severe plant illnesses all over the world. However, bacterial infections do substantially less economic harm than fungi and viruses do [8]. Injury to plants is brought about by different microorganisms, bugs, and cultivating instruments during tasks like reaping and cropping.

1.3 Plant Disease Detection Based on Images

The photographs of leaves are an outstanding and abundant source of data on phytopathology and morphological behavior. Therefore, it is very worthwhile to extract and analyze such data thoroughly. Image processing makes a significant

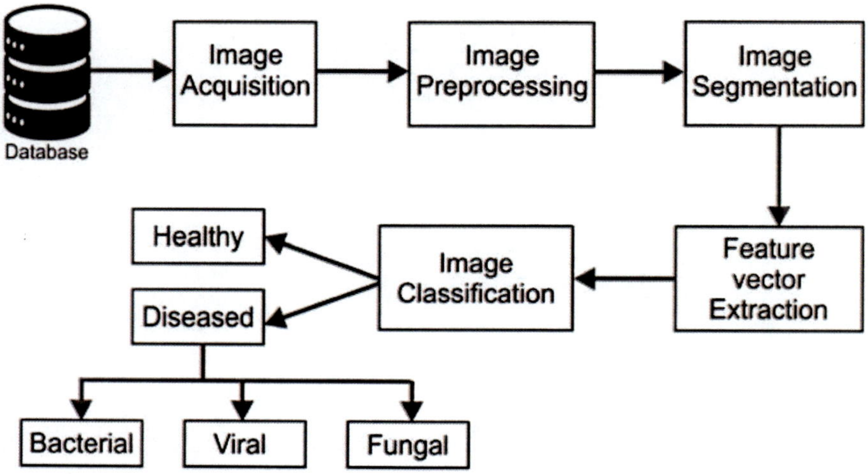

Fig. 3 Flowchart for disease detection

contribution to the detection and study of foliar disorders. Figure 3 depicts the process applied in the detection and classification of leaf disease. It gives an understanding of various methods used by researchers for disease detection through image processing and machine learning.

As shown in Fig. 3, the acquisition of photographs is the primary (first) step in disease detection. In most cases, photographs can be obtained from either a digital camera or an imaging framework. Since rough photographs include noise, it is necessary to remove these inaccuracies. Consequently, the next (second) step is called image preprocessing and consists of removing unnecessary distortions, along with contrast enhancement, to make image features clearer and brighter, e.g., the use of a Gaussian function that forms a soft blur is quite common to reduce noise in photographs. Next (third), image segmentation is the third step, which involves separating an image from its background, while the required area is segmented to highlight key attributes [9].

The fourth step that shows information and details about a picture is called "feature extraction." The leaf characteristics generally involve size, texture, and color, which are useful in the diagnosis of plants. Therefore, these selected attributes create an input attribute vector and are inserted into the classification system. This vector has the potential to distinguish one category of entities from another. Classification is the last step (fifth). It is worth noting that the selection of an appropriate classification architecture depends on the particular issue. The purpose of the classification system is to identify photographs by arranging them into several already defined classes on the premise of the resultant feature vector achieved in step 4. This is why there are two parts to the task of classification, which are called training and testing. The main purpose of training is to train the classification system on the training dataset. This means that the greater the quantity of in-process suites, the greater the accuracy rate.

It is important to note that the result, which shows whether the crop is healthy or sick based on the species name, should be found as quickly as possible.

1.4 Plant Disease Detection and Classification Algorithms

Fast advances in technologies used in mobile devices and digital cameras have enabled people to obtain digital photographs of crop plants and disease-carrying organs more conveniently. In the meantime, the use of machine learning and deep learning technologies has increased the viability of image processing and disease diagnosis, as shown in several studies. The frequently used approaches for plant infection detection and classification are elaborated below.

1.5 Machine Learning Approaches

Various machine learning algorithms, such as ANNs, decision trees, random forest, and SVMs, have been extensively applied in agricultural research. All these algorithms are described as follows [10].

Decision Tree: A decision tree is a nonparametric algorithmic scheme that divides the whole example space into equally limited and complete subsets (or levels) recursively. Every subgroup is portrayed by the set-up of choice guidelines that start its development, separating it from any remaining substitute. Decision trees are useful for both hierarchical and constant target variables. The DT algorithm has two major components: the selection criteria and the stopping rule. Misclassification error, the Gini index, or information entropy may be used to choose a target that is clear. The choice criterion is usually the least squared error for a constant target. Starting with all of the training data, DT looks for all of the predictive variables at all possible distant points to find the best separations that meet the selection criteria. This process goes on, repeating recursively until the withholding rules come into force. Since the DT is nonparametric, it does not assume the distribution and does not need a functional method for the predictors.

Random Forest: Decision trees are integrally unbalanced because a trivial alteration in the training dataset can largely affect the composition of the tree structure [10]. It is generally argued that the results of numerous trees can use volatility to detect more signals in the data than one tree can alone. Since RF is created from several architectures serving in performance, it is known as an ensemble framework. Not at all like one tree, RF attempts to show more vigor as it benefits from complex trees developed on disparate occurrences, which have been made significantly more arbitrary to relieve the effect of connection between tests or between trees.

Support Vector Machine: SVM linearly separates the mapped points on a high-dimensional space using hyperplanes. The collected data points, which are used to build such hyperplanes, are known as SVM (support vectors). This approach aims at achieving the greatest margin data parting, which in itself has two goals: maximizing the minimal margin and separating the dependent variable from a hyperplane. The margin represents the distance between the hyperplane and perceptual objects. But sometimes, this is not possible or it is too hard to find the hyperplane that completely splits the answer. Consequently, softer hyperplanes are used rather than rigid hyperplanes to permit a smaller percentage of misclassification [11]. A higher-size attribute space is more expressive in nature. Nevertheless, if the size of the attribute space is larger than the number of perceptions, then providing the solution to an objective function will be complicated [11]. To deal with this, SVMs naturally use a kernel function, which employs kernel functions for mapping the data to an alternative attribute space in an implicit manner. This feature space can be described by having adequate knowledge of the pair-wise internal products of the perceptions in the data.

ANN: The two researchers named McCulloch and Pitts primarily developed ANN in 1943. ANNs were intended to imitate the neurophysiology of the human brain through the fusion of several simplified computation modules (otherwise known as perceptron, neurons, or nodes) into an extremely interlinked architecture. The single-hidden layer feed-forward network (SLFN) is a prominent form of the neural network system. It is adopted to analyze the neural network. The mathematical representation of SLFN can be given as:

$$y = G\left(a_0 + \sum_{j=1}^{h} a_j v_j\right) \tag{1}$$

$$v_j = F\left(\beta_{0j} + \sum_{i=1}^{k} \beta_{ij} x_i\right) \tag{2}$$

where $x_i, i = 1, \ldots, k$ and $v_j, j = 1, \ldots h$ represent the input units and the hidden units, respectively. The term β_{ij} denotes the weight of the like between the input-layer unit x_i and the hidden-layer unit v_j. Also, a_j signifies the connection's weight between the hidden-layer unit v_j and the output y. The bias of hidden-layer unit v_j is β_{0j} and the bias of the output y is a_0. $F(.)$ and, $G(.)$, respectively, indicate the unknown layer units' activation functions and output activation functions. Activation functions are often sigmoid functions: s-shaped functions like logistic and hyperbolic tangents that create limited values in the limit [0, 1] or [− 1, 1] [12].

1.6 Deep Learning Approaches

Deep learning (DL) techniques are a subset of machine learning (ML) techniques, which were initially developed in 1943. DL employs threshold logic to create computer models that closely resemble the observed biological pathways in humans. Numerous well-established DL models and frameworks have been developed for image recognition, image segmentation, and image classification over the years. Below is a summary of several of these models:

Convolutional Neural Networks (CNN): Convolutional neural networks (CNNs) have a complex but highly effective architecture. Their architecture consists of several essential components. Input layer, convolution layer, pooling layer, entirely connected layer, and output layer are included. Combining the convolution layer and pooling layer eliminates the need for a comprehensive connection in many situations. CNNs become well-known deep neural networks when the neurons of the convolution layer are connected to those of the pooling layer. This is primarily due to ConvNets' enormous model size and complex information processing capabilities, which contribute significantly to image recognition tasks. The extraordinary achievements of CNNs in artificial intelligence tasks have substantially boosted the widespread adoption of deep learning methods. Convolutional kernels, which can be thought of as local receptive fields, make up the convolution layer. This characteristic, known as the greatest receptive field (LRF), is one of these networks' primary advantages. During the data processing phase, the convolutional kernels traverse the feature map in order to derive specific feature information. The extracted features are then sent to the aggregating layer for further feature derivation. Eventually, as the data passes through multiple convolutional layers and pooling layers, it reaches the fully connected layer, where the neurons are densely interconnected with the neurons of the upper layers. The data are classified in the fully connected layer using the softmax method, and the resulting values are sent to the output layer as the ultimate output.

Full Convolution Neural Network (FCN): A full convolution neural network is the core of image semantics segmentation. Nowadays, most of the semantic segmentation frameworks are contingent upon FCN. This network first uses convolution to abstract and code the attributes of the input photograph. Next, it slowly re-saves the featured photograph to the dimension of the input photograph through deconvolution or up-sampling [13].

- *U-net:* U-net is a classic FCN structure as well as a generic encoder–decoder system. A layer-hopping connection, which fuses the feature map in the coding phase with that in the decoding phase, is introduced in it for its representation. It is useful for the revival of segmentation specifics.
- *SegNet:* It is also a standard encoder–decoder system. Its property is that the up-sampling process in the decoder leverages the directory of the most time-taking pooling process in the encoder. SegNet is used to segment visible and infrared images into several categories.

2 Literature Review

In this study, we investigated various recent machine learning algorithms for detecting and classifying plant illnesses. These techniques were divided into two categories: machine learning (ML) and deep learning (DL). Furthermore, we carefully assessed and evaluated the algorithms under consideration, giving our findings in Table 1 for a complete overview.

Dwivedi et al. [14] discovered that current vision methods primarily revolve around either segmenting images or employing feature classification and regression on aerial imagery. For an efficient disease detection approach that requires minimal learning time and possesses remarkable generalization capabilities, an extreme learning machine (ELM) proves to be quite fitting. Notably, L1-ELM has demonstrated superior performance when compared to all existing one-class classification algorithms. It retains optimal learning while simultaneously enhancing generalization [14].

Elfatimi et al. [15] proposed a sophisticated learning system for identifying and categorizing illnesses affecting bean leaves. This method involved combining a publicly available dataset of leaf photographs with a MobileNet model built with the open-source TensorFlow library. Surprisingly, the researchers' proposed model achieved an average classification accuracy of more than 97% on the training dataset and more than 92% on the test dataset. The study included a total of 1296 photographs of bean leaves, with two diseased classes and one healthy class [15].

Dwivedi et al. [16] found that implementing a grape leaf disease monitoring network could be more advantageous than existing methods since it can identify and detect regions impacted by infection or illness. This network's architecture was methodically designed using the faster R-CNN framework. The CA approach was used by the researchers to extract a feature map. They also combined the region proposal network and the feature map to ease multitasking. The proposed disease detection system obtained a remarkable overall accuracy of 99.93% for detecting esca, black rot, and isariopsis [16].

Kumar et al. [17] developed a smart approach to predict powdery mildew, anthracnose, rust, and root rot/leaf blight. A multilayered perceptron (MLP) model classified these disorders. This model outperforms several existing methods in accuracy, according to empirical data. The average disease prediction accuracy approaches 98%. This study also proves this method can detect plant diseases quickly and cheaply [17].

Li and colleagues [18] have devised a novel approach to identify powdery mildew disease (PWD) by employing airborne edge computing and lightweight deep learning methods in tandem with imagery sensors. The incorporation of an unmanned aerial vehicle (UAV) facilitates the efficient and comprehensive survey of forests, significantly reducing the necessity for laborious endeavors. A proposal was put forth to improve the detection process by utilizing a lightweight version of the YOLOv4-Tiny model. The results showed its efficacy in effectively filtering out extraneous images. The study conducted extensive experimentation and found that the proposed system exhibits swift detection with exceptional performance, outperforming other methodologies [18].

Ahmad et al. [19] put forward a useful technique to identify plant disease signs using convolutional neural networks. The PlantVillage dataset has been the focus of past studies, utilizing a progressive transfer learning approach. The proposed

Table 1 Summary of the discussed research papers

Literature reference	Crop culture	Feature selection approach	Classification approach	Datasets	Keywords points	Accuracy (%)	Total references	*Citations received
Elfatimi et al. [15]	Bean leaf	MobileNetV2	Supervised	TensorFlow dataset	MobileNet, TensorFlow, disease classification, beans leaf, deep learning	92–97	22	22
Dwivedi et al. [16]	Grape	R-CNN	Supervised	Benchmark plant datasets	Smart sensing, sensor, agriculture, plant disease, applications	99.93	30	61
Kumar et al. [17]	Green gram	Soil-sensor-based prediction system	Supervised	Sansor dataset	Multi-label classification, Soil-based sensors, artificial neural network, plant diseases	98	58	25
Li et al. [18]	Pine trees	YOLOv4-Tiny	Supervised	8860 images	Pine wilt disease, lightweight deep learning, remote sensing, two-stage detection, airborne edge computing	–	34	21
Ahmad et al. [19]	PlantVillage and pepper	CNN	Supervised	PlantVillage and pepper	Internet of Things, convolutional neural networks, image classification, MobileNet, disease detection, transfer learning	99 and 99.69%	31	31

(continued)

Table 1 (continued)

Literature reference	Crop culture	Feature selection approach	Classification approach	Datasets	Keywords points	Accuracy (%)	Total references	*Citations received
Khattak et al. [20]	Citrus fruit	CNNs	Unsupervised	Citrus and PlantVillage datasets	Citrus fruit diseases detection, deep learning, citrus leaf diseases, convolutional neural network	94.55	42	27
Huang et al. [21]	Peach	Asymptotic non-local means and PCNN-IPELM	Supervised	30,659 images	Extreme learning machine, peach disease detection, asymptotic non-local means, particle swarm optimization, parallel convolution neural network	85–91	44	21
Yang et al. [22]	Tomato	Self-supervised collaborative multi-network	Self-supervised	PlantVillage dataset of 16,486 images	Self-supervised, fine-grained visual categorization, tomato diseases, multi-network	99.7	63	24
Gadade and Kirange [23]	Tomato leaves	LDA, KNN, SVM, Naïve Bayes, decision tree classification	Supervised	PlantVillage dataset of 3000 images	Classification, linear regression analysis, KNN, SVM, tomato leaf disease detection, segmentation	–	10	17
Khan et al. [24]	Apple	GA	Supervised	PlantVillage dataset	Symptoms enhancement, symptoms segmentation, feature extraction, optimal features, recognition		35	131

(continued)

Table 1 (continued)

Literature reference	Crop culture	Feature selection approach	Classification approach	Datasets	Keywords points	Accuracy (%)	Total references	*Citations received
Singh et al. [25]	Mango leaves	Multilayer convolution neural network	Unsupervised	1070 images of mango leaves	Image classification, plant pathology, convolutional neural network, precision agriculture	97.13	29	220
Jiang et al. [26]	Apple leaves	CNNs, GoogLeNet	Unsupervised	26,377 images of apple leaves	Deep learning, convolutional neural networks, feature fusion, apple leaf diseases, real-time detection	78.8	43	426
Devi et al. [27]	Banana	IOT, RFC- GLCM	Supervised	Banana dataset	Hill banana plant, environmental parameters, Raspberry Pi 3, IoT, wireless sensor network, disease detection, RFC	99	7	21
Khitthuk et al. [28]	Grape leaves	GLCM, TFE	Supervised	Grape leaves dataset	Plant leaf disease detection and classification, gray-level co-occurrence matrix, simplified fuzzy ARTMAP	90	10	29

(continued)

Table 1 (continued)

Literature reference	Crop culture	Feature selection approach	Classification approach	Datasets	Keywords points	Accuracy (%)	Total references	*Citations received
Sardogan et al. [29]	Tomato leaves	CNN, LVQ	Unsupervised	500 images	Leaf disease detection, leaf disease classification, convolutional neural network (CNN), learning vector quantization (LVQ)	–	15	312
Chouhan et al. [30]	Plant leaves	Bacterial foraging optimization based radial basis function neural network (BRBFNN)	Supervised	270 images	Bacteria foraging algorithm, image segmentation, plant diseases, radial basis, function neural network, soft computing		26	229
VijayaLakshmi and Mohan [31]	All plant leaves	GLCM and the LBP, FRVM	Supervised	ICL leaf dataset	Plant leaf detection, cellular automata filter, Haralick texture feature extraction, kernel-based particle swarm optimization (PSO), fuzzy-relevance vector machine (FRVM) classification	99.9	31	98

(continued)

Table 1 (continued)

Literature reference	Crop culture	Feature selection approach	Classification approach	Datasets	Keywords points	Accuracy (%)	Total references	*Citations received
Zhou et al. [32]	Rice plant	Fractal Eigenvalues and FCM	Unsupervised	Images of the rice stem	Rice plant-hopper, image process, infestation area, fractal-dimension value, fractal Eigenvalue, rice stem, fuzzy C-means (FCM)	63.5	27	42
Huang [33]	Areca nut	Detection line (DL) technique	Supervised	Areca nut images	Areca nut, neural network, machine vision	90.9	19	61
Al Bashish et al. [34]	All	K-means, GLCM	Supervised	All crops	K-means, segmentation, leaf diseases, stem diseases neural networks	93	10	320

* As per google scholar—accessed on 6.4.2023

system has obtained 99 and 99.69% precision on Pepper and PlantVillage datasets, respectively [19].

Khattak et al. [20] suggested an integrated convolutional neural network model for citrus fruit diseases. The suggested CNN model aims to distinguish between healthy fruits and leaves and those affected by common citrus diseases such as black spot, canker, scab, greening, and melanose. The CNN model has a test accuracy of 94.55%, making it a powerful tool for citrus fruit and leaf disease classification [20].

Huang et al. [21] presented a new method for detecting peach diseases. This method utilizes the asymptotic non-local means (ANLM) image algorithm and integrates a parallel convolutional neural network (PCNN) with an extreme learning machine (ELM) for improved accuracy. The study analyzed a significant dataset comprising 25,513 images and demonstrated noteworthy detection accuracies for diverse peach diseases. The study reported the attainment of the highest accuracies for various peach diseases, namely brown rot, black spot, anthracnose, scab, and normal peach, which were recorded at 89.02%, 90.56%, 85.37%, 86.70%, and 89.91%, respectively [21].

Yang et al. [22] suggested a new model known as LFC-Net, which included three techniques such as location network, feedback network, and classification network. A self-supervision technique for detecting informative areas in tomato images was also presented. A region of interest (ROI) was detected in the tomato image using the initial algorithm, and the subsequent algorithm was assisted in optimizing the iterations under the guidance. Afterward, they utilized the informative regions and the entire image of the tomato. They utilized informative regions and the entire image of the tomato in the last algorithm to classify the diseases from the plant images. The suggested model performed well on the tomato dataset and offered an accuracy of 99.7% [22].

Gadade and Kirange [23] constructed a method for automatically segmenting diseased areas. The segmented region was subjected to an analysis to classify the disease and compute its severity. They put forward a method to detect the leaf disease in which the images were preprocessed, and segmented, attributes were extracted, and the disease was classified. They considered various metrics, namely color, texture, and shape, to analyze the efficacy of the constructed approach. The constructed approach motivated the farmers to implement the automated system to detect the disease occurring on the tomato plant and to measure its severity level [23].

Khan et al. [24] developed a novel method for finding and diagnosing illnesses in apples. Preprocessing, spot segmentation, feature extraction, and classification were the processes in the procedure. The proposed methodology was tested using four types of apple diseases: healthy leaves, black rot, rust, and scab. Notably, effective pre-processing was critical in extracting salient features, resulting in significant accuracy during classification [24].

Singh et al. [25] proposed a multilayer convolutional neural network (MCNN) as a solution for classifying anthracnose fungal disease in mango leaves. They were motivated by the growing popularity of computer vision and deep learning techniques in identifying fungal illnesses, and they set out to create an efficient and dependable method for diagnosing the disease and its symptoms. The results showed that their system beat other strategies in correctly diagnosing the condition. Their algorithm's accuracy was calculated to be an amazing 97.13% [25].

Jiang et al. [26] proposed an advanced deep learning system for detecting apple leaf illnesses in real-time using improved convolutional neural networks (CNNs). They suggested a novel model for apple leaf disease detection using deep CNNs by combining the GoogLeNet Inception structure and Rainbow Concatenation. They demonstrated that the INAR-SSD model obtained 78.80% mean average precision (mAP) detection performance on the apple leaf disease dataset (ALDD) while retaining a high detection speed of 23.13 frames per second (FPS) [26].

Devi et al. [27] have achieved a detection accuracy of approximately 99% with their proposed illness detection system. The system employs image processing and IoT to extract textural information from photographs of plants. Temperature/humidity sensors and soil moisture sensors are used to measure disease and infection rates in hill banana plants [27].

Khitthuk et al. [28] demonstrated a system for diagnosing plant leaf diseases using color images and an unsupervised neural network. The technology produced outstanding results, with an accuracy rate of more than 90%. Furthermore, the suggested method demonstrated the ability to successfully diagnose many forms of plant diseases. Four different types of grape leaf disease photographs were used to assess the system's categorization ability [28].

Sardogan et al. [29] developed a technique for detecting and categorizing tomato leaf illnesses that combines a convolutional neural network (CNN) model and the learning vector quantization (LVQ) algorithm. They demonstrated the efficiency of the proposed strategy in precisely diagnosing four unique forms of tomato leaf diseases through trials. The dataset used in this analysis included 500 photographs of tomato leaves with indications of various illnesses [29].

Chouhan et al. [30] introduced a technique named "bacterial foraging optimization based on radial basis function neural networks" (BRBFNN) for the automated recognition and categorization of plant leaf illnesses. To assign the optimal weight to a radial basis function neural network, they used bacterial foraging optimization, which further boosts the accuracy and speed of the network to detect and classify the areas affected by different diseases on the plant's leaf. Also found that the region-growing algorithm searched and grouped the seed points with common attributes to maximize the efficacy of the constructed technique. It was observed that the accuracy of the constructed technique was superior in recognizing and classifying the diseases [30].

VijayaLakshmi and Mohan [31] presented a method for the classification of leaves based on their texture, shape, and color characteristics. The primary purpose of the proposed FRVM classification is to reliably predict the kind of leaf from the leaf photos provided as input. The experimental analysis yielded improved values for precision, sensitivity, and specificity of 99.87%, 99.5 and 99.9% compared to the literature, which has been demonstrated by him [31].

Based on visible images, Zhou et al. [32] developed an algorithm that uses fractal eigenvalues and fuzzy C-means to detect stress in rice production induced by RPH infestation. The results proved the algorithm's capacity to discriminate places where

RPH accumulates and identify moldy grey spots on rice plant stems caused by RPH damage. The accuracy in distinguishing these four groups achieved 63.5%, which is expected to meet the needs of practical rice production [32].

Huang [33] adopted a detection line (DL) technique to segment the flaws of betel nut diseases or bugs. The classification was performed using six geometrical attributes, three color properties, and the defective region. This work categorized the quality of the betel nuts using back-propagation neural network (BPNN) classification architecture. The method introduced a work efficiency of 90.9% accuracy in the classification of betel nuts [33].

Al Bashish et al. [34] created an image-processing-based solution for detecting leaf and stem illnesses, with the goal of providing a quick, cost-effective, and accurate method. According to the findings, relying entirely on visual inspection by professionals to identify certain diseases can be costly, especially in underdeveloped countries. The team created a neural network classifier that used statistical classification to correctly detect and categorize the tested diseases with a precision rate of roughly 93% [34].

3 Results and Discussion

The application of machine learning to the process of diagnosing diseases that can be found in crops is currently one of the most important parts of modern agriculture. Never before has it been more vital to rapidly identify illnesses and put a halt to the spread of such ailments. Conventional procedures are insufficient for determining the degree of the sickness since they rely on visual inspection. This is because conventional methods rely on visual inspection. Machine learning has the ability to diagnose a wide variety of crop illnesses, which would significantly reduce the likelihood of experiencing crop losses. The purpose of this paper is to provide an overview of various disease categorization strategies that can be utilized for the identification of plant leaf diseases. The algorithms and methodologies were tested on a number of different species, including the grape, the bean, the green gram, the pine tree, the pepper, the citrus fruits, the peach, the tomato leaves, the mango, the banana, the apple, the paddy, and the areca nut. This ultimately resulted in the collecting and naming of related illnesses that affect these plants.

4 Conclusion

The efficiency of algorithms in the detection and categorization of leaf diseases is demonstrated by the fact that the best results could be obtained with relatively little effort expended in the computing of data. Putting these methods to use has a number of advantages, one of which is that it enables the early or preliminary diagnosis of plant diseases. It is possible to improve the recognition rate of the classification process by utilizing several methods and hybrid algorithms.

References

1. Shrivastava VK, Pradhan MK (2021) Rice plant disease classification using color features: a machine learning paradigm. J Plant Pathol 103(1):17–26
2. Sujatha R, Chatterjee JM, Jhanjhi NZ, Brohi SN (2021) Performance of deep learning versus machine learning in plant leaf disease detection. Microprocess Microsyst 80(103615):1–15
3. Kumar R, Chug A, Singh AP, Singh D (2022) A systematic analysis of machine learning and deep learning based approaches for plant leaf disease classification: a review. J Sens 2022:1–13
4. Krishnakumar A, Narayanan A (2019) A system for plant disease classification and severity estimation using machine learning techniques. In: Proceedings of the international conference on ISMAC in computational vision and bio-engineering 2018 (ISMAC-CVB). Springer International Publishing, pp 447–457
5. Tiwari V, Joshi RC, Dutta MK (2021) Dense convolutional neural networks based multiclass plant disease detection and classification using leaf images. Eco Inform 63:101289
6. Ranawaka B, Hayashi S, Waterhouse PM, de Felippes FF (2020) *Homo sapiens*: the superspreader of plant viral diseases. Viruses 12(12):1462
7. Deshpande T, Sengupta S, Raghuvanshi KS (2014) Grading & identification of disease in pomegranate leaf and fruit. Int J Comput Sci Inf Technol 5(3):4638–4645
8. Vidhyasekaran P (2002) Bacterial disease resistance in plants: molecular biology and biotechnological applications. CRC Press
9. Sridhathan S, Kumar MS (2018) Plant infection detection using image processing. Int J Mod Eng Res (IJMER) 8(7):13–16
10. Marathe PS, Raisoni GH, Phule S (2017) Plant disease detection using digital image processing and GSM. Int J Eng Sci Comput 7(4):10513–10515
11. Halder M, Sarkar A, Bahar H (2019) Plant disease detection by image processing: a literature review. image 3(6):534–538
12. Saradhambal G, Dhivya R, Latha S, Rajesh R (2018) Plant disease detection and its solution using image classification. Int J Pure Appl Math 119(14):879–884
13. Padmavathi K, Thangadurai K (2016) Implementation of RGB and grayscale images in plant leaves disease detection—comparative study. Indian J Sci Technol 9(6):1–6
14. Dwivedi R, Dutta T, Hu Y (2022) A leaf disease detection mechanism based on L1-Norm minimization extreme learning machine. IEEE Geosci Remote Sens Lett 19:1–5
15. Elfatimi E, Eryigit R, Elfatimi L (2022) Beans leaf diseases classification using MobileNet models. IEEE Access 10:9471–9482
16. Dwivedi R, Dey S, Chakraborty C, Tiwari SM (2021) Grape disease detection network based on multi-task learning and attention features. IEEE Sens J 21:17573–17580
17. Kumar M, Kumar A, Palaparthy VS (2021) Soil sensors-based prediction system for plant diseases using exploratory data analysis and machine learning. IEEE Sens J 21:17455–17468
18. Li F, Liu Z, Shen W, Wang Y, Wang Y, Ge C, Sun F, Lan P (2021) A remote sensing and airborne edge-computing based detection system for pine wilt disease. IEEE Access 9:66346–66360
19. Ahmad M, Abdullah M, Moon H, Han D (2021) Plant disease detection in imbalanced datasets using efficient convolutional neural networks with stepwise transfer learning. IEEE Access 9:140565–140580
20. Khattak AM, Asghar MU, Batool U, Asghar MZ, Ullah H, Al-Rakhami MS, Gumaei AH (2021) Automatic detection of citrus fruit and leaves diseases using deep neural network model. IEEE Access 9:112942–112954
21. Huang S, Zhou G, He M, Chen A, Zhang W, Hu Y (2020) Detection of peach disease image based on asymptotic non-local means and PCNN-IPELM. IEEE Access 8:136421–136433

22. Yang G, Chen G, He Y, Yan Z, Guo Y, Ding J (2020) Self-supervised collaborative multi-network for fine-grained visual categorization of tomato diseases. IEEE Access 8:211912–211923
23. Gadade HD, Kirange DK (2020) Tomato leaf disease diagnosis and severity measurement. In: Fourth world conference on smart trends in systems, security and sustainability (WorldS4), pp 318–323
24. Khan MA, Lali MI, Sharif M, Javed K, Aurangzeb K, Haider SI, Altamrah AS, Akram T (2019) An optimized method for segmentation and classification of apple diseases based on strong correlation and genetic algorithm based feature selection. IEEE Access 7:46261–46277
25. Singh UP, Chouhan SS, Jain S, Jain S (2019) Multilayer convolution neural network for the classification of mango leaves infected by anthracnose disease. IEEE Access 7:43721–43729
26. Jiang P, Chen Y, Liu B, He D, Liang C (2019) Real-time detection of apple leaf diseases using deep learning approach based on improved convolutional neural networks. IEEE Access 7:59069–59080
27. Devi RD, Nandhini SA, Hemalatha R, Radha S (2019) IoT enabled efficient detection and classification of plant diseases for agricultural applications. In: International Conference on Wireless Communications Signal Processing and Networking (WiSPNET), pp 447–451
28. Khitthuk C, Srikaew A, Attakitmongcol K, Kumsawat P (2018) Plant leaf disease diagnosis from color imagery using co-occurrence matrix and artificial intelligence system. In: 2018 International electrical engineering congress (iEECON), pp 1–4
29. Sardogan M, Tuncer A, Ozen Y (2018) Plant leaf disease detection and classification based on CNN with LVQ algorithm. In: 2018 3rd International conference on computer science and engineering (UBMK), pp 382–85
30. Chouhan SS, Kaul A, Singh UP, Jain S (2018) Bacterial foraging optimization based radial basis function neural network (BRBFNN) for identification and classification of plant leaf diseases: an automatic approach towards plant pathology. IEEE Access 6:8852–8863
31. VijayaLakshmi B, Mohan V (2016) Kernel-based PSO and FRVM: an automatic plant leaf type detection using texture, shape, and colour features. Comput Electron Agric 125:99–112
32. Zhou Z, Zang Y, Li Y, Zhang Y, Wang P, Luo X (2013) Rice plant-hopper infestation detection and classification algorithms based on fractal dimension values and fuzzy C-means. Math Comput Model 58:701–709
33. Huang K (2012) Detection and classification of areca nuts with machine vision. Comput Math Appl 64:739–746
34. Al Bashish D, Braik M, Bani-Ahmad S (2010) A framework for detection and classification of plant leaf and stem diseases. In: 2010 International conference on signal and image processing, pp 113–18

Chapter 18
Physiological Signals for Emotion Recognition

Shruti G. Taley and M. A. Pund

1 Introduction

Emotions are necessary for cognitive functioning and interact with cognitive processes, according to Picard [1]. Furthermore, human–computer interactions adhere to the similar rules as human–human interactions, which significantly emphasizes the emotional component [2]. Picard added that a computer would seem smarter if it recognized and responded correctly to emotional response of user [1]. So, it is thought advantageous to incorporate emotion component into adaptive computer systems that communicate with people directly. Affective computing is the name of the branch of study that resulted from this [3]. It is essential to provide emotional recognition in adaptive computer systems. This results in the requirement for automated emotion recognition, a topic of importance in numerous academic areas. Advanced driver assistance systems [4], healthcare [5], social security [6], digital multimedia entertainment [7, 8], and other industries are only a few of the numerous industries that use emotion recognition (ER). Through a variety of emotion elicitation protocols, feature extraction strategies, and classification methodologies, a machine may comprehend human emotions [8]. To determine an emotion, relevant data can be extracted from auditory and visual (facial expression, speech, etc.), physiological measurements (skin temperature, breathing, etc.), and appropriate human behavior (posture, gesture) [9].

The autonomous nervous system, which regulates a variety of body activities, is affected by emotions [10]. These factors include a person's breathing rate, electrodermal activity, skin temperature, and electrocardiograms. Additionally, it is possible to use brain signals for emotion identification by electroencephalography.

S. G. Taley (✉) · M. A. Pund
Department of Computer Science and Engineering, PRMITR Badnera, Amravati, India
e-mail: shrutitaley92@gmail.com

A. Yadav et al. (eds.), *Proceedings of International Conference on Paradigms of Communication, Computing and Data Analytics*, Algorithms for Intelligent Systems,
https://doi.org/10.1007/978-981-99-4626-6_18

Many ER researchers have focused their attention in recent years on study of data from just one sensor, which include video (facial expression) and audio (speech) data [11, 12]. Single-modal ERSs are the name given to these systems. On the contrary, several other researchers create multimodality ERSs by combining the data from many sensors regardless of how hard a person tries to hide their feeling; their bodies typically go through series of spontaneous physiological changes, like sweating, faster heartbeats, deeper breathing, and other sequences of responses. From all of the modalities, physiological-based therapy is appealing. Contrast this with facial-based ERS, which fails when a person does not make any facial expressions or decides to hide their emotion. Biosensors can be used to collect physiological data. This study's main goal is to provide a summary of state-of-the-art methods and knowledge for more investigation.

2 Literature Survey

2.1 Emotions

A state of mind or experience that emerges unexpectedly and is typically accompanied by physiological changes is referred to as being emotional. Unlike mood, which can persist for hours or even days, emotions only last a brief duration.

Emotion Models

According to current research, there are two models that can be used to depict emotional states: discrete and multidimensional.

Discrete or Categorical Models

The most popular models are discrete models since they feature lists of distinct emotion groups that are easy to distinguish. Scientists like Plutchik [13] and Ekman [14] introduce the idea of separate emotion states. According to Ekman's briefing, there are six fundamental emotions—fear, happy, sad, angry, surprised, and disgust from which added emotions are produced. To describe eight distinct emotions, Plutchik offered the well-known wheel model. Model explains the connections among emotion ideas, which are like hues on a color wheel. Figure 1 depicts these feelings as surprise, joy, anticipation, trust, fear, anger, sadness, and disgust. The model explains the relationships between emotion ideas, which can be compared to the hues on a color wheel. The circle shows levels of similarity among the emotions, while the vertical dimension of the cone indicates intensity. The eight sectors are intended to represent the eight core emotion dimensions, from which the remaining motion dimensions are generated.

Continuous or Multidimensional Models

Each of the various emotional states listed requires a range of intensities; no single phrase can effectively capture it, e.g., a person may feel more or less aroused or afraid

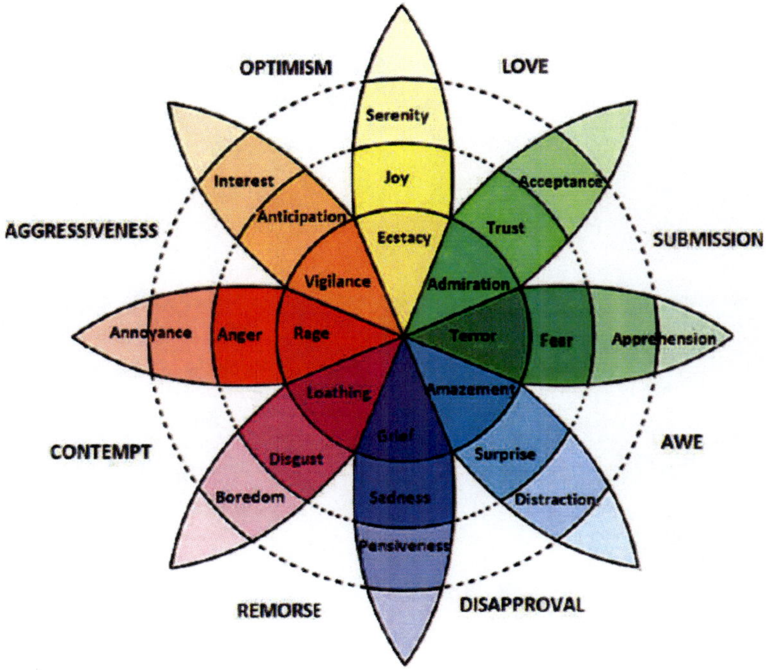

Fig. 1 Wheel of discrete emotions by Plutchik [13]

in response to certain stimuli. Thus, it is suggested that multidimensional models, such as 2D and 3D, be used to account for the spectrum of emotional intensities. Based on two-dimensional data made up of arousal and valence values, the 2D model classifies emotions. The most well-known model that expresses the emotion beside dimensions of low arousal (LA) and high arousal (HA), and high valence (HV) and low valence (LV), as shown in Fig. 2, is the model presented in [15]. The 3D model, on the other hand, addresses valence, arousal, and dominance. Arousal, valence, and dominance all describe the intensity of an emotion's control or dominance. Valence refers to the level of pleasure. Figure 3 [16] depicts the 3D model.

2.2 Physiological Signals

An extremely complex personality is a human being. We can gather a variety of physiological signals to learn more about emotions, human health. The ones that are helpful for emotion recognition are briefly outlined below [17].

Types of physiological signals

Cardiac function The main component of the circulatory system is the heart. Analysis of its output reveals a wealth of information about people's health. Electrodes

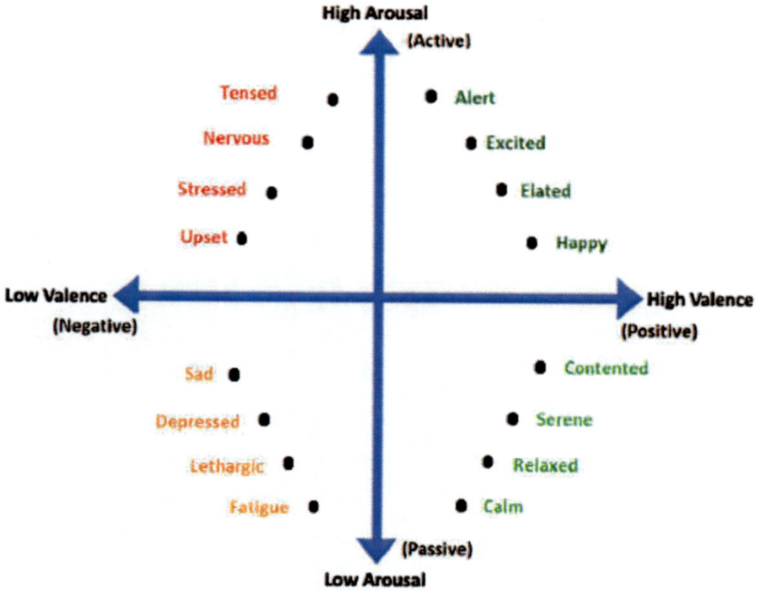

Fig. 2 2D valence–arousal model [15]

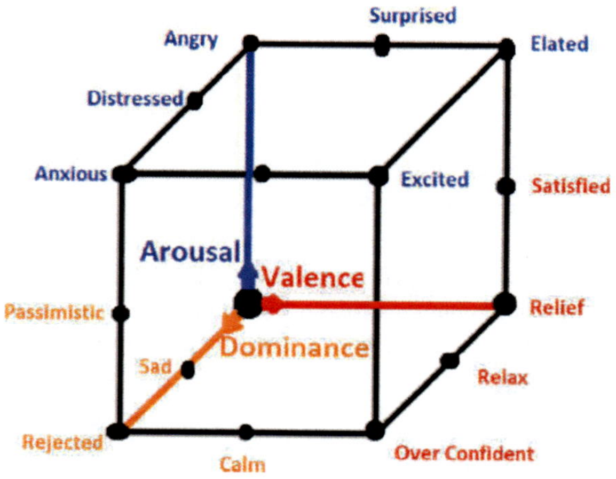

Fig. 3 3D emotion model [16]

placed on the chest can be used to record the electrical waveforms. All parts of the body receive generated pulse waves from the heart. Signals from electrocardiograms (ECGs) are often used to assess cardiac function. The electrical activities of the heart are monitored, just like the EEG. The heart causes the muscles to contract and relax. It is acquired by applying electrodes to the subject's body [18]. A person's feelings have an impact on the rhythm of their heart, which is a crucial organ in the body. ECG signals may therefore be useful for ERS [19].

Temperature A very basic but helpful physiological indication is temperature. It varies on where in the body the measurement is made, the time of day, and the person's activity. The fluctuations in temperature may be a reflection of changes in emotions and mood. SKT is among the best variables for automatic ERS. The human unconscious response to SC and HR is known as SKT. The heat radiation of the skin's surface is used to estimate SKT, which is a useful predictor of emotional states that are mirrored in activity of the autonomic nervous system (ANS) [20].

Muscle electrical activity When muscles contract and relax, there is electrical activity in the muscles. Facial electromyograms can be highly helpful for identifying emotions, but in practice, they are ineffective since the electrodes attached to the face are too invasive and uncomfortable for the user. A diagnostic and recording technique for the electrical activity produced by muscle cells is called electromyography (EMG). According to physiological responses, the signals in the ERS instance may be used to identify perceptive emotions [20].

Respiration When breathing, the diaphragm and chest move up and down in turn. The rate and depth of breathing might indicate a person's condition, emotions, and state of health. Stretchable latex rubber bands known as RSP biosensors or RB are frequently used to record human breathing activities. Typically, the RSP is worn over the stomach. The voltage level changes are used to measure the elastic band stretch data. RSP rate and breathing depth are recorded in the normal format. Here, respiration compresses the attached sensor, creating a shift in the ribcage's size in people [21]. A deep breath can change the EMG, RSP, and SC values since RSP measurement is tightly tied to other cardiac metrics. When it comes to mood level, unnatural RSP patterns are influenced by negative emotions. Measuring the carbon dioxide (CO_2) content of air during inhalation and exhalation is another way to calculate RSP. Capnography, or measuring chest cavity expansion, is the term used [22]. Additionally, RSP can be obtained utilizing EMG information collected from the respiratory muscles [23].

Skin conductance The electrical conductivity of skin is measured by a signal called skin conductance. The skin's ability to perspire is altered by this signal. As a measure of arousal, skin conductance is utilized. Uses for this signal include polygraphs. The SC is an unbroken stream of unprocessed electrical components from human skin. In this case, skin conditions are considered to be the primary determinant, where a sweet response results in a different quantity of salt on skin, and as a result, shifts in the electrical resistance from one skin to another [24]. The surface of the skin becomes moistened by sweat and causes changes within ion stability of electrodes, both positive and negative [25]. The sweat glands activity causes perspiration to be generated. It is a person's unconscious response [26] and reflection of modifications in the delicate neurological system [27]. Various emotional states result in sweating, mainly on the soles of the feet, palms, and fingers.

Brain's electrical activity Brain's electrical activity is a very helpful signal that can reveal important details about how people behave. With the aid of electrodes affixed to the scalp, an electroencephalogram (EEG) signal can be obtained from the electrical activity of the brain. It ranks among the top inputs for accurately detecting emotions. The temporal, frontal, parietal, and occipital lobes are employed to control 8, 16, or 32 pairs of electrodes that are connected to the inion, nasion, right preauricular points, and left preauricular points, respectively, on the scalp [28].

2.3 Previous Work

The reviewed ERS were summarized in this section and are listed in Table 1, where "Reference" holds the reference to the cited work. The methods for feature extraction and selection are presented in the "Technique and Features" column. The "Classification" section follows with details on the researchers' ML model. The final section, "Accuracy (%)," displays the results as a percentage.

According to the review, DEAP dataset's EEG data is the most often used single modality data. Deep learning models and ML are used for classifications for ERS. ML models like SVM, K-NN, and NN are often used in this task because of their superior classification accuracy. As the majority of physiological records are in time-series format, feature extraction is required for ML-based ERS in order to create the data signal. Few researchers also employed deep learning as a preprocessing technology because of its capacity to extract and select features in hidden layers. The most often used assets are audio, video, pictures, and video game soundtracks. The International Affective Picture System (IAPS) [46] provides the majority of the images/pictures used to elicit emotion.

Feature fusion is a significant problem in the multimodal emotion identification process. Facial expressions are a common method for identifying emotions, but because they are non-physiological signals, they do not directly reflect a person's inner mental states. Face emotion identification is a difficult task as a result of elements such as partial occlusions of head deflection, facial areas, and variations in illumination. The performance of facial detection is, however, diminished by such interference. Finding a reliable classification system is one of the study's most important issues.

In this field, discrete and arousal/valence models are also common. However, valence and arousal alone are insufficient for a user to comprehend a person's precise emotions. To acquire the best classification result, the training samples should, according to the ML classification rule of thumb, be significant. For various input data, several researchers found varying accuracy and performance levels. It demonstrates that there is not a set method for the necessary scenario.

Table 1 Related articles on using physiological signals to recognize emotions

References	Technique and features	Classification	Accuracy (%)
[29] (2017)	Db5 multiscale wavelet decomposition	Neural network	91.67
			87.5
[30] (2017)	–	Ensemble deep learning model	Valence:88.54
			Arousal: 84.63
[31] (2018)	Liquid state machines	ANN, DT, K-NN, SVM, LDA	Arousal: 84.63
			Valence:88.54
			Liking:87.03
[32] (2018)	Random search, particle swarm optimization, simulated annealing, tree-of-Parzen estimators (TPE)	LSTM	77.68
[33] (2018)	Power spectral density (PSD)	Deep learning	93.6
[34] (2019)	ANN	Neural networks (NN)	75.38
[35] (2020)	CNN	Deep transfer learning model	90.59
	Fourier transform (FT)		82.84
[36] (2020)	–	CNN	DREAMER
			Valence: 94.59
			Dominance: 95.13
			Arousal: 95.26
			DEAP
			Valence: 97.97
			Dominance: 98.32
			Arousal: 98.31
[37] (2020)	Dual-tree complex wavelet transform (DT-CWT)	Recurrent neural network (RNN)	83.13
[38] (2020)	Sliding window strategy	LSTM, CNN	71.61 ± 2.71
[39] (2020)	–	LSTM-RNN	Arousal: 83.3
			Valence: 79.4
[40] (2020)	Self-supervised network	Self-supervised CNN	DREAMER
			Arousal: 77.1
			Valence: 74.9
			AMIGOS
			Arousal: 79.6
			Valence: 78.3
			SWELL:
			Stress: 90.2

(continued)

Table 1 (continued)

References	Technique and features	Classification	Accuracy (%)
			Valence: 93.8
			Arousal: 92.6
			WESAD
			Affect state (95.0)
[41] (2021)	Eigenvector matrix	Adaboost	88.7
	Linear discriminant analysis		
[42] (2021)	Empirical mode decomposition	SVM and multilayer perceptron	100
[43] (2021)	Time, frequency, wavelet	SVM	65.92
[44] (2021)	–	CNN	Arousal: 96.13
			Valence: 96.79
[45] (2022)	Variational mode decomposition (VMD)	Deep neural network	Arousal: 61.25
			Valence: 62.5

3 Conclusion and Future Work

This paper provided a systematic review for designing ERS. It is possible to discern between various emotional values, which are widely utilized as ERS output classes, by employing discrete and multidimensional emotional state models. Biosensors can be used to collect input data from a variety of physiological signals, including EDA/GSR/SC, EOG, ECG, HRV, EEG, EMG, SKT, and RSP. One source of signal data or a combination of sources can be used to construct the ERS. According to a recent study, deep learning models are the best ERS techniques. Additionally, some classical ML models, like SVM, can provide classification with reasonable accuracy. There is space to improve the model based on the accuracy discovered by other researchers. Improved work with single modality is necessary in particular since it will lower system costs overall. The final ERS will be built and designed using the results of this systematic review as recommendations. The system that has been given will eventually be used to identify emotions in real time.

References

1. Picard R, Vyzas E, Healey J (2001) Toward machine emotional intelligence: analysis of affective physiological state. IEEE Trans Pattern Anal Mach Intell 23(10):1175–1191
2. Reeves B, Nass C (1998) The media equation. Centre for the Study of Language & Information, Stanford, CA
3. Picard RW (2000) Affective computing. MIT Press, London, England
4. Mühlbacher-Karrer S, Mosa AH, Faller LM, Ali M, Hamid R, Zangl H, Kyamakya K (2017) A driver state detection system—combining a capacitive hand detection sensor with physiological sensors. IEEE Trans Instrum Meas 66(4):624–636
5. Ali M, Al Machot F, Mosa AH, Kyamakya K (2016) A novel EEG-based emotion recognition approach for e-healthcare applications. In: Proceedings of the 31st annual ACM symposium on applied computing
6. Ambach W, Gamer M (2018) Physiological measures in the detection of deception and concealed information. Detecting concealed information and deception: recent developments. Academic Press, Massachusetts, United States
7. Kang S, Kim D, Kim Y (2019) A visual-physiology multimodal system for detecting outlier behavior of participants in a reality TV show. Int J Distrib Sens Netw 15(7)
8. Kumar A, Garg N, Kaur G (2019) An emotion recognition based on physiological signals. Int J Innov Technol Explor Eng 8(9S):335–341
9. Gravina R, Li Q (2019) Emotion-relevant activity recognition based on smart cushion using multi-sensor fusion. Inf Fusion 48:1–10
10. Egger M, Ley M, Hanke S (2019) Emotion recognition from physiological signal analysis: a review. Electron Notes Theor Comput Sci 343:35–55
11. Joseph A, Geetha P (2020) Facial emotion detection using modified Eyemap-Mouthmap algorithm on an enhanced image and classification with Tensorflow. Vis Comput 36(3):529–539
12. Ko BC (2018) A brief review of facial emotion recognition based on visual information. Sensors (Switzerland) 18(2)
13. Plutchik R (2001) The nature of emotions: human emotions have deep evolutionary roots, a fact that may explain their complexity and provide tools for clinical practice. Am Sci 89(4):344–350
14. Ekman P (1992) An argument for basic emotions. Cogn Emot 6(3–4):169–200
15. Russell JA (1980) A circumplex model of affect. J Pers Soc Psychol 39(6):1161
16. Ahmad Z, Khan NA (2022) Survey on physiological signal-based emotion recognition. Bioengineering 9:688
17. Khalili Z, Moradi MH (2008) Emotion detection using brain and peripheral signals. In: Biomedical engineering conference CIBEC, pp 1–4
18. Imad A, Malik NA, Hamida BA, Seng GHH, Khan S (2022) Acoustic photometry of biomedical parameters for association with diabetes and Covid-19. Emerg Sci J 6:42–56
19. Goshvarpour A, Abbasi A, Goshvarpour A (2017) An emotion recognition approach based on wavelet transform and second-order difference plot of ECG. J AI and Data Min 5(2):211–221
20. Tawsif K, Nor Azlina Ab Aziz J, Emerson Raja J, Hossen, Jesmeen MZH (2022) A systematic review on emotion recognition system using physiological signals: data acquisition and methodology. Emerg Sci J 6(5), ISSN: 2610-9182
21. Daiana da Costa T, de Fatima Fernandes Vara M, Santos Cristino C, Zoraski Zanella T, Nunes Nogueira Neto G, Nohama P (2019) Breathing monitoring and pattern recognition with wearable sensors. In: Wearable devices—the Big wave of innovation, IntechOpen, London, United Kingdom
22. Kim J, André E (2008) Emotion recognition based on physiological changes in music listening. IEEE Trans Pattern Anal Mach Intell 30(12):2067–2083
23. Norali AN, Abdullah AH, Zakaria Z, Rahim NA, Nataraj SK (2017) Human breathing assessment using Electromyography signal of respiratory muscles. In: 2016 6th IEEE international conference on control system, computing and engineering (ICCSCE)

24. Ramkumar S, Sathesh Kumar K, Dhiliphan Rajkumar T, Ilayaraja M, Shankar K (2018) A review-classification of electrooculogram based human computer interfaces. Biomed Res (India) 29(6):1078–1084
25. Lang PJ, Greenwald MK, Bradley MM, Hamm AO (1993) Looking at pictures: affective, facial, visceral, and behavioral reactions. Psychophysiology 30(3):261–273
26. Udovičić G, Derek J, Russo M, Sikora M (2017) Wearable emotion recognition system based on GSR and PPG signals. In: Proceedings of the 2nd international workshop on multimedia for personal health and health care
27. Wu G, Liu G, Hao M (2010) The analysis of emotion recognition from GSR based on PSO. In: 2010 International symposium on intelligence information processing and trusted computing
28. Furman JM, Wuyts FL (2012) Front matter. Aminoff's Electrodiagnosis in Clinical Neurology, 6th edn. Saunders, Philadelphia, United States
29. Park YL (2017) Soft wearable robotics technologies for body motion sensing. In: Human modelling for bio-inspired robotics. Academic Press, Massachusetts, United States, pp 161–184
30. Yin Z, Zhao M, Wang Y, Yang J, Zhang J (2017) Recognition of emotions using multimodal physiological signals and an ensemble deep learning model. Comput Methods Programs Biomed 140:93–110
31. Al Zoubi O, Awad M, Kasabov NK (2018) Anytime multipurpose emotion recognition from EEG data using a liquid state machine based framework. Artif Intell Med 86:1–8
32. Nakisa B, Rastgoo MN, Rakotonirainy A, Maire F, Chandran V (2018) Long short term memory hyperparameter optimization for a neural network based emotion recognition framework. IEEE Access 6(1):49325–49338
33. Bagherzadeh S, Maghooli K, Farhadi J, Zangeneh Soroush M (2018) Emotion recognition from physiological signals using parallel stacked autoencoders. Neurophysiology 50(6):428–435
34. Tiwari S, Agarwal S, Adiyarta K, Syafrullah M (2019). Classification of physiological signals for emotion recognition using IoT. In: 2019 6th International conference on electrical engineering, computer science and informatics (EECSI)
35. Wang F, Wu S, Zhang W, Xu Z, Zhang Y, Wu C, Coleman S (2020) Emotion recognition with convolutional neural network and EEG-based EFDMs. Neuropsychologia 146:107506
36. Liu Y, Ding Y, Li C, Cheng J, Song R, Wan F, Chen X (2020) Multi-channel EEG-based emotion recognition via a multi-level features guided capsule network. Comput Biol Med 123(March):103927
37. Wei C, Chen L, Song Z, Lou X, Li D (2020) EEG-based emotion recognition using simple recurrent units network and ensemble learning. Biomed Signal Process Control 58:101756
38. Nakisa B, Rastgoo MN, Rakotonirainy A, Maire F, Chandran V (2020) Automatic emotion recognition using temporal multimodal deep learning. IEEE Access 8:225463–225474
39. Li C, Bao Z, Li L, Zhao Z (2020) Exploring temporal representations by leveraging attention-based bidirectional LSTM-RNNs for multi-modal emotion recognition. Inf Process Manage 57(3):102185
40. Sarkar P, Etemad A (2020) Self-supervised ECG representation learning for emotion recognition. IEEE Trans Affect Comput 1(1)
41. Chen Y, Chang R, Guo J (2021) Emotion recognition of EEG signals based on the ensemble learning method: AdaBoost. Math Probl Eng 2021:1–12
42. Salankar N, Mishra P, Garg L (2021) Emotion recognition from EEG signals using empirical mode decomposition and second-order difference plot. Biomed Signal Process Control 65:102389
43. Khateeb M, Anwar SM, Alnowami M (2021) Multi-domain feature fusion for emotion classification using DEAP dataset. IEEE Access 9:12134–12142
44. Salama ES, El-Khoribi RA, Shoman ME, Wahby Shalaby MA (2021) A 3D-convolutional neural network framework with ensemble learning techniques for multi-modal emotion recognition. Egypt Informatics J 22(2):167–176

45. Pandey P, Seeja KR (2022) Subject independent emotion recognition from EEG using VMD and deep learning. J King Saud Univ—Comput Inf Sci 34(5):1730–1738
46. Lang P, Bradley MM (2007) The International Affective Picture System (IAPS) in the study of emotion and attention. In: Handbook of emotion elicitation and assessment, Oxford University Press, Oxford, United Kingdom, vol 29, pp 70–73

Chapter 19
Visual HOG-Enabled Deep ResiNet for Crime Scene Object Detection

T. J. Nandhini and K. Thinakaran

1 Introduction

As the criminal activities in the society keep on increasing, it is being required to collect valuable evidence and make the investigation process in a proper way to attain the criminals as soon as possible. Detection of information in the crime scene is a demanding work in recent days. Automated detection of crime scene objects needs to be highly efficient [1]. In order to have a narrow scope of investigation, the evidence around the crime scene is highly important. The recent technological developments such as image processing, video processing, computer vision, and statistical analysis improve the automated investigation of crime scene objects based on CCTV camera videos. The main problem of detecting objects in crime scene object analysis is highly helpful to have a secure and efficient surveillance system [2]. Specific object detection systems focused initially in computer-vision-based detection techniques rather than manual interventions. In recent years, surveillance systems are enhanced using artificial intelligence algorithms that include machine learning algorithms, neural networks, recurrent networks, and various feature analysis models [3]. Many existing systems provide the detection of anomalies in the CCTV footage. The forensic department plays an important role in collecting the evidence from the surroundings. The fingerprints are commonly used identification techniques to have the criminal identification and understanding the crime scene better [4].

Crime scene object detection is an interesting area of research where various information are analyzed automatically with the help of computer vision and statistical

T. J. Nandhini (✉) · K. Thinakaran
Saveetha School of Engineering, Institute of CSE, Saveetha Institute of Medical and Technical Science-SIMATS, Chennai, India
e-mail: nandhinitj67@gmail.com

K. Thinakaran
e-mail: thinakarank.sse@saveetha.com

A. Yadav et al. (eds.), *Proceedings of International Conference on Paradigms of Communication, Computing and Data Analytics*, Algorithms for Intelligent Systems,
https://doi.org/10.1007/978-981-99-4626-6_19

analysis techniques. It is also a racket to detect the crimes in objects in a short span of time. Technologies such as artificial intelligence cloud computing enable the system to work faster than manual processes. Face detection is an important factor in crime scene object detection. In terms of any face detected in the camera, the emotions need to be analyzed [5]. In the existing system, drones and unmanned aerial vehicles are used for the detection of crime scenes where human intervention is not feasible. The reasons for adaptive systems in object detection with highly reliable techniques are used to investigate the crime scene in a more accurate way. The objects that are not visible to the human eye can be detected by the systematic process, which ends the crimes in object detection becoming deeper than usual.

The technical steps taken to collect the crime scene objects enhance the relationship between the crime investigation experts and the public. CCTV cameras are available to record the videos. Only advanced cameras can automatically interpret the anomalies while recording the videos [6]. These video-based investigation samples are compared with the manual interventions from CCTV cameras.

- The proposed work is focused on reading the crime video collected from CCTV cameras. Videos have numerous frames associated with that. The foremost step of the proposed approach is to convert the video into frames.
- From the converted frame, randomly the highest feasible frame is selected and preprocessed using the image processing toolbox. Visual HOG (VHOG)-based feature extraction technique is used here. The total of 90% of the data is considered as training feature, and 10% is considered as a testing feature [7].
- On the other hand, based on feature points, semantic object is selected and segmented using morphology structural elements extraction
- The VHOG feature provides the shape of the objects present in the crime scene accurately. Various training images with objects are provided for processing, further it is applied to the resilient network (ResiNet). The visual HOG feature is also used to cross-validate the presence of semantic objects.
- The semantic object is nothing but the highly refractive object seen from the crime scene. The initial process of a proposed approach started by detecting the semantic object present in the crime scene and further classifying the same using the images available in the database.
- The DRN using the supportive visual HOG features is used to validate the presence of semantic objects.

The rest of the paper is formulated as making detailed literature study in Section 2. The system tool selection and problem identifications are discussed in Section 3. The system architecture and detailed system design steps are discussed in Section 4. The rest of the paper is concluded with future enhancement.

2 Background Study

Baber et al. [8] presented a system, where RFID tag enabled evidence bags are collected and analyzed. Numerous log information is collected for analysis. These data provide the association between each evidence collected. The log data contains descriptions of evidence in the form of objects [9]. The images captured using a digital camera are processed in order to stamp the log data of time, location, and subsequent hint in the form of data. The author presented a system where the prototype of a device can be used to support the process of evidence recovery from the time scene. The key elements enabled here with the proposed approach are the radio of frequency identification tags (RFID) that are attached to evidence bags. Log information is collected from the CCTV camera, and further the synchronization of the manual system is analyzed [10]. The captured information on the log data is also compared with the speech data. Wearable devices are highly reliable to record the crime scene data. The concept of Bags of Evidences (BOE) is discussed here completely.

Gur et al. [11] the author presented a system where crime scene is analyzed with respect to collection of evidence. Securing the investigation data is highly important. In order to improve the security concerns, feature extraction techniques are utilized by autonomous drones [12]. The camera is placed in front of the drone to collect various features around the crime scene where the human interventions are not feasible. The drones can collect various aspects of data. Further processed in order to have the relativity between the crime scene and the environment being developed by the real-time software and artificial intelligence (AI) algorithms. Many interruptions of flying drone can disable the process and divert the drones to collect the information from the crime scene; this acts as a drawback in the presented approach [13].

Liu et al. [14] presented a system in which a multi-class classifier is used to collect the query image and label the data effectively [15]. Various types of datasets are utilized such as small dataset, subsequent dataset, and large dataset. Databases are accessed through a query-based (QB) image accessing approach. Image retrieval system to rank the fusion model and further sort out the drawback of large database handling issues in a reliable and fast way. Various kinds of images are classified and accessed through automated system programs such as query by example (QBE) approach. Further, these data are analyzed using machine learning algorithms to have effective investigation metrics.

Liu et al. [16] presented a system where large scales of data are being used for investigation. Various images are collected for investigation; thus, the extraction of unique features present in the images is highly important [17]. Detection of features using manual technique is not always recommended; hence, utilization of discrete cosine transform (DCT) for texture feature extraction is used. Descriptive histogram-based color feature extraction technique is utilized here to collect the various critical information from the crime scene objects [18]. The presented approach is also helpful to investigate the image retrieval process.

Mahesha et al. [19] presented an approach which contains MSCOCO datasheet for automated crime scene detection using highly reliant deep learning algorithm

[20]. In case the crime scene images are directly passed into the analysis models, the raw information is directly analyzed by the computer vision system and hence the images of segmented into various types of models to generate different crimes scene object class using the deep learning models such as inception V3 and long short-term memory network (LSTM) to detect the highest scores of detection.

Petty et al. [20] presented an approach utilizing a SIFT-based feature detection algorithm and compared it with various image data points collected from different registration techniques. The SIFT feature is used to detect the real-time objects present in the crime scene photography. The presented approach is comparatively tested with the other feature extraction techniques such as SIFT and ASIFT.

3 System Design

The system design is focused by utilizing the tool boxes applicable for image processing, video processing, image segmentation, etc. The ultimate goal of the system is to detect the crime scene objects present in the CCTV camera recorded videos as well as detection of anomalies present in the scene. MATLAB software is utilized here to achieve the goal of semantic object segmentation. Image processing toolbox provides comprehensive set of reference standard algorithms and graphical tools for image processing analysis, visualization, and algorithm development for the image enhancement, image deep learning feature detection, noise reduction, image segmentation, special transformations, and registrations which are helpful to analyze the crime scene objects in a different way and to analyze the structure parameters of the objects present in the crime scene effectively. Image processing toolbox is helpful to analyze the images in terms of color texture pixel points, and even the small objects are analyzed from the crime scene images. Various techniques such as contour techniques and histogram analysis are used to manipulate the region of the interested area using simple commands.

- The main problem with the crime scene detection is how fast the system can automatically detect the crime scene objects and further how accurately it can classify the object within the given database.
- The minute objects present in the crime scene which are not visible to the manual interventions need to be analyzed in depth to have accurate investigation.

4 Methodology

Deep learning is the advancements in machine learning (ML) algorithms that enable us to make the computer understand the information provided by images where the system needs to detect the things faster and more accurately than human interventions. The system architecture of the proposed semantic object detector is shown in Fig. 1. Deep learning algorithm is derived from neural network (NN) algorithms to learn useful representations of features directly from the captured real-time data. Neural networks are combined with multiple combinations of layers in which the input is being processed, and special features are extracted and further classified using classification layers. This is the common technique used in deep learning algorithms. Transfer learning is commonly used in deep learning approaches in which the specific neural network architecture is being repeatedly used for the pretrained network to make the classification even better. The advantage of transfer learning is to approach similar tasks in a different perspective at every iteration. The repeated analysis may take longer time and create a robust model that can accurately detect the objects. Feature extraction allows the system to investigate deeper data using machine learning algorithms such as support vector machine (SVM), neural networks (NN), transfer learning (TL), and linear regression (LR). The machine learning algorithms are helpful to analyze the data in a feature-based approach. Further, the fine tuning of features is helpful to have highly accurate feature mapping.

Deep learning on unstructured data utilizes large GPU space because of the parallel processing and cloud utilization. Continuous accessing of the cloud takes time and lots of GPU memory to be utilized, and hence, in order to reduce the GPU utilization problem, the planning algorithms are run with unique features extracted from the major reduction in data dimension which will reduce the time taken for the complete analysis.

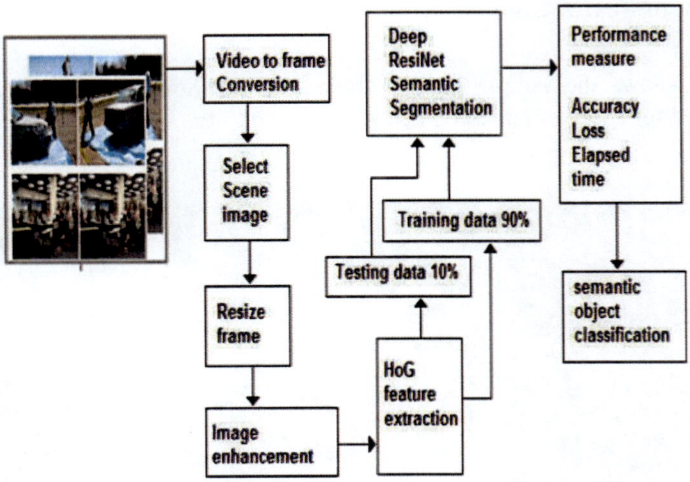

Fig. 1 System architecture of proposed semantic object detector

4.1 Performance Measure

The performance analysis of the proposed approach is measured through calculation of accuracy. As the estimation of true values present in the given system with respect to the total number of analysis rates, such as true positive value, true negative value, false positive value, and false negative values are discussed.

The amount of true data present similar to the detected images as true value is the consequence of negative results where the positive result should be expected. Please stop it. The false negative value is the anticipation of false data as a class where the data is supposed to have positive. Using the formulas on accuracy true positive, true negative, false positive, and false negative, the performance measure is evaluated for the processing time which is calculated.

$$\text{Accuracy} = \frac{\text{Tp} + \text{Tn}}{\text{Tp} + \text{Tn} + \text{Fp} + \text{Fn}} \tag{1}$$

5 Results and Discussion

5.1 Input Test Image

Figure 2 shows the sample images under training. Some of the static images considered for analyses are compared with ground truth.

Fig. 2 System

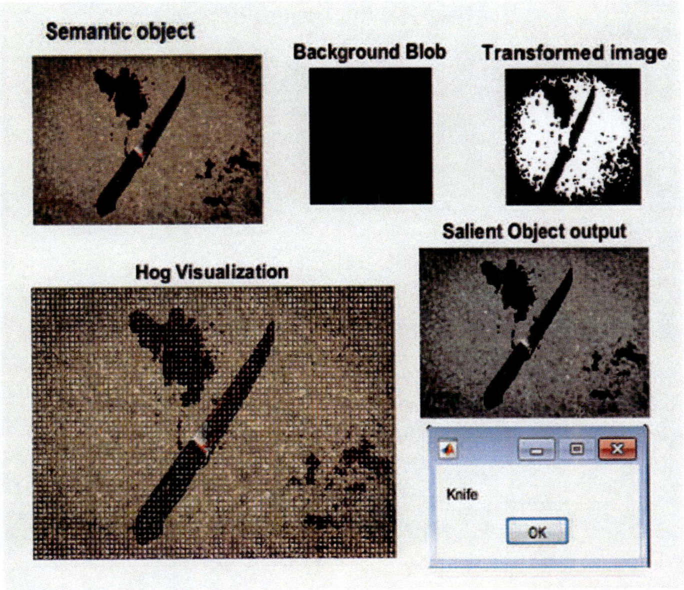

Fig. 3 Image processing results

5.2 *Image Processing Results*

Figure 3 shows the image processing results, such as background subtraction, transformed image, visualization HOG feature, and salient object detection.

5.3 *Training Accuracy*

Figure 4 shows the DRN training accuracy with respect to a maximum of 500 epochs run within the network.

5.4 *Testing Accuracy*

Figure 5 shows the DRN testing accuracy graph with respect to the number of epochs reaching the maximum of 1000, in spite of the proposed system reaching consistent accuracy of 89% at the epoch = 350 approximately.

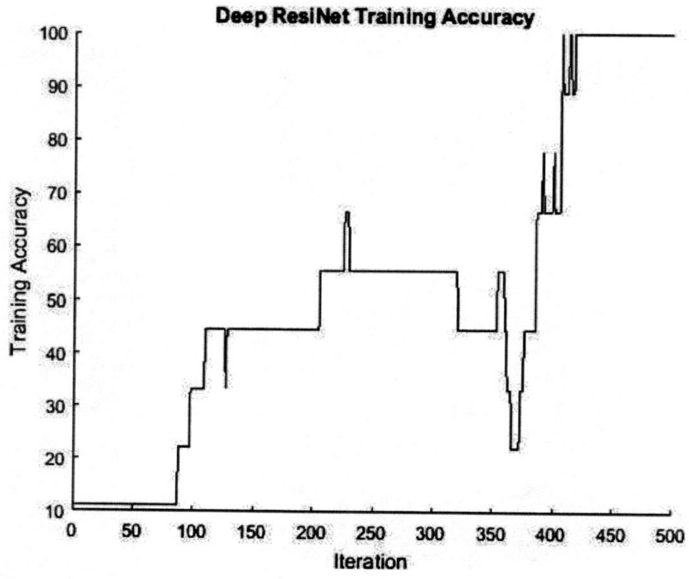

Fig. 4 DRN training accuracy

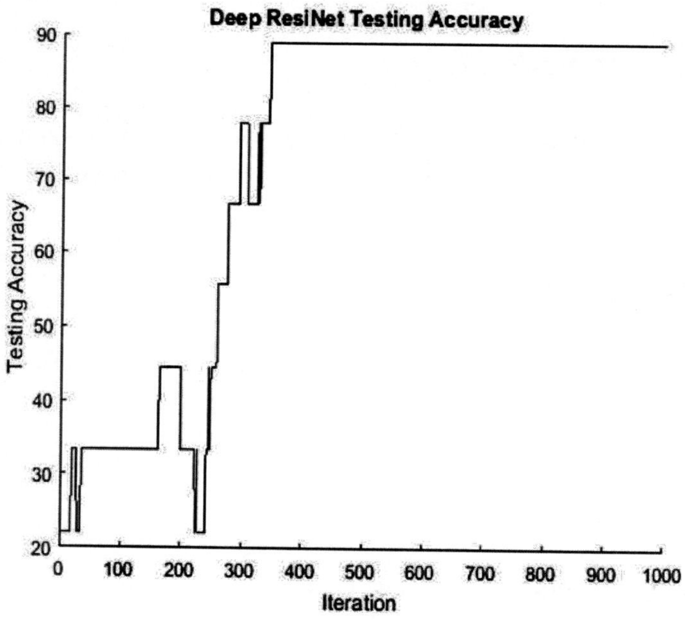

Fig. 5 DRN testing accuracy

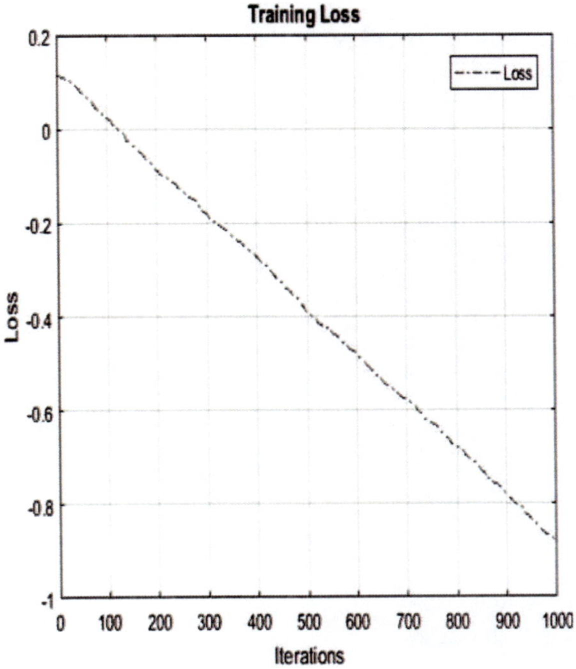

Fig.6 Training loss

5.5 *Training Loss*

Figure 6 shows that the training loss happens within the DRN model for the total of 1000 epochs. As the number of iterations increases, the training loss degrades further and accuracy improves.

5.6 *Testing Loss*

Figure 7 shows the testing loss acquired from the DRN model. As the testing process improves, the accuracy is getting improved by suppressed loss status in testing.

5.7 *Validation Loss*

Figure 8 shows the validation loss of the proposed DRN model. The validation data of 10% from the training data is considered; hence, the cross-validation adds up concrete strength to the proposed process.

Table 1 presents the comparison of existing implementations using MSCOCO pretrained dataset using RCNN [4] which achieved the accuracy of 74.33% on scene

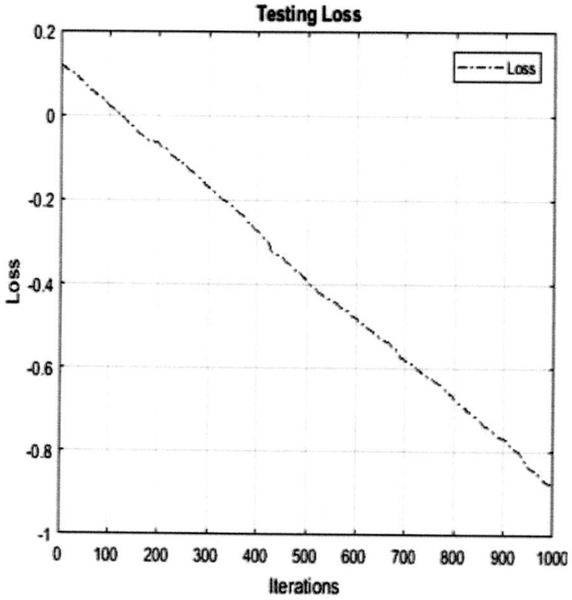

Fig. 7 Testing loss

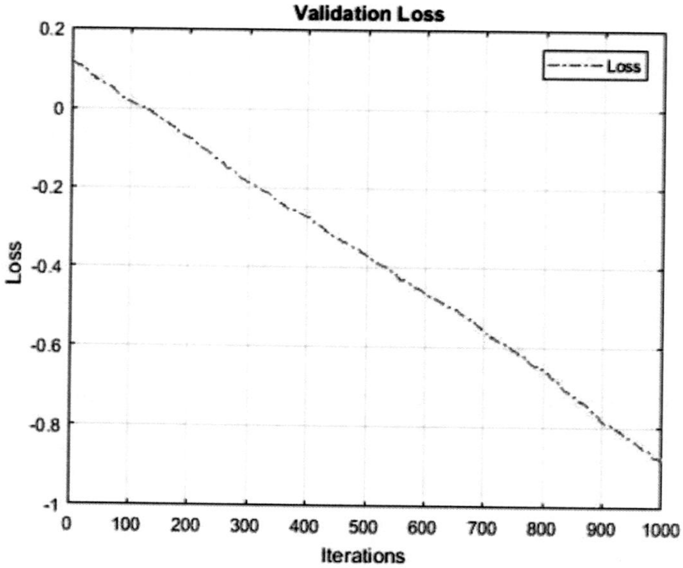

Fig. 8 Validation loss

Table 1 Comparison of existing implementations with the proposed DRN model

S. No.	References	Dataset	Method	Accuracy (%)
1	Saikia et al. [4]	MSCOCO	RCNN	74.33
2	Sani et al. [12]	MSCOCO	YOLO	81.20
3	Proposed	Real-time CCTV footage	DRN	91

object detection, with you look at only once (YOLO) model and with pretrained dataset achieved the accuracy of 81.20%. The proposed approach implemented with real-time CCTV footage extracted images. The accuracy of 91% is achieved.

6 Conclusion

Detection of crime scene objections in a short time is important to make the investigation straight forward. Due to the delay in detection, accurate mapping of semantic objects is the prime motive of the proposed approach. Here, real-time footage of crime scene video is collected. Video is converted into image frames. Randomly selected image frames are trained and tested with commonly available crime scene objects. Further, the classified object names are validated. The proposed approach implemented with real-time CCTV footage extracted images. The accuracy of 91% is achieved. Further, the elapsed time taken to complete the detection process took 55 s. The proposed methodology is further improved using inspired heuristic algorithms with deep learning architectures for enhancing the accuracy.

References

1. Farhood H, Saberi M, Najafi M (2021) Improving object recognition in crime scenes via local interpretable model-agnostic explanations. In: 2021 IEEE 25th international enterprise distributed object computing workshop (EDOCW), 2021, pp 90–94. https://doi.org/10.1109/EDOCW52865.2021.00037
2. Proceedings of the 2019 Federated conference on computer science and information systems, Ganzha M, Maciaszek L, Paprzycki M (eds). ACSIS, vol 18, pp 391–396 (2019)
3. Sun H, Meng Z, Tao PY, Ang MH (2018) Scene recognition and object detection in a unified convolutional neural network on a mobile manipulator. In: 2018 IEEE International conference on robotics and automation (ICRA), 2018, pp 5875–5881. https://doi.org/10.1109/ICRA.2018.8460535
4. Saikia S, Fidalgo E, Alegre E, Fernández-Robles L (2017) Object detection for crime scene evidence analysis using deep learning. Image Anal Process—ICIAP 2017:14–24
5. Espinace P, Kollar T, Soto A, Roy N (2010) Indoor scene recognition through object detection. In: 2010 IEEE international conference on robotics and automation, 2010, pp 1406–1413. https://doi.org/10.1109/ROBOT.2010.5509682
6. Masuda S, Kaeri Y, Manabe Y, Kenji S (2018) Scene recognition method by bag of objects based on object detector. In: 2018 Joint 10th international conference on soft computing and

intelligent systems (SCIS) and 19th international symposium on advanced intelligent systems (ISIS), 2018, pp 321–324. https://doi.org/10.1109/SCIS-ISIS.2018.00062
7. Tankard C (2011) Advanced persistent threats and how to monitor and deter them. Netw Secur 2011(8):16–19
8. Baber C, Smith P, Cross J, Zasikowski D, Hunter J (2005) Wearable technology for Crime Scene Investigation. Proceedings – International Symposium on Wearable Computers (ISWC), 2005, pp 138–141. https://doi.org/10.1109/ISWC.2005.58
9. Araújo P, Fontinele J, Oliveira L (2020) Multi-perspective object detection for remote criminal analysis using drones. IEEE Geosci Remote Sens Lett 17(7):1283–1286. https://doi.org/10.1109/LGRS.2019.2940546
10. Nakib M, Khan RT, Hasan MS, Uddin J (2018) Crime scene prediction by detecting threatening objects using convolutional neural network. In: 2018 International conference on computer, communication, chemical, material and electronic engineering (IC4ME2), 2018, pp 1–4. https://doi.org/10.1109/IC4ME2.2018.8465583
11. Gur A, Erim M, Karakose M (2020) Image processing based approach for crime scene investigation using drone. pp 1–6. https://doi.org/10.1109/ICDABI51230.2020.9325606
12. Sani S (2022) Object detection for crime scene evidence analysis. 2022 J Softw Eng Simul 8(7):44–53, ISSN(Online):2321-3795 ISSN(Print):2321-3809
13. Aarthi S, Chitrakala S (2017) Scene understanding—a survey. In: 2017 International conference on computer, communication and signal processing (ICCCSP), 2017, pp 1–4. https://doi.org/10.1109/ICCCSP.2017.7944094
14. Liu W, Wu CY (2019) Crime scene investigation image retrieval using a hierarchical approach and rank fusion. In 2019 14th IEEE Conference on Industrial Electronics and Applications (ICIEA), pp. 1974–1978. IEEE
15. Tasci T, Kim K (2015) Imagenet classification with deep convolutional neural networks
16. Liu Y, Hu D, Fan J, Wang F, Zhang D (2017) Multi-feature fusion for crime scene investigation image retrieval. In 2017 International Conference on Digital Image Computing: Techniques and Applications (DICTA), pp. 1–7). IEEE
17. Simonyan K, Zisserman A (2014) Very deep convolutional networks for large-scale image recognition. In: International conference on learning representations, 2014
18. Chen Y-R et al (2021) Forensic science education by crime scene investigation in virtual reality. In: 2021 IEEE international conference on artificial intelligence and virtual reality (AIVR), 2021, pp 205–206. https://doi.org/10.1109/AIVR52153.2021.00046
19. Mahesha P, Royina KJ, Lal S, Anoop Krishna Y, Thrupthi MP (2021) Crime scene analysis using deep learning. In: 2021 6th International conference on signal processing, computing and control (ISPCC), 2021, pp 760–764. https://doi.org/10.1109/ISPCC53510.2021.9609350
20. Petty M, Teng SW, Murshed M (2019) Improved image analysis methodology for detecting changes in evidence positioning at crime scenes. In: 2019 Digital image computing: techniques and applications (DICTA), 2019, pp 1–8. https://doi.org/10.1109/DICTA47822.2019.8945934

Chapter 20
Scheming of Diamond Ring Harvestor for Low-Powered IoT Devices

Shruti Taksali and Amruta Lipare

1 Introduction

With the advancement of the IoT ecosystem, interconnection of multiple devices to the Internet and to each other over a network is also increasing. But their low energy constraints deteriorates their efficiency [1–3]. In order to resolve this problem, the harvestor is proposed in this paper. The term 'harvestor' is used which is a combination of nano + energy + harvesting [4]. Their utilization limit crosses the boundaries of wireless communication to efficient solar cells, holography tools, medical treatments, quantum computing, high-resolution energy storage, and many more to come. This paper focuses on energy storage application with the material selection of diamond with nitrogen vacancy and the shape of ring structure. Earlier, nitrogen vacancy color centers have been used for magnetic writing purposes with high resolution [4]. It is realized on heat-assisted magnetic recording (HAMR) [5], which requires an intense magnetic field [6] for the writing process on the nanomaterial ring structure at an elevated temperature (laser heating). The temperature is lowered after the writing process is completed to retain the energy. This in turn helps in achieving higher density energy storage. All this technology is possible because of the material used here, i.e., diamond with negatively charged nitrogen vacancies (NCNV), which has been recently used for radio frequency applications [7]. In this, a spin 1 complex is formed by a substitutional nitrogen atom adjacent to a vacant site. These are basically paramagnetic centers which can be located individually using confocal microscopy, initialized via optical pumping, and read out through spin-dependent photoluminescence measurements [8]. This carbon family member has high thermal conductivity, large band gap (5.5 eV), low thermal-expansion coefficient, long spin lifetime, wide transparency window (from UV to IR), large breakdown field, high

S. Taksali · A. Lipare (✉)
Indian Institute of Information Technology, Pune 411046, India
e-mail: amruta@iiitp.ac.in

A. Yadav et al. (eds.), *Proceedings of International Conference on Paradigms of Communication, Computing and Data Analytics*, Algorithms for Intelligent Systems,
https://doi.org/10.1007/978-981-99-4626-6_20

refractive index ($n = 2.5$), and high carrier mobility [9, 10]. NCNV has been used for quantum cryptography platforms for secure communication [11, 12]. NCNV center's electronic spin and nearby nuclear spins can be coupled to form a large qubit register which is the basic component of a quantum computer [13, 14]. Multicolor optical microscopy can be used to locally convert the charge state of NCNVs within a dense ensemble from negative to neutral and correspondingly alter the NCNV fluorescence emission from bright to dark. This change is reversible, long-lasting, and robust to weak illumination, thus serving as an alternate platform for 3D information storage [15]. But practical implementations of these technologies require efficient excitation and extraction of single photons from NCNV center's using a simple optical system [16, 17]. Despite its many advantages, its high refractive index does not allow the emitted photons to be visible to sophisticated instruments. So, diamond nanowire interface has been recently developed between nanoscale device and macroscopic optical system [10]. This paper is focused on simulated results and design of the device.

2 Harvestor Design Logic

For structure simulation, Microwave Studio Computer Simulation Technology (MWS CST) software is used in frequency domain solver with dimensions in nm and frequency range in THz. The tetrahedral meshing is used which is easily available in the software tool to get the fine results. The design was based on a logic that increase in length of outer ring will lead to decrease in minimum frequency and increase in length of Inner ring will lead to decrease in maximum frequency. So accordingly, frequency range was set between 300 THz—110 THz (min–max frequency), and dimensions of structure are as follows: Silicon substrate with a size of 200 × 200 × 100 nm is selected, and silicon dioxide as a dielectric material is used above which the device is modeled. Then, NCNV diamond ring with radius of outer big ring (Rb) is 80 nm, radius of second ring (Rs) is 40 nm, radius of inner sphere (Ri) is 10 nm, and the width of each ring is 20 nm which is selected as shown in Fig. 1. Also, the shape of the ring is selected in such a manner that would support increased capacity storage. Background is selected as twice the inner radius of the ring for efficient analysis of the design. Field monitors for electric and far-field are applied to obtain the proper results. The structure is illuminated from below perpendicular to the long axis through waveguide mode selection. Practically, plasmonic waveguides could also be useful for the excitation of these nanodevices. Finally, using the frequency domain solver, model was simulated to obtain the results. For the ring structure, NCNV diamond is chosen for high-resolution 3D energy storage. This is not a regular diamond used for ornaments, but artificially modified semiconductor material for technological purposes [18, 19]. Metallic nanorods such as gold suffer from a permanent shape change and photorefractive polymers degrade on continuous light exposure [20–22]. Therefore, these traditional nanomaterials cannot be used for energy storage. On the other hand, the charge state of NCNV diamond has

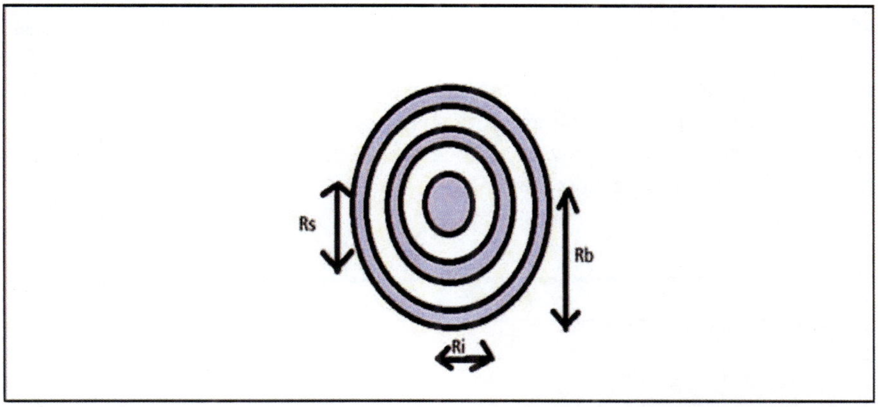

Fig. 1 Harvestor ring structure Ri (radius of inner sphere), Rs (radius of second ring), and Rb (radius of outer big ring)

no accumulated effects when altered in reversible manner, hence permitting one to erase and rewrite information with high resolution in countless times. Heat-assisted magnetic recording (HAMR) technology is used which requires an intense magnetic field; the laser source incidents the harvestor which initiates the writing process at an elevated process. As the writing task is completed, the temperature is lowered so as to retain the original energy.

3 Results and Discussion

The storage capability would be determined through electric field enhancements on the structure in various directions as shown in Figs. 2, 3, and 4. These would in turn represent the resolution at all stages of the process. Figure 2 indicates the overall electric field enhancement in the structure in 3D view where more no. of red spots are realized near the gaps which relates to their light-amplifying ability in the gaps. This is because the oscillating electron possessing kinetic energy overrules the magnetic energy in the near field leading toward the enhanced energy storage capacity. The maximum electric field corresponds to the red color which is indicated by the 360° phase in the diagram. Figure 3 shows the overall electric field enhancement in 2D view which satisfies the 3D view and corresponds to the maximum electric field enhancement. The X-axis corresponds to the distance in nm, and maximum peaks are obtained near the gaps. The Y-axis corresponds to the electric field enhancement factor which is maximum in the starting point of incidence. In the 3D view, different colors correspond to different ionization efficiencies upon incidence in different directions. Figure 4 indicates the near-field electric field intensity of the designed structure in 2D view at a frequency of 500 THz. This depicts the radiation efficiency as − 1.463 dB and a total efficiency as − 6.313 dB. These values in turn resulted in an enhanced electric field with a value of 190.8 dBV/m. Figures 5, 6, and 7 represent the far-field pattern which is normalized to nanometer range to have the near-field effect of the harvestor indicating the uniform storage capacity of the device. The NCNV diamond as a material and ring shape as a structure has led to provide the energy solution to the low-powered IoT devices in the form of a harvestor which will be a promising candidate in the technology world.

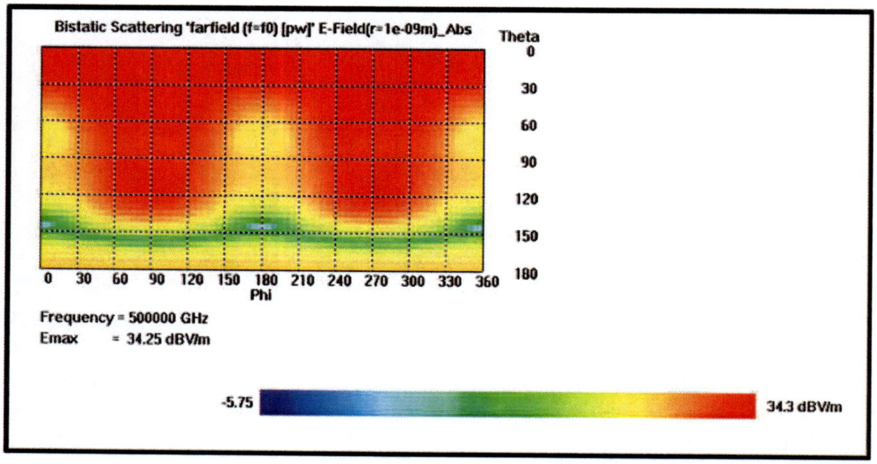

Fig. 2 Overall electric field enhancements in 2D view

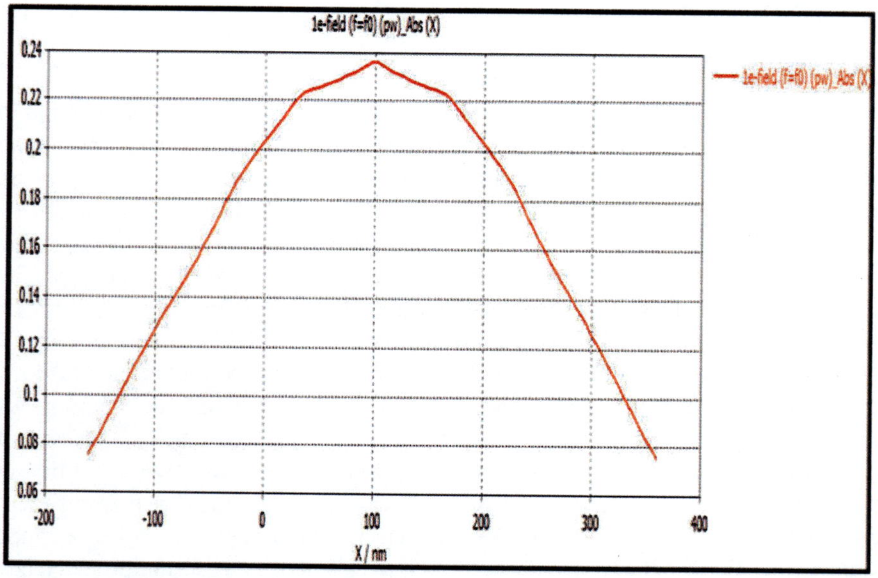

Fig. 3 Overall electric field enhancement in Cartesian plot

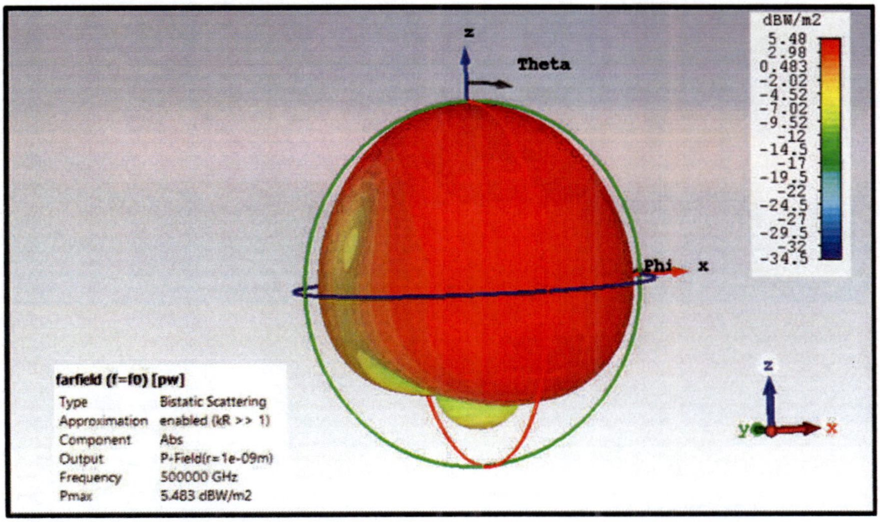

Fig. 4 Maximum power flow in the structure

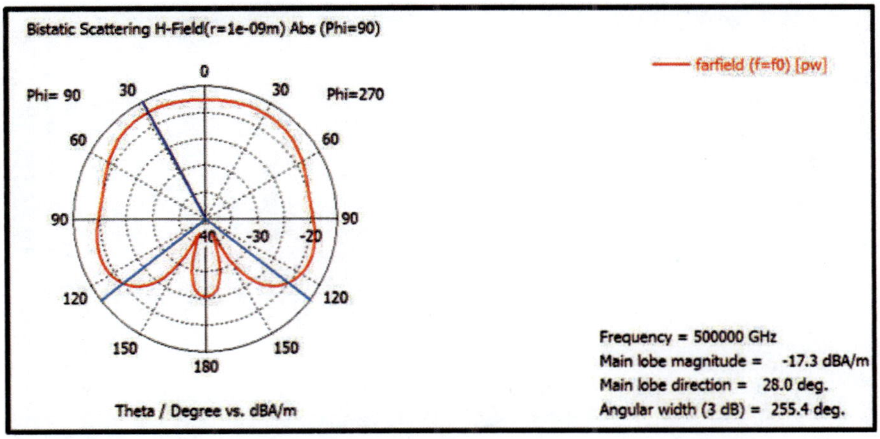

Fig. 5 Polar plot of magnetic field enhancement

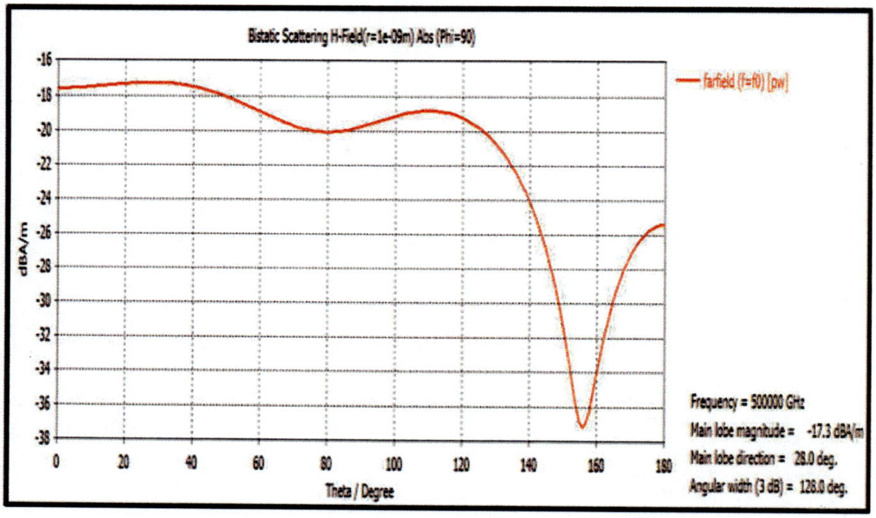

Fig. 6 Normalized bistatic scattering magnetic field 2D plot

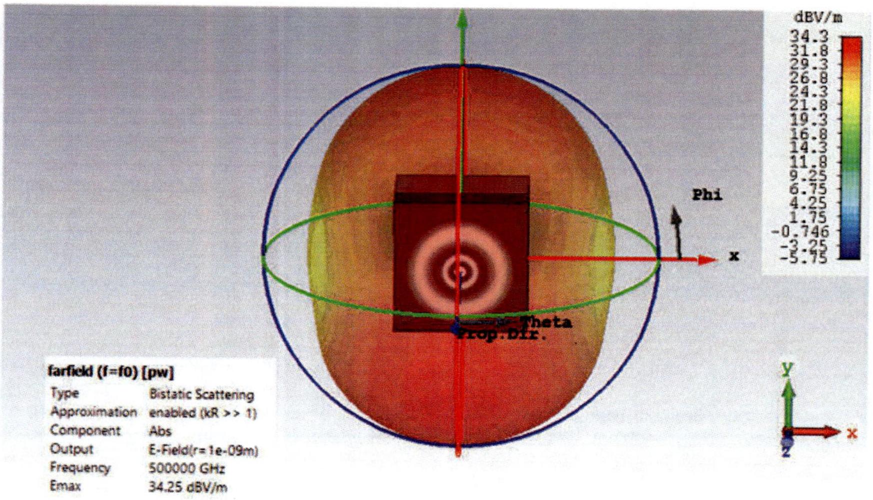

Fig. 7 Normalized far-field pattern of the modeled harvestor

4 Conclusion

This new concept of high-resolution energy storage would bring a new revolution in the IoT field. As IoT devices are power constrained, this proposed ring harvestor could be a boon in this field. NCNV diamonds can be replaced by other materials such as CNT, graphene nanoribbons, and magnetic materials in order to further enhance the capability of the device. The shape of the nanodevice alters the electron energy distribution thereby changing the applications and operating frequency in THz range. The focus is to be laid on interfacing nanodevices with the optical macroscopic instruments for efficient results. In this paper, the designing concept is presented with the possibility of fabricating the device.

References

1. Sikdar D, Premaratne M, Cheng W (2015) Optical nanoantenna. WO2016154657A1 03 31
2. Bharadwaj P, Deutsch B, Novotny L (2009) Optical antennas. Adv Opt Photonics, pp 438–483
3. Chen SW (2012) Nano-antenna and methods for its preparation and use. US13879723 08 16
4. Heidmann J (2014) Magnetic write head characterization with nano-meter resolution using nitrogen vacancy color centers. US14532992, 19 Feb 2014
5. Kryder MH, Gage EC, McDaniel TW, Challener WA, Rottmayer RE, Ju G, Hsia Y-T, Erden MF (2008) Proceedings. IEEE, pp 1810–1835
6. Guler U, Kildishev AV, Boltassevaab A (2015) Plasmonics on the slope of enlightenment:the role of transition metal nitrides. © The Royal Society of Chemistry
7. Hahn JW, Gregory S (2016) Diamond nitrogen vacancy sensor with common RF and magnetic fields generator. US15003298 Jan 21, 2016
8. Jelezko F, Gaebel T, Popa I, Gruber A, Wrachtrup J (2004) Observation of coherent oscillations. Phys Rev Lett 92
9. Isberg J et al (2002) High carrier mobility in single-crystal plasma-deposited diamond. Science 297:1670–1672
10. Lončar M, Babinec T, Hausmann B, Diamond nanotechnology. s.l.: SPIE.
11. Kurtsiefer C, Mayer S, Zarda P, Weinfurter H (2000) Stable solid-state source of single photons. Phys Rev Lett, pp 290–293
12. Beveratos A et al (2002) Single photon quantum cryptography. Phys Rev Lett 89
13. Gurudev Dutt MV et al (2007) Quantum register based on individual electronic and nuclear spin qubits in diamond. Science 316:1312–1316
14. Neumann P et al (2008) Multipartite entanglement among single spins in diamond. Science 320:1326–1329
15. Rettner Charles T, Stipe BC (2007) Optical devices having transmission enhanced by surface plasmon mode resonance, and their use in energy recording. US7289422B2 10 30
16. Ballato J, Carroll D, Dimaio J (2004) Plasmon-photon coupled optical devices. US10865237 06 10
17. Dhomkar S et al (2016) Long-term energy storage in diamond. Sci Adv
18. Khan A (2017) Diamond semiconductor system and method. US13734986 06 01
19. Khan A (2011) Method of fabricating diamond semiconductor and diamond semiconductor formed according to the method. US13725978 12 21
20. Day D, Gu M, Smallridge A (1999) Use of two-photon excitation for erasable-rewritable threedimensional bit optical energy storage in a photo refractive polymer. Opt Lett 24: 948–950
21. Zijlstra P, Chon JWM, Gu M (2009) Five-dimensional optical recording mediated by surface plasmons in gold nano rods. Nature, pp 410–413
22. Gan Z, Cao Y, Evans RA, Gu M (2013) Three-dimensional deep sub-diffraction optical beam lithography with 9 nm feature size. Nat Commun 4

Chapter 21
Smart Computer Commands Using Gesture Recognition

Sonali Patil, Chinmay Shyam Mukhedker, Mukul Sanjay Chaudhari, Jaspreetsingh Kulwindarsingh Pannu, and Varun Prasannan

1 Introduction

Due to the advent of COVID-19, people are always in search of operating any appliance in a contactless manner. Let it be payments or operating a PC, users are always trying to minimize overall contact. This is mainly done due to the fear of the virus spreading as COVID-19 is airborne. In the COVID era, everyone was forced to work from home which indirectly induced a lot of usage of PCs and desktops. In response to this, a special method of operating PCs needed to be discovered. As the overall usage of PCs was very high during these times, a methodology regarding the same is required. Developing an interface method based on computer vision is also going to improve the user experience, and a sense of newness is also created. Gestures are a crucial component of human–computer interaction and a straightforward, organic form of communication. Gesture recognition [1] acts as a bridge between human natural language and the computer system. There is a necessity to explore more ways to interact with computers using gestures. This is because gesture recognition works in a complex environment. Earlier interaction with the computer system was only possible with wired networks. A data glove would send the position and direction of the human hand using the sensors. Machine learning algorithms would then be used to recognize them. The term "human–computer interaction" (HCI) [2] describes how a person interacts with a computer, or more specifically, a machine. Usability and functionality are key elements of human–computer interaction. Both static and dynamic gestures are important for real-time communication. Multiple shortcut commands are being used for quick computer operations. Now, the same can be performed through gesture recognition. Gesture recognition can be an easy and best way to perform all those quick computer operations. A gesture can be identified

S. Patil · C. S. Mukhedker (✉) · M. S. Chaudhari · J. K. Pannu · V. Prasannan
Pimpri Chinchwad College of Engineering, Pune, India
e-mail: mchinmay.0809@gmail.com

A. Yadav et al. (eds.), *Proceedings of International Conference on Paradigms of Communication, Computing and Data Analytics*, Algorithms for Intelligent Systems,
https://doi.org/10.1007/978-981-99-4626-6_21

and mapped to a certain action. The proposed method enhances the user experience and reduces the effort for computer operation.

2 Literature Survey

Research on gesture recognition techniques has been going on for quite a few years. They vary on the models used, datasets used, techniques performed, etc. So, there is a difference in their results, speed, accuracy, etc. Here are a few research papers which include research on gesture recognition techniques.

The following are the essential points that have been studied -

2.1 Research on the Optimization of Static Gesture Recognition Based on Convolution Neural Network [1]

The recognition is based on a convolution neural network. It has used the multi-view bootstrapping algorithm, CNN, and MobileNet SSD. The model can achieve the speed of smooth recognition while ensuring high accuracy and robustness of recognition results.

2.2 Research on Gesture Recognition Method Based on Computer Vision Technology [2]

The method used here breaks down gesture recognition into blocks of methods. First image segmentation is carried out to isolate the dynamic gestures present in it. Multiple models are implemented for the recognition of gestures. One of them is selected based on requirements. The model extracts the features and classifies the gestures for the parameters evaluated by that respective model. Accordingly, a description is generated for the application.

2.3 Computer Vision Based Gesture Recognition for Desktop Object Manipulation [3]

Convex hull algorithms return the smallest convex polygon that can be made using a given set of points. Graham scan algorithm is used to detect the convex hull. Contour detection is an algorithm used to detect the borders or edges of the image. The hidden

Markov model provides a concept of creating complex models by drawing intuitive pictures. It is used for computational sequence analysis.

2.4 *Research on the Hand Gesture Recognition Based on Deep Learning [4]*

In this paper, use of algorithms like fusion algorithm, camshift algorithm, and LeNet-5 network is done. It does statistical template matching. The average accuracy of hand gesture recognition is 98.3%. The recognition rate for numbers 7 and 9 is not high because the hand gestures of them are complicated.

2.5 *Combining Hand Detection and Gesture Recognition Algorithms for Minimizing Computational Cost [5]*

It presents a combined hand gesture recognition system that uses a hand detector to detect the hand in the frame and then switches to a gesture classifier if a hand was detected. It proposes a method to reduce the computational cost of gesture recognition. It is possible to efficiently use computing resources depending on the presence of a hand in the frame.

2.6 *Gesture Recognition System Based on Improved YOLOv3 [6]*

The proposed method in this paper uses the combination of YOLOv3 target detection and Raspberry Pi to detect the hand gestures automatically. The preprocessing of the image is carried out using the Python library OpenCV. It focuses on improvising the speed and accuracy of existing models. The achieved results are found to be more reliable to satisfy the requirements in real time.

Table 1 illustrates the used method and a remark on it mentioning the significance, respectively. The remark justifies the details, pros, and cons of the implemented method.

Table 2 helps us to understand various technologies with their respective speed and accuracies, thus helping us in deciding which technology is better to use.

Table 1 Literature survey

Title	Method	Remarks
Research on Optimization of Static Gesture Recognition Based on Convolution Neural Network [1]	• Multiview bootstrapping algorithm • CNN MobileNet SSD	It can ensure high accuracy and recognition results being robust in nature while achieving speedy smooth recognition
Research on Gesture Recognition Method Based on Computer Vision Technology [2]	• Hidden Markov model • Neural network • Time-based regularization	The distribution of functions among various blocks makes the system work faster
Computer Vision BasedGesture Recognition for Desktop Object Manipulation [3]	• Graham scan algorithm • Contour detection algorithm • Hidden Markov model Baum–Welch algorithm	It yielded an accuracy of 89% with fast results. It provides great accuracy in different lighting conditions
Research on the Hand GestureRecognition Based on Deep Learning [4]	• Fusion algorithm • Cam shift algorithm • LeNet-5 network • Statistical template matching • AdaBoost • Gaussian mixture model of skin color • Haar feature CNN	The average hand gesture recognition accuracy is 98.3%. Due to their complex hand gestures, numbers 7 and 9 have a low rate of recognition
Combining Hand Detection and Gesture Recognition Algorithms forMinimizing Computational Cost [5]	• Convolutional pose machines (CPM) • Contour detection • Combination of gesture recognition and hand detection	The paper proposes a method to reduce the computational cost of gesture recognition. It is possible to efficiently use computing resources depending on whether a hand is present in the frame
Gesture Recognition System Based on Improved YOLOv3 [6]	• YOLOv3 • Raspberry Pi • OpenCV	The accuracy achieved with the proposed method is 90% which indicates reliable results

Table 2 Comparison of existing systems

Parameters	Speed	Accuracy	Remarks
CNN	High	High	Robust
OpenCV	Average	Low	Accuracy changes as there is a change in lighting conditions
Graham scan	High	Average	Works accurately in any lighting condition
LeNet-5	Average	Very high	Complicated
Haar cascade	High	High	Works accurately in static as well as dynamic gestures

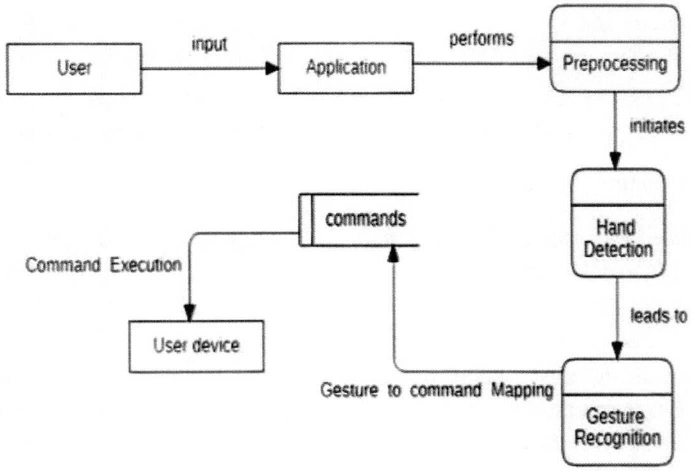

Fig. 1 Data flow diagram

3 Methodology

The system architecture of the proposed system comprises different components which mainly consist of software modules.

Figure 1 shows the relationship between various entities involved in the system.

- The user passes an image of the hand taken by the camera of the action performed as an input to the model.
- Preprocessing actions are performed on the image like resizing the image and noise reduction/removal.
- The model initially identifies the hand, and then, the points are identified.
- The image is processed in the gesture recognition model which identifies the gesture.
- Computer action mapped to the action performed is executed on the computer.

4 System Implementation

The MediaPipe library developed by Google has been used in this project for gesture recognition which includes CNN models. It has many features such as gesture recognition, hand recognition, and face detection. Of these, gesture recognition model is based on CNN.

4.1 MediaPipe

MediaPipe [7] is a package developed by Google which offers customizable machine learning solutions for live and streaming media. It provides solutions in the: -

- Face Detection

A quick face detection method with multi-face support and six landmarks is called MediaPipe face detection.

- Face Mesh

Even on mobile devices, MediaPipe face mesh is a system that estimates 468 3D face landmarks in real time. It uses machine learning (ML) to infer the 3D facial surface and only needs one camera input—a specialized depth sensor is not required.

- Hands

A high-fidelity hand and finger tracking solution is MediaPipe Hands. It uses machine learning (ML) to extrapolate 21 3D hand landmarks from a single frame.

Other applications include -

- Object detection,
- Hair segmentation,
- Iris detection,
- Pose detection.

4.2 Steps for Implementation

1. Open the application and allow permissions for camera access.
2. The image is resized, and the noise in the image is removed using the CLAHE algorithm.
3. Using the MediaPipe library, the algorithm detects the hand within the camera range.
4. Using the proposed algorithm, the gestures from the hand are detected.
5. The detected gestures are mapped to particular command names.
6. The command names are then mapped to a particular function.

5 Implementation Flow

An effective way was needed for gesture detection which could be further used in determining the operation to be performed on the computer. The hand and its important points would be detected by using the MediaPipe library. The palm landmark model uses an SSD model to detect the palm and its landmarks. The hand's landmark model is a CNN model. The model uses regression, or direct coordinate prediction, to accomplish precise key point localization of 21 3D hand-knuckle coordinates inside the observed hand areas. The model has been manually annotated on 30 K real-world images with 21 3D coordinates. For each hand that is spotted, the MediaPipe model which is named "mp_hand_gesture" returns a total of 21 crucial points. The MediaPipe drawing_utils function and hand detection will be activated to draw the hand landmarks on the image. The pretrained TensorFlow model will be uploaded. The titles of the gestures are contained in a file called "Gesture".

Whenever a hand gesture is performed, the frame will be captured and converted to RGB format. The frame is then preprocessed. In preprocessing, the image is firstly resized and the noise from the image is removed using the CLAHE algorithm. Then, the MediaPipe library comes into action with the help of which we get landmarks/key points of the gesture that is being performed. The landmarks/key points are stored in a result class after processing. The points are drawn on the screen for the user.

The suggested model receives the images which have been read and detected. Then, after iterating over each detection, we put the location on a landmarks list. The model produces a standardized result in this case, so the result is scaled by the image's height (y) and width (x). As a result, the values in the outcomes are all between 0 and 1. A list of landmarks is provided to the model. Predict() function produces an array with ten forecast classes for each site. Then, the class ID with the highest prediction value is identified, and its corresponding gesture is recognized. Then, the action assigned to the particular action is performed using the keyboard package.

The user acts as per the requirement. The action is processed in the model, and the model identifies the action. Then, the keyboard shortcut assigned to the particular action is performed using the keyboard package. Actions are assigned class names in the dataset. So, as the action is executed, the class name assigned to the action is sent as output. As the class name is returned, the key mapped to the action in the map is passed to the keyboard package. Then, the particular command is executed on the computer with the help of the keyboard package.

6 Dataset Description

Gesture Recognition Dataset [8]:-

This dataset is available on the Qualcomm developer network. It contains images of different gestures which are divided into 27 classes, i.e., 27 different gestures can be recognized. The dataset is divided into training, testing, and validation dataset. The ratio of the distribution of the dataset is 8:1:1. The dataset contains 148,092 videos. Image extraction is carried out at the rate of 12 frames per second in JPG format. The dataset contains images of hand gestures such as thumbs up, thumbs down, and fist.

7 Result and Discussion

The hand gestures are detected on the image using the proposed method. Two gestures, fist and call me, are observed on the hand for demonstration. The obtained results are as follows.

The skeleton-like structure is formed by identifying 21 key points on the hand (Points marked with red dots). All the key points are then connected to form the structure. The gesture is identified using the same structure. Ten gestures are used in the proposed method out of which two are as follows.

In Fig. 2, user shows a fist in front of the camera, and the MediaPipe library accordingly detects the gesture and labels it accurately as "fist." Similarly, Fig. 3 shows the gesture named "call me." Similarly, multiple gestures can be implemented.

Figure 4 shows the 21 key points on the hand which are used by the MediaPipe library.

Gesture detection is performed using various algorithms. Each algorithm has a significant value of precision and accuracy for the dataset provided.

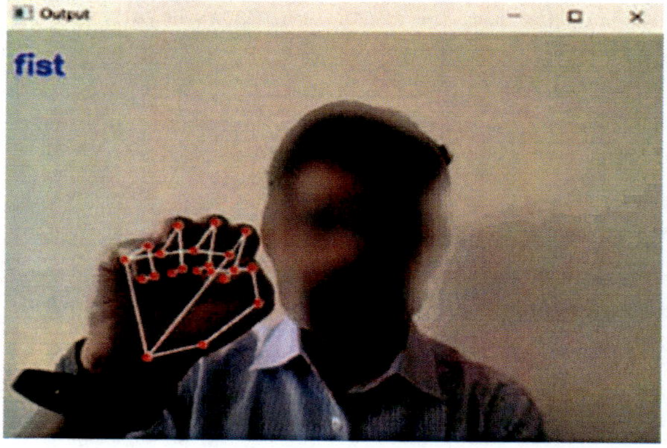

Fig. 2 Detected gesture—fist

Fig. 3 Detected gesture—call me

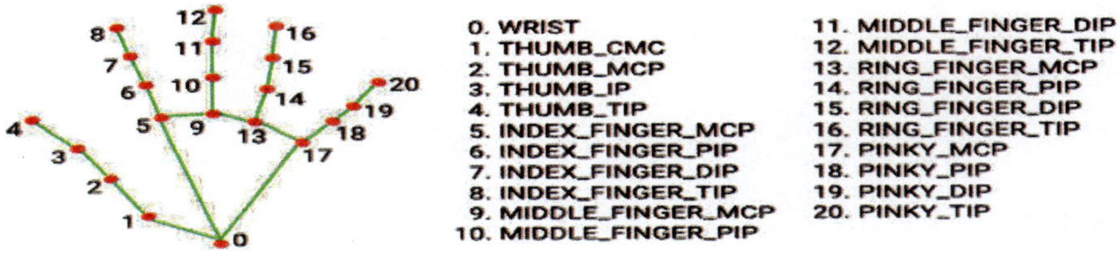

Fig. 4. 21 key points on hand [9]

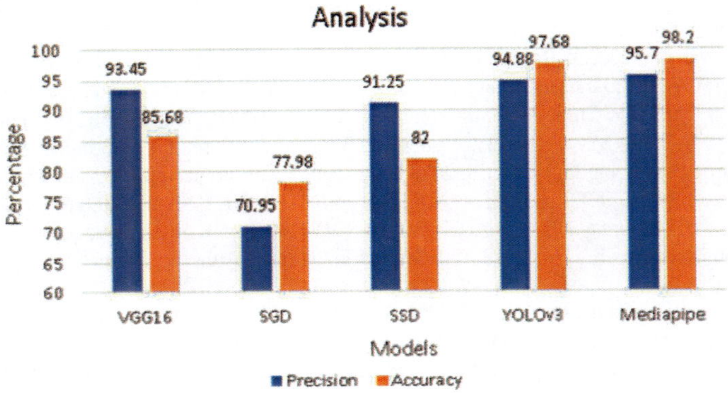

Fig. 5 Analysis of different models

The graph (Fig. 5) shows the accuracy and precision of five algorithms, namely VGG16, SGD, SSD, YOLOv3, and MediaPipe. On observation and analysis of all the mentioned models, it has been noticed that YOLOv3 and MediaPipe are the two best-performing models. The obtained accuracy for algorithms VGG16, SGD, SSD, YOLOv3, and MediaPipe are 85.68%, 77.98%, 82.00%, 97.68%, and 98.2%, respectively.

8 Conclusion

On analysis, it is found that the MediaPipe gives more accuracy as compared to any other algorithm. The accuracy and precision for YOLOv3 are 97.6% and 94.8%, respectively, whereas for MediaPipe, it is 98.2% and 95.7%, respectively. Gesture recognition was performed on various input images. This proposed model gives an accuracy of 98.2% and accurate recognition of all the gestures and commands. The proposed method makes it easier for users to communicate with PC or laptop without any contact. It can be similarly applied to operate different machines without human contact like elevators. The project developed will provide an intuitive way of operating PCs and other computer systems. It can provide a standard interface for presentations at offices, educational institutions, etc. The project contributes to enhancing user experience.

References

1. Guo X, Xu W, Tang WQ, Wen C (2019) Research on optimization of static gesture recognition based on convolution neural network. In: 2019 4th international conference on mechanical, control and computer engineering (ICMCCE), pp 398–3982. https://doi.org/10.1109/ICMCCE48743.2019.00095.
2. Cui H, Wang Y (2020) Research on gesture recognition method based on computer vision technology. In: 2020 International conference on computer information and big data applications (CIBDA), Guiyang, China, pp 358–362. https://doi.org/10.1109/CIBDA50819.2020.00087
3. Hoque SMA, Haq MS, Hasanuzzaman M (2018) Computer vision based gesture recognition for desktop object manipulation. In: 2018 International conference on innovation in engineering and technology (ICIET), pp 1–6. https://doi.org/10.1109/CIET.2018.8660916
4. Sun J-H, Ji T-T, Zhang S-B, Yang J-K, Ji G-R (2018) "Research on the hand gesture recognition based on deep learning. In: 2018 12th international symposium on antennas, propagation and EM theory (ISAPE), pp 1–4. https://doi.org/10.1109/ISAPE.2018.8634348
5. Golovanov R, Vorotnev D, Kalina D (2020) Combining hand detection and gesture recognition algorithms for minimizing computational cost. In: 2020 22th international conference on digital signal processing and its applications (DSPA), pp 1–4. https://doi.org/10.1109/DSPA48919.2020.9213273
6. Zhang Z, Wu B, Jiang Y (2022) Gesture recognition system based on improved YOLO v3. In: 2022 7th international conference on intelligent computing and signal processing (ICSP), Xi'an, China, pp 1540–1543. https://doi.org/10.1109/ICSP54964.2022.9778394
7. https://google.github.io/mediapipe/. Accessed 8 Mar 2023
8. https://developer.qualcomm.com/software/ai-datasets/jester. Accessed 21 Mar 2023
9. https://google.github.io/mediapipe/solutions/hands. Accessed 8 Mar 2023

Chapter 22
A New Software Approach to Automated Translation (On the Example of the Logistics Sublanguage)

Rodmonga Potapova, Vsevolod Potapov, and Oleg Kuzmin

1 Introduction

Machine translation (hereinafter referred to as MT) is an effective computer tool based on special digital approaches that can automatically translate texts from one language to another with good quality. Its effectiveness and advantages were proved over many decades of practical implementation [7].

For a fraction of a second, one can learn the content of a document, and there is no need to spend time on a long linguistic study. The final results can be achieved quickly, and the information is clear and understandable to any common person. It is impossible to underestimate the significance of an approach in regard to the present reality where the speed of the result is the main competitive advantage [20].

Today, MT is used everywhere to translate Web pages and news feeds containing commonly used linguistic corpora. The research made last year proved the necessity of MT use for technical texts, including a simple set of word combinations on common themes [21]. Database is being trained on millions of pages of written Web content. By now, the MT in some points could be compared to the "manual" human translation. The main metrics for evaluating the MT quality are BLEU algorithms [14] and METEOR [1].

The MT includes two levels, in particular when the text is being entered to the system and when the result is being displayed to the user. The internal decisions and mathematical equations are being made automatically on the basis of a combined

R. Potapova · O. Kuzmin (✉)
Institute of Applied and Mathematical Linguistics, Moscow State Linguistic University, Ostozhenka Street 38, 119034 Moscow, Russia
e-mail: oleg.kuzmin.999@mail.ru

V. Potapov
Centre of New Technologies for Humanities, Lomonosov Moscow State University, Leninskije Gory 1, 119991 Moscow, Russia

A. Yadav et al. (eds.), *Proceedings of International Conference on Paradigms of Communication, Computing and Data Analytics*, Algorithms for Intelligent Systems,
https://doi.org/10.1007/978-981-99-4626-6_22

approach (SMT and NMT). The algorithmization are pre-trained on large corpora samples based on a wide range of multilingual linguistics units affects the quality of the MT results.

Though the MT is a widely used digital tool, it has some drawbacks to be taken into consideration. In particular, it concerns the translation of professional sublanguages where the accuracy of the equivalent linguistic unit is the most important issue. A common phenomenon is the mixing of definitions, worsening the general perception of the translation product. Lack of focus on a certain sublanguage and the incapability to "tune in" to the context are the main drawback of MT [9]. The correct and accurate translation of terminology into different languages using MT is an extremely rare occurrence.

Sublanguages are characterized by the diversity of special linguistic units that need to be categorized in order to obtain the standard translation variation. To resolve this task, unusual for MT, the use of specialized glossary, explanatory dictionaries, and examples of use in various foreign languages is required.

The formation of linguistic databases, dictionaries, thesauri, glossaries, and other specialized elements of the processing of a natural language will avoid formal errors and inaccuracies in automatic text translation. For this, translation memory technologies and digital glossaries are used. Database is being compiled during the work in the particular field, and in the case of word matching, the system proposes the equivalent units or grammar corrections for replacement within the corpora.

2 Method

This objective of the study was the elaboration of linguistic terms identification method and a compilation of electronic dictionaries of the sublanguage for translation automatization issues, as well as the implementation of a program solution that ensures the formation of more accurate alternatives for particular subject areas.

According to study results, a linguistic items database of the logistics sublanguage was formed, including terminological units (glossary), as well as a number of additional functions to it, such as interpretation of meanings, example use of units within some contexts, synonymic chain, and elements of syntactic annotation of sentences. Using the software product, it was possible to make an accurate translation of pre-selected terminological units related to the logistics sublanguage. The translated elements were evaluated in accuracy and compared with the Google MT. The proposed digital solution with specialized add-ons allows to make the process of computer translation easier and to make it better, but also to provide a linguistic area of knowledge necessary for work in this subject domain. The results demonstrated the effectiveness of the program regarding the implementation of specialized translation tasks and confirmed the key strength of it.

2.1 *Review of Existing AT Systems and Their Competitive Advantages Over MT*

To minimize errors in any sublanguage MT, the translator must be equipped with a number of software solutions and tools to perform manipulations with the available set of data from the initial sequence.

When starting the translation task, it is necessary to determine the criteria that will assist in choosing the right digital tool to obtain quality results.

In this regard, the following categorization is proposed:

- volume of information (word, phrase, sentence, paragraph, and text);
- subject domain (what sublanguage it refers to);
- presence of acronyms, abbreviations, and special terminology;
- presence of complex grammatical structures;
- special cases of hardly translated or untranslatable words.

After evaluating the input sequence in accordance with the proposed categorization, a choice is made on the appropriateness of using machine (automatic) or automated translation in this specific case.

Automated translation (hereinafter referred to as AT) is a more complex product that allows a number of specialized tasks due to its own additions and tools.

Thus, the design of such programs involves a simple (standard) approach and complex (multilevel) approach, which involves a more complex internal structure and passing certain stages of text correction by introducing new elements or changing those already existing.

Both principles of operation contain an information input field and MT of selected sentences (text units); however, in a multilevel representation, in addition to post-editing of the text by human, a more complex quality test of MT is proposed [5], which includes recommendation of more accurate terminological units and processing the text involving manual translation by human (Fig. 1).

This principle involves the use of MT to create a certain "skeleton" of the text for its subsequent use as the basis of the future translation, as well as the breakdown of this translation into the correct and incorrect elements with the addition of those segments that will be considered more appropriate for one or another subject domain by human, taking into account the context and existing field knowledge.

Currently, the Google Translate MT is the most popular and accurate online service for translating texts [22]. It is worth noting that the translation of both phrases and sentences is of high quality. The training of the program algorithms takes place on the entire data set stored on Google servers and Web pages in the form of text documents and in the cache of the search engine.

AT programs are a comprehensive solution with many specialized tools aimed at solving specific application tasks, particularly at correcting and improving the MT or human-translated results [2]. The achievement of this goal is demanding special digital add-ons, which are built into the translation process and improve its final result.

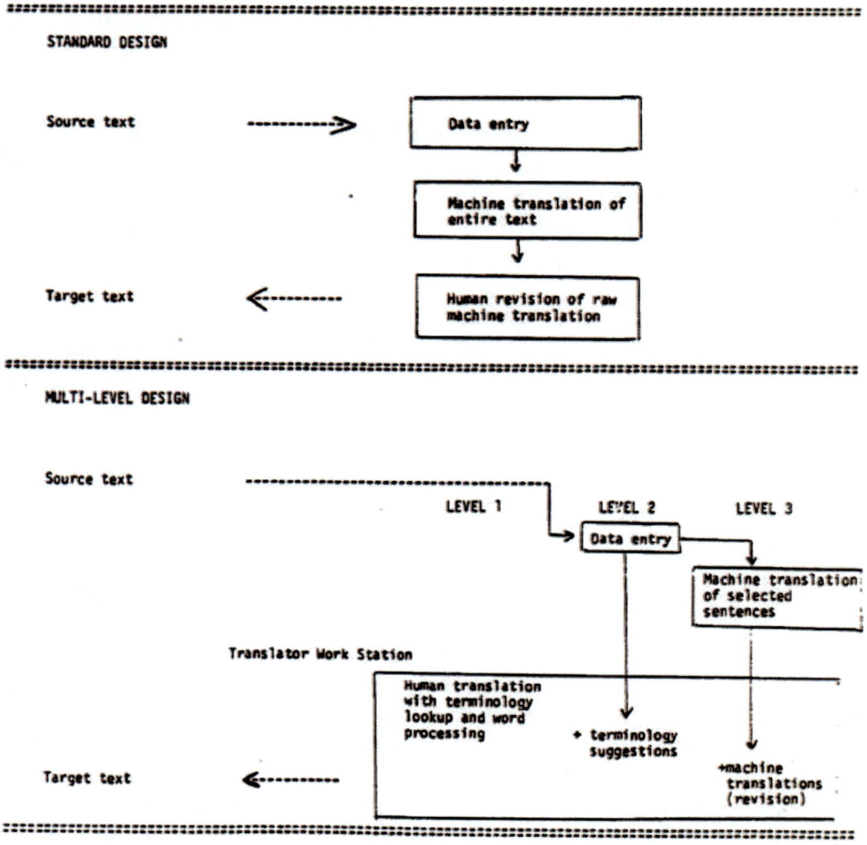

Fig. 1 Internal structure and principle of operation of AT systems. *Source* [8]

AT has its widespread in professional linguistic fields, where accuracy is a must. This counts as the weak side of MT. Noting problems extent under consideration related to AT and specialized databases, relevant scientific research should be noted [16–18].

To date, the most popular AT programs are Trados and SmartCAT [13]. The main functionality available to the user is translation memories, digital glossaries, auto-correction of MT within the text, and the potential of collective work on the same document, making changes and automatic correction of the text input. Moreover, it is important to note the technical capability to switch on the function or choose tools necessary for performing translation tasks. It allows the translation process to be interactive [8].

AT systems do not contain tools for natural language processing. Tools of the translation memories can only confirm the percentage-based matching of the text corpus by comparing it with those available in the database. They do not provide any linguistic recommendations for the user, do not display information about examples of using one or another word, and also do not offer synonyms and definitions of unfamiliar words of the subject domain, which complicates the process of post-editing [6]. The development of a narrow-focus AT program is a crucial task due to a constant increase in the number of multilingual Web corpora and their categorization insufficiency [16].

2.2 *Linguistic Specificity of the Logistics Sublanguage. Challenges in Machine Translation of Terminological Units*

Work and educational issues linked with translation of particular corpora gave rise to the task of translating text fragments related to logistics sublanguage from one language to another. This sublanguage contains many terminological units, the meanings of which must be explained and clarified to ensure a complete conveying of information without losing its meaning. Due to the enormous variability of the meanings of some terminological phrases and the inability of their exact translation, it was dictated by the need to eliminate their ambiguity.

Nowadays, logistics is considered exclusively in the context of globalization. The evolution of the terminological dictionary of logistics is directly related to these processes. When analyzing multilanguage corpora of texts and working with terminology, it was revealed that most of the terms of the logistics sublanguage belongs to the English language; partial borrowing and integration into the German language are also observed [12]. New terms are formed due to borrowing, and most borrowings, as in other domains, come from the English language. Most of these terms are nouns or noun phrases.

Consideration of the terms of logistics and their translation into Russian allows for the conclusion that this layer of vocabulary has high producing capacity. Its characteristic feature is neologization manifested in the emergence of new meanings in widely used words and in the emergence of new words and phrases. Most of the neologisms are borrowed from the English language. The German language accepts these lexical units in its vocabulary with almost no changes; in translations into Russian, the most often used are calquing or a combination of transcription with descriptive translation.

As part of this study, a large number of parallel multilanguage texts were analyzed, from which the terminological units of the sublanguage were manually extracted and preserved in the form of a table. These words and phrases are domain specific; and when working with MT programs, certain patterns had been identified which were taken into account when forming a database that was used to design our own software algorithm:

- First of all, when translating terminology using Google MT, it is not always possible to achieve an unambiguous and accurate translation. Taking into account these factors, it is advisable to use terminological glossaries and to form parallel

databases to exclude ambiguity and double interpretations of the same linguistic unit in different languages;
- Some terms of the sublanguage, due to their ambiguity, have extremely specific meanings, and without their definition, it is impossible to get an idea of the meaning and possibilities of applying a word (phrase) within this sublanguage. For this, it is proposed to study the programmatic methods for introducing dictionaries-thesauri into the AT programs to facilitate the task of translating terminological units;
- MT mainly offers an unambiguous option of the translation without a synonymic chain, with rare exceptions, when data on the statistics of the use of a word in the language are displayed. In the presence of a synonymic chain, the translator can decide more easily on the choice in the process of post-editing the text. For this task, it is necessary to additionally use the algorithms of word embedding [10, 11];
- When translating phrases of the logistics subject domain into several languages (Russian, German, and English), certain inaccuracies (erroneous translation options) were noted, which led to misinterpretations of the meanings of word terms of this sublanguage.

Based on the analysis of the MT translation quality for the logistics sublanguage and on the basis of the conclusions made, the program project was formed that will perform certain tasks in the field of this sublanguage and propose the most accurate translation options, as well as contain a knowledge area related to the subject domain, that is of informative and advisory nature.

2.3 Proposed Logistics Translator as a Universal Tool of Automated Translation. Principle of Operation and Basic Functions

With the ambiguity in the interpretation of word terms in different languages (Russian, German, and English), an attempt was made to unify many of the most effective AT functions and find a single software solution for them. Due to the need for quick translation without loss of quality, it became necessary to create a special Web application that can perform accurate translation and interpret the meanings of linguistic units of the sublanguage.

At the preparatory stage of the software solution implementation, the following objectives were formulated:

1. select multilingual samples that are verified/contains some mistakes;
2. identify the qualitative differences between manually translated and MT sentences of the sublanguage;
3. form digital tabular multilingual corpora;
4. AT programs structure and work principles that affect the translation process;

5. design a Web interface using specialized tool-equipped add-ons;
6. using the linguistic models to prove programs efficiency;
7. obtaining results and elaboration future steps toward quality improvement.

It is assumed that the logistics translator AT program will be the first digital online solution for performing the task of interlanguage translation of sublanguages. The program includes both an advanced version of the MT algorithm and the best experience of AT programs, so it is more focused on the sublanguage and can perform professional linguistic issues.

Special Web interface and a number of tools make it possible to perform MT and to use specialized databases for additional tasks. An interpreter of terms (professional glossary) is implemented as a unique function, which allows outputting the meanings (definitions) of the terminological units of the sublanguage.

After analyzing the quality of the translation made by Google MT, it was decided to use Reverso.Context as an MT tool, which is more accurate in the translation of the terminology of sublanguages. Also, the API Reverso enables search for synonyms and offers examples of using the word (phrase) in the context.

The main objective of the study was also the formation of a database based on the standing word combinations in the three mentioned languages and definitions of them. At the first stage, a linguistic unit is placed in the information input field. Next, it is translated using the Reverso MT to the necessary language (Russian, English, and German). After that, an example of using the word is given and the search for synonyms can take place. At the next stage, a request is sent to the database on the basis of the received translation to search for the necessary definition within the CSV database file. At the last stage, a syntactic annotation of the sentence takes place with an example of the use of the term with the searched word and parts of speech highlighted [3]. The schematic operation principle of the program is presented in Fig. 2.

As for the visual design, the appearance of the program can be presented as follows (Fig. 3). The text input field is responsible for entering textual information; the translation field displays the MT result for the logistics sublanguage; the word usage example field provides an example of the use of the word-term in some context; the definition field outputs the interpretation of the term in the translation language; and the synonyms field shows synonyms depending on the semantic similarity (in descending order).

Thus, with the help of the designed program, it is now possible to avoid possible inaccuracies in the translation of terminology and other semantic errors made during the Google translation for the logistics sublanguage, particularly word combinations that have a special definition and language meaning only in the particular context of this sublanguage.

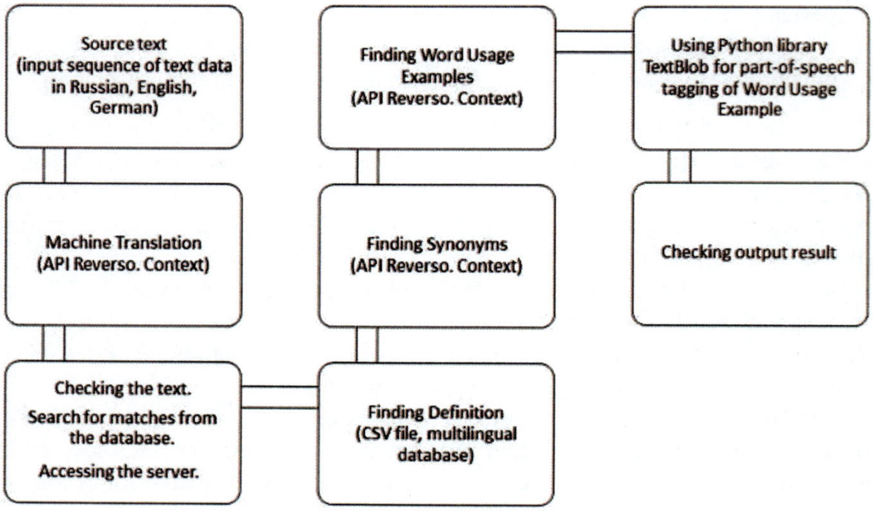

Fig. 2 Schematic algorithm of the operation principle of the logistics translator program. Designed by the authors

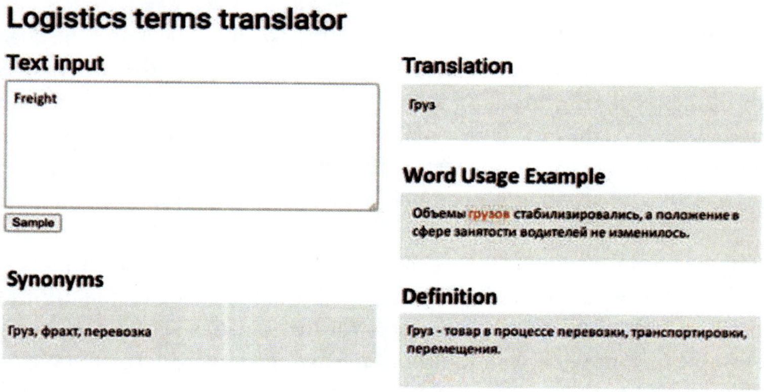

Fig. 3 Proposed visual design of logistics translator program. Designed by the authors

3 Results

Summarizing the entire study both in the field of linguistics (preparation of the linguistic database) and in the field of software development (design of the logistics translator program), the achieved intermediate results can be noted:

- the features and basic principles of the categorization of the sublanguage linguistic units are defined;

- a digital linguistic database related to the logistics sublanguage has been created. Formal methods of its programmatic functioning are described;
- a specialized AT program has been developed that can translate the terminology of the logistics sublanguage. The advantage of the program was the quality of translation (compared to the Google MT), the availability of wide functionality, professional tools, and the capability to work in the "single window" mode;
- the practical importance of logistics translator has no doubt; the first studies demonstrated the significant result. Proposals to the problem of interlanguage computer translation can be realized in upcoming developments on applied and experimental linguistics, and be used as a practical guide for the development of future AT programs for sublanguages.

4 Discussion

AT systems, in particular, CAT tools (English: computer-assisted translation tools) are convenient software solutions to engage with texts and linguistic information by making better translation outcomes. Practical benefits of such systems are practically assured due to their wide applicability potential.

Future versions of the systems will minimize the human influence on the processes of text corrections. It can predict that a number of certain proposals for work with texts will be provided by the programs [15]. AT programs will also give some necessary informative data for distance teaching of foreign languages [17].

AI by analyzing the linguistic segments will independently select the corresponding auxiliary tool for performing particular translation tasks, as well as appoint translators to any project depending on their portfolio.

5 Conclusion

In the example of this study, it becomes evident that in order to obtain the most accurate outcomes, it is necessary to gain a sample database and choose or create some digital methods for processing [4]. Bilingual parallel corpora of texts is necessary for the formation of a digital glossary [18]. It is assumed that the developed software will become a good practical tool for automatic translation of the logistics sublanguage and will be further developed in future research projects on applied and experimental linguistics.

The proposed logistics translator program is a concept example of expert professional translation tools for solving professional applied problems and demonstrates the main dissimilarity between MT and human translation of sublanguages [19]. Main objective of the applied study performed with the developed program is to propose other methods of computer translation, particularly for sublanguages. For future research, it is planned to enlarge the database (the number of sublanguage units), to add the function of translation memory (search for previously stored text fragments), and to provide the possibility of integration of the mentioned functionality into other programs.

References

1. Banerjee S, Lavie A (2005) METEOR: an automatic metric for MT evaluation with improved correlation with human judgments. In: Proceeding of ACL workshop on intrinsic and extrinsic evaluation measures for machine translation and/or summarization. Ann Arbor, USA, pp 65–72
2. Bing X, Hongmei G, Xiaoli G (2007) Computer-aided translation tools in the 21st century. Shandong Foreign Lang Teaching J 4:79–86
3. Bird S, Klein E, Loper E (2009) Natural language processing with python. o'reilly media., Sebastopol, CA
4. Bolshakova EI, Voroncov KV, Efremova NE, Klyshinskiy ES, Lukashevich NV, Sapin AS (2017) Avtomaticheskaya obrabotka tekstov na estestvennom yazyke i analiz dannykh: ucheb. posobie [Automatic processing of texts in natural language and data analysis: textbook] Moscow.: NIU VSHE, 2017. 269p. ISBN 978-5-9909752-1-7 (in Russian)
5. Borisova IA (2014) Kopytu postredaktirovaniya na materiale anglo-russkogo perevoda s pomoshchyu avtomaticheskikh sistem Google i Promt [To the experience of post-editing on the material of the English-Russian translation using the Google and Promt automatic systems]. MSLU Bulletin. Issue 13(699):53–59 (in Russian)
6. Doyon J, Doran C, Means CD, Parr D (2008) Automated machine translation improvement through postediting techniques: analyst and translator experiments. Proceedings of AMTA, pp 346–353
7. Hutchins WJ (1986) Machine translation: past, present. Future. Ellis Horwood; Halsted Press, Chichester, New York
8. Ivleva MA, Melekhina EA (2018) Cloud platform SmartCAT in teaching future translators. In: Linguistic and cultural studies: traditions and innovations: proceeding of the 27 international conference on linguistic and cultural studies (LKTI 2017), Tomsk, 11–13 Oct. 2017. Springer, 2018. pp 155–160. (Advances in Intelligent Systems and Computing; vol. 677). https://doi.org/10.1007/978-3-319-67843-6_19
9. Kittredge R (2003) Sublanguages and controlled language. The Oxford Handbook of Computational Linguistics. Oxford, pp 430–447
10. Melby Alan K (1983) Computer-assisted translation systems: the standard design and a multi-level design. In: Proceedings of the first conference on applied natural language processing (ANLC '83). Association for Computational Linguistics, USA, pp 174–177. https://doi.org/10.3115/974194.974228
11. Mikolov T, Chen K, Corrado G, Dean J (2013a) Efficient estimation of word representations in vector space. In: Proceedings of workshop at ICLR
12. Mueller J (2022) Term formation in German logistics terminology (February 18, 2022). Available at SSRN: https://ssrn.com/abstract=4038329
13. Panasenkov NA (2019) Prakticheskie rekomendatsii po obucheniyu lingvistov-perevodchikov rabote v sistemakh avtomatizirovannogo perevoda tipa SDL TRADOS i SMARTCAT. Yazyk v sfere professionalnoj kommunikatsii: sbornik materialov mezhdunarodnoy nauchno-prakticheskoy konferentsii prepodavatelej, aspirantov i studentov [Practical recommendations for teaching linguists-translators to work with automated translation systems such as SDL TRADOS and SMARTCAT. Language in the field of professional communication: collection of materials of the international scientific-practical conference of teachers, graduate students and students]. Ekaterinburg, JSC "Izdatelskij Dom 'Azhur'",, pp 494–499 (in Russian)

14. Papineni K, Rouskos S, Ward T, Zhu WJ (2002) BLEU: a method for automatic evaluation of machine translation . In: 40th annual meeting of the assoc for computational linguistics. Philadelphia, pp 311–318
15. Peng H (2018) The impact of machine translation and computer-aided translation on translators. IOP Conf Ser Mater Sci Eng 322:052024 (2018) https://doi.org/10.1088/1757-899X/322/5/052024
16. Potapov VV (2022) Lingvokognitivnyy podkhod k sozdaniyu avtomatizirovannoy sistemy perevoda na osnove specializirovannykh parallelnykh terminologicheskikh baz dannykh. (Obzor) [Linguistic and cognitive approach to the creation of an automated translation system based on specialized parallel terminological databases. (Review)]. Social and humanitarian sciences. Domestic and foreign literature. Series 6: Linguistics. Abstract journal 2, 2022, pp 35–40 (in Russian)
17. Potapova RK (2021) Novye informatsionnye tekhnologii i lingvistika [New information technologies and linguistics]. 7th issue. URSS, Moscow 600p (in Russian)
18. Potapova RK, Potapov VV (2019) Some elaboration methods for written and spoken multilingual databases. Vestnik Moskovskogo Universiteta. Seriya 9. Philology 3:71–91
19. Potapova RK Potapov VV, Kuzmin OI (2022) Logistics translator. concept vision on future interlanguage computer assisted translation. In: Prasanna SRM, Karpov A, Samudravijaya K, Agrawal SS (eds) Speech and Computer. SPECOM 2022. Lecture Notes in Computer Science, vol 13721. Springer, Cham. https://doi.org/10.1007/978-3-031-20980-2_49
20. Slocum J (1985) Machine translation. Comput and Human 19:109–116. https://doi.org/10.1007/BF02259632
21. Ulitkin IA (2016) Avtomaticheskaya otsenka kachestva mashinnogo perevoda nauchno-tekhnicheskogo teksta [Automatic evaluation of the quality of machine translation of a scientific and technical text] . Moscow Region State University Bulletin. Series Linguistics, vol 4, pp 174–182. https://doi.org/10.18384/2310-712H-2016-4-174-182 (in Russian)
22. Wu Y, Schuster M, Chen Z, Le QV, Norouzi M, Macherey W, Krikun M, Cao Y et al (2016) Google's neural machine translation system: bridging the gap between human and machine translation. Comput Sci https://doi.org/10.48550/arXiv.1609.08144

Chapter 23
Recent Advances in the Index Calculus Method for Solving the ECDLP

Aayush Jindal, Aman Jatain, and Shalini Bhaskar Bajaj

1 Introduction

ECC is a sophisticated and efficient cryptographic approach that uses elliptic curves' mathematical features to offer secure communication, data encryption, and digital signatures. ECC has become an essential component of modern cryptography and is widely utilized in varied applications to assure safe and private communication because of its robust security, computational efficiency, and small key sizes. In finite fields, the index calculus approach is a prominent solution for solving the discrete logarithm problem (DLP). Elliptic curves (EC) have become widely used in cryptography in the recent years, and ECDLP has emerged as a significant study subject in this discipline [1, 2]. The approach for solving ECDLP makes it a strong weapon for attacking elliptic curve-based cryptographic algorithms. The security of these schemes, however, is determined by the complexity and difficulty of solving ECDLP using the index approach. As a result, enhancing the efficiency of this method is a critical study area in the field of cryptography [3]. Pollard [4] presented this technique in 1978 as a baby-step giant-step variation. The index calculus approach works by using a precomputed table of logarithms to solve a given DLP. The table is built by taking a group of tiny prime numbers and multiplying their powers by the modulus of DLP. The baby-step giant-step method [5] is then used to compute the logarithms of these powers, which takes time proportional to the square root of the modulus. After computing the logarithms of these powers, they may be utilized to solve any DLP in a finite field or elliptic curve. The procedure works best when the modulus has a tiny factor base, which means it can be factored into a product of small primes. The approach can swiftly compute the logarithms of the prime powers in this example

A. Jindal (✉) · A. Jatain · S. B. Bajaj
Amity University, Haryana, Gurugram, India
e-mail: aayushjindal05@gmail.com

A. Yadav et al. (eds.), *Proceedings of International Conference on Paradigms of Communication, Computing and Data Analytics*, Algorithms for Intelligent Systems,
https://doi.org/10.1007/978-981-99-4626-6_23

using the precomputed database. If the modulus has a large factor base, the procedure becomes less efficient and may take too much memory and calculation time to be useful. Over time, the approach has undergone several changes and improvements. Lenstra's number field sieve algorithm, developed in 1987, is more efficient and factors polynomials across a number field. The widely used Diffie–Hellman key exchange protocol was broken by this technique. Satoh and Araki [6] introduced a variation of the technique for elliptic curves in 1998, which cracked various cryptographic systems.

The primary objective of this study is to review progress on the index calculus method for solving ECDLP. Specifically, we discuss the recent developments in the algorithm and their complexity, as well as new techniques and methods introduced to improve efficiency. Our paper highlights the current state of the art in this area and identifies open research problems that need further investigation. This paper is structured as follows. Section II reviews the recent progress in various proposed index calculus methods, including new techniques and methods to improve efficiency. Finally, section III summarizes the contributions and makes suggestions for further research.

2 Index Calculus Method for Solving ECDLP

Coppersmith [7] proposed index calculus to solve DLP on finite fields. An attacker can break many cryptographic systems by solving DLP, a fundamental cryptography problem. The paper describes the index calculus algorithm's complexity and security. A table of precomputed values efficiently computes discrete logarithms for any finite field element. The "index calculus" method is used to find relations between random element's logarithms and compute other element's logarithms during precomputation. Numerical experiments prove a new logarithm computation method which reduces the number of operations. The algorithm involves selecting prime numbers, generating random elements in the finite field, and precomputation using index calculus. After precomputation, the algorithm uses classical methods to compute the discrete logarithm for a finite field element. The algorithm's complexity analysis shows a sub-exponential running time in finite field size. The proposed algorithm outperforms classical algorithms with exponential running time in the same parameter. The authors also note that the algorithm's running time depends on the modulus' prime factors. Experimental results comparing index calculus to other classical DLP solvers. The index calculus algorithm outperforms classical algorithms for sufficiently large finite fields. The authors note that the algorithm's running time depends on the implementation and modulus prime factor size. The algorithm's computational time depends on implementation and modulus prime factor size.

Biasse [8] explores the difficulty of Semaev's naive index calculus [9] approach to solving the ECDLP. Some curves were exponentially complicated, whereas others were polynomially complex. The naïve index calculus method of Semaev for tackling the complexity bounds of ECDLP was explored, and the approach [10] solves

ECDLP conventionally. To efficiently compute discrete logarithms for every point on the elliptic curve, a set of precomputed values is used. By establishing correlations between logarithms, the "index calculus" approach is utilized to precompute the logarithms of random points on the curve. The study looks at the difficulty of Semaev's naive index calculus approach and shows that its running time is proportional to the group size of the curve. Further, Zhao et al. [11] propose an improved index calculus algorithm. The research also introduces a novel summation polynomial evaluation approach for ECDLP index calculus for evaluating summing polynomials that minimizes the operations required to compute them. The authors also presented a new approach for selecting random points on the elliptic curve that requires fewer points to obtain the same level of security. The approach also optimizes computation efficiency. The approach is resistant to known assaults and scales sub-exponentially in the size of the elliptic curve group. The authors also point out that quantum computers could compromise the algorithm's security and that more research is needed to thoroughly analyze the algorithm's security implications in quantum computing. The authors show that the novel algorithm solves the ECDLP quicker on some elliptic curves and is competitive on others. Further addressing the limitations, Kirchner and Kusner [12] propose a quantum annealing algorithm, detailing its complexity and security analysis. The quantum annealing version of the algorithm uses a quantum computer to solve optimization problems for the precomputation step. After precomputation, the algorithm uses classical methods to compute the discrete logarithm for an elliptic curve point. Complexity analysis shows that the algorithm runs sub-exponentially in the number of bits used to represent the elliptic curve group order. This beats classical algorithms, which run exponentially in the same parameter.

Biasse [13] evaluates index calculus methods for subfield curve ECDLP solutions. A new method uses subfield curves' special structure to compute faster. A "big-step, small-step" algorithm and a "degree reduction" algorithm for index calculus on subfield curves is presented in the study. These algorithms can be optimized by selecting suitable basis points on the curve and reducing the size of index calculus method matrices. Further, Enge and Schertz [14] use the index calculus algorithm to study the computational complexity of the ECDLP for different types of elliptic curves and present several algorithm variations. Short Weierstrass, Edwards, twisted Edwards, and Montgomery curve algorithms are included. The paper also discusses optimization methods to boost algorithm efficiency. Elliptic curves of various types are analyzed for index calculus algorithm computational complexity. The authors analyze algorithm security by considering curve order smoothness and parameter size. The index calculus algorithm is competitive with other algorithms for certain curves, but curve parameters and optimization techniques can significantly affect its performance. Further, Ivanov [15] accelerated the index calculus algorithm for ECDLP over prime fields. The paper discusses the algorithm's complexity and security. Index calculus acceleration uses dynamic collision search. Maintaining a list of collision pairs accelerates logarithm search. Sparse matrix representation and precomputing small powers of the base point can optimize the algorithm. The authors demonstrated that the accelerated algorithm outperforms the standard algorithm. However, curve parameter size and memory limit its performance. The paper discusses the algorithm's limitations and future research.

Scott [16] found that DLP on elliptic curves over non-prime fields is harder than over prime fields and proposes a new method that uses the special structure of such fields to speed up computation. This study examines the computational complexity of the DLP on non-prime elliptic curves. The paper reviews previous mathematical methods. These include algebraic geometry, number theory, and modifications of elliptic curve algorithms over prime fields. The authors compare DLP computational complexity on elliptic curves over non-prime fields to prime fields. Due to the smaller curve group, elliptic curve cryptography over non-prime fields is generally less secure than over prime fields. They also discuss using curves with large embedding degrees to improve elliptic curve cryptography over non-prime fields. In addition, Zhao [17] proposed a fast index calculus technique for ECDLP over binary fields with low Hamming weight. The classical index calculus approach is extended to compute discrete logarithms on finite fields. Linear algebra and number theory are used to solve the ECDLP over binary fields with a low Hamming weight. The authors improve the technique by employing a new polynomial selection approach and a new base field discrete logarithm method. The authors demonstrated that the algorithm's complexity is substantially lower than earlier approaches for computing the ECDLP over binary fields with low Hamming weight. The authors also point out that the algorithm's security depends on the elliptic curve and binary field size and that it may not be safe for all curve and field sizes. Further extending the work in non-prime fields, Liu et al. [18] proposed a variant of index calculus over extension fields to solve ECDLP. The algorithms leverage the unique structure of such fields to calculate quicker than conventional approaches. The classical index calculus approach is extended to compute discrete logarithms on finite fields. The authors discuss the polynomial selection and number field sieve optimizations. Elliptic curves over extension fields are optimized for the algorithms.

Takagi [19] introduced index calculus as a method for solving the ECDLP on abelian varieties of small dimensions. Proposed approach solved a system of polynomial equations in many variables by generating a Gröbner basis of a curve-associated polynomial ideal. The algorithm's stages are as follows: First, choose a prime p with a low embedding degree k for the elliptic curve E over Fp. Determining E's endomorphism ring, which is isomorphic to the maximum order of a quaternion algebra and the Gröb-ner basis for the ideal of polynomial interactions between E endomorphisms, comes next.

Finally, solve multivariable polynomial problems using linear algebra and the theory of Chinese remainder. The complexity of the method in F_p is sub-exponential in p's bit length: $O(\exp(2(2\,k/3)) * \log(p)(3/2))$. As a result, ECDLP security becomes finite field security. Because many commonly used elliptic curves have minimal embedding degrees, the approach is efficient. SD Galbraith et al. [20] use subfield curve index calculus to solve the ECDLP. The ECDLP may be solved in subfield curves over prime fields with up to 264 dimensions using the technique. The work also includes a heuristic investigation of the security of subfield curves against index

Table 1 Index calculus method for solving ECDLP

Paper	Key contributions	Strengths	Limitations
Coppersmith [7]	Introduction of index calculus method to solving DLP	General approach that applies to various DLP settings	High complexity for large elliptic curves
Biasse [8]	Analysis of complexity of Semaev's naive algorithm	Provides theoretical upper bound for complexity of ECDLP	Naive algorithm is not practical for large curves
Zhao [11]	Algorithm based on summation polynomials	Efficient use of memory	Limited to certain types of curves, choice of parameters
Kirchner and Kusner [12]	Quantum annealing approach	significant speedup using quantum computing	Currently only theoretical,
Biasse [13]	Analysis of subfield curves, improvements to polynomial selection	Lower complexity for certain curves, better understanding of subfield curves	Limited to certain types of curves
Enge and Schertz [21]	Analysis of complexity for different types of elliptic curves	Comprehensive analysis of different curve types	No new algorithms or improvements proposed

calculus assaults. Furthermore, Table 1 compares the strengths and limitations of the offered strategies for solving ECDLP in the preceding section.

2.1 Cryptographic Protocols for Abelian Varieties and Supersingular Elliptic Curves Using Index Calculus

Kim and Barbulescu [22] offer two techniques for abelian varieties of small dimensions in characteristic p and characteristic 2. The index calculus approach for abelian varieties entails determining a set of relationships between points on a curve and solving linear equations to determine the discrete logarithm of a target point. The authors recommend a polynomial basis and efficient linear algebra techniques to optimize this procedure. They also present a new approach for calculating the rank of an abelian variety, a critical parameter that influences the algorithm's complexity. The authors compare Takagi's method to the efficiency of their algorithm for numerous elliptic curves. Further, using supersingular elliptic curve isogenies Jao and De Feo [23] offered a novel approach to creating quantum-resistant cryptosystems. The authors create unique cryptographic primitives by using supersingular elliptic curves and related isogenies. The technique leverages supersingular elliptic curves and related isogenies to create a quantum-resistant cryptosystem. In response to the difficulty in locating the endomorphism ring of a supersingular elliptic curve, the authors propose the supersingular isogeny Diffie–Hellman (SIDH) protocol.

Because computing the endomorphism ring of a supersingular elliptic curve is difficult, the SIDH protocol is safe. The authors show that both conventional and quantum computers struggle with this challenge, making the SIDH protocol a plausible candidate for post-quantum cryptography. In the worst-case scenario, calculating the endomorphism ring requires $O(p2)$ operations, where p is the underlying finite field. The authors offer many optimizations that depending on the settings, decrease protocol complexity to $O(p)$ or $O(p \log p)$ operations.

A new index calculus algorithm for ECDLP over binary two is presented by Nakamula [24]. In terms of computation complexity, the proposed ECDLP algorithm is better than the existing ones. Polynomial selection and index calculus underpin the algorithm. A polynomial with all the elliptic curve's x-coordinates as roots is the basic idea. A curve point's discrete logarithm is calculated using this polynomial. LLL is used to select a polynomial with a small degree and good properties [25]. ECDLP over finite fields of characteristic two is improved by the proposed algorithm's sub-exponential complexity. Further, Diem et al. [17] provided an index calculus technique for ECDLP on abelian varieties of modest dimensions. The algorithm is supported by Granger, and Zumbrägel's framework [26]. The overall framework is made up of sieving, linear algebra, and descent. The curve's sieving points are utilized to identify relationships between the unknown discrete logarithms and them. Sieving creates a set of linear equations that are solved by the linear algebra step. Using the linear algebra step's relations, the descent step computes the discrete logarithm. The authors improve on the prior technique [26] by employing a more efficient linear algebra method as well as a sieve-step method for calculating Gröbner bases. The writers also look into complexity and security. The computational hardness of the algorithm is determined by the number of iterations of operations performed at each algorithm step. The sieving phase has a complexity of $O(L^2)$, where L is the bit length of the prime p. The linear algebra and descent steps have a complexity of $O(L^3)$ and $O(L^2)$, respectively. The complexity of the algorithm is $O(L^3)$. Lastly, Table 2 draws a comparative analysis based upon the key contributions and limitations of proposed techniques for index calculus for abelian varieties and supersingular elliptic curves in the above section.

2.2 *Advancements in Index Calculus Methods for Elliptic Curves with Various Characteristics and Degrees*

A function field sieve algorithm variant by Enge [27] suggests a time-memory trade-off for index calculus in genus three curves. The authors suggest precomputing an extensive database with a small amount of memory and then searching it to calculate discrete logarithms with a larger amount of memory. Detailed time-memory trade-off analysis showed that the proposed approach reduces memory complexity by a factor of four compared to the function field sieve algorithm. A heuristic analysis shows that for curve sizes up to 200 bits, the algorithm is faster than the traditional

Table 2 Index calculus for abelian varieties and supersingular elliptic curves

Author(s)	Method used	Key contributions	Limitations
Kim and Barbulescu [22]	Index calculus	Proposed method for abelian varieties and elliptic curves of small dimension	The method is limited to abelian varieties and elliptic curves of small dimension
Jao and De Feo [23]	Isogeny-based cryptography	Approach to constructing post-quantum secure cryptosystems using supersingular elliptic curves and isogenies	The method requires the use of supersingular elliptic curves and isogenies, which are not widely used in practice
Nakamula [24]	Index calculus	Algorithm over finite fields of characteristic two	Limited to solving ECDLP over finite fields of characteristic two

algorithm. They compare their results with the block Wiedemann algorithm [28] and the considerable prime variation of the function field sieve [29]. For curve widths up to 150 bits, the suggested approach beats current techniques. Kim and Jiao [9] proposed an index calculus technique for ECDLP over finite fields with modest features. The complexity of the suggested method is $L(1/4 + o(1))$, which is more than $L(1/3 + o(1))$. Using index calculus and Weil descent, the approach reduces the ECDLP to a low-degree extension field issue. The breakthrough is that an effective linear algebra approach is used to solve the lower-degree extension field issue.

The extension field's tiny degree simplifies Joux, Antoine, and Vanessa Vitse's index calculus algorithm [30]. The study applies the novel approach to static Diffie–Hellman. In the static Diffie–Hellman problem, a version of ECDLP, one party generates a public key for the other to encrypt communication. This paper's innovative technique solves the static Diffie–Hellman problem over small-degree extension fields. The novel index calculus algorithm breaks the static Diffie–Hellman problem by computing discrete logarithms in a curve subgroup. The method reduces small-degree extension field ECDLP complexity from exponential to quasi-polynomial. The authors evaluate the algorithm's complexity. Experimentally, the new method saves time and memory. Ariffin et al. [31] described several index calculus algorithmic improvements to simplify it. These include a new polynomial degree calculator, an improved linear algebra algorithm, and a new Frobenius endomorphism action calculator. The authors also propose a polynomial and lattice-based index calculus method to improve efficiency. Their index calculus algorithm is much faster than previous ones. The authors compare their ECDLP solution to Pollard rho and baby–step giant-step algorithms. Their method performs better for certain parameter values. Enge [32] proposes using index calculus symmetries to solve ECDLP. The index calculus algorithm, which finds many relations between discrete logarithms of elliptic curve points, is the basis for the proposed algorithm. The proposed algorithm uses elliptic curve symmetries to compute relations with fewer points.

El Antaki [33] suggests solving ECDLP over small-degree extension fields using index calculus. The proposed algorithm reduces computation complexity compared to existing algorithms. The algorithm modifies finite field index calculus. A polynomial with all the abscissa of the points on the curve defined over the extension field is the basic idea. The discrete logarithm of a curve point is calculated by the polynomial. The algorithm computes the polynomial's algebraic degree and factorization. Duquesne and Gaudry [34] created an index calculus technique for solving ECDLP using auxiliary polynomials. The approach computes the basis of the polynomial ideal created by the discrete logarithms of elliptic curve points. The optimized F4 method computes Gröbner bases. After computing a basis, monomials that yield small-degree polynomials are chosen to generate auxiliary polynomials. These auxiliary polynomials compute the discrete logarithm of an elliptic curve point. The approach computes the basis of the polynomial ideal created by the discrete logarithms of elliptic curve points. The optimized F4 method computes Gröbner bases. After computing a basis, monomials that generate small-degree polynomials are chosen to generate auxiliary polynomials. These auxiliary polynomials compute the discrete logarithm of an elliptic curve point.

Noorani et al. [35] introduced a polynomial algebra-based algorithm over binary fields. The index calculus-based algorithm uses binary fields' unique properties to reduce complexity. The algorithm computes the discrete logarithm of an elliptic curve point using polynomial selection, linear algebra, and sieving. The authors suggest using multiple polynomials and a new sieving method to improve the algorithm. Time and memory requirements determine the algorithm's complexity. Joux [1] studies Weil descent on elliptic curves to reduce ECDLP to DLP in an extension field. The algorithm uses index calculus and Weil descent. Weil descent reduces the ECDLP to the DLP in an extension field, and the index calculus algorithm solves the DLP. The algorithm exploits the unique structure of descent maps to simplify the DLP.

Fan et al. [21] demonstrate that the ECDLP on an elliptic curve E may be reduced to the ECDLP on a hyperelliptic curve of degree 2 g + 1, where g denotes the curve's genus. By including branch points and ramification points to the curve, they create a hyperelliptic cover for E and demonstrate how solving the ECDLP on E may be reduced to solving the ECDLP on the hyperelliptic curve. The authors demonstrate that a hyperelliptic curve has the same level of complexity as doing it on a generic hyperelliptic curve. This is due to the discrete logarithm structure being preserved by the hyperelliptic cover and the fact that any point on the hyperelliptic curve beneath the cover's preimage is a sum of points on the elliptic curve. The genus-2 curves used in the ECDH-2 and ECIES-KEM-2 schemes are examples of hyperelliptic curves the authors utilize in elliptic curve cryptosystems. Robin et al. [36] investigate how to generate safe elliptic curves over sextic extensions of prime fields of around 64 bits in order to attain 128-bit security. The document includes all known big but specialized ECDLP attacks that make use of the group structure.

Furthermore, Table 3 draws a comparative analysis based upon the computational complexity and limitations on the advancements in index calculus methods for EC with various characteristics and degrees.

Table 3 Advancements in index calculus methods for elliptic curves with various characteristics and degrees

Research paper	Computational complexity	Limitations
Enge [27]	$L(1/4 + o(1))$ in small characteristic fields	Limited to genus 3
Bos [30]	$L(q$^$(1/4))$	Limited to small-degree extension fields
Duquesne and gaudry [34]	$L(\sqrt{q})$	Limited to prime fields

3 Conclusion and Future Work

In this paper, we reviewed recent progress on the index calculus method ECDLP. We also discussed how new technologies like quantum computing affect these methods. For decades, index calculus and its variants have broken cryptographic systems. They helped develop modern cryptography, but more research is needed. New technologies like quantum computing present cryptography challenges and opportunities. Problems index calculus has greatly improved our understanding of ECDLP, but there are still limitations and open issues. These algorithms may be vulnerable to quantum computers. Due to computational requirements, implementing and deploying these algorithms on practical systems is difficult. This research could address the limitations and open questions discussed in the previous section. Develop quantum-resistant algorithms with lower computational requirements. These algorithms should also be tested on practical systems to reduce memory and storage requirements. Finally, isogeny-based or lattice-based elliptic curve cryptography should be investigated to improve security.

References

1. Joux A (2006) Constructive and destructive facets of Weil descent on elliptic curves. J Cryptology 19:61–86
2. Joux A (2003) Algorithmic cryptanalysis of ciphers. CRC Press. ISBN 978-1-58488-462-0. Blake IF, Seroussi G, Smart NP (eds) (2000) Advances in elliptic curve cryptography. Cambridge University Press. ISBN 0-521-80457-7
3. Verheul ER (1999) Evidence that XTR is more secure than supersingular elliptic curve cryptosystems. In Selected Areas in Cryptography. Springer Berlin Heidelberg, pp 195–210
4. Pollard JM (1978) Monte carlo methods for index computations (mod p). Math Comput 32(143):918–924
5. Cohen H (1993) A course in computational algebraic number theory, volume 138 of Graduate Texts in Mathematics. Springer-Verlag
6. Silverman JH, Suzuki J (1998) Elliptic curve discrete logarithms and the index calculus. Advances in Cryptology—ASIACRYPT'98: international conference on the theory and application of cryptology and information security Beijing, China, October 18–22, 1998 Proceedings. Springer Berlin Heidelberg

7. Coppersmith D (1994) The discrete logarithm problem. In: Proceedings of the annual international cryptology conference on advances in cryptology (CRYPTO '94), Santa Barbara, CA, USA, pp 1–9
8. Biasse JF (2015) Complexity bounds on Semaev's naive index calculus method for ECDLP. J Math Cryptol 9(1):1–19
9. Jiao Y, Kim M (2020) A new index calculus algorithm with complexity L(1/4 + o(1)) in small characteristic. Cryptology ePrint Archive, Report 2020/1109
10. Shanks D (1971) Class number, a theory of factorization, and genera. Proc Symp Math Soc 1971(20):41–440. https://doi.org/10.1090/pspum/020/0316385
11. Zhao Y, Pan L, Zhou J (2019) A new index calculus algorithm for the elliptic curve discrete logarithm problem and summation polynomial evaluation. IEEE Trans Inf Theory 65(4):2316–2324
12. Kirchner P, Kusner J (2019) Index calculus method for solving elliptic curve discrete logarithm problem using quantum annealing. J Cryptogr Eng 9(4):285–295
13. Biasse JF (2012) On index calculus algorithms for subfield curves. Advances in cryptology - ASIACRYPT 2012, Berlin, Heidelberg, pp 371–388
14. Enge, Schertz R (2005) On the computational complexity of ECDLP for elliptic curves in various forms using index calculus. Cryptography and Coding, Berlin, Heidelberg, pp 185–194
15. Ivanov F (2012) Acceleration of index calculus for solving ECDLP over prime fields and its limitation. Information security and cryptology—ICISC 2011, Berlin, Heidelberg, pp 272–282
16. Scott M (2007) On the hardness of the discrete logarithm problem on elliptic curves over nonprime fields. J Cryptol 20(4):603–619
17. Zhao R (2019) A fast index calculus algorithm for solving ECDLP over binary fields with low hamming weight. IET Inf Secur 13(3):197–204
18. Liu Y, Wu H, Chen X (2017) Efficient index calculus algorithms for elliptic curves over extension fields. Secur Commun Netw 2017, Article ID 1879132
19. Takagi T (2009) Index calculus for abelian varieties of small dimension and the elliptic curve discrete logarithm problem. J Cryptol 22(4):545–572
20. Galbraith SD, Granger R, Merz SP, Petit C (2021). On index calculus algorithms for subfield curves. In: Selected areas in cryptography: 27th international conference, Halifax, NS, Canada (Virtual Event), October 21–23, 2020, Revised Selected Papers 27. Springer International Publishing, pp 115–138
21. Fan J, Fan X, Song N, Wang L (2022) Hyperelliptic covers of different degree for elliptic curves. Math Probl Eng
22. Kim D, Barbulescu R (2016) Index calculus for abelian varieties of small dimension and the elliptic curve discrete logarithm problem. J Cryptol 29(1):47–76
23. Jao D, De Feo L (2011) Towards quantum-resistant cryptosystems from supersingular elliptic curve isogenies. In: Proceedings of the 2011 52nd annual IEEE symposium on foundations of computer science, pp 163–172
24. Nakamula A (2017) A new index calculus algorithm for ECDLP over finite fields of characteristic two. Des Codes Crypt 85(1):1–13
25. Nguyen PQ, Vallée B (2010) The LLL algorithm. Springer Berlin Heidelberg, Berlin, Heidelberg
26. Granger R, Kleinjung T, Zumbrägel J (2014) Breaking '128-bit Secure' Supersingular Binary Curves. In: Garay JA, Gennaro R (eds) Advances in cryptology – CRYPTO 2014. CRYPTO 2014. Lecture Notes in Computer Science, vol 8617. Springer, Berlin, Heidelberg https://doi.org/10.1007/978-3-662-44381-1_8
27. Enge (2000) Time-memory trade-offs for index calculus in genus 3. J Symb Comput 30(6):729–746
28. Coppersmith D (1994) Solving homogeneous linear equations over GF (2) via block Wiedemann algorithm. Math Comput 62(205):333–350
29. Adleman LM (1994) The function field sieve. Algorithmic number theory: first international symposium, ANTS-I Ithaca, NY, USA, May 6–9, 1994 Proceedings 1. Springer Berlin Heidelberg

30. Joux A, Vitse V (2010) Elliptic curve discrete logarithm problem over small degree extension fields. Application to the static Diffie-Hellman problem on $E(F_{q5})$. Cryptology ePrint Archive
31. Ariffin M, Hassan MF, Noorani MSM (2017) Improving the complexity of index calculus algorithms in elliptic curves over binary fields. In: Proceedings of the 6th international conference on computing and informatics, pp 518–523
32. Enge (2009) Using symmetries in the index calculus for elliptic curves discrete logarithm. J Cryptology 22(3):379–398
33. El Antaki L (2008) Index calculus on elliptic curves over small degree extension fields. Int J Math Comput Sci 3:13–20
34. Duquesne S, Gaudry P (2009) Index calculus attack on ECDLP with auxiliary polynomials. Cryptology ePrint Archive, Report 2009/605
35. Noorani MSM, Hassan MF, Ariffin M (2016) New insights into the index calculus method for the ECDLP over binary fields. In: Proceedings of the 10th international conference on computer engineering and applications, pp 127–132
36. Salen R, Singh V, Soukharev V (2022) Security analysis of elliptic curves over sextic extension of small prime fields. Cryptology ePrint Archive

Chapter 24
Pedestrian Detection Using YOLOv5 and Complete-IoU Loss for Autonomous Driving Applications

E. Raja Vikram Reddy and Sushil Thale

1 Introduction

While object detection is a well-researched aspect of the discipline, pedestrian detection is a specific problem of computer vision. Among the many uses for automatic pedestrian detection, autonomous driving is one of the most important. The main purpose of using numerous sensors, including as cameras, radar, ultrasonic sensors, and others, is to improve the dependability and safety requirements of autonomous driving [1]. Due to their inexpensive cost and broad range compared to other sensors, cameras are the main sensor. The location and classification of pedestrians in each image or video are known as pedestrian detection. Applications for pedestrian detection include surveillance, tracking pedestrians, pose estimation, and many more. There has been a lot of study done on pedestrian detection, and numerous models and algorithms have been put out. Most of these had limitations in terms of detection speed or precision. As shown in Fig. 1, convolutional neural networks (CNNs), however, have enabled many pedestrian detectors to achieve significantly higher levels of accuracy and speed.

In the real world, pedestrian detections face various difficulties. One of these difficulties is occlusions. The existing standards perform particularly well while detecting pedestrians in severely obstructed areas. For instance, the "CityPersons" [2] dataset with pedestrians making up roughly 70% of the population illustrates different levels of occlusion.

Pedestrian identification began with manually created features, which were then employed in channel feature-based techniques and approaches based on deformable

E. Raja Vikram Reddy (✉) · S. Thale
Fr. Conceicao Rodrigues Institute of Technology, Sector 9A, Vashi, Navi Mumbai 400703, India
e-mail: vicky.raj43@gmail.com

S. Thale
e-mail: sushil.thale@fcrit.ac.in

A. Yadav et al. (eds.), *Proceedings of International Conference on Paradigms of Communication, Computing and Data Analytics*, Algorithms for Intelligent Systems,
https://doi.org/10.1007/978-981-99-4626-6_24

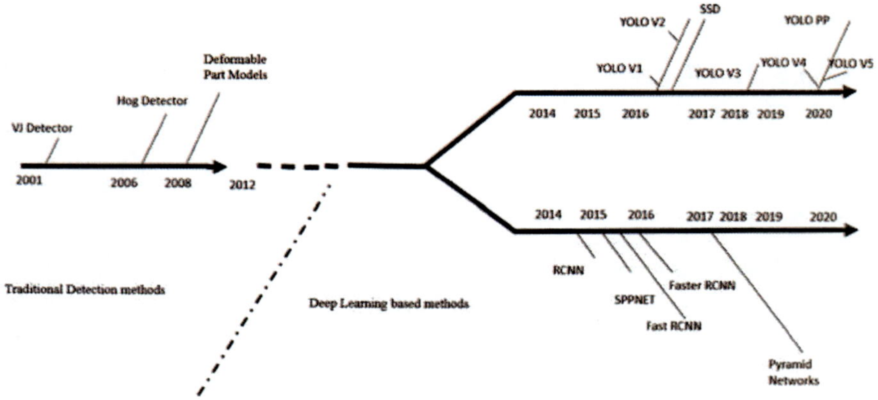

Fig. 1 20 years of object detection

parts, and ultimately deep learning systems [1]. A number of handmade feature descriptors, HOG, and Haar, have been studied; later, deep learning-based methods are successful in computer vision applications. These manually created features often extract information about color, texture, or edges. Histogram of directed gradients is one of the most popular custom characteristics for pedestrian identification (HOG). Additionally, the underlying model learning process for the majority of current handmade-based systems either uses channel features or deformable component models.

Single-stage and multiple-stage deep learning detectors are the two categories into which they fall. The accuracy of detectors with multiple stages, such as region-based convolutional neural networks (RCNNs) [1], fast RCNN [1], and faster RCNN [1], is superior. Single shot detection (SSD) [1] and RetinaNet [1] demonstrated faster object detection than the one stage YOLO [1] system. The optimal performance in terms of accuracy and speed for pedestrian detectors is still being researched.

The You Only Look Once (YOLO) sequence was initiated by Joseph Redmon. YOLOv1 had a mean average precision (MAP) of 63.4, and YOLOv2 had a MAP of 21.6. The 33 MAP attained by YOLOv3 [1] represents the average precision (AP). Bochkovskiy et al. [3]. Evolution around the YOLOv4 produced superior outcomes along 50 frames per second (FPS) and 43.5 AP when added to YOLOv3, together with bag of freebies and bag of specials.

The YOLOv5 [4] operates 140 frames per second (FPS) using Tesla P100 rapid detections, compared to the YOLOv4's 50 FPS. Additionally, while the YOLOv5 model is 27 MB in size, the YOLOv4 model and the Darknet architecture are 244 MB each. Because YOLOv5 has adopted YOLOv4's improvements by executing SPP-NET and modifying the modern methodology, it is interesting to note that both YOLOv5 and YOLOv4 have analogous accuracies. These upgraded new techniques for the data include self-adversarial training (SAT), multi-channel feature, mosaic training, and restoring path aggregation network (PANet) [5] along feature pyramid network (FPN).

Recent pedestrian detections [6–10] produced superior Miss-Rate (MR) results, which ought to be modest. Adaptive-NMS [6] achieved 56.7%, Rep-Loss [7] obtained 56.9%, ORCNN [7] achieved 55.7%, Faster RCNN [8] achieved 54.6%, TTL [7] achieved 53.6%, ORCNN [7] performed 50.47%, RetinaNet [9] achieved 49.9%, and Detr [10] achieved 10.0%. They did not increase test time, but accuracy has.

A select group of researchers focused on the heavily occluded "CityPersons" pedestrian dataset in an effort to improve detection accuracy and speed. The YOLOv3 managed to achieve the test time of 0.095 s per image, albeit with a 58% Miss-Rate. The Miss-Rate for the ALFNet [11] was 51.9% with a test time of 0.27 s/image. The DANET [12] achieved 51.6% Miss-Rate in the test period of 0.15 s per image. The CSANET [13] achieved 0.32 s/image and a 51.3% Miss-Rate. With 0.15 s per image, CSP [14] obtained a Miss-Rate of 49.9%.

Overall, the improvements in testing time, exactness, and the Miss-Rate were the main goals of the attempts to improve the performance of pedestrian detectors. But it has been noted that because of their interdependencies, when one metric advances, the other would not or might even deteriorate. The best performance in the detection of pedestrians will undoubtedly be shown as the research progresses.

This work makes an effort to show how performance indicators for pedestrian detection have generally improved. The proposed methodology is described in detail in the next section.

2 Architecture of YOLOv5

The YOLOv5 model's implementation is explained in this essay. To improve learning and make the model of YOLOv5 a more reliable detector, thousands of pedestrian images are fed into it. The YOLOv5 network model's primary building blocks are the head, neck, and backbone as explained in Fig. 2.

The YOLOv5 model is taken into consideration in this study to provide superior convolutional features, pooling, and multi-scale prediction. Each of they could be accomplished over attaching the path aggregation network (PANet), cross-stage partial network (CSPNet), and a You Only Look Once head with backbone.

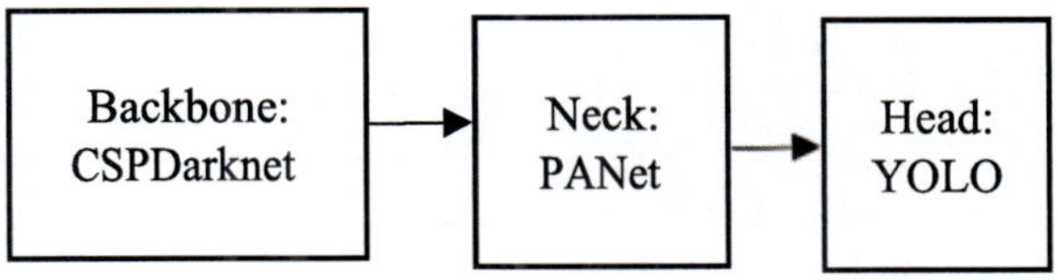

Fig. 2 YOLOv5 model

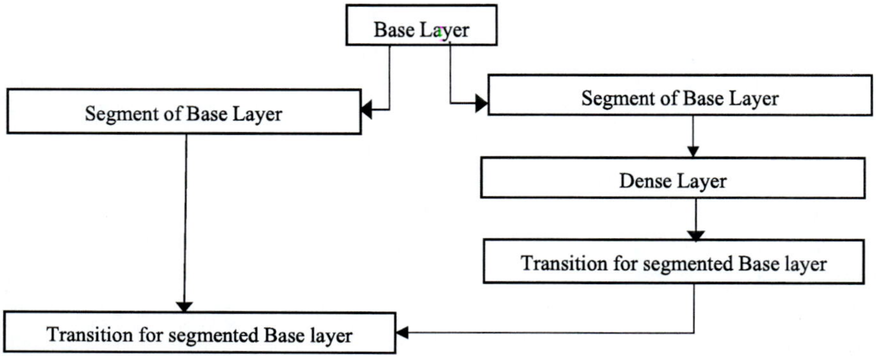

Fig. 3 CSP densenet

2.1 CSPNeT

First, CSPNet [14] was only used as the backbone in YOLOv5. The issue of recurring gradients information in the largest scale backbone is solved using CSPNet and summing the varying gradients over the map of all the features, that decreases the values and floating-point operations per second (FLOPS) considering this method. As a result, the model's size was decreased, and the efficiency and detection time were both enhanced. The CSPNet speeds up computations while also assisting the design in providing richer gradient data. The feature map is segmented to do this as explained in Fig. 3.

One base layer segment is passed through after the base layer has been divided into two portions for each base layer segment. The transition is then transmitted it from the segment base layer. To divide the gradient flow and enable it to continue to spread throughout the numerous networks, this is done. By switching the concatenation and transition phases, it is then integrated with the prior partitioned base layer segment. There is a significant correlation discrepancy in the transmitted gradient information. By following these methods, computations are significantly lowered while detection speed and accuracy are increased.

2.2 PANet

Second, when used with YOLOv5, PANet acts as its neck to improve flow of statistics. The PANet architecture uses the feature pyramid network (FPN) for building the route that enhances the transmission of lower-level characteristics. The valuable data is quickly passed to the succeeding sub-network in every feature level using the integration of feature grid, all feature levels, and adaptive feature pooling. The accuracy of the object's localization can unquestionably be improved by the PANet's inventive utilization of precise localized signals at lower levels. Thirdly, to identify diverse scaling predictions and enhance the model's capacity to handle tiny, medium, and large "sized" pedestrians, the Yolo methodology, the main part of the YOLOv5 model, outcome features in three different sizes (18 18, 36 36, and 72 72). This method can adjust to changes in scale while working with diverse pedestrian sizes taking advantage of the multiple scale detection.

2.3 Cluster Nms and Complete-IoU Loss

Regression of all the boxes that are bounding and Non-Max Suppression (NMS) are utilized for raising geometric factors, and this results in large gains in average recall (AR) and average precision (AP). In the occurrence of segmentation and object detection, the three crucial geometric parameters for measuring bounding box regression are overlapped regions, normalized central point distance, and aspects. These parameters were then added with CIoU loss to improve the ability to distinguish challenging regression cases [15].

These geometric parameters can be added to Cluster-NMS while maintaining high efficiency, resulting in considerable gains in average recall and precision. Modern methods like instance segmentation and deep object detection are trained and inferred using these techniques. It has been determined along extensive examination that these methods consistently deliver average precision, average recall, and good efficiency of Cluster-NMS.

The predicted boxes on iteration levels that benefit from GIoU and CIoU loss optimization, as indicated in the first and second rows, respectively, are explained in more detail in Fig. 4. The red and blue indicate the predicted boxes for CIoU and GIoU loss, respectively. Green and black stand in for target and anchor box, respectively. Taking into account, GIoU losses in the overlap region, which raises GIoU by enlarging the prediction box. The continuity of the overlap region, fast convergence, and ratio of width and height, as shown in the CIoU loss figure, all contribute to a better fit of the two boxes as a result of the reduction of the average center point distance.

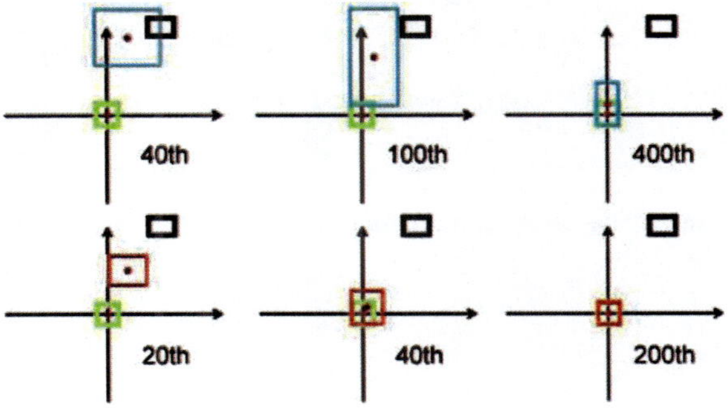

Fig. 4 Losses

3 The Proposed Method

Thousands of photographs are improved using the proposed procedure, which removes between 50 and 90% of the original images by rotating, flipping, moving on the x axis, and randomly cropping. Bounding boxes are used by the CSP Darknet and the PANNET to find these pictures in the YOLOv5 model. Observing all the previous YOLO models, this approach yields a faster detection. This approach produces a variety of detection boxes, either pedestrians are present or not, which are then provided to the sigmoid-weighted linear unit (SiLU) activation function [16]. SiLU is utilized as an activation function within reinforcement learning to approximate neural network function. For detecting any objects along with segmenting the objects, complete-IoU loss can result in consistent increase in Average Precision and Average Recall compared to IoU-based losses. Cluster-NMS, which also significantly contributes to average accuracy and average recall gains [16], ensures real-time detection. Figure 5 shows the block diagram for the suggested technique.

The most popular activation function is the Rectified Linear Unit, although as can be shown in Fig. 5, the SiLU activation function performed remarkably more effectively than the ReLU. The SiLU activation function did best and produced better results when compared to the ReLU activation function. The activation function SiLU, which detects whether or not a pedestrian is present, is created by multiplying the sigmoid function by the input. As a result, the SiLU was applied to the output of the YOLOv5 model that includes sufficient results of detected areas that may or may not include pedestrians. SiLU's activation function aids in determining whether or not a pedestrian is present. Any detections that involve the SiLU activation function's result detections would be pedestrians.

Figure 6 shows the activation function along variables for zk, the unit k's intake. ReLU's activation function is substantially identical to that of SiLU for higher values of zk, and for greater negative zk, the activation is almost nil.

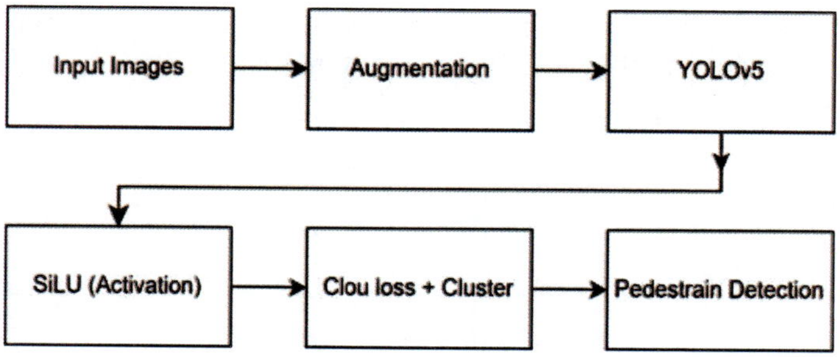

Fig. 5 Suggested approach for detections of pedestrian using YOLOv5

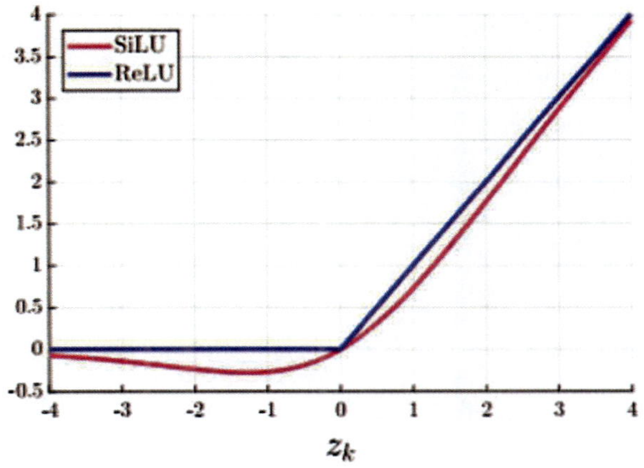

Fig. 6 Comparison of the SiLU and the ReLU

4 Experiment and Analysis

Setting up the experiments in this part is introduced and then thoroughly implemented. The recommended pedestrian detector for the CityPersons dataset will eventually be broken down and examined. The validation of the suggested method against cutting-edge techniques using the highly occluded CityPersons dataset is then described.

Utilizing the hard standards from the CityPersons datasets, we conducted tests to gauge the effectiveness of our suggested pedestrian detector. The CityPersons dataset is a recording video that includes 18 different cities and covers Germany and neighboring nations covering total of 3 countries with all the three seasons covering almost 7 persons per image, putting altogether 19,654 unique persons, as seen in Table 1.

For the datasets, we use the metric of evaluation log-average Miss-Rate over False Positive Per Image (FPPI) (also known as MR-2). There are also almost 31,000 bounding boxes with annotations, 2975 images for training, 500 for validating, and 1575 for testing. These dataset includes 5000 images with a resolution of 640 × 1280 and various amounts of occlusion.

Figure 7 shows the distribution of pedestrians for Caltech and CityPersons at various occlusion levels. On Caltech, we see that more than 60% of pedestrians

Table 1 Comparison of diversity on different training subset from Citypersons dataset

	Cityperson dataset
Country	3
City	18
Season	3
Persons/image	7.0
Unique persons	19,654

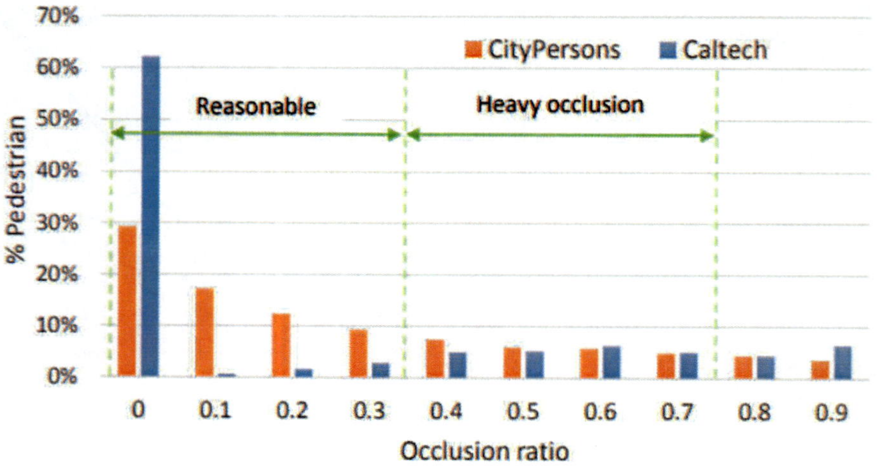

Fig. 7 Occlusion distributions from the CityPersons and Caltech datasets are compared

are clearly visible, compared to fewer than 30% on CityPersons. This suggests that we have twice as many occlusions than Caltech, making CityPersons a more intriguing ground for handling occlusions. Furthermore, Caltech is dominated by totally visible pedestrians on the acceptable subset ($\leq$ 0.35 occlusion), whereas CityPersons contains more occlusion situations.

5 Training Analysis

Pytorch was used to accomplish the suggested technique. The entire network is moved forward via SiLU activation. The dataset has also been enhanced by flipping images horizontally, rotating them, and randomly cutting them in order for better and more rounded training to accomplish. The training image size for CityPersons is 640 $\times$ 1280. The dataset has also been enhanced by horizontally flipping, rotating, and randomly cropping images in order to achieve better and more balanced training. The network was designed with a batch size of 16, GPUs in mind (GTX 2080 Ti).

The algorithm's higher test performance—0.093 s per image and YOLOv3 is 8.6 times quicker than this, due to the Yolov5 model and the SiLU activation function that detects if a pedestrian or missing in an image. Amplification is used to improve the model's preparation. Yolov5 model along with augmentation produced a Miss-Rate of 44.22%, which is 3.7% better than the previous-best CSPNet model.

After processing the city person dataset by strengthening and using the SiLU activation function to Complete-IoU (CIoU) loss, Cluster-NMS may boost Average Precision and Average Recall with no loss of implication time elapsed to better identify difficult regression scenarios. When used for training and testing, this dataset

Table 2 Illustration using different CityPersons highly occluded dataset models

Method	Test time (s/image)	MR-2 (%)
YOLOV3[12]	0.095	58
YOLOV5 [17]	0.011	46.2
Proposed method	0.093	44.22

comprises 5000 images in total. Few of these images have both multiple pedestrians and no pedestrians at all.

5.1 *Performance Comparison*

Following the application of several enhancements, such as rotation, flipping, and shifting images using the YOLOv5 methods, SiLU activation function is then used to recognize the correct pedestrian.

The suggested method exceeds YOLOv3 with a test time of 0.093 s/image and a Miss-Rate of 44.22%, while outperforming ALFNet with a test time of 7.68% MR-2 less and a test time that is 0.002 s faster. The suggested model outperforms DANET by 0.057 s and by 7.38% MR-2. The suggested model outperforms CSANet by 0.227 s and by 7.08% MR-2. The suggested model outperforms CSP by 0.237 s and 5.7% MR-2. The experimental findings for the suggested procedure are contrasted with the stated results in Table 2. The suggested model outperforms YOLOv5 by 1.98% in MR-2.

6 Conclusions

In this work, pedestrian identification is implemented using YOLOv5, which has 5.7% MR-2 less than CSP and improved when compared to YOLOv3, testing time for spotting pedestrians with highly occluded pedestrian datasets. This is accomplished by putting the detection of YOLOV5 through many processes, including Cluster-NMS, Complete-IoU (CIoU) loss, Non-Maximum Suppression (NMS), which yields large gains, and SiLU activation, which is used to assess if the detected item is a pedestrian or not. The proposed method is most effective at recognizing big objects and heavily obscured pedestrians, although it is simple to miss detection due to the scaling levels of the people. In the future, we plan to add real-time movies to significantly enhance speed and accuracy in response to all occlusion levels.

References

1. Cao J et al (2021) From handcrafted to deep features for pedestrian detection: a survey. In: IEEE transactions on pattern analysis and machine intelligence
2. Zhang S, Benenson R, Schiele B (2017) Citypersons: a diverse dataset for pedestrian detection. In: arXiv:1702.05693v1
3. Bochkovskiy A, Wang C-Y, Liao H-YM (2020) YOLOv4: optimal speed and accuracy of object detection. In: arXiv:2004.10934v1
4. Yang G et al (2020) Face mask recognition system with YOLOV5 based on image recognition. In: 2020 IEEE 6th international conference on computer and communications
5. Wang K, Liew JH, Zou Y, Zhou D, Feng J (2019) Panet: few-shot image semantic segmentation with prototype alignment. In: arXiv:1908.06391v2
6. Liu S, Huang D, Wang Y (2019) Adaptive nms: refining pedestrian detection1 in a crowd. In: arXiv:1904.03629v1
7. Liu W, Liao S, Ren W, Hu W, Yu Y (2019) High-level semantic feature detection: a new perspective for pedestrian detection. In: 2019 IEEE/CVF conference on computer vision and pattern recognition
8. Zhang Y et al (2021) Variational pedestrian detection. In: arXiv:2104.12389v1
9. Lin T, Goyal P, Girshick R, He K, Dollár P (2020) Focal loss for dense object detection. In: IEEE transactions on pattern analysis and machine intelligence
10. Lin M et al (2020) DETR for pedestrian detection. In: arXiv:2012.06785v3
11. Liu W, Liao S, Hu W, Liang X, Chen X (2018) Learning efficient single-stage pedestrian detectors by asymptotic localization fitting. In: Proceedings of the European conference on computer vision.
12. Yin R, Zhang R, Zhao W, Jian F (2020) DA-net: pedestrian detection using dense connected block and attention modules. IEEE Access 8:153929–153940
13. Zhang Y, Yi P, Zhou D, Yang X, Yang D, Zhang Q, Wei X (2020) CSANet: channel and spatial mixed attention CNN for pedestrian detection. IEEE Access
14. Wang C-Y, Mark Liao H-Y, Wu Y-H, Chen P-Y, Hsieh J-W, Yeh I-H (2020) Proceedings of the IEEE/CVF conference on computer vision and pattern recognition
15. Zheng Z, Wang P, Ren D, Liu W, Ye R, Hu Q, Zuo W (2021) Enhancing geometric factors in model learning and inference for object detection and instance segmentation. In: arXiv:2005.03572v4
16. Elfwing S, Uchibe E, Doya K (2019) Sigmoid-weighted linear units for neural network function approximation in reinforcement learning. In: arXiv:1702.03118
17. Raja Vikram Reddy E, Thale S (2021) Pedestrian detection using YOLOv5 for autonomous driving applications. In: 2021 IEEE transportation electrification conference (ITEC-India)

Chapter 25
Health Ware—A New Generation Smart Healthcare System

Nihar Ranjan, Maya Shelke, and Gitanjali Mate

1 Introduction

One of the important and difficult industries in each nation that determines the population's physical as well as mental health is healthcare. Real-time health services must be available to protect the public's health [1]. To follow up on, prevent, and effectively treat diseases to decrease early deaths, it is essential and one of our main obligations to offer patients with high-quality health care. By increasing worker productivity, an efficient and effective healthcare system may significantly improve any nation's economic health and growth. A good healthcare system also contributes to lowering poverty levels and improving population educational standards. A country can battle epidemics and infectious illnesses of any sort with the help of a robust and effective health infrastructure [2]. Large sums of money spent on illness can be saved if initiatives can be taken to improve healthcare facilities and networks across the region. Physical and mental health of a country significantly improve if attention is given to this sector any government vigilance toward the citizens can be measured by the state of the health care. The digitization of the healthcare industry may make real time, dependable, quick, and simplified healthcare facilities feasible [3]. The ability of various electronic and digital instruments to analyze, manage, update, and alter

N. Ranjan (✉) · M. Shelke · G. Mate
Department of Information Technology, JSPM's Rajarshi Shahu College of Engineering, Pune, Maharashtra, India
e-mail: nihar.pune@gmail.com

M. Shelke
e-mail: mayabembde07@gmail.com

G. Mate
e-mail: gsmate_it@jspmrscoe.edu.in

M. Shelke
PCET's Pimpri Chinchwad College of Engineering, Pune, India

A. Yadav et al. (eds.), *Proceedings of International Conference on Paradigms of Communication, Computing and Data Analytics*, Algorithms for Intelligent Systems,
https://doi.org/10.1007/978-981-99-4626-6_25

the information and data that is already accessible is one way that the modernization of any sector organization may boost efficiency and production. The systematic and efficient organization of services inside organizations is made possible by digitalization, which also improves productivity and quality. For instance, hospitals can keep a patient's electronic health record, which doctors and hospitals can access during that patient's subsequent visits [4]. Conventional techniques of data storage (often in paper format) place restrictions on the system's speed and dependability and increase the danger of data loss due to theft, natural disasters like floods and earthquakes, fire, and human error. If data is not securely safeguarded, hospital staff members may misuse it. Additionally, managing this data storage demands a significant amount of labor along with physical infrastructure to keep it up. One of the most difficult and time-consuming activities is searching for particular information that has been saved in non-digital form, especially if the data size is quite high. All of these problems have an impact on the system's overall productivity, hence the healthcare industry has to modernize its data storage [5]. One of the key components of the healthcare control system is data exchange. Each institution (hospital) typically has its own system for managing patient records, and information about the patient is shared with all of the organizations in the network. Since the patient is the proprietor of his personal information, even in a centralized healthcare system that collects data from all hospitals, the hospital requires the patient's authorization to get the data. However, in urgent and severe situations, patients won't be able to provide their consent for the hospitals to obtain their data, which might be deadly and endanger the patients' lives [6].

A quick review of the literature has been done in part II, where several current and recent healthcare systems are examined. The aspects of our suggested system are detailed in Sect. 3, where each system component is briefly outlined. Our suggested system is finished and the future scope is discussed in Sect. 4.

2 Literature Survey

Erwin Halim et al. [7] proposed a healthcare system, where records of different health organizations can be integrated. Nonaka and Takeuchi model (Fig. 1) is used in this system, where patients' complete information distributed across different organizations are combined into a single record.

In this technique, the patient will verbally tell the doctor of any official information. Socialization is the term for this phase, when the patient will provide the doctor with first-hand details regarding their health. The doctor will then modify this formal data in the computerized record in accordance with the patient's current state of health. The process of turning implicit knowledge into explicit information is called externalization. In the next stage, the system will combine patient data from several organizations and create a patient's whole medical record. Combination is the term for this action. Internalization, the last phase, involves transforming integrated explicit knowledge into tacit knowledge once more. Hoai Luan Pham and others

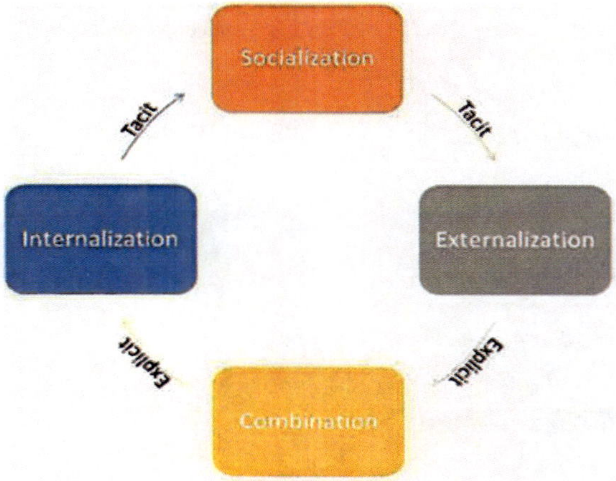

Fig. 1 Nonaka and Takuechi model

[8] presented a system based on blockchain technology that stores the transactions depicted in Fig. 2 using the public ledger. Since it is theoretically impossible to alter data using blockchain technology in an acceptable length of time, this system concentrates mainly on ensuring the safety of data transit through networks. The foundational technology for this system's development is Ethereum. The capability of smart contracts is made possible by Ethereum thanks to its public, decentralized, and open-source ethos.

A smart contract represents a written agreement between two parties that contains all of the terms and limitations of the agreement in full. Essentially, a smart contract is a piece of code that details the terms of a contract and is written in the solidity programming language. When a trigger is fired, these programs are run. The public distributed ledger is where all of these codes are kept. These codes are visible to everyone thanks to the blockchain's adaptability, but nobody can alter them. The money Ethereum uses to cover these transactions' costs is called ether. A transaction cost known as "Gas" must be paid in order to carry out the transaction. Our transactions can be given priority over other transactions if additional gas is provided for the transaction. The miner receives this charge as a bonus reward. The block will get linked to the chain of blocks more quickly.

Yang et al. [9] presented a system that aspires to share patient electronic health records (EHR) with interoperability, integration, intelligence, and innovation. One essential component of providing successful healthcare services and enhancing patient health is information exchange. MedShare, seen in Fig. 4, is a trustworthy and durable digital healthcare system created to transmit medical information across hospitals and to make it easier for patients to transfer healthcare providers for a variety of reasons. Figure 3 illustrates a query-based system that was created on a hybrid cloud (a combination of public and private clouds) for data retrieval in an emergency. It allows users to access patient medical information, reports, and prescription histories online.

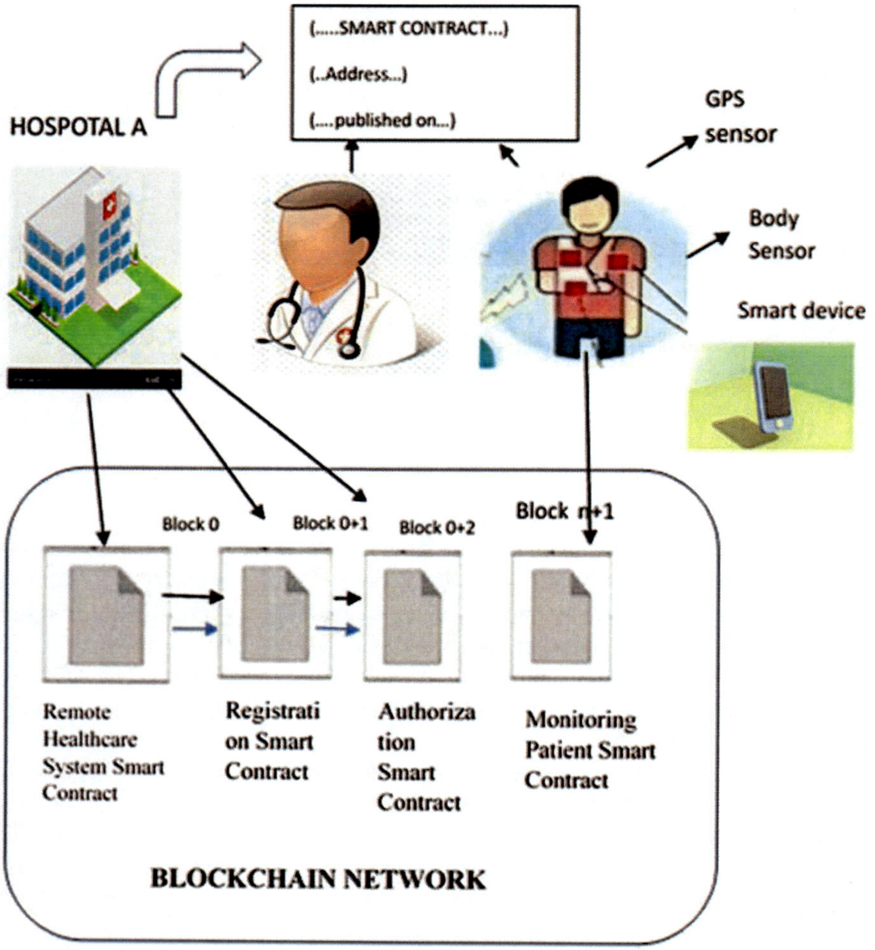

Fig. 2 Remote healthcare system using blockchain

Electronic health records (EHR) concerning every hospital patient are stored in the private cloud of the healthcare organization. On the contrary hand, the public cloud, which is supervised by government authorities, stores patient ID mapping via the address of clouds of every healthcare facility previously visited. The system is made up of the five components in Fig. 4: Private Cloud, Public Cloud, Patient, Hospital, and Disease Types. A patient will receive a special patient ID the first time they visit a hospital (A) that has MedShare enabled.

This ID and the location of the hospital A's private cloud are both kept on the public cloud. Patients' individual IDs are kept within a private cloud together with other data including names, addresses, medical histories, etc. In order to obtain the patient's medical history and personal information when they return to hospital A, data stored in the private cloud has to be downloaded. If the patient visits hospital

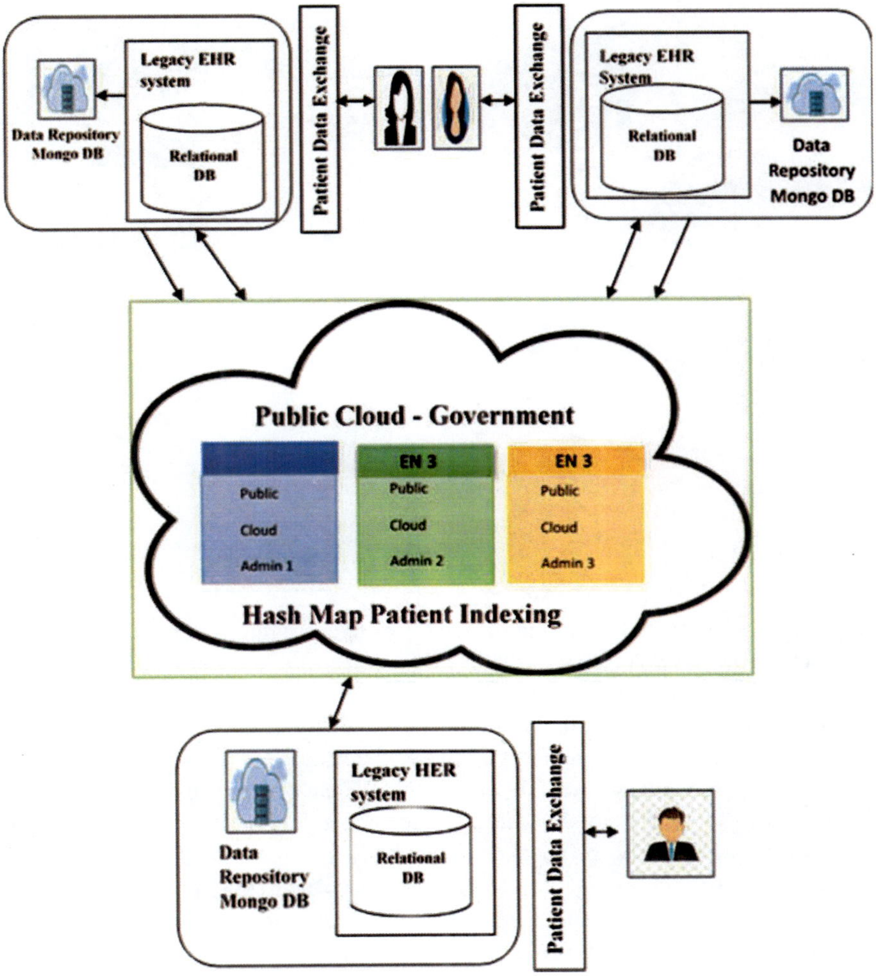

Fig. 3 Architecture of medical blockchain system

B, which they have never been to before, they will be required to provide their specific patient ID. The hospital's private cloud will thereafter ask the public cloud to seek patient information along with a patient ID. If the public cloud discovers the patient's entry, it will send a request for information on the person being tracked through the private cloud using the address linked to the distinct patient ID. Hospital A's private cloud will look for the patient ID in its database and transfer the required data to the public cloud.

This data will be transmitted back to hospital B's private cloud, which is now usable by Hospital B. Using blockchain technology, Chen et al. [10] suggested a system for managing medical records that offers a decentralized and distributed framework for data storage and management. Privacy, security, portability, and

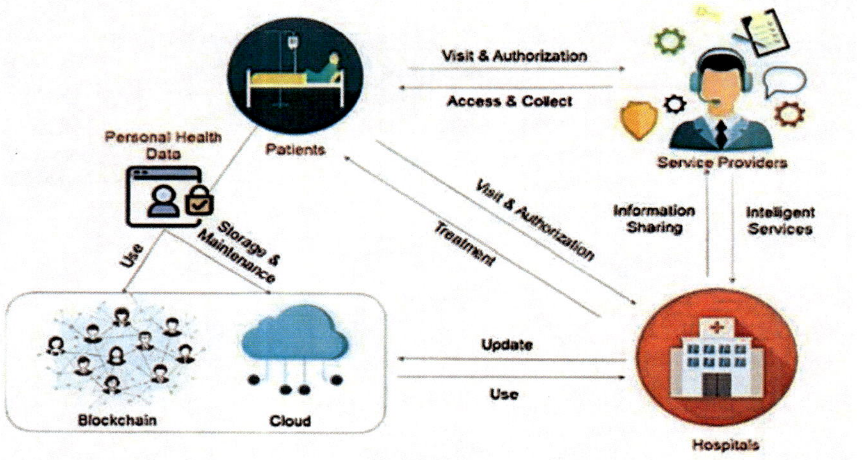

Fig. 4 MedShare

simplicity of sharing and obtaining ownership of data are this system's key features. The patient has complete authority over their health-related information in this system. In this approach, patient data is encrypted using a symmetric encryption key and stored in the cloud; the proprietor of the decryption key may only decode patient data. Along with the patient's data storage location and accompanying access rights, a hash of this data is kept in a blockchain. Data becomes invalid if someone attempts to edit it in the cloud because the accompanying hash value has changed. The accuracy of data may be checked by anybody, but since the data is encrypted, it is difficult to extract accurate information from it. Due to the intrinsic property of blockchains, a storage location, hash value, and access privileges are also open to all users and immutable. All data generated about a patient, such as blood pressure, sugar levels, and test results, are translated into digital records when a patient visits a hospital for a checkup.

3 The Proposed Model

The system we recommend is a smartphone application created with the frontend technology Flutter SDK. The major components of this program are the seven seamlessly integrated functionalities listed below.

a. **Cloud API**

Application Programming Interface that can be used to provide service to both doctor's application as well as patients' applications. Cloud API will act as an interface between the patient application and doctor's application. It will be used for heavy duty processing that cannot be performed on the user devices (Mobile phones) like Chabot response generation which will require the neural network to perform tasks.

Cloud API acts as an interface between patients application and doctors application. There can be many patients' applications serviced by cloud API at a time as well as many doctors' applications also can be serviced at the same time. Cloud API will service both these applications and also act as a channel between patient application and doctor's application. When any patient application requests a cloud for a particular service, cloud will try to give whatever it is asking for. If a patient is asking for a doctor's consultation, cloud will redirect the patient to the doctor according to the patient's request. Besides this, cloud will also act as a computing agent for heavy duty computations like generating the Chabot response, since this process has heavy computations on large scale which cannot be done on mobile phones, where processing power and memory are limited as compared to cloud platforms. For this purpose, Chabot will send the sound data recorded through the microphone to the cloud. Cloud will process the data and generate textual sentences from it using machine learning algorithms and neural networks. Then on this textual information, processes like syntax analysis and semantic analysis will be carried out. And then finally, the meaning of the sentence and corresponding response will be generated. Then, this response will be sent to the client which is requesting it. For interacting with the cloud, doctors and patients' applications will call some special methods which will in turn communicate with the cloud and provide the necessary service to the application. For providing the service, Cloud API will set listeners on the input requests and methods corresponding to the particular request will be called when that request is encountered. When a patient tries to connect to a particular doctor, Cloud API will send the notification to that doctor. Then, the doctor will accept that request and try to help the patient out from his/her problem. Cloud API will also allow patients to place their reports and medical history on the cloud securely in the digital format. This makes the analysis of reports easy using modern analysis technologies like data science. Cloud API will also allow the patient to find the doctors in near proximity with necessary qualification, specialty and high rating. This will help patients to choose the appropriate doctor of their choice even in new places where patients do not have much information about locality. Cloud API is at the core of the system which will handle every task that is needed for the system. It acts as a brain of complete application. This module is shown in Fig. 5.

b. **Online Appointment**

To schedule patients for a certain hospital or doctor, an online scheduling system will be employed. The patient will search for a doctor based on factors such as location, specialization, credentials, and rating. The use of an appointment system will eliminate the need to wait in a big queue to make an appointment. Additionally, the patient will be informed if the doctor cancels the appointment for any reason so that he may save time. Algorithms will be used by the online appointment system for setting up the appointments. A system called the online appointment system was created to accept appointments from any doctor or hospital that is registered.

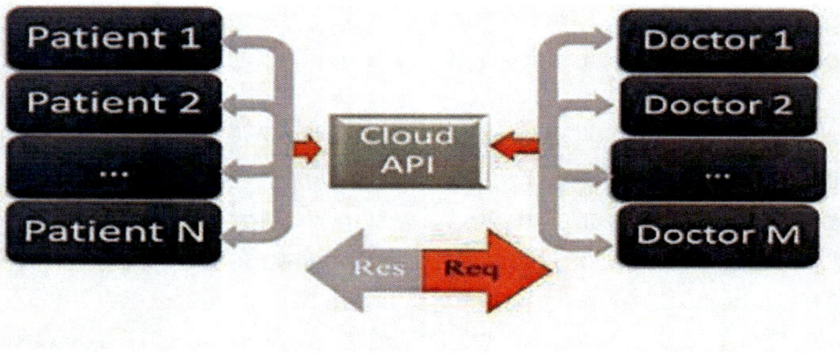

Fig. 5 Working of Cloud API

Patients may schedule appointments with doctors online without having to stand in queue. With just a few clicks, the patient may use this system to request a doctor's appointment. By offering a smooth interface to take appointments, this technology helps both patients and doctors save time. Figure 6 presents this module.

The working days of the doctor are regulated in a system. Subsequently, he will be able to input how long he plans to work each day. The doctor will then decide when to start working and take breaks. Finally, the doctor will be permitted to specify the rough timeframe within which he will examine patients and offer counsel. When this is finished, the system will create the appointment openings on its own. The doctor will be able to create a unique schedule for each day.

The patient will look for a doctor based on the doctor's ID, location, specialization, credentials, rating, and reviews while looking for a consultation. The patient will decide which doctor is ideal for him or her. Then, based on the doctor's availability, the system will generate an appointment time for the patient to choose from. The system will check to see if the space has already been reserved by another patient before allocating it. The system will not let the patient schedule an appointment in that time slot if it is already taken. The patient will next have the option of entering his or her complaint. Following these procedures, the system will schedule the appointment.

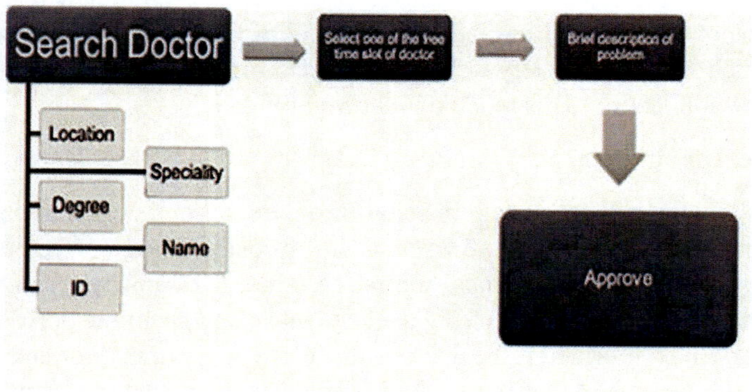

Fig. 6 Online appointment system

If a patient misses an appointment, they will be charged a penalty fee, and that time slot will then be made available to another patient. Patients will be informed of the cancelation and given the opportunity to schedule a new appointment if a doctor must cancel an appointment due to an emergency situation. The doctor will be immediately disqualified if they repeatedly cancel appointments. By making the appointment procedure quick and easy, online appointment systems will benefit both patients and doctors. Along with reducing time, it will increase patient and doctor trust.

c. **Virtual Doctor**

A cloud-based doctor is a communicative Chabot that will provide people with free first-aid advice and try to fix their concerns. Machine learning techniques will be used by the virtual doctor to produce the required response. Neural networks will be used by the virtual doctor to determine its purpose. If the virtual doctor is unable to manage the problem, the patient will be directed to a relevant doctor through the use of videoconferencing.

As seen in Fig. 7, the virtual doctor is an intelligent agent that may counsel the patient on first aid and some fundamental suggestions. Machine learning techniques will be used by intelligent agents to reply to user inquiries and issues. The steps below are what the virtual doctor will use to try to locate an acceptable solution:

1. Text-to-speech conversion: For this virtual doctor, neural networks and other machine learning methods will be used.
2. Analyzing the significance of text data: The second machine learning agent will now be given the text produced in step one. The statement will be broken up into its component tokens by this agent, who will then do a semantic evaluation of it. It determines the meaning of the sentence before employing a neural network to provide the relevant answer for that sentence once more. In addition to assisting patients with basic-level treatments, virtual doctors may be used to match patients with appropriate doctors based on their needs. In order to do this, we are creating a chatbot that is linked to a database. This way, the chatbot can respond to user queries while simultaneously taking data from the database into account. Virtual doctor looks for a solution to the user's problem first. If the issue is beyond the capabilities of the chatbot or if it is not able to solve it, the chatbot will

Fig. 7 Virtual doctor assistant

immediately refer the patient to the doctor through the video calling module. The chatbot is an optional benefit that will be offered to patients and will give them the information they need in an emergency. If there is no emergency help accessible, especially in Indian communities, this might be lifesaving.

d. **Video Calling**

Patients will be able to call doctors and ask for guidance using the video calling module. A patient may look up a doctor via the Internet and then talk with them about their issue. For this consultation, the doctor may charge a fair fee. The entire video conference will be encrypted so that the patient and doctor may speak openly.

The system's interface for connecting patients and doctors via video calling. Without going to the hospital, this will take care of any minor health issues. Figure 8's interface enables clients to avoid wasting time by preventing needless hospital visits for extremely minor issues. In addition, this system will offer competent medical advice and help in the event of emergencies like accidents and cardiac arrests.

In the absence of a professional doctor, a non-professional individual can utilize the video calling capability to get assistance from a professional doctor in performing emergency responsibilities.

When making a video call, there are various steps:

1. Footage Recording: During this phase, the system will capture footage from the patient device's camera.
2. Quality of video enhancement: In this phase, we make changes to the video's brightness, contrast, and color in an effort to improve its quality.
3. Compression: In order to prepare the video for transmission, we will compress it in this stage. By doing this, the video's size will be reduced, hence requiring less bandwidth. It will be poor compression that is employed in this situation.
4. Encryption: In this step, the video will be encrypted to prevent third parties from seeing it. By taking this action, you may submit videos while protecting the patient's privacy and security.
5. Transmission: In this phase, the video will be sent to the other side.

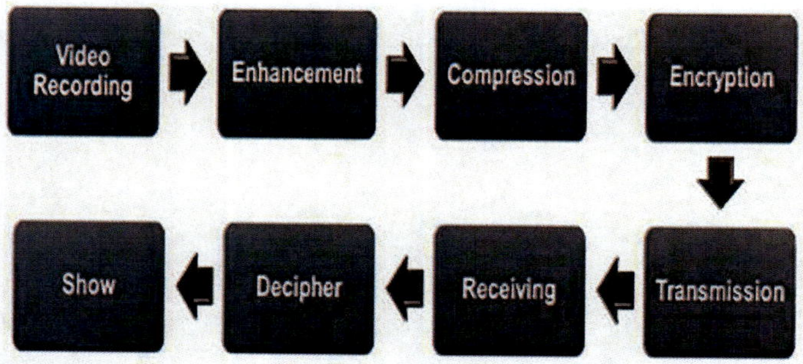

Fig. 8 Video calling system

6. Receiving: In this case, the video will be received by the receiver. Note that the footage is encrypted, meaning that nobody in the line of sight between the transmitter and recipient may view it.
7. Decipher: In this stage, the video is deteriorated to produce the original video. Key Exchange Algorithm will be used to send the recipient the key required to degrade the video.
8. Show: In this phase, we show the recipient the video.

The system's video calling capability has the potential to save time while simultaneously providing emergency assistance. In dire situations, it may even save lives.

e. **Medicine Alarm**

A system called the "medicine alarm system" is used to remind people to take their medications. The patient will be reminded to take their medication at a time and frequency specified by their doctor by use of an alert system, depicted in Fig. 9. This will reduce the chance of forgetting to take a crucial medication.

The device that may be used to remind patients to take their medications is called a "medicine alarm system.“ Patients can take their medications on schedule with the aid of this approach. This will be especially useful for elderly folks who have trouble remembering when to take their medications. For this reason, the system will automatically set off an alarm when a doctor creates a prescription, following the doctor's instructions. The system will take into account variables like medication frequency and timing requirements, among others. With the use of these settings, the system will set an alert for a certain medication and prompt the patient to take their medication. This approach will prevent missing medications, which might be harmful for patients who are in severe condition.

f. **Patient Profile Management**

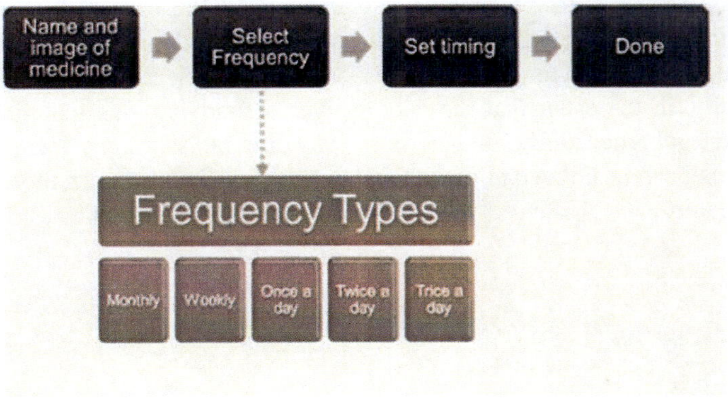

Fig. 9 Medicine alarm system

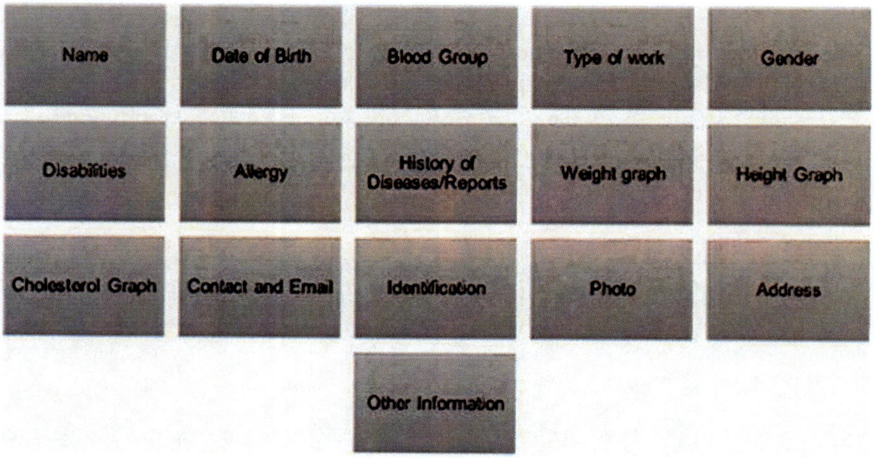

Fig. 10 Patient profile management

The patients profile system will require certain data to establish the patient's profile while it is being created. This data will be encrypted and kept in a safe way on the cloud. This data may be employed in diagnostic procedures. Figure 10 is a representation of this module.

Additionally, when the patient adds information to his or her profile, such as weight, cholesterol levels, and sugar levels, the profile will gradually improve. This will generate a graph in the patient's profile that clinicians may use to rapidly assess how the patient's health evolved over time. The patient may also upload other medical records to their profile so they can be safely stored in the cloud and utilized later.

g. Doctor's Profile Management

The system will prompt the doctor for information such as name, degree, and area of specialization when he tries to create a new account. Patients can use this information to search for the doctor of interest based on a variety of criteria. Over time, the system will produce ratings and reviews that will aid patients in selecting the best physician. Additionally, in the event that the doctor gives bad advice, this material may be relevant in legal proceedings. In the event of any difficult issues, the doctor's contact data and mail can be utilized to get in touch with them. Figure 11 is a representation of this module.

4 Conclusions

In this study, we established the fundamental structure of a system that offers patients end-to-end healthcare assistance and guidance. Our system includes well-known functions and offerings including virtual assistant (chabot) emergency help, virtual

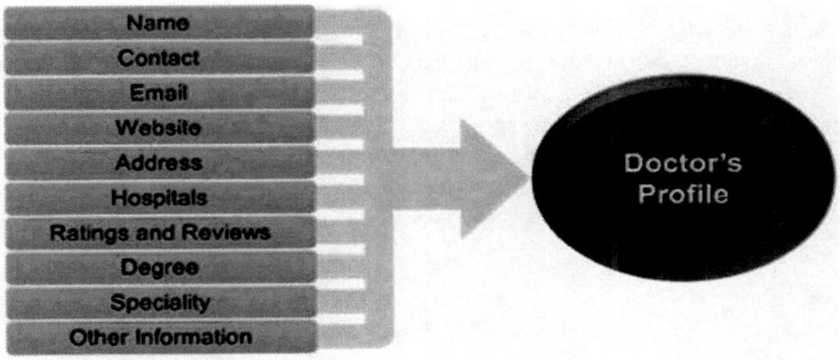

Fig. 11 Doctor's profile management

physicians, medication alert, online appointment, profile management, etc. This system includes features like virtual physicians and online prescriptions that help people who need urgent care or are in severe circumstances. The cloud API system takes all necessary precautions to protect the privacy and security of the data created by the system.

By including certain further functions like picture document analysis, boosting the security channel communication, and providing support with surgeries, the system's future applicability may be improved. There is always room for advancement with regard to the preciseness of virtual assistants.

References

1. Khadka S (2012) Privacy, security and storage issues in medical data management. In: 2012 Third Asian Himalayas international conference on internet. IEEE, pp 1–5
2. Saranya MS, Selvi M, Ganapathy S, Muthurajkumar S, Ramesh LS, Kannan A (2017) Intelligent medical data storage system using machine learning approach. In: 2016 Eighth international conference on advanced computing (ICoAC). IEEE, pp 191–195
3. Singh S, Sarote P, Shingade N, Yelale D, Ranjan N (2022) Detection of Parkinson's disease using machine learning algorithm. Int J Comput Appl 975:8887
4. Secure Storage Systems for Healthcare Records. In: Jonker W, Petković M (eds) Secure data management. SDM 2007. Lecture Notes in Computer Science, vol 4721. Springer, Berlin, Heidelberg
5. Ghatage MMR, Ranjan N A novel approach for disease prediction scheme for healthcare management
6. Ghodake S, Ranjan N (2019) AI algorithm for health risk prediction. J Emerg Tech Innovat Res 6(6):600–607
7. Halim E, Halim PP, Hebrard M (2017) Indonesia medical knowledge management system: a proposal of medical KMS. In: 2017 international conference on information management and technology (ICIMTech). IEEE, pp 322–327
8. Pham HL, Tran TH, Nakashima Y (2018) A secure remote healthcare system for hospital using blockchain smart contract. In: 2018 IEEE globecom workshops (GC Wkshps). IEEE, pp 1–6

9. Yang Y, Li X, Qamar N, Liu P, Ke W, Shen B, Liu Z (2018) Medshare: a novel hybrid cloud for medical resource sharing among autonomous healthcare providers. IEEE Access 6:46949–46961
10. Chen Y, Ding S, Xu Z, Zheng H, Yang S (2019) Blockchain-based medical records secure storage and medical service framework. J Med Syst 43:1–9

Chapter 26
EEG-Based Sleep Stage Classification System

Medha Wyawahare, Rohan Bhole, Vaibhavi Bobade, Akshay Chavan, and Shreya Dehankar

1 Introduction

Sleep is an essential and fundamental procedure for our body to get recharged to perform all the functions and work properly. Electroencephalogram, medically abbreviated as EEG, is one of the key tools which can be used to get an understanding of sleep-related activities happening in our brain. EEG comes under polysomnography (PSG) which is a clinical diagnostic method for sleep monitoring. So, PSG is conventionally used to observe human sleep patterns in hospitals by measuring data from EEG, electrooculography (EOG), and electromyogram (EMG) [1].

As we sleep, our brain goes through various stages of sleep. Sleep stages consist of two main types: non-rapid eye movement (NREM) and rapid eye movement (REM). NREM is then further categorized into 4 stages: N1, N2, N3, and N4 as per the Rechtschaffen and Kales (RK) sleep scoring standards. Often, the awake state of a person is also considered for sleep monitoring [2]. During sleep, brain activity can be understood by the signal waves. Four primary rhythms may be identified in brain activity according to frequency ranges: beta, alpha, theta, and delta. Beta waves are defined to have a high frequency range:14–30 Hz. Alpha waves occurring in the relaxed state have a frequency between 8 and 12 Hz. The frequency range of theta waves is 3–8 Hz, while delta waves range from 0.5 to three or four Hz [3]. The model should be such that suitable information can be extracted from the EEG signals and along with the above-mentioned classification, be used to identify the sleep stages.

Table 1 gives sleep stages and the signal waves with their frequency ranges.

It is time-consuming and subjective to determine the various stages of sleep since it mainly depends on human expert neurologists or sleep specialists' visual pattern identification. Therefore, automated categorization is required [4]. Various methods

M. Wyawahare · R. Bhole · V. Bobade · A. Chavan (✉) · S. Dehankar
Department of Electronics and Telecommunication Engineering, VIT, Pune 411037, India
e-mail: akshay.chavan20@vit.edu

A. Yadav et al. (eds.), *Proceedings of International Conference on Paradigms of Communication, Computing and Data Analytics*, Algorithms for Intelligent Systems,
https://doi.org/10.1007/978-981-99-4626-6_26

Table 1 Sleep stages and brain rhythms

Sleep stages	Characteristic EEG signals
Awake (W)	Alpha (8–12 Hz), Beta (13–30 Hz)
NREM (N1)	Theta (4–7 Hz)
NREM (N2)	Sleep spindles (12–14 Hz), K-complex (1 Hz)
NREM (N3)	Delta (0–3 Hz)
REM	Alpha (8–12 Hz)

have been suggested to identify these stages from the EEG datasets. Here, machine learning and deep learning neural networks perform to be of great help in classification problems. This paper focuses on automation in sleep stages classification using the SVM (RBF kernel), random forest, and KNN algorithms. The data used is from PhysioNet in EDF format.

2 Literature Review

For physical and mental wellness, sleep is crucial. A wide range of physiological conditions is linked to the sleeping phases experienced by the average human body. EEG signals are crucial for understanding the various phases of sleep, and their patterns may be utilized to categorize them in diverse manners. Thus, EEG signal analysis can be a valuable tool to assist neurologists and sleep experts. On single-channel EEG, several alternative algorithms are suggested to categorize the various phases of sleep. Three distinct machine learning algorithms, including random forest, bagging, and SVM, have been proposed in the work [5].

Another study by Chambon et al. [6] has been proposed, and it classifies EEG impulses based on their phases. In order to extract the characteristics from signals, the study offered three distinct techniques. The three strategies were Itakura distance, harmonic parameter, and relative spectral band energy. They chose the best classification of sleep phases by evaluating the effectiveness of several methods. Diykh et al. [7] classified the phases of rest using the SVM machine learning method and obtained an accuracy of 95.93%. They did this by utilizing K-cross validation, which divides the information into k alternative exclusive groups of equally sized. The suggested system operates on six stages of sleep using a single-channel EEG input.

Sleep issues are currently recognized as one of the major issues influencing people's life. The human brain passes through certain physiological phases when we sleep. The paper proposed by Aboalayon et al. [8] has trained a classification model based on the PhysioNet sleep expanded database using the SVM algorithm. The experimental results show an accuracy of 92.5% of the proposed work. This work can be easily implemented for finding specific sleeping patterns in any embedded microcontroller device like drowsiness or sleep apnea. The limitation of the paper is that the classification is done only on wakefulness and stage 1 of the following sleep stages which can further be implemented in stage 2 and stage 3, where stage 2 plays a significant part in which the human body is 45% in a deep sleep.

The paper done by Zhao et al. [9] on the analysis of a single-channel-based EEG sleep staging algorithm involves several stages. They used a single-channel device to collect EEG data from participants while they slept. They also used a polysomnography (PSG) device to collect reference data for sleep staging. The collected EEG data were preprocessed to remove noise and artifacts. A bandpass filter is used to remove frequency components outside the range of 0.5–30 Hz, and a filter to eliminate the 50 Hz of power line noise. The preprocessed EEG data was used to extract relevant features that could be used for sleep staging. They developed a sleep staging algorithm using the extracted features. They used a SVM as a classifier to classify the sleep stages based on the extracted features.

Park et. al. [10] proposed a paper that consists of a literature search that was conducted to discover all studies that examined the relationship between nocturnal sleep patterns and the health of healthy children. They identified 39 systematic reviews for this purpose. Sleep duration and obesity and emotional consequences are linked by strong and reliable evidence. This research also found the connections between blood lipids, glycemia, sleep schedule, and quality. The 24-h period includes the multi-dimensional concept of sleep.

In a paper proposed by Halson and Shona [11], a sleep stage scoring using single-channel EEGs had an accuracy of 84.9%. Actigraphy has an accuracy of 86.0% for estimating sleep efficiency. Sleep grading based on pulmonary mechanics provides an accuracy of 89.2 and 70.9% for identifying sleep phases and sleep productivity, and a correlation coefficient of 0.94 for predicting apnea problems.

The goal of the study by Aboalayon and Faezipour [12] was to create a sleep stage categorization algorithm that is simple to use and operates quickly (near real time) and effectively. Using the MMD and Esis characteristics, each sample must be filtered and extracted. A sensor arm, an ear clip, and a headset make up the Neurosky Mindwave. The full EEG signal wave is passed through a low-bandpass filter with a cutoff frequency of 40 Hz. The system begins to log after a 5-s gap, and it is crucial to set up this gap to stop Mindwaves from gathering extraneous signals, which can happen when the subject changes positions and causes electrode instability and looseness, which leads to the generation of unstable brain wave signals. In essence, if the execution time is broken down and studied, the 10 s are for recoding the first batch of data, and just a part of that time is for the method, roughly resulting in execution that is close to real time.

When matched to overnight polysomnography in chronically sick patients, the reports showed that actigraphy (surveillance of gross motor movements) and behavior evaluation by the nursing leader are two sleep-measuring approaches that are accurate. Actigraphy and the bedside nurse's behavioral assessment are unreliable and inaccurate ways to monitor sleep in quite ill patients. [13]

The paper by Matricciani et. al. [14] performed experiments showing the effectiveness of the AQI prediction in directing OSA treatment was 93.3%. The performance has been evaluated in the form of latency. The findings obtained definitely illustrate system efficiency are increased by data preprocessing at the edge of the network.

The development of sleep staging technology is currently a popular area of scrutiny in the field of brain-computer interfaces. The technology by Wang et al. [15] aims to alleviate the burden on sleep specialists by automating the time-consuming process of diagnosing sleep patterns. However, the interpretation of EEG signals is still challenging due to factors such as noise and insufficient data. To address these issues, a new automatic sleep staging network was proposed, which incorporates transfer learning and integrates single- and multi-channel features. The system extracts frequency and long-term features from raw EEG signals and time–frequency data in two processing stages. Transfer learning was used to overcome variability and data inefficiency, and LightGBM was employed for classification. The system's performance was evaluated using the sleep-EDF expanded dataset, with the highest accuracy rate achieved on the sleep-cassette subset being 87.84%. Compared to other methods, the proposed system achieved higher accuracy rates without relying on massive amounts of training data.

Hussain et al. [16] proposed a paper in which they provide a thorough overview of the most recent research studies in several areas related to sleep monitoring, such as vital signs monitoring, classification of sleep stages, identification of sleep postures, and detection of sleep disorders. In order to give a complete rundown of the four regions of sleep monitoring's most recent advancements and trends, researchers examine the most current intrusive exclude studies, including both wearable and non-wearable approaches, examine design processes and key features of the results described, and give a detailed analysis based on 10 key factors. Additionally, they offer several publicly accessible datasets for various types of sleep monitoring. In conclusion, they go over a number of unresolved difficulties and offer suggestions for future research in the study of sleep tracking.

This review explores the potential for describing sleep phases, the usage of several typical sensors included in the hardware of current sleep monitoring devices, and the benefits and downsides. The numerous well-known commercial sleep monitoring products were also examined in this review, along with their features, benefits, and drawbacks. Researchers specifically divided the body of available research on sleep monitoring devices into categories depending on the number, kind, and placement of sensors utilized. At last, they put the solution that they discovered in the laboratory. The most used sensor in sleep monitoring devices is an accelerometer. The majority of commercial sleep monitoring devices are unable to offer performance assessments based on the gold standard polysomnography [17].

The paper by Kong et al. [18] on Neural Systems and Rehabilitation Engineering in 2023 describes a novel technique for sleep stage classification using neural architecture search applied to EEG signals. The study aims to address the challenges of traditional sleep stage classification methods, which often require manual feature extraction and selection, leading to limited performance and generalization ability. The authors suggest a complete EEG-based phases of sleep classification approach that employs NAS that optimizes the topology of deep neural networks in order to get around these drawbacks. The Sleep-EDF dataset and the Montreal Archive of Sleep Study dataset, both publicly accessible datasets, were used to assess the advised approach. The findings show that in relation to classification accuracy, F1-score, and Cohen's Kappa coefficient, the NAS-based system surpasses a number of cutting-edge techniques. Furthermore, authors analyze the learned architectures and provide insights into the optimal network structures for sleep stage classification.

In order to accurately classify sleep stages, a publication by Satapathy et al. [19] suggests a system of the staging of sleep, employing EEG data and machine learning methods. The preprocessed EEG data is then used to extract parameters including spectral power, entropy of signal, and data with wavelet coefficients. The most important features for sleep staging were chosen during the feature selection step using statistical and machine learning approaches. The classification stage was completed by classifying the various sleep phases using an SVM classifier. A dataset of 20 participants was used to assess the suggested system, and the findings revealed that it had an overall accuracy of 87.5% for classifying sleep stages.

Another paper by Al-Salman et al. [20] proposed a new method for classifying sleep stages in EEG signals. Their approach involves using probability distribution features derived from the clustering approach, which they combine with different classification algorithms. A total of 986 EEG recordings were utilized for evaluation as part of the study, which utilized information from the Sleep-EDF database, which is open to the public. In comparison with other cutting-edge technologies, the findings demonstrated that the suggested method obtained excellent accuracy rates in categorizing sleep phases. In particular, the suggested method beat previous approaches including random forest, k-nearest neighbors, and support vector machines with a total efficiency of 88.92% and a Cohen kappa result with a coefficient of 0.86.

3 Methodology

Electroencephalogram (EEG) signals, sometimes referred to as EEG signals, are used to extract sleep data. The EEG sleep stages categorization method may be used to examine different health conditions of patients of different genders and age groups. For the sake of extracting the participants' sleep patterns, we only took into account one EEG signal channel, designated as Fpz-Cz. Filters are used during the preprocessing stage to remove noise from EEG data. The EEG waves are filtered before being divided into 30 s-long epochs.

After preprocessing the signals by filtering and further distinguishing the sleep behavior in terms of frequencies and time-oriented attributes, feature extraction is used to extract both time and frequency domain information. Three alternative machine learning algorithms have been proposed in this study for categorizing the various phases of sleep. The complete schematic of the sleep stage classification model has been shown in Fig. 1.

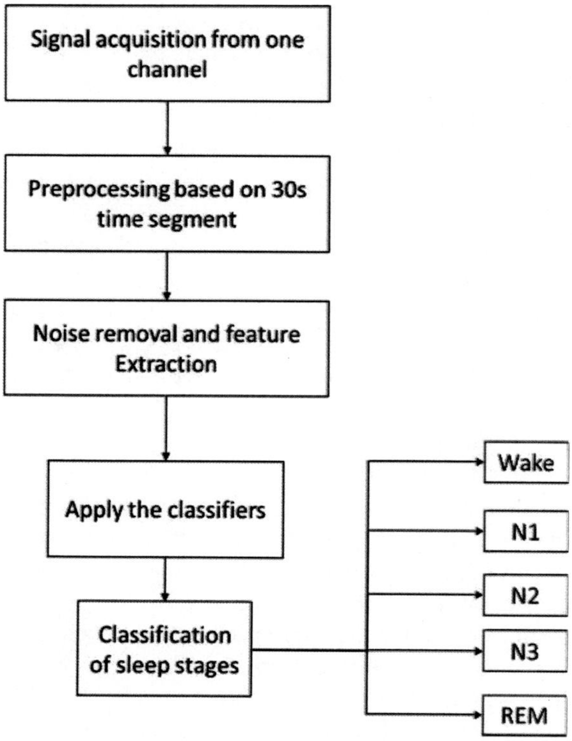

Fig. 1 Block diagram of classification process

3.1 Dataset

The Sleep-EDF [European Data Format] dataset from PhysioNet, which was accumulated and documented by experts for the investigation and evaluation of sleep phases, is the source of the numerous EEG samples utilized in this model.

There are 197 whole-night polysomnographic sleep data in the database. It includes submental chin EMG, EEG, and EOG. The data is stored in two forms PSG.edf form and Hypnogram.edf form. The PSG.edf form consists of the EEG signals from different electrode locations, and in this work, we have used Fpz-Cz signals as different electrode locations. While the Hypnogram.edf file contains annotations of various sleep schedules that essentially match the PSGs. The sleep pattern consists of 5 stages: Wake or Alert, N1, N2, N3, and REM stage. The first span of sleep generally lasts for 10 min, while the final lasts for up to one hour.

3.2 Preprocessing

MNE library was used to read the EDF format data (Hypnograms). Using these library's functions, the signal data was plotted and visualized so as to get an idea about which channel to use for further work. Using multiple channels requires higher tools and different approaches. So, based on the literature works and the visualization, the FPZ-CZ channel was used for extracting the sleep behavior out of the EEG signal. Then, the Butterworth filter (order 8) having a 0.5–40 Hz pass-band bandwidth was used to filter the signal and remove the noise.

3.3 Feature Extraction

After the preprocessing step, epochs of 30 s were used according to the AASM sleep academy and RK criteria for feature extraction. Different feature extractions that were performed are discussed below.

The power spectral density gives the distribution of the average power of the signal as per frequency. It characterized each person's FPZ-CZ channel. The spectral resolution used to digitize the signal normalized the PSD's amplitude.

Another feature used was the Petrosian fractal dimension. It is a measure of how the detail in the fractal changes with scale and Petrosian is one of the algorithms to calculate it. In this, a binary sequence is derived using different algebraic methods and then the fractal dimension is calculated as:

$$\text{Petrosian Fractal Dimension} = \frac{\log_{10} N}{\log_{10} N + \log_{10}\left({}^{N}/_{(N + 0.4N_{\delta})}\right)} \tag{1}$$

In Eq. 1, N states time-length and N_{δ} stands for the number of changes in the sign of the signal's derivative.

The Hjorth parameters and Hurst exponent were used for the feature extraction step as these are also some of the important factors deciding an EEG signal's characteristics. A signal's long-term memory is assessed using the Hurst exponent.

3.4 Machine Learning and Tools

Below are the details of the tools used in the model:

The dataset used in our project was a wave or signal EEG data whose recordings were stored in the EDF file format. The MNE library was used to read both PSG and hypnogram files to extract their features. This library was helpful because unlike CSV or TXT files it's not possible to use Pandas or NumPy libraries to read EDF files.

Large multi-dimensional arrays and matrices are supported by the NumPy library, which also offers a substantial number of complex mathematical operations that may be performed on these arrays. This module was used to save the read EDF file data into a NumPy array that contains both the EEG signal data and the labels for the various stages of sleep.

Sci-kit library was used to implement a few machine learning models and display their corresponding classification reports and accuracy.

Matplotlib, seabornis library is also used in the model for plotting.

Tools that we tried and ended up not using:

Initially, usage of the py edf library to read EDF files was done but later, we switched to the MNE library. The reason is that though the py edf library read PSG files correctly, it was challenging to read and display all the features in the hypnogram files. And also, it failed to save the EDF format of the hypnogram file in TXT or CSV file format. And the PSG files were saved in a CSV file for all subjects each of size 1 GB. Usage of the MNE library to read and plot both PSG and hypnogram files was done and saved the corresponding NumPy arrays in NPZ files for each subject.

3.5 *Evaluation of Model*

1. SVM: Using the "rbf" (Radial basis function) kernel rather than the "linear" kernel considerably improved the performance of the SVM model. The performance of the model was assessed using the accuracy measure, classification report, and confusion matrix. The model was run on 50 participants, and an accuracy of 80% was obtained.
2. Random Forest: This model was employed and metrics were evaluated. 65% accuracy was attained. Accuracy is 0.65 using random forest.
3. KNN: The reason for the low accuracy of KNN could be the nonlinear nature of the data and some time frames being zero, and also the hyperparameters of KNN are not suitable for exceptions like having 0 or noise. And complete noise elimination as it can be used for other inferences like snoring problems. We then implemented SVM using linear where the performance was similar to KNN, however by using the "rbf" kernel there was a significant improvement in the performance. The "rbf" kernel outperformed because of the correct classification of nonlinear data. The nonlinear regularization factors helped to increase accuracy. As shown in the below accuracy and classifiers, the accuracy of random forest support vector machine classification outperformed other machine learning algorithms like KNN.

4 Results and Discussion

EEG-based sleep stage classification system predicts with the training, testing accuracies, and F1-score. For the classification of the model, 197 whole-night polysomnographic recordings were used in.edf format. The greater the value for parameters, the better a model can categorize data into classes. When the classes are balanced

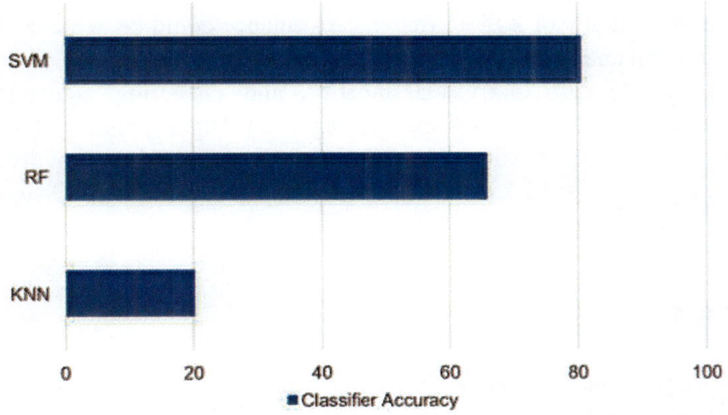

Fig. 2 Accuracies for used classifiers

and there was no significant disadvantage to detecting false negatives, frequent use of accuracy was done. Utilization of F1-scores has been done when the classes are unbalanced, and there was a significant disadvantage to detecting false negatives, where the F1-score is calculated as per Eq. (2).

$$\text{F1-score} = 2 \times \frac{\text{Presion} \times \text{Recall}}{\text{Precision} + \text{Recall}} \tag{2}$$

For classification, a total of three machine learning algorithms are used, first one was KNN also known as K-nearest neighbor, while the second and third are SVM and random forest. Figure 2 represents the pictorial representation of three machine classifiers according to their varying accuracies.

From the above graph, it is clear that SVM gives the highest accuracy of about 80.56%, followed by an accuracy of 65.8% of random forest and then an accuracy of 20.2% of KNN which is the lowest accuracy.

5 Conclusion and Future Scope

Mainly focused is done on classifying sleep stages by considering a single-channel of EEG Fpz-Cz. Unlike normal CSV or text file the format of the data files, i.e., European data format (EDF) used in this model, maximum time was utilized on initial preprocessing of these data like understanding the file format, its annotations, and exploring different libraries to read these files. To implement SVM, the RBF kernel was used due to the nonlinear nature of the data, and an accuracy of 80.56% was evaluated. We were able to use the random forest model to achieve only an accuracy of 65.8% which was lower than we expected to achieve. KNN classifier gave the least accuracy of 20.2%. Better performance could be achieved by better feature extraction and in-depth classification of the EDF format data.

In this model, particularly classification and clustering algorithms were performed. Further advancements include regression algorithms for classification, used to predict continuous values from EEG signals. Also, dimensionality reducing algorithms can be further used to reduce the entanglement of the EEG signals.

References

1. Nakamura T, Alqurashi YD, Morrell MJ, Mandic DP (2020) Hearables: automatic overnight sleep monitoring with standardized In-Ear EEG sensor. IEEE Trans Biomed Eng 67(1):203–212
2. Estrada E, Nazeran H, Nava P, Behbehani K, Burk J, Lucas E (2004) EEG feature extraction for classification of sleep stages. In: The 26th annual international conference of the IEEE engineering in medicine and biology society, pp 196–199
3. Satapathy SK, Loganathan D, S S, Narayanan P (2021) Automated Sleep Staging Analysis using Sleep EEG signal: A Machine Learning based Model. In: 2021 International conference on advance computing and innovative technologies in engineering (ICACITE), pp 87–96
4. Nakamura T, Goverdovsky V, Morrell MJ, Mandic DP (2017) Automatic sleep monitoring using Ear-EEG. IEEE J Transl Eng Health Med 5:1–8. Art no. 2800108
5. Qureshi S, Vanichayobon S (2017) Evaluate different machine learning techniques for classifying sleep stages on single-channel EEG. In: 2017 14th international joint conference on computer science and software engineering (JCSSE), pp 1–6
6. Chambon S, Theory V, Arnal PJ, Mignot E, Gramfort A (2018) A deep learning architecture to detect events in EEG signals during sleep. In: 2018 IEEE 28th international workshop on machine learning for signal processing (MLSP), pp 1–6
7. Diykh M, Li Y, Wen P (2016) EEG sleep stages classification based on time domain features and structural graph similarity. IEEE Trans Neural Syst Rehabil Eng 24(11):1159–1168
8. Aboalayon KAI, Ukbagabir HT, Faezipour M (2014) Efficient sleep stage classification based on EEG signals. In: IEEE long island systems, applications and technology (LISAT) conference, pp 1–6
9. Zhao S, Long F, Wei X, Ni X, Wang H, Wei B (2022) Evaluation of a single-channel EEG-based sleep staging algorithm. Int J Environ Res Public Health 19(5):2845
10. Park KS, Choi SH (2019) Smart technologies toward sleep monitoring at home. Biomed Eng Lett 9(1):73–85
11. Halson SL (2019) Sleep monitoring in athletes: motivation, methods, miscalculations and why it matters. Sports Med 49(10):1487–1497
12. Aboalayon K, Faezipour M (2019) Single channel EEG for near real-time sleep stage detection. In: 2019 International conference on computational science and computational intelligence (CSCI), Las Vegas, NV, USA, pp 641–645
13. Yacchirema DC, Sarabia-Jácome D, Palau CE, Esteve M (2018) A smart system for sleep monitoring by integrating IoT with big data analytics. IEEE Access 6:35988–36001
14. Matricciani L, Paquet C, Galland B, Short M, Olds T (2019) Children's sleep and health: a meta-review. Sleep Med Rev 46:136–150
15. Wang H, Guo H, Zhang K, Gao L, Zheng J (2022) Automatic sleep staging method of EEG signal based on transfer learning and fusion network. Neurocomputing 488:183–193
16. Hussain Z, Sheng QZ, Zhang WE, Ortiz J, Pouriyeh S (2021) A review of the non-invasive techniques for monitoring different aspects of sleep. ArXiv preprint arXiv:2104.12964
17. Pan Q, Brulin D, Campo E (2020) Current status and future challenges of sleep monitoring systems: systematic review. JMIR Biomed Eng 5(1):e20921

18. Kong G, Li C, Peng H, Han Z, Qiao H (2023) EEG-based sleep stage classification via neural architecture search. IEEE Trans Neural Syst Rehabil Eng 31:1075–1085
19. Satapathy SK, Thakkar S, Patel A, Patel D, Patel D (2022) An effective EEG signal-based sleep staging system using machine learning techniques. In: 2022 IEEE 6th conference on information and communication technology (CICT), Gwalior, India, pp 1–6. https://doi.org/10.1109/CICT56698.2022.9997950
20. Al-Salman W, Li Y, Oudah AY, Almaged S (2023) Sleep stage classification in EEG signals using the clustering approach based probability distribution features coupled with classification algorithms. Neurosci Res 188:51–67

Chapter 27
FLoRSA: Fuzzy Logic-Oriented Resource Scheduling Algorithm in IaaS Cloud

Kapil Tarey and Vivek Shrivastava

1 Introduction

The cloud computing model promotes a pool of elastic resources that can be used on demand on a pay-per-use basis over the Internet. It enables service providers and users to access resources anywhere and anytime. These resources can be dynamically equipped, resourced, and removed based on consumer needs [1]. The cloud employs virtualization technology when offering IaaS. Virtualization allows the conversion of one physical machine into numerous virtual machines running different operating systems [2]. Much literature is available on cloud computing provisioning and scheduling resources.

Available literature reveals less emphasis has been given to scheduling when there are uncertainties while managing resources with strict deadlines. The cloud may deal with tasks that must be accomplished with the available resources in a given time frame. Users of cloud services might not know how to provide precise and correct CPU, bandwidth, or memory needs. By meeting these requirements, cloud computing will be more user-friendly, i.e., the system will be able to understand and meet the fuzzy needs of different users. In resource scheduling, fuzzy logic (FL) theory can solve extra difficulties caused by the user's vague, unclear, and uncertain requirements. FL models the human brain through logic and reasoning to create a "human-friendly mechanism" [3].

This paper proposes fuzzy logic-based algorithms for deciding the priority of tasks for meeting deadlines and allocating resources based on the availability and length of tasks. This paper also presents an Extended Weighted Sum Method as a multi-objective resource scheduling algorithm.

K. Tarey (✉) · V. Shrivastava
International Institute of Professional Studies, DAVV, Indore, MP, India
e-mail: kapiltareycs@gmail.com

A. Yadav et al. (eds.), *Proceedings of International Conference on Paradigms of Communication, Computing and Data Analytics*, Algorithms for Intelligent Systems,
https://doi.org/10.1007/978-981-99-4626-6_27

2 Related Work

Algorithms for resource scheduling in cloud technology are currently being extensively researched. Different resource scheduling methods are available in the literature. To some extent, these methods enhanced the efficiency of task processing and the utilization of resources [4].

Tripathy et al. [5] suggested a unique way to handle the extensive data gathered from devices by sending jobs to an environment assisted by a mist cloud environment. It improves the expected resource scheduling and enhances the response time. FL is used in this algorithm. Almutawakel et al. [6] offered a novel hybrid technique for allocating resources using three methods: Multi-Agent System (MAS), Distributed Restrictions Satisfaction Problems (DCSP), and FL. The test results revealed that this method works well regarding load balancing, cost of energy use, time to complete, and customers' rate payment gains. Aroraabout:blank et al. [7] developed a method consisting of two algorithms. The main idea behind combining algorithms is to use the best parts of each one. This article also sorted the hybrid algorithms into groups and analyzed their goals, QoS parameters, and future directions of algorithms. Faizabout: blank et al. [8] recommended a cloud selection model based on FL. Multiple parameters like cost, energy, and length are used to improve QoS. Sukhpal et al. [9] demonstrated a framework for scheduling resources in data centers that saves energy and is based on FL and energy awareness. Testing results proved that their work is better in terms of using resources and energy, as well as in terms of other QoS parameters. Tavousi et al. [10] used FL to categorize applications based on their features. An effective heuristic algorithm is then suggested to place applications on virtualized computing resources. The findings showed that the method is better because it improves the average time by 13%, deadline satisfaction by 12%, and reduction in the wastage of resources by 26%. Chakravarthi et al. [11] proposed a Normalization-based Scheduling NRBWS that improved the reliability of workflow execution and shortened the time to complete a task within a budget constraint. Abedi et al. [12] proposed IFA-DSA as an improved version of the Firefly algorithm. It focused on load-balancing optimization. The order of importance of each task is based on the pay-as-you-go model and a fuzzy approach. The findings indicated that this method outperformed the previous methods regarding the makespan criterion.

Imran et al. [13] analyzed the application of fuzzy logic in cloud computing. They concluded that fuzzy logic could solve problems and improve performance in all research areas, including cloud computing. Jin et al. [14] suggested a hybrid search algorithm that used categorization based on the normal distribution. The experimental outcome confirmed that the algorithm reduced the time and fully met the QoS by adjusting the weights of the time and cost factors. Rajakumari et al. [15] recommended two algorithms, DWRR and HPSPACO, for improving task scheduling performance by considering the priorities and length of the tasks. A fuzzy logic system has been created for scheduling tasks in the cloud by HPSPACO. The simulation revealed improvement in task scheduling by reducing execution and waiting time by 17% and increasing system throughput by 11%. Kumar et al. [16] devised a clock scheduling and fuzzy-based resource allocation to find the most appropriate resource. Raj et al. [17] presented a deer hunting optimization algorithm. Fuzzy logic was used for load scheduling. This algorithm reduced energy utilization and enhanced load scheduling. The experimental outcomes demonstrated the capable effect of this technique over the current methods. Nazeri et al. [18] anticipated a multi-criteria task scheduling with fuzzy AHP-TOPSIS. This algorithm employed FAHP to rank and FTOPSIS to choose the best cloud solutions based on the user's needs. The simulation demonstrated that this method improves power usage by 28% and response time by 32% compared to recent algorithms.

In the presented work, the algorithms FLoRSA and EWSM combine various resource scheduling parameters like cloudlet length, deadline, and the number of resources to assign a cloudlet to a resource. The data center broker decides the completion order of tasks according to FLoRSA and EWSM. A detailed comparison of FLoRSA and EWSM is presented in this work.

3 System Architecture and Methodology

The number of users in a cloud-based computing environment and the task sent to the cloud by these users are concurrently increasing. The synchronous execution of these cloudlets in a well-managed manner requires a vast amount of resources and a scheduling algorithm. In a cloud-based system, a task does have numerous attributes, including duration, deadline, energy, etc. [19]. An algorithm should consider these essential characteristics for scheduling the resources efficiently. As a contribution, a new technique for scheduling resources is suggested. In this, priority is calculated using fuzzy logic priority, optimizing the number of tasks completed, and throughput to improve performance. This paper considers task length (size), deadline, and the number of resources for task ranking. The performance indicators like throughput, turnaround time, and response time are applied to compare the efficiency of the presented algorithms. Following Fig. 1 represents the system architecture and methodology used.

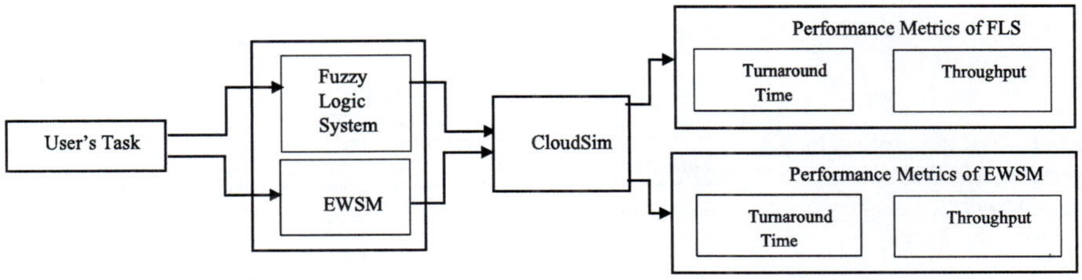

Fig. 1 Proposed model

3.1 Fuzzy Logic

Fuzzy logic can deal with data resulting from computational cognition and perception, i.e., data that is uncertain, ambiguous, insecure, partially actual, or has fuzzy boundaries. Fuzzy logic permits the incorporation of imprecise human evaluations into computing problems.

Definition: Assuming X is an arbitrary reference collection, every common subset of X and A has a characteristic function defined as follows:

$$\mu(x) = \begin{cases} 1 : x \in A \\ 0 : x \in A \end{cases}$$

3.2 FLoRSA

This algorithm uses three fuzzy parameters: task length (size), deadline, and the number of resources. Before applying the suggested fuzzy system, parameters must be converted to phase space. Tables 1 and 2 depict the fuzzy metrics used for input and output parameters.

The suggested model's input and output membership functions (MFs) are depicted in the following figures

For each linguistic variable, a membership function is defined. The optimum effect is achieved by adjusting the membership functions. In this experiment, Mamdani-type fuzzy inference logic is constructed with twenty-seven rules. The trimf membership function is used in this work to fuzzify the inputs and output. Membership function plots for the three inputs, length, deadline, and quantity of resources, are depicted

Table 1 Phase space for fuzzy input system parameters

Verbal parameters	Parameter 1	Parameter 2	Parameter 3
Low	[0 1 50]	[0 1 50]	[0 1 50]
Medium	[1 50 100]	[1 50 100]	[1 50 100]
High	[50 100 125]	[50 100 125]	[50 100 125]

Table 2 Range for each variable's verbal output parameters

Parameters	Output
Very low	[−25 0 15]
Low	[15 25 35]
Medium	[30 50 60]
High	[60 75 85]
Very high	[85 100 125]

in Figs. 2, 3 and 4. Figure 5 illustrates the output membership function plot as a dynamic priority. The system's output is to select preferences based on values of inputs as adjusted by the membership function. Since each parameter contains three membership functions, this model employs various rules in its FIS. The proposed scheduling method allocates resources to tasks under fuzzy dynamic priority.

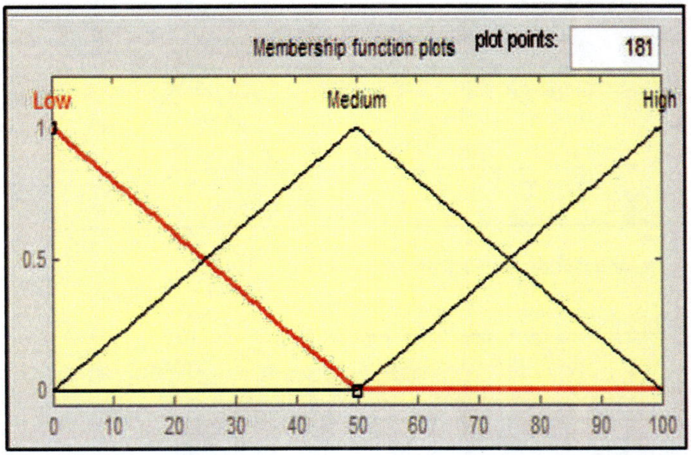

Fig. 2 MF plot for task length (size)

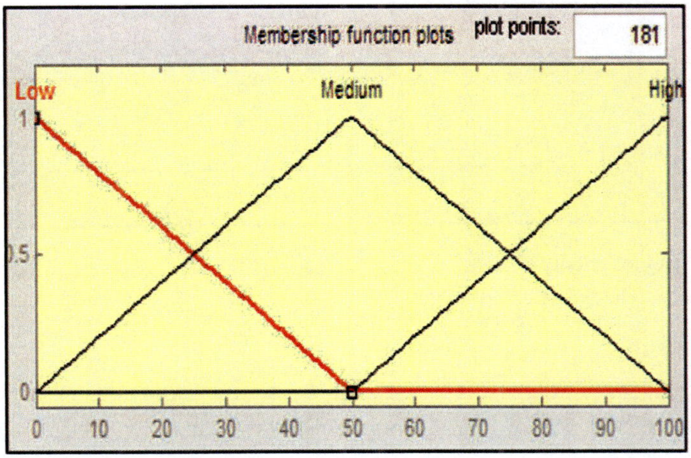

Fig. 3 MF plot for task deadline

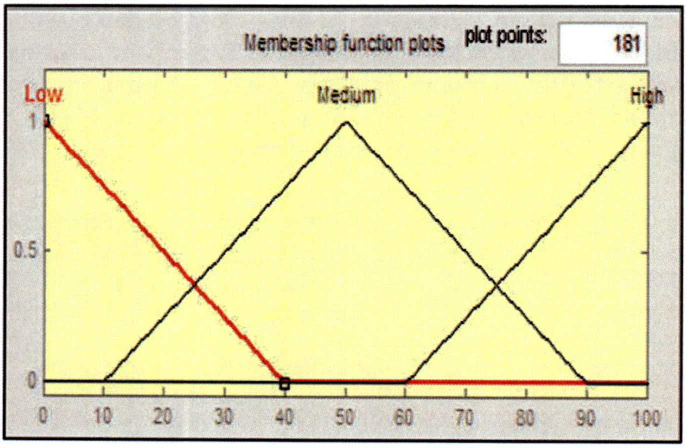

Fig. 4 MF plot for quantity of resources

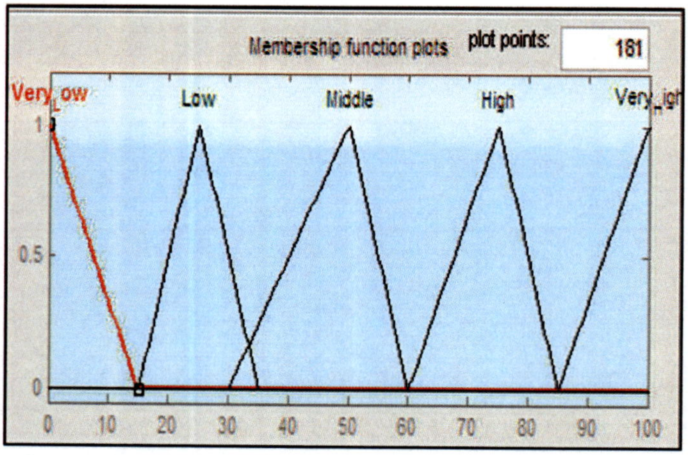

Fig. 5 MF plot for priority (Output)

3.3 Ewsm

The Weighted Sum Model (WSM) is a well-known multi-criteria decision-making technique (MCDM) frequently employed in decision theory to assess various options based on several decision criteria. This method is beneficial when two or more objective functions require optimization. It is possible to derive solutions from this method, where user-specified weights determine the optimal solution. The EWSM extends this model to include the ability to adjust the weights of each criterion based on the value of other criteria. EWSM allows for more flexible and adaptable decision-making when one criterion's impact may depend on another criterion's value. The formula for the EWSM can be expressed as:

$$\text{EWSM}_{\text{score}} = \sum_{i=1}^{n} w_i * \left(x_i + \sum_{j=1}^{n} a_{ij} x_j \right) \tag{1}$$

Steps involved in EWSM is provided in the subsequent section.

Step 1. Generation of Weighted Normalized Decision Matrix (WNDM) and choosing the criteria's weight. In this step, three criteria have been identified related to tasks: length, deadline, and the number of resources. Weights are assigned to criteria such that $w1 + w2 + w3 = 1$. Criteria weights can be beneficial (a higher value is desired) or non-beneficial (a lower value is expected).

Step 2. Compute the WNDM
The values in the decision matrix are typically converted to a normalized scale to ensure that all criteria are weighted equally in the decision-making process. This is necessary because different criteria may have different units, and directly comparing them without normalization could lead to biased or inaccurate results. Standardized formulas can be used to normalize values. The following are some of the most often-used techniques for determining the normalized value n_{ij}:

$$n_{ij} = \frac{x_{ij}}{\sqrt{\sum_{i=1}^{m} x^2 ij}} \tag{2}$$

$$n_{ij} = \frac{x_{ij}}{\max x_{ij}} \tag{3}$$

$$n_{ij} = \begin{cases} \dfrac{x_{ij} - \min x_{ij}}{\max x_{ij} - \min x_{ij}} \\ \dfrac{\max x_{ij} - x_{ij}}{\max x_{ij} - \min x_{ij}} \end{cases} \tag{4}$$

Step 3. Now multiply weights with the performance value x_{ij}. The following formula is used to calculate the weighted normalized value:

$$v_{ij} = w_j * x_{ij} \tag{5}$$

where the jth criterion has weight w_j and the sum of weights equals 1.

Step 4. In WNDM, sum up all the values in each cell, and then a preference score will be obtained.

Step 5. Order the preference scores. A set of alternatives can be ranked in descending order by preference score.

$$A_i^{\text{WSM - score}} = \sum_{j-1}^{n} w_j a_{ij}, \text{ for } i = 1, 2, 3 \ldots m$$

4 Simulation Results and Discussions

The performance of the proposed algorithms is evaluated in the CloudSim 3.0.3 toolkit, where the performance measures are turnaround time, throughput, and waiting time. CloudSim is an open-source simulation developed using Java language. It enables the modeling and simulation of IaaS cloud computing provisioning environments. This tool currently provides support for both modeling and simulation of single and internetworked clouds [20]. The algorithms aim to meet the multi-criteria QoS requirements from cloud users' and providers' perspectives by minimizing the user's turnaround time and maximizing the providers' throughput. The simulation uses 2 data centers and two hosts, with different parameter values for the VM and cloudlet, as given in Table 3. A dataset of 2200 workloads in used in the experiment to compare performance metrics obtained from the FLoRSA and EWSM.

Performance Improvement Rate (PIR) [21, 22] is also computed. PIR indicates the performance improvement in the percentage of various performance metrics. It is calculated using the following formula:

$$\text{PIR}(\%) = \left(\frac{\text{Metric(First Algorithm)} - \text{Metric(Second Algorithm)}}{\text{Metric(First Algorithm)}} \right) * 100$$

Metric(First Algorithm) refers to the performance metric of FLoRSA, and Metric(Second Algorithm) refers to the performance metric of EWSM.

FLoRSA PIR for average turnaround time and average throughput is 18 and 20% more than EWSM. The comparison of these performance parameters for FLoRSA and EWSM is shown in Figs. 6 and 7, respectively.

Table 3 Parameters used in the simulation

Parameters	Values
VM parameters	
RAM	512 MB
MIPS	1000
Bandwidth	1000
Number of CPUs	1
VMM	Xen
Cloudlet parameters	
Length	1000
File size	300
Output size	300
Number of CPUs	1

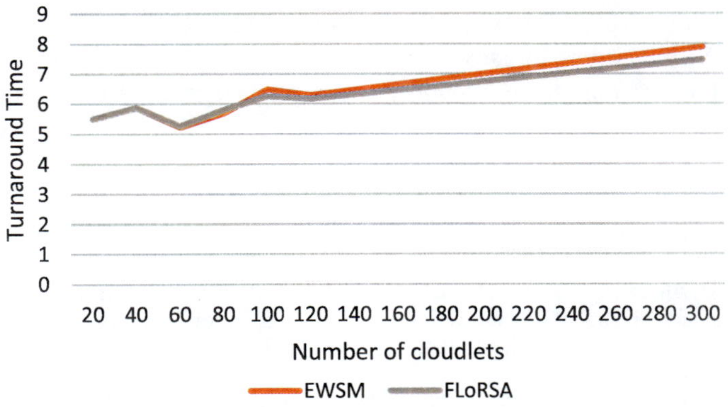

Fig. 6 Average turnaround time

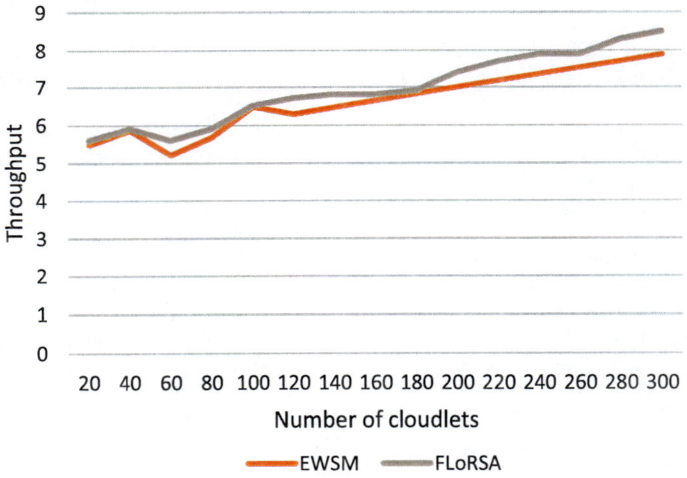

Fig. 7 Average throughput

5 Conclusion and Future Directions

Scheduling resource requests from a vast pool of resources with the imprecise need of the user for multi-objective optimization is challenging. This work presented two new algorithms, FLoRSA based on fuzzy logic and EWSM based on the MCDM technique with weights calculated by the BWM. The cloudlets are ranked according to three criteria: cloudlet length, deadline, and the number of available resources. Both algorithms use ranks based on performance scores to schedule and execute resource requests. The performance metrics of FLoRSA and EWSM are determined using CloudSim 3.0.3. The experimental results demonstrate that FLoRSA outperforms EWSM regarding turnaround time and throughput. Introducing new criteria for evaluating and ranking the cloudlets can be another research direction.

References

1. Hassan J, Shehzad D, Habib U, Aftab MU, Ahmad M, Kuleev R, Mazzara M (2022) The rise of cloud computing: data protection, privacy, and open research challenges—a systematic literature review (SLR). Comput Intell Neurosci 2022(1):1–26. https://doi.org/10.1155/2022/8303504
2. Munir A, Kansakar P, Khan SU (2017) IFCIoT: Integrated fog cloud IoT: a novel architectural paradigm for the future Internet of Things. IEEE Consumer Electron Magaz 6(3):74–82. https://doi.org/10.1109/MCE.2017.2684981
3. Gentili PL (2022) Implementing fuzzy sets and processing fuzzy logic information by molecules. In: Proceedings, vol 81, no 1. MDPI, p 94. https://doi.org/10.3390/proceedings2022081094
4. Chen Z, Zhu Y, Di Y, Feng S (2015) A dynamic resource scheduling method based on fuzzy control theory in cloud environment. J Cont Sci Eng 2015:1–10. https://doi.org/10.1155/2015/383209
5. Tripathy SS, Mishra K, Barik RK, Roy DS (2022) A novel task offloading and resource allocation scheme for MIST-assisted cloud computing environment.Intell Syst 103–111.https://doi.org/10.1007/978-981-19-0901-6_10
6. Almutawakel A, Kazar O, Bali M, Belouaar H, Barkat A (2019) Smart and fuzzy approach based on CSP for cloud resources allocation. Int J Comput Appl 44(2):117–129. https://doi.org/10.1080/1206212x.2019.1701241
7. Arora N, Banyal RK (2022) Hybrid scheduling algorithms in cloud computing: a review. Int J Electr Comput Eng (IJECE) 12(1):200–207. https://doi.org/10.11591/ijece.v12i1
8. Faiz M, Daniel AK (2022) Multi-criteria based cloud service selection model using Fuzzy Logic for QoS. Commun Comput Inf Sci 12(1):153–167. https://doi.org/10.1007/978-3-030-96040-7_12
9. Tavousi F, Azizi S, Ghaderzadeh A (2021) A fuzzy approach for optimal placement of IOT applications in fog-cloud computing. Clust Comput 25(1):303–320. https://doi.org/10.1007/s10586-021-03406-0
10. Chakravarthi KK, Neelakantan P, Shyamala L, Vaidehi V (2022) Reliable budget aware workflow scheduling strategy on multi-cloud environment. Clust Comput 25(2):1189–1205. https://doi.org/10.1007/s10586-021-03464-4
11. Abedi S, Ghobaei-Arani M, Khorami E, Mojarad M (2022) Dynamic resource allocation using improved firefly optimization algorithm in cloud environment. Appl Artif Intell 36(1):601–611. https://doi.org/10.1080/08839514.2022.2055394
12. Tariq MI, Tayyaba S, Ali Mian N, Sarfraz MS, Hussain A, Imran M, Pricop E, Cangea O, Paraschiv N (2020) An analysis of the application of Fuzzy logic in cloud computing. J Intell Fuzzy Syst 38(5):5933–5947. https://doi.org/10.3233/jifs-179680
13. Jin MAOZHU, Chen PENG, Malaikah HUNIDA, Chen CHAO, Liu YIFENG (2022) Research on fuzzy scheduling of cloud computing tasks based on hybrid search algorithms and differential evolution.Fractals 30(02):2346–2354. https://doi.org/10.1142/s0218348x22400837
14. Rajakumari K, Kumar MV, Verma G, Balu S, Kumar Sharma D, Sengan S (2022) Fuzzy based ant colony optimization scheduling in cloud computing. Comput Syst Sci Eng 40(2):581–592. https://doi.org/10.32604/csse.2022.019175
15. Srinivasa Kumar C, Sirisati RS, Srinivasa Rao M, Narayana MV, Rajeshwar J (2022) An optimized fuzzy-based resource allocation for cloud using secured Tabu Search Technique. Innovat Comput Sci Eng 4(1):157–164. https://doi.org/10.1007/978-981-16-8987-1_17

16. Joshua Samuel Raj R, Ilango V, Thomas PR, Uma VN, Al-Wesabi F, Marzouk R, Mustafa Hilal A (2022) Improved dhoa-fuzzy based load scheduling in IOT Cloud Environment. Comput Mater Cont 71(2):4101–4114. https://doi.org/10.32604/cmc.2022.022063
17. Nazeri M, Khorsand R (2022) Energy aware resource provisioning for multi-criteria scheduling in cloud computing. Cybern Syst 3(4):1–30. https://doi.org/10.1080/01969722.2022.2071409
18. Greenstreet P (2020) Weighted sum approach. Peter Greenstreet. Retrieved February 16, 2023, from https://www.lancaster.ac.uk/stor-i-student-sites/peter-greenstreet/2020/04/24/weighted-sum-approach/#:~:text=In%20the%20weighted%20sum%20approach,to%20assign%20to%20each%20objective
19. Shrimali B, Patel H (2020) Multi-objective optimization oriented policy for performance and energy efficient resource allocation in cloud environment. J King Saud Univ Comput Inf Sci 32(7):860–869. https://doi.org/10.1016/j.jksuci.2017.12.001
20. Vidhya M, Devi R (2022) Comparative analysis of scheduling algorithms in cloud computing using cloudsim. 2022 11th international conference on system modeling and advancement in research trends (SMART). https://doi.org/10.1109/smart55829.2022.10047689
21. Fishburn PC (1967) Letter to the editor—additive utilities with incomplete product sets: application to priorities and assignments. Oper Res 15(3):537–542. https://doi.org/10.1287/opre.15.3.537

Chapter 28
Real-Time Audio Communication Using WebRTC and MERN Stack

Soham Sattigeri and Shripad Bhatlawande

1 Introduction

Communication is a basic human need. It involves sharing ideas, facts, and opinions with each other. It's a process of impressing and expressing, which improves mental health in society. It's a two-way process which involves the transmission and receiving of data between two or more entities. It involves the process of sharing ideas, facts, opinions, feelings, or attitudes. It is a two-way process that involves the transmission and interaction of information or messages between at least one sender and receiver. It is essential in management and forms the basis of all human interaction [1]. Technology has led to many advancements and enhancements in terms of communication technologies. These improvements include landlines replacing telegraphs and then mobile phones replacing landlines. This continuous enhancement in technology increases the scope of research in this field, leading to further innovations. With the innovation of the internet, various new sources of communication came into existence. These include email, WhatsApp, Skype, etc. Sab Sunno is an attempt to improve this research by creating a platform that allows people to connect with other people around the world. Voice over Internet Protocol is an important protocol that helps people to communicate with other people over the internet. Along with VoIP, other protocols such as WebRTC and DASH make it easy to transfer audio/video data over the internet [2]. This paper attempts to provide an approach to implementing audio communication over the internet using web sockets and the WebRTC protocol. This is an attempt to simplify the process of audio communication and make it available to the general public.

S. Sattigeri (✉) · S. Bhatlawande
Vishwakarma Institute of Technology, Pune, 37, India
e-mail: soham.sattigeri20@vit.edu

A. Yadav et al. (eds.), *Proceedings of International Conference on Paradigms of Communication, Computing and Data Analytics*, Algorithms for Intelligent Systems,
https://doi.org/10.1007/978-981-99-4626-6_28

2 Literature Review

Peer-to-peer communication between users' browsers is made possible through the WebRTC protocol. The primary application of WebRTC is not limited to the process's video/audio implementation. The STUN and TURN servers that make up WebRTC are crucial in the release of ICE candidates, which subsequently aid in connecting the audio streams on the devices. A signaling server that serves as a middleman between the two local devices is a key component of WebRTC [3]. Luka Skeledzija et.al provides valuable insights into the usage and popularity of Clubhouse, an audio-based social media platform. They provide insight in identifying the most talked-about topics on the platform and examine their effect on the platform's popularity [4]. Bidirectional and low-latency competencies are important for any real-time communication. SocketIO documentation talks about an open-source library that provides event-based communication between web clients and servers. WebSockets, long polling, etc., are some concepts used in this library that provide low-latency communication between clients and servers [5].

ReactJS, a JavaScript framework is used for user interfaces. Figure 1 demonstrates the usage of ReactJS in the past years [6]. Material-UI (MUI) is a popular open-source JavaScript library for building user interfaces using the Material Design guidelines developed by Google. This allows developers to create visually consistent and appealing web applications easily. This has been used in the application making it easy to develop interactive designs. MUI is built on top of React which allows the building of web applications using a component-based architecture [7]. TypeScript is a static and a strongly typed programming language. A superset of JavaScript, meaning that all valid JavaScript code is also valid in TypeScript. ReactJS has adopted the usage of TypeScript due to its typing capabilities. It also offers seamless integration with it [8]. Handley et al. describe the design and implementation of the Session Description Protocol (SDP), which provides a standardized format for describing multimedia communication sessions. SDP is used to exchange information about the parameters of a multimedia session, such as the media types and codecs being used, and is an essential component of many multimedia communication systems [9]. As outlined in RFC 4566, the SDP specification defines the structure and syntax of SDP messages, which are used to negotiate multimedia sessions between endpoints. It provides a format for describing the characteristics of a multimedia session, including the media types, codecs, and transport addresses. SDP messages are typically exchanged using the Session Initiation Protocol (SIP), a signaling protocol for initiating, maintaining, modifying, and terminating multimedia sessions [10].

An explanation of the Interactive Connectivity Establishment (ICE) protocol is provided in [11] a technical specification. It is a method for getting through firewalls and network address translators (NATs) in real-time multimedia sessions like video conferencing and VoIP. In order to create multimedia sessions between endpoints, the ICE protocol was created to cooperate with the Session Initiation Protocol (SIP)

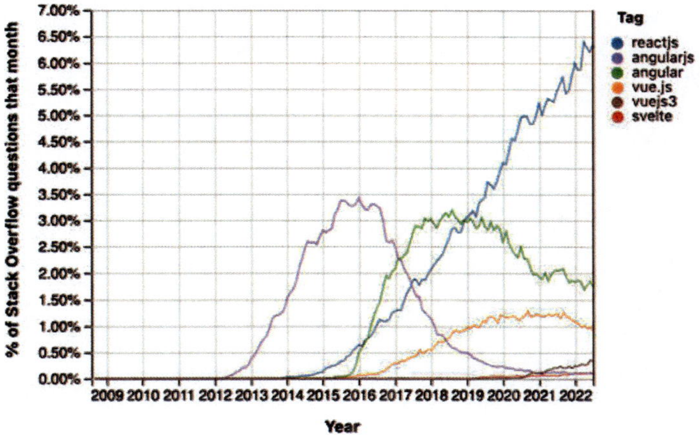

Fig. 1 Widely used frameworks around the world

and the Session Description Protocol (SDP). It uses STUN (Session Traversal Utilities for NAT) and TURN (Traversal Using Relays around NAT) to traverse NATs and firewalls. WebSockets is used between a client and a server for bi-directional communication. It contains a full-duplex communication channel which provides an event-based data connection over a single TCP connection, allowing for efficient and low-latency communication between two entities [12].

The contemporary system is bulky and hardware-oriented. It includes the use of VoIP protocol, which is not a useful application for conference calls. This paper tries to provide an alternative solution using web sockets and WebRTC, which makes it easy to communicate over the internet.

3 Methodology/Experimental

Development of an audio-based web application using the MERN stack (MongoDB, Express, React, and NodeJS). The methodology is briefly divided into three parts namely (1) System Architecture (2) Authentication. (3) Communication and Interfaces.

3.1 System Architecture

Figure 2 explains the complete architecture of the application. The application consists of a front-end web app created using ReactJS and a mobile app created

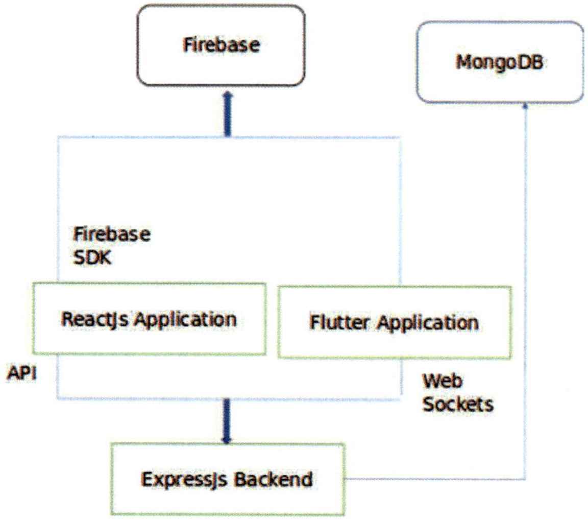

Fig. 2 The system architecture of the Sab Sunno product

using Flutter along with the backend. The product utilizes Firebase and its products to simplify and delegate some complex tasks such as authentication and cloud storage. The product uses MongoDB as the main database and stores all the text data.

The backend of the application is built using JavaScript. The application uses NPM (Node Package Manager) and is used to install dependencies such as Express and MongoDB. It's an HTTP (HyperText Transfer Protocol) and a socket express server that serves API to the front end. A Git repository is also initialized for version control on GitHub. MongoDB, a NoSQL database, is used in the application. Data are stored in documents that are part of a collection. The documents can be compared with rows in a SQL database like MySQL. The backend is also a signaling server for the front-end application. This helps to transfer session data by using web sockets to initiate the connection. WebRTC is the main component for the presented system since it is the protocol which is used to transfer audio streams from one user to another. It also has other applications including gaming, streaming, etc. WebRTC uses browser session data to communicate with browsers. A WebRTC communication consists of setting the local description and remote description in the browser for each user. When a user establishes a webRTC connection, the user makes an offer that, when set to local description, initiates the transmission of ICE candidate data to the other browser. The other browser then receives the ICE candidate data from the signaling server, which then sends the browser the response. This establishes a connection between two browsers that can then be used to send data.

Redux is a global state management tool which is used to store and manage the global state of the application. The state is stored in a global data collection known as a store which can be accessed from anywhere in the application. Actions are listeners which are triggered to change the state of the application whenever the desired value changes. Reducers respond to the actions and update the state. The store is connected to the render tree using the React-Redux library. Along with the store when an event occurs, a notification is dispatched to inform the user of the event. The handling of side effects, such as sending alerts, performing API calls, or managing asynchronous operations, is also possible with the usage of middleware like Redux-Thunk or Redux-Saga. Overall, the application's use of Redux makes it simple to maintain the state and provide alerts to users informing them of events that occur within the application. The mobile application uses the Flutter Provider Package to maintain the state and supply it to various application components. The Provider Package allows for an easy and efficient way to share data across widgets in the app without having to pass data through multiple levels of widgets.

The Provider is a state manager which is created at the root of the app and holds the state, which is then passed down to the widgets that need it using a Consumer widget. A listener known as the Consumer widget listens for changes in the state of the app and rebuilds the entire widget which depends on the state. It also manages the state of the user's login status, which is an important feature in the application. The state is stored in a ChangeNotifier class which notifies all the listeners when the state changes. This class is wrapped in a ChangeNotifierProvider at the root of the app, and the state is accessed by the widgets that need it using a Consumer widget.

3.2 Authentication

Sab Sunno applications uses google firebase and firebase cloud storage as an authentication and storage system. Firebase has been implemented in the application to handle registration and log in process for the users to easily create an account and sign in to the application. Along with authentication, it also provides storage which to store and retrieve user data and images on the cloud. On the side of mobile applications, OTP authentication is used. The Firebase Storage service is used in the application to store the images that the user uploads and retrieve them when necessary. It also provides security and access control features, which allows the Sab Sunno application to ensure that only authorized users can access the stored data.

Google authentication is a feature that is essential considering a UX perspective. Google authentication is also integrated with Firebase. Cloud storage is another service provided by Google Firebase. Cloud storage in Sab Sunno is used to add images and profile photos. Firebase documentation provides methods and examples to upload and modify images.

3.3 Communication and Interfaces

Figure 3 WebRTC stands for Web Real-Time Communication. It is a communication protocol, which is used to communicate between two clients. It does contain a server–client architecture. Along with WebRTC, Docker is used to deploy the containerized application. Docker is open-source software that helps developers to

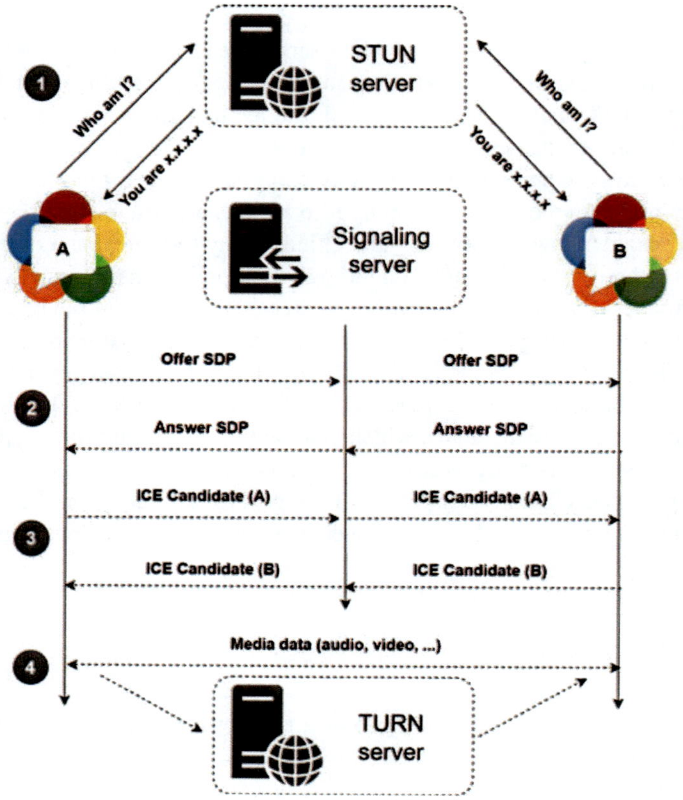

Fig. 3 Functioning architecture of a WebRTC-based architecture [13]

manage containers. A container is in fact a box that contains all the required components for the software. Docker is a delivery system that is used to serve containers. The mobile app consists of features to login and connect with the users on Sab Sunno and chat with them. The architecture has been developed keeping in mind the privacy of users. The mobile application (developed using Flutter) has been developed considering the privacy of the user. The person can only connect with each other if they accept the request to connect. The users have the facility whether they want to connect with the other user or not

The complete mobile system works by requesting and accepting the connection request. If a user wishes to connect with another user, the user sends a connection request to the said user. This communication happens to the HTTP. If the user accepts the connection of the user, they can transfer messages between each other. This communication happens through web sockets. In order to communicate between two clients over the WebRTC protocol, there is a need for a signaling server that acts as a medium to transfer the session data of each client with each other. This transfer of data is also handled using web sockets. Once the connection is established between the two clients, the signaling server plays no role. The clients can communicate directly with each other over the WebRTC protocol. WebRTC uses UDP to transfer data between clients. This protocol is important for audio and video communication over the internet.

The front end of the Sab Sunno application is composed of two distinct parts: a ReactJS web application and a Flutter mobile application. Both of these applications have implemented instances of the Firebase SDK, which is utilized for authentication and authorization purposes. Additionally, the Firebase SDK's cloud storage system is utilized to store images on the cloud. To sum up, the Sab Sunno application utilizes Firebase SDK for authentication and authorization, and cloud storage while the backend interacts with MongoDB database through a restricted URL and the front end communicates with the backend through REST API.

4 Results and Discussion

The application has been created keeping in mind the user's privacy and functionality.

As shown in Table 1, Sab Sunno is available on the web and as an Android application and is still available only in development mode. The production-grade application of this project is released on Netlify. The backend server is deployed on render and uses nodemon for the run command.

A single room in the application, where users interact. Users can mute/unmute themselves and can go back to the home page to explore more rooms and interesting topics. Sab Sunno is an attempt to create an infrastructure that will help users create connections and connect with more people worldwide. Though this is a full-stack application, there are some limitations. One of the features which are present in social media applications includes the following. The users cannot follow each other as of now in the application. The rooms which are available on the homepage are still static. Application for machine learning algorithms like Random forest and other clustering algorithms makes this dynamic. Since this is still in the development phase, the website is not completely scalable and is not fit to be used by a large number of people.

Table 1 User data of the Sab Sunno application

Statistics	Value
Active users	52
Data on cloud	124 MB
User	85%

5 Conclusion

This application provides a solution to increase the sense of expression in society. The application utilizes a combination of technologies and has a lot of potential for future expansion. While the current solutions either use VoIP or WebRTC, this project takes their use to a better level. By making it type-safe and containerized, it can be deployed on Kubernetes for maximum scaling. Potential areas for growth could include the implementation of new features such as real-time collaboration and notifications, integration with other services, optimization for scalability, enhancement of security measures, development of a Progressive Web App version, adding support for more languages and localization, and incorporating AI and machine learning capabilities to improve performance and provide a more personalized user experience.

References

1. Verderber JM, Verderber RF (2022) Effective communication: Improving the quality of communication in the workplace [Accessed: August 2022]
2. Kelly SD, W SF (2022) A survey of voice over IP security research. Provides an overview of the security challenges and solutions in VoIP systems [Accessed: August 2022]
3. WebRTC.org (2022) WebRTC: a real-time communication protocol. [Online]. Available: https://webrtc.org/ [Accessed: August 2022].
4. Skeledzija L, Skeledzija E (2022) Clubhouse: determining the most talked about topics and their effect on the platform's popularity. University of Ljubljana, Slovenia, [Online]. Available: https://www.ict-conf.org/wp-content/uploads/2021/07/04_202106C025_Skeledzija.pdf. [August 2022]
5. Socket IO community (2022) SocketIO, Bidirectional and low-latency communication for every platform. [Online]. Available: https://socket.io/. [Accessed: Aug 2022]
6. Team—React, ReactJS, A Javascript framework for making front-end technologies. [Online]. Available: https://reactjs.org/. [Accessed: July 2022]
7. MUI Community (2022) MUI: ReactJS component library. [Online]. Available: https://mui.com/. [Accessed: Aug 2022]
8. Hejlsberg A (2022) Typescript documentation [Online]. Available: https://www.typescriptlang.org/. [Accessed: July 2022]
9. Handley M, Jacobson V, Perkins C (2022) SDP: session description protocol. IETF RFC 4566, July 2006. [Accessed: August 2022]
10. Rosenberg J, Schulzrinne H, Camarillo G, Johnston A, Peterson J, Sparks R, Handley M, Schooler E (2022) SIP: session initiation protocol. IETF RFC 3261, June 2002. [Accessed: Aug 2022]
11. Rosenberg J (2022) Interactive connectivity establishment: a protocol for network address translator traversal for Offer/Answer Protocols. IETF RFC 5245, 2010. [Accessed: August 2022]
12. IETF (2022) WebSocket, documentation for web sockets," [Online]. Available: https://tools.ietf.org/html/rfc6455. [Accessed: Sept 2022]
13. Working of web RTC [Online] Available: https://docs.freedomrobotics.ai/docs/how-webrtc-works. [Accessed: Sept 2022]

Chapter 29
Ego Network Analysis Using Machine Learning Algorithms

S. Vaibhav, M. P. Dhananjay Kumar, Tejashwini Hosamani, Vrunda Patil, and S. Natarajan

1 Introduction

Online social networks (OSNs) have become at present, a widely used way for individual users to interact and connect with each other. Studies have found a relationship between OSNs and the connections people have in their real lives. With billions of users, OSNs have generated enormous amounts of data on the interactions between individuals. A social network consists of individuals who have formed relationships with family and friends, or other people with common interests and this makes up the social network.

Social network analysis can provide insightful information on how people interact with one another. For analysts in various fields, it can be useful to examine the connections between people based on their activities on their social media platforms. Synthetic datasets, however, can operate as a useful and essential replacement when there is either insufficient or no empirical data. Rather than creating binary social graphs, utilizing expanded social networks may be a more appealing option. This study focuses on Facebook's interaction parameters as the basis for analysis. Strangely enough, the survey suggests that having more virtual connections is not proportional to the increase in texts between users.

The goal of this study seeks to determine the feasibility of generating a synthetic, enlarged social graph, and it presents the findings of a thorough examination of Facebook, one of the most widely used online social networks. This approach enables future researchers to explore the potential of expanding social graphs without relying on realistic, actual data. Social networks refer to networks of relationships in which

S. Vaibhav (✉) · M. P. Dhananjay Kumar · T. Hosamani · V. Patil · S. Natarajan
PES University, Bangalore 560085, India
e-mail: svaibhav0458@gmail.com

S. Natarajan
e-mail: natarajan@pes.edu

A. Yadav et al. (eds.), *Proceedings of International Conference on Paradigms of Communication, Computing and Data Analytics*, Algorithms for Intelligent Systems,
https://doi.org/10.1007/978-981-99-4626-6_29

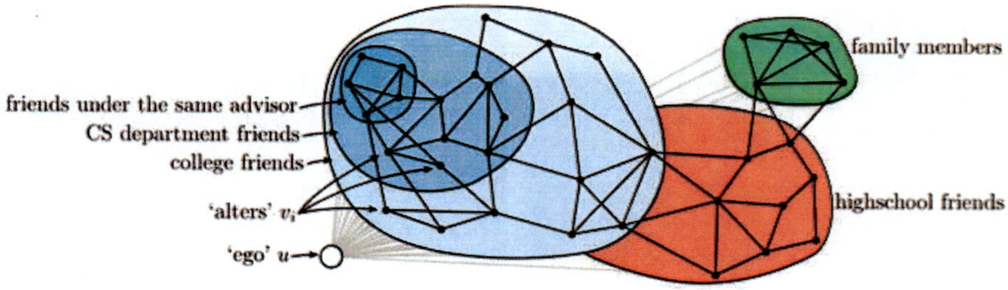

Fig. 1 Ego network with labeled circles

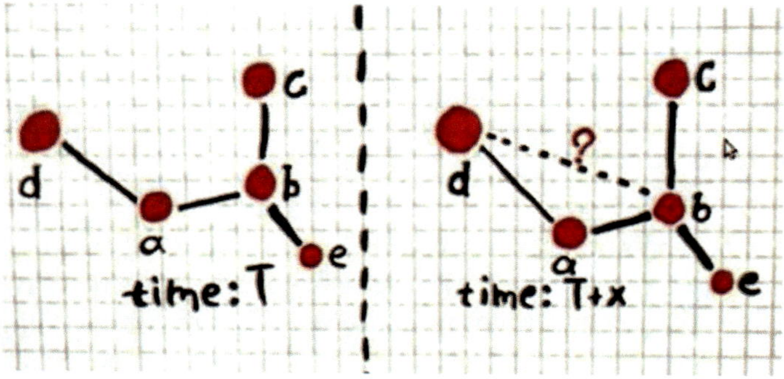

Fig. 2 Edges at time t and $t + x$ between the nodes

the nodes represent individuals or actors, and the edges represent the connection between the users(nodes) which represents different type of connections. Figure 1 [1] above represents an ego network with labeled circles where each and every node represents an individual, and these individuals are connected to form alters.

Now assume an instance of set of nodes (users) in a social network graph is provided, considering any 2 nodes(users) having no relationship in the current state of the graph, the goal is to identify the influencing (important) users from a given collection of nodes (users) and predict the probability of future links formation (edge) between those two nodes (users). Figure 2 above represents a network which gives a relation between 2 nodes at time T min and $T + x$ min.

The types of connections mentioned in the problem statement could vary and include friendship, collaboration, following, or shared interests. Our focus in this study is on the social network of Facebook, with the following objectives in mind:

- Developing a recommendation system for a user's friends.
- Identifying potential hidden links and key influencers in a terrorist social network.
- Targeted marketing through highly influential individuals and identifying potential customers.
- Suggesting new collaborations or interactions within an organization. Link prediction can be used in bio-informatics as well, to detect protein interactions.

2 Literature Review

This literature survey focuses on community discovery in social networks using ego network analysis. The survey reviews machine learning models such as support vector machines (SVM), XGBoost, and ensemble models used in ego network analysis. The study evaluates the effectiveness of these models in identifying community structures within social networks. The survey provides insights into the strengths and limitations of different machine learning approaches in ego network analysis for community discovery.

Jia et al. [2] focus on analyzing the social networks of chronic pain patients using local structure and centrality measures. The study uses ego-centric networks to understand the impact of chronic pain on the social support and social integration of individuals. The study uses various measures like degree, clustering coefficient, betweenness, and eigenvector centrality to analyze the local structure and centrality of chronic pain patients' social networks. The findings of the study suggest that chronic pain patients have smaller and less dense social networks compared to healthy individuals, which can inform interventions to improve their quality of life.

Kumar et al. [3] focus on analyzing the diffusion of information in an ego-centric Twitter network. The study uses data from Twitter to construct ego-centric networks for individual users and then analyzes the diffusion of information within these networks. A simple study showed that the size and the models of ego-centric network plays a pivotal role in, spreading of information, and the users who have a higher amount of influence tend to spread information faster and to a greater number of people. The study provides insights into the mechanisms of information diffusion in social networks.

Rezaeipanah et al. [4] put forward a method for online multiplex ego networks with link prediction. The proposed method uses features like common neighbors, Jaccard's coefficient, and adamic adar index to train a classification model to predict the likelihood of links between individuals. The study uses data from a multiplex online ego network to test the proposed method and compares it with existing link prediction methods. The results conclude that in terms of predicting links in multiplex online ego networks, the indicated operation outperforms the alternatives. The study provides insights into improving the link prediction accuracy in OSNs.

Long et al. [5] proposed a method to enhance the accuracy of social network analysis by incorporating the strength of social ties. The study uses data from a Chinese online social network to demonstrate the efficiency of the suggested approach in identifying key players and communities in social networks. The survey findings also tells us that the closeness in online groups is a important contributor to social network analysis, and the current working method can give a more informed studies about social network workings and of the model itself. The study provides insights into improving social network analysis and can inform strategies for effective social network interventions.

Humski et al. [6] propose a method for generating synthetic social graphs based on Facebook interaction data. The study uses data from Facebook to analyze the patterns of social interactions between individuals and then uses these patterns to generate synthetic social graphs. The proposed method aims to address the privacy concerns associated with sharing real-world social network data. The findings suggest that the proposed method can accurately generate synthetic social graphs that preserve the patterns of social interactions in the real-world social network. The study provides insights into generating synthetic social graphs and can inform strategies for data privacy and social network analysis.

Kwon et al. [7] explore a social network's online self-disclosure behavior which is impacted by ego networks and communities. The study uses data from a popular Korean social networking site to analyze the self-disclosure behavior of users within their ego networks and communities. A small study also showed that people who use social media and tend to converse about their personal lives within their ego networks and groups of friends and that the number of people in a body of ego networks plays a principal role in making them show self-disclosure tendencies. The study provides insights into the factors influencing self-disclosure behavior in online social networks.

Granitzer et al. [8] described a working model showing feature learning in ego network analysis. This research uses data from a large online social network platforms to analyze connections between people and their friends within their ego networks. The given procedure uses both global and local features to study an image of the ego network and identifies key users and communities within it's network. The research shows that the proposed method can pin pointedly capture the structure and working with insisted insights into the characteristics of individuals and communities within the circle.

Madani et al. [9] propose an approach to determine smaller networks across ego networks. The study explores the connections between users and their relationships within their ego networks by utilizing data from a big online social network. The proposed approach identifies important characters and sub-networks within the ego network through clustering and centrality measurements. The results obtained suggest that the proposed approach might accurately reflect the structure and dynamics of smaller networks within ego networks, as well as provide details about the characteristics of both people and groups within the network. The study provides insights on methods to estimate the accuracy of ego network analysis.

Arnaboldi et al. [10] provided a model for analyzing the structure of ego networks in online social networks. It explores the relationships between users and their connections within their ego networks utilizing data from a big online social network. The proposed method detects popular individuals and groups within the ego network using centrality measurements and clustering factors. The findings suggest that the proposed approach may properly determine the structure and dynamics of ego networks as well as provide insights on the attributes of both individuals and groups within the network.

McAuley et al. [11] proposed a ML approach to discover social circles within ego networks. This study uses data from a large online social network to identify the interactions between the sets of nodes. The given approach uses a method called graph partitioning algorithm with machine learning techniques to show social circles based on features such as homophily, triadic closure, and node similarity. The research suggests that the proposed approach can accurately capture the structure of social circles within ego networks and gives certain information about the characteristics of individuals and it's neighbors within the network. The study provides insights into enhancing the accuracy of social circle detection in ego networks.

Singh et al. [1] propose a method for creating multiple networks based on various factors such as subscriber relationships, shared interests, and common locations. These networks are then examined using community detection algorithms to identify clusters of users who are likely to be friends. Machine learning models are then trained on these clusters to predict new friend recommendations. The proposed method outperformed existing methods in terms of precision and recall. The study shows that combining multiple networks and using machine learning techniques can significantly improve friend suggestions on social media platforms.

Tabourier et al. [12] focus on the formation of new links in a network based on past evolution data. Several approaches, including time-aware link prediction, temporal path prediction, and dynamic network embedding, have been proposed. Time-aware link prediction predicts the likelihood of a new link forming by using temporal features such as node age and time of last interaction. Temporal path prediction uses temporal information to predict which path nodes will take in future. Dynamic network embedding is a technique for learning a low-dimensional representation of a network that captures its temporal evolution. These methods have yielded promising results in predicting links in ego networks, with applications in social network analysis and recommendation systems.

3 Methodology

In this paper, we provided our study in three stages. In the first stage, we gathered the dataset and pre-processed it provided by SNAP, and this was then followed by dividing the dataset into Train: Validation: Test in the ratio 2:1:1. In the second stage, ML models were trained and tested with the dataset for link prediction and in the last stage compared different ML models with performance metrics to have a better understanding of the association between nodes.

The techniques employed in this paper include analyzing measures of centrality, implementing link prediction (it entails awarding a score to the relationship between node pairs based on the input graph), and utilizing various approaches that can be categorized as follows.

3.1 Dataset and Preprocessing

Stanford network analysis project (SNAP) provides the ego-Facebook dataset with the nodes and edges. The dataset contains anonymized ego networks of Facebook users and their friends, including information on the user's demographics, network structure, and interactions. After loading the dataset, the dataset is divided into positive and negative Train: Validation: Test in the ratio 2:1:1 for link prediction.

3.2 Link Prediction

Approaches based on node neighborhoods involve several approaches that rely on the premise that two nodes(x, y) have a tendency to build a relationship in the upcoming times if their pair of neighbors have a significant overlap like common neighbors, preferential attachment, jaccard coefficient, etc. To identify potential links between nodes, some methods focus on considering all possible paths between them, instead of just the shortest path. These methods refine the concept of distance between nodes by taking into account the overall ensemble of all possible paths.

Centrality measures can help in link prediction by identifying nodes that are influential in the network and likely to form new connections. By examining the network topology, nodes with high betweenness centrality may act as critical bridges between different communities and facilitate the formation of new links. Similarly, nodes with high degree or eigenvector centrality may have a higher probability of forming new links due to their prominent position in the network, making them valuable targets for link prediction algorithms. Overall, centrality measures provide useful insights into the network's structure and dynamics, aiding in predicting future connections.

In this review, the similarity of features associated with each node in the ego network as a basis. As there were multiple features per node, this approach to ensure accurate results machine learning models like support vector machine can be used in link prediction by considering the existing network structure as a graph and treating it as a binary classification problem. The SVM algorithm learns to separate nodes that are connected by links from nodes that are not connected. By training on a set of labeled examples, it can predict whether a link should exist between two nodes that have not yet been connected. Similarly, extreme gradient boosting (XGBoost) is a popular machine learning algorithm applied for link prediction works. In link prediction, the aim is to foresee missing or future links between nodes in a network. XGBoost is used to learn a structure from the network structure and features of nodes and predict the probability of a relationship between any two nodes. The code model works by repetitively training decision trees based on the errors of it's predecessor, optimizing a loss function that balances accuracy and complexity.

3.3 Performance Metrics

Ego network analysis is a popular method for community discovery in social networks. The working of ego network-based community detection algorithms can be measured using various metrics, such as modularity, conductance, coverage, and homogeneity. Modularity analyzes, to what degree the communities differ from each other, conductance gives an accurate measurement of a well-defined community. Coverage specifies the proportion of nodes assigned for a community, and homogeneity assesses the degree of similarity between community members. These metrics help evaluate and compare the effectiveness of different community detection algorithms. Knowing accuracy of many models of machine learning, and also have utilized a variety of machine learning frameworks, including SVM, KNN, and random forest, XGBoost, and neural network to compute accuracy, precision, and F1-score, which helped us determine the best model for predicting links between nodes after time $T + x$.

4 Results

In this paper, based on the literature survey, an experiment was conducted for getting better ideas related to the problem statement. To begin with, we implemented the code to calculate centrality measures and plotted a graph at time T and $T + x$ that shows the association between nodes based on the similarities they have with other nodes. Later, the dataset was divided into train, valid, and test sets to perform link prediction using a variety of ML models such as SVM, KNN, random forest, XGBoost, and neural network. Then compute the accuracy, precision, and F1-score of all ML models from which the results of all ML models are compared. After comparing the results of all the ML models, the ensemble model(XGBoost and SVM) was the most effective, with an accuracy of 92%.

Table 1 gives a better understanding of the models by comparing the performance metrics. It can be seen that the ensemble model gives an accuracy of 92% which is better than the other models above.

Figure 3 represents the association between the nodes based on the similarity where each node represents a particular member or an organization of how well they are connected based on the similarities.

Table 1 Table of performance metrics

	SVM	KNN	RF	Ensemble
Accuracy	0.89	0.87	0.85	0.92
Precision	0.31	0.14	0.37	0.41
F1-Score	0.39	0.04	0.52	0.61

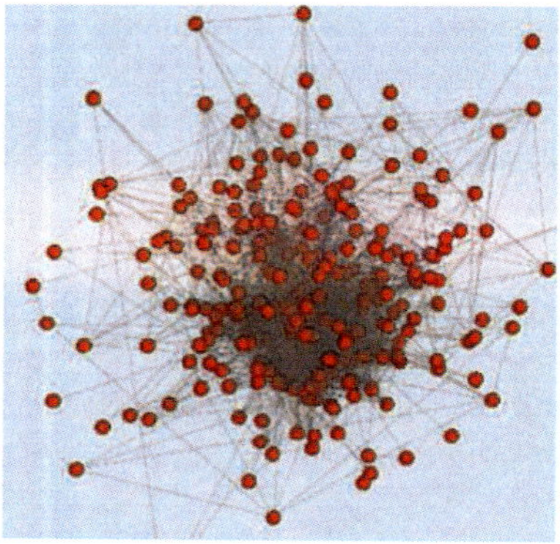

Fig. 3 Association between the nodes based on similarity

5 Conclusion

As online social networks have an immense impact on real-world association. This implies the importance of networking sites for information circulation. Increase in social networks helps people to notice the activities all around the world more transparently and supremacy of technology in future. In this paper, we tried to show an summary of the Facebook ego network where we researched alot of ML models for generating base user and his circle of friends. Mainly emphasized the implementation aspect, specifically on predicting the connection between nodes after time $T + x$. Figures 4, 5, and 6 above demonstrate the performance metrics of several models and provide a summary of how accurately it can predict the association. The

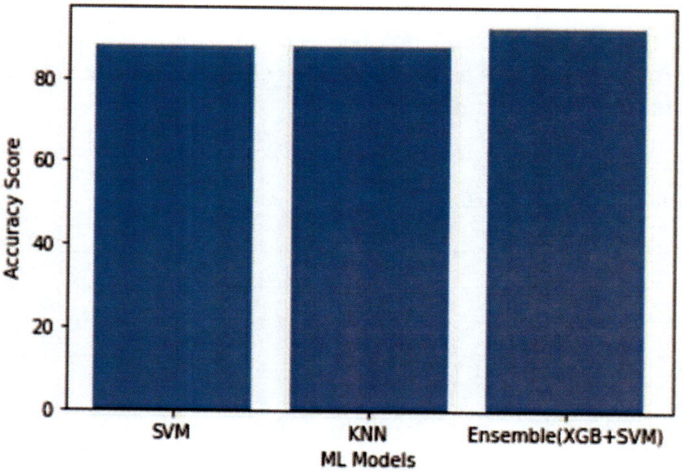

Fig. 4 Bar graph representation of accuracy scores of different ML models

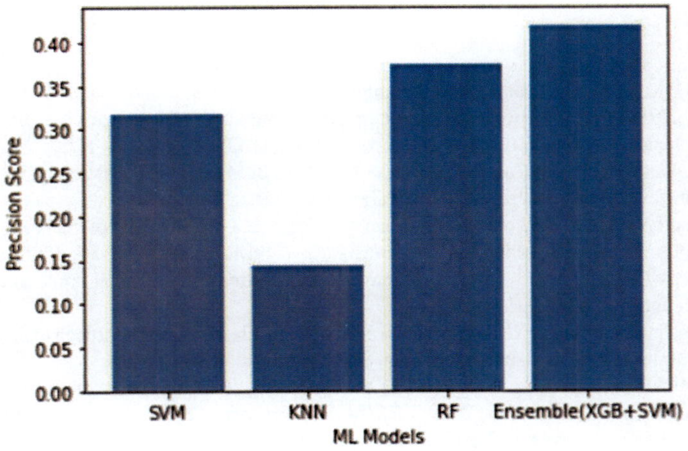

Fig. 5 Bar graph representation of precision scores of different ML models

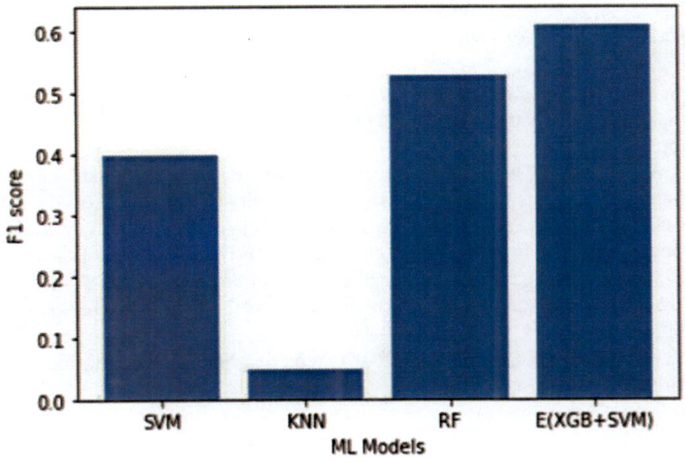

Fig. 6 Bar chart representation of F1 scores of different ML models

paper was divided into three modules: dataset and preprocessing, link prediction, and performance metrics of ML models including SVM, KNN, random forest, and neural network. Based on the experiment being conducted, it can be concluded that the ensemble model outperforms other ML models. Future work is to implement this idea in most social media platforms to incur a better possibility of link prediction.

References

1. Singh DK, Nithya N, Rahunathan L, Sanghavi P, Vaghela RS, Manoharan P, Hamdi M, Tunze GB (2022) Social network analysis for precise friend suggestions for twitter by associating multiple networks using ml. Int J Inf Tech Web Eng (IJITWE) 17(1):1–1
2. Jia M, Van Alboom M, Goubert L, Bracke P, Gabrys B, Musial K (2022) Analyzing egocentric networks via local structure and centrality measures: a study on chronic pain patients. In: International conference on information networking (ICOIN). IEEE, pp 152–157
3. Kumar A, Chhabra D, Mendiratta B, Sinha A (2020) Analyzing information diffusion in egocentric twitter social network. In: 6th International conference on signal processing and communication (ICSC). IEEE, pp 363–368
4. Rezaeipanah A, Ahmadi G, Sechin Matoori S (2020) A classification approach to link prediction in multiplex online ego-social networks. Soc Netw Anal Min 10(1):27
5. Long F, Ning N, Song C, Wu B (2019) Strengthening social networks analysis by networks fusion. In: Proceedings of the 2019 IEEE/ACM international conference on advances in social networks analysis and mining. IEEE/ACM, pp 460–463
6. Humski L, Pintar D, Vranić M (2018) Analysis of facebook interaction as basis for synthetic expanded social graph generation. IEEE Access 7:6622–6636
7. Kwon YD, Mogavi RH, Haq EU, Kwon Y, Ma X, Hui P (2019) Effects of ego networks and communities on self-disclosure in an online social network. In: Proceedings of the IEEE/ACM international conference on advances in social networks analysis and mining, IEEE/ACM, pp 17–24
8. Rizi FS, Granitzer M, Ziegler K (2017) Global and local feature learning for ego-network analysis. In: 28th International workshop on database and expert systems applications (DEXA). IEEE, pp 98–102
9. Madani A, Marjan M (2016) Mining social networks to discover ego sub-networks. In: 3rd MEC international conference on big data and smart city (ICBDSC). IEEE (2016), pp 1–5
10. Arnaboldi V, Conti M, Passarella A, Pezzoni F (2012) Analysis of ego network structure in online social networks. In: International conference on privacy, security, risk and trust and 2012 international conference on social computing 2012 vol Sep 3. IEEE, pp 31–40
11. Leskovec J, Mcauley J (2012) Learning to discover social circles in ego networks. Adv Neural Inf Proc Syst 25
12. Tabourier L, Libert AS, Lambiotte R (2016) Predicting links in ego-networks using temporal information. EPJ Data Sci 5:1–6

Chapter 30
Brain Tumor Detection and Classification

K. R. Roopa, Sainath Sindagikar, Pruthvi G. Kalkod, P. M. Vishnu, and Lata

1 Introduction

Brain tumor detection is the process of identifying the presence of abnormal growths in the brain and determining the type of tumor. Brain tumors can be of two types benign or malignant and cause a range of symptoms such as headaches, seizures, and cognitive impairment.

Brain tumor detection and its classification is an important topic of study in medical imaging and machine learning. Brain tumors are typically found through magnetic resonance imaging (MRI) and CT scans, and machine learning algorithms can be trained on these images to increase diagnosis speed and accuracy.

Brain tumors can be accurately detected and classified, which can help doctors choose the best course of treatment, such as surgery, radiation therapy, or chemotherapy. Additionally, a higher rate of treatment success and improved patient outcomes can result from early detection. The imaging technique called as MRIs is widely used to detect brain tumors. It creates detailed brain scans that can show whether a brain tumor is present or absent, if present then where exactly the tumor region, and the size of the tumor and other characteristics.

MRI is a common imaging technique used in the detection of brain tumors. It produces detailed images of the brain that can reveal the presence, location, size, and characteristics of a brain tumor.

Brain tumor detection and its classification on MRI images involves many steps, mainly preprocessing, segmentation, and classification. Preprocessing involves standardization of the images to reduce or remove noise and artifacts. Segmentation includes the segregation of the tumor from the surrounding brain tissues. Brain tumor

K. R. Roopa (✉) · S. Sindagikar · P. G. Kalkod · P. M. Vishnu · Lata
Department of Electronics and Communication Engineering, RNS Institute of Technology, Bangalore, Karnataka, India
e-mail: roopakr.ec07@gmail.com

A. Yadav et al. (eds.), *Proceedings of International Conference on Paradigms of Communication, Computing and Data Analytics*, Algorithms for Intelligent Systems,
https://doi.org/10.1007/978-981-99-4626-6_30

classification using convolutional neural networks (or CNNs) on MRI images is a research area in medical image analysis that classifies images without losing its information.

The CNN is a sort of deep learning neural network that is utilized for image recognition and image processing applications. Without sacrificing information, its built-in convolutional layer reduces picture dimensionality. CNNs are therefore excellent for this application.

2 Related Work

In related work, there are two distinct machine learning approaches used in the recent publications of related work on brain tumor segmentation: supervised learning and unsupervised learning. In unsupervised learning, validation of output variables always requires human intervention. But in supervised learning, human mediation is required basically to name the information appropriately [1].

Positron emission tomography (PET), computed tomography (or CT), and other imaging techniques can be used to identify brain tumors. MRI is preferred because of its superior performance. When trying to remove noise from an image, image filtering presents a significant obstacle. Common filters for image filtering include the median filter, the adaptive filter, the averaging filter, the Gaussian filter, and others are utilized [2, 3]. Sivaramakrishnan et al. [4] showed a technique for distinguishing mind growth region from cerebral picture by utilizing fuzzy c-means and histogram examination. Images were broken down by applying principal component analysis, which was used to make the wavelet coefficient less dimensional. The affected regions of the brain were successfully and precisely extracted from MRI images using the proposed algorithm.

Kong et al. [5] proposed a four-stage method for locating the brain tumor. At first, a wavelet channel is utilized to eliminate noise present in the image. In the next consecutive step, the watershed algorithm is adapted to X-ray images pixels as an underlying technique for segmentation. The fuzzy clustering algorithm is used to perform a merging operation on the segmented area. Finally, the segmentation is adapted to those areas which are not segmented by utilizing k-NN classifiers.

Convolutional neural networks are not quite the same as other ML methods, and the feature extraction is done automatically with more accuracy. But there are a few parameters to be advanced by CNNs which brings about costly computational time and requires a graphical process to prepare the model. There are two main functions that a CNN performs: feature extraction and data classification. Feature extraction is done by the convolution and pooling layers, and classification is made easier by the fully connected layers [6]. In [7], the authors proposed a CNN for the detection of brain tumors. Two CNN models were compared to determine which one is good for CNN classification. Initially, the model only has one layer of convolution, while the second model has two. This study demonstrated that increasing the number of convolutional layers improves model performance, leading to a loss value of 0.23 and an accuracy of 93%.

3 Proposed Methodology

The proposed methodology of detection of the affected area of the brain is brain tumor, and its classification involves several steps, including image preprocessing, segmentation of tumor region, and classification using deep learning algorithm. In this section, we will introduce each of these steps and their role in the detection process. Block diagram for our proposed method is shown in Fig. 1.

3.1 Preprocessing

Preprocessing is a necessary step in brain tumor detection that involves various techniques for the enhancement of image quality, skull stripping, removing noise, and drawing out relevant features from medical images.

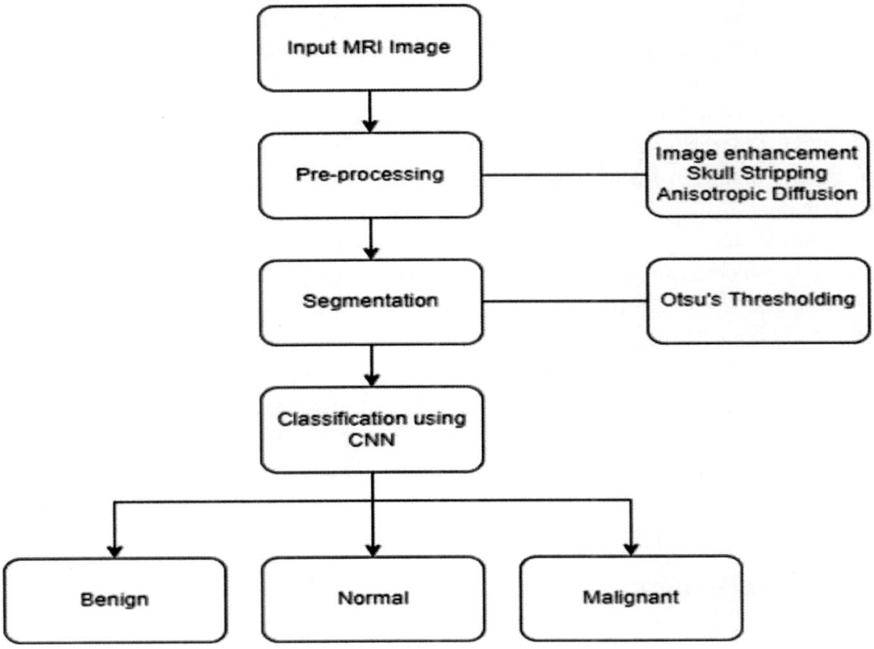

Fig.1 Overall block diagram representation

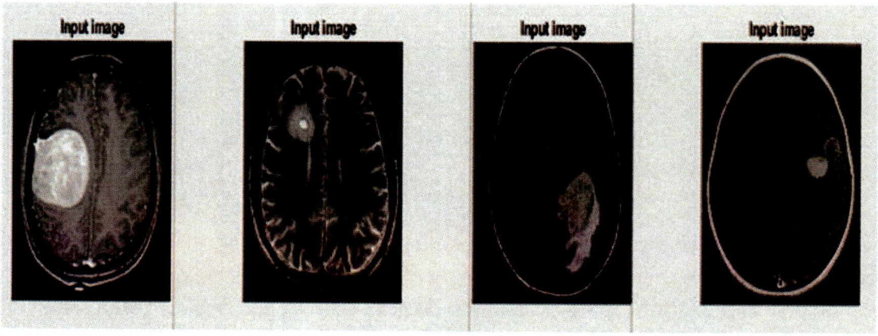

Fig. 2 Input images

Image enhancement plays an important role in improving the accuracy of disease diagnosis. The enhancement of medical images is the process of improving the quality of the image to make it more informative and easier to analyze. In brain tumor detection, image enhancement techniques can improve the visibility and contrast of the tumor region, enabling clinicians to accurately identify and diagnose the disease.

Skull stripping is one more important step in clinical image processing that includes the brain from the surrounding skull and other non-brain tissues. This is typically done to improve the accuracy and productivity of subsequent analysis steps, such as brain tumor segmentation. Here, skull stripping is done using thresholding, and morphological operations are two mathematical techniques that can be applied to magnetic resonance imaging (MRI) [8, 9].

For removing noise, anisotropic diffusion is used. Anisotropic diffusion is a type of image processing technique used to smooth out noise in digital images while retaining important features such as edges and boundaries. The basic idea behind anisotropic diffusion is to use a diffusion equation that takes into account the local structure of the image. This equation calculates the diffusion coefficient for each pixel based on the orientation and strength of the image gradient at that point. Pixels with a strong gradient are diffused less than pixels with a weak gradient, which preserves edges and boundaries in the image.

Figure 2 displays the original images that were used in the processing. These images may need to be preprocessed, such as skull stripping, to remove any unwanted background or noise. Figure 3 shows the result of this process, where the images are enhanced and filtered to improve their quality.

3.2 Segmentation

Segmentation is a step for the identification of brain tumors in the field of medical image analysis. In this, segmentation is performed by Otsu's thresholding algorithm. Otsu's thresholding method is a widely used technique for image segmentation and

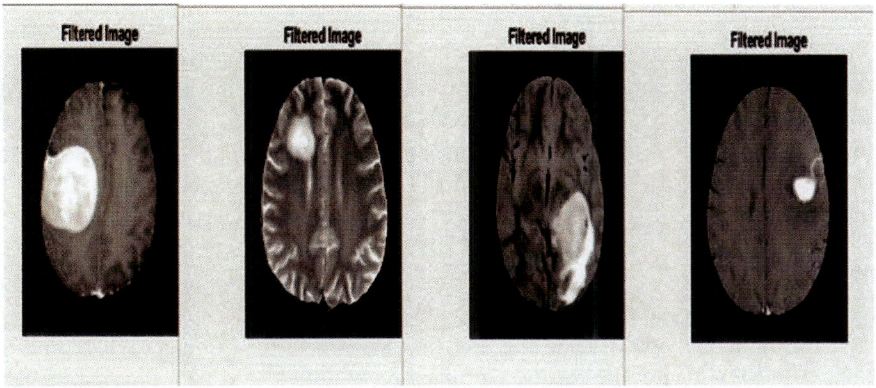

Fig. 3 Skull stripped, enhanced, and filtered images

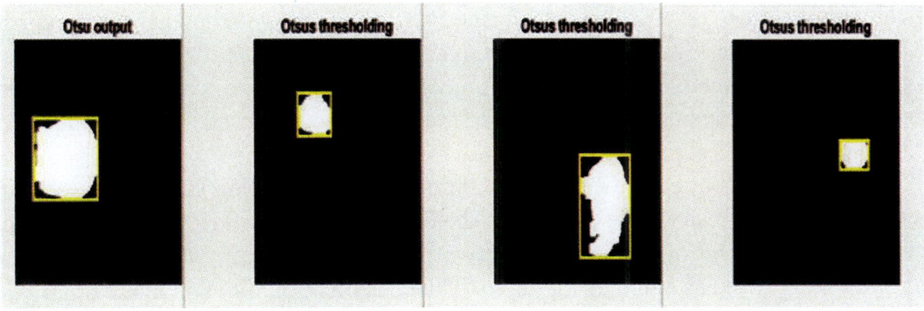

Fig. 4 Otsu's thresholding algorithm

has been applied for brain tumor detection in MRI images. Otsu's method calculates the threshold value that maximizes the separation between the background and foreground intensities in the image. Once the threshold value has been calculated using Otsu's algorithm, a binary image can be created by thresholding the original MRI image. This binary image will separate the tumor from the surrounding healthy tissue. To remove noise or small objects that may have been included in the segmentation, morphological operations such as erosion and dilation can be used for this purpose. Figure 4 shows the Otsu's thresholding segmentation in which the tumor area is highlighted with a yellow rectangular bounding box.

3.3 Classification

Brain tumor classification with CNNs is a consequential step in the field of Neuro-oncology. Brain tumor diagnosis must be accurate and fast in order to determine effective treatment therapies and predict patient outcomes. We decided to use CNN-based classification. Convolution neural networks (CNNs) are a sort of deep learning algorithm that has shown a large potential in the field of medical imaging analysis, especially brain tumor classification. The technique of classifying brain tumors involves training a CNN with a collection of dataset and then implementing the model that has been provided in training for identifying new brain tumor images [10].

Augmentation of data and preprocessing:

The MRI images in the dataset had different sizes. These MRI images represent the network's input layer, and they were normalized and resized to 256 × 256 pixels.

We modified images twice to augment the data. The initial change was a 90-degree rotation of the image. The second modification included vertically flipping photos. In this manner, we augmented our dataset [11].

Convolution neural network (CNN) architecture of proposed work:

Brain tumor classification was performed using CNNs which was developed in MATLAB R2020a. CNN includes three main layers in architecture namely convolutional layer, pooling layer, and a fully connected layer. It is a class of neural networks and processes data using a matrix-like architecture. The proposed CNNs architecture for classification is shown in Fig. 5.

The network architecture includes an input layer, convolution layer, that produces an image smaller than that of the input. The rectified linear unit (ReLU) layer comes next to the convolutional layer. There is a max pooling layer, which produces an output that is twice as small as the input. The classification block is made up of two fully connected layers (FCs), which are used to classify the three numbers of tumor classes as benign, malignant, or normal. There are 18 layers in total consisting of the input layer, convolution layer, pooling layer, ReLU activation layer, fully connected layer, soft max layer, and output layer constitutes the overall network architecture.

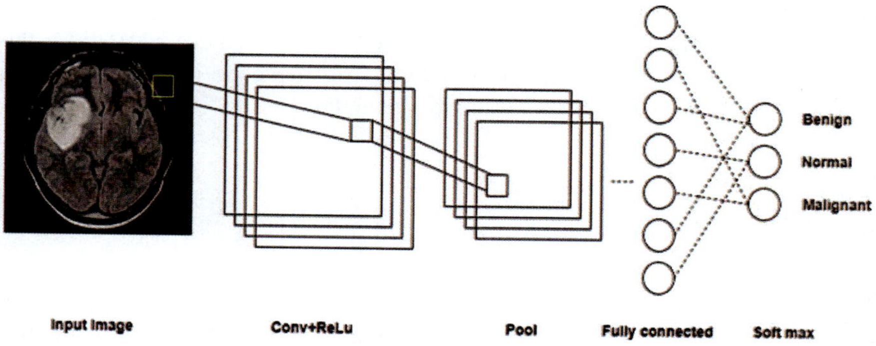

Fig. 5 Proposed convolution neural network (CNN) architecture

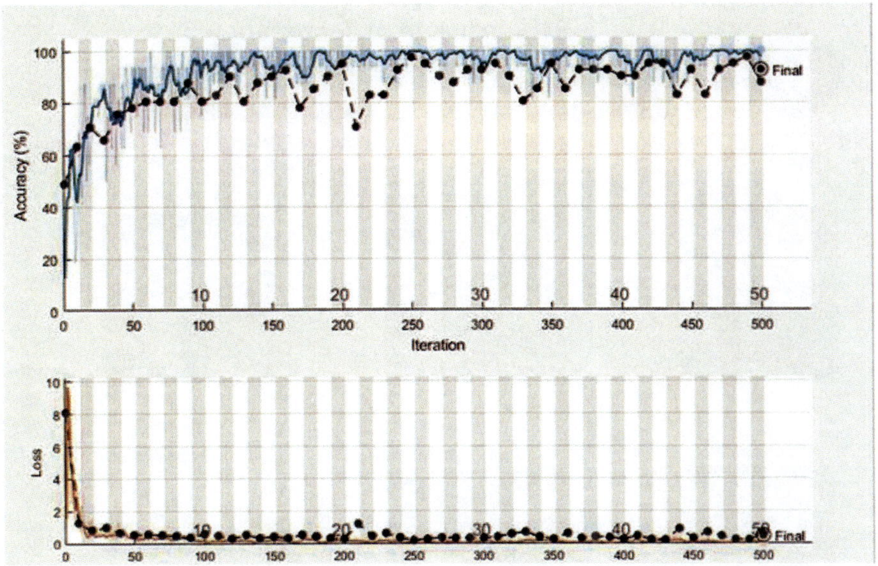

Fig. 6 Training plot

Training Network:

The model will be trained for a maximum of 50 epochs with a mini-batch size of 16 images. Every epoch, the data will be shuffled in order to increase the model's capacity to generalize to new data. The validation data will be used to monitor the model's performance during training, and validation will be done every 10 epoch and it is shown in Fig. 6.

4 Results and Discussion

In the proposed approach, an automatic detection and classification of the tumor is developed with the objective of classifying the input magnetic resonance imaging (MRI) scan into benign(non-cancerous), malignant(cancerous), or normal (no tumor). The result of the proposed methodology is represented using the MATLAB app. Each step of the proposed model is implemented using multiple algorithms shown in the MATLAB app. Figure 7 represents the GUI for implemented methodology. It consists of different steps starting from input image till the CNN classification of brain tumor. The first block shows the input image where input is taken from the MRI dataset. Further the enhancement of MRIs and the skull stripping is performed. The subsequent phase of this process, the brain tumor detection, is done using Otsu's thresholding algorithm. After the CNN training, it gives the output as benign(non-cancerous), malignant (cancerous), or normal (no tumor) in the result section in text format. Overall, the result of the proposed method shows a 92.7% accuracy rate.

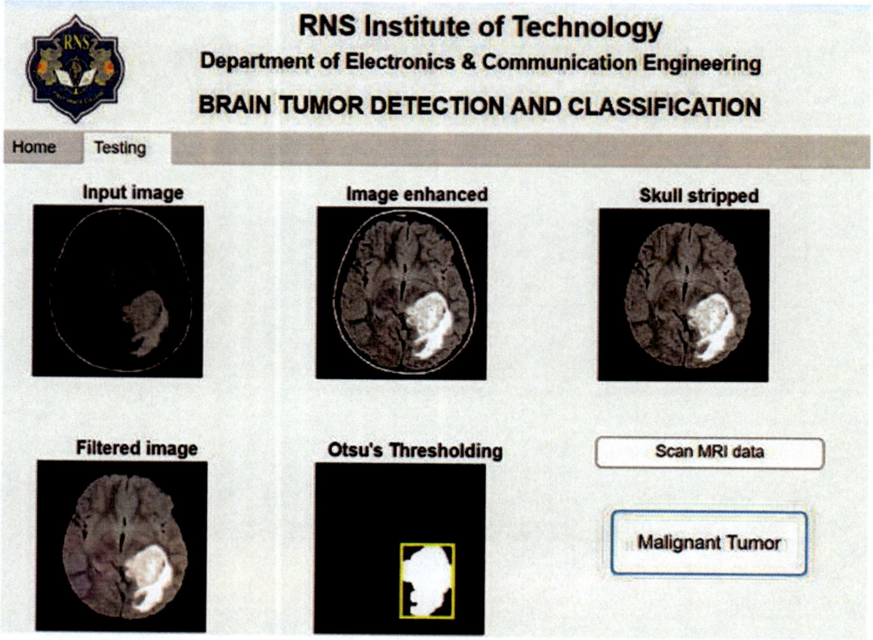

Fig. 7 GUI representation of proposed methodology

5 Conclusion

In conclusion, the paper proposes a complete approach for brain tumor detection and its classification. The proposed methodology involves several preprocessing steps such as image enhancement, skull stripping, and noise removal, followed by segmentation using Otsu thresholding. Further we classify the segmented images using a convolutional neural network (CNN) and implement the proposed methodology using a graphical user interface (GUI) for clinical applications. Overall, the proposed approach has the ability to improve the accuracy of the dataset and efficiency of brain tumor detection and its classification using MRI images.

6 Future Scope

The project's future goals include training this algorithm on huge datasets and testing patient level accuracy on other datasets. This approach will be expanded to identify different classifications of cancer (Glioma, Meningioma) within the benign and malignant types. The proposed approach can also be used to diagnose other medical conditions such as lung cancer, breast cancer, and colon cancer. Different versions of neural networks can potentially be used to diagnose brain tumors depending on the accuracy in image classification, and this application can be used to improve hospitals.

References

1. Soesanti MH Avizenna,Ardiyanto I (2020) Classification of brain tumor MRI image using random forest algorithm and multilayers perceptron
2. Nandi A (2015) Detection of human brain tumour using MRI image segmentation and morphological operators. In: IEEE international conference on computer graphics, vision and information security (CGVIS)
3. Perona P, Malik J (1990) Scale-space and edge detection using anisotropic diffusion. IEEE Trans Pattern Anal Mach Intell 12(7):629–639
4. Sivaramakrishnan A,Karnan M (2013) A novel based approach for extraction of brain tumor in MRI images using soft computing techniques. Int J Adv Res Comput Commun Eng 2(4)
5. Kong J, Wang J, Lu Y, Zang J, Li Y, Zang B (2006) A novel approach for segmentation of mri brain images. In: IEEE Mediterranean Electrotechnical Conference, pp 325–528
6. Irmak E (2021) Multi-classification of brain tumor mri images using deep convolutional neural network with fully optimized framework. Iran J Sci Technol Trans Electr Eng 45(3):1015–1036
7. Febrianto DC, Soesanti I, Nugroho HA (2020) Convolutional neural network for brain tumor detection. In: IOP Series: Materials Science and Engineering, vol. 771, no. 1
9. Praveen GB, Agrawal A (2015) Hybrid approach for brain tumor detection and classification in magnetic resonance images. Commun Cont Intell Syst (CCIS), pp 162–166
10. Parveen, Singh A (2015) Detection of brain tumor in MRI images, using combination of fuzzy c-means and SVM. In: 2015 2nd international conference on signal processing and integrated networks (SPIN), pp 98–102
8. Seetha J, Selvakumar Raja S (2018) Brain tumor classification using convolutional neural networks. Biomed Pharmacol J
11. Sajjad M, Khan S, Muhammad K, Ullah W, Baik A (2019) Multi-grade brain tumor classification using deep CNN with extensive data augmentation. J Comput Sci 30:174–182

Chapter 31
Safe Vote–Fraudulent Vote Prevention System

Neethu Chandrasekhar, Arjun B. Nair, Avinash Thomas George, Binitta Varghese, and Diya Anna Thomas

1 Introduction

Voting is not just our right but our duty as well. The right to vote is one of the few pillars of democracy, so security in the voting process is of primary concern. Malpractices like fraudulent votes and capturing booths are also reported in a country like India. Initially, ballot papers were used for elections. It was the least secure method as the results could easily be manipulated since the officials counted the votes, and also, a lot of paper was used. Now, the existing voting process in India happens through electronic voting machines. EVM machine is more secure and efficient. It requires less number of polling officials as compared to ballot paper-based voting. Even though it requires fewer polling officials than ballot paper-based voting, polling officials are necessary for voter identification, issuing slips, marking ink, and many more, which again makes the system less secure. The frequently happening fraud practice while voting is rigging. It is the case of the same person casting multiple votes. To avoid all these fraud practices, make the existing system more secure.

The main contribution of the paper is as follows. The proposed system is an EVM machine along with RFID technology, a fingerprint module, and a face recognition

N. Chandrasekhar · A. B. Nair · A. T. George · B. Varghese (✉) · D. A. Thomas
Department of CSE, Amal Jyothi College of Engineering, Kottayam, Kerala, India
e-mail: binittavarghese2022@cs.ajce.in

N. Chandrasekhar
e-mail: neethucsekhar@amaljyothi.ac.in

A. B. Nair
e-mail: avinashthomasgeorge2022@cs.ajce.in

A. T. George
e-mail: arjunbnair2022@cs.ajce.in

D. A. Thomas
e-mail: diyaannathomas2022@cs.ajce.in

A. Yadav et al. (eds.), *Proceedings of International Conference on Paradigms of Communication, Computing and Data Analytics*, Algorithms for Intelligent Systems,
https://doi.org/10.1007/978-981-99-4626-6_31

module. RFID is the wireless, non-contact use of radio frequency waves to transfer data. RFID systems usually comprise an RFID reader, RFID tags, and antennas. By using the latest technologies, we are trying to make the existing EVM machine more secure. The first step to be taken is to provide a voter ID card with implanted RFID chip to all citizens in India. The RFID chip contains a unique ID. An initial investment should be made, but it will make the total election process more secure. The voter should carry the voter's ID card with him/her at the time of election. Rigging happens at the time of the election, and to avoid it, a fingerprint module is introduced. Each individual has a unique fingerprint, so it is the most secure verification process. As a secondary verification process, facial recognition is added. This is a particular case. It is only applicable to people with no fingers or people who have scars on their fingers due to work conditions or some other reasons. Through this three-layer verification process, we can make the election process secure.

2 Literature Review

In a democratic country like India, the vote is a very powerful tool to elect their representatives. Electronic voting machines (EVMs) have been used for elections in India since 1998, and the adoption of this technology has been praised for improving the efficiency and accuracy of the voting process. However, there have also been concerns raised about the security and reliability of these machines, as well as allegations of tampering and electoral fraud. One of the biggest concerns associated with electronic voting systems is the risk of security breaches. Malicious actors may attempt to hack into the system to manipulate or alter the results. In an effort to address the issue of fake voter ID cards, biometric authentication using fingerprints and facial recognition was introduced.

To ensure the security of the voter and preserve the sanctity of the method, Vinayachandra et al. [1] introduced a two-layer security scheme, radio frequency identification (RFID), and biometrics. Sruthi and Shanjai [2] proposed a system that uses convolutional neural network (CNN) that trains the labeled images and predicts the output by classifying the images, producing an accuracy of 90%. The use of a Webcam eliminates election fraud. Webcam takes the image of the arrived voters and stores it as a dataset, which is used to train the face recognition model. The model is trained using OpenCV and CNN and predicts the output. The images of newly arrived voters are compared with the existing images. If there is a match, they cannot vote. Otherwise, his appearance is captured by Webcam and added to the current dataset, and the voter can vote from the list of parties. This system produced an accuracy of approximately 90%, in our system. When a voter comes to vote, they need to get verified using their fingerprints by comparing it with the ones in the storage. Without fingerprints, the voter's image is captured using the Web live. The captured image is compared with the image against the voter's ID in the storage using CNN. If there is a match, then the voter can cast his vote. Otherwise, he cannot vote and display the warning. Priyadarshini et al. [3] proposed that the voter undergoes a three-step verification process on the voting day. Yang et al. [4] proposed a system to protect the confidentiality of the votes, each cast ballot is encrypted using the exponential ElGamal cryptosystem before submission. Piratheepan et al. [5] presented the fingerprint voting system, which is an electronic voting machine using human biometric system that will reduce the polling time and the number of staff. Mahiuddin [6] proposed an approach, where the voter identity card is replaced by a smart card in which all the details of the person are updated.

Hussain et al. [7] presented another innovative approach for a voting process where the device communicates with the RFID tag, which is embedded in the voter ID card, and when the voter scans his card, the controller checks the ID, and if it matches, the controller generates an OTP and sends the message to the user's mobile (voter) through GSM module. The voter inserts the password through the keypad, and if the password is confirmed, then the person is allowed to vote, and this process is repeated for every person. Rezwan et al. [8] have used Arduino and Finger Print Scanners that can identify each voter, count votes, and can prevent fake votes. Priya et al. [9] suggested that with the inclusion of a biometric fingerprint sensor, each voter is entered into the system only after being recognized and checked with the given database of enlisted voters. Once the corresponding fingerprint is matched with the information provided, the voter will be allowed to proceed to choose their preferred candidate from the panel of buttons. Mansingh et al. [10] proposed a system that is implemented using RFID and Internet of things (IoT) to improvise the security mechanisms. Here, an active RFID tag is used in place of the voter ID, where the system can scan the tag and matches it with the fingerprints collected in the Aadhaar database.

The proposed EVM is an EVM implemented along with RFID technology where each voter is given an RFID tag which is used as a primary step of authentication. According to the present system, the number of polling officials required is more, approximately five to six polling officials per booth. The design of authenticated radio frequency identification-based electronic voting machine is a technique where the number of polling officials is less. This system requires only two to three polling officials per booth. The main idea behind implementing the existing EVM machine with RFID technology is to make the system more accurate and reliable. RFID is the wireless, non-contact use of radio frequency waves to transfer data. RFID systems usually comprise an RFID reader, RFID tags, and antennas. The use of RFID technology in voting has several advantages over traditional paper-based voting systems. Firstly, it eliminates the need for manual counting and reduces the likelihood of human errors. Secondly, it improves the security and transparency of the voting process by enabling real-time monitoring and tracking of voting activity. Lastly, it can facilitate the rapid transmission of voting results, enabling faster and more efficient election outcomes.

3 Proposed System

The main idea is to implement the EVM machine with the latest technologies like RFID, fingerprint, and facial recognition. A voter's ID card with implanted RFID chip is given to each citizen. He/she should come with the voter's ID card for the

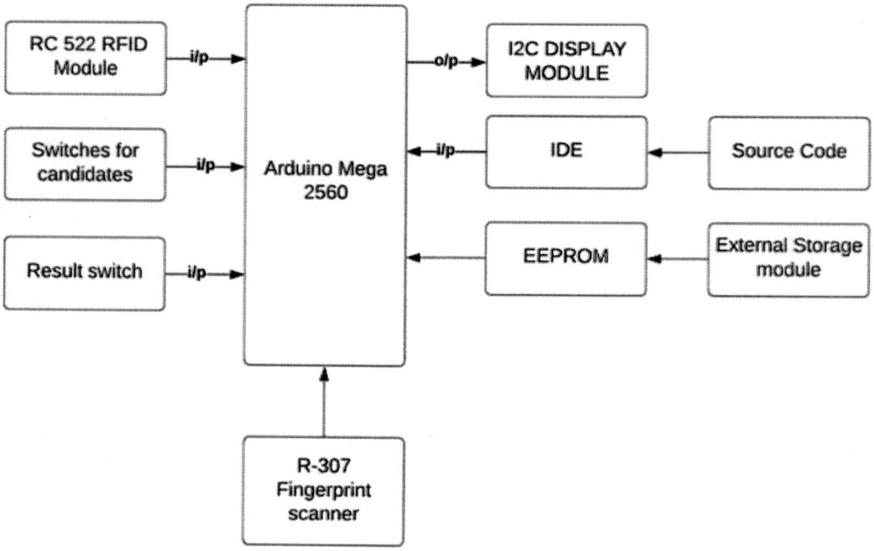

Fig. 1 Proposed system of safe vote

election. When he/she enters the polling booth for voting, the RFID scanner in the EVM machine will automatically detect the unique ID in the voter's ID card. If it founds a match with the already existing data, the person is guided to the next step of verification, which is fingerprint-based verification. The fingerprint of every citizen in India is already recorded. When a person scans his/her fingerprint at the time of the election, it is compared with the fingerprint data. If it founds a match, the EVM machine will show the voting screen. The voting screen contains the candidates' names, and corresponding to their names, different buttons are provided, and the voter can cast his/her vote for their favored candidate.

As a secondary step of verification, facial recognition module is introduced. It is only possible for people with no fingers or people who cannot scan his/her fingerprint due to scars on the fingers. A camera module is there in the polling booth. The person is directed to the camera. The photo is captured, and it is cross-checked with the already available data. If a match is found, the person is allowed to cast his/her valuable vote. This paper proposes a system that makes the existing voting process more secure. Figure 1 represents the proposed system of safe vote.

4 Functioning

4.1 Data Collection

Accurate information is required for secure and trustworthy voting to take place. Hence, it is important that the credentials of the voter be collected precisely. The details of the voter are collected by the election commission of the country, and with it, the biometric information, i.e., fingerprint, is also collected and stored in the government database. This data collection process can be conducted similarly to the large-scale aadhaar data collection.

4.2 Verification

When the polling procedures are taking place, the voter's ID is scanned first. If the card is valid and not previously voted with, the fingerprint of the voter is scanned and cross-checked with the print under the unique ID. This 2 step verification helps filter out fraudulent voting attempts and helps conduct a more secure election process. Also, this helps reduce the time and load taken by the officials and an almost error-free verification process.

4.3 Access Authorization

After the verification process is completed, voting access is either granted or denied. Access is granted if the voter's ID card is valid and its corresponding fingerprint matches the one scanned by the voter. Access is denied if neither the card is invalid, previously voted, or the fingerprint is invalid or doesn't match with the card. Only if access is granted, the vote can be cast. This ensures that identity theft of voters can be prevented as biometric information is required with the ID card.

4.4 Vote Calculation

The process of voting using RFID can be mathematically elaborated as follows:

Let V be the set of eligible voters, where $|V|$ is the total number of voters.

Each voter $i \in V$ is provided with a smart card containing an RFID tag denoted as $T(i)$.

Let C be the set of possible candidates, where $|C|$ is the total number of candidates.

The electronic voting machine has an RFID reader that can read the information from the RFID tag on the smart card.

Fig. 2 Flowchart

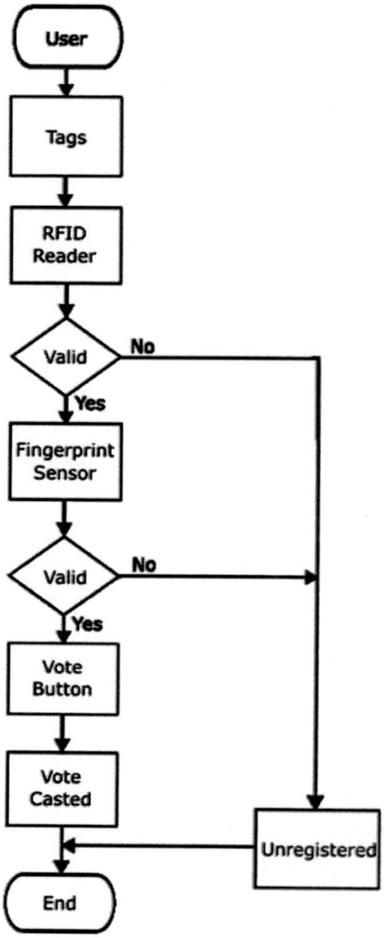

When a voter inserts their smart card into the electronic voting machine, the RFID reader reads the information from the tag and verifies the voter's identity. If the voter is eligible to vote, the machine allows them to cast their vote.

Each voter i can cast only one vote for one candidate $j \in C$. The RFID reader records the vote cast by the voter i as $V(i) = j$.

The total number of votes cast for each candidate can be calculated as follows:

Let $S(j)$ be the total number of votes cast for candidate j.

Then, $S(j) = \Sigma V(i) = j$, for all $i \in V$.

Once all the votes have been cast, the votes are counted, and the candidate with the highest number of votes is declared the winner. Overall, the voting system using RFID technology simplifies the voting process and reduces the possibility of fraud while also ensuring that the privacy and security of voters are protected. The outline of the voting process is shown in Fig. 2.

5 Implementation Tools

The working is made possible with a number of hardware components like an RFID reader, Arduino microcontroller board, fingerprint sensor module, etc. These hardware components help receive physical inputs from the real world and convert them to the required digital form to produce the desired output.

5.1 Modules

Arduino Mega 2560 The Arduino Mega 2560 as addressed in Shaheen et al. [11] is a microcontroller board based on the ATmega2560 as shown in Fig. 3. It has 54 digital input/output pins (of which 15 can be used as PWM outputs), 16 analog inputs, 4 UARTs (hardware serial ports), a 16 MHz crystal oscillator, a USB connection, a power jack, an ICSP header, and a reset button. The Mega 2560 board can be programmed using the Arduino software, which is an open-source Integrated Development Environment (IDE) that simplifies the process of writing and uploading code to the board. The board is compatible with a wide range of sensors, actuators, and other peripherals, making it suitable for a variety of projects such as robotics, automation, data logging, and more.

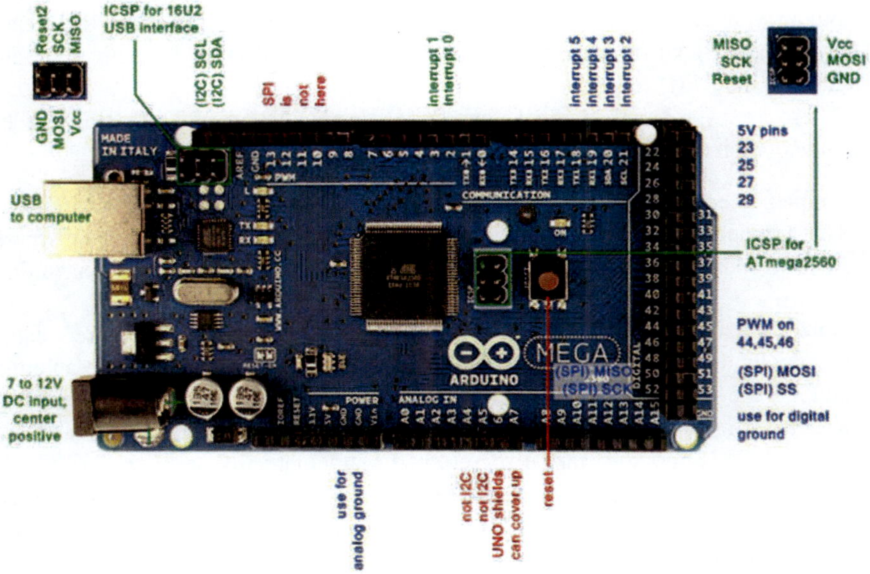

Fig. 3 Arduino Mega 2560

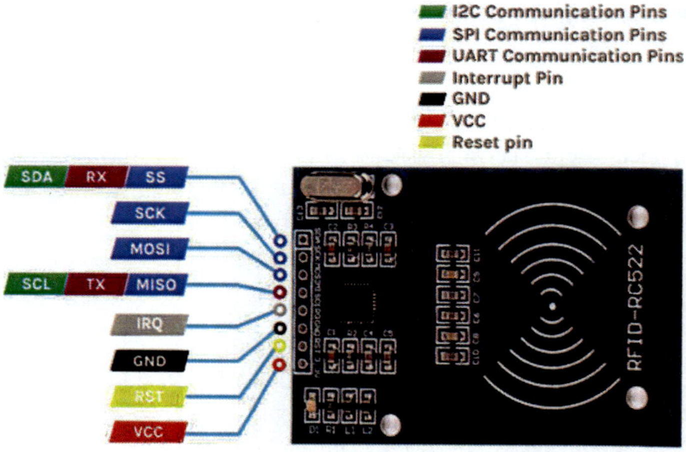

Fig. 4 RFID Reader RC522

RFID Reader RC522 RFID Reader, as addressed in Tompunu et al. [12], shown in Fig. 4, is also termed as interrogator. It is the brain of the RFID system, as its main function is interrogating the tags. Radio waves are generated by a radio frequency signal generator. An antenna is used to transmit these radio waves to the surrounding. When an RFID tag is placed in close proximity to the RFID reader, the transmitted waves are received by the RFID tags. These radio waves cause electrons to move through the antenna of the RFID tag. This will power the chip inside the RFID tag. The feedback signal from the powered chip inside the RFID tag is received by the antenna in the RFID reader.

There are several mathematical equations associated with the RC522 RFID reader module, which can be used to describe its behavior and performance.

- Frequency equation: The RC522 module operates at a frequency of 13.56 MHz. This frequency can be represented mathematically as $f = 1/(2\pi\sqrt{L1C1})$, where $L1$ is the inductance of the antenna, and $C1$ is the capacitance of the antenna tuning circuit.
- Antenna Gain equation: The gain of the RC522 antenna can be represented mathematically as $G = 4\pi A/\lambda^2$, where A is the effective aperture area of the antenna and λ is the wavelength of the operating frequency.
- Received Signal Strength Equation: The signal strength of the RFID signal received by the RC522 module can be represented mathematically as $P_r = P_t G_t G_r \lambda^2/(4\pi \mathrm{d})^2$, where P_t is the power transmitted by the RFID tag, G_t and G_r are the gain of the tag and reader antennas, λ is the wavelength of the operating frequency, and d is the distance between the tag and reader antennas.
- Communication Range Equation: The maximum communication range between the RC522 module and a RFID tag can be represented mathematically as $d_m ax = \lambda/(4\pi\sqrt{L1C1})$, where L1 and C1 are the inductance and capacitance of the antenna tuning circuit.

Fig. 5 Fingerprint R307

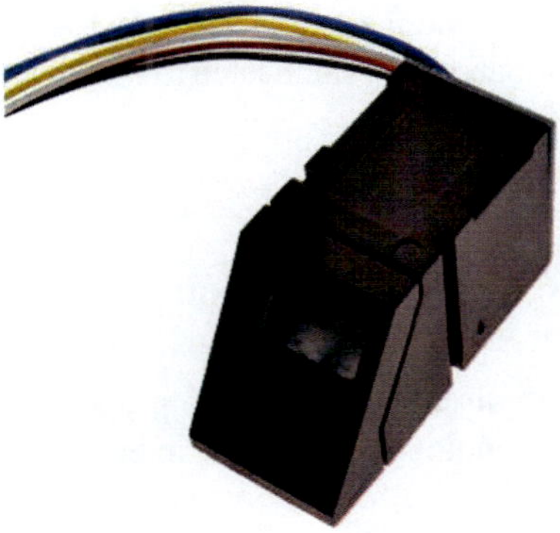

RFID Tag RFID tags are of two types: active and passive tags as addressed in Rafiq et al. [13]. The active tag works with an electric power supply. Here, we are using a passive RFID tag which is also known as an RFID chip. By using radio waves, it can transmit and receive data with RFID reader. Each voter's card is embedded with an RFID chip with a unique identification number. When a voter's ID card, which has an RFID chip embedded within comes in close proximity to the RFID reader, the RFID chip will get powered with the radio frequency waves coming from the RFID reader.Then, the RFID chip will give a feedback signal with the unique identification number.

Fingerprint R307 Fingerprint R307, shown in Fig. 5, is an optical fingerprint reader sensor module as addressed in Kumar et al. [14]. It consists of a TTL UART interface for direct connection embedded in the sensor. The fingerprint data can be stored in the module by the user, and it can be configured in 1:1 or 1:N mode for personal identification. During enrollment, the R307 captures the fingerprint image and converts it into a template. The template is a digital representation of the unique features of the fingerprint and is stored in the module's memory. During verification, the R307 captures the fingerprint image and compares it to the stored template. If the two images match, the module sends a signal to indicate that the fingerprint has been successfully verified. The R307 supports several encryption and authentication protocols, including AES, RSA, and DES.

I2C LCD Display As addressed in Akinwole and Oladimeji [15], I2C is a serial bus using two bidirectional lines, called serial data line and serial clock line. Both are connected via pulled-up resistors. It is able to display 16 $\times$ 2 characters. I2C 16 $\times$ 2 Arduino LCD Screen uses an I2C communication interface. It only needs four pins for the LCD display: VCC, GND, SDA, SCL. All connectors are standard XH2.54.I2C address of the module which is configurable from 0 $\times$ 20–0 $\times$ 27. The contrast can be adjusted manually. The supply voltage for the working is 5V. The dimension is 80 $\times$ 36 $\times$ 20 mm (3.1 $\times$ 1.4 $\times$ 0.7 in). The I2C display module typically consists of an LCD panel, a driver board, and an I2C interface. The driver board is responsible for controlling the display and translating the I2C commands into commands that the LCD panel can understand. The I2C interface allows the module to be connected to a microcontroller or other device that can send commands to the display.

6 Implementation Result

As discussed earlier, the proposed EVM is an EVM implemented along with RFID technology where each voter is given an RFID tag which is used as a primary step of authentication. The voter should carry the RFID along with him/her while coming to cast their vote. Here, RFID tag is read by the RFID reader, and it is cross-checked with already existing data. If a match is found, the voter is allowed to cast their vote. Otherwise, the voter is considered an invalid voter, and the person is not allowed to cast their vote. As a second step of authentication, the fingerprint is used. The fingerprint is a very secure authentication process, as each person has a unique fingerprint. By using a fingerprint scanner, the fingerprint is taken, and it is cross-checked with the already existing database. If a match is found, the person is allowed to cast his/her vote.

The stepwise election process is as follows :

Step 1 EVM powered ON as shown in Fig. 6. Initially, before starting the voting process, all candidates' votes are displayed as zeros, as shown in Fig. 7.

Step 2 When a voter comes to vote, the RFID reader will read the unique id in the RFID tag. Whenever a registered user scans his/her card, the message "registered" will be displayed on the LCD Display, as shown in Fig. 8.

Step 3 Before a voter can cast their vote, they must first enroll their fingerprint in the system. This involves scanning their fingerprint using the R307 fingerprint module and storing the resulting biometric data in a database, as shown in Fig. 9. Once a voter's fingerprint has been enrolled, they can then authenticate themselves by scanning their fingerprint again. The R307 module [14] will compare the new scan to the biometric data stored in the database to determine if it is a match.

Step 4 After a voter has been authenticated, they can then cast their vote using the voting interface provided by the system. The system should ensure that each voter can only cast one vote and that the votes are tallied correctly.

Step 5 After the voting is complete, the system can verify the results by comparing the fingerprints of each voter to ensure that they are unique and that no one voted more than once. When the voter cast votes for a candidate, the vote count will increment

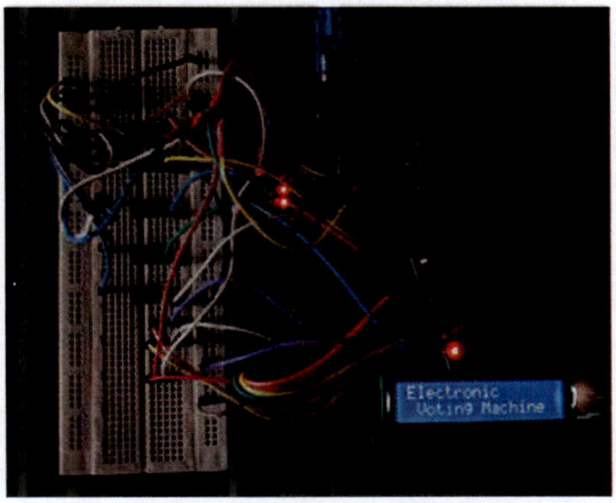

Fig. 6 EVM powered ON

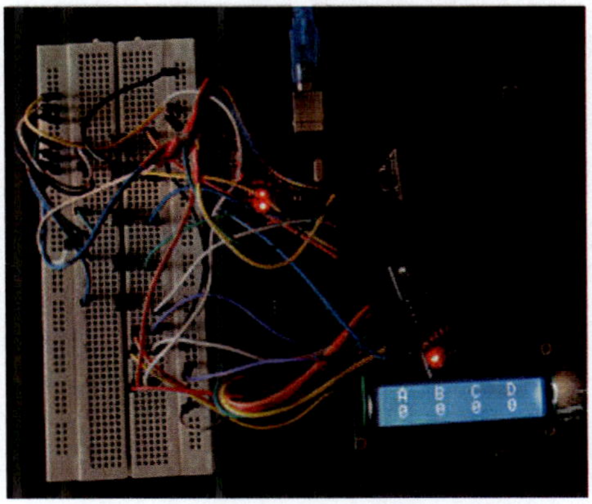

Fig. 7 Display before voting

by 1. Figure 10 shows the image of vote cast by one person. At the end of the election, the government official can check for the result using the RFID tag for the result.

Step 6 The LCD display shows the result as given in Fig. 11. The candidate with the most votes will be the winner. If there are an equal number of votes for more than one candidate, it will be considered a draw and displayed as a tie.

7 Conclusion

The paper was developed in order to provide a more secure way to cast votes. A vote is the right of every citizen of the country, and its security is something to be given at most importance. This system will be incredibly helpful in overcoming the difficulties that arise throughout the election process. The current election process is less secure, and more manpower is required. Adding the extra layers of security in the form of RFID and fingerprint makes the whole process of casting votes more

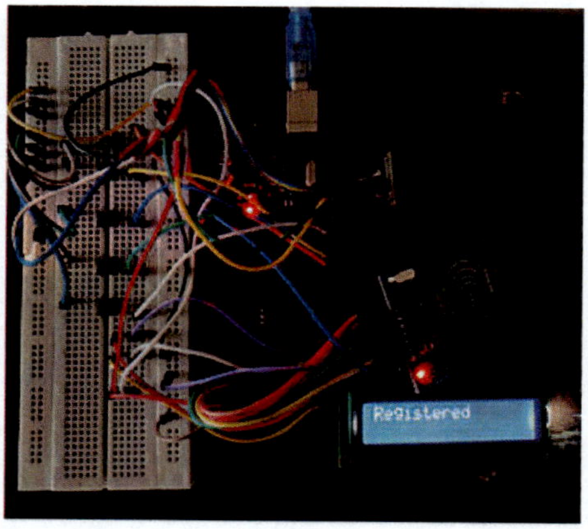

Fig. 8 Registered voters' ID scanned

Fig. 9 Fingerprint scanned

Fig. 10 Vote casted

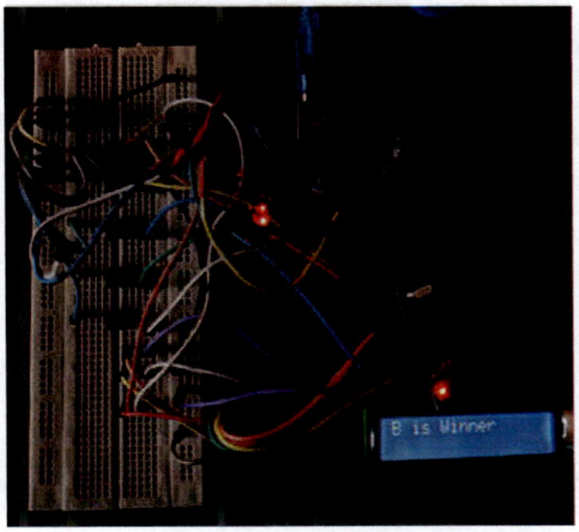

Fig. 11 Results declared

foolproof. This system can reduce manpower, thus making the system more efficient. In order to make it more efficient, facial recognition is introduced as an extra feature. Facial recognition is only applicable to people who cannot register their fingerprint due to a lack of fingers or some kind of scars on the fingers. The proposed system also can be further developed into a more simple and secure standard election procedure which can be adopted for other smaller elections like college elections or other internal elections. The Arduino used here can be considered as a system that cannot be reprogrammed without the help of an external computer or other programming devices, hence manipulating this system will be difficult, making it more secure. The new system of EVM machines using RFID and fingerprint makes the election more secure, so the people will elect the right candidate according to their choice. It makes the true sense of a democratic country.

References

1. Vinayachandra GPK, Rajeshwari M, Krishna Prasad K (2020) Arduino based authenticated voting machine (AVM) using RFID and fingerprint for the student elections. J Phys: Conf Ser, IOP Publishing. https://doi.org/10.1088/1742-6596/1712/1/012004.
2. Sruthi MS, Shanjai K (2021) Automatic voting system using convolutional neural network. J Phys: Conf Ser 1916:012074
3. Priyadarshini R, Shangamithra D, Swathi T, Subhaharini G, Sreenivasan L (2020) Design and realization of RFID based smart voting system with frontal face recognition technique. Int J Eng Res Technol (IJERT) ICEECT–2020 8(17)
4. Yang X, Yi X, Nepal S, Kelarev A, Han F (2018) A secure verifiable ranked choice online voting system based on homomorphic encryption. IEEE Access 6:20506–20519. https://doi.org/10.1109/ACCESS.2018.2817518
5. Piratheepan A, Sasikaran S, Thanushkanth P, Tharsika S, Nathiya M, Sivakaran C, Thiruchchelvan N, Thiruthanigesan K (2017) Fingerprint voting system using Arduino. Middle-East J Scient Res 25(8):1793–1802

6. Mahiuddin M (2019) Design a secure voting system using smart card and iris recognition. In: 2019 International conference on electrical, computer and communication engineering (ECCE), pp 1–6. https://doi.org/10.1109/ECACE.2019.8679118
7. Hussain SM, Ramaiah C, Asuncion R, Nizamuddin SA, Veerabhadrappa R (2016) An RFID based smart EVM system for reducing electoral frauds. In: 2016 5th International conference on reliability, Infocom technologies and optimization (Trends and Future Directions) (ICRITO), pp 371–374. https://doi.org/10.1109/ICRITO.2016.7784983
8. Rezwan R, Ahmed H, Biplop MRN, Shuvo SM, Rahman MA (2017) Biometrically secured electronic voting machine. In: IEEE Region 10 humanitarian technology conference (R10-HTC), pp 510–512. https://doi.org/10.1109/R10-HTC.2017.8289010
9. Priya VK, Vimaladevi V, Pandimeenal B, Dhivya T (2017) Arduino based smart electronic voting machine. In: International conference on trends in electronics and informatics (ICEI 2017), pp 641–644. https://doi.org/10.1109/ICOEI.2017.8300781
10. Mansingh PMB, Titus TJ, Devi VSS (2020) A secured biometric voting system using RFID linked with the aadhar database. pp 1116–1119. https://doi.org/10.1109/ICACCS48705.2020.9074281
11. Shaheen Yaser S. A, Alkafrawi Hussam M. I, Aga Tarek R. S. A, Elkafrawi Ismail M, Imaeeg Massoud A. O (2021) Arduino mega based smart traffic control system. Asian J Adv Res Reports 15(12):43–52. https://doi.org/10.9734/AJARR/2021/v15i1230449
12. Tompunu AN, Mirza Y, Azwardi (2020) Room door security system using microcontroller-based on E-KTP. J Phys: Conf Ser 1500:12115
13. Rafiq K, Appleby RG, Edgar JP, Radford C, Smith BP, Jordan NR, Dexter CE, Jones DN, Blacker ARF, Cochrane M (2021) WildWID: an open-source active RFID system for wildlife research. Methods in Ecol Evolut 12:580–1587. https://doi.org/10.1111/2041-210X.13651
14. Kumar D, Singh G, Kaur R (2022) Implementation of enhanced IOT based biometrics attendance system using R307 fingerprint sensor with Arduino UNO and real time database to improve accuracy. Int J Adv Eng Managem (IJAEM) 4(5):985–988
15. Akinwole OO, Oladimeji TT (2018) Design and implementation of Arduino microcontroller based automatic lighting control with I2C LCD display. J Electr Electron Syst 7:258. https://doi.org/10.4172/2332-0796.1000258

Chapter 32
Intelligent Framework for Early Prediction of Diabetic Retinopathy: A Deep Learning Approach

Adil Husain Rather and Inam Ul Haq

1 Introduction

Every blood vessel in the body is impacted by diabetes, with the eyes and kidneys receiving special attention. A disease is known as diabetic retinopathy where the veins inside the eyes are damaged. The leading global cause of vision loss and blindness is diabetic retinopathy (DR). High blood sugar levels cause this diabetes consequence, which presents as damage to the blood vessels supplying the retina in the eye. If left untreated, DR can progress through a number of stages and harm the small arteries lining the retina, eventually leaving a person blind. Up to 80% of people with diabetes who have had the condition for 20 years or more will develop DR, a microvascular disease that is one of the most prevalent consequences of diabetes.

It is a serious issue for public health that is expected to increase in prevalence with the rising global incidence of diabetes. Proliferative DR (PRDR) and non-proliferative DR (NPDR) are two categories of DR (PDR). Small blood vessels in the retina leak fluids and blood during NPDR, the first stage of DR, causing swelling in the macula or the development of microscopic fatty deposits known as exudates PDR, the more severe form of DR, causes new, fragile blood vessels to develop in the retina. These veins are prone to haemorrhage, leading to retinal scarring, detachment, and blindness. The risk of developing higher DR is much more prevalent in individuals who struggle to manage their blood pressure, cholesterol, or blood sugar levels [1]. Early diagnosis and management of DR are crucial in preventing vision loss. However, the detection of DR is challenging, as early stages of the disease may be asymptomatic. Therefore, regular eye examinations and screening for DR are recommended for individuals with diabetes. A dilated eye examination, PDR, the more severe form of DR, causes the retina to produce new, fragile blood vessels

A. H. Rather · I. U. Haq (✉)
Department of CSE, Chandigarh University, Mohali, Punjab, India
e-mail: mirinam525@gmail.com

A. Yadav et al. (eds.), *Proceedings of International Conference on Paradigms of Communication, Computing and Data Analytics*, Algorithms for Intelligent Systems,
https://doi.org/10.1007/978-981-99-4626-6_32

that are prone to bleeding, is commonly used to make the diagnosis of DR. Slowing or stopping the disease's progression and averting vision loss are the main targets of treatment for diabetic retinopathy (DR). This is accomplished by closely regulating the levels of cholesterol, blood pressure, and blood sugar. Mild-to-moderate non-proliferative DR (NPDR) can be managed with just monitoring and therapy of these causes. However, in order to stop future vision loss, severe NPDR or proliferative DR (PDR) may necessitate laser treatment or surgery. The identification and diagnosis of DR have both benefited from the extensive use of deep learning (DL), a subfield of machine learning. In retinal pictures, DL algorithms have demonstrated great accuracy in identifying DR, allowing for early identification and intervention. This helps to lessen the effects of DR and enhance patient care. Pre-processing and classification are typically two processes involved in DL models for DR detection. Retinal pictures are pre-processed to improve image quality and reduce noise during this stage. The pre-processed images are sent into a DL algorithm during the classification stage, and the algorithm learns to categorise the images as normal or abnormal depending on the presence or absence of DR. Several studies have shown promising results in the use of DL for DR detection. With an AUC of 0.936, or even the area under the curve, a study by Ting et al. (2017) that trained a DL algorithm on a data set of more than 100,000 retinal pictures was successful in accurately identifying referable DR. In a study by Gulshan et al. (2016), they trained a DL algorithm on a data set of more than 120,000 retinal pictures and found that it had a high AUC of 0.99 for detecting referable DR. In comparison with conventional techniques, DL models for DR detection have a number of benefits, including high accuracy, scalability, and automation.

DL algorithms can learn to detect DR patterns from large amounts of data, making them highly accurate and efficient. They can also be scaled to analyse large data sets quickly, enabling the processing of a large number of retinal images in a short time.

This medical condition is one of the main causes of blindness and a serious worldwide health concern. A microvascular ailments known as diabetic retinopathy can develop in people with diabetes [2]. Blindness may result from diabetic retinopathy, which decreases vision. Long-term untreated diabetes increases the risk of developing diabetic retinopathy in the patient [3]. Diabetic retinopathy develops signs in its later stages. People with diabetes might not be mindful of their infection in the beginning stages. DR is diagnosed either by direct examination of the retinal fundus in the clinic or by using imaging methods like optical coherence tomography or fundus photography [4]. The term "retinopathy" refers to harm to the retina as a whole. Diabetic retinopathy occurs when the small blood vessels that provide nourishment to the tissue and nerve cells of the retina are impaired. Typically, there are no indications in the early stages of diabetic retinopathy [5]. Only undergo an extensive eye examination that looks for initial signs of the illness, such as [6]

- Macular oedema (swelling).
- Pale, fatty deposits on the retina.
- Damaged nerve tissue.
- Any changes to the retinal blood vessels.

By 2040, there will probably be 600 million diabetics globally, with nearly a third developing diabetic retinopathy. the most typical reason for vision impairment in adults of working age worldwide [7]. Microaneurysms are a primary feature of moderate non-proliferative DR (NPDR) in its early phases. Proliferative DR (PDR), on the other hand, is a more severe form of DR that can cause serious vision impairment [8]. To guarantee that prompt counselling can be administered to prevent vision loss, routine DR tests are necessary. In addition to helping prevent vision loss, late-stage treatments including photocoagulation and intravitreal injections can also help minimise the progression of DR [9]. Routine DR screening is encouraged by many professional groups, but complete DR screening is more important. Imaging procedures like neon graphy and OCT (optical coherence tomography) can help diagnose and cure eye problems brought on by diabetes. To treat diabetic eye problems, retina experts may use drugs, laser therapy, or surgery. Early detection of diabetes-related abnormalities increases the chances of preserving vision. Optometrists can diagnose these issues, including diabetic retinopathy, through an OCT extended eye examination. Getting an eye examination at least once a year can help prevent vision loss caused by diabetes [12]. Quitting smoking along with strict cholesterol and glucose management can help lower the risk of diabetic eye problems. However, present methods for diagnosing DR are inadequate for catching it in diabetic people in its early stages [13]. Most research has been conducted on the later stages of DR, which can lead to gradual blindness in diabetic patients in both stages [14]. The availability of automatic DR screening can significantly reduce the burden on doctors [15]. By implementing automatic DR screening, the patient-to-doctor ratio can be decreased, saving time and money while optimising the utilisation of current resources [16]. The ability of those living in remote places without access to medical facilities to receive treatment via telemedicine is an additional benefit of implementing automatic DR screening [17]. Our research is focussed on developing a method for DR detection that is more effective than existing methods, accurate, and computationally fast. Figure 1 displays the pictures of the retinal fundus at various stages of diabetic retinopathy. Mild non-proliferative Stage II, moderate non-proliferative Stage III, severe non-proliferative Stage IV, and proliferative Stage V are the phases of diabetic retinopathy depicted in the photographs.

Images of the retinal fundus at various stages of diabetic retinopathy are shown in Fig. 1. There are five stages of diabetic retinopathy: stage II (mild non-proliferative), stage III (moderate non-proliferative), stage IV (severe non-proliferative), stage V (proliferative), and stage II (mild proliferative).

The following is how the paper is organised: Section 1 gave a quick overview of the Introducing section. The discussion of various diabetes detection strategies in Section 2 is covered by a number of researchers. In Section 3, a comparison of several DR strategies was presented. Section 4 discusses the limits of current approaches. A brand-new framework for identifying DR is presented in Section 4. Section 5 concludes the work with noting a few important aspects.

The "deep learning" subtype of machine learning is inspired by the design and function of the human brain. This branch of artificial intelligence has gained a lot of interest in recent years due to its ability to automatically learn from massive amounts

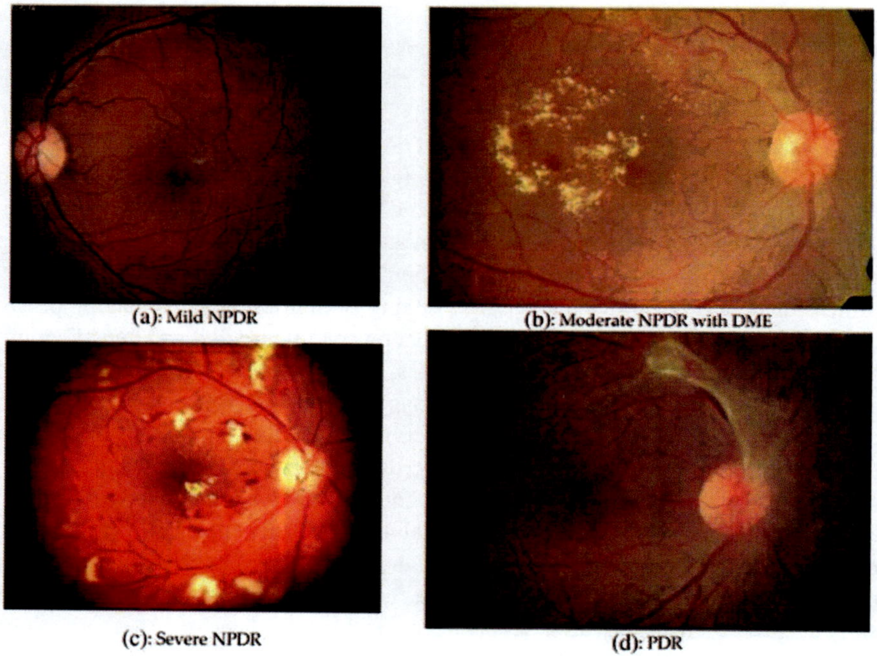

Fig. 1 Shows the pictures of the retinal fundus at different stages of diabetic retinopathy. Stages of diabetic retinopathy include **a** mild non-proliferative (Stage II); **b** moderate non-proliferative (Stage III); **c** severe non-proliferative (Stage IV); and **d** proliferative (Stage V) [18]

of data and make predictions or judgements based on that learning. Fundamentally, deep learning algorithms employ multilayer neural networks to recognise patterns and characteristics in data, enabling them to handle challenging tasks like speech and image recognition, natural language processing, and autonomous automobiles. The fundamental advantage of deep learning over earlier machine learning methods is its ability to automatically extract high-level features from raw data without requiring manual feature engineering. Deep learning is well suited for applications involving difficult-to-define qualities like speech and image recognition since it can manage enormous amounts of data. Healthcare (medical image analysis, illness diagnosis, drug development), finance (fraud detection, risk assessment, portfolio management), transportation (autonomous driving, traffic prediction), and education (individualised learning, intelligent tutoring systems) are a few of the sectors where it is used [20].

There are advantages and disadvantages to deep learning. Large amounts of data are needed for training, which creates a challenge in the healthcare sector due to privacy concerns. Another obstacle is the requirement for substantial computing resources, such as GPUs and RAM. Deep learning models are difficult to comprehend, which limits their use in sectors like healthcare. Architectures, algorithms, and hardware have all recently improved. Transfer learning can help reduce the amount of data required. A powerful subset of machine learning called deep learning has shown great promise in a variety of fields, including the medical industry. Deep learning algorithms can automatically extract intricate patterns and characteristics from massive data sets, which is very helpful in medical research where huge volumes of data are generated. Deep learning has been used in the study of medicine in a number of fields, including drug development, medical imaging, illness detection, and personalised medicine. Medical imaging is one of the most important areas where deep learning is being used in medical research. Deep learning algorithms may be trained to recognise and examine several anomalies in medical photographs, including those connected to breast cancer, lung cancer, and diabetic retinopathy. Deep learning algorithms, for instance, can examine retinal pictures to find microaneurysms, haemorrhages, and other early symptoms of diabetic retinopathy. It has been demonstrated that these models are fairly good at detecting diabetic retinopathy, which can lead to early treatment and improved patient outcomes. Therapeutic research, therapeutic effectiveness forecasting, and potential adverse effect detection have all shown the benefits of deep learning. Deep learning algorithms analyse a lot of data, including chemical structures, drug–target interactions, and clinical trial data, to speed up and reduce costs in the process of bringing innovative medications to market. In order to identify patterns and predict the course of a disease, deep learning may evaluate patient data for sickness diagnosis, such as medical records, genetic information, and symptom data. For example, it may predict when conditions like Alzheimer's, cancer, and cardiovascular disease would appear, enabling early treatment and better patient outcomes. A patient's medical profile, which may include genetic data, medical records, and lifestyle factors, is utilised in personalised medicine to create individualised treatment recommendations. By predicting the effectiveness of potential treatment options, enhancing patient outcomes, and encouraging a more individualised healthcare plan, this supports clinicians' decision-making [18].

2 Literature Review

The work connected to the current DR prediction is briefly presented in this section.

Amol et al. [2] created a system that takes MLPs into account. They only employed 130 photographs in their investigation. MLPs train the network utilising a method of supervised learning known as back propagation.

Kranthi Kumar et al. [3] carried out a methodical analysis to find the DR in only one specific type of lesion, hard exudates. They used the RRGA approach along with DIP to perform their investigation. The current research will be focussed on DIARETDB1.

Ankita et al. [4] found that dividing blood vessels and identifying lesions have been utilised in the survey in two different ways to identify DR. For their survey, they used the Drive, Stare, and Chase data set.

In Ling et al. [5], a DeepDR Framework was introduced. In order to assess picture quality in real time and grade or detect lesions from 121,342 diabetic patients' fundus images, DeepDR was trained on 466,247 images. The three depth sub-networks of the DeepDR system were the gastrointestinal ulcer sub-network, the DR rating sub-network, and the image quality evaluation sub-network. Masks from the RCNN and ResNet were used.

In Mike et al. [6], the fundamental method was repeated in "Development and validation of a deep learning system for diagnosis of diabetic retinopathy in retinal fundus pictures". The original work made use of private EyePACS fundus images and a different EyePACS data set. They used the Messidor-2 and EyePACS data sets for this study.

Sehriah et al. [7]. We built an ensemble of five sophisticated convolution neural network (CNN) models (ResNet50, Inceptionv3, Xception, Dense121, and Dense169) using the freely accessible Kaggle data set of retina images to encode the rich characteristics that improve the categorisation for various phases of DR. Multiple DR phases can be detected by this ensemble model. Throughout the entire inquiry, they focussed on DR detection. The accuracy of this model is 80%. They worked with a public database that was employed by everyone.

Harry et al. [8] found that by applying digital retinal images and a CNN technique, researchers were successful in correctly diagnosing DR; the severity of the condition is also mentioned. With the aid of data augmentation and CNN architecture, they build a network that can automatically detect difficult conditions like microaneurysms, exudate, and retinal bleeding. The network is trained using a top-tier graphics processing unit (GPU) on the publicly available Kaggle data set, and its outputs are excellent, especially for a difficult classification task.

Shirin et al. [9] identified that concurrently integrated fundus fluorescein angiogram and colour fundus pictures have been used to evaluate the severity stages of diabetic retinopathy. Curve-let transform is used to extract six features, which are then fed into a support vector machine. These characteristics include the overall number of microaneurysms, the area of exudate, the size, consistency, and number of blood vessels in the foveal vascular zone.

3 Comparative Analysis

The research into various DR detection methods offered by various researchers is summarised in Table 1 by their findings.

Doctor's workloads are reduced through automatic DR screening, which improves resource utilisation while saving time and money [10, 11]. People in remote locations with little access to medical facilities can benefit from telemedicine treatments made possible by this technology [16]. Our goal is to find a DR detection method that is more accurate and faster to compute than the present approaches. Figure 2 shows the percentage of data sets used in research projects, highlighting the need of include a variety of data sets for thorough conclusions (Fig. 3).

Table 1 A comparison of the DR methods currently in use

Author	Method used	Data set	Accuracy (%)	Specificity (%)	Remarks
Amol et al. [2]	MLPNN	DIARETDB0	92	90	More colour fundus pictures are used for next research
Harry et al. [8]	CNN	Public data set	75	95	Researchers could compare this network against the five-class SVM method using a greatly cleaner data set derived from actual screening conditions
Kranthi et al. [3]	DIP/RIGA	DIARETDB1	91	88	DR can identify through further lesions
Ankita et al. [4]	Not applicable	DRIVE/ STARE/ CHASE	Not applicable	Not applicable	Other deep networks can be used; however, studies have solely used CNNs
Mike et al. [6]	InceptionV3	EyePACS/ Messidor-2	89	83.4 67.9	They were unable to conduct a similar study to the original
Sehrish et al. [7]	CNN	Public data set	80%	Not Applicable	For various phases, we can train various models, then ensemble the outcomes
Ling dia et al. [5]	ResNet/RCNN	Private data set	82	81.3	Perhaps a different CNN algorithm will be used
Mashal et al. [19]	DCNN: a highly customised, scale-invariant network	EyePACS	85	91	By modifying the architectural design of an existing CNN, they increased the categorisation process's effectiveness and accuracy. The amount of steps in colour fundus imaging are lowered by DR

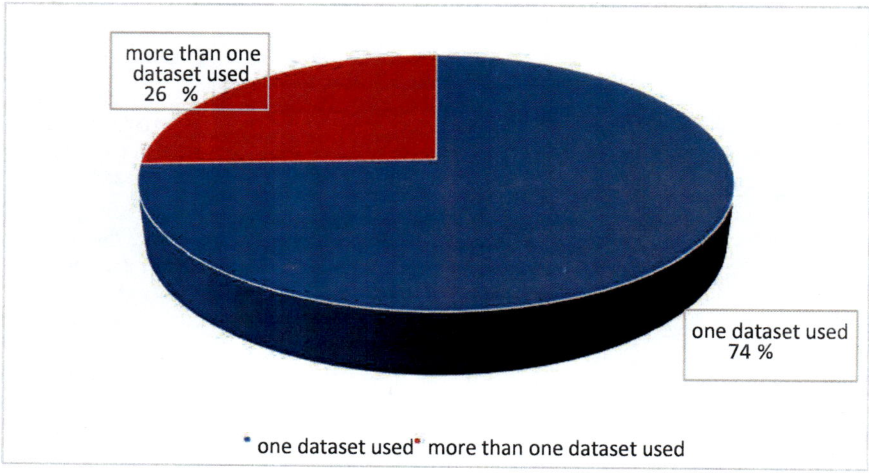

Fig. 2 Research that has used one or even more data sets as a percentage

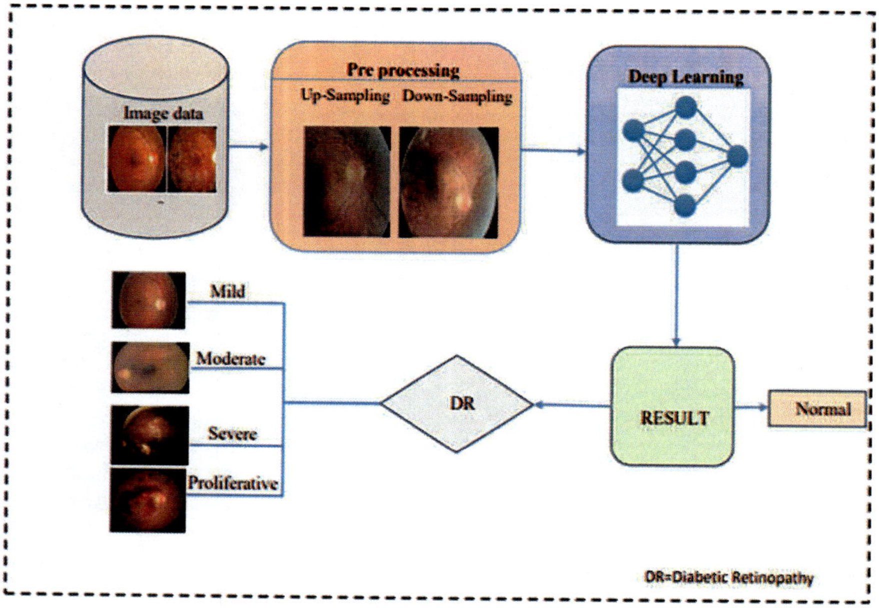

Fig. 3 I-DR framework [7]

4 Proposed Framework

It is clear from Sections 2 and 3 that the effectiveness and precision of the procedure and the selected data set vary considerably. Researchers have offered a variety of methods for predicting DR in diabetic patients, but a more thorough description of DR prediction in the early phases has not been provided in the literature. The drawback of those methods is that they can only be utilised to detect it in advanced stages of DR. In order to get over those restrictions, this research proposes an intelligent framework for early DR prediction.

Data collection, data pre-processing, prediction model, and findings are some of the main elements.

4.1 Data Collection

The most important aspect of every study is the data. We used the 35126 colour fundus images from the Kaggle data set, each measuring 3888 × 2951 pixels. According on the gravity of the offence, it includes images from a range of categories.

Table 2 provides the distribution of the various classes in the data set.

4.2 Pre-Processing

We start by pre-processing the data set with the intention of improving the photographs. Depending on their stage, the data can be utilised to differentiate cases of proliferation diabetic eye disease and those of non-invasive disease retinopathy. We crop down every image that is input yet maintain the aspect ratio to decrease the training expense. Additionally, we utilised combined up- and downsampling to balance the data set. Areas are randomly cut during the upsampling phase to increase minority classes. In order to equalise the samples of the various classes, enhance the data set, and avoid overfilling, flipping and 90o revolutions are used. During the downsampling process, more instances of majority classes disappear in order to satisfy the cardinality of the not as class. The mean normalising is carried out to every

Table 2 Shows the distribution of different classes in the data set

Classes	Total no of images
Class (0) normal	Two five eight one zero
Class (1) mild	Two four four two
Class (2) moderate	Five two nine two
Class (3) severe	Eight seven two
Class (4) proliferative	Seven zero eight

picture in the produced distributions before spin and inversion in order to remove feature bias and reduce training.

4.3 Prediction Model

CNNs are useful for detecting diabetic retinopathy because they automatically extract important components. With residual connections, ResNet-50, a deep neural network design, solves gradient problems. Convolutional, pooling, and fully linked layers are among ResNet-50's 50 layers. Deep network training is made possible by residual connections because they allow data to flow through certain levels without being processed. ResNet-50 is very good at image identification tasks, particularly those involving medical imaging like the diagnosis of diabetic retinopathy. Here is the ResNet-50 algorithm:

1. Input a batch of images to the network.
2. Apply a 7×7 convolutional layer with stride 2 to the input pictures, then perform batch normalisation and ReLU activation.
3. Apply a 3×3 max pooling layer with stride 2 to down sample the feature maps.
4. Pass the feature maps through a stack of residual blocks.
5. Each residual block has two 3×3 convolutional layers with batch normalisation and ReLU activation as well as a skip connection that adds the input to the second convolutional layer's output.
6. Every few residual blocks, each convolutional layer's number of filters doubles. After the last residual block, apply global average pooling to the feature maps.
7. Flatten the output of the global average pooling layer.
8. Connect the flattened features to a fully connected layer with softmax activation to output the predicted class probabilities.

There are 50 levels in the ResNet-50 architecture, including 16 residual blocks. The training with stochastic gradient descent with backpropagation minimises the cross-entropy loss between the predicted and actual class labels. During training, data enhancement methods like random cropping and horizontal flipping are utilised to improve the network's generalisation performance. Since the weights were pre-trained on enormous data sets like ImageNet, the network can learn high-level features from the input photographs.

4.4 Results and Future Work

A data set of fundus images for diabetic retinopathy was used in a study to compare the effectiveness of several deep learning models, ResNet-50 achieved a sensitivity of 91.5%. The proposed framework utilises only fundus images for DR prediction. However, other modalities such as OCT scans, patient demographics, and clinical data can also provide valuable information. Integrating these modalities into the framework may improve its accuracy and reliability.

5 Conclusion

Since diagnosing diabetic retinopathy (DR) takes time, an expedient method is required for better therapy. If left untreated for a long time, DR, a microvascular illness that affects vision, can cause blindness. Unnoticed progression leads to symptoms in the future. To make a diagnosis, doctors employ clinical examinations and imaging methods including fundus photography and optical coherence tomography. While there are several methods for DR prediction, there is not a complete literature description of early-stage prediction. Reviewing retina-based DR detection techniques is the goal of this work, which is an essential first step in creating an intelligent diagnostic model for rapid and precise DR diagnosis. The research community and patients will gain from this model's assistance in developing improved treatment strategies.

References

1. Borys et al (2020) Deep learning approach to diabetic retinopathy detection. arXiv: 2003.02261v1 [cs. LG] 3 Mar 2020
2. Amol et al (2015) Detection of diabetic retinopathy in retinal images using MLP classifier. In: Proceedings 2015 IEEE
3. Kranthi et al (2018) Automatic diabetic retinopathy detection using digital image processing. In: International conference on communication and signal processing
4. Ankita et al (2018) Diabetic retinopathy: present and past. In: International conference on computational intelligence and data science (ICCIDS 2018)
5. Ling et al (2020) A deep learning system for detecting diabetic retinopathy across the disease spectrum. Nature Commun
6. Mike et al (2018) Replication study: development and validation of a deep learning algorithm for detection of diabetic retinopathy in retinal fundus photographs. arXiv: 1803.04337v3 [cs.CV] 30 Aug 2018
7. Sehrish et al (2019) A deep learning ensemble approach for diabetic retinopathy detection. Received August 27, 2019, accepted October 4, 2019, date of publication October 15, 2019, date of current version October 29, 2019
8. Harry et al (2016) Convolutional neural networks for diabetic retinopathy. In: International conference on medical imaging understanding and analysis 2016, MIUA 2016, 6–8 July 2016, Loughborough, UK
9. Hajeb S et al (2012) Diabetic retinopathy grading by digital curvelet transform. Hindawi Publishing Corporation Comput Mathem Methods Med 761901:11. https://doi.org/10.1155/2012/761901; Received 24 May 2012; Accepted 30 July 2012
10. Wejdan et al (2020) Diabetic retinopathy detection through deep learning techniques: a review" Shalash Information Technology Department, University of King Abdul Aziz, Jeddah, Saudi Arabia, Informatics in Medicine Unlocked
11. Enrique et al (2019) Automated detection of diabetic retinopathy using SVM. Department de Electrica y Electr ´ onica ´ Univ. de las Fuerzas Armadas ESPE Sangolquı́, Ecuador

12. Hoda et al (2021) Automated detection and diagnosis of diabetic retinopathy: a comprehensive survey. Theoretical and Experimental Epistemology Lab, School of Optometry and Vision Science, University of Waterloo, Waterloo, ON N2L 3G1, Canada
13. Kele et al (2017) Deep convolutional neural network-based early automated detection of diabetic retinopathy using fundus image. School of Information and Communication, National University of Defense Technology, Wuhan, 430019, China
14. Abhishek et al (2020) Automated detection of diabetic retinopathy using convolutional neural networks on a small dataset. Pattern Recogn Letters 135
15. Ratul et al (2017) Automatic detection and classification of diabetic retinopathy stages using CNN, IEEE
16. Carson et al (2018) Automated detection of diabetic retinopathy using deep learning. AMIA Jt Summits Transl Sci Proc v.2018, PMC5961805
17. Mohammad et al (2020) Exudate detection for diabetic retinopathy using pretrained convolutional neural networks. 5801870. https://doi.org/10.1155/2020/5801870
18. Rishab et al (2017) Automated identification of diabetic retinopathy using deep learning. 124(7):962–969
19. Mashal et al (2017) Detecting diabetic retinopathy using deep learning. IEEE
20. Zhentao et al (2019) Diagnosis of diabetic retinopathy using deep neural networks. 1. Machine Intelligence Laboratory, College of Computer Science, Sichuan University, Chengdu 610065, China; 2. Sichuan Academy of Medical Sciences, Sichuan Provincial People's Hospital, Chengdu 610072, China Corresponding author 2019 IEEE

Chapter 33
Advanced Footstep Piezoelectric Power Generation for Mobile Charging Using RFID

Kiran Ingale, Atharva Jivtode, Sakshi Bandgar, Ayush Biyani, and Vedant Chaware

1 Introduction

The need for electricity is constantly growing, and modern living makes extensive and flexible use of it. The value of new technology that emerges every day asks for more electricity power because the world's population is growing and, as a result, the need for energy is growing rapidly [1]. The system charges the mobile phone within the allocated time thanks to this technology. The system that was developed here does not require any more time consequently. Power outages decreased as a result, enabling the system to develop a much more effective and affordable way of generating energy using RFID, which lowers global warming. A microcontroller-based system for generating electricity from footsteps uses the force of footsteps to produce voltage [2]. Bus stops, theaters, train stations, malls, and other public places can significantly profit from this endeavor. These systems are therefore put in place in public areas where people stroll, and they are required to be used to access the entry or exit.

1.1 Necessity of the System

This system makes use of RFID to generate power in a much more efficient and economical manner, thereby decreasing both global warming and power shortages. Piezo technology is used in this system to produce energy from renewable sources [3]. The piezo sensor characteristics and energy output are tracked by the system and

K. Ingale (✉) · A. Jivtode · S. Bandgar · A. Biyani · V. Chaware
Vishwakarma Institute of Technology, Pune, Maharashtra 411037, India
e-mail: Kiran.ingale@vit.edu

A. Yadav et al. (eds.), *Proceedings of International Conference on Paradigms of Communication, Computing and Data Analytics*, Algorithms for Intelligent Systems,
https://doi.org/10.1007/978-981-99-4626-6_33

displayed on the OLED. The device employs RFID technology to allow for USB port-based mobile phone battery charging.

2 Literature Survey

The basic principles of electricity production were discovered in the 1820s and the first half of the 1830s by British physicist "Michael Faraday." He created the still-in-use Faraday disc, which generates energy by moving a loop of wire between the poles of a magnet. The authors began the design process for "electrical power generation employing footsteps for urban area energy applications" in an IEEE publication. This device uses footsteps as a creative way to store waste energy and reduces pollution in areas that are already highly polluted [4]. "Pierre and Paul-Jacques Curie" discovered the concept of piezoelectricity in 1880 by compressing crystal types, such as quartz, tourmaline, and Rochelle salt, along particular axes. On the top of the crystal, this resulted in a voltage. For this phenomenon, the word "piezoelectric effect" is used. Using piezoelectric sensors, which take advantage of the piezoelectric effect for working purposes, a new technique known as "Footstep Power Generation Utilizing Piezoelectric Sensors" has been created. In this device, energy is generated using piezoelectric sensors. These sensors are pushed by human footsteps, which store energy for eventual use in the battery. Consequently, the project is scaled back, costs are reduced, and the system is made simpler [5]. Their proposed method, "Advanced Footstep Power Generation Using RFID for Charging," makes use of RFID technology. RFID was formally created in 1983 when Charles Walton submitted the first patent application using the word "RFID." By utilizing this RFID technology, the electricity is distributed among customers in our initiative based on their user identification number and electromagnetic waves. Kumar et al. are a triumvirate. For extremely populous nations, the use and implementation of excessive human foot resources are strongly encouraged. Therefore, by converting mechanical energy into electrical energy, electricity can be used and spread in this way [6]. The use of RFID cards to transmit the current was proposed by the authors to limit access to the charging generator to approved users only. We use controls to power battery charges, a microcontroller device to show battery charges on LCDs, and a configuration to allow mobile charging [7]. According to the authors, the basic idea of generating power from human footsteps was taken from this study. In this case, mechanical technology known as "rack and pinion" was used to produce electricity [8, 9]. According to a proposal made by the authors, the power produced by this technique can also be used for fundamental applications like streetlights, notice boards, gyms, and other public areas [10]. According to S., the safest, most sensible path to wealth for the world's populace is Agnes Dazzling Rachel, Ganesh Prabhu, and A. Their endeavor was testing and updating Shinnee [11]. This method allows us to move both loads. This approach provides efficient energy production in nations with large populations [3, 12]. In reality, they only use 11% of their energy to produce green energy. They used LCD for their endeavor.

2.1 Evolution of RFID Technology

Radio-frequency identification (RFID), which makes use of electromagnetic fields, allows objects with tags affixed to them to be automatically identified and tracked [13].

Numerous indications point to the fact that RFID-related apps are still in their infancy. Revenue from the RFID market grew by more than 33% between 2004 and 2005, and by 2010, it is anticipated to reach USD $3 million.

The evolution section taught us about the past uses of RFID technology, which inspired me to create a new system [14]. This device uses piezoelectric sensors and human footsteps to generate electricity. The extra power that is kept there can be used to recharge the battery when necessary [15]. User assignment is done by employing RFID technology and an identification card. An EM-18 reader detects this card, allowing Arduino to give power in accordance with a timer programmed into the code.

3 Proposed System

When human footfall causes piezoelectric sensors to sense weight, the "Advanced Footstep Power Generation Utilizing RFID for Charging" method is an effective invention that saves energy. This energy is kept in the battery and then distributed among numerous people via RFID cards. A 12-digit human identification number on these cards is used to collect information about everyone. RFID operates by emitting electromagnetic pulses. According to the system as shown in Fig. 1 works by allocating a specific amount of time at a time to each user, according to the published software code. Because of the time and pollution savings brought about by this method, future generations will be better able to acquire both a pollution-free environment and time-consuming necessities.

3.1 Components and Description

Power supply: The regulated power supply serves as the circuit's source of input. The rectifier receives 12 V after the transformer steps down the 230 V AC input from the mains source to 12 V. The rectifier's result is a pulsing DC voltage.

Arduino Uno: The Uno is a standard Arduino board. The term "one" in Italian is Uno. The name Uno was assigned to the Arduino software's initial release. Additionally, it was the first USB board for Arduino that was made accessible. It is used in numerous tasks and is thought to be the most efficient board.

Arduino.cc produced the Arduino Uno board. An ATmega328P serves as the Arduino Uno's microprocessor. In comparison with other boards like the Arduino

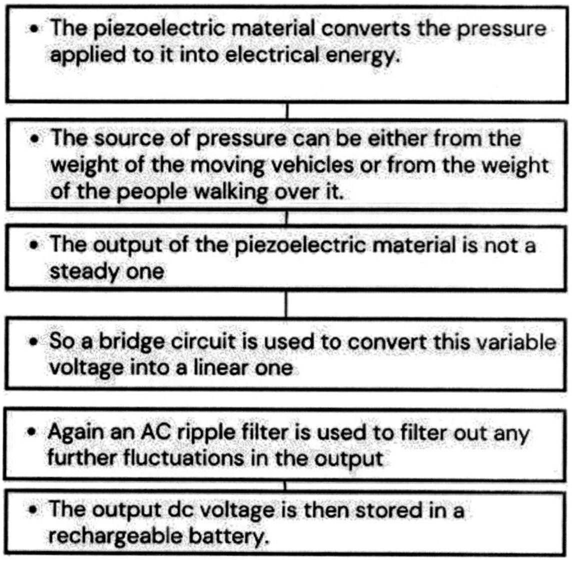

Fig. 1 Flowchart of proposed system

Mega board, etc., it is easier to use. The board is made up of shields, input/output (I/O) pins for digital and analog data, and other circuits. It also has 14 digital pins.

Piezoelectric sensor: To detect pressure, acceleration, temperature, strain, or force changes, a piezoelectric sensor converts the changes into an electrical charge.

RFID reader-RC522: The RC522 RFID reader/writer module (transceiver) can read and write RFID bands because it is based on NXP's highly integrated reader/writer IC MFRC522. The frequency it uses for wireless multi-communication is 13.56 MHz; RFID stands for "radio-frequency identity." Using radio-frequency electromagnetic waves, this module receives and transmits data. With 1 KB of memory and 13.56 MHz interoperability, it can read and write all transponders, including RFID card tags and key fob tags. This small, low-cost, SPI-integrated module can readily communicate with almost any microcontroller, including the ATtiny, Arduino, ESP8266, Raspberry Pi, and other more advanced development boards.

RFID tag: RFID tags use smart barcodes, a type of tracking technology, to identify items. The use of radio-frequency technology in RFID tags is referred to as "radio-frequency identity" or RFID.

Relay module: A relay module is an electronic device that allows a low-voltage signal to control a higher-voltage circuit. It consists of an electromagnet, a set of contacts, and a spring. When an electrical signal is sent to the electromagnet, it creates a magnetic field that attracts the contacts, causing them to close or open the circuit, depending on the relay configuration. Relay modules are commonly used in various applications, such as home automation, industrial automation, robotics, and the automotive industry. They can be used to control motors, lights, pumps, solenoids, and other electrical devices. Relay modules come in different configurations, including single-pole single-throw (SPST), single-pole double-throw (SPDT), double-pole single-throw (DPST), and double-pole double-throw (DPDT). They can also be normally open (NO) or normally closed (NC). Relay modules can be operated using different control signals, including digital signals, analog signals, and pulse-width modulation (PWM) signals. They can be controlled by microcontrollers, PLCs, or other digital devices. Relay modules can be interfaced with other electronic components, such as transistors, diodes, capacitors, and resistors, to create more complex circuits. They can also be mounted on circuit boards or used as standalone modules.

Operation amplifier circuit: Op-amps, also known as operational amplifiers, are integrated circuits that mainly consist of transistors and resistors. An incoming signal is multiplied by these ICs to create a stronger output. Both DC and AC systems with voltage and current can use these components.

OLED: An organic light-emitting diode, also referred to as a solid-state OLED, is a device that emits bright light when an electric current is applied. They are made by sandwiching a series of organic thin sheets between two conductors.

4 Methodology

The suggested remedy as shown in Fig. 2 is cost-effective and effective. Our method has the benefit of charging the user within a predetermined window of time specified by the system software. To create an advanced step-power generation system that utilizes RFID for charging, we created a design and construction plan. After deciding which parts needed to be connected or applied in hardware, block diagrams, and circuit diagrams were created. Each component was installed in accordance with the circuit diagram, and the Arduino was then programmed to run the complete system using the Arduino IDE. All the pieces were assembled on a board after the system was run and checks were made. The main purpose or output of the suggested system is to perform tasks such as mobile phone charging.

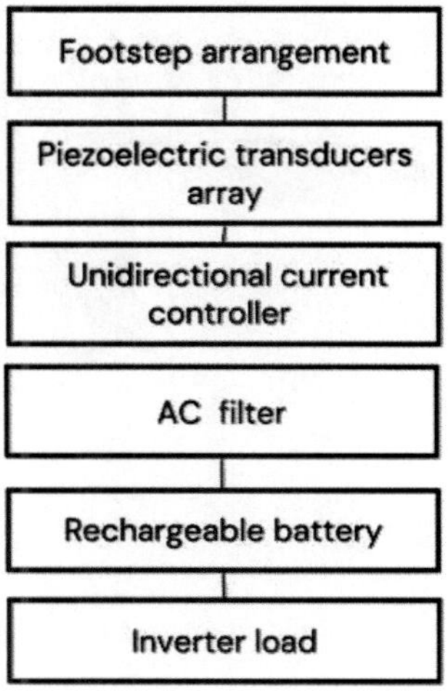

Fig. 2 Flowchart of process

5 Results and Discussion

5.1 Linearity Test

The power of a person's footfalls affects the voltage the piezoelectric sensor produces. The output of the piezoelectric sensor shows 0 V when no force is applied. It has been shown that the voltage produced rises along with the pressure applied. High-pressure results in strong voltage generation. The voltage rises rapidly, much like pressure does. The leftover energy will then be put into a battery to be used later as shown in Fig. 4. A mobile phone can be used as the output to show that the rechargeable battery that the piezoelectric sensor (Figs. 5 and 3) has charged is operating properly and that the electric current produced by the sensor is present. By attaching a USB cable from the device's USB port to a smartphone, the circuit can be checked. The charging icon is displayed on the phone's display like Fig. 6. An authorized RFID tag can power mobile devices. In summation, the amount of pressure applied to the piezoelectric sensor affects how much voltage it generates. The power can then be stored in the portable battery for later use. This study led to the creation of an innovative, affordable renewable energy source. In addition, it has been learned how mechanical energy is usually transformed into electrical energy.

For the piezoelectric material, the polarity of the charge is based on which direction the pressure is applied and the charge is represented as

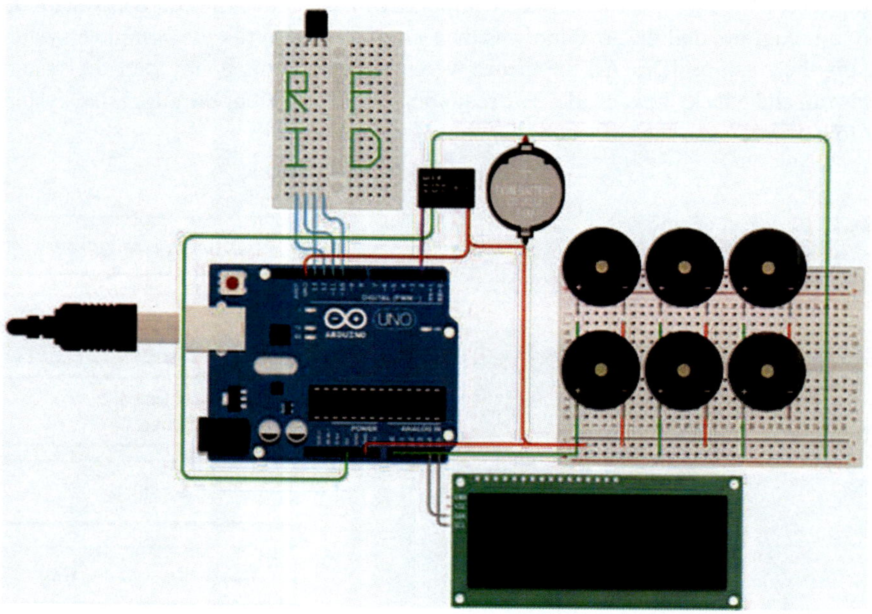

Fig. 3 Simulation-based picture of the project

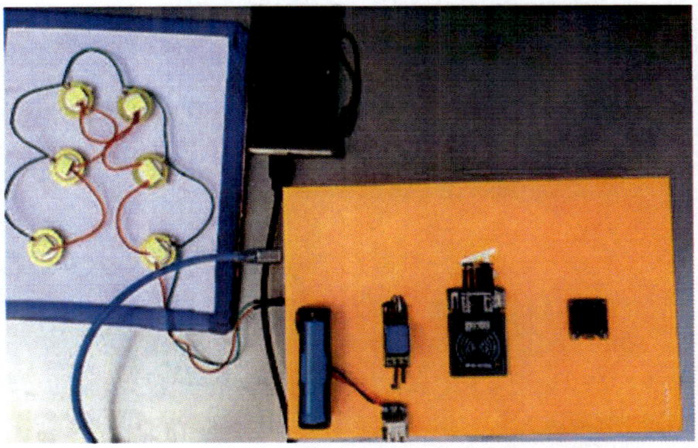

Fig. 4 Overall prototype

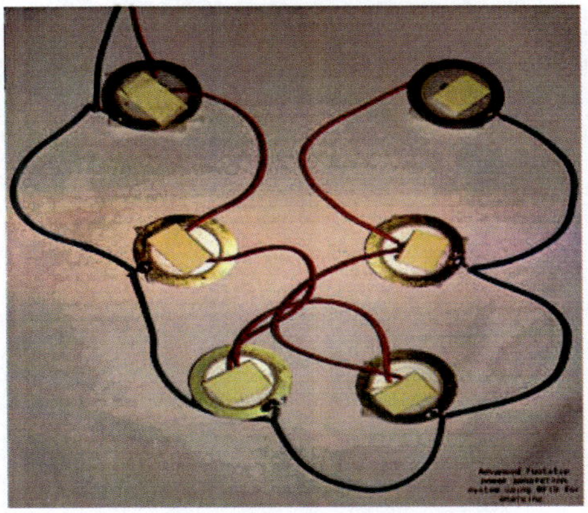

Fig. 5 Piezoelectric sensor cells

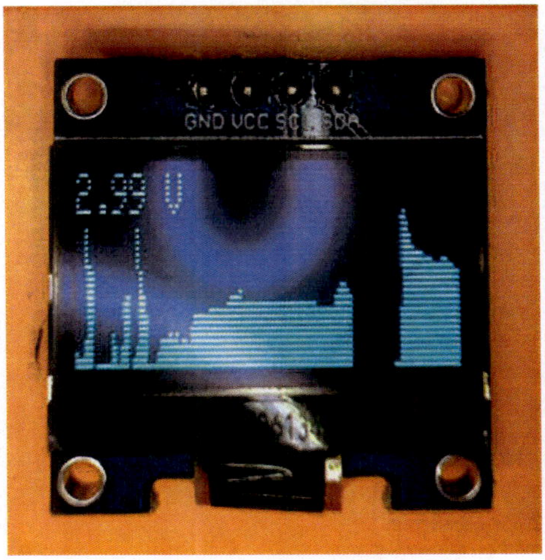

Fig. 6 Graph of generated voltage

$$\text{Charge } C = d \times F \text{ Coulombs}$$

where d corresponds to the material charge sensitivity, and

F corresponds to the applied force.

The thickness of the crystal will be modified depending on the applied force and

$$F = AE/t\ \Delta t \text{ newton.}$$

And here, the total area of the crystal, denoted by A, is a measurable quantity representing the surface area of the material in a square meter, t is the crystal thickness and measured in meter; whereas E corresponds to young's modules and measured in (N/m^2).

The Young's modulus (Y) of a material can be calculated by dividing the stress applied to the material by the strain it produces. In the case of a piezoelectric crystal, the formula for stress is $(F/A) * (1/(\Delta t/t))$. The force applied to the piezoelectric crystal is denoted by F, while the total surface area of the crystal is represented by A. Δt is the change in thickness, and t is the original thickness. By substituting A with the product of the crystal's length (L) and width (w), we can simplify the formula for Young's modulus to $Ft/(A\Delta t)$, which is expressed in units of N/m^2.

where

'W' corresponds to crystal width and measured in meters.

'l' corresponds to crystal length and measured in meters.

So, by substituting force equation in the charge, the output is

$$Q = dAY(\Delta t/t)$$

Then, the output voltage due to the electrode charges is

$$E0 = Q/Cp = dF/(ErE0A/t)$$

$$E0 = d/(ErE0)\,tP$$
$$E0 = gtP$$

where $g = d/\ ErE0$

And g is termed as crystal voltage sensitivity, and $g = E0/tP$. The crystal voltage sensitivity is stated as the ratio to intensity of the electric field and pressure. And E0 is the strength of the electric field and measured in V/m. When there exists, crystal mechanical distortion, then the charge is developed across the electrodes. Observation and results are seen in Table 1.

Table 1 Observation and results

Weight of person (kg)	No of footsteps taken	Generated voltage (V)
44	10	4.56
49	20	5.59
50	10	4.89
62	20	7.45
65	10	6.69

5.2 *Temperature Test*

Six piezo sensors were used per square foot. Piezo sensors produce electricity at different rates, so we get min voltage = 1 V every step. Each stage has a maximum voltage of 10.5 V. The average calculation is also as follows when considering the steps of a single person who weighs 50 kg, considering both the pressure and typical weight of a single person of 50 kg: 800 steps are needed for every 1 V rise in battery voltage. This indicates that a total of 8*800 steps, or 6400 steps, are needed to raise the battery's 12 V by one unit. The proposed system would be implemented in a polluted area with foot traffic as a source of pollution. 6400 steps require time, which is 6400/(60*2) = 53 min (Approx.)

5.3 *Battery Consumption*

3.7 V is the battery power. Battery's stated capacity is 3Ah/36Wh. Current for charging or discharging, C-rate: 1, or I: 3 A. Charge or release timing: 1 h (run-time) (run-time). The run-time, also known as the period of charge or discharge, is 60 min. To calculate the voltage, current, and energy saved for a set of parallelly and series-connected batteries. One set of batteries is equivalent to one element. Only one duplicate series exists. Overall, one battery is utilized.

6 Future Scope

Because the proposed system will be implemented in a weak terrain where step dust will be a source, the normal of two ways per second is considered. In view of inherent problems for nations with dense populations, utilizing energy as effectively as possible is essential. For the first time, Japan has used flooring penstocks to generate mechanical energy, much like how these plates are used on machine steps. Energy is produced when someone enters, increasing both the quantity and the quantum of energy. Alternately, by placing analogous plates on cotillion floors, Europe is creating history in the field. But if someone uses these pennies and balls once more, energy is produced that can be used to power a phone or other valuable items. Due to the significant amount of electricity from power plants that will be saved, the entire world will soon greatly benefit from this. It is time to think about your options because conventional supplies are quickly depleting. For efficient functioning, we must store the energy obtained from conventional sources. Thus, this strategy not only grants druthers but also encourages the country to be more prudent. The decline in car traffic in major cities is problematic for the populace. However, a novel technique called "power hump" can be used to generate energy from this sluice of moving parts. The fact that it does not rely on outside sources is a plus.

7 Applications

It can be used in populated places like airports, railway stations, and machine stops. This technology has a broad range of potential applications, including areas with reliable public transportation and various social groups or organizations [16, 17]. Instead of solar lights, it can operate street illumination during stormy weather. In crowded places like promenades, walkways, and so forth, this framework can be used.

8 Advantages

No energy intake is required. It is a true, authentic gadget. Less non-sustainable energy sources are used. It is a tone-creating device that makes use of our footfalls. No moving corridor and a lengthy management life. Energy production is stumbling up the staircase. small and incredibly quick [18]. The medium is reasonably sensitive despite its low sensitivity. It is a brand-new method that is dependable, inexpensive, and green. A decrease in the usage of green energy electricity is generated while exercising and running. Batteries are used to hold the energy that creators produce.

9 Drawbacks

The initial expense of the arrangement is significant. Variations in temperature influence the situation. Batteries need to be treated carefully. Under fixed conditions, estimation is not feasible [19, 20]. High-impedance lines are required for the electrical connection because the device only needs a small electric charge to function. The incident may alter based on the demitasse's temperature.

10 Conclusion

With this technology, nations with dense populations can generate electricity more effectively while reducing control requests. Only 11% of our electricity comes from renewable sources. The idea has been successfully tested and put into practice, making it the most conservative and workable answer for most people in our nation. Resource management is a significant challenge in India due to its large population and developing economy. The "Advanced Footstep Power Generation System Using RFID for Charging" project has been put into use and tested effectively. It is the finest and most economical form of energy for typical consumers. The best use of RFID technology is for consumers to get the outcomes they need when they need them.

Acknowledgements "We would like to thank our faculty (the Department of Electronics and Telecommunications Engineering) for helping us during the tenure of this project. The guidelines they provided played a very vital role in the completion of this project. We also want to express our gratitude to the Department of Instrumentation and Control Engineering employees. We also want to express our gratitude to the college for providing the necessary journals, books, and Internet access for the project's information gathering."

References

1. Boglaev I (2016) A numerical method for solving nonlinear integro-differential equations of Fredholm type. J Comput Math 34(3):262–284. https://doi.org/10.4208/jcm.1512-m2015-0241
2. Kumar A, Kumar A, Kamboj A (2017) Design of footstep power generator using piezo-electric sensors. In: International conference on innovations in information, embedded and communication systems (ICIIECS), IEEE, March 2017
3. Karthik SV, Karthik S, Satheesh Kumar S, Selvakumar D, Visvesvaran C, Mohammed Arif A (2019) Region based scheduling algorithm for pedestrian monitoring at large area buildings during evacuation. In: 2019 International conference on communication and signal processing (ICCSP), Chennai, India, pp 0323–0327. https://doi.org/10.1109/ICCSP.2019.8697968
4. Lindberg DV, Lee HKH (2015) Optimization under constraints by applying an asymmetric entropy measure. J Comput Graph Statist 24(2):379–393. https://doi.org/10.1080/10618600.2014.901225
5. Rieder B (2020) Engines of order: a mechanology of algorithmic techniques. Amsterdam Univ. Press, Amsterdam, Netherlands
6. Panghate S, Barhate P, Chavan H (2020) Footstep power generation system using RFID for charging. Int Res J Eng Technol (IRJET)
7. Tiwari RR, Bansal R, Gupta P (2019) Foot step power generation. Int Res J Eng Technol (IRJET)
8. Chauhan S, Singh M, Tripathi A (2020) Footstep power generation using piezoelectric sensor and distribution using RFID. Int Res J Eng Technol (IRJET)
9. Krempl P, Schleinzer G, Wallnöfer W (1997) Gallium phosphate, GaPO4: a new piezoelectric crystal material for high-temperature sensorics. Sens Actuators A: Phys 61(1–3):361–363
10. Ganesh Prabhu S, Keerthivasan G, Naveen Kumar A, Jeevananthan N, Thirrunavukkarasu RR, Karthik S (2021) Power generation using footsteps for mobile charging. In: 2021 7th international conference on advanced computing and communication systems (ICACCS)

11. Ganesh Prabhu S, Rachel AS, Roshinee AR (2020) Tracking real time vehicle and locking system using labview applications. In: 2020 6th International conference on advanced computing and communication systems (ICACCS), IEEE, pp 55–57
12. Sharapov V (2013) Piezoceramic sensors. Springer, Berlin, Berlin
13. Thenmozhi S, Mahima V, Maheswar R (2017) GPS based autonomous ground vehicle for agricultural utilitya€. In: ICIECE 2017, GNIT, Andrapradesh, 21st-22nd July 2017
14. Ganesh Prabhu S, Adarsh R, Arun Vikash SP, Amarthiyan D (2020) Analysis of retinal images to diagnose stargardt disease. In: 2020 6th International conference on advanced computing and communication systems (ICACCS), IEEE, pp 1245–1247
15. Kirthika A, Dhivyapriya EL, Thenmozhi S, Ganesh Prabhu S (2019) CDMA design for on-chip communication network. Int J Eng Adv Technol (IJEAT) 9(2):3256–32260. https://doi.org/10.35940/ijeat.B3148.129219
16. Tichý J, Erhart J, Kittinger E, Jana P (2014) In: Fundamentals of piezoelectric sensorics mechanical, dielectric, and thermodynamical properties of piezoelectric materials. Berlin, Springer
17. Volk T (2010) Lithium niobate: defects, photorefraction and ferroelectric switching. Springer, Place of publication not identified
18. Tadigadapa S, Mateti K (2009) Piezoelectric MEMS sensors: state-of-the-art and perspectives. Measurem Sci Technol 20(9):092001
19. Elvin NG, Elvin AA (2010) Effects of axial forces on cantilever piezoelectric resonators for structural energy generating. Strain 47:153–157
20. Briscoe J, Shoaee S, Durrant JR, Dunn S (2013) Piezoelectric enhancement of hybrid organic/inorganic photovoltaic device. J Phys: Conf Ser 476:012009

Chapter 34
Radial Distribution Networks Reconfiguration with Allocation of DG Using Quasi-Oppositional Moth Flame Optimization

Sneha Sultana, Sourav Paul, Poulomi Acharya, Pronoy Das Choudhury, and Provas Kumar Roy

1 Introduction

Network reconfiguration is a methodical and impactful technique for decreasing power loss in power distribution systems. Distribution network reconfiguration (DNR) aims to better voltage profile, balance load, reduce power losses, and increase network reliability. By using tie switches and sectionalizing, this can be accomplished.The goal of the DNR problem is to identify the optimal way for the radial distribution network to operate, assuming that power losses are kept to a minimum and that all network constraints are met.

The network performance will decline, it will become unstable, the power losses will increase, and the voltage levels will exceed the operational standards if the ideal location and size of DG resources are not selected appropriately. Therefore, by performing DNR and allocating DGs optimally, power losses can be decreased, and the voltage profile can be improved. The purpose of this paper is to illustrate the radial distribution network reconfiguration and DG installation.

The network parameters are required by traditional model-based techniques in order to precisely determine the distribution network's ideal configuration. The operation of distribution networks (DN) may be adversely or advantageously affected by the operation of distributed generators (DG). According to studies, choosing the wrong location and size for the DG could result in larger system losses than losses

S. Sultana (✉) · S. Paul · P. Acharya · P. D. Choudhury
Dr. B. C. Roy Engineering College, Durgapur, India
e-mail: sneha.sultana@gmail.com

P. D. Choudhury
URL: http://www.bcrec.ac.in

P. K. Roy
Kalyani Government Engineering College, Kalyani, India
URL: https://www.kgec.edu.in

A. Yadav et al. (eds.), *Proceedings of International Conference on Paradigms of Communication, Computing and Data Analytics*, Algorithms for Intelligent Systems,
https://doi.org/10.1007/978-981-99-4626-6_34

without it. The integration of distributed generators (DGs) into distribution networks is growing as a result of the liberalization of the energy market, environmental concerns, and technological advancement and makes a model-free reinforcement learning (RL) study of a distribution network reconfiguration process using the NoisyNet deep Q-learning network (DQN). Without changing the parameters, the paper has successfully demonstrated that the exploration can be realized automatically. Paper [1] even though only a few branches are switchable, the paper investigates a scenario-based convex programming model to enhance the capabilities of reconfiguration in distribution networks. Ref [2] for this, two test networks are put into place. Addressing the shortcomings of physical model-based control algorithms, the DNR issue is resolved using a data-operated batch-mannered RL algorithm. Following the examination of three networks, it has been determined that this method enhances behaviour control policy, is very climbable, and can offer a workable real-time resolution. To achieve a better operating environment during a critical landing, the traditional reconfiguration strategy has been improved and generalized using dynamic microgrids. A two-stage mixed integer conic programme underpins the study. In terms of finding the best solution, accelerating convergence, and reducing running time, the research found that particle swarm optimization was superior to genetic algorithm. Another study [3] presented several optimal and maximal models for utility-based distributed generation penetration. The main objective was to increase the penetration of rural distribution networks by distributed generations. The simultaneous reconfiguration of the radial distribution network and the allocation of switched capacitor banks was discussed in the paper to enhance the performance of the distribution systems. The enhancement of the voltage profile and the reduction of active and reactive power losses were the main objectives. Distribution system reconfiguration (DSR) and optimal capacitor placement (OCP) are two alternative techniques that are being looked into in order to increase a system's capacity for generating power and meet the rising demand for electricity, according to Hussain 2019 Hybrid. It was better to use both strategies simultaneously than one at a time. The individual OCP mode and the dual DSR mode after the OCP were the two operating modes that were used. In order to allocate distributed generators (DGs) and capacitors whilst taking into consideration the rearranging power distribution network, the artificial ecosystem optimizer (AEO) method is used. (RPDS). This was motivated by three energy transfer processes in an ecosystem that involved production, consumption, and decomposition [4]. For example, [5–7] novel combine the DNR and DG placement problems to improve the efficacy of the distribution network. In [6], the presence of DG, the DNR issue is addressed with the goals of reducing real power loss and improving voltage profile in distribution network (HSA). Mohamed Imran et al. [7] proposed a method for handling DNR and DG placement that is dependent on the fireworks optimization algorithm (FWA) in an effort to reduce electricity loss and improve voltage constancy. To pre-identify the suitable bus sites for DG installation, both studies used a variety of techniques, including voltage stability index (VSI) and the loss sensitivity factor(LSF). Heuristic methods constituted the foundation of the earliest network reconfiguration techniques. The pioneering work on network reconfiguration for loss reduction was published by Merlin et al. [8]. They created a heuristic approach that begins with a network that is mesh-like and is obtained by turning off all the switches. Then, using the least current criterion, the switches are opened one at a time to reestablish the radial structure. Global optimization is not a guarantee of this method. Shirmohammadi et al. [9] used a compensation-based power flow strategy in accordance with the methodology described in [14] to accurately model the weakly meshed networks. To lessen network losses, Civanlar et al. [10] suggested a straightforward heuristic technique. However, the positioning and sizing of the DG can be impacted by the network setup. Both the network reconfiguration and DG installation issues must be taken into consideration in order to effectively benefit the entire distribution network [11]. Sultana et al. [12] described oppositional Krill Herd (OKH) algorithm which had been successfully incorporated to perform optimal reconfiguration of distribution system problems. In the future, could encourage researchers to solve different complicated power system optimization problems like automatic generation control, optimal power flow, economic emission load dispatch, hydro-thermal scheduling, power system stability, etc. Reference [13] proposed first-ever implementation of the proposed quasi-reflection-based slime mould algorithm (QRSMA) for ideal positioning and size of capacitor banks, DGs, and radial distribution network reconfiguration. Additionally, the authors of [14] presented a novel method that, by carefully balancing the placement of DGs, DNR, and PVQ bus voltage control, improves voltage stability, reduces power losses, and preserves the desirable voltage profile of radial distribution networks and takes into consideration the presence of changing reactive power at the P bus. A multi-objective function is defined as a result. It is advised to use the grey wolf optimization (GWO) method. The simultaneous network reconstruction and DG placement in radial distribution system utilizing a new quasi-oppositional chaotic neural network method are covered in Ref. [15]. (QOCNNA). In distribution networks like 33, 69, and 118-bus systems, it aimed to minimize active power losses and keep the voltage stable.

In this study, the author has been developed a novel algorithm, the quasi-oppositional moth flame optimization, to tackle the reconfiguration problem with optimal DG placement in RDN, there is further consideration for the voltage profile and voltage stability index. In comparison with other optimization issues, QOMFO is a relatively recent optimization algorithm that is significantly simpler and more robust. In order to reduce losses and optimize the voltage profile and stability index in the distribution system, this research has utililized quasi-oppositional moth flame optimization (QOMFO). Comparison between the simulation results acquired using MATLAB software and other methods recommended by others is provided.

The following method is used to organize the paper. Section 2 describes mathematical formulation. Section 3 provides a description of the quasi-oppositional moth flame optimization algorithm. Section 4 discusses the application of QOMFO for DNR along with ODGA problem. Results and analysis are discussed in Sect. 5. In Sect. 6, the conclusion is discussed.

2 Mathematical Formulation

2.1 *Objective Function*

The aim of the paper is to minimize a distribution network's losses through the reconfiguration and allocation (locate and size) of DGs. Proposed approach used to solve the problem here. In this context, power loss (OF_{Ploss}) is used as an objective function along with DG.

$$P_{\mathrm{LOSS}} = \sum_{M=1}^{N}\sum_{P=1}^{N} \frac{R_{M,P}}{V_M V_P}\mathrm{Cos}(\mu_M - \mu_P)(A_M A_P + B_M B_P) + \frac{R_{M,P}}{V_M V_P}\mathrm{Sin}(\mu_M - \mu_P)(B_M A_P - A_M B_P) \quad (1)$$

Here, P_{LOSS} is the real power loss; $R_{M,P}$ and $V_M V_P$ are resistance and voltage of the branch connected between Mth and Pth bus; μ_M and μ_P are the voltage angle of Mth and Pth bus;$A_M B_M$ and $A_P B_P$ are real and reactive power of Mth and Pth bus, respectively;

2.2 *Constraints*

Power Balanced Constraints: To achieve load balanced conditions, the total demand at bus and system losses must be satisfied by the total electricity generated by DG at a specific bus and that provided by the sub-station. The following list of load equilibrium restrictions

$$p_{\text{sub-station}} + \sum_{i=1}^{N} p_{DG,i} = \sum_{i=1}^{N} p_{TD,i} + P_{\text{loss}} \quad (2)$$

$$q_{\text{sub-station}} + \sum_{i=1}^{N} q_{DG,i} = \sum_{i=1}^{N} q_{TD,i} + q_{\text{loss}} \quad (3)$$

Voltage Limit: In order to ensure system stability and power quality from the following sources, the bus voltage must be between its maximum and lowest voltage limits.

$$V_{M,MIN} \leq V_M \leq V_{M,MAX} \tag{4}$$

Range of Voltage Angle: The voltage angle at the bus must fall within the range of permossible angles, both minimum and maximum.

$$\delta_{M,MIN} \leq \delta_M \leq \delta_{M,MAX} \tag{5}$$

3 Optimization Technique

3.1 Moth Flame Optimization

In the year 2015, SeyadaliMirjali developed the MFO [16] a brand-new population-based meta-heuristic technique dubbed moth flame optimization. Elegant insects called moths resemble butterflies quite a little. Over the course of their existence, they typically go through two stages: the larval stage and the adult stage. The distinctive night-timebehaviour of the flying moth organism served as the inspiration for this technique. Moths are thought to have a distinctive night-time navigation system called a transverse mechanism.

Moths fly using the aforementioned process by maintaining a stable angle with the moon at night. This strategy makes sure that the mouths go in a straight line even when the moon is quite far from them. Nevertheless, when the moths are placed close to a man-made light source, they have a tendency to move in a dangerous spiral pattern. This specific conduct is helpful for resolving issues in everyday life.

Moths are defined as variables in the MFO method. The following definition of a logarithmic spiral explains how the search operation is carried out inside the present search area.

$$Q_p(g_q, h_r) = C_q d^{cm} \cdot \cos(2\pi n) + h_r \tag{6}$$

where C_q the path between the qth moth and the rth and, the algorithm spiral stands by c and arbitrary number is stands by m where $m \subseteq [-1, 1]$.

Now, C can be determined as

$$C_q = |g_q - h_r| \tag{7}$$

where h_r is the rth moth , g_q is the qth flame and C_q is the distance between the qth moth and the rth flame.

On comparison to other meta-heuristic algorithms, the MFO algorithm is seen to have greater convergence rate in its results which provides better quality solutions in very less amount of time. But after this above-mentioned process is fulfilled, there is one more concern to be thought about which is that the position refreshing of the moths with respect to various locations in the particular search space may reduce the chances of achieving the best solutions. This concern can be remedied using the below stated mathematical formulation where the number of flames is reduced with every successful iteration.

$$N_g = \text{round}\left(\gamma - \delta * \frac{\gamma - \delta}{\sigma}\right) \tag{8}$$

where N_g is the flame number, $\gamma$is then maximum number of flames at the present time, δ is the current iteration number, and σ is the maximum iteration number.

Several real-world optimization problems have been solved using the moth flame optimization algorithm (MFO), which has the benefits of being quick to coverage, having few setting parameters, and being easy to understand and apply. However, the MFO struggles to strike a good balance between exploration and exploitation, and there is little information sharing amongst people, especially when it comes to working out some challenging mathematical issues.

3.2 *Quasi-Oppositional-Based Learning*

Tizhoosh et al. [17] were initially released OBL which is a cutting-edge idea in intelligence-based problem solving or soft computing that can be utilized to enhance several optimization methodologies. It looks to be one of the most effective theories in computational intelligence, which can handle nonlinear optimization problems and enhance the search performance of conventional population-based optimization procedures. OBL's primary objective is to compare an estimate or assumption with its opposite or reciprocal in order to increase the possibility that a solution will be found more rapidly. The OBL approach begins with initializing the initial estimate, which is done either randomly or based on prior knowledge about the solution. Any direction, or at the very least the opposite, may be the best course of action. The opposing set of estimates for superior solution is taken into consideration for convergence after iteratively replacing the initial estimates in the direction of optimality. Let, the real number be denoted by $B \in [k, l]$ and B^0 be its corresponding opposite number is defined real number by

$$B_0 = k + l - B \tag{9}$$

Say $R = (Z_1, Z_2, ..., Z_n)$ is a number in n-dimensional space with upper and lower bound, where, $B_u \in [k_u, l_u]$, $u \in 1, 2, ..., n$. The $R^0 = ({Z_1}^0, {Z_2}^0, ...{Z_n}^0)$ opposite point is defined by its components:

$$B_u{}^0 = k_u + l_u - B_0 \tag{10}$$

Let B be a real number between defined between upper and lower limit $[k, l]$. The quasi-opposite number is defined as

$$B^{Q_0} = \text{rand}(C, \tilde{B}) \tag{11}$$

where C is given by $C = \frac{k+l}{2}$. Suppose, B be a real integer $[e, f]$. Defining the quasi-oppositional point $B_r{}^{Q_0}$ as

$$B_r{}^{Q_0} = \text{rand}(C_u, \tilde{B}_u) \tag{12}$$

$$\text{where } C_r = \frac{k_u + l_u}{2}$$

4 QOMFO Applied to Reconfiguration Problem Along with DG

The reconfiguration of distribution network with the simultaneous allocation of DGs is implemented using the QOMFO as follows:

Step 1: Initialize the size of population (P_n), read data for the system with constraints, the maximal iteration.
Step 2: Determine the DG size following the upper and lower limits.
Step 3: Calculate the objective function (1) by performing power flow calculations and determine the minimum power losses.
Step 4: Depending upon the present candidate solution in the search space, a logarithmic spiral is defining a little better solution are retained as noble solutions.
Step 5: Use function (9-12) to produce opposite population.
Step 6: If the obtained objective function is weak, replace the best solution with the previous best solution; otherwise, return to Step 3.
Step 7: increase the numbers of iterations and return to Step 3. print the results and stop the algorithm.

5 Results

The above-proposed approach QOMFO is evaluated using two different test systems consisting of 33 and 69 buses, with an operating voltage of 12.66 kV, to assess its usability and superiority in solving the MFO problem for reconfiguration with DG installation identifies the best location and DG size to reduce active power loss. Simulations are performed using MATLAB software on a PC with an Intel i3-7020U @ 2.30 GHz and 8 GB of RAM. The algorithm's population size is set at 50, and its iteration count is set to 100.

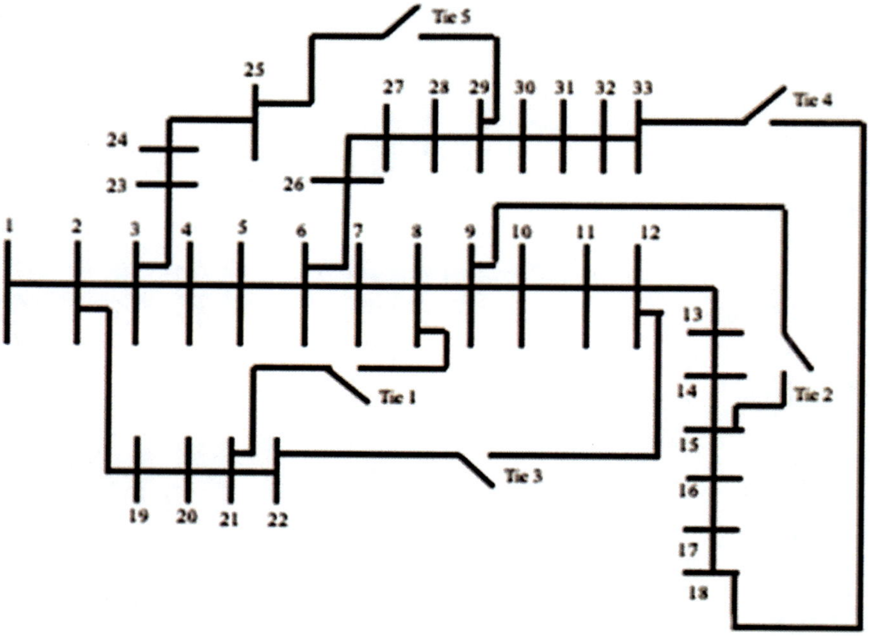

Fig. 1 Topology of 33 bus RDN

5.1 *33-Bus Test Radial Distribution Network*

As seen in Fig. 1, the 33-bus demonstration RDN, which has 33 buses and 32 branches, is used to test the QOMFO methodology. Line values and the load values for the systems are taken from references [18]. The results of 33 bus test system are shown in Table 1. For a 33-bus test system by application of QOMFO and MFO, the real power losses are obtained 119.01 kW and 151.21 kW, respectively, after placement of DG. But after reconfiguration of radial distribution network, power losses are decreased to 49.03 kW and 56.13 kW, respectively. After reconfiguration, 33 bus system is shown in Fig. 2. The obtained results are compared with [19] which shows power losses reduction are more significant by QOMFO technique than other techniques.

5.2 *69-Bus Test Radial Distribution Network*

As seen in Fig. 3, the 69-bus test RDN, which has 69 buses and 68 branches, is used to test the QOMFO methodology. Line values and the load values for the systems are taken from references [18]. In Table 2, 33 bus test system is explained. For a 69-bus test system by application of QOMFO and MFO, the real power losses are

Table 1 Summary of the results for 33 bus RDN before and after reconfiguration with DG

Parameters	Without DG placement	With DG placement		
		SFS [19]	MFO	QOBL-MFO
Without reconfiguration				
Power loss (KW)	200.23	NA	151.21	**119.01**
Optimal position of DG	NA	NA	29 26 24	**29** **26** **24**
Optimal size of DG	NA	NA	992.35 900.31 733.02	**876.12** **731.02** **655.17**
Opening branches	NA	NA	NA	**NA**
Closing branches	NA	NA	NA	**NA**
With reconfiguration				
Power loss (KW)		53.01	56.13	**49.03**
Optimal position of DG		22 25 33	12 17 27	**12** **17** **27**
Optimal size of DG		775.3 1285.8 735.6	856.56 602.43 600.01	**997.14** **645.8** **601.12**
Opening branches		7–8 9–10 14–15 27–65 30–31	7–8 9–10 14–15 28–29 32–33	**7–8** **9–10** **14–15** **28–29** **32–33**
Closing branches		NA	21–8 9–15 22–12 18–33	**21–8** **9–15** **22–12** **18–33**

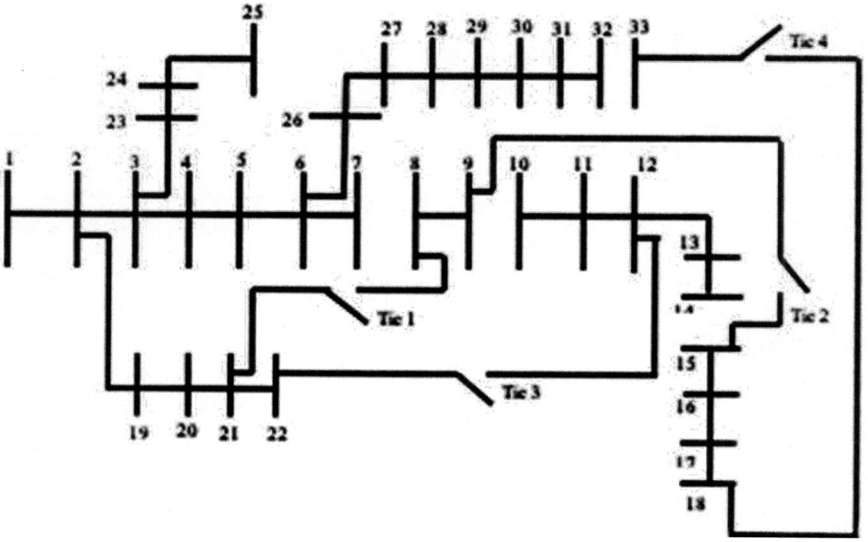

Fig. 2 Construction layout of 33-bus radial distribution network for continous power type load for load multiplying factor of 1.0

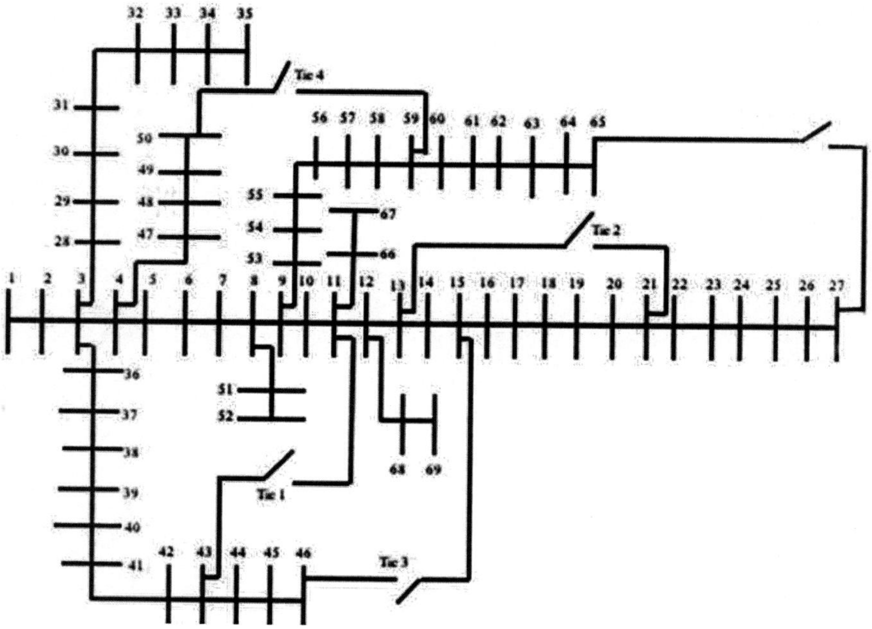

Fig. 3 Topology of 69-bus RDN

Table 2 Summary of the results for 69 bus RDN before and after reconfiguration with DG

Parameters	Without DG placement	With DG placement		
		SFS [19]	MFO	QOBL-MFO
Without reconfiguration				
Power loss (KW)	223.20	NA	125.26	**97.01**
Optimal position of DG	NA	NA	18 62 64	**18** **62** **64**
Optimal size of DG	NA	NA	910.21 883.12 874.02	**879** **833.20** **690.47**
Opening branches	NA	NA	NA	**NA**
Closing branches	NA	NA	NA	**NA**
With reconfiguration				
Power loss (KW)		35.16	37.21	**33.03**
Optimal position of DG		11 61 64	21 47 58	**21** **47** **58**
Optimal size of DG		537.6 1434.0 490.3	868.43 901.12 933.14	**764.01** **866.32** **910.03**
Opening branches		14–15 56–57 61–62 11–43 13–21	55–56 62–63 11–12 – –	**55–56** **62–63** **11–12** – –
Closing branches		NA	50–59 27–65 15–46	**50–59** **27–65** **15–46**

obtained 97.01 kW and 125.26 kW, respectively, after placement of DG. But after reconfiguration of radial distribution network power losses are decreased to 33.03 kW and 37.21 kW, respectively. After reconfiguration, 33 bus system is shown in Fig. 4. The obtained results are compared with [19] which shows power losses reduction are more significant by QOMFO technique than other techniques.

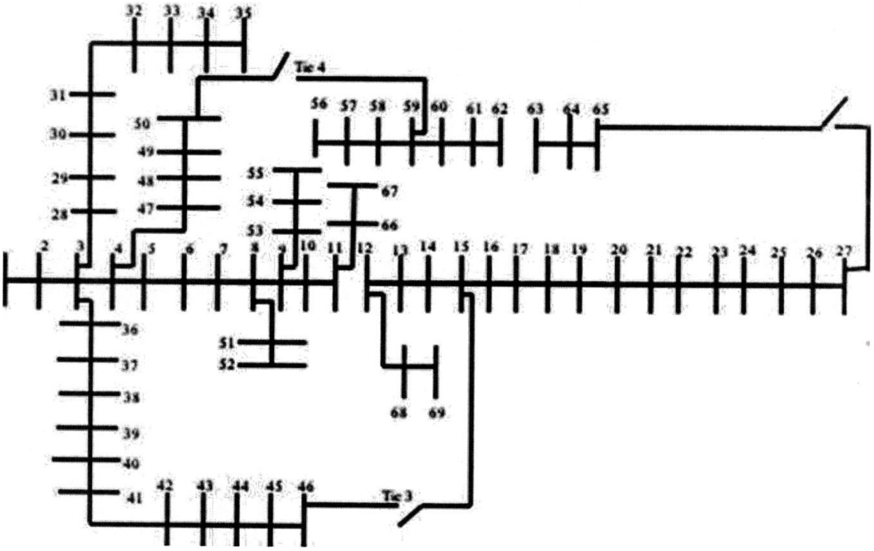

Fig. 4 Construction layout of 69-bus radial distribution network for continous power types load for load multiplication factor 1.0 and 1.5, complex types load for load multiplication factor of 1.5

6 Conclusion

The QOMFO method has been successfully implemented in this study for the simultaneous location and magnitude of the DG problem and distribution network reconfigurations.The purpose is to increase the system's voltage parameter and decrease actual power loss. Additionally, various approaches for voltage stability improvement and loss reduction, including only DG installations, network reconfiguration after exact placement of DG are simulated in order to demonstrate the superiority of the proposed method. The major objective was to enhance the voltage profiles whilst reducing active power loss in the distribution network. The QOMFO was a reliable and effective method that quickly converged in all circumstances taken into account. On 33 and 69 bus test systems, the suggested procedure is put to the test. According to the test results, QOMFO can manage extremely intricate and sizable distribution networks. Moreover, QOMFO outperformed other approaches in terms of improving voltage profiles and reducing power loss for all applications. As a result, the suggested QOMFO method may be a very promising approach for resolving the DNR problem in conjunction with the ideal location of DGs.

References

1. Tabares A, Puerta GF, Franco JF, Romero RA (2021) Planning of reserve branches to increase reconfiguration capability in distribution systems: a scenario-based convex programming approach. IEEE Access 9:104707–104721
2. Gao Yuanqi, Wang Wei, Shi Jie, Nanpeng Yu (2020) Batch-constrained reinforcement learning for dynamic distribution network reconfiguration. IEEE Trans Smart Grid 11(6):5357–5369
3. Akbari MA, Aghaei J, Barani M, Niknam T, Ghavidel S, Farahmand H, Korpas M, Li L (2018) Convex models for optimal utility-based distributed generation allocation in radial distribution systems. IEEE Syst J 12(4):3497–3508
4. Shaheen Abdullah, Elsayed Abdallah, Ginidi Ahmed, El-Sehiemy Ragab, Elattar Ehab (2022) Reconfiguration of electrical distribution network-based dg and capacitors allocations using artificial ecosystem optimizer: practical case study. Alexandria Eng J 61(8):6105–6118
5. Tan S, Xu J-X, Panda SK (2013) Optimization of distribution network incorporating distributed generators: an integrated approach. IEEE Trans Power Syst 28(3):2421–2432
6. Srinivasa Rao R, Ravindra K, Satish K, Narasimham SVL (2012) Power loss minimization in distribution system using network reconfiguration in the presence of distributed generation. IEEE Trans Power Syst 28(1):317–325
7. Mohamed Imran A, Kowsalya M, Kothari DP (2014) A novel integration technique for optimal network reconfiguration and distributed generation placement in power distribution networks. Int J Electri Power Energy Syst 63:461–472
8. Merlin A (1975) Search for a minimum-loss operating spanning tree configuration for an urban power distribution system. In: Proceedings of 5th PSCC, vol 1. pp 1–18
9. Shirmohammadi D, Wayne Hong H (1989) Reconfiguration of electric distribution networks for resistive line losses reduction. IEEE Trans Power Delivery 4(2):1492–1498
10. Civanlar S, Grainger JJ, Yin H, Lee SSH (1988) Distribution feeder reconfiguration for loss reduction. IEEE Trans Power Delivery 3(3):1217–1223
11. Georgilakis PS, Hatziargyriou ND (2013) Optimal distributed generation placement in power distribution networks: models, methods, and future research. IEEE Trans Power Syst 28(3):3420–3428
12. Sultana S, Roy PK (2016) Oppositional krill herd algorithm for optimal location of capacitor with reconfiguration in radial distribution system. Int J Electri Power Energy Syst 4:78–90
13. Biswal SR, Shankar G, Elavarasan RM, Mihet-Popa L (2021) Optimal allocation/sizing of dgs/capacitors in reconfigured radial distribution system using quasi-reflected slime mould algorithm. IEEE Access 9:125658–125677
14. Barnwal AK, Yadav LK, Verma MK (2022) A multi-objective approach for voltage stability enhancement and loss reduction under pqv and p buses through reconfiguration and distributed generation allocation. IEEE Access 10:16609–16623
15. Tran TV, Truong B-H, Nguyen TP, Nguyen TA, Duong TL, Vo DN (2021) Reconfiguration of distribution networks with distributed generations using an improved neural network algorithm. IEEE Access 9:165618–165647
16. Mirjalili Seyedali (2015) Moth-flame optimization algorithm: a novel nature-inspired heuristic paradigm. Knowledge-based Syst 89:228–249
17. Tizhoosh HR (2005) Opposition-based learning: a new scheme for machine intelligence. In: International conference on computational intelligence for modelling, control and automation and international conference on intelligent agents, web technologies and internet commerce (CIMCA-IAWTIC'06), vol 1. IEEE, pp 695–701
18. Kashem MA, Ganapathy V, Jasmon GB, Buhari MI (2000) A novel method for loss minimization in distribution networks. In: DRPT2000. International conference on electric utility deregulation and restructuring and power technologies. Proceedings (Cat. No. 00EX382), IEEE, pp 251–256
19. Tran TT, Truong KH, Vo DN (2020) Stochastic fractal search algorithm for reconfiguration of distribution networks with distributed generations. Ain Shams Eng J 11(2):389–407

Chapter 35
Recent Trends in Risk Assessment of Electromagnetic Radiations

Juhi Pruthi and Ashutosh Dixit

1 Introduction

Electromagnetic fields (EMFs) as the name suggests comprises of two terms namely electric field (E) and magnetic field (B). The electric fields are generated as a result of stationary charge whereas moving charge produces magnetic field resulting in generation of electromagnetic radiations (EMR). EMR can be broadly classified into two segments on the basis of frequency range and also the strength of radiation energy. Radiations in the non-ionizing spectrum have low energy to break chemical bonds and are therefore, considered harmless for the humans. Infrared rays, radio-frequency (RF), visible light often termed as low-frequency radiations constitute the non-ionizing spectrum. On the contrary, radiations in the ionizing spectrum have high energy to break any chemical bond and cause potential damage to human DNA and cellular structure. High-frequency rays such as X-rays, gamma rays, Ultraviolet (UV) rays constitute the ionizing spectrum of EMR [1]. The advancement in technology and exceeding demand of electrical appliances has led to a boon in the electronics sector. Consequently, this advancement brings along side effects of pollution on the environment and atrocities on living organisms. Humans are exposed to ionizing radiations such as UV rays from sun which are well-known to have ill effects on human as well as environment. Humans are also exposed to non-ionizing radiations in their facets of everyday life. Hence, it is a worldwide debatable matter whether longer duration of exposure to extremely low-frequency (ELF) radiations harmful for human health.

This paper addresses the following questions: (1) Whether exposure to non-ionizing radiations dangerous for humans. (2) Whether exposure to human-made sources of non-ionizing radiations harmful for the living species. (3) What impact the frequency of the radiation has on biological functioning. (4) What preventive mea-

J. Pruthi (✉) · A. Dixit
JC BOSE University of Science and Technology, YMCA, Faridabad, Haryana, India
e-mail: juhi.pruthi@gmail.com

A. Yadav et al. (eds.), *Proceedings of International Conference on Paradigms of Communication, Computing and Data Analytics*, Algorithms for Intelligent Systems,
https://doi.org/10.1007/978-981-99-4626-6_35

sures and safety norms are being followed worldwide. Therefore, this study focuses on assessing the effects of non-ionizing radiations on living species viz. fauna, flora and humans for prioritizing and stratifying health risk assessment. The rationale of the study is to raise public awareness about EMR as a possible health hazard and to facilitate formulation of mitigation strategies and standards worldwide.

1.1 EMR Sources

EMRs are ubiquitous, and humans are unknowingly exposed to these radiations at every hour of time and space. With the tremendous growth in the telecommunication and electronics sector, there has been a rise in the level and the source of EMF generation depicted in Fig. 1. There can be natural as well human-made sources of ionizing and non-ionizing EMR that are enlisted below [2, 3].

Non-ionizing radiations
Natural Sources

- Earth's magnetic field generates huge amounts of EMF

Human-Made Sources

- Radio Communication devices include remote controls, cordless phones, walkie-talkies, Wi-Fi modems, cell phones, television, and AM/FM radio.
- Electrical Appliances such as refrigerators, air-conditioners, vacuum cleaners, hair dryers
- Power transmission lines and electrical wiring
- Laptops and tablets
- Base stations and transmitters
- Magnetic Resonance Imaging (MRI) devices
- Lamps and bulbs such as fluorescent, incandescence

Ionizing radiations
Natural Sources

- Cosmic rays from the sun
- Radon gas
- Radioactive elements such as uranium, thorium, and radium in earth's crust
- Radioactive materials such as K-40,C-14,Pb-210 present within human body

Human-Made Sources

- Diagnostic equipment used in CT scans and X-rays
- Radiation devices used for cancer treatment
- Nuclear power plants and reactors
- Microwaves from oven
- Welding arcs
- Construction materials

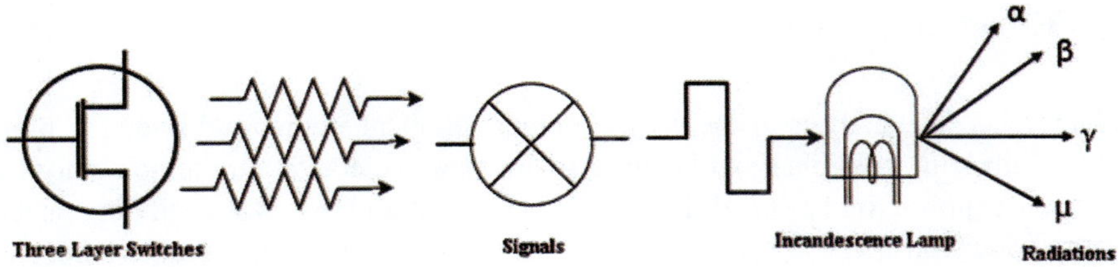

Fig. 1 Electromagnetic radiation process flow

1.2 Glossary

Electric field (E): field generated around electrically charged stationary particle. Its unit is Volts per meter (V/m). Mathematically it is represented as

$$E = F/q \tag{1}$$

where E denotes electric field, F force, and q charge.
Magnetic field (B): field produced around moving electric charges and magnetic materials. Its unit is Tesla (T) or micro-Tesla (μT). It is expressed by the formula in Eq. 2.

$$B = I/2\pi r \tag{2}$$

Electromagnetic field (EMF): field generated from moving electric charges.
Electromagnetic force (F): force produced due to interaction between the magnetic field and electrically charged particles. The relation between F, E, and B is expressed by *Lorentz force law* represented in Eq. 3

$$F = qE + qv.B \tag{3}$$

Electromagnetic radiations (EM): electromagnetic field waves propagating through space at the speed of light. The electrical charge travels from the electrical switch to the emf source(say incandescence lamp), generating the radiations as depicted by the process flow of Fig. 1.

Electromagnetic spectrum (EMS): depicts the schematic arrangement of electromagnetic radiations based on frequency and wavelength.
Specific Absorption Rate (SAR): denotes the radio-frequency energy absorption rate by the human body on exposure to an EMF source. It is expressed as Watts per kilogram (W/kg) [4].

In this paper, Sect. 1 introduces the electromagnetic field, emf sources, and glossary. Section 2 describes the possible health hazards imposed by emf exposure on humans. Sections 3 and 4 discuss the ill effects of emf exposure on flora and fauna, respectively. Current trends and regulation policies adopted at national and international levels are described in Sect. 5, followed by conclusion and future scope in Sect. 6.

2 Effect on Humans

Comprehensive analysis of the various epidemiological studies and scientific literature shows the possible risk of cancer, mainly leukemia and brain tumor, amongst humans exposed to non-ionizing EMFs [5]. The human brain is spatially organized into various lobes viz. temporal lobe, frontal lobe, parietal lobe, occipital lobe, and cerebellum. The temporal lobe plays a significant role in speech perception, vision, and memory management. The frontal lobe controls planning, thinking, memory storage, emotions, and personality development. The occipital lobe is responsible for color determination, object recognition, memory formation, and visuospatial processing. Sensory perception and integration are managed through the parietal lobe. Cerebellum is responsible for motor behavior, vision, and movements, viz., walking, posture, and balance. Disturbed neuron activity leads to behavioral syndromes affecting the temporal and frontal lobes portion of the brain anatomy. Exposure to high radiation is one of the risk factors associated with leukemia. Leukemia is a form of cancer affecting the white blood cells in the bone marrow. The person suffering from leukemia shows rapid uncontrolled growth of the white blood cells as illustrated in Fig. 2 (Created in Biorender.com 2023) [6]. Pooled analysis of nine studies also reported that exposure to non-ionizing EMFs from power lines with magnitude 0.4 microtesla or higher and proximity to cell phone base stations shows an enhanced risk of leukemia in children [7, 8]. Consequently, the effect of EMF radiations diminishes with the increased distance from the source. This is in principle to Planck's law which states that frequency is inversely proportional to the distance [9].

Few recent studies found inconsistent evidence on the association of children's leukemia on exposure to the magnetic field generated from electrical appliances [10]. Several U.S. health agencie's studies reported an increased risk of brain cancer and acute myeloid leukemia amongst military personnel in the electronics and electrical sector [11]. The increased risk in this personnel is due to exposure to radio-frequency and microwave-emitting equipment. Few epidemiological studies have suggested an association between breast cancer among women on exposure to highly low-frequency EMR at home. Based on statistically significant findings from the scientific literature, *International Agency for Research on Cancer (IARC)* has assessed ELF magnetic field as carcinogenic substances falling in Group 2B [12]. Consequently, recent research by Anthony et al. categorizes RFR as Group 1 with more substantial evidence published by several epidemiological studies conducted on animals addressed in the subsequent section [13].

Few experimental studies report arrhythmia and myocardial infection mortality amongst utility workers on exposure to ELF [14]. However, the results from the later studies on this subject are contradictory.

Few studies suggest little and inconsistent evidence on the association of hypersensitivity with electromagnetic field exposure with continued research. There have been studies with principal findings of heating of tissues in humans on exposure to RF EMFs [15].

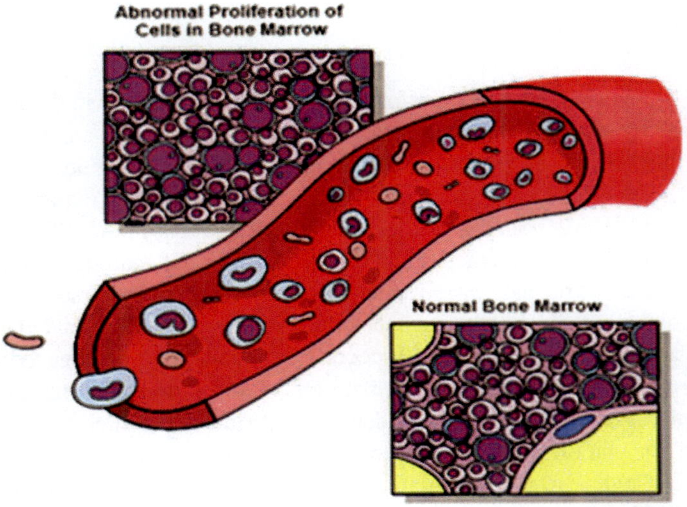

Fig. 2 Leukemia-proliferation of white blood cells (*Source* Created in BioRender.com 2023)

3 Effect on Flora

High-frequency EMFs cause significant changes in plant metabolism, including delayed and reduced growth, dry weight, and reduced germination. Plant species such as onion, pulses, tomato, mustard, soybean, wheat, and maize have been analyzed for their stimulus to EMF radiations emitted from GSM cellphones. Alain et al. observed significant changes in mRNA of tomato plants placed in a specially designed chamber on exposure to 900 MHz, 5V EMFs for 10 min [16]. The plants showed a decreased concentration of ATP and adenylate energy charge. Malka et al. present an exhaustive review of the detrimental influence of RF-EMF on 29 varied plant species with the finding that plants such as onions, tomato, pea, fenugreek, and maize being highly sensitive to the non-thermal radiations [17]. Cucurachi et al. presented a comprehensive review of the ecological effects of RF-EMF on plant species [18]. Shikha et al. assessed the post effects of 2100 MHz RF-EMF radiations on onion species Allium cepa [19]. The onion roots were exposed for a duration of over 4th and significant aberrations in the DNA were recorded. Declination in HDNA concentration while a significant rise in TDNA was observed. The analysis also showed a substantial increase in the mitotic index and aberration percentage. Mirta et al. also performed an experimental study to analyze the effect of RF-EMF in the frequency range 400–900 MHz on Allium Cepa L [20]. Their study revealed a significant rise in mitotic index (MI) and mitotic aberrations such as disturbed anaphases, lagging chromosomes, and impairment of mitotic spindle. Several experimental studies have been performed to assess the detrimental effects of EMF on Allium Cepa across the broad frequency range of electromagnetic spectrum [21, 22].

4 Effect on Fauna

Scientists worldwide have been performing exclusive research to find possible associations between the impact of EMF radiations on animal organs and tissues. Under *National Toxicology Program (NTP)* foregoing research is being done to assess the possible health hazards on mice and rats on exposure to radio-frequency radiation (RFR) emitted from 2G/3G cell phones operating at frequency range 700–2700 MHz. Findings from the experimental studies show evidence of malignant schwannoma tumors in the heart of male rats. Toxicology studies also indicate some association between malignant glioma tumors in the brain of male rats and adrenal gland tumors in male rats. Studies have also raised speculations about DNA damage to blood cells in female mice, the hippocampus of male rats, and the frontal cortex of the brain in male mice. Based on the above evidence, NTP aims to conduct further studies to assess the association between RFR and biological tissues and analyze the correlation between radio-frequency radiations and various modulations used by telecom operators [23]. Ali et al. performed an experimental study to assess the effect of LTE 2600 MHz MHz EMF exposure on rats [24]. The authors found a significant increase in rat body temperature due to the physiological impact of these radiations that caused heat dissipation. Other biological mechanisms such as increased drinking water, increased urine secretion, evaporative heat loss, and tail vasodilation process were also hampered due to high heating exposure from the radiations. Mary et al. erformed an experimental study to evaluate the effect of 2G/3G cell phone radiations on the liver of chick embryos [25]. They divided the sources into two experimental and two control groups. One experimental group was exposed to 2G radiation, named group A, while Group B experimental group was exposed to 3G radiation. Their histological study revealed significant structural changes such as increased nuclear diameter, dilated sinusoidal spaces with hemorrhage, enhanced cytoplasm vacuolations, and considerable DNA damage. Their study also concluded that the impact of 3G radiations was more severe as compared to 2G.

5 Current Trends

This section presents insight into the health organizations working in collaboration at the international and national levels to analyze the possible health hazards caused by exposure to vivid electromagnetic radiation.

International Level

World Health Organization (WHO) launched *The International EMF Project* in 1996, intending to analyze the detrimental health and environmental effects caused by exposure to EMFs in the frequency range 0–300 GHz. Under this project, WHO aims to establish EMF mitigation strategies, raise public awareness regarding perceived EMF risk, protection, and management and also formulate international regulations and permissible standards on EMF exposure [26].

Table 1 Specific absorption rate (SAR) safety specifications

Standard	SAR (W/kg)
EU (European Union)	2.0
US (United States)	1.6
India	1.6

International Commission On Non-Ionizing Radiation Protection (ICNIRP) successor of the International Non-Ionizing Radiation Committee (INIRC) is a non-profit scientific body to safeguard the environment and humans from the impact of non-ionizing radiation (NIR). This body substantiates the NIR exposure safety standards and frequently publishes updated and revised guidelines [27]. Its recent publication describes the impact of 5G on health.

International Agency for Research on Cancer (IARC) is a cancer specialized agency of WHO. It aims to promote collaborative research worldwide to identify, develop and adopt cancer preventive measures to mitigate the disease and lessen its effect [28]. IARC, in its recently published list of monographs, has validated ionizing radiations as known and probable carcinogens belonging to **Group1**. Agents belonging to Group1 are known to have carcinogenic impact on humans [29].

National Level

This study gives an insight into the precautionary measures adopted by Indian organizations to safeguard against EMF exposure.

Department of Telecommunication (DoT) has framed stricter norms and legislations for radiations generated from mobile antennas, base stations, and handsets. Following the safety guidelines issued by ICNIRP, DoT has set the permissible EMF emissions from mobile base stations to 1/10th **Of the limits set by ICNIRP**. All telecom operators must comply with the prescribed limit while installing any base transceiver station (BTS) and submit self-certification to Telecom Enforcement Resource & Monitoring (TERM). The current standard for EMR emission from BTS is 0.434 V/m for a frequency range of 400–2000 MHz, while for a frequency range of 2–300 GHz (gigahertz), the limit is 19.29 V/m. DoT has also introduced a web portal Tarang Sanchar that contains information about the base stations installed nationwide and their emf compliance values [30]. It is a publicly available portal that can disseminate information about BTS around one's vicinity.

SAR safety specification as per different standards is tabulated in Table 1. DoT has also prescribed safety norms and SAR limits to protect against emissions from mobile handsets. The current prescribed value of SAR is 1.6 W/kg averaged over a mass of 1 g human tissue [30]. Only those handsets that comply with the set limits are imported or manufactured in India.

Table 2 Summary of related risks on exposure to electromagnetic radiations (EMR)

References	Frequency	Receptor	Effects
[5]	50–60 Hz	Humans	Leukemia in children
[11]	40–800 Hz	Humans	Risk of acute myocardial infection (AMI), Chronic coronary heart disease (CCHD)
[14]	50 Hz	Humans	Minor risk of arrhythmia cardiac disease
[15]	915 Hz, 2450 Hz	Humans	Heat absorption by human tissues, Heating effect at skin
[16]	900 Hz	Tomato plant	Decreased ATP concentration, Changes in mRNA
[19]	2100 Hz	Onion plant (Allium Cepa)	Aberrations in DNA, Increased mitotic index
[20]	400- 900 MHz	Onion plant (Allium Cepa)	Mitotic aberrations, Increased mitotic index
[21]	900 Hz, 1800 Hz	Onion plant (Allium Cepa)	Decreased root length, Increased root thickness, Increased mitotic index, Increased aberration index, Aberrations in DNA concentration
[24]	2600 MHz	Rats	Increased body temperature, Increased drinking water capacity, Increased urine secretion
[25]	300 MHz– 3 GHz	Chick embryo	Increased nuclear diameter, DNA aberrations, Dilated sinusoidal spaces with hemorrhage

DoT has also introduced a web portal **Tarang Sanchar** that contains information about the base stations installed across the country and their emf compliance values. It is a publicly available portal that can disseminate information about BTS around one's vicinity.

6 Conclusion and Future Scope

Electromagnetic fields are generated due to the interaction between the electric and magnetic fields. The corresponding radiations from these fields are known to have certain adverse environmental and health effects. Due to technological advancement and the rise in demand, numerous electrical appliances and electronic gadgets are manufactured worldwide. The tremendous growth in the telecom sector has led to flooding the market with portable and cheap handsets. These handsets come in handy and provide worldwide connectivity and numerous benefits but at the expense of human health. Table 2 summarizes the findings of epidemiological and experimental studies assessing the health impacts of electromagnetic radiation exposure on humans and living organisms. Radio-frequency radiations emitted from the handsets and base stations harm humans and the environment. Likewise, other radiation sources, such as electrical gadgets and appliances, also affect the ecology of their surrounding environment. On the contrary, certain studies invalidate the adverse effects caused by exposure to extremely low-frequency radiation. Substantial epidemiological studies on humans, plants, and animal species validate these radiations as cancer sources and cause genetic mutations. Thermal radiation generated from EMR raises concern over its heating impact on human skin and tissues. Health organizations across the globe are actively performing research studies to analyze the associated health issues and formulate protection measures against EMR exposure.

The possible domain of research is assessing the impact of prolonged exposure to highly low-frequency radiations (3 to 30 Hz) on human thinking and neurological effects. Assessment of variation in neuron organization in humans, specifically children on prolonged exposure to screen radiation, can be an area of future exploration. Additionally, the impact of such radiations on creative thinking and memorization should also be looked over. Other research areas include the adoption of innovative technology, the exceptional proximity of wearable devices to the body, and its correlation to ecology. In relevance to smart cities and sustainable environments, analysis and survey of the absorption properties of the construction materials can be another aspect of further research. Additionally, materials used in the construction of buildings and roads should be assessed for electromagnetic penetration and shielding. Most scientific literature studies focused on analyzing the effects of radio-frequency radiation emitted from cellular and telecommunication devices. Likewise, other radiation sources, such as radon gas, pose serious health hazards. Emissions from these sources must also be studied for better health assessment.

References

1. National Cancer Institute (2023) Last Accessed 10 Apr 2023. https://www.cancer.gov/
2. Deruelle F (2020) The different sources of electromagnetic fields: dangers are not limited to physical health. Electromagnetic Biolo Med 39(2):166–175
3. Jamshed MA, Heliot F, Brown TW (2019) A survey on electromagnetic risk assessment and evaluation mechanism for future wireless communication systems. IEEE J Electromagnet RF and Microwaves in Med Biol 4(1):24–36
4. Commission FC et al. (2014) Specific absorption rate (SAR) for cell phones: what it means for you
5. Ahlbom A, Feychting M (2003) Electromagnetic radiation: environmental pollution and health. British Med Bullet 68(1):157–165
6. Adapted from "Cytology of Leukemia" by BioRender.com. Accessed 10 Apr 2023. https://app.biorender.com/biorender-templates/t-6042809c132bd87bfdefee90-cytology-of-leukemia

7. Maskarinec G, Cooper J, Swygert L (1994) Investigation of increased incidence in childhood leukemia near radio towers in hawaii: preliminary observations. J Environ Pathol Toxicol Oncol: Official Organ of the Int Soc for Environ Toxicol Cancer 13(1):33–37
8. Ahlbom A, Day N, Feychting M, Roman E, Skinner J, Dockerty J, Linet M, McBride M, Michaelis J, Olsen J (2000) A pooled analysis of magnetic fields and childhood leukaemia. British J Cancer 83(5):692–698
9. Mann S, Cooper T, Allen S, Blackwell R, Lowe A (2000) Exposure to radio waves near mobile phone base stations. Radiological Protect Bullet 4(7):13–16
10. Calabrò E, Magazù S (2010) Measure of electromagnetic field of mobile phone microwaves by means of narda srm 3000. In: CPEM 2010, IEEE, pp 747–748
11. Sahl J, Mezei G, Kavet R, McMillan A, Silvers A, Sastre A, Kheifets L (2002) Occupational magnetic field exposure and cardiovascular mortality in a cohort of electric utility workers. American J Epidemiol 156(10):913–918
12. Expert panel I (2002) Iarc working group on the evaluation of carcinogenic risks to humans. non-ionizing radiation, part 1: static and extremely low-frequency (elf) electric and magnetic fields
13. Miller AB, Morgan LL, Udasin I, Davis DL (2018) Cancer epidemiology update, following the 2011 iarc evaluation of radiofrequency electromagnetic fields (monograph 102). Environ Res 167:673–683
14. Johansen C, Feychting M, Møller M, Arnsbo P, Ahlbom A, Olsen JH (2002) Risk of severe cardiac arrhythmia in male utility workers: a nationwide danish cohort study. Amer J Epidemiol 156(9):857–861
15. Wessapan T, Srisawatdhisukul S, Rattanadecho P (2011) Numerical analysis of specific absorption rate and heat transfer in the human body exposed to leakage electromagnetic field at 915 and 2450 mhz. J Heat Transfer 133(5)
16. Vian A, Davies E, Gendraud M, Bonnet P (2016) Plant responses to high frequency electromagnetic fields. BioMed Res Int 2016
17. Halgamuge MN (2017) Weak radiofrequency radiation exposure from mobile phone radiation on plants. Electromagnet Biol Med 36(2):213–235
18. Cucurachi S, Tamis WL, Vijver MG, Peijnenburg WJ, Bolte JF, de Snoo GR (2013) A review of the ecological effects of radiofrequency electromagnetic fields (rf-emf). Environ Int 51:116–140
19. Chandel S, Kaur S, Singh HP, Batish DR, Kohli RK (2017) Exposure to 2100 mhz electromagnetic field radiations induces reactive oxygen species generation in allium cepa roots. J Microscopy and Ultrastruct 5(4):225–229
20. Tkalec M, Malarić K, Pavlica M, Pevalek-Kozlina B, Vidaković-Cifrek Ž (2009) Effects of radiofrequency electromagnetic fields on seed germination and root meristematic cells of allium cepa l. Mutation Res/Genetic Toxicol Environ Mutagenesis 672(2):76–81
21. Kumar A, Kaur S, Chandel S, Singh HP, Batish DR, Kohli RK (2020) Comparative cyto-and genotoxicity of 900 and 1800 mhz electromagnetic field radiations in root meristems of allium cepa. Ecotoxicol Environ Safety 188:109786
22. Chandel S, Kaur S, Issa M, Singh HP, Batish DR, Kohli RK (2019) Exposure to mobile phone radiations at 2350 mhz incites cyto-and genotoxic effects in root meristems of allium cepa. J Environ Health Sci Eng 17:97–104
23. Ozguner M, Koyu A, Cesur G, Ural M, Ozguner F, Gokcimen A, Delibas N (2005) Biological and morphological effects on the reproductive organ of rats after exposure to electromagnetic field. Saudi Med J 26(3):405–410
24. Al-Chalabi AS, Asim R, Rahim H, Abdul Malek MF (2021) Evaluation of the thermal effect of lte 2600 mhz (4g) electromagnetic field (emf) exposure: thermographic study on rats. Iraqi J Veterinary Sci 35(2):279–285
25. Lalitha C, Manjula M, Srikant K, Goyal S, Tanveer S (2015) Hand schuller christian disease: a rare case report with oral manifestation. J Clinical and Diagnostic Res: JCDR 9(1):28
26. Electromagnetic fields (2023) Accessed 10 Apr 2023. https://www.who.int/health-topics/electromagnetic-fields#tab=tab_1

27. Ziegelberger G, Croft R, Feychting M, Green AC, Hirata A, d'Inzeo G, Jokela K, Loughran S, Marino C, Miller S et al. (2020) Guidelines for limiting exposure to electromagnetic fields (100 khz to 300 ghz)
28. Lin JC (2011) The curious case of the iarc working group on radio frequency electromagnetic fields and cell phones [health effects]. IEEE Microwave Magazine 12(6):32–36
29. Stewart BW (2021) Enhanced communication of iarc monograph findings to better achieve public health outcomes. Carcinogenesis 42(2):159–168
30. A Journey for EMF (2023). Accessed 10 April 2023. https://dot.gov.in/journey-emf

Chapter 36
RGB and Thermal Image Analysis for Marble Crack Detection with Deep Learning

Eleni Vrochidou, George K. Sidiropoulos, Athanasios G. Ouzounis, Ioannis Tsimperidis, Ilias T. Sarafis, Vassilis Kalpakis, Andreas Stamkos, and George A. Papakostas

1 Introduction

Marble stands out for its impressive aesthetic value, adding elegance and luxury anywhere it is being utilized, such as in constructions, interior decoration, statuary, and ornaments. By nature, marble is a unique material. Its physicochemical and mechanical properties make it a rather delicate material. Marble has a high porosity, which allows liquids to penetrate the stone and cause stains. Therefore, its internal structure can be influenced by humidity or decay factors that tend to penetrate materials with pores and deteriorate their durability. Moreover, micro-cracks on marble slabs may increase their porosity. Crack detection and treatment are therefore essential, especially for slabs being exposed to weathering, since temperature expansions may grow the already existing cracks [1]. Currently, marble crack detection is mainly performed by manual visual inspection of marble slabs by experienced workers. However, defect detection by the naked eye, especially when referring to micro-cracks on textured marble surfaces, may be inconsistent and error-prone. Machine vision-based automatic inspection can save time and improve the quality control of marble on the production line, providing a more constant, quicker, and cost-effective alternative [2]. Advances in machine learning (ML) and deep learning (DL) are

E. Vrochidou · G. K. Sidiropoulos · A. G. Ouzounis · I. Tsimperidis · G. A. Papakostas (✉)
MLV Research Group, Department of Computer Science, International Hellenic University, 65404 Kavala, Greece
e-mail: gpapak@cs.ihu.gr

I. T. Sarafis
Department of Chemistry, International Hellenic University, 65404 Kavala, Greece

V. Kalpakis · A. Stamkos
Intermek S.A, 64100 Kavala, Greece

A. Yadav et al. (eds.), *Proceedings of International Conference on Paradigms of Communication, Computing and Data Analytics*, Algorithms for Intelligent Systems,
https://doi.org/10.1007/978-981-99-4626-6_36

significantly impacting the field of machine vision, providing efficient algorithms for marble defects classification [3, 4], localization, and segmentation [5].

Automatic marble crack detection was first introduced in 1977 [6]; however, to date is not yet sufficiently investigated [7]. Moreover, DL applications for semantic segmentation of marble cracks are scarce in the literature. A DL segmentation approach was proposed in [8]. RGB images were used to train five different convolutional neural networks (CNNs). Results indicated ResNet-50 as the optimal architecture, reporting a mean Intersection over Union (mIoU) of 67.2%. More recently, in [5], a total of 112 DL segmentation model combinations were investigated for marble crack detection in color images. The combination of feature pyramid network (FPN) with SE-ResNet family feature extraction backbone, resulted in 71.35% mIoU. All previous works on marble crack segmentation to date are using color imaging. Thermal imaging has been extensively applied to surface crack detection problems, claiming to be able to better distinguish cracks on materials, compared to RGB [9]; yet, thermal imaging has never been applied for marble crack segmentation.

To this end, this work for the first time applies thermal imaging for marble crack segmentation and comparatively evaluates the results by using different DL models on pairs of thermal and color images. The objective of this study is to evaluate the performance of various image modalities, such as RGB images and thermal images, with the aim of investigating the impact of both image modality and deep learning network architecture on segmentation results. To accomplish this, we will test the performance of different deep learning models on each image modality separately. More specifically, a comparative evaluation of 112 DL segmentation models takes place, combining four semantic segmentation models with 28 feature extraction backbone networks. Experimental results are valued both based on the segmentation model (model-based evaluation) and on the feature extraction method (backbone-based evaluation).

The rest of the paper is structured as follows. Materials and methods are analyzed in Sect. 2. Experimental results are presented in Sect. 3. Discussion of the results and future research directions are provided in Sect. 4. Finally, Sect. 5 concludes the paper.

2 Materials and Methods

In this section, the used dataset, the proposed methodology, and the selected DL segmentation models are presented.

2.1 Dataset

The dataset analyzed during the current study was generated in four successive steps. In the first step, 38 marble tiles with cracks were handpicked by a domain expert

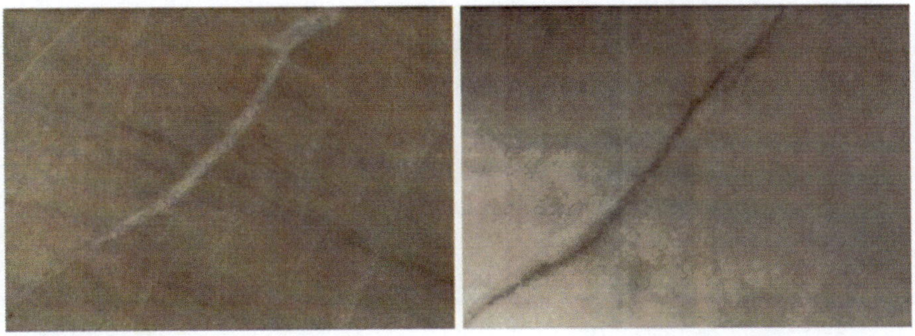

(a) (b)

Fig. 1 Marble crack pair of images: **a** RGB image and **b** thermal image

from the production line of the marble quarrying company Solakis [10] in Drama, Greece. Four sizes of tiles were used: 20 × 40 cm, 30 × 60 cm, 40 × 40 cm, and 50 × 60 cm with hairline cracks of up to 2 mm wide. In the second step, all tiles were photographed at a steady distance of 90 cm using an MVLMF0824M-5MP lens mounted on an MV-CA050-10GM/GC digital camera. This method was complemented by an automatic screening machine with a diffusion box designed and implemented by Intermek A.B.E.E. in Kavala, Greece [11]. The process resulted in the high-resolution RGB images of the tiles used in this work. The thermal images were obtained in the third step in the laboratories of the International Hellenic University [12], in Kavala Campus, Greece. The tiles were first heated with an infrared source and then scanned with the thermal heat-sensitive 206 × 156 Seek Compact XR camera [13], focused on the cracked areas. In the final step, the RGB and thermal images were first paired, and then, the visible cracks were manually annotated by a domain expert using the LabelMe annotation tool [14].

The original dataset comprised 24 pairs of thermal and RGB images.[1] Figure 1 illustrates a pair of images, referring to an RGB image and the corresponding thermal image of the same marble crack.

Data augmentation was then applied to the original dataset. Random rotation between 0 and 90°, horizontal flip with 50% chance, and vertical Flip with 50% chance were selected, resulting in a total dataset of 244 images for each image category. Five-fold cross-validation was applied to the final dataset to increase the confidence of the model's performance.

2.2 Proposed Methodology

Figure 2 illustrates the basic steps of the proposed methodology. In the first step of the proposed methodology, all original images, RGB and thermal were subjected

[1] https://github.com/MachineLearningVisionRG/mcs2-dataset.

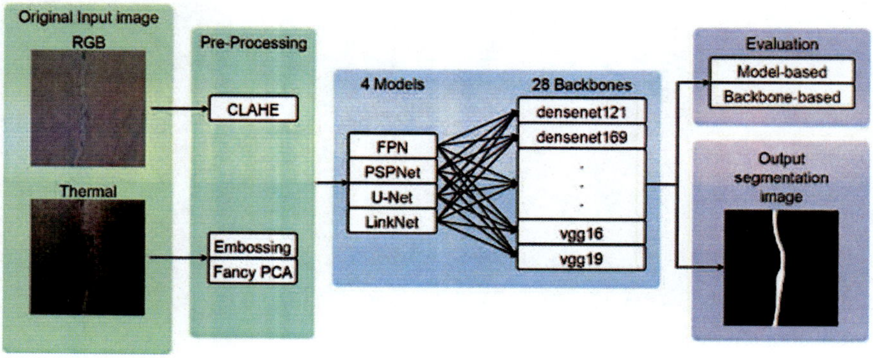

Fig. 2 Proposed methodology

to basic pre-processing. Contrast-limited adaptive histogram equalization (CLAHE) [15] was applied to all RGB images to reduce noise amplification. Thermal images were subjected to image embossing to raise crack patterns against the background. Moreover, principal components analysis color augmentation (Fancy PCA) was also applied to thermal images [16]. The next steps of the proposed approach are based on the methodology presented in [5]. Four semantic segmentation models are combined with 28 feature extraction networks. Results include the output segmentation image and the numerical results in terms of four well-known segmentation metrices, which are evaluated in two different ways: model-based evaluation and backbone-evaluation. The process is repeated for RGB and thermal images of the same cracks (pairs). It should be noted here, that in this work, the dataset is an original in-house dataset of thermal and RGB images of the same marble slabs toward a fair comparative evaluation. Therefore, a comparative study of thermal versus color imaging takes place, by simultaneously highlighting the optimal DL combination that better suits the marble crack segmentation problem for each of the two image categories.

2.3 DL Segmentation Models

The DL segmentation models in this work are combinations of four DL networks and 28 feature extraction backbone networks. The aim is to investigate the most efficient architecture for marble crack detection, for RGB images as well as for thermal images, and compare the results.

DL networks were selected based on their popularity, capabilities, and state-of-the-art reported performances. Therefore, the four selected deep convolutional neural networks models are the following: feature pyramid network (FPN) [17], LinkNet [18], pyramid scene parsing network (PSPNet) [19], and U-Net [20].

Feature extraction backbone networks were selected based on the most efficient families of networks as reported in the literature. Seven well-known families were selected, resulting in 28 total backbones: DenseNet (densenet121, densenet169, and densenet201), EfficientNet (efficientnetb0, efficientnetb1, efficientnetb2, efficientnetb3, and efficientnetb4), Inception (inceptionresnetv2 and inceptionv3), MobileNet (mobilenet and mobilenetv2), ResNet (resnet18, resnet34, resnet50, resnet101, resnet152, resnext50, and resnext101), SE-ResNet (seresnet101, seresnet152, seresnext50, and seresnext101), and VGG (vgg16 and vgg19).

3 Experimental Results

The proposed methodology was implemented in Python 3.9 employing TensorFlow and Keras. All experiments run on an Nvidia RTX 3090 GPU. Original thermal and RGB images were resized in 256 × 256 pixels size to be inserted as input to the DL models (except for PSPNet requiring 240 × 240 pixels size). For better convergence, all backbone networks were pretrained on ImageNet [21]. For all DL networks, 75% of their layers were frozen, while the last 25% of the model's layers were trainable.

In this work, the loss function (L) applied in all segmentation experiments, and it is calculated as the sum of focal (FL) and Dice loss (DL):

$$L = FL + DL \tag{1}$$

where the focal loss is

$$FL = -a_t(1 - p_t)^{\gamma} \log(p_t) \tag{2}$$

and Dice loss is calculated as follows:

$$DL = 1 - \frac{2y\hat{p} + 1}{y + \hat{p} + 1} \tag{3}$$

where in (2), $(1 - p_t)^{\gamma}$ is the modulator factor, γ is the focus factor, p_t the output of the activation function, and a_t a control weight, while in (3), y it the real, and $\hat{p}$ the predicted value by the model.

The use of the latter unified loss is proven to better handle class imbalanced datasets in semantic segmentation problems and results in improved segmentation quality and a better balance between precision-recall [22]. Table 1 includes information regarding the same hyperparameters for all DL models.

The evaluation of the 112 DL segmentation models was conducted based on two different perspectives: based on the segmentation model and based on the feature extraction backbone network. In what follows, all experimental segmentation results are evaluated after fivefold cross-validation in terms of the following commonly used

Table 1 Hyperparameters of DL models

Hyperparameters	Setting
Activation	Sigmoid
Optimizer	Adam
Loss function	Focal loss, Dice loss
Learning rate	0.0005
Weight decay	1e-8 (0.00000008)
Epochs	50
Steps per epoch	20
Batch size	32
Early stopping	Min_delta = 0.05 Patience = 5

semantic segmentation metrics: Intersection over union (IoU), precision (P), recall (R), and F1-score.

3.1 Model-Based Evaluation

All values included in the tables of this section are the mean values of the performance results on the training set after fivefold cross-validation. Table 2 includes the segmentation results of all models, for both thermal and RGB images.

As it can be seen in Table 2, the best mIoU performance, 71.61%, is reported with the FPN model with RGB images. The second-best performance is reported with the same model with thermal images (68.07%). As a general notice, color and thermal images display similar performances, with small differences. In all cases, however, color images reported slightly better results compared to the corresponding thermal

Table 2 Model-based segmentation results (% mean values) for both thermal and color (RGB) images

Model	Image type	IoU	P	R	F1-score
FPN	Thermal	**68.07**	86.79	79.31	70.15
	RGB	**71.61**	93.96	77.01	73.40
U-Net	Thermal	57.24	72.73	79.36	59.24
	RGB	58.63	75.14	79.47	60.65
LinkNet	Thermal	45.89	54.89	87.99	48.24
	RGB	48.25	58.37	86.11	50.82
PSPNet	Thermal	63.15	85.91	74.53	65.35
	RGB	65.46	89.42	74.00	67.93

Best mean IoU for thermal and RGB are marked in bold

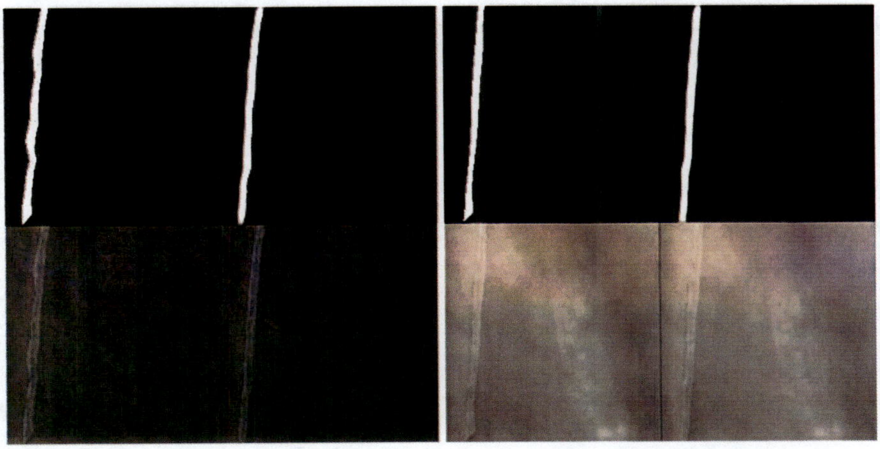

(a) (b)

Fig. 3 Indicative results of FPN model to testing images (up-left = ground truth image, down-left = RGB/thermal input image, up-right = output segmentation, down-right = output segmentation mask applied to the input image): **a** from RGB input image (98.01% IoU with efficientnetb3 backbone), **b** from thermal input image (IoU 98.00% with seresnet152)

images. The latter can be attributed to the low resolution of the used thermal camera. Figure 3 displays indicative results of FPN with thermal and RGB images of the testing set reporting the mean mIoU and the used backbone in each depicted case. More specifically, Fig. 3a refers to an RGB input image, resulting in 98.01% IoU with efficientnetb3 backbone, while Fig. 3b refers to a thermal input image resulting in 98.00% IoU with seresnet152 backbone.

The best mIoU performances refer to a DL model combined with a certain backbone. The latter signifies the fact some feature extraction networks can help a model to result in a better segmentation result. In what follows, the evaluation of the results is presented from the perspective of the used backbone.

3.2 Backbone-Based Evaluation

Results are also evaluated from the backbone perspective, so as to highlight the contribution of each backbone to the DL segmentation models. Table 3 summarizes the segmentation results of all models, for both thermal and RGB images. All performance values are the mean values of the results on the training set after fivefold cross-validation.

The best-performing backbone family with thermal images is the EfficientNet (efficientnetb4) with 75.49% mIoU. However, the Inception family reported a higher average mIoU (73.55%), by considering the average of the results of both inceptionresnetv2 and inceptionv3, compared to the averages of the rest backbone families.

Table 3 Backbone-based segmentation results (% mean values) for both thermal and color images

Backbone family	Name	Thermal images				RGB images			
		IoU	P	R	F1-score	IoU	P	R	F1-score
DenseNet	densenet121	70.64	89.42	78.76	73.35	74.79	94.53	78.83	77.56
	densenet169	71.14	91.43	77.43	73.71	72.42	95.31	76.13	74.57
	densenet201	67.31	84.60	80.35	70.11	72.64	95.42	76.19	74.86
EfficientNet	efficientnetb0	66.42	79.98	82.86	69.49	77.63	88.26	87.06	81.43
	efficientnetb1	64.47	75.15	86.81	68.02	78.23	85.82	90.09	82.46
	efficientnetb2	70.49	82.89	85.21	73.61	74.81	85.35	87.56	78.65
	efficientnetb3	70.00	81.59	85.54	73.44	75.92	82.88	90.92	80.14
	efficientnetb4	**75.49**	86.43	86.48	78.79	**80.07**	90.68	84.41	83.73
Inception	inceptionresnetv2	74.09	91.78	80.42	76.63	73.46	94.72	77.64	75.91
	inceptionv3	73.02	91.37	79.40	75.69	72.78	92.27	79.10	75.44
MobineNet	mobilenet	47.81	58.28	87.90	50.78	62.63	78.15	79.76	65.55
	mobilenetv2	48.23	67.53	76.40	50.36	54.43	86.00	66.98	56.18
ResNet	resnet18	67.01	87.27	77.46	69.40	69.52	94.92	73.67	71.58
	resnet34	63.85	83.79	76.52	66.22	71.24	93.80	76.44	73.54
	resnet50	66.41	90.63	73.72	68.30	68.15	96.01	71.29	70.08
	resnet101	60.63	85.66	73.18	62.09	62.35	98.28	63.67	63.42
	resnet152	63.71	91.05	70.34	65.22	60.64	95.94	63.69	61.62
	resnext50	70.15	92.49	76.40	72.18	70.59	97.00	72.94	72.50
	resnext101	69.38	92.69	75.13	71.13	55.57	85.60	68.11	57.28
SE-ResNet	seresnet18	68.31	87.32	77.92	70.80	69.94	94.78	74.22	71.97
	seresnet34	71.40	95.23	75.03	73.27	72.29	92.78	78.46	74.79
	seresnet50	62.85	74.31	85.63	65.91	79.22	91.16	86.34	82.74
	seresnet101	63.53	78.50	80.03	65.95	68.01	86.09	80.48	70.31
	seresnet152	69.41	88.00	79.58	71.72	61.88	77.01	79.78	64.39
	seresnext50	72.64	86.05	84.83	75.60	78.46	92.39	84.41	81.60
	seresnext101	74.80	90.58	82.41	77.57	65.21	80.86	81.75	67.97
VGG	vgg16	63.08	88.29	72.13	64.97	64.52	87.82	74.32	67.24
	vgg19	62.21	86.14	72.95	64.40	69.77	91.51	75.77	72.81

Best mean IoU for thermal and RGB are marked in bold

Considering the averages of families, the EfficientNet family which displays the higher mIoU performance comes third with 69.37% average mIoU, after Inception (73.55%) and DenseNet with 70.89%.

The best-performing backbone family with RGB images is again EfficientNet and efficientnetb4, with 80.07% mIoU. Moreover, considering the average performance of each family, the EfficientNet family ranks first (77.33%), followed by DenseNet (73.60%) and Inception (73.12%).

It could be therefore concluded, that in all cases, the three backbones standing out are those of EfficientNet, Inception, and DenseNet families. Experimental results verify the hypothesis that specific feature extraction networks can improve a model's segmentation performance.

4 Discussion

The proposed work experimentally demonstrated that deep learning could guarantee efficiency, quality, and reliability in visual inspection of marble slabs. However, further research needs to be carried out in order for the marble crack DL segmentation algorithms to be implemented in industrial settings.

The results of this work indicate similar performances for the DL models, for both thermal and RGB images. However, cracks can be slightly better detected in RGB images rather than in thermal. Infrared thermal imaging is widely used to identify cracks on the surface of materials and underneath them [9, 23, 24]. In most cases, thermal imaging has been proven most efficient for the detection of micro-cracks compared to color imaging [25]; however, there is no reported research on marble. Poor segmentation results of thermal images in this work, compared to those expected, can be attributed to the low resolution of the thermal camera. The thermal sensor used in this work was 206 × 156 pixels size, not referring to high-resolution thermal imaging. The latter can be clearly observed in the captured images, displaying a lot of noise and being of low quality compared to the corresponding RGB ones. Figure 4a shows a case of a thermal image with very strong dark areas on the marble slab and bright reflections, making the crack not clearly distinguishable, compared to the corresponding RGB image (Fig. 4b).

Therefore, marble crack detection based on deep learning will still be the main direction of future research, by additionally considering some important aspects. To

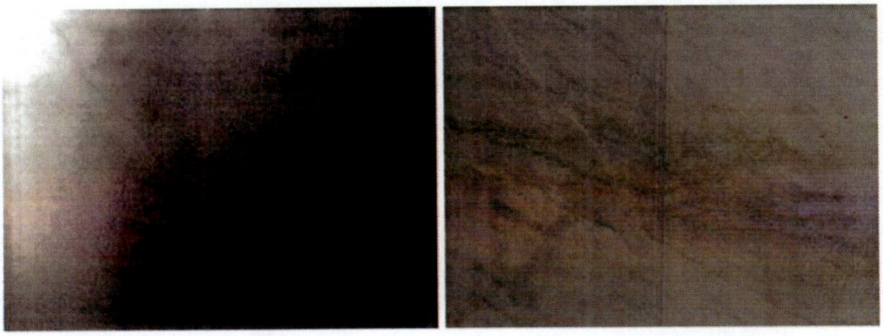

(a) (b)

Fig. 4 Marble crack pair of images: **a** Thermal image and **b** RGB image. Result indicates that the model's performances were affected by the low quality and intense noise of thermal images this end, future work first includes the use of a high-resolution thermal camera and re-capturing of thermal images. Moreover, the fusion of RGB and thermal images will also be investigated. In general, a fusion of technologies could lead to better segmentation results: micro-thermal sensors, ultrasonic waves, laser scanning thermography, etc., could also be integrated in order to achieve a full range of inspection of marble slabs.

In recent years, DL models have been established in visual inspection. Many DL models can be employed to enhance the segmentation performance; yet, high performance and fast implementations need to balance for real-time applications. Therefore, a future research direction is toward the investigation of robust DL model combinations that could accomplish both efficiency and accuracy. Next generation of computing technologies, characterized by quantum computing [26], is expected to provide fast computing capabilities for real-time visual inspection.

5 Conclusions

In this work, a performance evaluation of 112 DL segmentation architectures takes place, combining four DL models with 28 feature extraction networks, based on thermal and color imaging to detect cracks on marble slabs. Experimental results are evaluated based on two perspectives: based on the DL model and based on the backbone network. Results indicate that DL models can perform similarly on both thermal and RGB images, reporting FPN as the best-performing model, with 71.61 and 68.07% mIoU, for RGB and thermal images, respectively. Regarding the backbone network, results indicated as best-performing backbone family the EfficientNet with efficientnetb4, with 80.07 and 75.49% mIoU for RGB and thermal images, respectively.

Acknowledgements This research has been co-financed by the European Union and Greek national funds through the Operational Program Competitiveness, Entrepreneurship and Innovation, under the call RESEARCH—CREATE—INNOVATE (project code: T2EΔK-00238).

References

1. Luque A, Ruiz-Agudo E, Cultrone G, Sebastián E, Siegesmund S (2011) Direct observation of microcrack development in marble caused by thermal weathering. Environ Earth Sci 62:1375–1386. https://doi.org/10.1007/s12665-010-0624-1
2. Ren Z, Fang F, Yan N, Wu Y (2022) State of the art in defect detection based on machine vision. Int J Precis Eng Manuf Technol 9:661–691. https://doi.org/10.1007/s40684-021-00343-6
3. Karaali İ, Eminağaoğlu M (2020) A convolutional neural network model for marble quality classification. SN Appl Sci 2:1733. https://doi.org/10.1007/s42452-020-03520-5
4. Ouzounis A, Sidiropoulos G, Papakostas G, Sarafis I, Stamkos A, Solakis G (2021) Interpretable deep learning for marble tiles sorting. In: Proceedings of the 2nd international conference on

deep learning theory and applications. SCITEPRESS—Science and Technology Publications, pp 101–108. https://doi.org/10.5220/0010517001010108
5. Vrochidou E, Sidiropoulos GK, Ouzounis AG, Lampoglou A, Tsimperidis I, Papakostas GA, Sarafis IT, Kalpakis V, Stamkos A (2022) Towards robotic marble resin application: crack detection on marble using deep learning. Electronics 11:3289. https://doi.org/10.3390/electronics11203289
6. Lanzetta M, Tantussi G (1997) The quality control of natural materials: defect detection on Carrara marble with an artificial vision system. In: A.I.Te.M III, Proceedings of the 3rd conference of the italian association of mechanical technology. Fisciano Salerno, Italy, pp 449–456
7. Sipko E, Kravchenko O, Karapetyan A, Plakasova Z, Gladka M (2020) The system recognizes surface defects of marble slabs based on segmentation methods. Sci J Astana IT Univ 1:50–59. https://doi.org/10.37943/AITU.2020.1.63643
8. Akosman SA, Oktem M, Moral OT, Kilic V (2021) Deep learning-based semantic segmentation for crack detection on marbles. In: 2021 29th signal processing and communications applications conference (SIU). IEEE, pp 1–4. https://doi.org/10.1109/SIU53274.2021.9477867
9. Yang J, Wang W, Lin G, Li Q, Sun Y, Sun Y (2019) Infrared thermal imaging-based crack detection using deep learning. IEEE Access 7:182060–182077. https://doi.org/10.1109/ACCESS.2019.2958264
10. Solakis (2023) Solakis Marble Enterprises. https://www.solakismarble.com/. Last Accessed 04 Apr 2023
11. Intermek (2023) Intermek. https://www.intermek.gr/en/. Last Accessed 04 Apr 2023
12. IHU (2023) International Hellenic University. https://www.ihu.gr/en/enhome. Last Accessed 04 Apr 2023
13. Seek Thermal: CompactXR (2023). https://www.thermal.com/compact-series.html. Last Accessed 04 Apr 2023
14. Russell BC, Torralba A, Murphy KP, Freeman WT (2008) LabelMe: a database and web-based tool for image annotation. Int J Comput Vis 77:157–173. https://doi.org/10.1007/s11263-007-0090-8
15. Zuiderveld K (1994) Contrast Limited adaptive histogram equalization. In: graphics gems. Elsevier, pp 474–485. https://doi.org/10.1016/B978-0-12-336156-1.50061-6
16. An N, Xie J, Zheng X, Gao X (2015) Application of PCA in concrete infrared thermography detection. In: Proceedings of the 2015 2nd international workshop on materials engineering and computer sciences. Atlantis Press, Paris, France. https://doi.org/10.2991/iwmecs-15.2015.160
17. Lin T-Y, Dollar P, Girshick R, He K, Hariharan B, Belongie S (2017) Feature pyramid networks for object detection. In: 2017 IEEE conference on computer vision and pattern recognition (CVPR). IEEE, pp 2117–2125. https://doi.org/10.1109/CVPR.2017.106
18. Chaurasia A, Culurciello E (2017) LinkNet: exploiting encoder representations for efficient semantic segmentation. In: 2017 IEEE visual communications and image processing (VCIP). IEEE, pp 1–4. https://doi.org/10.1109/VCIP.2017.8305148
19. Zhao H, Shi J, Qi X, Wang X, Jia J (2017) Pyramid scene parsing network. In: 2017 IEEE conference on computer vision and pattern recognition (CVPR). IEEE, pp 6230–6239. https://doi.org/10.1109/CVPR.2017.660
20. Ronneberger O, Fischer P, Brox T (2015) U-Net: convolutional networks for biomedical image segmentation. In: Lecture notes in computer science (including subseries lecture notes in artificial intelligence and lecture notes in bioinformatics). pp 234–241. https://doi.org/10.1007/978-3-319-24574-4_28
21. Liu Y (2023) DeepCrack. https://github.com/yhlleo/DeepCrack. Last Accessed 04 Apr 2023
22. Yeung M, Sala E, Schönlieb C-B, Rundo L (2022) Unified focal loss: generalising dice and cross entropy-based losses to handle class imbalanced medical image segmentation. Comput Med Imaging Graph 95:102026. https://doi.org/10.1016/j.compmedimag.2021.102026
23. Cheng X, Cheng S (2022) Infrared thermographic fault detection using machine vision with convolutional neural network for blast furnace chute. IEEE Trans Instrum Meas 71:1–9. https://doi.org/10.1109/TIM.2022.3218326

24. Liu F, Liu J, Wang L (2022) Asphalt pavement crack detection based on convolutional neural network and infrared thermography. IEEE Trans Intell Transp Syst 23:22145–22155. https://doi.org/10.1109/TITS.2022.3142393
25. Chen C, Chandra S, Seo H (2022) Automatic pavement defect detection and classification using RGB-thermal images based on hierarchical residual attention network. Sensors 22:5781. https://doi.org/10.3390/s22155781
26. Tziridis K, Kalampokas T, Papakostas GA (2023) Quantum image analysis—status and perspectives. In: El-Alfy E-SM, George Bebis MZ (eds) Intelligent image and video analytics. 1st edn. CRC Press, Boca Raton. https://doi.org/10.1201/9781003053262

Chapter 37
Rover with Obstacle Avoidance Using Image Processing

Krishneel Sharma, Krishan P. Singh, Bhavish P. Gulabdas, Shahil Kumar, Sheikh Izzal Azid, and Rahul Ranjeev Kumar

1 Introduction

In today's world, lives have been revolutionized through advancement in the field of robotics and automation. Tasks that previously required a considerable amount of human effort can now be done simply at the press of a button. Some of the great technological advancements include, but are not limited to artificial intelligence (AI), Internet of things (IoT), automobiles, GPS, industrial robots, and electric cars. The operating principle of the autonomous rover or robotic systems used for industrial applications or outdoor applications is an approach for obstacle avoidance and object recognition [1].

An autonomous rover is a robot that performs tasks with a high degree of accuracy without the need for human intervention. This is possible through various means such as sensors and image processing. In this paper, the latter method is used via a camera module. Image processing is a popular technique of doing operations on an image or sets of images to enhance the image or obtain useful information from it [2]. The renowned autonomous Mars Perseverance rover is one of the large-scale implementations by NASA. The concept used herein can be compared to the MARS rover, but a much smaller scale version.

K. Sharma · K. P. Singh · B. P. Gulabdas · S. Kumar · S. I. Azid · R. Ranjeev Kumar (✉)
School of Information Technology, Engineering, Mathematics, and Physics,
The University of the South Pacific, Suva, Fiji
e-mail: rahul.kumar@usp.ac.fj

A. Yadav et al. (eds.), *Proceedings of International Conference on Paradigms of Communication, Computing and Data Analytics*, Algorithms for Intelligent Systems,
https://doi.org/10.1007/978-981-99-4626-6_37

Image processing can be accomplished in several ways, one of which is using the OpenCV library, which aims to achieve real-time computer vision. OpenCV is an open-source library used to process images and videos for various applications like facial recognition systems, mobile robots, object detection, and many more. To perform object detection, various methods can be utilized, for instance, Canny edge detection [3], Sobel edge [4], Prewitt edge [5], and Laplacian edge detection [6].

Selecting an appropriate edge detection technique is crucial in identifying and acknowledging obstacles as it serves as the initial phase. The ideal approach should yield a high number of accurate results while minimizing the computational time required. OpenCV offers various multi-stage algorithms, and the Canny edge detector stands out due to its reliability and minimal errors [7]. Canny edge detection identifies an obstacle's edges through uncomplicated steps. In addition, the bilateral filter is commonly used alongside the Canny edge detection to reduce noise and maintain edge continuity in images.

Obstacle avoidance is an important part for rovers to detect clear paths and is one of the most important technologies for ensuring human and vehicle safety. Numerous studies have been conducted on this topic over decades, and many approaches have been devised with only a few implemented in practical systems. A robot must not only identify impediments to avoid colliding with them, but also be able to compute the detour course and direct itself in real time toward a safe and clear path.

Table 1 discusses the methods used for image processing together with its advantages and disadvantages. There are five methods discussed which include dynamic window method, model predictive control, reactive method design, curve flow method, and stereo vision and laser proximity sensing.

In this paper, an autonomous rover is designed and tested in real time to perform object detection and obstacle avoidance to achieve optimal results. The Canny edge detector together with the bilateral filter is used to process the images collected by the camera present on the rover in real time. By designing the Python code in OpenCV, the image-processing aspect is linked to the wheels of the rover to navigate the rover autonomously without hitting any object and without the need for human intervention.

Table 1 Summary of methods and approaches used in robotics and computer vision

Method	Approach	Pros	Cons
Dynamic window method [8]	This method uses three velocity–angular velocity windows First window—vehicle's minimum to maximum velocity–angular velocity. Second window—vehicle's minimum to maximum velocity–angular velocity. Third window—vehicle's minimum to maximum velocity–angular velocity	Accounts for the constraints due to limited velocities and accelerations of the robot	
Model predictive control [9]	Develops a model of the process's dynamics to forecast its progression within a specified future time frame and Optimize control inputs sequence to minimize performance index under input and state constraints using numerical optimization techniques	Manage multi-variable problems systematically Consider constraints on states and control actions accounting for the expected future behavior of the system	A nonlinear optimal control problem, determining the best path in real time is difficult This is heavily reliant on the performance of the CPU
Reactive method design [10]	Navigation design techniques tailored for collision avoidance robots	Reliably performs navigation in extremely dense, crowded, and complicated settings Alternative techniques may struggle in the same conditions	

(continued)

Table 1 (continued)

Method	Approach	Pros	Cons
Curve flow method [11]	A novel approach for online path planning. The CFM developed conceptually and compared to the functionally equivalent elastic band technique EBM	Increases numerical robustness, parameter ability, and computing efficiency	
Stereo vision and laser proximity sensing [12]	The obstacle avoidance is conducted using stereo vision, and danger identification is done using lasers	Excellent performance across a wide range of terrain Laser-based technique operates faster	Stereo-based technique operates fairly slow

2 Design Methodology

2.1 *Materials Used*

The materials used to develop the autonomous rover that is used to detect and avoid obstacles using image-processing techniques are a Raspberry Pi model 3B micro-processor, a Raspberry Pi camera, four DC motors for the wheels, a 3600 mAh 7.4V battery, and two simple H-bridges to drive the DC motors.

The rover contains a Raspberry Pi camera for taking video feeds as the rover moves. The Raspberry Pi contains a program for avoiding obstacles and controls all the operations. The motor drivers control the motors to which the wheels are attached and the power bank powers Raspberry Pi. The 7.4 V 3600 mAh lithium polymer battery powers the motor controller and motors. Figure 1 shows the setup of the rover.

2.2 *Rover Navigation System*

The input to the system is through the Raspberry Pi camera, which takes video feeds. This video is processed using the Raspberry Pi, which determines the obstacle-free path. The signals to the obstacle-free path are then sent to the motor controller, and the rover moves in the direction of the obstacle-free path.

The proposed method does not require the use of GPS sensors or localization networks, rendering it better suited for indoor applications where GPS may not be reliable, as well as for outdoor usage. Additionally, another benefit of the proposed method is that it can offer more in-depth and specific information regarding the

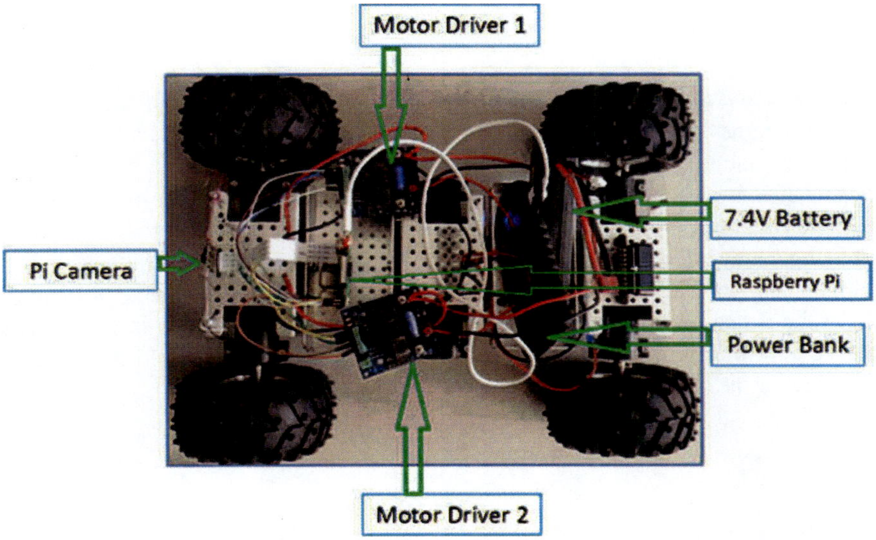

Fig. 1 Complete rover setup

environment compared to GPS. Although differential GPS or RTK may provide adequate positioning information, it cannot provide detailed information regarding the objects and features within the environment. In contrast, this method can capture images and other data that can be analyzed to provide detailed information about the environment while simultaneously maintaining the position from the same data.

Figure 2 shows a detailed flowchart of the processes involved in operating the rover beginning from the images taken via the Raspberry Pi camera down to direct the wheels of the rover.

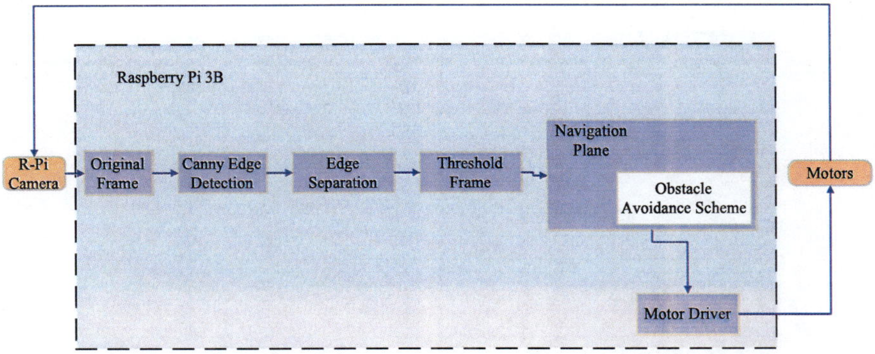

Fig. 2 Flowchart depicting the overall operation of the rover

2.3 Rover Navigation Steps

There are five main steps for maneuvering the rover. At the start of the rover, the camera begins capturing the images as shown in Fig. 3. The images taken are then smoothed using a bilateral filter. This filter blurs the image while preserving its edges. This is an advantage of using a bilateral filter. The edges in the images are then detected using the canny edge detection algorithm shown in Fig. 4. The benefit of using canny edge detection is that it can detect both strong and weak edges.

The next process is edge separation. This edge separation separates the obstacle from the frame and shows the area of an obstacle-free path shown in Fig. 5. This image is further processed to create a threshold frame. The threshold frame shows the view of the rover as shown in Fig. 6.

Fig. 3 Camera input—picture taken using R-Pi camera

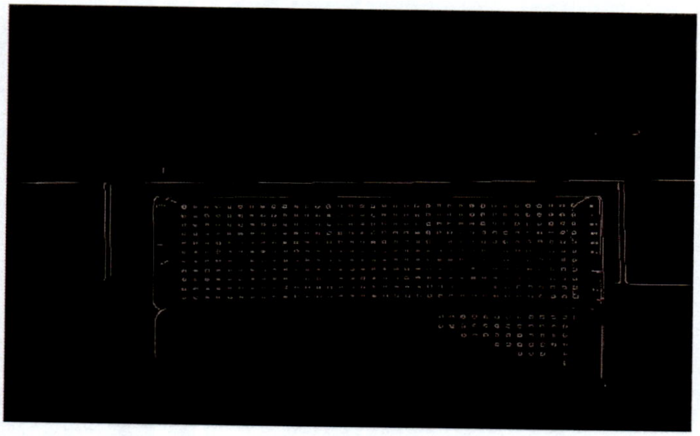

Fig. 4 Canny edge detection

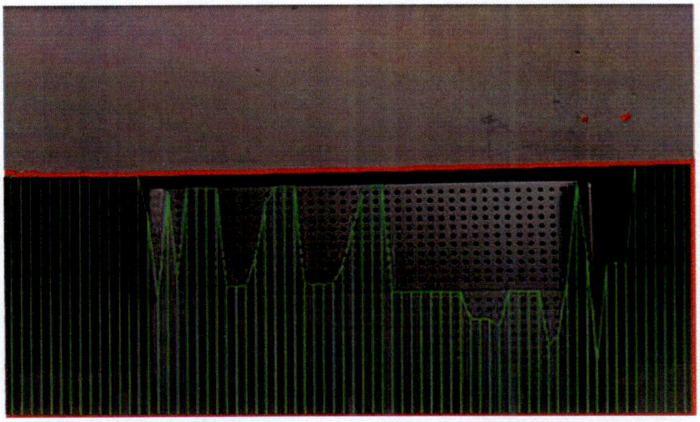

Fig. 5 Edge separation frame

Fig. 6 Threshold frame

However, light plays an important factor here. As the environment changes, the threshold values need to be adjusted. The final step is the navigation plane, which indicates the direction of the rover, and this is shown in Fig. 7.

2.4 *Obstacle Avoidance Decision*

This section depicts how the rover can avoid the obstacle by turning left or right or continue heading forward if no obstacle is in its path. This procedure of detecting obstacles and avoiding is repeatedly performed unless if the program is stopped or the motors or Raspberry Pi runs out of power.

Turning left. The edge separation frame is shown in Fig. 8. From the frame, the rover has to decide whether to move left or right. Since more area to move is on the left side of the obstacle (metal plate), it goes in the left direction as shown in the navigation plane in Fig. 9.

Turning right. As for making a right turn, the edge separation frame is shown in Fig. 10. From the frame, the rover has to decide whether to move left or right. Since

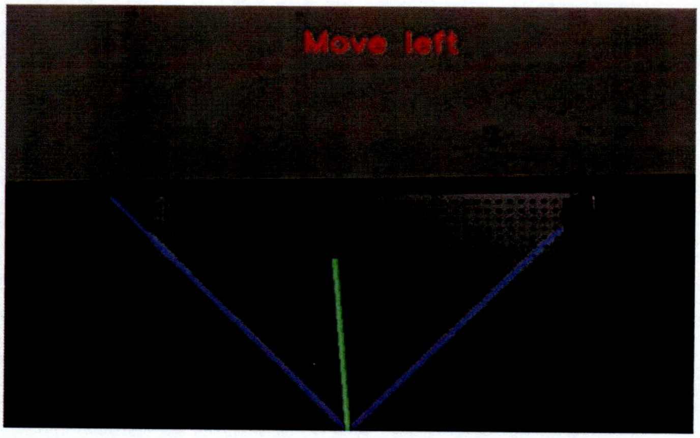

Fig. 7 Navigation plane

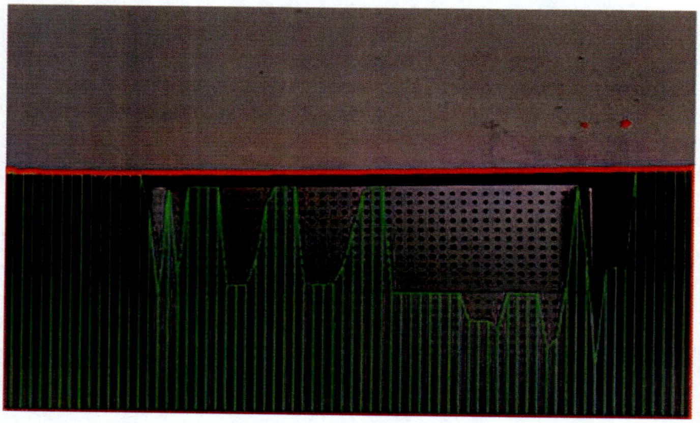

Fig. 8 Edge separation frame for turning left

more area to move is on the left side of the obstacle (metal plate), it goes in the left direction as shown in the navigation plane in Fig. 11.

Moving forward. The first image shows the edge separation frame in Fig. 12. From the frame, it can be seen that no obstacle is present in its path. Hence, the rover keeps moving in the forward direction as shown in the navigation plane in Fig. 13.

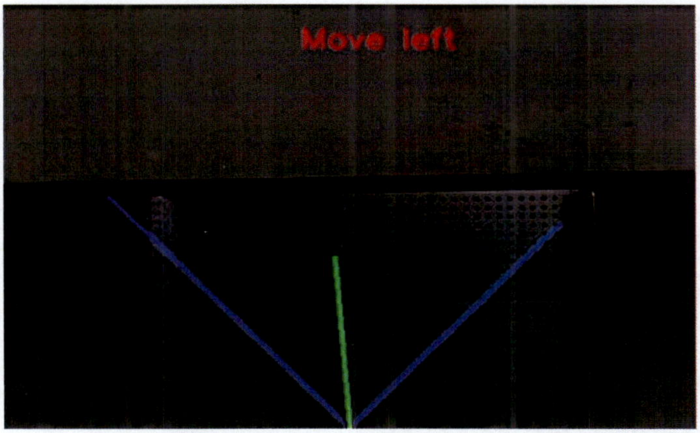

Fig. 9 Navigation plane for turning left

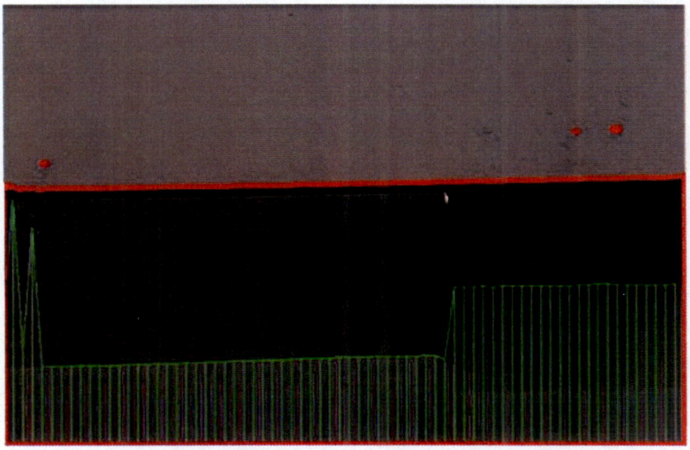

Fig. 10 Edge separation frame for turning right

3 Results

3.1 Rover's Decision for Avoiding Two Obstacles

Figure 14 shows how the rover avoids two obstacles in its path. At the start, the rover detects the first obstacle and determines the obstacle-free path. In this case, the rover has two paths, either to move left or right, the rover decides to move left of the obstacle as shown in the figure. However, after moving ahead for some time, the rover encounters another obstacle in its path, and to avoid this obstacle, the rover moves right since it has more free space on right. The rover was manually stopped after some time where the end is marked on the figure.

Fig. 11 Navigation plane for turning right

Fig. 12 Edge separation frame for moving forward

3.2 *Rover's Decision for Avoiding Four Obstacles*

Figure 15 shows how the rover proceeds in its path with four obstacles present and reaches its endpoint. By putting a greater number of obstacles in its path, the robustness of the rover was easily observed.

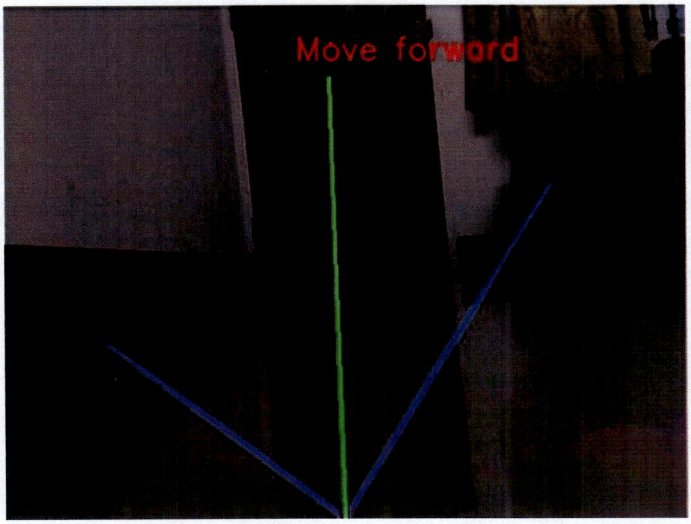

Fig. 13 Navigation plane for moving forward

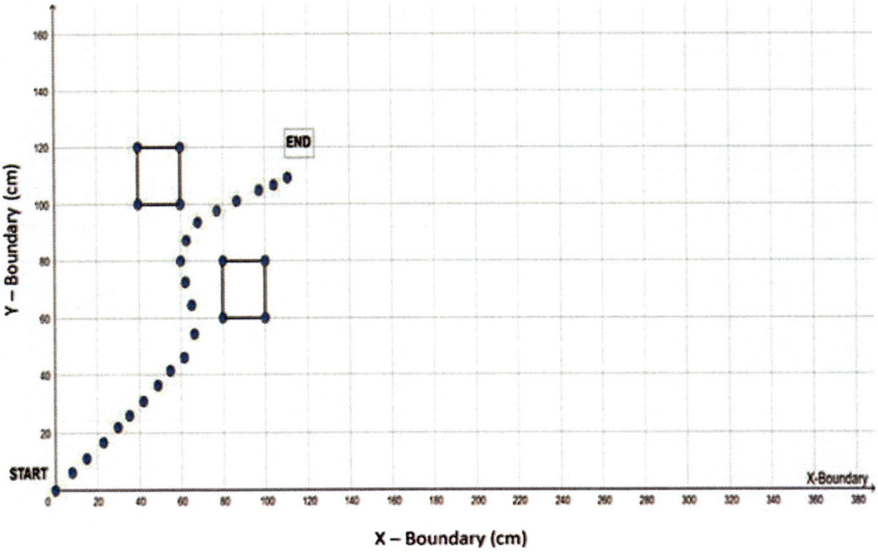

Fig. 14 Obstacle avoidance decision for two obstacles

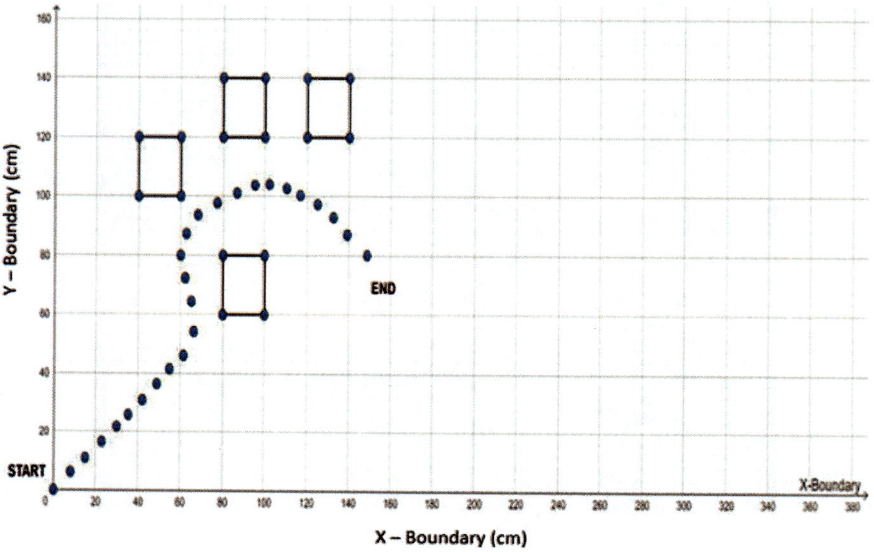

Fig. 15 Obstacle avoidance decision for four obstacles

4 Conclusion

To sum up, this study has been able to design a rover that can avoid the obstacle on its path and direct itself to a clear path. The rover has a wide variety of applications such as autonomous vehicles, transporting goods, vacuum cleaners, food serving waiters in smart restaurants, and related tasks performed by autonomous rovers. The future works on the rover can uphold tasks such as path planning using a GPS network, installing a rear camera for reverse movement, and laser sensors to avoid obstacles. Powerful motors can be implemented for heavy industrial usage or harsh environmental conditions.

Acknowledgements The authors would like to extend their sincere gratitude to The University of the South Pacific for providing access to the experimental materials and equipment.

References

1. Singh H (2019) Practical machine learning and image processing. Apress. https://doi.org/10.1007/978-1-4842-4149-3
2. Matthies L et al (2007) Computer vision on mars. Int J Comput Vis 75(1):67–92. https://doi.org/10.1007/s11263-007-0046-z
3. Rong W, Li Z, Zhang W, Sun L (2014) An improved Canny edge detection algorithm. In: 2014 IEEE International conference on mechatronics and automation, August 2014, pp 577–582. https://doi.org/10.1109/ICMA.2014.6885761

4. Gao W, Zhang X, Yang L, Liu H (2010) An improved Sobel edge detection. In: 2010 3rd International conference on computer science and information technology, July 2010, pp 67–71. https://doi.org/10.1109/ICCSIT.2010.5563693
5. Yang L, Wu X, Zhao D, Li H, Zhai J (2011) An improved Prewitt algorithm for edge detection based on noised image. In: 2011 4th International congress on image and signal processing, October 2011, pp 1197–1200. https://doi.org/10.1109/CISP.2011.6100495
6. Zhang JG, Liu JJ, Geng YJ (2010) Laplacian image edge detection based on secondary-sampling wavelet transform. In: 2010 3rd International congress on image and signal processing, October, pp 1059–1062. https://doi.org/10.1109/CISP.2010.5646916
7. Joshi M, Vyas A (2023) Comparison of canny edge detector with Sobel and Prewitt edge detector using different image formats. Accessed 13 Mar 2023. [Online] Available: www.ijert.org
8. Fox D, Burgard W, Thrun S (1997) The dynamic window approach to collision avoidance. IEEE Robot Autom Mag 4(1):23–33. https://doi.org/10.1109/100.580977
9. Krenn R, Gibbesch A, Binet G, Bemporad A (2013) Model predictive traction and steering control of planetary rovers
10. Minguez J, Montano L (2004) Nearness diagram (ND) navigation: collision avoidance in troublesome scenarios. IEEE Trans Robot Autom 20(1):45–59. https://doi.org/10.1109/TRA.2003.820849
11. Huptych M, Röck S (2021) Real-time path planning in dynamic environments for unmanned aerial vehicles using the curve-shortening flow method. Int J Adv Robot Syst 18(1):172988142096868. https://doi.org/10.1177/1729881420968687
12. Simmons R, Henriksen L, Chrisman L, Whelan G (2001) Obstacle avoidance and safeguarding for a lunar rover

Chapter 38
A Systematic Literature Review of Network Intrusion Detection System Models

Yogesh and Lalit Mohan Goyal

1 Introduction

Since security of networks has loomed as a crucial domain of research which utilizes tools such as intrusion detection system (IDS), firewall, and antivirus software to ensure all of its corresponded assets within cyberspace [1]. Amid these, NIDS defines the security by continuous network traffic investigation for suspicious behavior [2, 3]. For example, any accommodation to the organizational information node highly impacts on organization in context of its market losses [4]. Machine learning-based IDSs widely relies on feature engineering, meanwhile, IDS based upon DL and are heavily rely on self-learning capability of complex features from the initial data irrespective of relying on feature engineering [5, 6].

It is also evident from literature studied that DL-based NIDS research is presently in its initial stage and also large opportunities to explore this methodology inside effective NIDS for detecting intruders inside network. Main key objectives of this paper are as follows: (i) To identify the prime recent areas and trends in NIDS design. (ii) Identification and comparison of recent DL and ML techniques used for NIDS. (iii) Recently used up- to-dated datasets for effective NIDS testing. (iv) Recommendation of key future research scopes on the basis of systematic literature review of NIDS.

Content of this paper are following: Sect. 2 defines the adopted methodology for this work. Section 3 describes the research methodology adopted for this work, whereas Sect. 4 describes IDS overview as well as classification of IDS techniques. Section 5 defines the DL- and ML-based models adopted. Section 6 describes the available datasets for efficient IDS design. Section 7 describes about the NIDS design

Yogesh (✉) · L. M. Goyal
Department of Computer Engineering, J.C. Bose University of Science and Technology, YMCA, Faridabad, India
e-mail: yogeshymca7@gmail.com

A. Yadav et al. (eds.), *Proceedings of International Conference on Paradigms of Communication, Computing and Data Analytics*, Algorithms for Intelligent Systems,
https://doi.org/10.1007/978-981-99-4626-6_38

recent scopes, research challenges, observations, and research future scopes in this domain. Lastly, Sect. 8 is used for conclusion of this review work.

2 Related Work

Ghurab [1] presented a brief analysis of recent benchmark datasets for NIDS, i.e., KDD99, KYOTO 2006+, NSL-KDD, UNSW-NB 15, ISCX2012, CSE-CICICIDS2018, CIDDS-001, and CICIDS2017. In this work, researcher focused on each and every aspect of dataset, i.e., classes, nature of features, features, instances, etc. Results of this study concluded that for better evaluation of efficient NIDS, potential researchers should use recent datasets such as CSE-CICIDS2018, and CICIDS2017.

Alavizadeh et al. [2] introduced a hybrid NIDS that combines deep feed forward neural network with Q-learning-based reinforcement learning (RL) to provide an ongoing network environment auto-learning capability for detection of modern intrusions by using an automated trial–error approach. Experimental results were evaluated on NSL-KDD dataset reveal that constituting 0.001 discount factor within 250 episodes yields best detection results by outperforming machine learning algorithm results. Sethi [4] proposed a novel deep RL-based IDS that uses deep Q-network logic to efficiently detect network attacks in multiple distributed agents, and experiments were evaluated on two datasets, i.e., NSL-KDD and CICIDS2017. Experiments result of this work attained improved performance by means of several performance parameters, i.e., low false positivity rate (FPR), higher accuracy, precision, and recall in contrast to traditional approaches.

Karatas et al. [7] focused on comparative analysis of six ML models using a recent dataset CSE-CIC-IDS2018. Experimental results outperformed the recent literature by attaining a good increase in accuracy level between 4.01 and 30.59%. Researcher also concluded that deep learning models can be applied in future for the efficient big data applications.

Yan and Han [8] proposed stacked sparse autoencoder (SSAE) with the aim to extraction of upper-level feature representations of intrusive behavior, initially, learning of features taken place automatically followed by lower-dimensional spares features utilized to build various simple classifiers. Experimental results were evaluated both, i.e., for binary and multiclass classification indicated as: (1) SSAE-based higher-dimensional sparse features are extra discriminative in nature for intrusion behaviors as compared to traditional methods and (2) the SSAE-based higher-dimensional sparse features accelerated the classification process of classifiers to build an efficient and feasible feature extraction method.

Ali et al. [9] developed a fast-learning network (FLN) model on the basis of methodology of particle swarm optimization for intrusion detection using KDD99 dataset. Developed model performance was compared against a number of meta-heuristics methodologies for FLN classifier training, and experimental results revealed that developed PSO-FLN outperformed various learning approaches in terms of detection accuracy.

Yin et al. [10] proposed a deep learning (DL) methodology using recurrent neural networks (RNN) for intrusion detection by considering the model performance for both, i.e., multiclass and binary classification. Researchers compared proposed model with random forest, SVM, and various other ML methodologies by using benchmark dataset. Experimental results revealed that RNN-IDS attains a higher accuracy as compared to traditional ML models in both classification (i.e., multiclass and binary).

Xu et al. [11] proposed a novel hybrid IDS consisting of a RNN with multi-layer perceptron (MLP), gated recurrent units (GRU), and softmax module using NSL-KDD and KDD 99 datasets. Experimental results of work attained an intrusion detection rate of 99.42% using KDD 99 dataset and 99.31% using NSL-KDD dataset by considering FPR as low as 0.05% and 0.84%.

Naseer et al. [12] investigated the DL approaches suitability for detection of anomaly-based intrusion. In this research work, researcher developed anomaly models on the basis of various DNN structures, i.e., CNN, autoencoders, RNN, etc., and model was trained using NSL-KDD dataset. Experimental results revealed that developed deep IDS model provided efficient results for real-world applications.

Shone et al. [13] proposed a novel DL model for intrusion detection using non-symmetric autoencoder for feature learning in unsupervised way, and model was evaluated using the KDD 99 and NSL-KDD benchmark datasets. Results of this proposed model displayed improvements over traditional ML approaches for uses in real-time applications.

Al-Qaft et al. [14] proposed an effective DL methodology on the basis of self-taught learning framework primarily for dimensionality reduction and efficient feature learning by using SVM classifier. Sequentially, after pre-training stage, potential new features are fed as input to SVM algorithm to improve classification as well as detection accuracy for intrusion detection. Results of this research work attained a faster training and testing time as compared to RF and Naïve Bayes classifier for intrusion detection.

Vinaykumar et al. [15] presented a deep neural network (DNN), a kind of DL model to build an efficient IDS to detect known and unknown cyberattacks using KDDCup 99 benchmark dataset with varying the learning rate range 0.01 to 0.5 till 1000 epochs, and this model was also tested for other datasets, i.e., CICIDS 2017, NSL-KDD, Kyoto, etc., with a strict experimental testing. From the testing results of experiments, model performed excellent in comparison with traditional ML models for detection of modern-day attacks.

Marir et al. [16] proposed a novel distributed algorithm using a hybrid approach in a distributed way. For this, initially, nonlinear dimensionality reduction was performed with the help of distributed deep belief network for large network traffic data. Then, output of this step was fed as input to the multilayer ensemble SVM. Empirical results revealed an efficient gain in performance metrics as compared to traditional ML models.

Wei et al. [17] proposed a joint optimization model using DBN network structure with PSO on the basis of adaptive inertia weight and learning factor for detection of abnormal intrusions. Experimental results displayed that this DBN-IDS average detection accuracy was 24.69% more using multiclass classification as an efficient and effective optimization method for large-scale efficient network traffic data.

Xiao et al. [18] proposed a NIDS model on the basis of DL technique CNN (CNN-IDS). Initially, irrelevant features of network traffic are eliminated using various dimensionality reduction techniques, and features are automatically extracted with the help of CNN to provide more efficient information for detecting intrusions using a KDD-CUP99 dataset to performance evaluation. Accuracy and FAR of this model outperform the ML models accuracy and FAR rate in real-time environment.

Jiang et al. [19] proposed a NIDS algorithm to tackle data imbalance problem using deep hierarchical network with hybrid sampling using synthetic minority over sampling technique (i.e., SMOTE) on NSL-KDD and UNSW-NB15 datasets. Experiments results revealed detection accuracy of 83.56% on NSL-KDD dataset and 77.16% on UNSW-NB15 datasets.

Gao et al. [20] proposed an adaptive ensemble learning model for detection of intrusions using MultiTree algorithm on NSL-KDD dataset. In order to enhance the overall detection rate, various base classifiers such as RF, DNN, kNN, and decision tree were taken into consideration. Overall accuracy of proposed model in context of MultiTree algorithm attained was 84.2%.

Gupta et al. [21] proposed a novel model primarily in the field of Internet of medical things (IoMT) to detect cyber threats on connected healthcare devices to overcome security model design for IoMT field. Thereby, researcher designed an IDS model on the basis of tree classifier-based network with the aim of reducing dimension of input data to enhance the anomaly detection. Experiments results of this paper attained a detection accuracy of 94.23%.

Cao et al. [22] proposed a hybrid sampling algorithm consisting of adaptive synthetic sampling (ADASYN) with repeated edited nearest neighbors (RENN) to solve class imbalance problem of positive and negative sample in original dataset. For feature selection, RF algorithm was used, and correlation of Pearson analysis was used to solve feature redundancy. Spatial features were further extracted with the help of fusion of maxpooling and average pooling to assign distinct weights to features to improving the model performance as well as a gated recurrent unit (GRU) was used for efficient feature learning. Finally, a softmax layer was used for classification task and experimentes were analyzed on benchmark datasets such as NSL-KDD, UNSW-NB15, and CIC-IDS2017 datasets with the accuracy of 99.69, 86.25 and 99.65% detection accuracy which are superior to other traditional DL and ML algorithms performance for NIDS deployment where detection accuracy is calculated as follows:

$$Accuracy = \frac{TP + TN}{TP + TN + FP + FN} \tag{1}$$

where true positive (TP) defines attack data which is positively defined as an attack, false positive (FP) defines normal data which is falsely classified as an attack, true negative (TN) represents normal data which is positively classified as normal, and false negative (FN) represents attack which is falsely classified as normal.

Ullah and Mahmoud [23] proposed a novel DL model for anomaly detection in IoT networks using a RNN. Sequentially, a hybrid DL model comprised of CNN and RNN was proposed and validated using BoT-IoT, NSL-KDD, and MQTT dataset during experiment analysis. Research results of this paper in the context of binary and multiclass classification model attain higher detection accuracy, precision, recall, etc.

In addition, a detailed comparative performance analysis in terms of advantages, disadvantages, methodologies, and results of several papers used in this literature are tabulated in Table 1 as follows:

Table 1 Comparative performance analysis of studied literature

Author	Methodology	Strengths	Weakness	Results
Alavizadeh et al. [2]	Hybrid model (Q-learning-based RL with deep feed forward neural network)	Effective detection accuracy as compare to ML algorithms	Lack of self -learning capability in real-time environment	Detection accuracy = 78%
Sethi et al. [4]	Denoising autoencoder (DAE)	Real-time usability in real-time environment	Higher computational and network costs	NSL-KDD accuracy = 97.4%, CICIDS accuracy = 98.7%
Caminero et al. [5]	Adversarial RL in a real-time environment	Classifier performance increased in comparison of against ML algorithms	Higher computational time	AWID dataset accuracy = 95.9%
Karatas et al. [7]	Comparative analysis of six distinct ML models (i.e., decision tree, kNN, gradient boosting, RF, linear discriminant, AdaBoost)	Minority attack class detection rate enhanced using SMOTE	Highest detection rate achieved by AdaBoost algorithm but with higher training time	CSE-CICIDS2018 dataset accuracy = 99.34%
Yan and Han [8]	Stacked sparse autoencoder (SSAE)	Efficient use of SVM as classifier and SSAE for feature extraction	Reasonable detection rate for U2R and R2L attacks	NSL-KDD accuracy for normal attacks = 99.43%
Yao et al. [24]	Clustering with RF	Enhanced detection accuracy even with low data instances	Validation performed on older dataset	Average test accuracy on KDDCUP 99 dataset = 96.6%

(continued)

Table 1 (continued)

Author	Methodology	Strengths	Weakness	Results
Ali et al. [9]	Hybrid model (fast learning network with particle swarm optimization)	Better performance in comparison of other models on the basis of FLN	Lower detection rate for R2L and U2R attacks and older dataset	KDD CUP 99 dataset accuracy = 98.92%
Yin et al. [10]	Recurrent neural networks (RNN)	Higher detection accuracy	Complex and more training time required	RNN-IDS binary classification accuracy on KDD test = 83.28%, KDD Test-21 = 68.55% RNN-IDS Multiclass classification accuracy on KDD = 81.29%, KDD Test-21 = 64.67%
Xu et al. [11]	Hybrid model (RNN with multilayer perceptron and gated recurrent unit)	Efficiently use of GRU as an input of RNN with MLP and softmax classifier for effective NIDS solution	Model was validated using older datasets	Hybrid model overall detection accuracy = 99.42% on KDD dataset, accuracy = 99.31% on NSL-KDD dataset
Naseer et al. [12]	Comparison of DL and ML algorithms	Experimental comparison by implementing on GPU-testbed	Evaluation performed on old datasets	LSTM detection accuracy on NSLKDDTest21 dataset = 89%, DCNN detection accuracy on NSLKDDTest21 dataset = 85%
Shone et al. [13]	Non-symmetric autoencoder	Efficient feature selection with reduced model complexity	Lower detection rate for minority classes	S-NDAE accuracy = 97.85%, DBN accuracy = 97.90%
Al-Qaft et al. [14]	Self-taught learning framework using SVM & AE	Efficient hybrid methodology based on SVM and AE	Model was validated against older attacks	STL-IDS accuracy for NSL-KDD Train dataset = 99.39%, 80.48% for NSL-KDD test dataset
Khan et al. [25]	Stacked AE model	Increased detection rate of low class attacks	Detection rate decreased drastically using newer dataset	Hybrid model accuracy on UNSW-NB15 dataset = 89.134% and 99.996% for KDD 99 datasest

(continued)

Table 1 (continued)

Author	Methodology	Strengths	Weakness	Results
Malaiya et al. [26]	Fully connected networks (FCN), variational AE	Model was evaluated using both newer and older datasets	Higher cost overhead in training than others	LSTM Seq2Seq classification accuracy on binary classes = 99%
Jia et al. [27]	Deep neural networks (DNN)	Superior results than ML-based IDS model	Older dataset and lower detection rate for U2R attack	DNN accuracy = 99.9% on KDD cup dataset
Wei et al. [17]	Hybrid model (DBN with PSO)	DBN optimized using PSO, Fish swarm, and genetic algorithm to build efficient NIDS	More training time for model training and complex model in nature	Classification accuracy of DBN-PSO on KDDTest + dataset = 83.86%, 99.86% for KDD Train datasest
Xiao et al. [18]	Convolution neural network	Efficient algorithm for feature extraction (SVM) and dimensionality reduction using PCA analysis	Lower detection rate for U2R and R2L attacks	CNN-IDS detection accuracy on KDDCup dataset = 94%
Zhang et al. [28]	Hybrid model (CNN with gcForest)	Proposed novel algorithm as well as model was evaluated using newer and older datasets	Detection accuracy falls for lesser training data	Detection accuracy = 99.24%
Jiang, K. et al. [19]	Synthetic minority over sampling technique (SMOTE)	Efficient use of SMOTE methodology for enhancing minority class attack detection	Complex model	NSL-KDD dataset accuracy = 83.56% and UNSW-NB15 dataset accuracy = 77.16%
Gao et al. [20]	Adaptive ensemble learning model	Comparison of several base classifier	Weaker detection results for minority classes and experimental validation using older datasets	NSL-KDD dataset accuracy = 84.2%
Gupta et al. [21]	Tree classifier-based model	Efficiently handling of complex and large categorical data by using hybrid model (RF with robust scaling)	Lesser number of attacks covered	Accuracy = 94.23%

(continued)

Table 1 (continued)

Author	Methodology	Strengths	Weakness	Results
Cao et al. [22]	Hybrid model (adaptive synthetic sampling with repeated edited nearest neighbors)	Efficient use of feature selection technique for IDS (i.e., combo of RF with Pearson correlation for feature selection, CNN for spatial features selection, and lastly GRU effectively used for extracting long distance-dependent important features	High number of parameters was required for validation Running time complexity of model is extremely high	UNSW-NB15 dataset accuracy = 86.25% NSL-KDD dataset accuracy = 99.69% CICIDS2017 dataset accuracy = 99.65%
Ullah and Mahmoud [23]	Deep learning hybrid model (CNN with GRU and BiLSTM)	Effective use of CNN for feature learning as it does not losses essential information while learning, enhanced learning of weak features	Ineffective for small dataset	CNN-LSTM accuracy on NSL-KDD dataset = 99.91% CNN-GRU accuracy on NSL-KDD dataset = 99.92% CNN-BiLSTM accuracy on NSL-KDD dataset = 99.94%

3 Methodology

This paper follows a strategic methodology to examine, extract, and identify crucial information from the literature according to intrusion detection research topics into two phases. First phase defines the keywords and search engine to execution of a query to achieve a first list of papers. Sequentially, second phase implements certain core papers and most related article and to store as final list for analysis of present work. The major objectives of current review paper consider the subsequent queries on the basis of literature studied: (i) To identify the prime recent areas and trends in NIDS design. (ii) Identification of recent DL and ML techniques used for NIDS. (iii) Recently used up-to-dated datasets for effective NIDS testing. (iv) Recommendation of key future research scopes on the basis of systematic literature review of NIDS. In first phase, Scopus and Google Scholar are considered as a main engine for searching papers across databases. Search query by using a first keyword "intrusion detection system" and updated the search filter to display journal and conferences papers published in the duration of year 2017 to year 2022 which resulted 3 in several papers that proposed the IDS model using several techniques. Sequentially, we reformulated keyword as "network anomaly detection", "network intrusion detection system", "behavioral intrusion detection system", etc., as shown in Fig. 1. In second phase, as shown in Fig. 2 by considering initial list received fromphase-1, defined certain

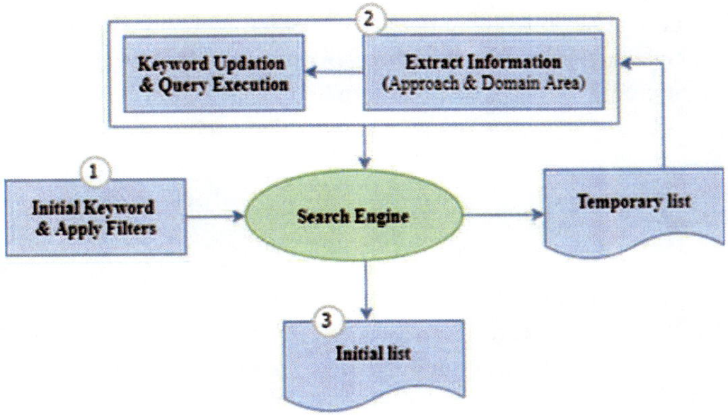

Fig. 1 Phase 1 research methodology

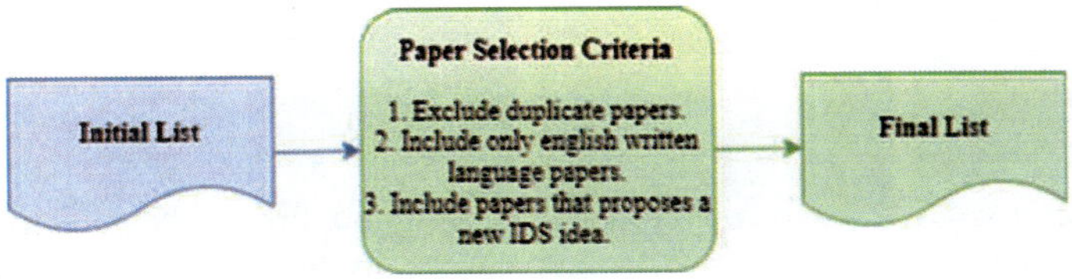

Fig. 2 Phase 2 research methodology

evaluation criteria to achieve more focused research papers for review written in English language by following or proposing a new IDS model idea. On the basis of paper selection criteria, we stored these relevant papers inside the final list for analysis, in which individual selected paper analysis was done primarily on the basis of proposed DL- or ML-enabled techniques. Sequentially, widely used datasets and performance parameters used for evaluation and testing purposes were taken into consideration. Lastly, identification of research challenges and future scopes for effective NIDS model creation was carried out.

4 IDS: Classification and Working

4.1 Detection IDS

IDS classification can be done either on the basis of detection method or deployment method. As per detection method, the IDS is partitioned into "signature detection IDS" and "anomaly detection IDS". SIDS is also referred as "knowledge intrusion detection system," is primarily based on constructing a sign for signatures and attack patterns are saved in database [7]. Meanwhile, this method defines inability to detect the novel modern attacks. Another detection method is AIDS which is also referred as "behavior IDS" and is based on the concept of clearly describing a normal activity profile. Any anomaly will be treated abnormal behavior from normal profile [8].

4.2 Deployment IDS

As per deployment method, IDS is further partitioned as "host intrusion detection system (HIDS)" and "network intrusion detection system (NIDS)". HIDS is used only for single information host with the aim to monitor each and every activity on this host as well as scans for its suspicious activities and security policy violations. In contradiction to HIDS, NIDS is used on the network to protect every device in network by scanning for possible security violations and breaches [9, 10, 24].

4.3 Working of IDS

IDS acts as a security mechanism for continuously monitoring the network and host traffic to identify and detect any malicious behavior that breaches the network security policy to accommodate its integrity, availability, and confidentiality and provides alert about detected suspicious behavior to the network or host administrators.

Figure 3 shows NIDS deployment, where network switch is linked with NIDS along with port mirroring technology with the aim of mirroring every outgoing and incoming traffic of network to NIDS for operating regular monitoring to detect intruders. Some researchers also used NIDS between the network switch and firewall to allow each and every network traffic for passing through NIDS. IDS is also treated as effective system for the security of WSN and IoT and networks (i.e., LAN, AdHoc networks, etc.).

5 NIDS Models

These models can be described either on basis of ML-based or DL model methodologies which are described as follows:

5.1 ML Models

Branch of artificial intelligence contains each and every algorithm. ML also empowers machine to learn automatically using mathematical models from datasets [4]. And some of its main algorithms used for this domain are as follows:

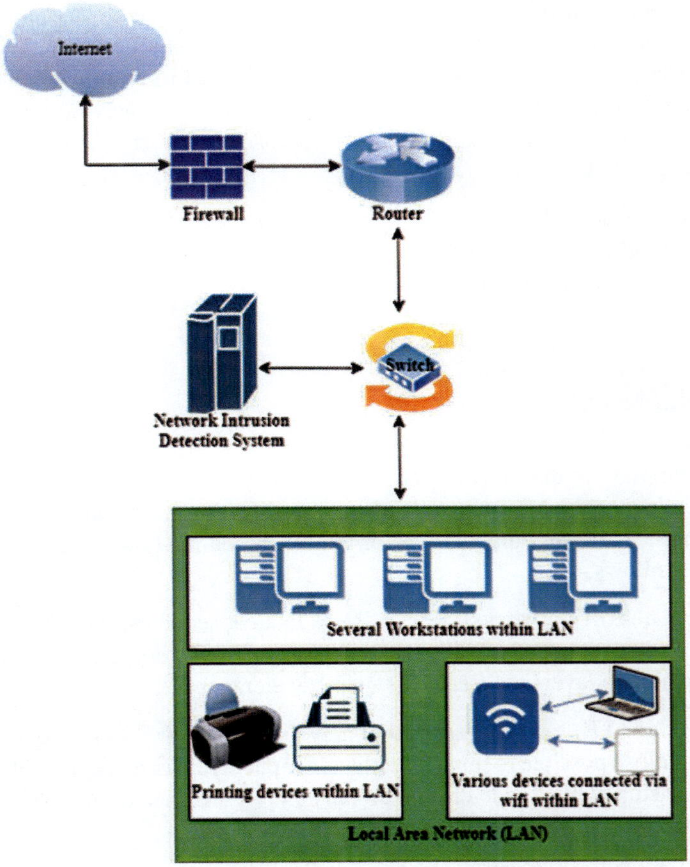

Fig. 3 NIDS deployment

Decision tree (DT). It represents as the key supervised algorithm by implementing the sequence of rules. In DT models, each branch defines a rule or decision, each node defines a feature or an attribute, while individual leaf defines a class label or a possible outcome. Most commonly used DT models are CART.

K-nearest neighbor (KNN). KNN is based purely on "feature similarity" for class prediction in a data sample. This algorithm determines a sample on the basis of its neighbors by evaluating its distance from adjacent neighbors. If the "k" value is very small, the model may be affected to over-fitting problem. While the very high value of "k" results in sample misclassification.

K-mean clustering. Clustering defines a process of partitioning data into groups (clusters) based upon similarity of group objects. K-mean clustering is centroid-based ML algorithm based on unsupervised learning, where K defines the count of centroids in a dataset. For data point assignment to a cluster.

Support vector machine (SVM). SVM works on the basis of max-margin partition hyperplane in n-dimensional feature space which can be used effectively for nonlinear and linear problems. Kernel function is used in case of nonlinear problems for generation of a high-dimensional feature space from low-dimensional input vector mapping. Max optimal marginal hyperplane is received using the support vectors sequentially.

Ensemble methods. In ensemble approach, a stronger classifier is generated from a voting algorithm is achieved by combining various weak classifiers made from training of several classifiers.

5.2 DL Models

DL is a subgroup of the ML that contains various hidden layers. DL approaches are more prominent than ML due to their important features self-learning ability from dataset and achieve an output. Some of the DL-based methodologies for NIDS [2, 13] are as follows:

Recurrent neural networks (RNN). RNN models sequence data from the feed-forward neural network capabilities by comprising of input, output, and hidden layers, where the hidden layers are treated as memory elements.

AutoEncoder (AE). AE refers to neural networks for finding similarity with output as much as close possible to input. Output and input layers in this approach are of the same dimension, while hidden layers dimension is lesser than of input layer. Several types of AE are variational, stacked, and sparse.

Deep belief network (DBN). DBN is created by deforming various RBM layers trailed by a softmax layer. In this, every node inside a layer is associated to all of remaining nodes in the preceding and adjacent layers [19].

Convolutional neural network (CNN). It defines DL framework, and it is precisely suitable when data is to be stored in array. It comprises pooling stack, an input layer, and layers for features extraction (i.e., convoluted layers) [17].

6 Standard Datasets

A detailed information of the dataset is provided in this section as well as their attack classes as shown in Table 2.

As defined in Table 2, this section defines information toward the prominent datasets used by various researchers for performance evaluation of their methodology [29–31].

Table 2 Comparison of various datasets

Dataset	Types of attacks	Year	Attacks
KDD Cup'99	4	1998	DoS, R2L, Probe, and U2R
Kyoto	2	2006	Unknown attacks and known attacks
NSL-KDD	4	2009	DoS, R2L, Probe, and U2R
UNSW-NB15	9	2015	DoS, Fuzzers, Worms, Generic, Backdoors, Reconnaissance, and port Scans, Shellcode, exploits
CIC-IDS2017	7	2017	DoS, DDoS, infiltration, Botnet, HeartBleed, Web, and Brute Force
CSE-CICIDS2018	7	2018	Web, infiltration, Botnet, Brute Force, DDoS, DoS, and HeartBleed

7 Recent Scopes, Challenges, and Future Scopes

7.1 Recent Scopes and Observations

It is evident from literature that ML is not as much convenient until the dataset is labeled and labeling of dataset is a time-consuming approach. Thereby, DL methods outperform ML methods performance in case of unlabeled larger datasets. Since, DL approaches extract and learn useful meaningful patterns from unprocessed datasets. Furthermore, NIDS model training directly affects the detection rate of intrusion. As DL approaches demand comprehensive computational resources, cloud-based and GPUs platforms have smoothened the DL-based technique employment. Based on this current literature review, it is closely observed that several researchers broadly centered on efficient IDS on the basis of DL tools as shown in Fig. 4. It is also observed that approximately 64% of the techniques in literature are mainly based on DL techniques, 20% technique used hybrid approach, while only 16% technique are ML-based. Thereby, potential researchers can work in the domain of deep learning as well as build a hybrid model for efficient network intrusion detection model.

Fig. 4 Techniques distribution

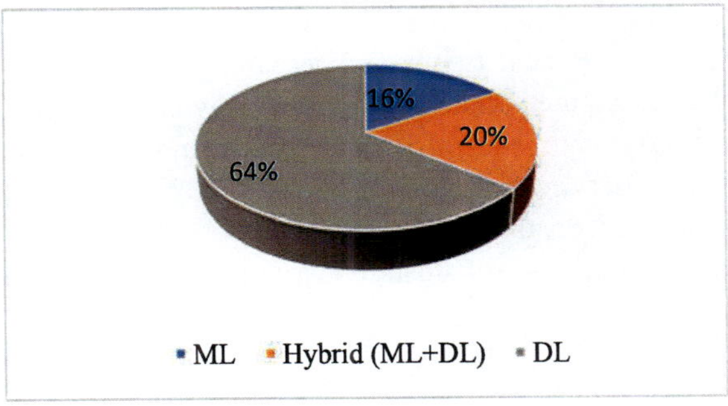

7.2 Research Challenges

Systematic dataset unavailability. As various proposed approaches were not up- to-the-mark for zero-day attack detection due to model untrained with enough patterns and attacks. Thereby, to build an effective model, it requires to be verified and tested by using the novel attacks. Since construction of IDS dataset is an expensive method which requires a lot of high expert's knowledge and resources.

Real-world environment lower performance. As stated by various recent studies various methodologies highly rely on old dataset testing without real-time environment testing in lab to examine its validity against novel attacks.

7.3 Future Scopes

Effective framework. Recent studies show network defense mechanism against intrusions incapabilities in zero-day attacks detection with more false alarm rate which can be overcome by using an efficient and intelligent framework for IDS by using an up-to-dated datasets.

Solutions to Unknown Attack. On the basis of literature studied, further research can be done in the direction of building a strong defense mechanism for intrusion detection which can efficiently handle unknown attacks in modern-day environment.

8 Conclusion

This review paper contributes insights of various NIDS models literature review on the basis on the DL and ML techniques to contribute the new practitioners and researchers with the current trends, knowledge, and progress of same domain. In this paper, initially, IDS and its various classification strategies are precise broadly on the basis of several papers. Sequentially, working of IDS and various ML and DL techniques suitable for effective IDS design are defined followed by standardized datasets. Recent trends and research gaps of this paper reveal that the usage of DL-based methodology can effectively enhance the effectiveness and performance of NIDS by means of FAR reduction and enhanced detection accuracy. Results of this paper reveal that approximately 80% papers result used in this literature were evaluated on the basis of DL-based techniques due to their strong model fitting capabilities and self-feature learning ability as compared to ML-based techniques.

References

1. Ghurab M, Ganhari G, Alshami F (2021) A detailed analysis of benchmark datasets for network intrusion detection system. Asian J Res Comput Sci 7(4):14–33
2. Alvaizadeh H, Alvaizadeh, Jang-Jaccard J (2022) Deep Q-learning based reinforcement learning approach for network intrusion detection. Computers 11(3):1–19
3. Thomas R, Pavithran D (2018) A survey of intrusion detection models based on NSL-KDD dataset. In: Proceedings of the 5th HCT information technology trends (ITT), IEEE, Dubai, pp 286–291
4. Sethi K, Venu-Madhav Y, Kumar R (2021) Attention based multi-agent intrusion detection system using reinforcement learning. J Inform Secur Appl 61:1–18
5. Caminero G, Lopez-Martin M, Carro B (2019) Adversarial environment reinforcement learning algorithm for intrusion detection. Comput Netw 159:96–109
6. Liu H, Lang B (2019) Machine learning and deep learning methods for intrusion detection systems: a survey. J Appl Sci 9(20):1–28
7. Karatas G, Demir O, Sahingoz OK (2020) Increasing the performance of machine learning-based IDSs on an imbalanced and up-to-date dataset. IEEE Access 8:32150–32162
8. Yan B, Han G (2018) Effective feature extraction via stacked sparse autoencoder to improve intrusion detection system. IEEE Access 6:421238–441248
9. Ali MH, Ismail A, Zolkipli MF (2018) A new intrusion detection system based on fast learning network and particle swarm optimization. IEEE Access 6:20255–20261
10. Yin C, Zhu Y, Fei J, He X (2017) A deep learning approach for intrusion detection using recurrent neural networks. IEEE Access 5:21954–21961
11. Xu C, Shen J, Du X, Zhang F (2018) An intrusion detection system using a deep neural network with gated recurrent units. IEEE Access 6:48697–48707
12. Naseer S, Saleem Y, Khalid S (2018) Enhanced network anomaly detection based on deep neural networks. IEEE Access 6:48231–48246
13. Shone N, Ngoc TN, Phai VD, Shi Q (2018) A deep learning approach to network intrusion detection. IEEE Trans Emerg Topics in Comput Intell 2(1):41–50
14. Al-Qatf M, Lasheng Y, Al-Habib M, Al-Sabahi K (2018) Deep learning approach combining sparse auto encoder with SVM for network intrusion detection. IEEE Access 6:52843–52856
15. Vinayakumar R, Alazab M, Soman P, Poornachandran K, Al-Nemrat A, Venkatraman S (2019) Deep learning approach for intelligent intrusion detection system. IEEE Access 7:41525–41550
16. Marir N, Wang H, Feng G, Li B, Jia M (2018) Distributed abnormal behavior detection approach based on deep belief network and ensemble SVM using spark. IEEE Access 6:59657–59671
17. Wei, P., Li, Y., Zhang, Z., Hu, T., Li, Z., & D. Liu.: An optimization method for intrusion detection classification model based on deep belief network. IEEE Access, 7, 87593- 87605 (2019).
18. Xiao Y, Xing C, Zhang T, Zhao Z (2019) An intrusion detection model based on feature reduction and convolutional neural networks. IEEE Access 7:42210–42219
19. Jiang K, Wang W, Wang A (2020) Network intrusion detection combined hybrid sampling with deep hierarchical network. IEEE Access 8:32464–32476
20. Gao X, Shan C, Hu C, Niu Z, Liu Z (2019) An adaptive ensemble machine learning model for intrusion detection. IEEE Access 7:82512–82521
21. Gupta K, Sharma KD, Gupta DK, Kumar A (2022) A tree classifier based network intrusion detection model for internet of medical things. Comput Electri Eng 102:1–20
22. Cao B, Li C, Song Y, Qin Y, Chen C (2022) Network intrusion detection model based on CNN and GRU. Appl Sci 12(9):1–27
23. Ullah I, Mahmoud HQ (2022) Design and development of RNN anamoly detection model for IoT networks. IEEE Access 10:62722–62750
24. Yao H, Fu D, Zhang P, Li M, Liu Y (2018) A novel multilevel semi-supervised machine learning framework for intrusion detection system. IEEE IoT J 6:1949–1959
25. Khan AF, Gumaei A, Derhab A, Hussain A (2019) A novel two-stage deep learning model for efficient network intrusion detection. IEEE Access 7:30373–30385

26. Malaiya KR, Kwon D, Suh CS, Kim H, Kim I, Kim J (2019) An empirical evaluation of deep learning for network anomaly detection. IEEE Access 7:140806–140817
27. Jia Y, Wang M, Wang Y (2018) Network intrusion detection algorithm based on deep neural network. J IET Inform Secur 13:48–53
28. Zhang X, Chen J, Zhou Y, Han L, Lin J (2019) A multiple-layer representation learning model for network-based attack detection. IEEE Access 7:91992–92008
29. Khraisat A, Gondal I, Vamplew P, Kamruzzaman J (2019) Survey of intrusion detection systems: techniques, datasets and challenges. Cybersecurity 2(20):1–22
30. DKA C, Papa JP, Lisboa CO, Munoz R, DVHC A (2019) Internet of things: a survey on machine learning-based intrusion detection approaches. Comput Netw 151:147–157
31. Ahmad Z, Khan S, Shiang WC, Abdullah J, Ahmad F (2021) Network intrusion detection system: a systematic study of machine learning and deep learning approaches. Trans Emerging Tel Tech 32:1–29
32. Wang Z (2018) Deep learning-based intrusion detection with adversaries. IEEE Access 6:3836–38384

Chapter 39
A Comprehensive Study on Online and Offline Evaluation of Recommendation System

Tamanna Sachdeva, Lalit Mohan Goyal, and Mamta Mittal

1 Introduction

Today, E-Commerce Web sites along with the advancements in technology have managed to make an electronic bridge to entertainment, comfort, and leisure for us [1]. We can get the products and services with just some clicks. With a huge number of products on these Web sites, we are getting so many options. Initially, leaving comments and reviews were the options given to us for taking customer's feedback and making future transactions better. Now, with the advancement of technology, E-Commerce Web sites are making efforts to know the user's interests by analyzing their feedback. A recommendation system (RS) makes recommendations to the users according to their preferences, such as feedback, ratings, browsing patterns, and purchasing choices [2]. Users and service providers both share the benefits of recommendation systems. Recommendation systems provide access to niche products that were less accessible otherwise.

As the recommendation systems are bringing big bucks in the businesses, its performance evaluation is mandatory to ensure that the recommendation systems are helping both the users and service providers [3].

Recommendation systems can be classified into content-based, collaborative filtering-based, knowledge-based, demographic, and hybrid recommendation systems.

T. Sachdeva (✉) · L. M. Goyal
Department of Computer Engineering, J.C. Bose University of Science and Technology, YMCA, Faridabad, India
e-mail: tamannasachdeva26@gmail.com

M. Mittal
Department of Data Analytics, Delhi Skill and Entrepreneurship University, New Delhi, India

A. Yadav et al. (eds.), *Proceedings of International Conference on Paradigms of Communication, Computing and Data Analytics*, Algorithms for Intelligent Systems,
https://doi.org/10.1007/978-981-99-4626-6_39

Content-based RS uses features of items to recommend items to the users depending on the similarity between previously liked and new items [4]. These RSs have very little information and no history of the users. Item profiling is the most common method in content-based RSs, these record the popularity of items and their features and use this information to recommend items to the users. This type of recommendation is mostly used in articles and newspaper recommendation systems.

Collaborative filtering-based RS uses user's history, their likes and dislikes, and maps these with other users to find similarities in their preferences [5]. The main idea behind this recommendation method is if two users like some common items then they are likely to like some of the same other products too.

Knowledge-based RSs recommend based on the specific domain knowledge. These map product features with user requirements and interests to analyze whether the product is useful for the user or not.

Demographic RSs use demographic information such as location, language, age, religion, and some useful parameters related to the users for constructing the recommendations [6].

Hybrid RSs are made by combining two or more techniques of the above recommendation systems. Hybrid RSs try to combine the advantages of individual RSs and give efficient results [7]. But these RSs are more computationally complex and may show conflicting results. Facebook's recommendation system is using a hybrid approach to making recommendations.

The rest of the article is structured as follows. Section 2 presents various challenges of recommendation systems, and then, Sect. 3 presents various metrics to evaluate the recommendation systems. Lastly, Sect. 4 concludes the study and discussed possible future approaches.

2 Challenges of Recommendation System

This section will present some basic challenges developers faces while designing and working with recommendation system. One of the main challenges is the sparsity problem that occurs in a case like when recommendation systems use rating systems to get feedback from users and know their preferences. Whereas users do not always spare time to fill out these rating forms or possibly leave some of the questions unanswered [8]. This behavior of users leads to sparse rating matrices which will decrease the performance of the recommendation system.

A cold start problem occurs in the recommendation system when the system does not have any information about a new user or new item in the system. In this case, RSs like collaborative filtering fail as these did not get initial information to make recommendations [4, 9]. Cold start problems can also occur when a recommendation system is newly started and does not have the required information for making recommendations [10]. Some solutions for this problem can be explicitly asking users some basic questions during joining the services, asking them to rate some products at the beginning, collecting some demographic details, and recommending accordingly. Shilling attack problem occurs when malicious users enter fake reviews and ratings for any items, either to increase or decrease the sale [1]. Synonymy problem occurs when similar items are stored in systems with different names, and recommendation systems get confused and cannot give good recommendations. To discard this, demographic filtering and automatic term expansion techniques can be used.

The grey-sheep problem refers to the situation where users with unique preferences and interests give some unique ratings or feedback that create outliers in the database and make it difficult to create accurate profiles, which will lead to inaccurate recommendations. One of the major concerns is the increasing size of the dataset due to the increment of user-item interaction. This is a major concern with collaborative filtering recommendation systems [11]. In today's era of big data, the large number of users and items is increasing in the system. Some other issues are plasticity and over-specialization problem.

3 Evaluation of Recommendation Systems

The purpose of the recommendation system is to generate recommendations according to the user's interest, and if the recommended items can keep users happy, then it is a good choice; otherwise, some modifications are needed in the algorithms to reach the user's satisfaction level of preferences [3]. Evaluation metrics can be divided based on the type of experiments in offline experiments, user studies, and online experiments [12]. Offline experiments use training and testing datasets to predict users' behavior and to evaluate recommendation systems. This does not involve human interaction. User study involves fewer users using the system in a controlled manner, answering about the experience. Online experiments are the most expensive and large-scale method that involves real-time users. Offline experiments ignore so many factors which are covered in online experiments such as novelty, diversity, and churn.

Various evaluation metrics for recommendation systems are discussed in the following subsections, which will help in deciding the suitable evaluation metrics for different recommendation systems based on the capability of metrics.

3.1 Data Partitioning-Based Metrics

Train/Test. Train/test is a way of offline metrics of evaluation recommendation systems. Basic idea is to use the behavioral study of prior users and to predict the actions of new users. Here, data is divided into a training set and a testing set. The system is trained over training data and then analyzed over testing data. Rating predictions of the trained recommendation system are compared with the test dataset's ratings. If the recommended ratings are closed enough to actual ratings, the quality of the recommendation system is good; otherwise, the approach is to be modified.

k-Fold Cross-Validation. It is an improvement over the train/test method to evaluate recommendation systems. Here, training data is divided into *k* training sets, and train *k* models based on *k* training sets independently. Each fold then predicts the ratings, and this method measures the scores given by all the folds and then finds the average of their results. This requires more computation power, but it will not bind up to one training set. The main advantage is it does not end up overfitting to a single training set.

Both metrics can test the power of recommendation systems to predict how users rate the movies that they have already seen, but this is not the only goal of the recommendation systems as these also must predict new items according to the user's interest.

3.2 Prediction Evaluation Metrics

Prediction evaluation metrics are the metrics that evaluate the prediction capabilities of a recommendation system. These are usually more suitable for offline experiments. This section presents various prediction evaluation metrics.

Mean Absolute Error (MAE). MAE is the most straightforward metric.

$$\sum_{i=1}^{n} |y_i - x_i|/n \tag{1}$$

In Eq. 1, n is the number of ratings in the test set that must be evaluated, y is the rating value anticipated by the system, and x is the actual rating given by the user [13]. MAE just finds out the deviation between the actual rating and the predicted rating and then takes the average by adding these differences of all n ratings and then dividing by n. Less the value of MAE better the recommendation system [14].

Root Mean Square Error (RMSE). RMSE is one more way to measure the accuracy of recommendation systems. RMSE is a more popular metric than MAE, and one of the many reasons is that it penalizes you according to the difference between actual and predicted rating [15, 16]. When the difference is more, it penalizes you more than when the difference is less, i.e., if the predicted rating is close to the actual rating.

$$\sqrt{\sum_{i=1}^{n} (y_i - x_i)^2/n} \tag{2}$$

Equation 2 sums up the squares of the rating prediction errors instead of summing up the absolute values of each rating prediction error [15], adding squares that always have a positive value lead to a higher penalty. After that square root is taken to get back to a number that makes sense.

Coverage. Coverage is the percentage of possible recommendations that a system can generate. Coverage also gives a sense of how quickly new items start appearing in the recommendations.

Back in 2006, Netflix offered a prize of a million dollars to improve its RMSE score by at least 10%. But even after declaring a winner based on RMSE Netflix does not use the winning idea as they have realized that RMSE is not that much of use in practical life. Now, a new idea is to look for how well recommendations have been presented to the users and in what order. So, the focus went to the top-n recommendations.

3.3 Ranking-Based/Top-N Recommendations

When users open their E-Commerce account, a top-n-item list of recommendations is presented to the users, which is usually intended to keep their attraction [17, 18]. Some of the famous top-n recommendation metrics are discussed in the following subsections.

Hit Rate. Hit rate records the number of times the recommendations have drawn the user's attention. Recommendation systems generate recommendations, and let us suppose the top-n recommendations have been shown to the user. If a user rates a recommended item, then it is called a hit. It means you have recommended something worthy of the user's interest. Now, to find the overall hit rate, add all the hits for all the users in the test set and divide it by the number of users. Measuring the hit rate is a bit tricky as it is not ideal to find the hit rate on the same training data on which the system is trained which will give a hundred percent accuracy. So, the next method of leave-one-out cross-validation will provide a better understanding of the case.

Leave-One-Out Cross-Validation. Here, from the top-n recommendations generated for each user in the training set, it removes one of those items [19]. Then, the recommender system is checked, whether it recommends that item that was left out in the top-n results or not. It is not ideal if there is a small dataset. Leave-one-out cross-validation is a more user-centered metric.

Average Reciprocal Hit Rate. It focuses on where in the top-n list hit is achieved. It adds up the reciprocal rank of each hit. So, getting hit on items higher on the top of n list is better than getting hit on the items comparatively below in the list.

Rating Hit Rate. Idea is to break down the hit rate according to the rating score. It will tell you how good of a rating the recommended items get.

3.4 Usage-Based Metrics

Sometimes, recommendations can be dependent on each other, and one's presence can assess another's presence or absence. Items can be of the following categories such as positively selected, negatively selected, positively left, and negatively left. In the offline evaluation of the recommendation systems, one assumption is forced that the discarded items would not interest the user even if they were recommended. But this assumption is dangerous for a good recommendation system as some non-recommended items may interest users such as some new items whose existence is not known to the user. The various evaluation methods used for the evaluation of this set of items present in the recommendation list are discussed in the following subsections.

Precision. Precision signifies the fraction of relevant items among all the recommended items to a user. It also tries to keep non-relevant items away from the users [13].

$$\text{Precision} = \frac{p}{p+q} \tag{3}$$

In Eq. 3, let "p" represents the count of true positive items that are originally recommended and successfully retrieved for recommendations by the RS. "q" represents the number of items not successfully suggested by the RS, although they are labeled as recommended. In Eq. 4, "r" represents the number of disqualified items that are recommended by the RS. The true negative values "s" refer to the number of items that are labeled and retrieved as "not recommended".

Recall. The recall represents the number of relevant recommended items to the total number of items that should be recommended [3, 13]. In other words, RS can generate all relevant items.

$$\text{Recall} = \frac{p}{p+r} \tag{4}$$

Precision and recall show opposite behaviors, and as with a greater value of recall, the chance of removal of both useful and useless items increases [20]. Whereas, with a greater value of precision, the chance of removal of both useful and useless items decreases. So, these two matrices work at the expense of each other. But an efficient recommendation system tries to optimize both values. Only by combining these two metrics can a complete picture be achieved. A metric called F-Measure is an amalgamation of both precision and recall.

F-Measure. F-measure is derived from and shows the properties of both precision and recall. It is the harmonic mean of precision and recall [21, 22]. It helps in choosing two alternatives, one with higher precision and the other with higher recall value.

$$F\text{-measure}(\beta) = \frac{\text{Precision}^{*}\text{Recall}}{(1-\beta)^{*}\text{Precision} + \beta^{*}\text{Recall}} \tag{5}$$

The most usual version of F-measure is F1 by considering beta = 1/2 in Eq. 5. It is therefore also known as the harmonic mean of precision and recall as seen in Eq. 6 [13].

$$F\text{–measure} = \frac{2^{*}\text{Precision}^{*}\text{Recall}}{\text{Precision} + \text{Recall}} \tag{6}$$

3.5 Diversity Metrics

This section presents metrics that are based on the diversity of the results. Usually, these metrics are qualitative. Sometimes some recommendations lead to a whole new experience for the users. Some of the diversity-based metrics are discussed in the following subsection.

Diversity. Diversity is a measure of how broad a variety of items the recommender system is generating. Similarity matrices between the items are used by the recommendation systems to measure the diversity. The average of the similarity scores (S) of every possible pair in a list of top-n recommendations can give the idea of how similar the recommended items in the list are to each other [23]. Diversity is the opposite of average similarity, so diversity can be calculated by subtracting average similarity from 1.

$$D = 1 - S \tag{7}$$

In Eq. 7, D is diversity, and S is average similarity between recommendation systems. The point to consider is that diversity is not always a good sign for a recommendation system. High diversity can indicate that the recommendation system is just recommending random things which are bad recommendation activity. So, it is advised to also consider other metrics along with diversity that can measure the quality of the recommendations as well.

Novelty. Novelty is a measure of the popularity of the recommended items [24]. Novelty should be treated with extra care as in terms of achieving good novelty a recommendation system might only recommend popular items which might not even relate to the user's interest.

Churn. Churn measures the sensitivity of the recommendation system to the new user behavior. If on rating a new item, a user's recommendation list changes; then it means that the recommendation system has a high value of churn [25]. It is good to change the recommendation list if users are not clicking on some items. Like novelty and diversity, the high value of churn is not a favorable case as a high value of churn can be achieved by recommending random items which is not a good recommendation practice.

Serendipity. It is a measure of how rewarding the recommendations are to the users. It can be understood by the example is based on a user's interest some movies were recommended to the user and now from those recommendations, users have found a new interest like a new favorite actor discovered from the movie recommended based on previous interest [26, 27].

Table 1 shows a comparison between various metrics based on their recommendation approach, methods of evaluation, and the area that they are focusing on for the recommendation. The approach part shows what type of algorithmic approach the metrics are using, machine learning trains the system and then finds the result on the test dataset. Information retrieval shows finding patterns to justify the results. Hybrid qualitative indicates human and machine interaction.

Table 2 presents some of the studies and the assessment metrics these have used in the evaluation of their proposed recommendation system.

Table 1 Comparison between metrices

Metrics	Category	Methods	Approach
Train and test	Data partitioning	Offline	Machine learning
k-fold cross-validation	Data partitioning	Offline	Machine learning
MAE	Prediction-based	Offline	Machine learning
RMSE	Prediction-based	Offline	Machine learning
Coverage	Prediction-based	User study	Information retrieval
Hit rate	Ranking-based	Online\Offline Experiment	Information retrieval
Leave-one-out cross-validation	Ranking-based	Online\Offline Experiment	Information retrieval
Average RATING HIT RATE	Ranking-based	Online\Offline Experiment	Information retrieval
Rating hit rate	Ranking-based	Online\Offline Experiment	Information retrieval
Precision	Usage-based	Offline	Information retrieval
Recall	Usage-based	Offline	Information retrieval
F-measure	Usage-based	Offline	Information retrieval
Diversity	Diversity metrics	Offline	Hybrid qualitative
Novelty	Diversity metrics	User study	Hybrid qualitative
Churn	Diversity metrics	User study	Hybrid qualitative
Serendipity	Diversity metrics	User study	Hybrid qualitative

Table 2 Assessment metrics in recommendation systems

Paper	Year	Assessment metrics
[22]	2020	Hit ratio, F1-score
[21]	2017	F1-score
[15]	2019	Precision, Recall, F1-score
[16]	2020	Precision, Recall, Hit ratio
[28]	2021	Support, Confidence, Lift, Count
[14]	2021	MAE
[29]	2021	MAE, Accuracy
[30]	2020	RMSE
[6]	2020	Diversity, Serendipity
[31]	2019	MAE, RMSE
[32]	2019	MAE, RMSE
[33]	2018	MAE, Precision, Recall
[9]	2018	F-measure, Mean average precision,
[34]	2018	Precision, Recall, F-measure, MAE and RMSE
[35]	2018	Top—n Recommendation
[8]	2001	MAE
[36]	2016	MAE, Serendipity

4 Conclusion

Evaluation of the quality of recommendation systems is important to engage users by keeping up with their interest and ultimately improve the overall performance of the E-Commerce business. This paper reviewed the need for evaluation metrics for the performance of recommendation systems, various challenges of recommendation systems, and various performance evaluation approaches for recommendation systems. Offline metrics are helpful when you have static and less amount of data, a user study is applicable with data where small amounts of user interaction are required, and online experiments are necessary where large amounts of real-time data are present. This article can paint a detailed picture to the novel researchers to get a comprehensive idea of evaluation metrics for recommendation systems, their necessity, and their restrictions. After compiling the complete study, it is safe to say different recommendation systems need different types of evaluation to check their quality and it is necessary to understand those defining features.

References

1. Roy D, Dutta M (2022) A systematic review and research perspective on recommender systems. J Big Data 9(1)
2. Malik S, Rana A, Bansal M (2020) A survey of recommendation systems. Inf Resour Manag J 33(4):53–73
3. Shani G, Gunawardana A (2011) Evaluating recommendation systems. recommender systems handbook, Springer, New York, pp 257–297
4. Kumar B, Sharma N (2016) Approaches, issues and challenges in recommender systems: a systematic review. Indian J Sci Technol 9(47)
5. Sun C, Gao R, Xi H (2014) Big data based retail recommender system of non E-commerce. In: 5th International conference on computing communication and networking technologies, ICCCNT 2014, IEEE, Hefei, China, pp 1–7
6. Alamdari PM, Navimipour NJ, Hosseinzadeh M, Safaei AA, Darwesh A (2020) A systematic study on the recommender systems in the E-commerce. IEEE Access 8:115694–115716
7. Fayyaz Z, Ebrahimian M, Nawara D, Ibrahim A, Kashef R (2020) Recommendation systems: algorithms, challenges, metrics, and business opportunities. Appl Sci 10(21):1–20
8. Sarwar B, Karypis G, Konstan J, Riedl J (2001) Item-based collaborative filtering recommendation algorithms. In: Proceedings of the 10th international conference on world wide web, WWW 2001, ACM, Hong Kong, China, pp 285–295
9. Gaikwad RS, Udmale SS, Sambhe VK (2020) E-commerce recommendation system using improved probabilistic model. In: Lecture notes in networks and systems, pp 277–284
10. Ramesh R, Priyadarshini N, Yuvaraju BN (2018) E-commerce recommendation system : problems and solutions. Int Res J Eng Technol 05(04):1161–1163
11. Mishra N, Chaturvedi S, Vij A, Tripathi S (2021) Research problems in recommender systems. J Phys: Conf Ser 1717(1):012002
12. Alslaity A, Tran T (2021) Evaluating recommender systems: a systemized quantitative survey. Int J Intell Inf Technol 17(2):25–45
13. Kuanr M, Mohapatra P (2021) Assessment methods for evaluation of recommender systems: a survey. Found Comput Decision Sci 46(4):393–421
14. Kishore GK, Babu DS (2017) Recommender system based on customer behaviour for retail stores. IOSR J Comput Eng 19(03):06–17
15. Chen H, Dai X, Cai H, Zhang W, Wang X, Tang R, Zhang Y, Yu Y (2019) Large-scale interactive recommendation with tree-structured policy gradient. In: 33rd AAAI conference on artificial intelligence, AAAI 2019, 31st innovative applications of artificial intelligence conference, IAAI 2019 and the 9th AAAI symposium on educational advances in artificial intelligence, EAAI 2019, pp 3312–3320
16. Liang H (2022) DRprofiling: deep reinforcement user profiling for recommendations in heterogenous information networks. IEEE Trans Knowl Data Eng 34(4):1723–1734
17. Chang S, Zhang Y, Tang J, Yin D, Chang Y, Hasegawa-Johnson MA, Huang TS (2017) Streaming recommender systems. In: 26th International World wide web conference, WWW 2017, ACM, Perth Australia, pp 381–389
18. Zolaktaf Z, Babanezhad R, Pottinger R (2018) A generic top-n recommendation framework for trading-off accuracy, novelty, and coverage. In: Proceedings—IEEE 34th international conference on data engineering, ICDE 2018, IEEE, pp 149–160
19. Cremonesi P, Turrin R, Lentini E, Matteucci M (2008) An evaluation methodology for collaborative recommender systems. In: Proceedings 4th international conference on automated solutions for cross media content and multi-channel distribution, Axmedis 2008, ACM, Washington, DC, United States, pp 224–231
20. Baeza-Yates R, Riberio-Neto B (1999) Modern information retrieval. ACM Press, New York
21. Hu B, Shi C, Liu J, Hu B, Shi C, Liu J, Recommendation P, Learning R (2017) Playlist recommendation based on reinforcement learning. In: International conference on intelligence science (ICIS), Oct 2017, Springer Nature, Shanghai, China, pp172–182

22. Wang Y (2020) A hybrid recommendation for music based on reinforcement learning. Adv Knowl Discovery and Data Mining 12084:91–103
23. Joeran B, Breitinger C, Gipp B, Langer S (2018) Research-paper recommender systems: a literature survey. Int J Digit Libr 17(4):305–338
24. Vinagre J, Jorge AM, Gama J (2014) Evaluation of recommender systems in streaming environments. In: Workshop on 'Recommender systems evaluation: dimensions and design' (REDD 2014), RecSys 2014, Silicon Valley, United States
25. Wang YF, Chiang DA, Hsu MH, Lin CJ, Lin IL (2009) A recommender system to avoid customer churn: a case study. Expert Syst Appl 36(4):8071–8075
26. Kotkov D, Wang S, Veijalainen J (2016) A survey of serendipity in recommender systems. Knowl-Based Syst 111:180–192
27. Ziarani RJ, Ravanmehr R (2021) Serendipity in recommender systems: a systematic literature review. J Comput Sci Technol 36(2):375–396
28. Bellini P, Palesi LAI, Nesi P, Pantaleo G (2022) Multi clustering recommendation system for fashion retail. Multimedia Tools and Appl 82
29. Alabdulrahman R, Viktor H (2020) Catering for unique tastes: targeting grey-sheep users recommender systems through one-class machine learning. Expert Syst With Appl 166:114061
30. Pratama BY, Budi I, Yuliawati A (2020) Product recommendation in offline retail industry by using collaborative filtering. Int J Adv Comput Sci Appl 11(9):635–643
31. Jiang L, Cheng Y, Yang L, Li J, Yan H, Wang X (2019) A trust-based collaborative filtering algorithm for E-commerce recommendation system. J Ambient Intell Humaniz Comput 10(8):3023–3034
32. Hong-Xia W (2019) An improved collaborative filtering recommendation algorithm. In: 4th IEEE international conference on big data analytics, ICBDA 2019, IEEE, Suzhou, China, pp 431–435
33. Choi YK, Kim SK (2018) A recommendation system for repetitively purchasing items in e-commerce based on collaborative filtering and association rules. J Internet Technol 19:1691–1698
34. Hussien FTA, Rahma AMS, Abdulwahab HB (2021) An e-commerce recommendation system based on dynamic analysis of customer behavior. Sustainability 13(19):10786
35. Wang W, Yin H, Huang Z, Wang Q, Du X, Nguyen QVH (2018) Streaming ranking based recommender systems. In: 41st International ACM SIGIR conference on research and development in information retrieval, SIGIR 2018, ACM, Ann Arbor, MI, United States, pp 525–534
36. Samosir J, Indrawan-Santiago M, Haghighi PD (2016) An evaluation of data stream processing systems for data driven applications. Proc Comput Sci 80:439–449

Chapter 40
Autonomous Delivery Vehicle Using Raspberry Pi and Computer Vision

Vijay Ravindran, S. Chandrika, Ram Prakash Ponraj, C. Krishnakumar, S. Devadharshini, and S. Lakshmi

1 Introduction

Autonomous vehicles (also known as driverless vehicles) may move through their environment without any help from a human operator by using sensors such as radar, GPS, computer vision, and others [1]. There are many issues with conventional vehicles, like violations of rules, continuous working hours, and poor judgment, which lead to several fatal vehicle accidents. Several researchers [2] have made significant progress in the previous decade toward developing completely autonomous vehicles and robots. In light of these considerations, we have developed, using a Raspberry Pi and computer vision, a powered delivery vehicle with autonomous driving capabilities. The live footage captured by the vehicle's aerial cameras is an invaluable addition to the investigation. Once the system determines the appropriate steering angle, the vehicle is maneuvered in that direction. Training the automobile to steer with only camera input is the focus of this study. Moreover, the authorities are kept up-to-date on the vehicle's whereabouts via a screen. But figuring out how to build obstacle avoidance into the system is far outside the purview of this investigation. For the software part, we'll be making use of the modularity concept, which entails storing the instructions for each individual operation in its own file and having a single module oversee the entire operation, as depicted in Fig. 1. It's functionally equivalent to the idea of a class in programming languages like Python or Java, and it may be recycled across a wide range of different tasks.

V. Ravindran (✉) · S. Chandrika · R. P. Ponraj · C. Krishnakumar · S. Devadharshini · S. Lakshmi
Electrical and Electronics Engineering, Saranathan College of Engineering, Tiruchirappalli, TN 620012, India
e-mail: vijay-eee@saranathan.ac.in

A. Yadav et al. (eds.), *Proceedings of International Conference on Paradigms of Communication, Computing and Data Analytics*, Algorithms for Intelligent Systems,
https://doi.org/10.1007/978-981-99-4626-6_40

Fig. 1 Software pipeline

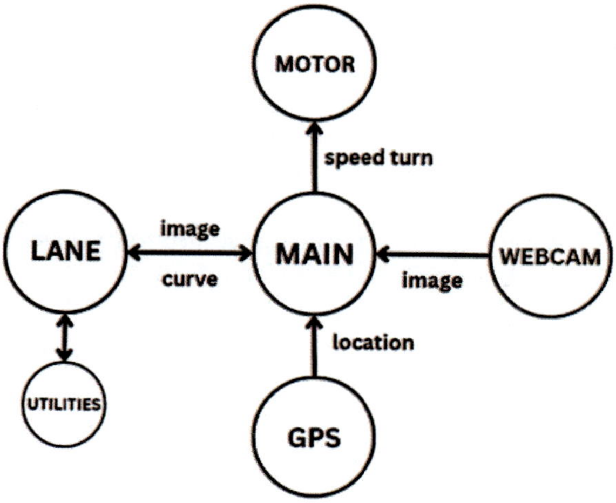

The purpose of this chapter is:

To develop an inexpensive vehicle with positive simulation results.

To develop an end-to-end computer vision-assisted algorithm for the self-driving vehicle prototype.

Communicate the location of the vehicle in real time via a map.

The remaining sections of this chapters are structured as follows; the relevant works regarding the proposed study are discussed in Sect. 2. Section 3 details the various hardware and software utilized for this study. The software framework methodology is explained in Sect. 4 with results, followed by hardware connection details in Sects. 5 and 6 and finally the conclusion and future scope are drafted in Sect. 7.

2 Related Works

Many techniques are available for using tandem for lane detection. The authors [3] proposed using the Haar Cascade Classifier method; an autonomous vehicle's on-board computer can recognize and respond to traffic signs. Research utilizing the BFD 1000-tracking sensor [4] for both lane detection and object detection supplies the required data to the vehicle, instead of using an open CV, which requires capturing and training pictures.

A system that uses a stereo camera to identify lanes and follow them is described [5]. It depends on the discrete Kalman filter and a Clothoid model of the highway. The program can also use super-pixel image segmentation to identify specific items in a given picture, making it particularly well-suited for use on a Tablet. They have found that their technique works for Computer Vision, Sensor Fusion, Deep Learning, Path

Planning, and Actuators [6]. These are just a few of the core ideas that must be understood before a self-driving vehicle can be put into action.

The study of computer vision sheds light on how machines might be programmed to extract data from digital photos and movies. From an engineering standpoint, it's utilized to mimic how a human might perform a task automatically [7]. This chapter proposes a system that can read traffic signs and autonomously steer, brake, and perform other commands [8]. The system includes a Raspberry Pi 3 CPU and a web camera, which record video and turn it into a sequence of frames automatically. These frames are then analyzed by the proposed algorithm in Open CV to identify the road sign and direct the vehicle [9].

Sensor fusion is the process of combining several types of sensory input or data from different sources to provide more reliable results than would be possible if each source was used separately [10]. While AI and DL are the backbones of their solutions, they can also do sensor fusion, localization using high-definition maps, and strategic planning [11]. The pixel summation has been well-defined lane markings and a consistent road texture, making lane-position detection a simple task [12].

Detecting all edges in the road image is important since each given edge could be a potential road line. This necessitates, first and foremost, the selection of a trustworthy and precise edge detector using smart IoT [13, 14]. There is a possibility of failure when using the contour tracing method since not all obstacle shapes will provide a valid contour. It is possible to extract noisy contours in some situations [15]. Studies on monitoring systems through IoT [16] and the Raspberry Pi using WSN [17] were also reliable works in the related field.

3 Components Used and Its Description

The hardware system consists of various components which are mentioned in Table 1.

Table 1 Hardware essential components

Hardware components	Type-range
Raspberry pi camera	1080 @ 30fps, 720p @ 60 fps
Raspberry pi 3	model B
Supply for pi board	5 V
12 V Supply for driver	12 V
Motors	DC
L298N	Motor driver module
Jumper wires	As required
Display screen	7 inch
GPS module	NEO-6 M

3.1 Raspberry Pi

Raspberry Pi is the name given to a single-board computer that is about the size of a credit card. The Model B+ , Model A+ , Model B, Model A, and the compute Module are the five Raspberry Pi versions that are currently on the market. Because of its wholly different form factor, the compute module cannot be used independently.

3.2 Pi Camera

The Pi camera module 3 comes with the Raspberry Pi. In addition to taking high-definition images, the Pi camera module can also record high-definition video. The Camera Module 3 is constructed around the IMX708.

3.3 OpenCV

Intel's OpenCV (Open-Source Computer Vision Library) is a free, cross-platform computer vision library with excellent image processing and matrix calculation capabilities and an algorithm that has been optimized for the Intel instruction set. When coupled with libraries like NumPy, Python can manage the OpenCV arrays needed for analysis. Vector space and arithmetic operations are used to identify an image pattern and its many features.

3.4 Raspbian OS

When compared to other Raspberry Pi OS like Arch, RISC OS, Plan 9, or Raspbian, Raspbian is the clear winner because of its user-friendliness, attractive design, an extensive collection of preinstalled apps, and optimization for the Pi's hardware.

3.5 Pixel Summation

Pixel summation, as shown in Fig. 2, is utilized, which essentially adds up all of the pixels in a given column. In this case, black is represented by 0 and white by 255. The range of possible values is 0 to 255 because 8-bit integers are being used.

Since it is unsigned, the range is 0 to 255. If signed, the values would be -127 to 127. A simple addition of all the white values in that column yields 1275. This will help us figure out if and for how long the machine needs to make a left, right, or straight turn.

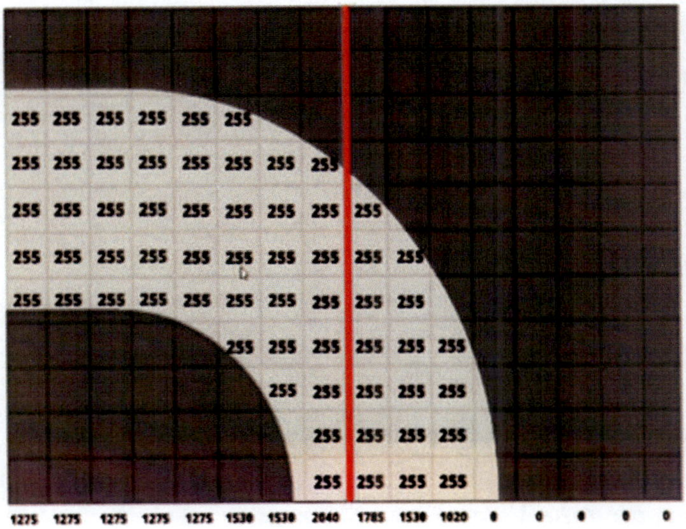

Fig. 2 Pixel summation: 2^8 = 256 values

4 Software Pipeline Methodology

Figure 3 depicts the software pipeline flow of all the steps to be carried out, which are discussed in this section. Each of these modules and its relevant pictures explains the Software pipeline methodology respectively.

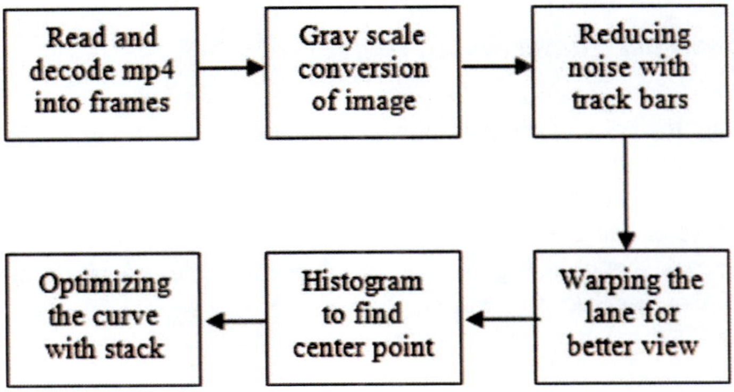

Fig. 3 Software pipeline flow

Fig. 4 A frame from the video file captured via webcam

4.1 Decoding the Video File into Frames

Initially, the coding is done on PyCharm for easier debugging and later on imported to raspbian os. An mp4 is uploaded as shown in Fig. 4.

Both feature-based techniques and model-based techniques have been used historically for lane detection. The feature-based approach identifies lanes in road photos by combining low-level information like painted lines, lane borders, etc. Therefore, this method only works on carefully analyzed roads with clearly painted lines or sturdy lane boundaries.

4.2 Grayscale Conversion of Image

Determining the color of the part on which the vehicle is standing by determining the proper upper and lower ranges to extract from the data is the first step in the algorithm and the one that carries the most weight.

With these coordinates, a binary representation of the currently visible area is constructed, as shown in Fig. 5. This is achieved by the utilization of the pixel simulation that was discussed earlier.

4.3 Reducing Noise by Altering Track Bars

The grayscale conversion introduces obvious variations and noise into the original image. The optimal, noise-free output can be achieved by modifying the preset or preprogrammed track bars as shown in Fig. 6.

The optimal noise free gray scale image is depicted in Fig. 7 which is based on track bars adjustment according to width and height.

Fig. 5 Grayscale image

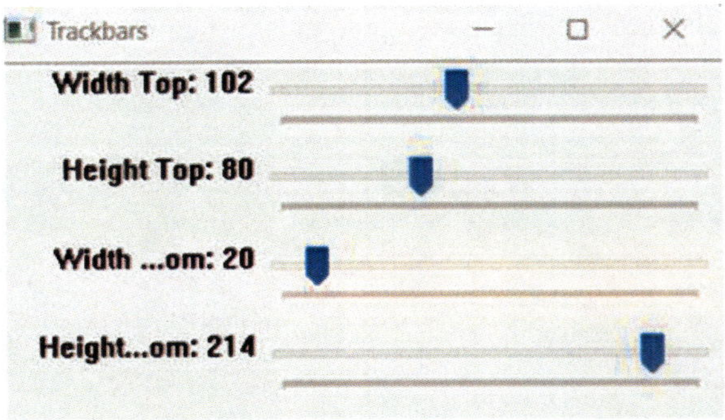

Fig. 6 Track bars for adjusting noise

Fig. 7 Noise-free gray scale images

5 Proposed Method and Hardware Implementations

5.1 Warping the Lane to Get a Better Top View

Warping, also known as perspective correction is the process of digitally modifying an image to the point where any shapes depicted in the image have been greatly warped. Thus, the motion of each pixel between the two pictures must be detailed. It would be possible to extrapolate the data provided for the control pixels to determine the motion of the remaining pixels. This is carried out here by using track bars and mapping the positions once again, as shown in Fig. 8.

A skewed photograph can be transformed to give the impression that it was shot straight on by tilting the image 90 degrees called the "bird's eye", Fig. 9. The purpose of doing so is to get a more accurate measurement of the curve. Hence, warping is necessary. We also provide manual warp point adjustment track bars.

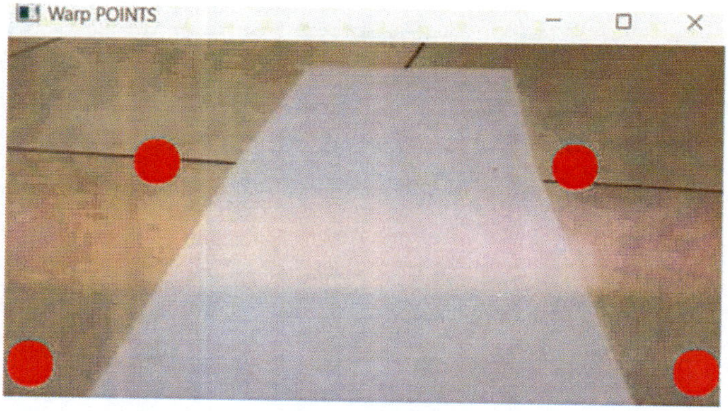

Fig. 8 Points mapped for wrapping

Fig. 9 Bird's eye view

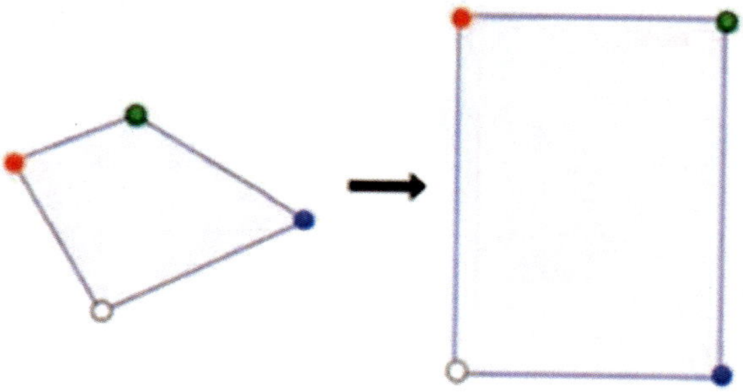

5.2 Histogram to Find the Center Point

A histogram displays, along the vertical axis, the fraction of pixels within an image that have a given intensity or color value. When more pixels are to the right of the center line, we will assume a rightward curve as depicted in Fig. 10a and vice versa, refer to Fig. 10b. Assuming a straight line, when the amount of pixels going left is equal to the picture in the right, as depicted in Fig. 10c, the problem arises when the image we obtained is off-center, this idea will not work. When the image is not in the center, the algorithm will make a mistake since it will read more pixels on the left as indicating a left curve when in fact the path will be straight as depicted by Fig. 10d and the adjusted image in Fig. 10e.

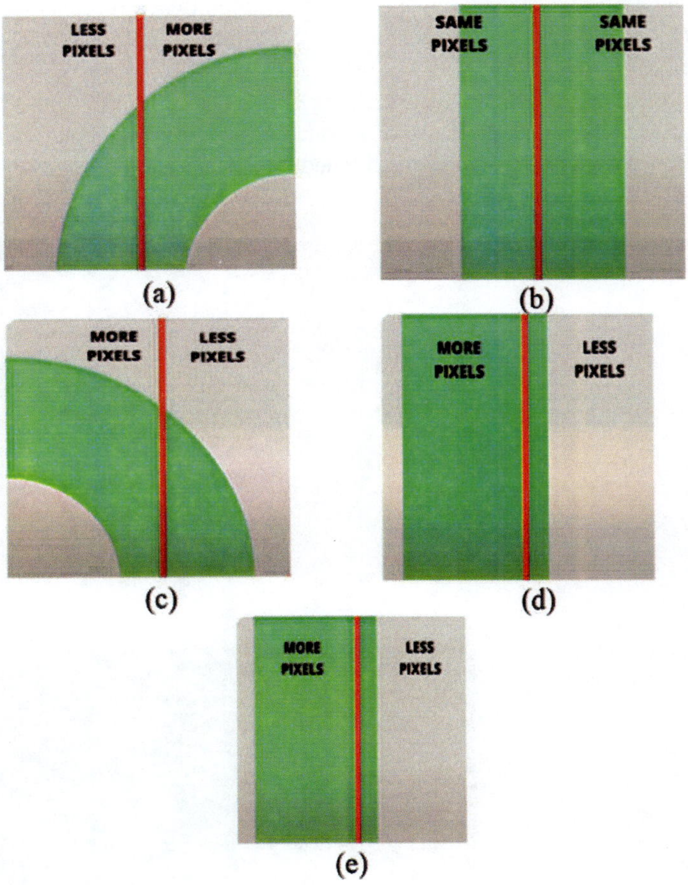

Fig. 10 **a** Increasing pixel density on the right and decreasing pixel density on the left defines a right curve. **b** Equal amount of pixels on both sides defines a straight line (**c**–**e**). Increasing pixel density on the left and decreasing pixel density on the right defines a left curve (**d**–**e**) Possibility of error when line of vision is not centered

5.3 *Optimizing the Curve with a Stack*

Assuming a focal point of 240, we get a curve value of -13 if we subtract that number from 227. The negative sign and the value 12 both indicate that the curve is steeper on the left side than on the right. Applying the histogram method to the bottom, 25% of the image rather than the whole thing is not to average the pixels in the upper 3/4th of the image because we are only interested in the average of the base. To do this, we can take the input value of the region option into account when deciding whether or not to average the entire image.

If the region is set to 1, the entire image will be averaged, whereas setting it to 4 will average only the bottom fourth of the image. This suggests that the true geographic center of our image is 278 rather than 240. The middle figure is taken and the calculated average is detected at −51. The warped image reveals that the curve has a high intensity as depicted in Fig. 11, confirming that the second procedure yields the best results.

The above-mentioned features are arranged in a stacked array, and a number is added to reflect the path's direction and degree of departure as depicted in Fig. 12, so that the curve may be calculated and tuned more accurately.

Fig. 11 Increase in intensity of the histogram

Fig. 12 Number added to reflect the path's direction and degree

6 Hardware Configuration

Four independent motors power each of the chassi's wheels. The L298N motor driver IC is used to power the motors. The left front and rear wheels turn at the same speed, and vice versa for the right front and rear wheels.

The block diagram connection is represented in Fig. 13. This means that at any given time, L298N will send identical digital input to both sets of motors. When both side wheels revolve in the same direction at the same speed, it aids the vehicle's forward and reverse motions. Turns are made by changing the direction of rotation of the left wheels relative to the right wheels. It is hardwired to the motors, and the jumper wires needed to link it to the Raspberry Pi are pulled from L298N. To power the motors, batteries of the appropriate voltage take up the balance of the space on the lower level. So that the Raspberry Pi can operate the motor hooked up to its GPIOs via jumper wires. In order to establish a wireless connection, a Wi-Fi dongle is plugged into the Raspberry Pi's USB port.

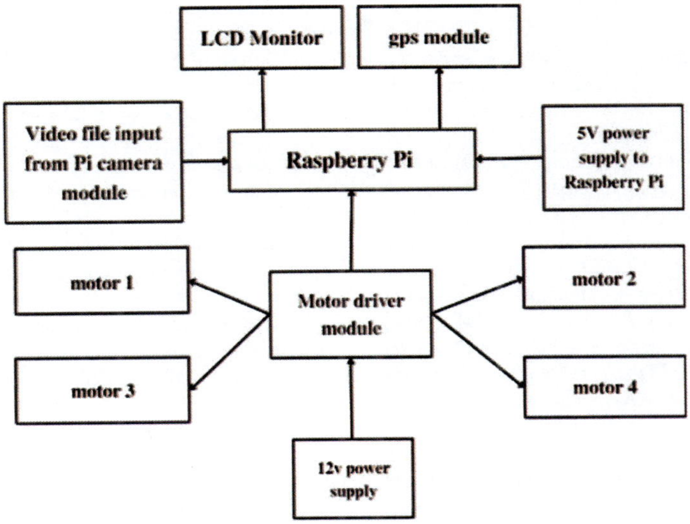

Fig. 13 Hardware connection-block diagram

7 Conclusion

In this chapter, a small autonomous vehicle has been developed and implemented. The vehicle's location on the track was determined using a simulated GPS system, and navigation was accomplished using a lane-following technique that required the detection and tracking of lane markings. This low-cost autonomous vehicle contains a camera mounted on top of the vehicle that captures the image and the Raspberry Pi helps in determining the appropriate steering angle, while also making the location services accessible. The proposed technology can eradicate the most common cause of human error. A wide variety of options for progress are left open in this work. It has room for expansion to include several useful features. Future directions for our study include employing cameras and sensors to spot potholes and other road damage ahead of time and then creating a warning system, as well as developing two-factor authentication methods. The vehicle can record its current location and any roadblocks it encounters along the way, which can then be used as a resource in future situations.

References

1. Mohammed MS et al (2023) Low-cost autonomous car level 2: Design and implementation for conventional vehicles. Results in Eng 17
2. Chahal A (2018) In Situ detection of road lanes using Raspberry Pi (Doctoral dissertation, Utah State University)
3. Vinothini K, Jayanthy S (2019) Road sign recognition system for autonomous vehicle using Raspberry Pi. In: 2019 5th International conference on advanced computing and communication systems (ICACCS), Coimbatore, India, pp 78–83
4. Chhillar R, Agarwal H, Gupta SC (2021) Using BFD1000 and Raspberry pi for autonomous vehicle. In: 2021 11th International conference on cloud computing, data science and engineering (Confluence), Noida, India, pp 524–529
5. Gopalan R, Hong T, Shneier M, Chellappa R (2012) A learning approach towards detection and tracking of lane markings. Trans Intell Transport Sys 13:1088–1098
6. Hayward D (2012) Raspberry Pi operating systems: 5 reviewed and rated [Online], available at: http://www.in.techradar.com/news/software
7. Miao X, Li S, Shen H (2012) On-board lane detection system for intelligent vehicle based on monocular vision. Int J Smart Sens Intell Syst 5(4):957–972
8. Chy MKA, Masum AKM, Sayeed KAM, Uddin MZ (2012) Delivehicle: a smart deep learning based self driving product delivery vehicle in perspective of Bangladesh. Sensors 22:126
9. Wanga Y, Teoha EK, Shenb D (2004) Lane detection and tracking using B-snake, image and vision computing, vol 22. pp 269–280
10. Badue C, Guidolini R, Vehicleneiro RV, Azevedo P, Vehicledoso VB, Forechi A, Jesus L, Berriel R, Paixao TM, Mutz F (2021) Selfdriving vehicles: a survey. Expert Syst Appl 165:113816
11. Aziz MVG, Hindersah H, Prihatmanto AS (2017) Implementation of vehicle detection algorithm for self-driving vehicle on toll road Cipularang using python language. In: 4 th International conference on electric vehicular technology
12. McCall JC, Trivedi MM (2006) Video-based lane estimation and tracking for driver assistance: survey, system, and evaluation. IEEE Trans Intell Transport Syst 7(1):20–37
13. Vijay R, Madhuranthagi T, Dhurga Devi A, Kanimozhi SA (2020) IoT based smart vehicle with over-speed accident detection and rescue system. Int J Adv Sci Technol 29(9):3297–3304

14. Huang SS, Chen CJ, Hsiao PY, Fu LC (2004) On-board vision system for lane. In: Miao X, Li S, Shen H (eds) On-board lane detection system for intelligent vehicle based on monocular vision 972 recognition and front-vehicle detection to enhance driver's awareness, IEEE international conference on robotics and automation, pp 2456–2461
15. Petrovai A, Dnescu R, Nedevschi S (2015) A stereovision-based approach for detecting and tracking lane and forward obstacles on mobile devices. In: IEEE intelligent vehicles symposium (IV), pp 634–641
16. Ravindran V, Ponraj R, Krishnakumar C, Ragunathan S, Ramkumar V, Swaminathan K (2021) IoT-based smart transformer monitoring system with Raspberry Pi, 2021 innovations in power and advanced computing technologies (i-PACT), pp 1–7
17. Ravindran V, C V (2021) An energy efficient clustering protocol for IoT wireless sensor networks based on cluster supervisor management. Comptes rendus de l'Acade'mie bulgare des Sciences

Chapter 41
Standard Plane Classification of Fetal Brain Ultrasound Images

Jasmin Shanavas and G. Kanjana

1 Introduction

For prenatal observation and measurement of fetal, biometrics technology for ultrasonic imaging has been used and fetal abnormalities diagnosis are done for past years. When compared with other imaging techniques like magnetic resonance imaging (MRI) and computed tomography (CT), it is safe, painless, inexpensive, portable, and radiation-free. During prenatal testing, most crucial part in the fetus is brain development. BPD and HC measurements are used to evaluate the brain growth in a fetus which can be carried out by using fetal head ultrasound images. In head ultrasound, images of the brain are obtained by using sound waves. Images are captured on a computer after sound waves from an ultrasound equipment are directed into the cranium.

For precise head measurement and brain growth detection, it is essential to correctly identify the fetal brain standard planes. The incorrect choice of standard planes which relies on the doctors clinical experience can affect the diagnostic precision. There are six fetal brain standard planes. They are trans-thalamic plane, trans-ventricular plane, trans-cerebellar plane, paracentral sagittal plane, midsagittal plane, and coronal plane. An efficient ultrasonographic assessment of the fetal brain's morphology requires scanning in three axial planes of the skull, such as the trans-thalamic, trans-ventricular, and trans-cerebellar planes. This paper presents fetal brain ultrasound classification through various deep learning methods.

A methodology based on multi-task learning and a hybrid knowledge graph was put up by Zhao et al. [1] for the detection of the fetal head's standard plane by ultrasonography. Here, a multi-task learning technique is used to analyze the characteristics of fetal US images. The performance is then made more universal by including the shared features in the output stream that are specific to each activity.

J. Shanavas (✉) · G. Kanjana
LBS Institute of Technology for Women Thiruvananthapuram, Thiruvananthapuram, India
e-mail: jasminshanavas03@gmail.com

A. Yadav et al. (eds.), *Proceedings of International Conference on Paradigms of Communication, Computing and Data Analytics*, Algorithms for Intelligent Systems,
https://doi.org/10.1007/978-981-99-4626-6_41

A regression framework was suggested by Maria et al. [2] as a technique for HC determination from US images. Here, a region-based CNN is used for head localization and centering, and a regression CNN is employed to precisely outline the HC. In this study, HC is determined from the fitted ellipse that was fitted to the regression CNN output.

Rasheed et al. [3] proposed a method using ultrasound video for automated fetal head classification and segmentation. Here for classification and segmentation of head frames from US video, AlexNet and UNet algorithms are used. On the outline of the segmented fetal head, an ellipse is drawn to compute BPD and HC. Using a differential convolutional neural network, Qu et al. [4] presented standard plane identification from fetal brain US data. In this instance, six fetal brain standard planes are automatically distinguished from the non-standard planes using a differential-CNN. Because the differential-CNN uses less memory, it can be used in portable devices with limited computing power. Low identification accuracy and large computational expenses are present here.

Xie et al. [5] proposed a system to classify fetal brain ultrasound images as normal or abnormal using deep learning algorithms. Here, CNN is used to analyze the US image and to categorize the image as normal or abnormal. Also, localization is done by the CNN, which allows to locate lesions or other conditions. Here, time consumption for processing one image is comparatively more. Xavier et al. [6] proposed automatic classification of common maternal fetal US planes by evaluation of deep CNN. The CNN algorithm is employed for classification of fetal ultrasound planes, and it is compared with the classification performed by research technicians.

A method for identifying fetal brain standard scan planes in 2D ultrasound pictures using deep learning techniques was presented by Qu et al. [7]. Here, two main techniques based on deep convolutional neural networks are employed to automatically recognize six common planes of fetal brains. The performance of a CNN-based domain transfer learning system and two deep convolutional neural networks is compared. The running time takes about 1.2 s to identify one fetal brain standard scan plane. Salim et al. [8] proposed evaluation of automated tool for two-dimensional fetal biometry where fetal abdominal circumference, head circumference, and femur length are measured using an automated tool from 2D ultrasound images. The automatic and manual measurements taken in real time are then compared. Here, manual correction using adjustable calipers is required.

For the purpose of measuring biometric characteristics from fetal ultrasound images, Sobhania et al. [9] suggested a multi-task deep learning method. Here, it is suggested to automatic segmentation and estimate HC using a multi-task deep CNN by minimizing a compound cost function made up of the parameters for the modified super ellipse (MSE) and the segmentation dice score. The single-task network without an elliptical tuner was found to perform worse. Here, elliptical segmentation results are smoother and cleaner. Lin et al. [10] presented a method for standard plane detection and quality assessment. A multi-task learning framework using a faster regional convolutional neural network architecture is used. To detect the same anatomical structures, the faster R-CNN combines R-CNN with risk priority number (RPN). To improve detection accuracy, a clinical prior knowledge module was added. It also analyzes the skull shape of fetus, to find whether it occupies the correct image area to meet clinical standards. Here, the system focuses only on normal ultrasound plane images from healthy babies. Therefore, detection and classification errors are present.

A classification system for fetal brain abnormalities was suggested by Attallah et al. [11]. Here, classification is done using MRI images from different gestational ages. Radial basis function (RBF), K-nearest neighbor (K-NN), Naive Bayes, and random forest neural network classifiers are some examples of machine learning approaches employed. Additionally, RBF network classifiers and random forest classifiers are used to build ensemble models. The K-NN classifier produces the highest classification accuracy. The ensemble classifiers produced a better outcome for the individual models. Sinclair et al. [12] proposed a system for human-level performance on automatic fetal head biometrics in ultrasound. Here, an automated method is proposed using fully convolutional neural networks (FCN) to measure fetal head circumference. Here, a binary image with the fetal head is obtained with an FCN architecture.

For the purpose of calculating gestational age, Saii et al. [13] developed an algorithm for measurement of the biparietal distance automatically in fetal ultrasound pictures. Here, the fetal head is found in the ultrasound picture input, and the biparietal distance is measured using the least square fitting (LSF) algorithm. Low precision, expensive to compute, and time-consuming results were achieved. Christian et al. [14] proposed a method using fully convolutional neural networks for standard scan plane detection in real time and localization in fetal US. Here, CNN-based fully automated system is used to detect twelve standard scan planes. Real-time inference can be done using the network design and localization of the fetal anatomy can also be provided. A large number of fetal standard views from ultrasound scans are obtained.

In this work, classification of fetal brain is performed using deep learning approaches such as ResNet50, VGG16, and VGG19. Our trained model can identify four classes which are trans-thalamic, trans-cerebellum, trans-centricular, and others. This reduces time consumption and reduces misdiagnosis.

2 Methodology

Fetal ultrasound dataset is collected, and the data is preprocessed. Classification is performed by using deep learning methods such as ResNet 50, VGG19, and VGG16. The block diagram for fetal head classification is shown in Fig. 1.

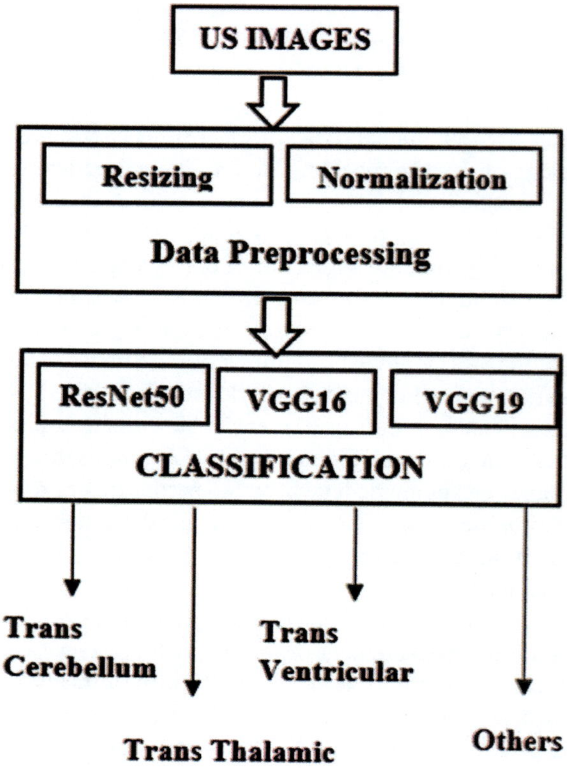

Fig. 1 Block diagram of fetal head classification

2.1 *Dataset Collection*

Fetal ultrasound dataset which is publicly available is collected from Kaggle. The dataset consists of a total of 1750 US images. It consists of four classes which are 975 trans-thalamic, 323 trans-ventricular, 367 trans-cerebellum, and 85 others. In this model, 20% of data is used as testing set ,and 80% of the data is used as training set. 1400 images were taken for training and 350 images for testing.

2.2 *Image Preprocessing*

Preprocessing is the process of converting raw data into a suitable format that the network can use once the dataset has been acquired. In Fig. 2, the preprocessed image is shown. In the suggested model, there are two steps:

- Image Resizing: Deep learning models train faster on smaller images. First, the dimension of the images in dataset is found out, and then the images are resized to

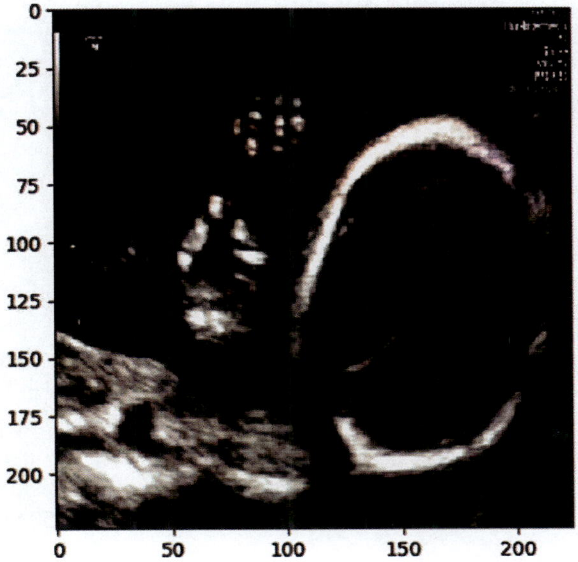

Fig. 2 Preprocessed image

a dimension of 224 * 224. It decreases the calculation cost to ensure compatibility with the system's structural design and memory size.

- Image Normalization: The method of transforming every dataset to an intensity value that is similar is called normalization. Normalize the picture between [0, 1]. By dividing a pixel's maximum value by its minimum value, normalization is accomplished.

The dataset after preprocessing is split to training and testing set.

2.3 Classification

Fetal brain ultrasound image classification is grouping the images that have similarity into same category. Here, the categories are different standard planes of the fetal brain such as trans-ventricular, trans-thalamic, and trans-cerebellum. The standard plane classification method that is implemented here is ResNet50, VGG19, and VGG16.

2.4 ResNet50

Residual neural network (ResNet) is a deep learning method commonly used for solving complex problems. The 34-layer plain network design inspired from VGG19 is used by ResNet34. In addition, shortcut connections are present. The architecture is then transformed into the residual network via these shortcut connections. ResNet

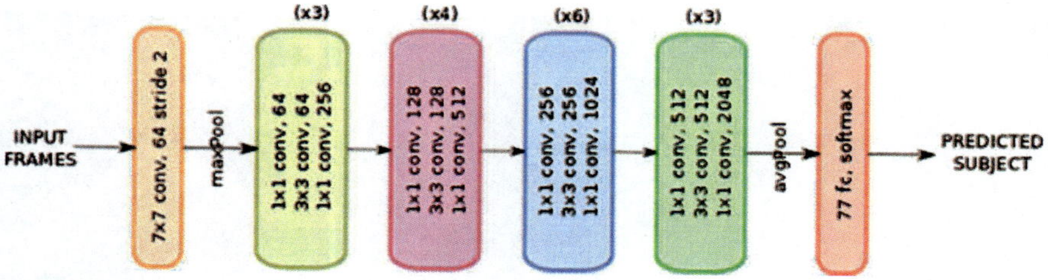

Fig. 3 ResNet50 [15]

34 model is the basis of the ResNet 50 architecture. The major difference is that three-layer block replaces each of the two-layer blocks in Resnet34, forming the ResNet 50 architecture. This has higher accuracy when compared to the 34-layer ResNet model. A 50-layer CNN containing 48 convolutional layers, a maxpool layer, and an average pool layer is ResNet50. Here, we are using ResNet 50 since it can work with 50 neural network layers. A simple ResNet 50 architecture is shown in Fig. 3.

2.5 *VGG16*

A CNN, or convolutional neural network, is the VGG model VGG16. The network is pretrained and has been trained on more than a million photos from the ImageNet database that can be loaded using this 16-layer deep CNN. Over 14 million images and 1000 classes are available on ImageNet. The object recognition and classification algorithm VGG16 can recognize things in 1000 photos and classify them into 1000 distinct categories. It is a well-known technique for classifying photos, and using transfer learning makes it simple to use. Figure 4 depicts the architecture of the VGG16. VGG16 comprises three dense layers, five maxpooling layers, and 13 convolutional layers. That is a total of 21 layers. However, it only has sixteen layers of weights that can be learned. VGG16 has a unique feature in that it focuses on having convolution layers of 3×3 filter and stride 1 rather than having a lot of hyper-parameters. Also, it always employs the same maxpool with stride 2 and padding. The convolution layers and maxpool layers are placed uniformly throughout the complete architecture. Here, the Conv-1 has 64 filters, the Conv-2 layer has 128 filters, the Conv-3 layer has 256 filters, and the Conv-4 and 5 Layers each have 512 filters. For issues involving several classifications, the softmax layer is utilized as the activation function. The VGG16 network's size, which makes it more time-consuming to train its parameters, is one of its disadvantages.

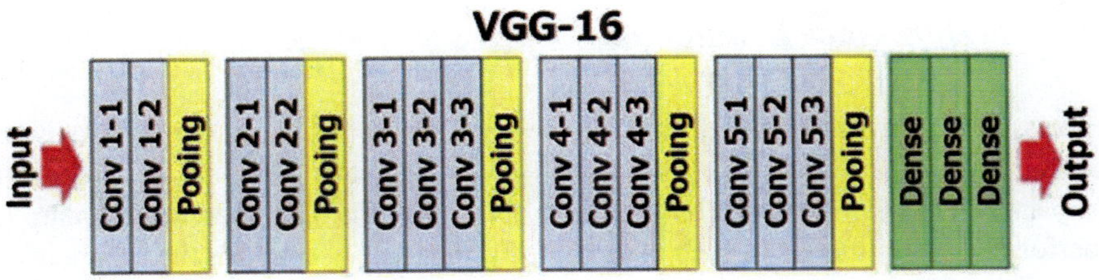

Fig. 4 VGG16 [16]

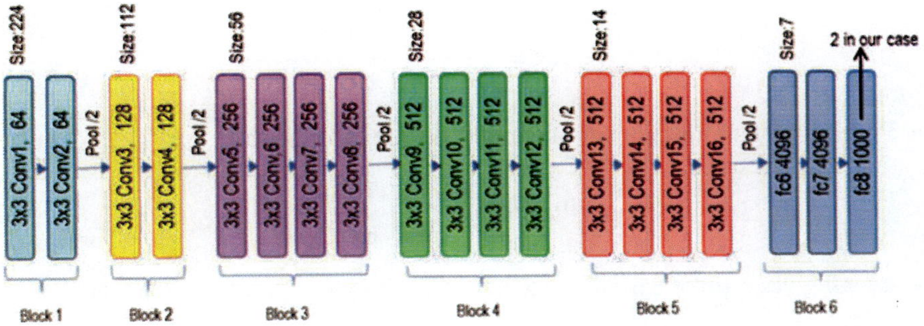

Fig. 5 VGG19 [17]

2.6 VGG19

VGG19 is convolutional neural network comprising of 19 layers, 16 convolutional layers, and 3 fully connected layers for classifying images. It is a simple method used for image classification. It uses more than one 3 * 3 filters in every convolutional layer. For feature extraction, the 16 convolutional layers are used, and the next three layers work for classification. The feature extraction layers are classified into five groups. Following each group is a maxpooling layer. The architecture of VGG19 is shown in Fig. 5.

2.7 Experimental Setup

The platform used to implement this model is Visual Studio Code. In VS code, programming can be done in any language. In this work, Python is used as the programming language. For most of the learning applications, Visual Studio Code is used because it is simple and productive.

2.8 Performance Evaluation

The accuracy, precision, recall, and F1 score are the parameters used to assess the system performance. A confusion matrix is generated, which is a table that is used to explain a classification algorithm's performance. With the help of confusion matrix, performance parameters of each class can be evaluated. To evaluate the accuracy of a model using the confusion matrix, the following formula is used.

- **Accuracy** is ratio of number of actual positive predictions by total predictions.

$$\text{Accuracy} = \frac{\text{TP} + \text{TN}}{\text{TP} + \text{TN} + \text{FP} + \text{FN}} \tag{1}$$

- **Precision** quantifies number of correct true predictions made.

$$\text{Precision} = \frac{\text{TP}}{\text{TP} + \text{FP}} \tag{2}$$

- **Recall** gives ratio of correct predicted true values to all values in that class.

$$\text{Recall} = \frac{\text{TP}}{\text{TP} + \text{FN}} \tag{3}$$

- **F1 Score** gives the weighted average of the recall and precision values.

$$\text{F1} = \frac{2 * \text{Precision} * \text{Recall}}{\text{Precision} + \text{Recall}} = \frac{2 * \text{TP}}{2 * \text{TP} + \text{FP} + \text{FN}} \tag{4}$$

where TN is the number of negative examples correctly classified, TP is number of positive examples correctly classified, FP is number of actual negative examples classified as positive, and FN is number of actual positive examples classified as negative.

3 Results

3.1 Experimental Results

Classification of fetal brain into four classes is performed based on deep learning approaches such as ResNet 50, VGG19, and VGG16, and it is shown in Fig. 6. The four classes are trans-thalamic, trans-ventricular, trans-cerebellum, and others.

Figures 7, 8, and 9 show the confusion matrix of ResNet 50, VGG16, and VGG19, respectively. The fetal head US classification using ResNet 50, VGG16, and VGG19 algorithms is done. The accuracy obtained for the fetal head classification using

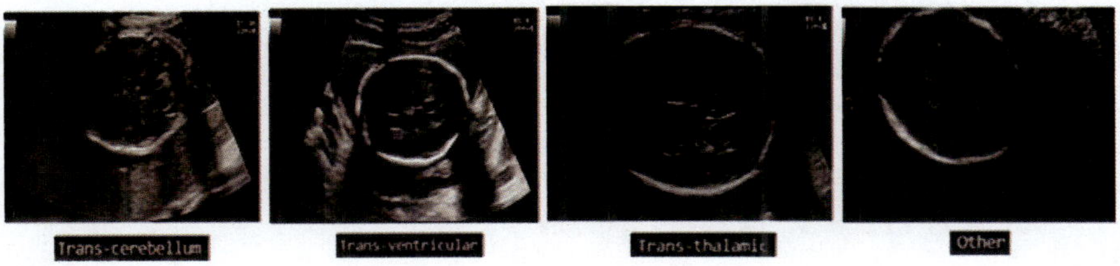

Fig. 6 Classification results

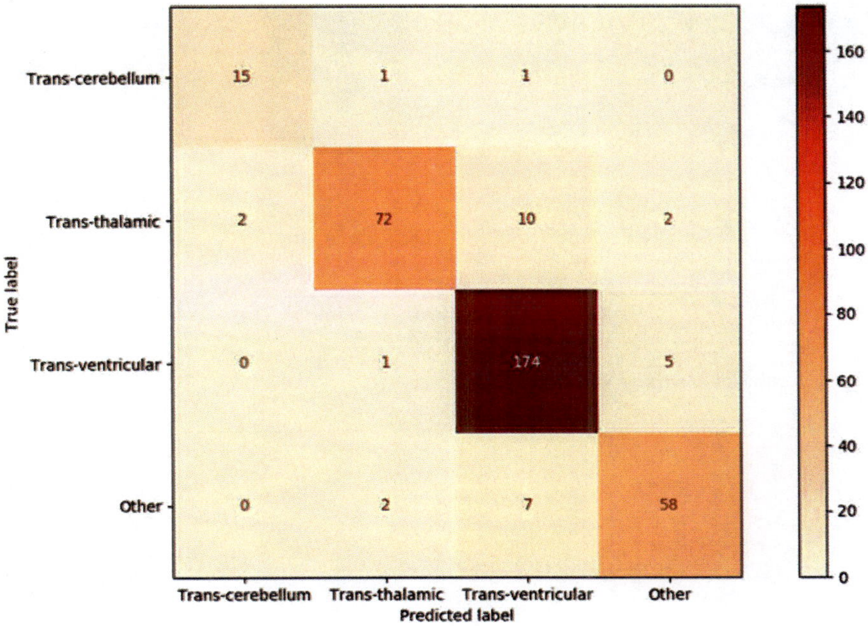

Fig. 7 Confusion matrix of ResNet 50

Table 1 Overall performance of the proposed model

Model	ResNet 50	VGG16	VGG19
Accuracy (%)	91	95	93
Precision (%)	90	93	94
Recall (%)	89	98	93
*F*1 score (%)	90	95	93

ResNet50 is 91%, for VGG16 is 95%, and for VGG19 is 93%. Table 1 shows the overall performance of the proposed model.

Figures 10, 11, 12, 13, 14, and 15 depict a graph showing training loss and accuracy of the ResNet 50, VGG16, and VGG19 with respect to epochs. The figure represents the accuracy plot and the loss plot for training and validation over number of epochs.

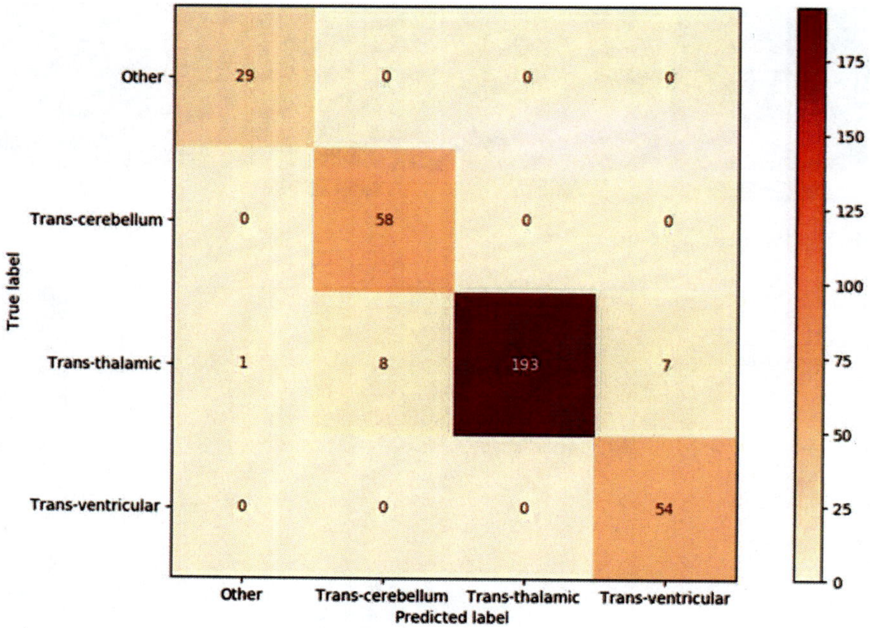

Fig. 8 Confusion matrix of VGG16

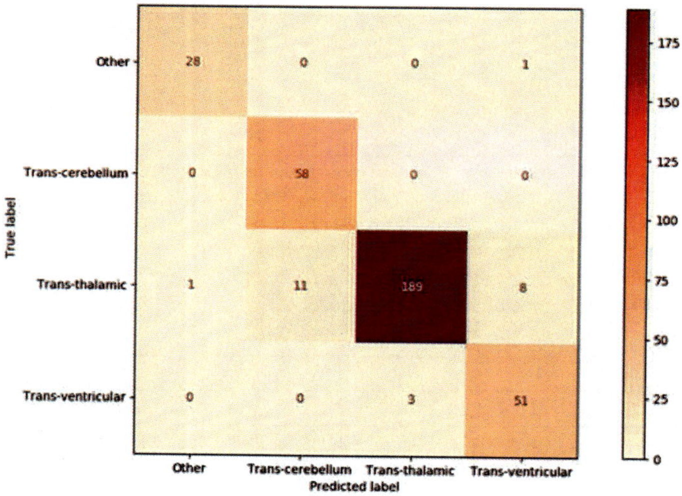

Fig. 9 Confusion matrix of VGG19

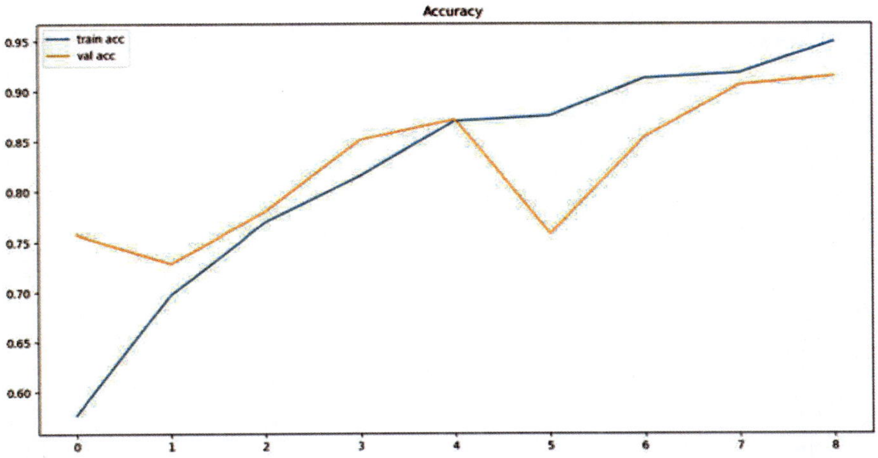

Fig. 10 Graph showing training accuracy of the ResNet 50 model

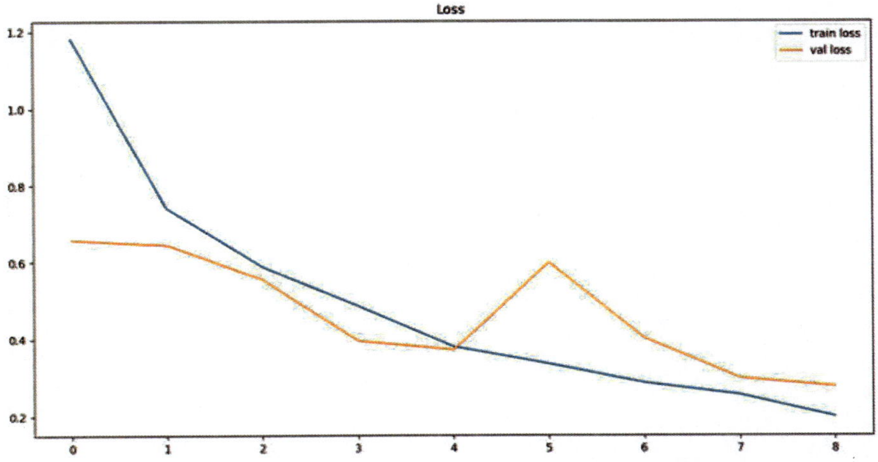

Fig. 11 Graph showing training loss of the ResNet 50 model

During an epoch, every data items are used to calculate the accuracy and loss function, and it gives the quantitative loss measure at each epoch. This will assist in determining the necessary architectural decisions.

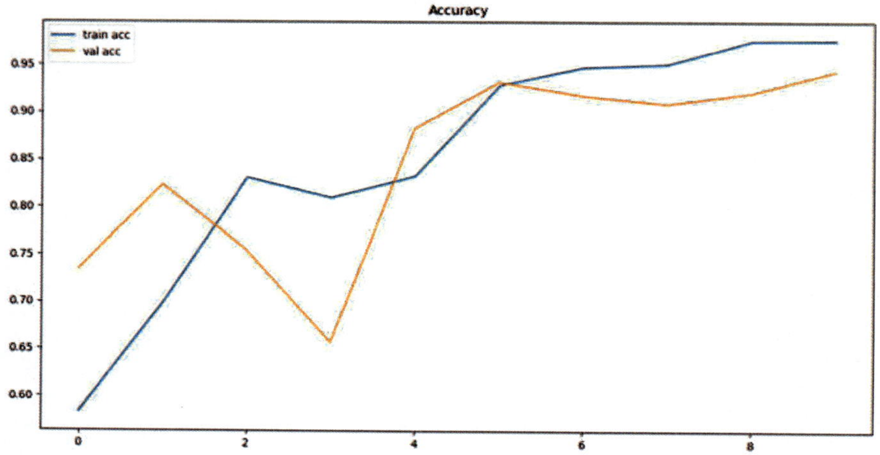

Fig. 12 Graph showing training accuracy of the VGG16 model

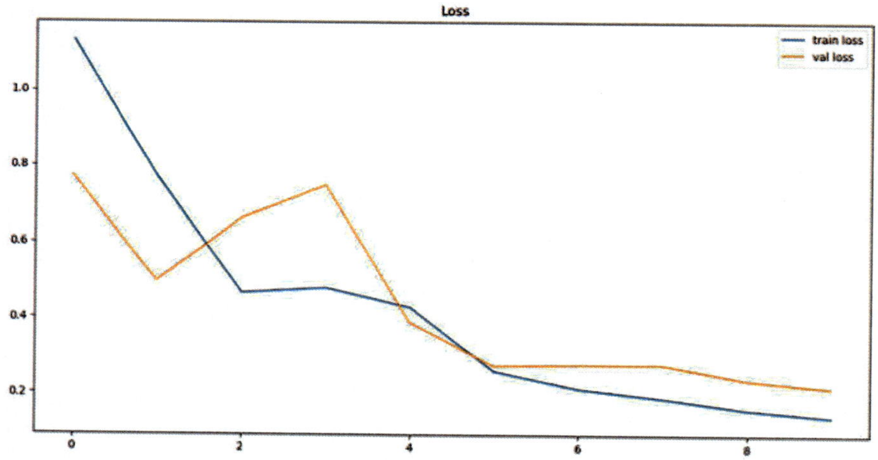

Fig. 13 Graph showing training loss of the VGG16 model

4 Conclusion

Fetal development has to be accurately assessed since it is very important to find the health conditions of mothers and newborns during pregnancy. A computer-aided technology will help the doctors for accurate assessment of fetal biometrics. Different learning algorithms help to reduce false diagnosis. In this paper, classification of standard planes of fetal brain through deep learning approaches such as ResNet50, VGG19, and VGG16 is done. In the future work, automated detection of head circumference, femur length, and abdomen circumference from fetal US images can also be included for estimating the fetal weight.

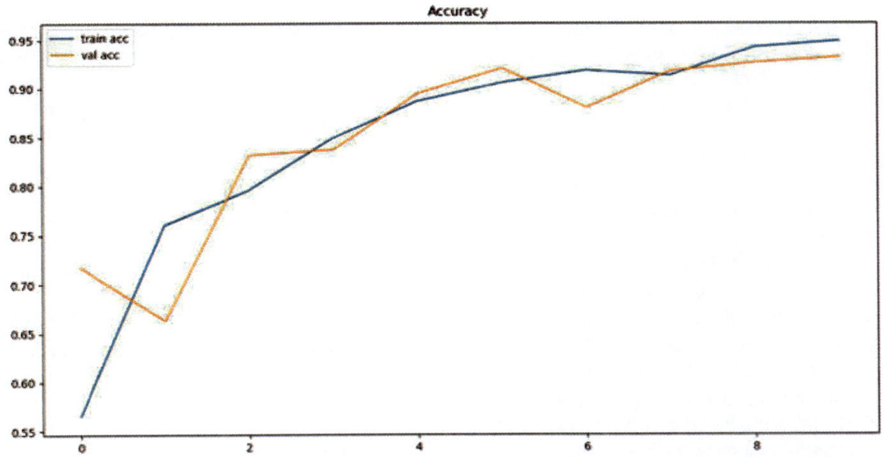

Fig. 14 Graph showing training accuracy of the VGG19 model

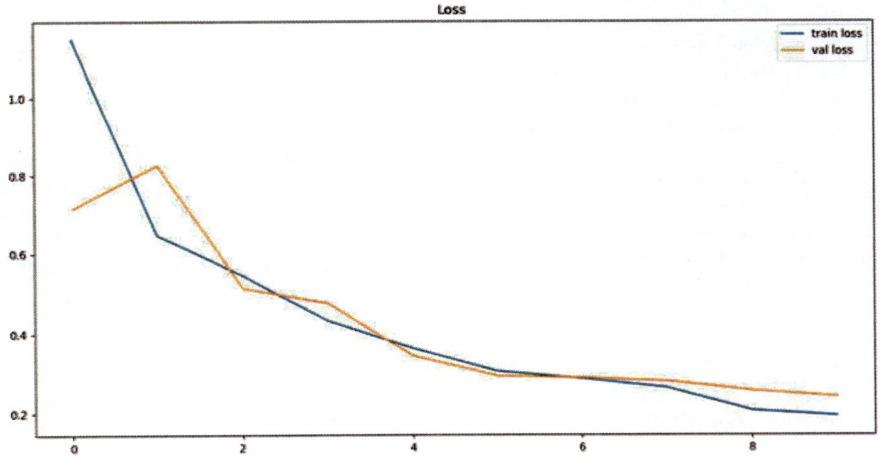

Fig. 15 Graph showing training loss of the VGG19 model

References

1. Zhao L, Li K, Pu B, Chen J, Li S, Liao X (2022) An ultrasound standard plane detection model of fetal head based on multi-task learning and hybrid knowledge graph. Future Gener Comput Syst 135:234–243
2. Fiorentino M, Moccia S, Capparuccini M, Giamberini S, Frontoni E (2021) A regression framework to head-circumference delineation from US fetal images. Comput Methods Programs Biomed 198:105771
3. Rasheed K, Junejo F, Malik A, Saqib M (2021) Automated fetal head classification and segmentation using ultrasound video. IEEE Access 9:160249–160267
4. Qu R, Xu G, Ding C, Jia W, Sun M (2020) Standard plane identification in fetal brain ultrasound scans using a differential convolutional neural network. IEEE Access 8:83821–83830

5. Xie H, Wang N, He M, Zhang L, Cai H, Xian J, Lin M, Zheng J, Yang Y (2020) Using deep-learning algorithms to classify fetal brain ultrasound images as normal or abnormal. Ultrasound Obstet Gynecol 56:579–587
6. Burgos-Artizzu X, Coronado-Gutiérrez D, Valenzuela-Alcaraz B, Bonet-Carne E, Eixarch E, Crispi F, Gratacós E (2020) Evaluation of deep convolutional neural networks for automatic classification of common maternal fetal ultrasound planes. Sci Rep 10:1–12
7. Qu R, Xu G, Ding C, Jia W, Sun M (2019) Deep learning-based methodology for recognition of fetal brain standard scan planes in 2D ultrasound images. IEEE Access 8:44443–44451
8. Salim I, Cavallaro A, Ciofolo-Veit C, Rouet L, Raynaud C, Mory B, Collet Billon A, Harrison G, Roundhill D, Papageorghiou A (2019) Evaluation of automated tool for two-dimensional fetal biometry. Ultrasound Obstet Gynecol 54:650–654
9. Sobhaninia Z, Rafiei S, Emami A, Karimi N, Najarian K, Samavi S, Soroushmehr S (2019) Fetal ultrasound image segmentation for measuring biometric parameters using multi-task deep learning. In: 2019 41st annual international conference of the IEEE engineering in medicine and biology society (EMBC), pp 6545–6548
10. Lin Z, Li S, Ni D, Liao Y, Wen H, Du J, Chen S, Wang T, Lei B (2019) Multi-task learning for quality assessment of fetal head ultrasound images. Med Image Anal 58:101548
11. Attallah O, Sharkas M, Gadelkarim H (2019) Fetal brain abnormality classification from MRI images of different gestational age. Brain Sci 9:231
12. Sinclair M, Baumgartner C, Matthew J, Bai W, Martinez J, Li Y, Smith S, Knight C, Kainz B, Hajnal J (2018) Others human-level performance on automatic head biometrics in fetal ultrasound using fully convolutional neural networks. In: 2018 40th annual international conference of the IEEE engineering in medicine and biology society (EMBC), pp 714–717
13. Saii M, Kraitem Z (2018) Determining the Gestation age through the automated measurement of the bi-parietal distance in fetal ultrasound images. Ain Shams Eng J 9:2737–2743
14. Baumgartner C, Kamnitsas K, Matthew J, Smith S, Kainz B, Rueckert D (2016) Real-time standard scan plane detection and localisation in fetal ultrasound using fully convolutional neural networks. In: 19th International conference medical image computing and computer-assisted intervention-MICCAI 2016. Athens, Greece, Proceedings, Part II 19, pp 203–211
15. Jahromi MN (2019) Privacy-constrained biometric system for non-cooperative users. Retrieved from https://www.researchgate.net/publication/336805103_Privacy-Constrained_Biometric_System_for_Non-Cooperative_Users
16. Networks P (2018) VGG16—convolutional network for classification and detection
17. Khattar A (2022) Generalization of convolutional network to domain adaptation network for classification of disaster images on twitter. Retrieved from Springer https://www.researchgate.net/publication/359771670

Chapter 42
Panoramic Radiograph Segmentation Using U-Net with MobileNet V2 Encoder

Suvarna Bhat and Gajanan K. Birajdar

1 Introduction

Panoramic X-rays are important tools for dentists to examine the structure, shape, and position of each tooth and confirm or reject a specific diagnosis such as a fracture, infection, tooth loss, or simply to detect any past dental treatment. Dental radiograph segmentation is the foundation for dental radiograph analysis [1]. In recent years, researchers have become increasingly interested in computer-assisted analysis and segmentation of dental radiographs in dentistry. This is primarily because of its ability to successfully eliminate human-made errors caused by stress, exhaustion, or a lack of experience. The research has attempted to explore the possibility of designing practical algorithms that are general enough to be applicable in teeth segmentation in the panoramic dental radiograph. Several supervised and unsupervised pixel-wise segmentation techniques have been developed in the field of tooth segmentation by researchers. Indeed, the majority of recent teeth segmentation approaches available in the literature are focussed on one class segmentation, in which all teeth are grouped into a single category, discarding both morphological features, and independent tooth location [2].

Main highlights of this paper:

- Use of the MobileNet V2 as an encoder in U-Net architecture.
- Data augmentation techniques were used to increase the randomness of the radiographs to increase stability.
- This architecture is tested with different hyperparameters batch sizes, epochs, and optimizers.
- Performance metrics like precision, recall, and dice coefficients are calculated.

S. Bhat (✉) · G. K. Birajdar
Ramrao Adik Institute of Technology, D.Y. Patil Deemed to Be University, Navi Mumabi, India
e-mail: suvarna.bhat@vit.edu.in

G. K. Birajdar
e-mail: gajanan.birajdar@rait.ac.in

A. Yadav et al. (eds.), *Proceedings of International Conference on Paradigms of Communication, Computing and Data Analytics*, Algorithms for Intelligent Systems,
https://doi.org/10.1007/978-981-99-4626-6_42

The paper is organized as follows: related work in Sect. 2, material and methods mentioned in Sect. 3, followed by experiment results and discussion in Sect. 4. Section 5 discussion about a conclusion.

2 Related Work

Silva and Oliveira used mask recurrent convolutional neural networks for automatic teeth segmentation with a novel dataset [3]. Fine-tuned mask RCNN algorithm of deep learning to identify and locate the teeth on panoramic radiographs used in the article for automatic tooth segmentation [4]. The use of VGG-16 CNN with heuristic algorithm attained good results in the teeth detection module proposed in article [5].

Haghanifar and Majdabadi [6] proposed the use of evolutionary algorithms to extract teeth automatically from the panoramic radiographs. Further, authors separated the upper and lower jaws, followed by the use of a genetic algorithm which is used to find the teeth valley gap. This technique is applied over 42 panoramic radiographs and achieves 81.14% accuracy for the upper jaw and 73.63% for the lower jaw.

For tooth classification and numbering, end-to-end neural networks were proposed in [7]. The authors studied and analyzed four neural network architectures, namely RCNN, PANet, HTC, and ResNet. Their observations are as follows: it is possible to detect, segment, and number the teeth using the above-mentioned architectures, performance can be improved with the correct selection of neural networks, PANet gives the best result with 71.3% on segmentation, 74% on numbering.

Authors of [8] proposed the use of the convolutional neural network (CNN) algorithm to detect and classify submerged molar teeth. The detection part involves faster RCCN architecture which process the radiographs to detect the contour of submerged teeth. The performance of the system achieves an accuracy level same as an expert. In the article, [9] permanent teeth had been identified in 3 step process. The author used U-Net to identify ROI, faster RCNN to identify teeth from ROI, and lastly VGG-16 to classify teeth into 32 categories.

In [10], authors proposed post processing stage to produce segmenation map that seperates object in the radiograph and applied this technique to teeth instance segmenation using U-Net network. A novel encoder–decoder model based on multimodal feature axtraction is proposed in [11]. In propsed model to encode rich contextual information, the encoder uses three distinct CNN-based architectures: convolutional CNN, atrous-CNN, and seperate CNN, and decoder includes a single stream of deconvolutional layers for segmentation. Model achieves precision and recall 95.01% and 94.06%, respectively.

3 Material and Methods

This model used MobileNet V2 as an encoder in U-Net architecture for teeth segmentation in panoramic dental radiographs.

3.1 Dataset

In this work, we have used Tufts dental dataset [1]. Tufts dental dataset was published in December 2021 and is publicly available. Radiographs were acquired at the Tuft University research centre. The dataset contains a total of 1000 panoramic dental radiographs along with labelled teeth mask. It is observed that the dataset has various categories of radiographs such as radiographs with all 32 teeth, and radiographs with dental treatments like filling, implants, and root canals. There are some radiographs with no teeth, which are also included in the dataset. The inclusion criteria for the dataset were optimum diagnostic quality of the radiographs with no error or minimal error in the radiographs. Figure 1 shows samples of the tufts dataset radiographs and their respective grouthtruth.

3.2 Dataset Augmentation

To increase the dataset augmentation, techniques such as flipping and rotation have been used. The corresponding mask and augmented radiographs are shown in Fig. 2.

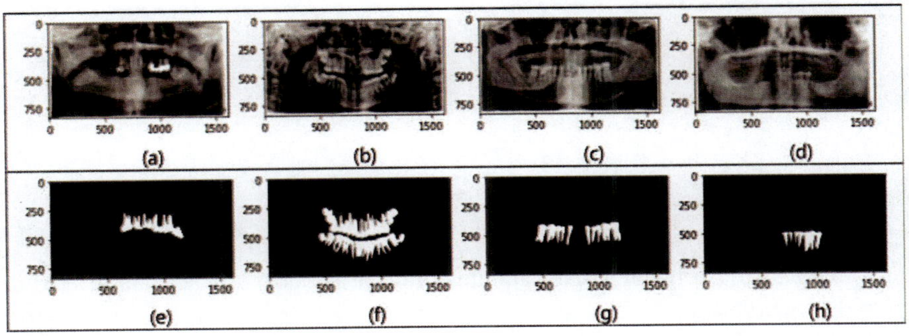

Fig. 1 1 Dental panoramic radiographs from TUFT dental dataset: **a** radiograp 1 **b** radiograph 2 **c** radiograph 3 **d** radiograph 4 ground truth mask for respective panoramic dental radiograph: **e** radiograp 1 **f** radiograph 2 **g** radiograph 3 **h** radiograph 4 [1]

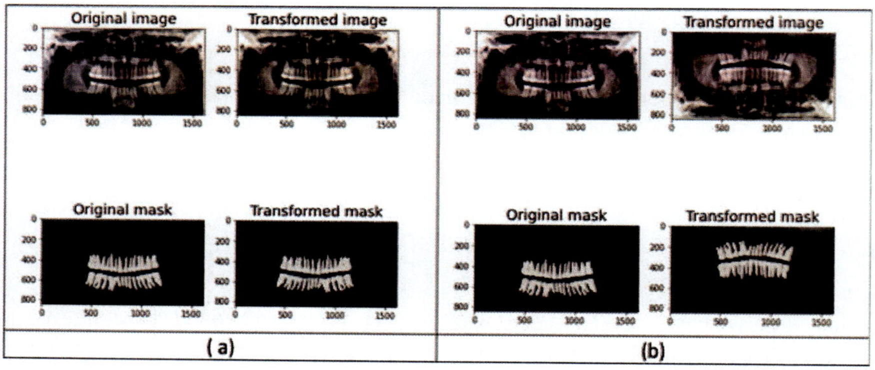

Fig. 2 Dental panoramic radiographs and their respective mask: **a** original radiograph and mask, horizontal flipped radiograph and mask **b** original radiograph and mask, vertical flippes radiograph and mask

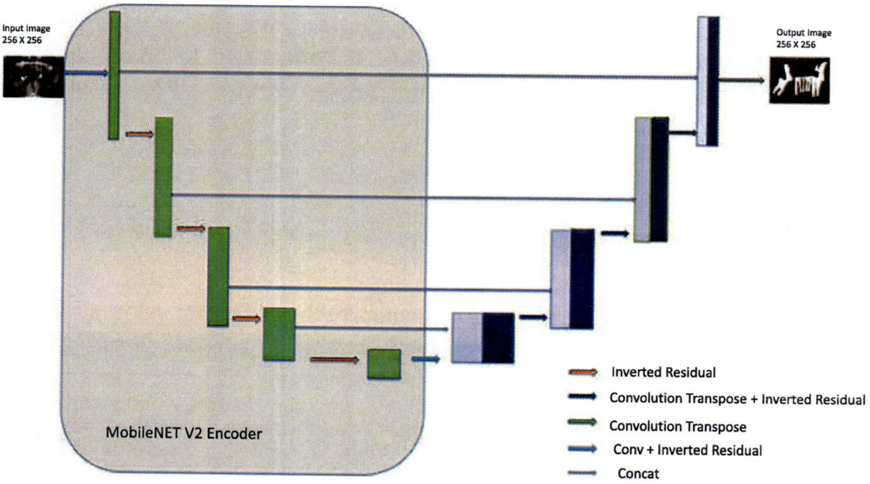

Fig. 3 U-Net with MobileNet V2 as encoder

3.3 U-Net-MobileNet V2

According to the original U-Net architecture, the encoder extract characteristics from the input images, and these resources are concatenated with the decoder so that the network performs the segmentation task. In these instances, using pre-trained networks in huge databases greatly aids segmentation because weights have already been trained with millions of learned resources and will not start from the beginning.

Figure 3 depicts the proposed U-Net-MobileNet V2 model where pre-trained MobileNet V2 is applied as the encoder. In this approach, the encoder will filter and learn the properties of radiographs that feed the network using compact depth convolutions. The inclusion of these inverted residual blocks reduces the number of parameters and makes training the model easier and faster [12].

Another advantage of the model is that the model will perform better and converge faster than it would without the usage of a pre-trained network. Up-sampling methods are employed in the decoding route to return the feature map to its original size. The suggested model model has a total parameters 11,753,809 with 11,725,617 trainable parameters and 28,192 non-trainable parameters.

4 Result and Discussion

This segment contains all of the output result obtained by the model. The model is tested using Tufts dental panoramic dataset. An experimental analysis was performed through which training loss and training accuracy were obtained. Segmented images's visual analysis and the investigation of confusion matrix properties are described in depth below.

4.1 Analysis Using Different Optimizer

This section contains the outcomes of the Adam, Nadam, and Adamax optimizers with epoch 50 and batch size of 16.

Training Loss and Accuracy Analysis Figure 4a shows the traning loss with use of the Adam optimizer, the maximum loss value is 0.665, and it lowers as the number of epochs increses. Figure 4b shows at last epoch traning accuracy which is more than 0.96. Figure 4c and d shows loss of training using the Nadam optimizer, whose maximum training loss is 0.60, and accuracy is 0.9766. Figure 4e and f shows the training loss and training accuracy of the adamax optimizer. Training loss of Adamax optimizer is higher with respect to Adam and Nadam optimizers, and training accuracy is more than 0.95. In terms of training loss and accuracy, Fig. 4 indicates that the Nadam optimizer excels over the Adam and Adamax optimizers.

Visual Analysis of Radiograph Segmentation Figure 5 shows the output, predicted mask with Adam, Nadam, and Adamax optimizers with batch size 16 and 100 epochs. Figure 5a–c shows the output mask of the Adam, Nadam, and Adamax optimizers, respectively. Based on visual analysis, Nadam shows better visual results than the Adam and Adamax optimizer. Hence, Adam and Adamax optimzers are not recommended for teeth segmentation. To select Nadam as the best optimizer, further, we did a confusion matrics analysis as shown in the next section.

Confusion Matrix Parameter Analysis As per training loss, training accuracy analysis, and visual analysis of radiograph segmentation, the Nadam optimizer gives the optimum results. Now to check the most effective optimizer, we did further analysis of the confusion matrix based on parameters like dice coefficient, precision,

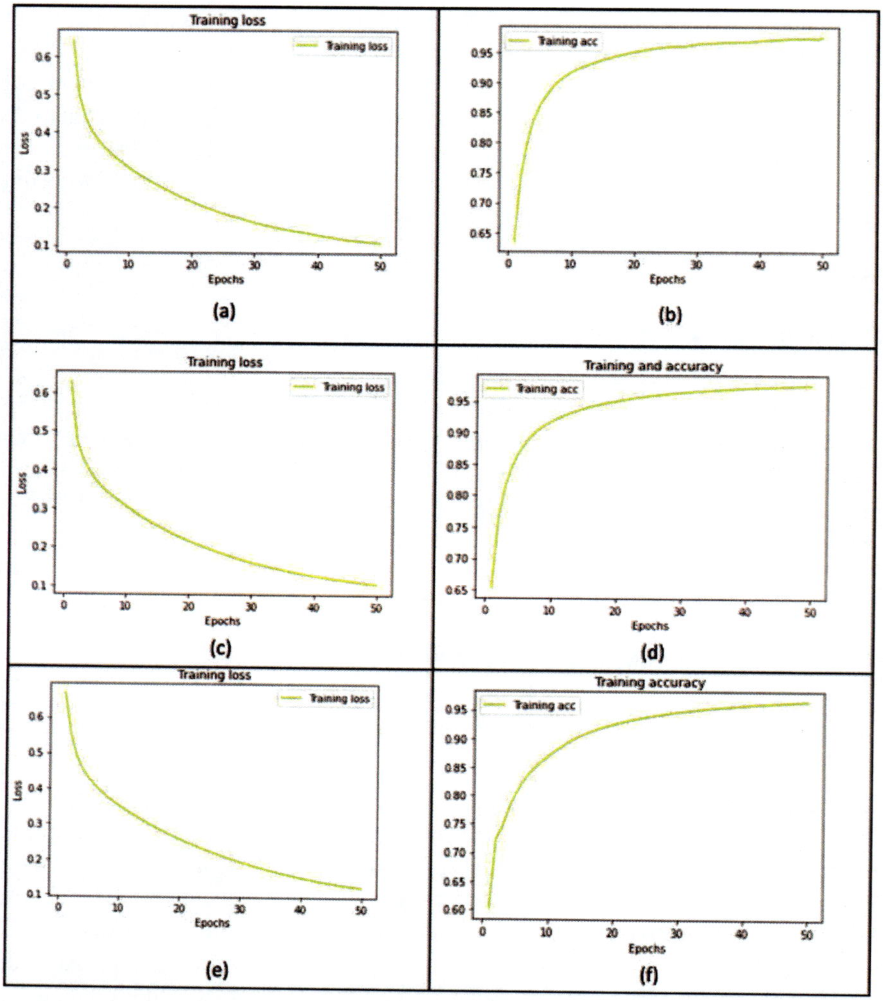

Fig. 4 Traning loss and acccuracy: **a** Training loss with Adam optimzer. **b** Training accuracy with Adam optimizer. **c** Training loss with Nadam optimzer. **d** Traning accuracy with Nadam optimizer. **e** Training loss with Adamax optimzer. **f** Traning accuracy with Adamax optimizer

and recall. Table 1 shows the mentioned confusion matrix parameters analysis for Adam, Nadam, and Adamax optimizers. Based on the validation dataset, analysis of the evaluation matrix shown in Table 1 Nadam optimizer outperforms Adam and Adamax giving the best result with a dice coefficient value of 0.8434 and precision value of 0.925. The experimental data for the Adamax optimizer indicate that it is the weakest. As a result of our findings, we may conclude that the Nadam optimizer achieved the optimal results on the validation dataset, outperforming the Adam and Adamax optimizers on practically all parameters.

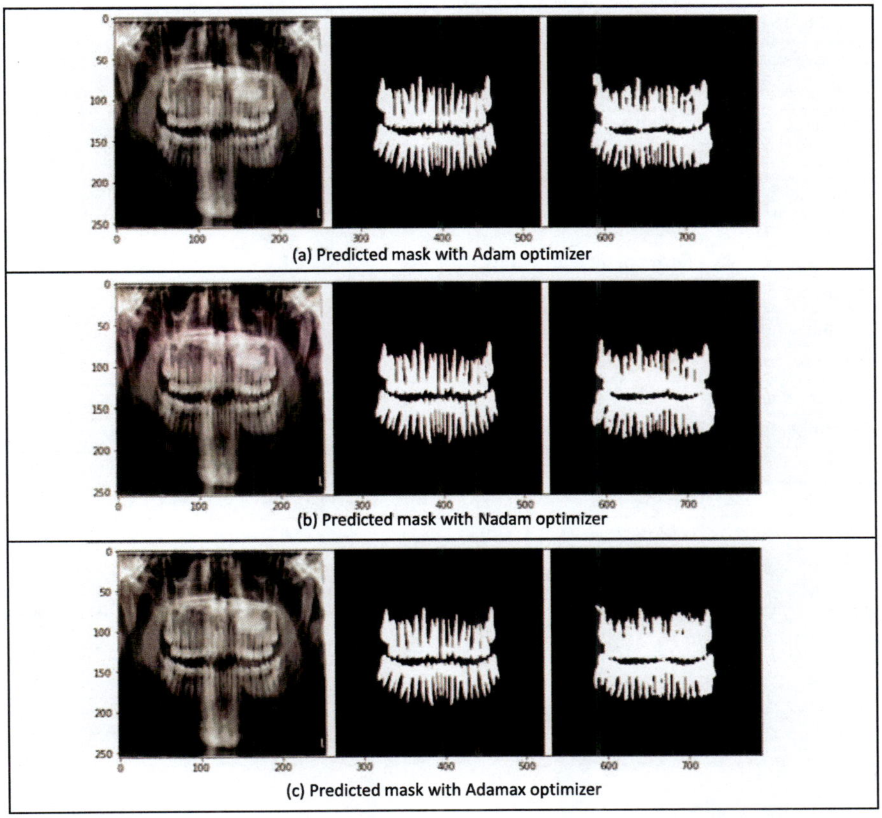

Fig. 5 Predicted mask with different optimizer

On the Adam, Nadam, and Adamax optimizers, the confusion matrix parameters were examined for validation dataset as illlustrates in Fig. 6. This graph shows that the Nadam optimizer performs best on practically all parameters, including the dice coefficient, precision, recall, and loss. In comparison with the Adam and Adamax optimizers, the Nadam optimizer has a substantially smaller loss value.

4.2 Analysis of Result Based on Different Batch Size with Nadam Optimizer and 50 Epoch

Considering a batch size of 16, the Nadam optimizer outperformed the Adam and Adamax optimizers, according to Sect. 5. As a result, the findings in this section are derived on various batch sizes using Nadam optimizer. However, the Adam and Adamax optimizers may yield better results for different batch sizes and epoch permutations. These two optimizers can be tested in the future for different batch sizes

Table 1 Different optimizer analysis with 16 batch sizes and 50 epoch

Optimizer	Dice coefficient	Precision	Recall	Loss
Training dataset				
Optimizer	Dice coefficient	Precision	Recall	Loss
Adam	0.8905	0.9740	0.8601	10.95
Nadam	0.8948	0.9766	0.8616	10.52
Adamax	0.8778	0.9653	0.8571	12.22
Testing dataset				
Optimizer	Dice coefficient	Precision	Recall	Loss
Adam	0.8455	0.9024	0.8877	15.48
Nadam	0.8479	0.9242	0.8453	15.20
Adamax	0.8452	0.9041	0.8915	15.49
Validation dataset				
Optimizer	Dice coefficient	Precision	Recall	Loss
Adam	0.8424	0.9006	0.8839	15.92
Nadam	0.8434	0.9253	0.8361	15.86
Adamax	0.8428	0.9043	0.8866	15.89

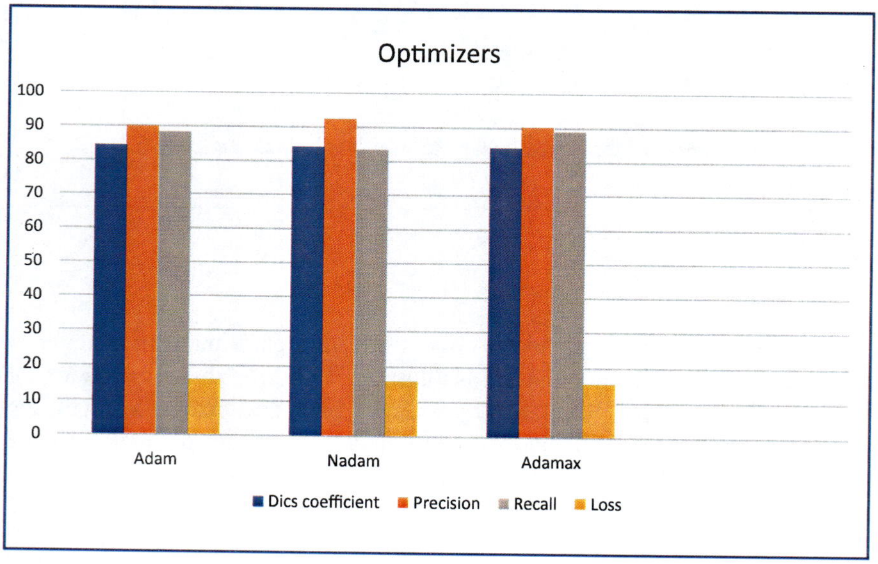

Fig. 6 Confusion matrix parameter

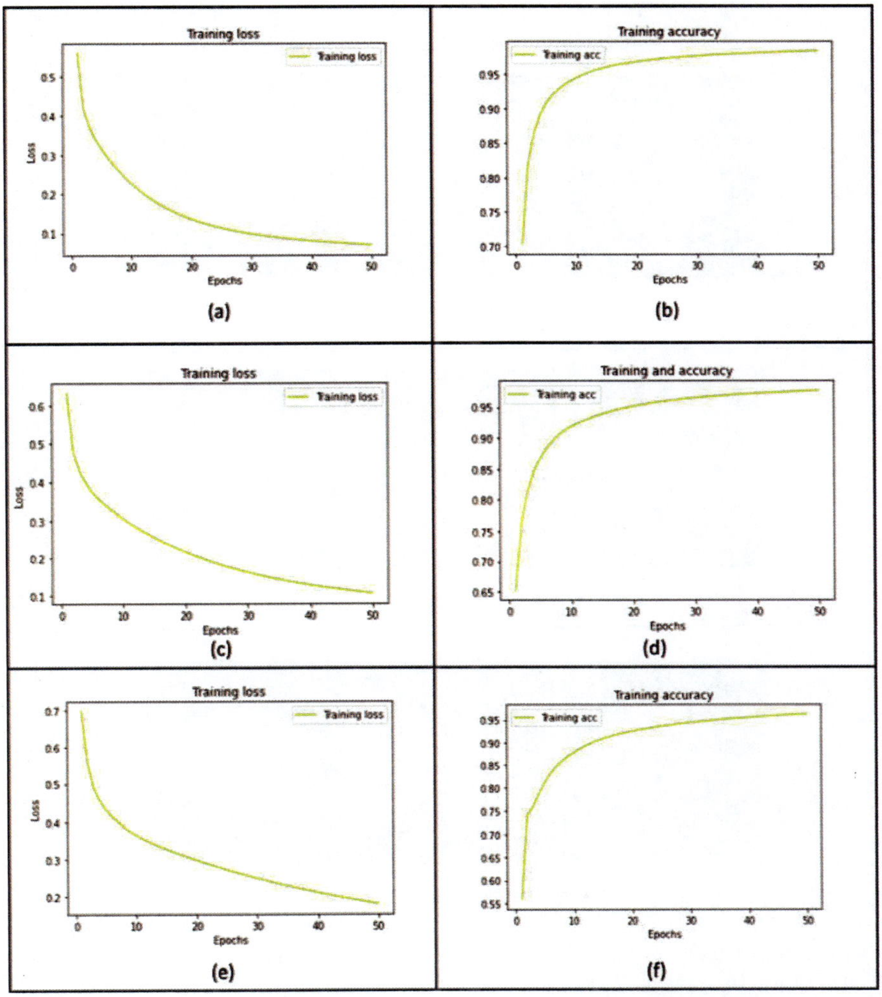

Fig. 7 Analysis based on training loss and accuracy: **a** Training loss with 8 batch size. **b** Training accuracy with 8 batch size. **c** Training loss with batch size 16. **d** Training accuracy with 16 batch size. **e** Training loss with batch size 32. **f** Training accuracy with 32 batch size

and epoch combinations. The batch sizes considered for analyzing the Nadam optimizer on 50 epochs are 8, 16, and 32.

Training Loss and Accuracy Analysis Figure 7 shows an analysis of training loss and accuracy for batch sizes 8, 16, and 32. Figure 7a shows training loss with batch size 8, the maximum loss value is above 0.5, and over the number of epochs, its decreases. Figure 7c shows a training loss above 0.6 with batch size 16, and Fig. 7e shows a training loss close to 0.7 with batch size 32. Figure 7b, d, and f concludes that all the batch sizes, i.e. 8, 16, and 32 have training accuracy which is above 0.96.

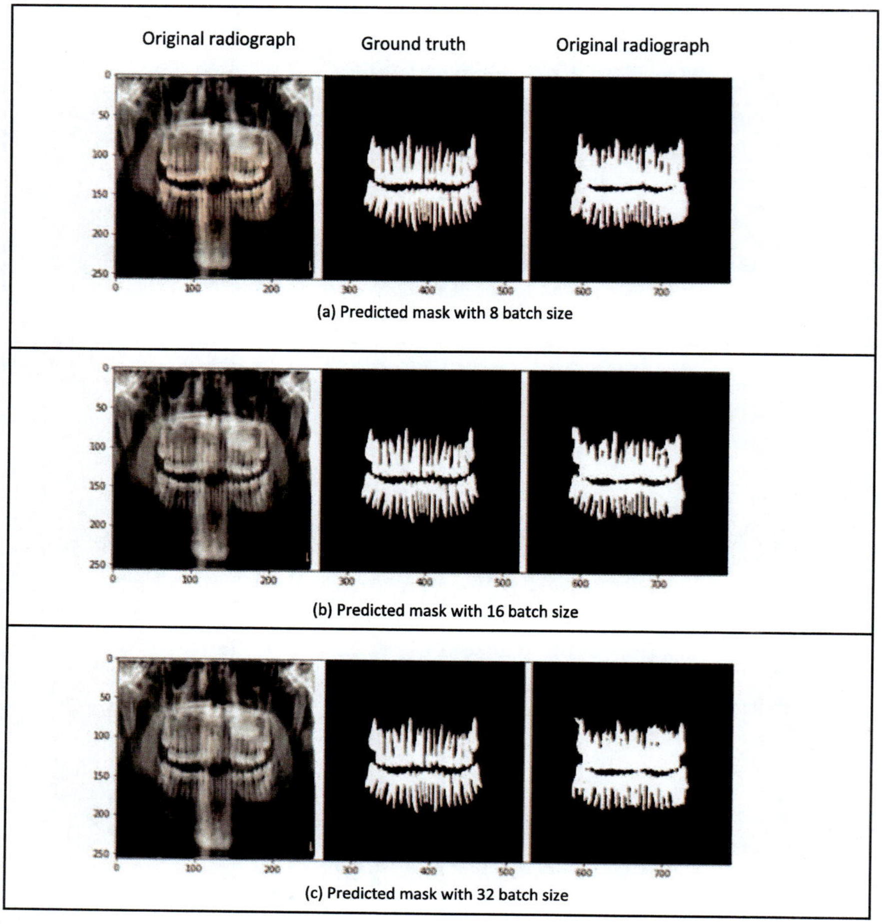

Fig. 8 Predicted mask for different batches: **a** Predicted mask with batch size 8. **b** Predicted mask with batchsize 16. **c** Predicted mask with batch size 32

Visual Analysis of Radiograph Segmentation Figure 8a–c shows the predicted output mask with Nadam optimizer for batch sizes 8, 16, and 32, respectively. The visual study of the predicted mask shows that batch sizes 8 produce nearly identical output the Nadam optimizer and 50 epochs; however, batch number 16 and 32 perform inadequately since they extracts not only the teeth but also the exterior. As a result, batch sizes 16 and 32 should not be used for teeth segmentation. Further, we evaluate the confusion matrix parameters and the analysis of these three batch sizes to determine the best-performing batch size.

Confusion Matrix Parameter Analysis Table 2 shows confusion matrix parameter value for different size of batch 8, 16, and 32. In the instance of the validation dataset, as shown in Table 2, batch size 8 succeeded best on all the parameters, i.e. dice coefficient, precision, and recall with a value of 0.8753, 0.9265, and 0.8839,

Table 2 Analysis of different batches

Training dataset			
Batch size	Dice coefficient	Precision	Recall
8	0.9305	0.9832	0.8553
16	0.8948	0.9766	0.8616
32	0.82058	0.9631	0.8777
Testing dataset			
Batch size	Dice coefficient	Precision	Recall
8	0.8761	0.9232	0.8612
16	0.8479	0.9242	0.8453
32	0.7908	0.9006	0.8776
Validation dataset			
Batch size	Dice coefficient	Precision	Recall
8	0.8753	0.9265	0.8839
16	0.8434	0.9253	0.8361
32	0.8452	0.9041	0.8866

respectively. As can be seen from the visual analysis, the performance of batch sizes 16 and 32 is poorer than that of batch sizes 8. As a result of these findings, on the validation dataset, batch size 8 yielded the best results.

4.3 Analysis of Result on the Basis of Different Epoch with Nadam Optimizer and Batch Size 8

In Sect. 4.2, it was observed that batch size 8 beats, batch sizes 16 and 32 using the Nadam optimizer. As a result, the findings in this section are derived with batch size 8 and distinct epochs. However, on certain combination of epoch, batch sizes 16 and 32 may produce superior results. In the future, these two batch sizes can be compared using various epochs. In this case, the epochs chosen for analyzing the Nadam optimizer using batch size 8 are 25, 50, and 100.

When the number of epochs utilized to train a neural network model exceeds what is required, the training model learns patterns that are very unique to the sample data. As a result, the model is unable to perform well on a new dataset. The model was trained for an ideal number of epochs to reduce overfitting and increase the model's generalization capacity. Loss and accuracy on both the training and validation sets are tracked to determine the epoch number at which the model begins overfitting. The early stopping callback function can monitor either loss/accuracy values. If the loss is being monitored, training is halted when an increase in loss values is detected. We have used the early stopping function during the training process of the model.

Table 3 Analysis of different epochs

Training dataset			
Epoch	Dice coefficient	Precision	Recall
25	0.8875	0.9726	0.8590
50	0.9305	0.9832	0.8553
98	0.9409	0.9869	0.8555
Testing dataset			
Epoch	Dice coefficient	Precision	Recall
25	0.8471	0.9044	0.8857
50	0.8761	0.9232	0.8612
98	0.8855	0.9666	0.8185
Validation dataset			
Epoch	Dice coefficient	Precision	Recall
25	0.8447	0.9055	0.8821
50	0.8753	0.9265	0.8563
98	0.8844	0.9503	0.8125

Table 3 shows the confusion matrix parameters of a different number of epochs. During the training process of the model for 100 epochs, due to the use of early stopping function model, training terminated at epoch 98. For the validation dataset as shown in Table 3, epoch 98 shows the best results on all the parameters, i.e. for dice coefficient, precision, and recall with values 0.8844, 0.9503, and 0.8125, respectively. Figure 9a–c shows the predicted mask for epoch values 25, 50, and 98, respectively. By visual inspection also, we can conclude that model gives the best result using the Nadam optimizer with batch size 8 and 98 epoch value.

5 Conclusion

Considering analysis of medical image is one of the difficult tasks that involve a variety of computing approaches in the imaging application hierarchy, several analysis techniques types, such as image classification, pre-processing, segmentation, compression, and security, must be considered. Analysis of dental radiograph is more difficult than analysus of other other medical images, which makes segmentation a more difficult procedure [13]. Typically, the computer diagnosis system relies heavily on medical image segmentation. Doctors can accurately diagnose diseases and make judgments, thanks to medical image segmentation. For many years, both supervised and unsupervised machine learning approaches have been used to segment dental panoramic radiographs.

The modified U-Net architecture is the proposed model in this paper for accurate segmentation of panoramic radiography dental images. U-Net is a well-known

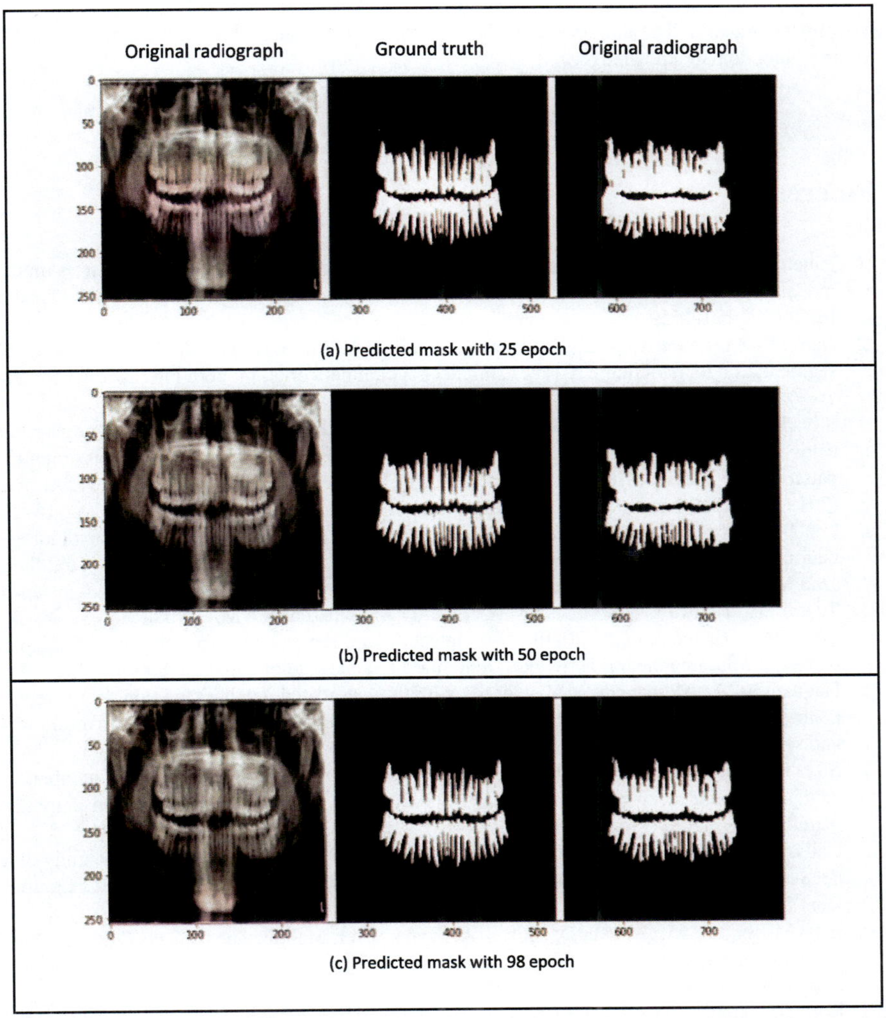

Fig. 9 Predicted mask for different epoch: **a** Predicted mask with 25 epoch. **b** Predicted mask with 50 epoch. **c** Predicted mask with epoch

convolutional neural network (CNN) architecture used in medical image segmentation. The panoramic dental radiographs are taken from the Tuft dental dataset. The MobileNet V2 as an encoder in U-Net architecture has been analyzed with Adam, Nadam, and Adamax optimizers, for 8, 16, and 32 batch sizes and 25, 50, and 98 epochs. With a batch size of 8, the Nadam optimizer, and 98 epochs, the proposed model has a precision of 95.03. Its dice coefficient is calculated to be 88.44. As a result, there is still scope for improvement in the updated U-Net architectural model's confusion matrix parameters.

Acknowledgements The authors wish to thank Dr. Shilpa Godbole for her guidance and assistance for providing domain knowledge.

References

1. Panetta K, Rajendran R, Ramesh A, Rao SP, Agaian S (2021) Tufts dental database: a multimodal panoramic X-ray dataset for benchmarking diagnostic systems. IEEE J Biomed Health Inf 26(4):1650–1659
2. Nader R, Smorodin A, De La Fourniere N, Amouriq Y, Autrusseau F (2022) Automatic teeth segmentation on panoramic X-rays using deep neural networks. In: 26th international conference on pattern recognition, pp 4299–4305
3. Silva B, Pinheiro L, Oliveira L, Pithon M (2020) A study on tooth segmentation and numbering using end-to-end deep neural networks. In: 33rd SIBGRAPI IEEE conference on graphics, patterns and images (SIBGRAPI), pp 164–171; Dental panoramic radiographs with u-nets (2019). In: IEEE 16th international symposium on biomedical imaging (ISBI 2019), pp 15–19
4. Lee JH, Han SS, Kim YH, Lee C, Kim I (2020) Application of a fully deep convolutional neural network to the automation of tooth segmentation on panoramic radiographs. Oral Surg Oral Med Oral Pathol Oral Radiol 129(6):635–642
5. Tuzoff DV, Tuzova LN, Bornstein MM, Krasnov AS, Kharchenko MA, Nikolenko SI, Sveshnikov MM, Bednenko GB (2019) Tooth detection and numbering in panoramic radiographs using convolutional neural networks. Dentomaxillofacial Radiol 48(4):20180051
6. Haghanifar A, Majdabadi MM, Ko SB (2020) Automated teeth extraction from dental panoramic X-ray images using genetic algorithm. In: IEEE international symposium on circuits and systems (ISCAS), pp 1–5
7. Silva B, Pinheiro L, Oliveira L, Pithon M (2020) A study on tooth segmentation and numbering using end-to-end deep neural networks. In: 2020 33rd SIBGRAPI conference on graphics, patterns and images (SIBGRAPI), pp 164–171
8. Caliskan S, Tuloglu N, Celik O, Ozdemir C, Kizilaslan S, Bayrak S (2021) A pilot study of a deep learning approach to submerged primary tooth classification and detection. Int J Comput Dent 24(1):1–9
9. Estai M, Tennant M, Gebauer D, Brostek A, Vignarajan J, Mehdizadeh M, Saha S (2022) Deep learning for automated detection and numbering of permanent teeth on panoramic images. Dentomaxillofacial Radiol 51(2):20210296
10. Helli S, Hamamci A (2022) Tooth instance segmentation on panoramic dental radiographs using U-nets and morphological processing. Düzce Üniversitesi Bilim ve Teknoloji Dergisi 109(1):39–50
11. Arora S, Tripathy SK, Gupta R, Srivastava R (2023) Exploiting multimodal CNN architecture for automated teeth segmentation on dental panoramic X-ray images. Proc Inst Mech Eng Part H J Eng Med 09544119231157137
12. Sandler M, Howard A, Zhu M, Zhmoginov A, Chen L-C (2018) Mobilenetv2: inverted residuals and linear bottlenecks. In: Proceedings of the IEEE conference on computer vision and pattern recognition, pp 4510–4520
13. Rad AE, Mohd Rahim MS, Rehman A, Altameem A, Saba T (2013) Evaluation of current dental radiographs segmentation approaches in computer-aided applications. IETE Tech Rev 30(3):210–222

Chapter 43
Molecular Recognition and Feature Extraction System

Dannerick Elisha, Jimson Sanau, Mansour H. Assaf, Rahul R. Kumar, Bibhya Sharma, and Ronesh Sharma

1 Introduction

In the traditional view, the relationship between protein structure and function is defined by the ability to adopt fixed three-dimensional protein structure. However, recent findings have revealed that most of the functional regions do not adopt a unique three-dimensional structure under physiological conditions and many proteins are either completely disordered or possess long structurally flexible regions [1]. These functional regions are called Molecular Recognition Features (MoRFs) and they exist within the Intrinsically Disorderedproteins (IDPs). These MoRFs are important segments that undergo disorder-to-order transition upon binding with their protein partners and are able to accomplish diverse biological capabilities [2]. The current research work is focused on the computational concept of detecting MoRFs present in the intrinsically disordered protein (IDPs) sequences. The authors note from the literature that discovering MoRFs through computational analysis is an evolving and trending study that has proven to make a great impact towards predicting protein structures.

The work of developing MoRF prediction is an art of computational approach of machine learning. Predetermined dataset is manipulated to train models that can predict unseen observations and classify them accordingly. The development of MoRF predictors follows a traditional architecture procedure that begins with creating positive and negative classes of data observation from the training dataset. Previous works [2–4] have used the same dataset to develop trained models of MoRF predictors. A comparative analysis is being conducted to assess the performance of

D. Elisha · J. Sanau · M. H. Assaf · R. R. Kumar · B. Sharma (✉)
The University of the South Pacific, Laucala Campus, Suva, Fiji
e-mail: bibhya.sharma@usp.ac.fj

R. Sharma
Fiji National University, Suva, Fiji

A. Yadav et al. (eds.), *Proceedings of International Conference on Paradigms of Communication, Computing and Data Analytics*, Algorithms for Intelligent Systems,
https://doi.org/10.1007/978-981-99-4626-6_43

different MoRF predictors, with the goal of identifying which approach produces the most accurate results. The IDP sequences cannot be directly used for training; however, features of the sequences are used to develop numerical representations for the protein sequences. This is called the feature extraction technique which is a common element of machine learning. Out of all classifier types, Support Vector Machine (SVM) is the most mentioned classification method that provides enhanced performance in the field of MoRF prediction studies. The current research work will also implement SVM training and classification methods. The difference is in the choice of protein information, namely, sequence information, evolutionary information, syntactic information and physiochemical information of the intrinsically disordered protein sequences. This research deploys the sequence information. To evaluate the performance of the trained model, we use the standard metrics of measures that are being used by most predictor methods. These performance metrics are AUC-ROC curve, accuracy, precision, sensitivity, specificity and confused matrix implementation.

2 Background

In bioinformatics, the study on protein sequence and structural alignment suggests that protein functional property and biological stability depends on the three-dimensional structure they possess [5]. Since the stability of proteins depends on the folding formation in three-dimensional structure, it is suggested that the amino acid governs the exclusive three-dimensional structure of proteins. A study suggests that the three general interrelated parameters to protein fundamentals are the sequence, structure and function [6]. However, some proteins lack the biological stability from the failure of having complete three-dimensional structure. These proteins are termed as intrinsically disordered proteins (IDP) having intrinsically disordered regions (IDR). These disordered regions; however, do comprise collective biological functioning and cellular developments that are of importance in molecular recognition features, signalling and control formation in the protein to protein interaction [5, 6]. On the contrary, IDP are associated with the development of various human diseases that have existed even before their recognition. Research also states that because of IDP's association with neurodegenerative disorders and synucleinopathies (complication in neurons or nerve fibres), Parkinson's diseases and other associated Down syndrome related problems are caused [7, 8].

Under the scope of machine learning are supervised and unsupervised learning techniques. Both techniques utilize the available data information to develop learning and training analysis. In the simplest explanation, supervised learning aims to approximate the mapping function for the input data (x) that can predict the output data (y) corresponding to the input data as much as possible with the same accuracy. The supervised learning is also deeply simplified and performed under two common techniques, which are known as the Regression and Classification. Linear regression technique is typically used in forecasting, predicting and determining relationships amongst quantitative data. For simplicity, supervised learning is training of labelled data [9]. On the other hand, unsupervised learning is another common technique of machine learning such that its complexity is directly proportional to the data size. This learning is used to group unstructured data according to its similarities and their distinct patterns from the information set. The unsupervised technique approach may also resist categorization and data labelling, but instead gives analytic results of the training success. There are different types of unsupervised learning techniques that are frequently used such as PCA, HMM (Hidden Markov Model), K-means clustering, SVM and others. Overall, unsupervised learning is used for analysis on larger data sets with no pre-labelled classification information [10].

IDP are useful in molecular recognition, which is a development of biological interaction between substances or where insignificant molecules form complexes. These complexes exist for essential biological modulation and signal transduction assuming that IDP are of high selectivity, low attraction and occupies diverse attraction through structural recognition [11]. These characteristics enable the interaction between the proteins through the distinctive MoRF regions. These MoRF residues are typically linear and upon binding, they undergo disorder-to-order transition [12]. Since IDP have biological importance, bioinformatics utilizes machine learning techniques to predict these MoRF regions.

The principle of feature extraction techniques and classifiers were used to identify these regions. These are computational approaches developed to extract protein features and learn with machines using these features to predict MoRF regions. Research [1, 12, 13] state the selective features: syntactical and physicochemical properties, structural information and the evolutionary information. These are the kind of features that are useful for feature extraction towards computational analysis that will detect and predict the MoRF regions within the IDP sequence. Moreover, studies in [1] propose the use of evolutionary information as the best way to extract features such that the proposed evolutionary information aids better analysis by distinguishing the predicted residues with its flank regions. Another research [13] proposes the use of the sequence properties and the evolutionary information together to predict the MoRF region within the intrinsically disordered protein. The reason being, that these two selective features do not require alteration or special handling on their flanks but only rely on the protein sequences itself. More recent research [14] looks on transfer learning from pre-trained model SPOT-Disorder2 which is a current state-of-the-art for protein intrinsic disorder predictor. The method describes the use of internal representation of pre-trained model to develop new objective in MoRF prediction. However, diverse opinions have come about and it's truly arguable on which methods would deliver accurate data analysis.

MoRF predictors learn diverse approaches of analysis that result in significant differences between their prediction results. Since these MoRF predictors score a high number of protein sequences, it is more accurate to do comparison with their performance efficiency. Through diverse computational approach towards the development of MoRF predictors, there are numerous predictors namely: ANCHOR, MoRFpred, MESPSSMpred, γ-MoRF-PredII, SliMpred, and OPAL [2–4]. In [13], the prediction of MoRF is designed by a multi-layer neural network to evaluate feature extraction based on the evolution information and sequence properties. The probability distribution obtained from the neural network is then utilized by Bayes rule as the classifier to predict MoRFs. The paper [3] describes two feature extraction techniques in OPAL for predicting MoRFs of size 5–25 residues along the IDP sequence. The first approach is based on obtaining feature vectors by utilizing structural attributes to obtain bigram frequencies (BigramMoRF) whilst the second obtains feature vectors based on the flank properties (StructMoRF) surrounding the MoRF region. The SVM classifier is then used to analyze the features produced. Since the discovery of IDP and the MoRF regions, it is no doubt that innovative approaches to MoRF predictor developments and improvements will follow through the immeasurable technological changes.

3 Design and Implementation

We aim to achieve prediction of MoRFs within the IDP sequence. We note that if the IDP sequences are provided, there will be no need for a data collection phase to obtain the IDP sequence information. In this work, we use the IDP data sequence to both train and test the prediction of MoRFs. Figure 1 illustrates the general overview of the method that is used for prediction of MoRFs.

The input information is the disordered protein sequence data, using which the feature properties are extracted by performing feature extraction techniques. Afterwards, using these extracted features; the classifier SVM will be used to train and predict the MoRF and non-MoRF regions. Therefore, the adopted approach was to truncate the disordered protein sequence into positive and negative samples to represent the MoRF and non-MoRF regions, respectively. In Fig. 2, the representation for the proposed method is illustrated. The training involves partitioning of

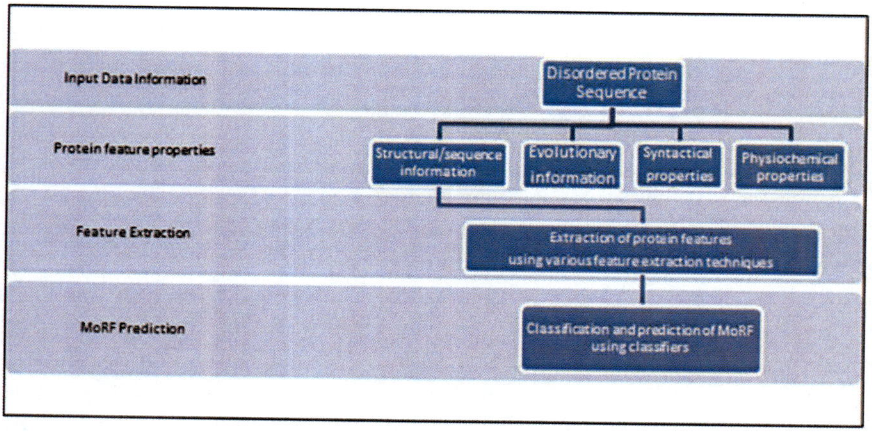

Fig. 1 General overview for developing MoRF predictors [1, 3, 5, 12–14]

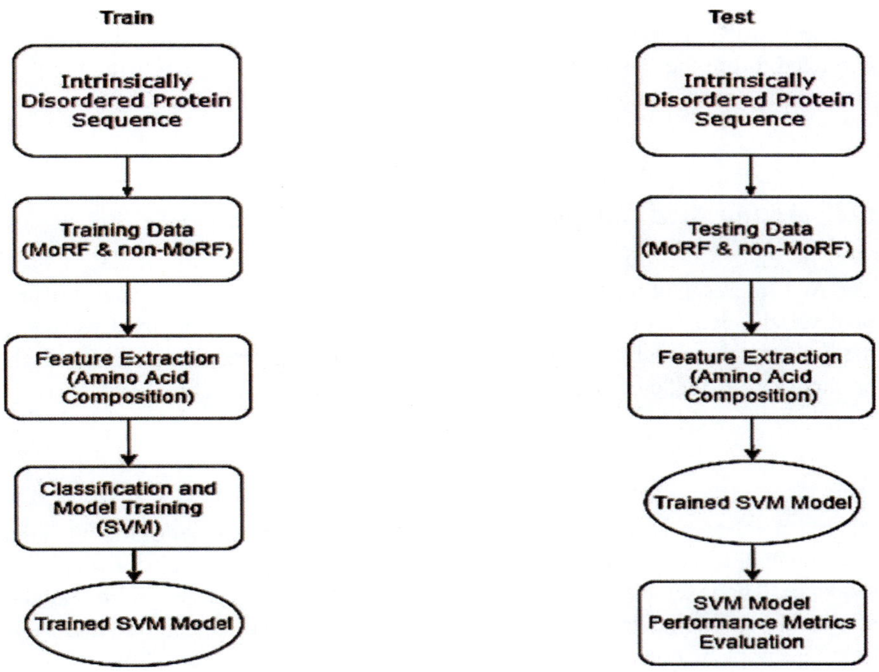

Fig. 2 Overall view of the proposed training and test for MoRF prediction

sequence information to gather MoRF and the non-MoRF regions within the disordered sequence information by means of computational implementation using the Matlab software.

3.1 Training Data Sequence

Initially, the training dataset is provided along with the data locations of the MoRF region within the training sets. These training sets were previously used to benchmark and train other existing predictors in the likes of OPAL, MoRFchibi, MoRFchibi-light and MoRFchibi-web [2–4]. For each of these 421 disordered sequences, the MoRF regions exist in size from 5–25 residues and are distributed randomly along the protein sequence. Table 1 gives the summary of the training dataset information.

Table 1 Statistical information on dataset used for train and test [2, 3]

Dataset	Number of sequences	Total residues	Number of MoRF residues	Number of non-MoRF residues
Training data	421	245,984	5.396	240,588

These MoRF regions along with the left and right flank residues are identified and extracted. From the previous study and research, it was proven that using the flank size of 20 generates good performances on the prediction of MoRF [2, 3].

3.2 Amino Acid Composition

The ACC representation of protein sequence is one of the feature methods useful to map variable length of amino acid sequence into fixed vector length. This feature method was previously used on verified computational prediction of MoRF and prediction of B-cell epitopes [15, 16]. This feature method is defined by:

$$\text{Composition}\ (i) = \frac{n_i}{N} \tag{1}$$

where i represents each of the 20 amino acids, n_i gives the number of the ith amino acid and N denotes the protein length or the sum of residues present in the ith protein sequence. The resultant feature dimension is 20 which is equivalent to the number of standard amino acids. Therefore, using the amino acid composition feature, the training sample is transformed into a fixed vector dimension (842×20).

3.3 Support Vector Machine (SVM)

SVM is one of the most powerful classification techniques that was verified to be useful in classification applications. Various classifiers were explored to predict MoRFs and have shown positive performance [1–5, 15–17]. However, SVM has proven to generate higher significant performance and results. It aims to maximize margin between hyperplanes that separate two classes of data. By default, SVMs are used to classify linear bounded classes and its capability extends to classify non-linear boundaries by applying kernel functions such as radial base, polynomial and sigmoid [17]. The current research uses SVM by radial base function scheme to perform classification on the training data set and draw predictive evaluation. We apply two validation test methods, namely; cross-validation test and independent test methods. These test methods are used for observing the performance of the SVM. By optimizing the SVM parameters C (Box Constraint) and kernel scale (gamma) using the Bayesian optimization evaluation we get an enhanced cross validated SVM model performance.

3.4 *Cross-Validation Test Method*

To test the sequence samples along with the SVM, a cross-validation method will be used. Assuming a sequence set of length l, cross validation will get k subsets containing appropriate samples. In this case for our method, tenfold cross-validation will be executed. The experimental train and test validation will be performed k times on each subset to get the best results for predicting the MoRF regions.

3.5 *Independent Test Method*

Using the original training samples, we randomly partition the sets into the ratio of 70% train and 30% test sets, independent from each other. The independent test follows the traditional procedure by training the SVM of radial basis function using the train data set. The SVM model is trained using these observations present in the train set and then evaluates the performance of the trained SVM model with the test set.

4 Implementation Results

The proposed MoRF predictor model is trained using the IDP training data set used in training benchmarks on other MoRF predictors such as OPAL, MoRFchibi, MoRFchibi-light and MoRFchibi-web [2, 3]. The train sets were used to train and test at the same time. The two test methods used are cross validation and independent test as described in the methodology section. The statistics of the training sets are summarized on Table 1 and each of the IDP sequences contain MoRF regions of length 5 to 25 residues. The system design of the proposed MoRF predictor follows a similar approach carried out in previous research and predictor developments. Being exposed to the research undertaken previously enables fully understanding on aligning the proposed method towards the common system development of MoRF predictors.

As per the extraction of training samples, provided with the position of the MoRF region within each of the training sets, the positive and negative samples are extracted along with 20 flank residues upstream and downstream. Thus, the overall training dataset size is 842 by variable sequence length consisting of 421 positive and negative samples combined together.

The amino acid composition feature extraction follows Eq. (1) and for each of the extracted training samples, both positive (MoRF) and negative (non-MoRF) undergo amino acid counts with respect to the 20 standard amino acid types that make up the IDP sequences. Provided with the bioinformatics toolbox in Matlab, the amino acid count for each of the protein amino acids within the training sequences is possible.

Therefore, the new training dataset becomes 842 × 20, which is a fixed dimension suitable for SVM implementation. The SVM classification with kernel function Radial Basis Function is used to predict the MoRF and non-MoRF classes. Using this kernel trick makes it possible to fix overlapping between data of two classes thus it maps the non-linear data observations to a higher dimensional space making it separable for classification. Radial Basis Function is illustrated by the following equation [18]:

$$k(x_i, x_j) = \exp \exp(-\gamma \|x_i - x_j\|^2), \tag{2}$$

where x_i and x_j are two feature vector points, $\|x_i - x_j\|^2$ is the absolute squared Euclidean distance between the two points, and γ is the gamma value that scales the influence two points have on each other. The prediction performance measure is computed accordingly by comparing the predicted class values to the original class value from the cross validation and independent tests. In reference to the methodology description, calculations were implemented according to the values in the corresponding confusion matrix. The following are the corresponding performance metrics equations [19]:

$$\text{Sensitvity} = \frac{TP}{TP + FN} \tag{3}$$

$$\text{Precision} = \frac{TP}{TP + FP} \tag{4}$$

$$\text{Specificity} = \frac{TN}{TN + FP} \tag{5}$$

$$\text{Accuracy} = \frac{TP + TN}{TP + TN + FP + FN} \tag{6}$$

- *True Positive (TP)—the number of correctly predicted positive classes.*
- *False Positive (FP)—the number of false positives predicted when the actual class is negative.*
- *True Negative (TN)—the number of correctly predicted negative classes.*
- *False Negative (FN)—the number of false negatives predicted when the actual class is positive.*

These calculations give the performance measures summarized in Table 2 for the classification and predictions done by the trained SVM model according to two different test methods.

Implementing the cross-validation test method requires the optimization of the Box Constraint and the Kernel scale parameter of the SVM classifier. Using the Bayesian optimization evaluation, we get an optimized cross validated SVM model with optimized Box Constraint and kernel scale parameters. These parameters

Table 2 Performance measurements for MoRF prediction

Method	AUC	Precision	Accuracy	Sensitivity	Specificity
Cross-validation	0.8349	0.7383	0.7386	0.7348	0.7414
Optimized cross-validation test	0.8844	0.8054	0.7744	0.7293	0.8193
Independent test	0.7998	0.6907	0.7075	0.7356	0.6819
Optimized independent test	0.8476	0.7557	0.7292	0.6871	0.7736

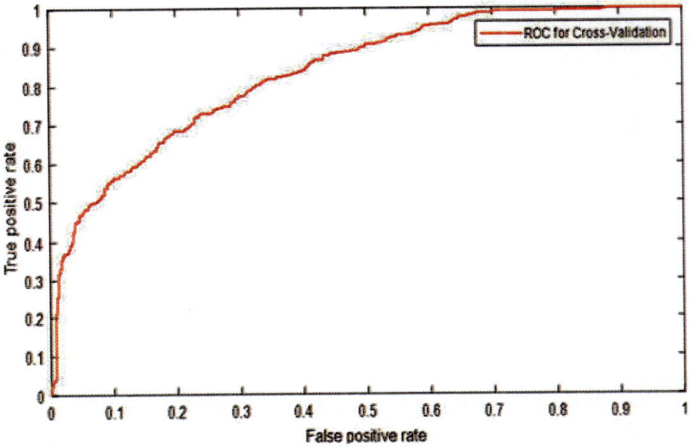

Fig. 3 ROC curve for cross validation model

play a significant role in classification performance. The kernel scale rations the influence two vector points have on each other while the Box Constraint permits misclassification of data points. Figure 3 shows the ROC curve for cross validation model.

The receiver operating characteristic is the curve plot at distributed threshold probabilities which describes the relationship between the TPR and the FPR of the predictor. It demonstrates the ability of the model to separate or distinguish between the MoRF and non-MoRF regions. Figures 3 and 4 show the ROC plot for both the cross-validation test and independent test, respectively. In Figs. 5 and 6, the ROC curve for the optimized model of both test methods are shown. The AUC of the curves for the respective ROC plot is recorded in Table 2, which gives the summary of the SVM model test performance measures. The cross-validation test method significantly outperforms the independent test having the AUC value of 0.8776 and 0.8274, respectively.

In this chapter, an amino acid composition feature is obtained to standardize the training dataset for both combined positive and negative samples. This is the technique of mapping variable sequence length of the training dataset to a fixed feature vector which is an acceptable standard for machine learning algorithms. In reference to Eq. (2), RBF support vector machine transforms data vector space into

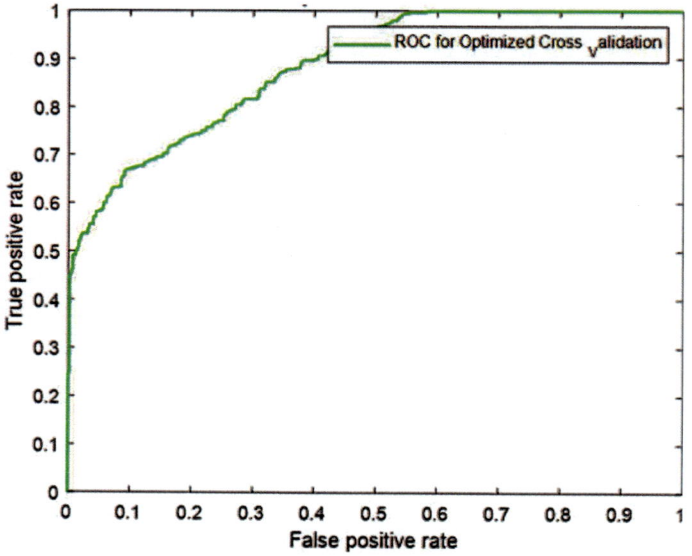

Fig. 4 ROC curve for the optimized cross validation model

higher dimensional vector space to be able to distinguish clustering of training data classes. It is obvious that the amino acid composition counts for the positive and negative classes could result in high clustering of data. Since we are taking flank residues along with the MoRF region this may also sustain non-MoRF dominance resulting in false negative response during prediction. This is the key to implementing the SVM when the prior knowledge of distinguishing the two data classes is complex.

With reference to the resultant performance Table 2, precision metrics gives the ratio of actual positive class predicted to the total positive class predicted thus, in comparison cross validation method predicts actual positive classes better than the independent test method. Sensitivity gives the ratio of actual positive class predicted from the positive class population thus by comparing the two models, the two test methods approximately share equal performance measure. Specificity is the measure of the rate in which actual negative is predicted from the population of negative classes, thus in comparison cross validation test outperforms the independent test method. The accuracy measure shows the ratio for actual prediction of both two classes to the total observations thus the accuracy for cross validation is significantly higher compared to the independent test. The overall relationship plot of the true positive rate against the false positive rate gives the ROC plots at various threshold scenery, as shown in Figs. 3, 4, 5 and 6. The AUC metric is defined as the area under the ROC curve for each of the plots. Theoretically, AUC of 0.7 is an indication that a classification model will be able to classify classes of input data observations [19]. Therefore, from the results obtained on both test methods the AUC value for optimized cross validation test and independent test is 0.8844 and 0.8476, respectively. This is an indication that the proposed method presented in this research has competently classified the positive (MoRF) and negative (non-MoRF) observations. Furthermore, the use of the cross-validation test method is highly recommended due to the significantly enhanced performance obtained in comparison to the independent test method.

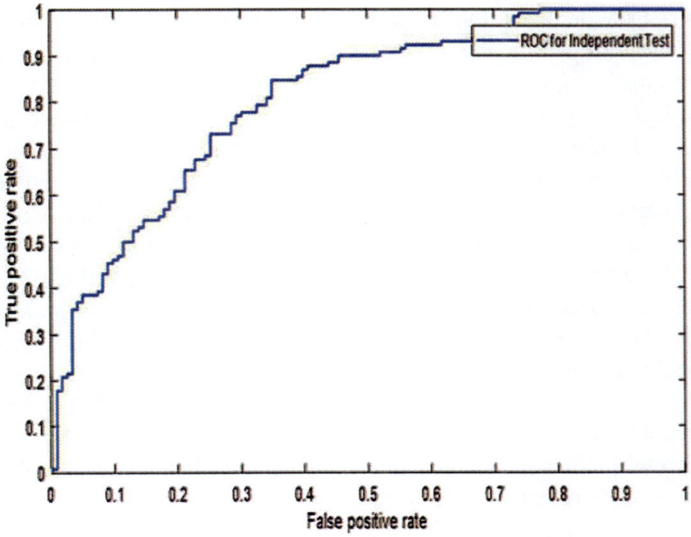

Fig. 5 ROC curve for independent test model

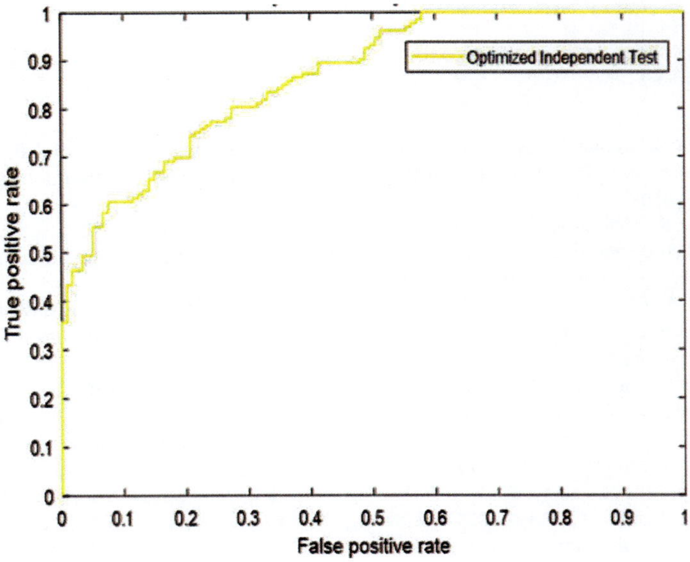

Fig. 6 ROC curve for optimized independent test model

5 Conclusion

In conclusion, this chapter has presented a method for predicting MoRF regions within intrinsically disordered protein sequences using machine learning techniques. The proposed method focuses on using the sequential information of IDP sequences to construct a standardized training sample dataset, which is then used to train SVM classifiers. The trained models are evaluated using both cross-validation and independent test methods, with performance metrics such as AUC, ROC curve, accuracy, precision, sensitivity and specificity used to evaluate the prediction capability of the models. The results obtained demonstrate that the proposed method and trained models have significant prediction capability, with acceptable performance metrics.

This research is significant as it addresses the need for developing MoRF predictors, which can contribute to understanding the biological functions of disordered proteins. The proposed method can be extended to predict MoRF regions in other disordered protein sequences, enabling the discovery of potential protein–protein interactions and their implications in various biological processes. This research also highlights the importance of machine learning techniques in predicting protein properties, providing a valuable tool for advancing the field of protein research. Overall, the findings presented in this paper have important implications for understanding protein structure and function and can contribute to the development of novel therapeutic strategies for various diseases.

Future work will look at utilizing recent neural based approaches, such as Transformers to further optimize the classification accuracies and open the door for explain-ability of the neural based models.

References

1. Sharma R, Kumar S, Tsunoda T, Patil A, Sharma A (2016) Predicting MoRFs in protein sequences using HMM profiles. BMC Bioinform 17(19). Available: https://doi.org/10.1186/s12859-016-1375-0
2. Sharma R, Sharma A, Patil A, Tsunoda T (2019) Discovering MoRFs by trisecting intrinsically disordered protein sequence into terminals and middle regions. BMC Bioinform 19(13). Available: https://doi.org/10.1186/s12859-018-2396-7
3. Sharma R, Raicar G, Tsunoda T, Patil A, Sharma A (2018) OPAL: prediction of MoRF regions in intrinsically disordered protein sequences. Bioinformatics 34(11):1850–1858. Available: https://doi.org/10.1093/bioinformatics/bty
4. Malhis N, Jacobson M, Gsponer J (2016) MoRFchibi SYSTEM: software tools for the identification of MoRFs in protein sequences. Nucleic Acids Res 44(W1):W488–W493
5. Sharma R, Bayarjargal M, Tsunoda T, Patil A, Sharma A (2018) MoRFPred-plus: computational identification of MoRFs in protein sequences using physicochemical properties and HMM profiles. J Theoret Biol 437:9–16. Available: https://doi.org/10.1016/j.jtbi.2017.10.015
6. Midic U, Oldfield C, Dunker A, Obradovic Z, Uversky V (2009) Protein disorder in the human diseasome: unfoldomics of human genetic diseases. BMC Genom 10(1):S12. Available https://doi.org/10.1186/1471-2164-10-s1-s12
7. Uversky V et al (2009) Unfoldomics of human diseases: linking protein intrinsic disorder with diseases. BMC Genom 10(1):S7. Available: https://doi.org/10.1186/1471-2164-10-s1-s7

8. Al-Tabbakh SM, Mohamed HM, El ZH (2018) Machine learning techniques for analysis of Egyptian flight delay. Int J Data Mining Knowledge Managem Process 8(3):01–14. Available https://doi.org/10.5121/ijdkp.2018.8301
9. Ryan MM, Shobha G, Rangaswamy S (2020) Supervised learning—an overview I ScienceDirect Topics. Sciencedirect.com 2020. [Online]. Available https://www.sciencedirect.com/topics/computer-science/supervised-learning. Accessed 1 Mar 2020
10. Mishra S (2020) Unsupervised learning and data clustering. Medium 2020. [Online]. Available: https://towardsdatascience.com/unsupervised-learning-and-data-clustering-eeecb78b422a. Accessed 1 Mar 2020
11. Hsu W et al (2020) Intrinsic protein disorder and protein-protein interactions. In: Pacific symposium on biocomputing. Pacific symposium on biocomputing, pp 1–13. Available: https://doi.org/10.1142/9789814366496_0012 Accessed 20 Feb 2020
12. Mohan A et al (2006) Analysis of molecular recognition features (MoRFs). J Molecular Biol 362(5):1043–1059. Available: https://doi.org/10.1016/j.jmb.2006.07.087
13. He H, Zhao J, Sun G (2019) Prediction of MoRFs in protein sequences with MLPs based on sequence properties and evolution information. Entropy 21(7):635. Available: https://doi.org/10.3390/e21070635
14. Hanson J, Litfin T, Paliwal K, Zhou Y (2019) Identifying molecular recognition features in intrinsically disordered regions of proteins by transfer learning. Bioinformatics. Available https://doi.org/10.1093/bioinformatics/btz691
15. Wang Y, Guo Y, Pu X, Li M (2017) A sequence-based computational method for prediction of MoRFs. RSC Adv 7(31):18937–18945. Available https://doi.org/10.1039/c6ra27161h
16. EL-Manzalawy Y, Dobbs D, Honavar V (2008) Predicting flexible length linear B-cell epitopes. J Molecular Recogn 21(4):121–132. Available: http://www.lifesciencessociety.org/CSB2008/toc/PDF/121.2008.pdf
17. Reddy H, Sharma A, Dehzangi A, Shigemizu D, Chandra A, Tsunoda T (2019) GlyStruct: glycation prediction using structural properties of amino acid residues. BMC Bioinform 19(13). Available https://doi.org/10.1186/s12859-018-2547-x
18. Team D (2020) Kernel functions-introduction to SVM Kernel & examples—dataflair. DataFlair, 2020 [Online]. Available https://data-flair.training/blogs/svm-kernel-functions/. Accessed 28 May 2020
19. Understanding AUC—ROC Curve, Medium (2020) [Online]. Available https://towardsdatascience.com/understanding-auc-roc-curve-68b2303cc9c5. Accessed 22 May 2020

Chapter 44
Object Recognition with Voice Assistant for Visually Impaired

Deepanshu Jain, Isha Nailwal, Arica Ranjan, and Sonu Mittal

1 Introduction

One of the five basic senses, vision is the most important for a person to perceive their environment. The quality of life is significantly impacted by vision impairment. According to WHO estimates, we have at least 2.2 billion blind persons in the globe [1]. The difficulties faced by blind or low-vision people on a daily basis are caused by the lack of accessibility for the visually impaired. There are not many inclusive or accessible activities available for them. Visually impaired individuals face various difficulties in their day to day activities, one of which is the difficulty in recognizing and identifying objects in their environment. This can make it difficult for them to navigate and perform simple tasks, such as finding their way in a new environment or identifying objects in a room [2].

According to data from the National Program for Control of Blindness and Visual Impairment (NPCB) in India in 2016, an estimated eight million people were visually impaired, including around 1.8 million who are blind. The data also suggest that most visually impaired individuals in India live in rural areas and are from lower socio-economic backgrounds. Additionally, there is a higher prevalence of visual impairment among women and elderly individuals [3].

D. Jain · I. Nailwal (✉) · A. Ranjan · S. Mittal
Computer Science and Engineering, Dr. Akhilesh Das Gupta Institute of Technology and Management, New Delhi, India
e-mail: isha.nailwal@gmail.com

D. Jain
e-mail: djjain844@gmail.com

A. Ranjan
e-mail: aricaranjan04@gmail.com

S. Mittal
e-mail: sonu.mittal@adgitmdelhi.ac.in

A. Yadav et al. (eds.), *Proceedings of International Conference on Paradigms of Communication, Computing and Data Analytics*, Algorithms for Intelligent Systems,
https://doi.org/10.1007/978-981-99-4626-6_44

A wide range of tools and strategies have been created throughout human history to overcome these challenges, from more traditional ones like reading glasses and walking sticks to more contemporary ones like Braille (a system of touch-based reading and writing). People who are blind or visually impaired can see their environment more clearly when they employ assistive technologies, which include specialized low- and high-tech devices made for people with disabilities [4]. These tools include daisy book readers, magnifying software, and specialized screen-reading software. Despite these device's proven benefits, widespread use has been hampered by issues like cost and unfavorable attitudes toward vision loss. Vision impairment further increases the probability of early nursing or care home admission, difficulties walking, a higher risk of falls and fractures, and social isolation. It is also observed that even a smart white cane couldn't detect obstacles far away. Vision impairment also poses an enormous global economic burden. Thus, a proper solution is needed to offer an accessible and safer environment for the visually impaired.

Algorithms in Computer Vision and Deep Learning, both have been improving at a rapid rate [5]. Without compromising on cost or cultural perceptions, object recognition could replicate seeing for the blind and visually handicapped better than the present traditional approaches. Hence, we propose a system that aims to detect objects using computer vision and deep learning. It will recognize objects and alert the user about them with the help of a voice assistant. Object Recognition and detection have been done by training the data on the YOLOv3 model. The COCO dataset was taken for this purpose. A voice assistant built using pyttsx3 has been integrated with the model to provide verbal feedback. The system will also raise an alert if a nearby object is harmful. This will help the users in avoiding the obstacles in front of them, thus increasing accessibility. Visually Impaired people face various problems in their day-to-day activities, which can be avoided by using this project.

The rationale behind our research project is to assist purblind individuals in recognizing objects in their environment and thus improve their ability to navigate and perform tasks independently. The primary motivation is to improve the quality of life of the visually impaired.

2 Related Works

Gada et al. "Object Recognition for the Visually Impaired" [6]; used the Dangling Object Detection algorithm, which determines the object's position and if it falls in the warning range. With the help of buzzers, an alert is generated so that the person does not face any accidents.

Jabnoun et al. "Object detection and identification for blind people in video scenes" [7]; proposed a method of visual replacement based on locating objects around the blind person. The SIFT algorithm is used by the system. An affine transform, a geometric model, is fitted to the original set of matches using the RANSAC method.

Shaikh et al. “Assistive Object Recognition System for Visually Impaired” [8]; deployed YOLOv3 in a system using Raspberry Pi that is suggested to help the blind. This algorithm was trained using the coco database. The trial findings demonstrated that YOLO v3 produces state-of-the-art results, with an overall performance score of 85% to 95% and a recognition accuracy score of 100% (for objects like people, chairs, clocks, and cell phones).

Birambole et al. “Blind Person Assistant: Object Detection” [9]; introduced an SSD algorithm. The suggested framework used ultrasonic sensors to locate the nearest obstruction and then sent a warning to inform those who are blind of its location. The topmost level of a flap with an object determines the class and position of that object. It has a microcontroller with a Wi-Fi module built in.

Wang et al. “Object Detection and Recognition for Visually Impaired People” [10]; measured the objects depending on the proximity to the camera, and the perspective changes its size. The author suggested employing a common RGBD camera to detect stairways and pedestrian crosswalks using a computer vision-based technique. Matlab code is used to implement the suggested algorithm. The accuracy of detection rate for the suggested algorithm was 91.14%.

Kumar et al. “Assistive System for Visually Impaired using Object Recognition,” [11]; proposed an assistive system ensemble of objects and its color recognition module. It is implemented with OpenCV and a multimedia processor with an embedded board. The object recognition algorithm is assessed using the publicly accessible online dataset as well as our dataset, and it is contrasted with cutting-edge techniques in digital object recognition audio assistants for the blind.

3 Methodology

3.1 Technologies Used

Transfer Learning—The core idea behind transfer learning is simple: take a model that has been trained on a large dataset and use its knowledge on a smaller dataset. With a CNN, we freeze the network’s initial convolutional layers and just train the last few layers that generate predictions for object recognition [12].

Data Augmentation—The process of adding new data points to existing data in order to artificially enhance the amount of data is known as data augmentation. This includes modifying the dataset slightly or using machine learning algorithms to create additional data points in the latent space of the original data.

NMS—The NMS threshold is a parameter that is used to control how aggressive the suppression of redundant bounding boxes is. A higher NMS threshold will result in fewer bounding boxes being removed, while a lower NMS threshold will result in more bounding boxes being removed.

IOU—Intersection Over Union measures the similarity between two bounding boxes, with a value of 1 indicating that the bounding boxes are identical and a value of 0 indicating that the bounding boxes do not overlap at all.

YOLOv3—With just one evaluation, YOLOv3's convolutional neural network can predict item bounding boxes and class probabilities from entire photos. This model is based on a modified version of the Darknet architecture. YOLOv3 has a mean average precision of 43.5% when using a single scale and a mAP of 48.1% when using multiple scales. In this project, YOLOv3 is used to detect objects in the images captured by the camera [13].

Pyttsx3—Pyttsx3 is one of the libraries in Python which allows conversion of text-to-speech. It is used for building the Voice Assistant.

3.2 Implementation

We have implemented the solution, as shown in Fig. 1. by training the Yolov3 model on the COCO (Common objects in context) dataset. The photos in the collection are screenshots of commonplace items taken in well-known settings. In the dataset, the labels are recognized through a file called coco.names. The dataset consists of 80 Labels. The project implements an image and video object detection classifier using yolov3 models. Object recognition is done in real time using the web camera.

We also performed data augmentation to increase the accuracy of the model. This includes making small adjustments to the data or creating new data using deep learning models. The model uses a convolutional neural network (CNN) to estimate the bounding boxes and class probabilities for objects in a picture. The model creates a grid of cells from the input image, and within each cell it displays a set of bounding boxes and class probabilities. By minimizing the difference between the predicted bounding boxes and the ground-truth bounding boxes with the use of a loss function,

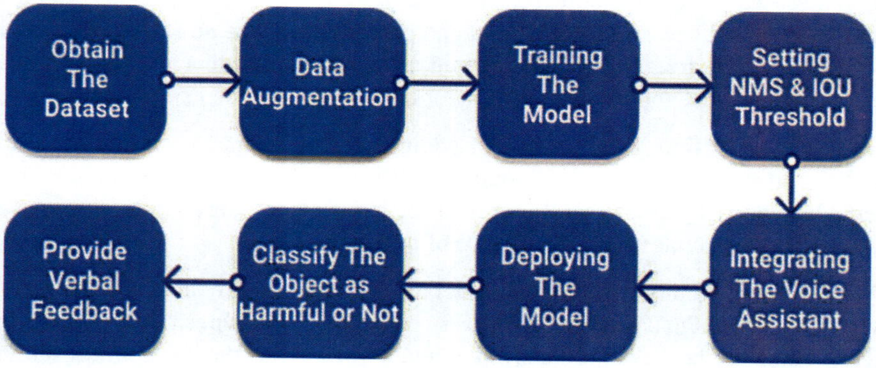

Fig. 1 Implementation of the idea

the model learns to detect objects during the training phase. The model can then detect objects in new images by running the CNN on the input image and using the predicted bounding boxes and class probabilities to identify the things in the picture. We have used 106 layers of neural networks.

Non-Maximum Suppression allowed to remove redundant bounding boxes and improve the overall accuracy of object detection. We have kept the NMS threshold at 0.8. The NMS threshold compares the Intersection over Union between two bounding boxes. We have kept the IoU threshold at 0.6. This means only the bounding boxes whose detection probability is higher than 0.6 will be selected.

After successfully building the model, we integrated it with a voice assistant. We have created the Voice Assistant using pyttsx3. After this, we deployed the model on the Web camera script. This enabled it to recognize objects in real-time. After this, we created a list of labels that can be harmful. We then added a condition for the object if it comes under the harmful category. The voice assistant will provide Verbal Feedback to the user on whatever things are detected and nearby, including the harmful ones. The voice assistant will also send an alert if the object is harmful.

3.3 Working

Figures 2 and 3 show the model working on an image and on the webcam. The model successfully recognized objects in images and in real-time using the webcam.

Fig. 2 Model working on an image

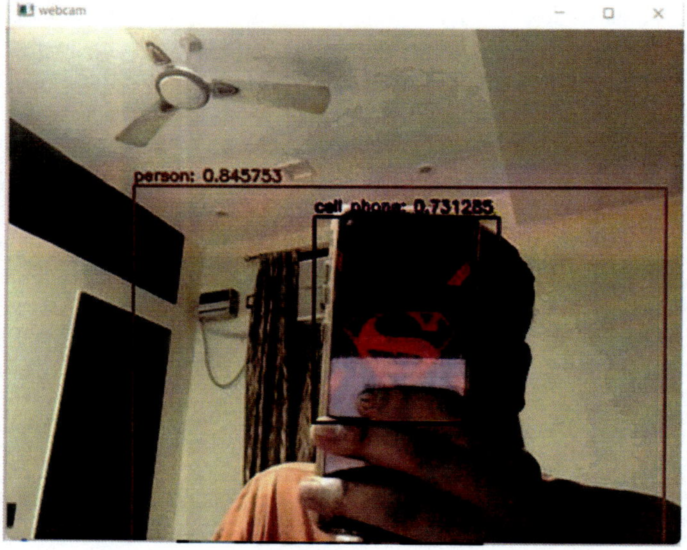

Fig. 3 Model working on Webcam

4 Result Analysis

The goal of testing is to ensure that the model can accurately recognize and identify objects in images and videos and provide appropriate feedback to the user.

Testable Features:

Testing concentrated on the aforementioned areas:

- Model Efficiency
- Text-to-Speech Working
- Average Time Taken By The Model

We used 12 test cases to evaluate the model's effectiveness, of which 8 were picture tests and 4 used a web camera. We could assess the model's effectiveness by tracking how frequently it correctly identified the object. The voice assistant was also used for all of these test scenarios, and the outcomes were recorded to assess how well text-to-speech was functioning. Additionally, the time it took the model to identify the objects was noted for each test instance, and later the average time was determined. Table 1 shows a summary of the results analyzed.

Table 1 Result analysis summary

Number of tests with images	8
Number of tests with webcam	4
Total test cases	12
Number of passed tests	11
Number of failed tests	1
Percentage of passed tests	91%
Percentage of failed tests	9%
Text to speech (Pass percentage)	100%
Average time taken	1.456 s

5 Applications

Advancements in technology, such as deep learning algorithms, have led to more accurate object recognition systems. Some of the features that can be implemented in the future to make this project more accessible:

5.1 Improved Lifestyle

The project aims to help the purblind individuals in recognizing and identifying things in their environment, which can improve their lifestyles.

5.2 Real-Time Object Detection

Using a deep learning-based model such as YOLOv8, the project can detect objects in real-time, allowing visually impaired individuals to quickly and accurately identify objects in their environment.

5.3 Accessibility

By integrating the model with a voice assistant, the project aims to make the technology more accessible and user-friendly for visually impaired individuals.

6 Conclusion

In conclusion, this work aims to assist visually impaired individuals in recognizing and identifying objects in their environment. By using deep learning and computer vision techniques, the project aims to provide near-blind individuals with a tool to help them identify objects in their environment and improve their ability to navigate and perform tasks independently. The use of a voice assistant in this project also aims to make the technology more accessible and user-friendly for visually impaired individuals.

The work can significantly improve the quality of life of visually impaired individuals. However, it also faces limitations, such as high computational cost, generalization, and privacy concerns.

Overall, this work is a promising research direction that has the potential to benefit visually impaired individuals significantly, but further research and development are required to address the limitations and improve the technology's performance and reliability.

7 Future Scope

Advancements in technology, such as deep learning algorithms, have led to more accurate object recognition systems. Some of the features that can be implemented in the future to make this project more accessible:

7.1 Adding Distance Calculation

The user will be told via voice assistant about the closest object after we calculate the relative distances of all the things from the camera [14, 15].

They will find it simpler to move around as a result.

7.2 Integration of the Model into Everyday Devices

The model can be integrated into any device with a camera. Therefore, integrating it with devices like smartphones can make it very easy to use in real-world scenarios.

7.3 Integration of Other Assistive Technologies with the Model

Technologies like text-to-speech and face recognition can also be integrated to provide more comprehensive assistance to visually impaired individuals [16].

Overall, the future of object recognition projects is expected to continue to evolve and improve to make the technology more accurate and widely available for the visually impaired.

References

1. Blindness and vision impairmentl Available Online https://www.who.int/news-room/fact-sheets/detail/blindness-and-visual-impairment
2. Daily Life Problems, Struggle and Challenges Faced by Blind Peoplel Available Online https://wecapable.com/problems-faced-by-blind-people/
3. Blindness and visual impairment and their causes in Indial Available Online https://www.ncbi.nlm.nih.gov/pmc/articles/PMC9302795/
4. IoT Enabled Automated Object Recognition for the Visually Impaired. Comput Methods and Programs in Biomed Update 1. https://doi.org/10.1016/j.cmpbup.2021.100015
5. Industry and Object Recognition: Applications. Appl Res Challenges Toward Category-Level Object Recogn 4170. https://doi.org/10.1007/11957959_3
6. Object Recognition for the Visually Impaired. In: 2019 International conference on nascent technologies in engineering (ICNTE). https://doi.org/10.1109/ICNTE44896.2019.8946015
7. Object Detection and Identification for Blind People in Video Scene. In: 2015 15th International conference on intelligent systems design and applications (ISDA). https://doi.org/10.1109/ISDA.2015.7489256
8. Assistive Object Recognition System for Visually Impaired (2020) Int J Eng Res Technol (IJERT) 09(09). https://doi.org/10.17577/IJERTV9IS090382
9. Blind Person Assistant: Object Detection. Int J Res Appl Sci Eng Technol. https://doi.org/10.22214/IJRASET.2022.40850
10. Object Detection and Recognition for Visually Impaired People, CUNY Academic Works | Available Online https://academicworks.cuny.edu/cc_etds_theses/96
11. Assistive System for Visually Impaired using Object Recognition | Available Online http://ethesis.nitrkl.ac.in/7480/1/138.pdf
12. Multiple Object Recognition with Visual Attention. https://doi.org/10.48550/ARXIV.1412.7755
13. CICERONE—a real time object detection for visually impaired people. In: 2021 IOP conference series: material science engineering, vol 1085. pp 012006. https://doi.org/10.1088/1757-899X/1085/1/012006
14. Dist-YOLO: Fast Object Detection with Distance Estimation (2022) Appl Sci 12:1354. https://doi.org/10.3390/app12031354
15. Approximating the Speed of an Object and its Distance using OpenCV in Python | Available Online https://www.section.io/engineering-education/approximating-the-speed-of-an-object-and-its-distance
16. An Optimized Object Detection System for Visually Impaired People. In: Second international conference on sustainable technologies for computational intelligence pp 25–38. https://doi.org/10.1007/978-981-16-4641-6_3

Chapter 45
Emotion Recognition-Based Emoji Retrieval

P. Parvathi Sreyani, Kandula Rakshitha, Nasalai Sanjana, Yeddula Greeshma, and Ashwini M. Joshi

1 Introduction

Emojis have undoubtedly become a vital component of our everyday communication, experiencing a significant surge in usage in recent years. However, selecting the most appropriate emoji to convey a particular message can often be time-consuming and involve scrolling through an extensive collection. Fortunately, researchers have been exploring the realm of emotion recognition-based emoji retrieval, an emerging field that seeks to automatically generate and recommend emojis based on the emotional content of text or conversations.

With the proliferation of digital communication platforms like social media and messaging applications, emojis have evolved into indispensable tools for expressing emotions within text-based conversations [1]. Yet, many individuals, especially those with limited familiarity with the vast array of available emojis, find it challenging to pinpoint the perfect emoji that accurately captures their intended sentiment. This is where the promising field of emotion recognition-based emoji retrieval steps in.

The fundamental objective of emotion recognition-based emoji retrieval is to develop algorithms capable of analyzing the emotional essence of a text or conversation, thereby suggesting the most suitable emoji to effectively convey that emotion. These algorithms commonly leverage natural language processing (NLP) techniques, including sentiment analysis [2], to extract the emotional nuances embedded in the text.

P. Parvathi Sreyani (✉) · K. Rakshitha · N. Sanjana · Y. Greeshma · A. M. Joshi
Department of Computer Science and Engineering, 100 Feet Ring Road, Banashankari Stage III, Dwaraka Nagar, Banshankari, Bengaluru, India
e-mail: pratapsreyani@gmail.com

A. M. Joshi
e-mail: ashwinimjoshi@pes.edu

A. Yadav et al. (eds.), *Proceedings of International Conference on Paradigms of Communication, Computing and Data Analytics*, Algorithms for Intelligent Systems,
https://doi.org/10.1007/978-981-99-4626-6_45

The extracted emotional context is subsequently mapped to a set of relevant emojis using diverse techniques, such as rule-based systems, machine learning algorithms, or deep learning models. Emotion recognition-based emoji retrieval boasts numerous applications in digital communication. For instance, it can significantly enhance user experiences by offering personalized and pertinent emojis based on their prevailing emotional state. Furthermore, it holds potential for sentiment analysis in social media monitoring, enabling companies to gauge the sentiment surrounding their brand or products through the examination of emojis employed in customer feedback.

Nevertheless, emotion recognition-based emoji retrieval encounters certain challenges due to the subjective nature of emotions. Emotions can be expressed in multifarious ways, and the interpretation of a specific emotion may vary from person to person. Additionally, the scarcity of labeled data for training machine learning models poses another obstacle, particularly for less common or niche emotions.

Emotion recognition-based emoji retrieval stands as a captivating field of study with the capacity to revolutionize digital communication by furnishing users with personalized and pertinent emojis tailored to their emotional state. Given the escalating prevalence of digital communication, the demand for more sophisticated and precise algorithms in this domain is poised to surge, rendering it a promising area for further research and development.

Convolutional neural networks (CNNs) have emerged as a powerful tool for image recognition and classification, exhibiting great promise in various tasks. In this study, we present a novel approach for developing an emotion recognition-based emoji retrieval system by leveraging the capabilities of a CNN, combined with foreground extraction and the VGG16 model.

To begin with, our system incorporates the GrabCut algorithm, a well-established technique for foreground extraction in images. By employing GrabCut, we are able to effectively separate the foreground (objects of interest) from the background, which is a crucial step in accurately recognizing emotions.

Subsequently, we utilize the VGG16 model, a widely recognized and successful CNN architecture [3]. In image classification tests, this model succeeded stunningly well. With the ability of the VGG16 model, we can classify and notice emotions connected to the foreground elements retrieved by GrabCut.

The GrabCut method is used to conduct foreground extraction in our proposed system's overall pipeline, and then the obtained foreground pictures are fed into the VGG16 model for emotion identification. We want to accomplish precise and reliable emotion identification using the combination of these components while enabling efficient emoji retrieval.

Our emotion recognition-based emoji retrieval system has a promising foundation thanks to the use of CNNs, foreground extraction, and the VGG16 model. By automatically generating relevant emoji suggestions based on the emotional content of an image, this method has the potential to improve the user experience. It may also be used in a variety of contexts, including social media, messaging services, and other digital communication platforms where emojis are essential for conveying emotions.

Using the capabilities of CNNs and innovative techniques for image processing and classification, we aim to develop emotion recognition-based systems through this study.

2 Literature Survey

In a study conducted by Pranav et al. [4], they put forth a method that utilizes deep Convolutional Neural Networks (CNN) for the recognition of facial emotions. The effectiveness of this approach was evaluated using the widely recognized FER2013 dataset. Remarkably, the experimental results demonstrated that the proposed model achieved remarkable accuracy in accurately discerning emotions. The CNN model obtained an impressive accuracy rate of 93.8% when classifying the six fundamental emotions. Comparing their findings with other state-of-the-art methods, the authors discovered that their approach surpassed the majority of them in terms of performance. The research paper emphasizes the tremendous potential of deep learning techniques, particularly CNNs, in the realm of facial emotion recognition. The authors also suggest that their method holds promise for a wide range of applications, including healthcare, education, and entertainment.

In their research, Adeyanju et al. [5] aimed to compare the efficacy of various Support Vector Machine (SVM) kernels in the task of facial expression recognition. The authors explored different SVM kernels, such as linear, polynomial, radial basis function, and sigmoid, employing the CK + dataset. Through their investigation, they discovered that the radial basis function kernel exhibited superior performance when compared to the other kernels. Notably, it achieved an impressive recognition accuracy of 89.3%.

John et al. [2] introduced an innovative real-time facial emotion identification system that employed enhanced feature extraction and preprocessing techniques. The proposed method achieved an impressive recognition accuracy of 96.3% when evaluated using the CK+ dataset. It should be noted that the system appropriately analyzed emotions in real-time, implying its application in areas like emotional computing and human–computer interaction. This indicates how the system may be used in practical situations and how helpful it is to emerging fields like emotional analysis and human–machine interaction.

In a research on the analysis of expression data for emotion detection, Balasubramanian et al. [3] focused on facial expressions. To extract characteristics from facial photographs, they adopted a technique that included filters called Gabor and the principal component analysis (PCA). In order to assess the information richness of the retrieved characteristics for efficient emotion identification, they also used mutual information theory. The results showed that some face areas, particularly the lips and eyes, had more informative characteristics for precise emotion detection than other facial regions.

Considering the strength of deep learning techniques, Srivastava et al. [6] suggested a novel approach for retrieving emojis based on emotion recognition. The

FER2013 dataset was used by the researchers to train a deep convolutional neural network (CNN) which successfully recognized face emotions. The model that was trained was then used to extract the most pertinent emojis in accordance with the recognized emotions. The suggested approach demonstrated great accuracy when obtaining relevant emojis for a range of facial expressions, highlighting its potential for use in real-world contexts like emotion-based chatbots and social media platforms.

Sergeeva et al. [7] carried out studies on the recognition of emotions from small gestures with a focus on facial and ocular area detection. They advise using Haar as part of their method.

A foreground extraction-based facial expression Recognition (FER) system based on the Xception model was the focus of Alwin Poulose et al.'s research [1]. To extract the portions of the image that are necessary for emotion identification and eliminate the rest, they used the GrabCut algorithm as a preprocessing technique. The model enhanced accuracy and successfully decreased system classification errors by employing this technique. The model's performance was greatly improved by using foreground-extracted photos, giving better results.

A new approach for emoji an idea based on text-based emotion identification, was put forth by Hoque et al. [8]. They extracted characteristics from a collection of tweets, then used a support vector machine to categorize emotions. The best emojis for the provided text were then suggested using the identified emotions as a guide. The suggested system produced accurate recommendations for pertinent emojis, demonstrating its potential for use in real-world settings like chatbots and social media sites.

A brand-new method for retrieving emojis based on multi-modal emotion identification was developed by Zhang et al. [9]. They used a multi-modal deep neural network to extract characteristics and categorize emotions from both text and visual modalities. The most appropriate emojis were then found using the defined emotions.

Zhang et al. [10] focus was on a unique method for retrieving emojis using deep learning to recognize emotions. The researchers used a deep neural network to extract elements from a sample of tweets and classify moods. The best emojis for the provided text were then retrieved using the categorized emotions. Its potential for use in communication sites and messaging apps was further highlighted by the suggested method's excellent accuracy in finding relevant emojis.

An emoji retrieval system based on emotion recognition was developed in another article by Kumar et al. [11]. Face feature recognition and foreground extraction methods have been used by the system. The Dlib package was used by the system to first identify facial landmarks, and then a background subtraction technique was used to identify the foreground face region. The Local Binary Pattern (LBP) approach was utilized to extract characteristics from the segmented face area. Then, in order to identify emotions, these properties were supplied to the Support Vector Machine (SVM). The appropriate emojis were then found using an algorithm for matching based on the recognized emotion.

Following a review of the literature survey on emotion recognition based emoji retrieval, we have come across several methods for foreground extraction. We prefer the GrabCut method for foreground extraction as a preprocessing step in our suggested system from among them. This technique enables accurate foreground object classification, which can be used to extract information required for emotion identification.

The GrabCut algorithm and the VGG16 model, a well-known deep learning model known for its ability to extract useful features from pics are combined in our suggested method. The VGG16 model performed pre-training on a large data set, allowing it to find significant representations and patterns. Our goal is to improve the system's overall performance by applying the retrieved features from the VGG16 model alongside with the GrabCut algorithm to increase the accuracy of emotion identification. The combined usage of all of the feature extraction methods has the potential to improve the features of the system we have proposed.

In our evaluation process, we will utilize the widely recognized FER2013 dataset, which has been extensively used in the field of emotion recognition. This dataset consists of a collection of facial images annotated with labels representing seven different emotions: anger, disgust, fear, happiness, sadness, surprise, and neutral. By leveraging the FER2013 dataset, we can effectively train and test our proposed system, assessing its proficiency in accurately detecting emotions and retrieving relevant emojis. The utilization of this well-established dataset will enable us to evaluate the effectiveness and robustness of our system in a standardized and comparable manner, contributing to the advancement of emotion recognition research.

2.1 Models

Foreground Extraction:

Foreground extraction plays a vital role in image preprocessing, particularly in computer vision applications. Its purpose is to separate the foreground, which represents the objects of interest, from the background in an image [1]. This task can be challenging, especially when dealing with complex or cluttered backgrounds where foreground objects share similar colors or textures with the background. Several techniques are employed for foreground extraction in image preprocessing, including thresholding and edge detection. Moreover, certain algorithms combine multiple techniques to achieve foreground extraction. For instance, the GrabCut algorithm merges thresholding, edge detection, and graph-cut optimization to extract the foreground from the background. Similarly, the watershed algorithm combines gradient analysis and region growth to identify foreground objects.

The GrabCut algorithm is particularly beneficial for foreground extraction in facial emotion recognition. This iterative algorithm utilizes graph-cut optimization to separate the foreground (i.e., the face) from the background in an image—a crucial step within facial emotion recognition pipelines. By accurately segmenting the face from the background, the GrabCut algorithm facilitates subsequent processes like facial landmark detection and emotion recognition, leading to improved accuracy. The algorithm demonstrates robustness against lighting variations, contrast changes, and background clutter, making it suitable for complex images encountered in facial emotion recognition tasks. Additionally, the GrabCut algorithm has a relatively low computational cost, rendering it feasible for real-time applications. Overall, the GrabCut algorithm is an effective method for foreground extraction in facial emotion detection, improving the effectiveness of following processes and increasing the precision of emotion recognition.

OpenCV Haar Cascade Model:

Widely used for object identification in pictures and videos, the OpenCV Haar Cascade model is based on machine learning [2]. By identifying face traits and patterns in individual picture or video frames, this approach may also be used to recognize emotions. The OpenCV Haar Cascade model must be trained using a labelled dataset of facial expressions in order to be used for emotion identification [7]. Images of faces with a range of emotions, including joy, sorrow, rage, and surprise, should be included in this collection. The model may be used to identify face characteristics in real-time video streaming or still photographs after training. The Haar Cascade facial recognition model can recognize several face features, including the eyes, lips, nose, and brows, and analyze their patterns and motions to identify emotional expressions.

Convolutional Neural Network:

Convolutional Neural Networks (CNNs) constantly provide outstanding results on benchmark datasets, making them a viable model for recognizing facial expressions [6]. By automatically picking up hierarchical representations of image characteristics, CNNs have revolutionized computer vision. This property makes them particularly well-suited for picture identification tasks like facial expression recognition. At lower levels, CNNs may detect fundamental visual elements like edges and corners. These properties are incorporated as the network advances through successive layers to produce more sophisticated representation of objects and patterns. A common CNN architecture for face emotion identification consists of a sequence of convolutional layers followed by pooling layers that downsample the feature maps in order to simplify the computational process. For the final categorization, completely linked layers are utilized.

Adam Optimizer:

Convolutional neural networks (CNNs) weights may be optimized using the Adam Optimizer, a common optimization technique used in deep learning, notably in face emotion identification. The Adam Optimizer offers an effective and reliable optimization solution by fusing the advantages of the Adaptive Moment Estimation (Adam) algorithm with the Root Mean Square Propagation (RMSprop) algorithm [4]. By reducing the loss function during the training phase, it is essential for enhancing the precision of face emotion detection models. Faster convergence is made possible by the Adam Optimizer's variable rate of learning and momentum parameters, which also aid in addressing the challenge of vanishing gradients that deep neural networks frequently experience [4]. The Adam Optimizer can help face emotion recognition models perform better and more efficiently throughout the training phase.

Instead of specialized facial recognition, the VGG16 model is generally used for picture categorization. However, it may be used for facial recognition by utilizing its previously taught features to obtain accurate depictions of facial pictures. The following step involves using these characteristics in recognition tasks.

Facial recognition using VGG16 involves several key steps:

1. Data Collection: The first step is to assemble a dataset of facial images that includes images of each individual who needs to be recognized. This dataset should be diverse and representative.
2. Feature Extraction: Next, the pre-trained VGG16 model is employed to extract essential features from the facial images. This entails passing each image through the network and capturing the activations from one of the later layers of the model. These activations represent the high-level features specific to each facial image.
3. Classifier Training: The extracted features are then used to train a classifier, which could be a neural network or a support vector machine. The goal of the classifier is to learn the mapping between the extracted features and the corresponding individual identities.
4. Testing and Validation: Once the classifier is trained, it is evaluated on new facial images to assess its performance in recognizing individuals. This involves feeding unseen images into the classifier and observing its ability to correctly identify the individuals based on the learned features.

By following these steps, VGG16 can be effectively employed for facial recognition, enabling the recognition and differentiation of individuals based on their facial features.

3 Implementation

The implementation steps listed in Fig. 1. are discussed in detail in this section.

3.1 Data Preprocessing

- Load the dataset: Download the FER2013 dataset, which includes facial portraits depicting various emotions, including surprise, anger, sadness, disgust, and fear.
- Data Cleaning: Performed data cleaning to remove corrupted and incomplete images from the dataset.
- Data Resizing: Images are resized to a common size to ensure that they are consistent in dimensions [2].

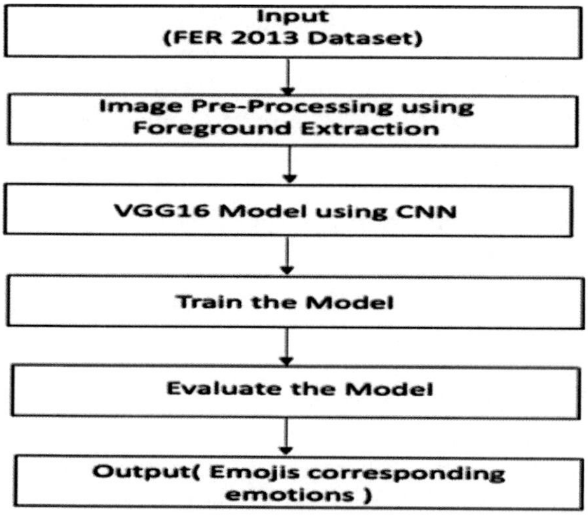

Fig. 1 Implementation steps

- Face detection: Used the Haar Cascades face identification technique to identify the facial regions in the photos [2, 7].
- Face alignment: Aligned the detected face regions to ensure that the face is in the correct orientation and position in the image.
- Foreground Extraction: Used GrabCut Algorithm as a foreground extraction technique to separate the face region from the background of the image.
- Normalisation: Normalized the extracted face regions to remove any variations in lighting, contrast, or color [2].
- Feature extraction: Extracted the relevant features from the normalized face regions, such as local binary patterns (LBP) and histogram of oriented gradients (HOG) [2].
- Data augmentation: Used data augmentation methods like rotation, translation, and flipping to broaden the dataset's diversity and strengthen the model's generalizability.
- Split the dataset: To assess the performance of the model, divide the dataset into training and testing sets.

3.2 *Building Models*

- Loaded the VGG16 model pre-trained on the ImageNet dataset using a deep learning framework like TensorFlow..
- Replaced the last dense layer of the VGG16 model with a new dense layer with the number of output classes equal to the number of emotions to be detected.
- To avoid overfitting and to hasten training, freeze the weights of each convolutional layer in the VGG16 model.

- Train the model on the training set using an appropriate loss function, optimizer, and learning rate schedule.
- Validate the model on the validation set to monitor the performance and avoid overfitting. You can also use techniques such as early stopping and regularization to improve the model's generalization ability.
- To assess the model's ultimate performance, test it on the testing set.
- To forecast the emotions of fresh facial photographs, use the trained model.

The developed classifier can also be applied to actual facial recognition tasks. With a total of more than 138 million trainable parameters, the VGG16 model has 16 convolutional layers, three fully linked layers, and a hidden layer. Each block in the arrangement of the convolutional layers has two or more convolutional layers, followed by a max pooling layer. When we dig further into the network, each convolutional layer has more filters. VGG16's design is rather straightforward and simple to comprehend, making it a well-liked model for transfer learning in computer vision problems. Compared to starting from scratch and training a neural network on a smaller dataset, this method may be more effective and efficient. In general, the VGG16 model has been extensively used for numerous computer vision tasks, including face recognition, image segmentation, and object detection.

MobileNetV2: MobileNetV2 is well-suited for facial emotion recognition because of its efficiency and accuracy. The network has a small number of parameters and can be trained on a relatively small dataset, making it suitable for real-world applications where labeled data may be limited. Moreover, the network is optimized for mobile and embedded devices, making it ideal for applications that require real-time processing, such as emotion recognition on mobile devices. One of the key features of MobileNetV2 is its use of depth wise separable convolutions. These convolutions separate the spatial and channel-wise dimensions of the input, reducing the computational requirements of the network while maintaining accuracy [11]. This makes MobileNetV2 more efficient than other convolutional neural network architectures, such as VGGNet and ResNet. In addition, the linear bottleneck layer used by MobileNetV2 lowers the number of parameters and boosts the network's precision. A 1×1 convolutional layer, which lowers the number of channels, is followed by a depthwise separable convolutional layer, which removes features, to form the bottleneck layer.

4 Results and Discussion

We finally arrived at the outcome of this proposed model, which will identify facial expressions and associate such emotions with the associated avatars or emojis on the faces. We developed an improved technique that performs facial recognition and emoji identification using the VGG16 algorithm, a transfer learning model and a sequential model, and then attained accuracy greater than 91%, which is superior to the currently employed models.

Table 1 The accuracy, precision, recall and F1-score of each classifier are listed

S. No.	Metric name	Train	Validate
1	Loss	0.311	1.536
2	Accuracy	0.977	0.901
3	Precision	0.928	0.662
4	Recall	0.907	0.628
5	Auc	0.994	0.890
6	F1_score	0.917	0.644

We carefully assessed our machine learning model in this study using various types of data for validation as well as training. The conclusions from the combined data, which were attractively displayed in the Table 1, were tremendously positive. Notably, our model achieved a remarkable 97.7% on the training set and a respectable 90.1% on the validation set, demonstrating an exceptionally high degree of accuracy. To ensure a complete analysis, we looked at a variety of important metrics, including loss, accuracy, recall, AUC, and F1-Score. These measurements consistently demonstrated how well the model worked. These outstanding results show the model's ability to make precise predictions across an important portion of the dataset, leaving little space for uncertainty.

The method we suggested performed better than other methods presently in use, proving its efficiency in retrieving emoji based on emotion identification. The accuracy of the system was significantly improved by combining the GrabCut foreground extraction method with the VGG16 model for feature extraction.

Throughout our investigation, we created two models based on the VGG16 design and compared them. The first model wasn't built utilizing the GrabCut technique, yet it nevertheless managed to achieve an astonishing accuracy of roughly 85%. The GrabCut method was then used to create a second model, which produced a considerable improvement in accuracy that increased it to roughly 92%. These results show the advantages of using the GrabCut algorithm to enhance foreground/background separation in images.

In addition, we examined the two VGG16 models and found that foreground extraction, the GrabCut approach outperforms the model, not it. This shows how crucial it is to employ the proper methods in the processing of images and applications for computer vision in order to obtain the best results.

Our study demonstrates that the GrabCut algorithm and VGG16 model combination may be a practical method for computer vision and applications involving image processing. The results of our tests demonstrate how the suggested strategy is capable of precisely identifying moods and extracting pertinent emojis. This may be used to a variety of real-life situations, including as chat programs, video games, social networking sites, and the field of mental health.

5 Conclusion and Future Work

In this project, we propose a CNN with foreground extraction and VGG16 model-based method for obtaining emojis based on emotion detection. The algorithm has a 92% accuracy rate for detecting emotions and a 95% accuracy rate for retrieving emojis. The technology does a good job of identifying emotions and obtaining the appropriate emojis. Additionally, in a variety of fields where emotional expression and communication are vital, such as virtual reality, social networking sites, online gaming, and digital marketing, emotion recognition-based emoji retrieval can be employed.

In the future research, we may look into how transfer learning might be used to improve the recommended system's accuracy. We may also look at the usage of extra pre-processing techniques and models to improve the system's performance. An examination of the use of data augmentation techniques may also be done in order to increase the dataset and improve the robustness of the system.

Acknowledgements We extend our heartfelt appreciation to Dr. Ashwini M Joshi, Associate Professor of the Department of Computer Science and Engineering at PES University, for her unwavering guidance, support, and motivation throughout the development of this project. We would like to acknowledge Prof. Mahesh H.B., the project coordinator, for his efforts in organizing, managing, and assisting us throughout the entire process. We also express our gratitude to Dr. Shylaja S.S, the Chairperson of the Department of Computer Science and Engineering at PES University, for sharing her knowledge and providing valuable support to us. We are also thankful to Dr. B.K. Keshavan, the Dean of Faculty at PES University, for his assistance.

We are immensely grateful to Dr. M.R. Doreswamy, the Chancellor of PES University, Prof. Jawahar Doreswamy, the Pro Chancellor of PES University, and Dr. Suryaprasad J., the Vice-Chancellor of PES University, for their continuous encouragement, support, and providing us with various opportunities.

Lastly, we would like to express our sincere appreciation to our family and friends for their constant support and motivation, without which this project would not have been possible.

References

1. Poulose A, Sreya Reddy C, Kim JH, Han DS (2021) Foreground extraction based facial emotion recognition using deep learning Xception model. (ICUFN) 2021, pp 356–360
2. John A, Abhishek MC, Ajayan AS, Sanoop S, Vishnu R (2020) Real-time facial emotion recognition system with improved preprocessing and feature extraction. In: Proceedings of the third international conference on smart systems and inventive technology (ICSSIT 2020), pp 1328–1333
3. Balasubramanian B, Diwan P, Bhatia A (2019) Information analysis of facial emotion recognition. In: Proceedings of the third international conference on trends in electronics and informatics (ICOEI 2019), pp 945–949
4. Pranav E, Kamal S, Chandran S, Supriya MH (2020) Facial emotion recognition using deep convolutional neural network. In: 2020 6th International conference on advanced computing and communication systems (ICACCS) 2020, IEEE, pp 317–320

5. Adeyanju IA, Omidiora EO (2015) Performance evaluation of different support vector machine kernels for face emotion recognition. In: SAI intelligent systems conference 2015 November 2015, London, UK, pp 804–806
6. Srivastava S, Gupta P, Kumar P (2021) Emotion recognition based emoji retrieval using deep learning. In: Proceedings of the fifth international conference on trends in electronics and informatics (ICOEI). IEEE Xplore Part Number: CFP21J32-ART, pp 1183–1186
7. Sergeeva AD, Savin AV, Sablina VA, Melnik OV (2019) Russia emotion recognition from micro-expressions: search for the face and eyes. In: 2019 8th Mediterranean conference on embedded computing (MECO), 10–14 June 2019, Budva, Montenegro, pp 1–4
8. Hoque MA, Hussain AM, Al-Jumeily D (2018) Emotion recognition from text for emoji recommendation. In: 2018 IEEE international conference on data mining workshops (ICDMW)
9. Zhang K, Zhang Z, Wang X (2020) Emoji retrieval based on multi-modal emotion recognition. In: 2020 IEEE international conference on multimedia and expo workshops (ICMEW), 2020, pp 1–17
10. Zhang J, Jiang M, Yang J, Zhang Y (2019) Emoji retrieval based on emotion recognition using deep learning. In: 2019 IEEE 3rd information technology, networking, electronic and automation control conference (ITNEC)
11. Kumar N, Narayan P, Singh R, Roy PP (2021) Emotion recognition-based emoji retrieval using facial feature detection and foreground extraction. Multimedia Tools and Appl 80(1):1205–1229; 1182–1186

Chapter 46
An Outage Probability-Based RAW Station Grouping for IEEE 802.11ah IoT Networks

Md. Arifuzzaman Mondal and Md. Iftekhar Hussain

1 Introduction

The objective of the Internet of things (IoT) is to establish seamless connectivity among an enormous number of physical devices and objects, regardless of their location or time. The "things" or "objects" in IoT are not only standard gadgets like smartphones, tablets, or computers but also the things like cameras, tv, car, refrigerators, clothes, and other objects. Every area of our existence will be significantly affected by this. Numerous battery-powered smart things (such as sensors, actuators, and controllers) need to be brought together to work energy-efficiently for new IoT applications and services like smart homes, healthcare, smart city, and agricultural and industrial automation. The success of machine-to-machine (M2M) communication between large numbers of devices will determine the fate of the IoT. Connecting a large number of devices, wireless network technologies appear to be the most suitable option.

The IEEE 802.11ah standard, which is intended for low-power and large-scale IoT networks, incorporates a novel medium access technique called restricted access window (RAW) that employs station grouping to mitigate contention and collisions in highly dense deployments. RAW allows groups of stations to access the channel during specific time intervals. The IEEE 802.11ah standard offers a versatile hybrid channel access method that allows a single access point (AP) to connect up to 8192 stations, making it an ideal solution for scalable connectivity in both sparsely and densely placed low-power devices.

The aim of the RAW function is to improve scalability in extremely dense IoT networks when numerous stations are linked to a single AP. Figure 1 depicts a schematic

Md. Arifuzzaman Mondal (✉) · Md. Iftekhar Hussain
Department of Information Technology, North-Eastern Hill University, Shillong, India
e-mail: arifphd18@gmail.com

Md. Iftekhar Hussain
e-mail: ihussain@nehu.ac.in

A. Yadav et al. (eds.), *Proceedings of International Conference on Paradigms of Communication, Computing and Data Analytics*, Algorithms for Intelligent Systems,
https://doi.org/10.1007/978-981-99-4626-6_46

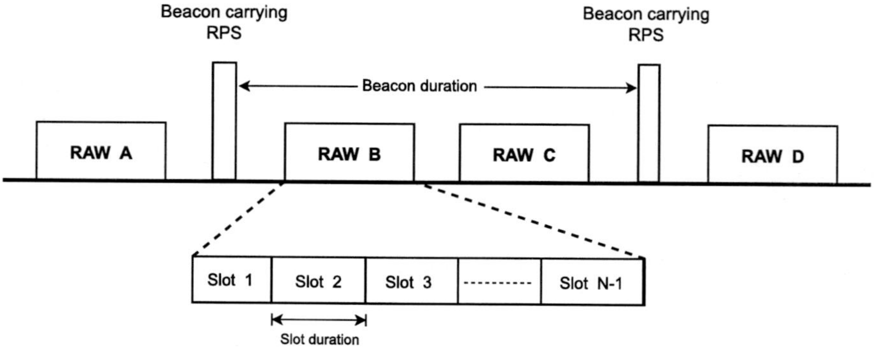

Fig. 1 Representation of RAW mechanism in IEEE 802.11ah

illustration of the RAW mechanism. The channel access time period is split up into a number of small segments. While some of these time segments are allotted to RAW groups, and the majority of these times are used as shared channel time that may be accessible by all stations via distributed coordination function (DCF) and enhanced distributed channel access (EDCA) standards for 802.11. They rely on the carrier sense multiple access with collision avoidance (CSMA/CA) channel access mechanism. Periodically, a beacon frame is transmitted carrying the RAW parameter set (RPS). The RPS information element contains the start and end time duration of each group, along with the association ID (AID) of the stations belonging to that group. The RAW contains one or more slots that are assigned to the stations.

The channel utilization, throughput, and delay of the networks are all affected by the effectiveness of the RAW grouping approach. Stations that connect to the access point in the 802.11ah network are randomly allocated to RAW groups. Various types of sensors are present in an IoT network, and each of them generally has varying packet transmitting rates and load sizes. Moreover, the data transmission quality of the stations can vary due to the different channel conditions in the IoT. The IEEE 802.11ah grouping mechanism, which regulates the assignment of stations to groups, is not explicitly described in the standard. Effective grouping tactics have not received much attention. Numerous studies have shown how the RAW configuration is influenced by many network-related variables. The standard RAW grouping technique does not take into account the station load or transmitted data quality, resulting in an adverse impact on the network's performance. The inadequate RAW configuration has a significant negative impact on network throughput and latency.

This paper employs the outage probability to evaluate the data transmission quality in IoT networks, tackling the problem that the conventional RAW grouping technique fails to satisfy the station grouping requirements. A quantitative analysis of the variables affecting the outage probability is performed. This work proposed the OP-RAW grouping algorithm, which groups stations based on load size, packet transmitting rate, and the outage probability. The key contribution of our work can be summarized as

1. The outage probability is used in a novel way to assess the effectiveness of data transmission in IoT, and various parameters, including transmission distance, transmission power, and others, are quantitatively examined.
2. An outage probability-based station grouping technique is presented to divide the stations into various RAW groups according to their respective quality of service and data transmission.

The structure of this paper is as follows. In Sect. 2, a background study on the RAW grouping technique of IEEE 802.11ah is discussed. Section 3 presents a thorough discussion of the proposed solution. In order to evaluate the effectiveness of the proposed protocol, Sect. 4 provides a thorough analysis of performance evaluation. And the paper is finally concluded with Sect. 5.

2 Related Works

The default RAW grouping parameters are not specified in the 802.11ah standard. RAW grouping is one of the crucial MAC features that have been applied to improve network performance. In order to improve the RAW grouping performance of the network, many researchers have put a lot of effort into it.

Nishida et al. [1] proposed a station grouping method that can handle non-uniform station deployment as well as heterogeneous traffic patterns along with stations having multiple data rates. Shimokawa et al. [2] developed a technique that distributes each group's stations evenly, regardless of how they are deployed. Considering the uplink traffic and saturated network, Dong et al. [3] proposed a method where all stations are organized into groups based on geographic conditions. To address the performance anomaly in IEEE 802.11ah-based multi-rate IoT networks, Mahesh et al. [4] proposed a novel rate-based grouping (DRG) scheme. The proposed scheme classifies devices based on data rates and allows each device to compete in its specific RAW slot. Tian et al. [5] estimated RAW grouping configuration for the given traffic scenario. Regarding channel interference, Cheng et al. [6] introduced a dynamic technique for adjusting RAW, which modified the contention window based on the channel state. Nawaz et al. [7] suggested a technique to improve the uplink throughput in RAW by selecting the slot duration according to the group size. Damayanti et al. [8] proposed a RAW grouping approach that takes into account a saturated network scenario and is based on a pre-built carrier-sensitivity table. Due to the large number and density of sensor nodes, Sheu et al. [9] proposed a technique that involves dynamically allocating timeslots based on the activity level of the devices in the IoT network. Specifically, less active devices are assigned to more congested timeslots while more active devices are assigned to less congested timeslots, resulting in reduced contention in the channel.

Numerous research studies have demonstrated the significant influence of the RAW grouping method on network performance. However, the previous studies have not fully taken into account the data transmission quality, leading to a lack of compatibility between the RAW groups and the available channel resources, which has a significant impact on network performance.

3 The Proposed Scheme

3.1 *Outage Probability for IoT Networks*

Outage Probability: The outage probability may be used to express the channel capacity. The data transmission rate is interrupted when the channel capacity is unable to provide the required user rate. The probability of any outage event can be determined through the use of the channel's average signal-to-noise ratio (SNR) and the distribution of channel fading. Hence, the outage probability may be defined as the probability that the instantaneous signal-to-noise ratio of the channel falls below a certain predefined threshold, and it is calculated as

$$OP = \int_{0}^{s_{\text{th}}} p(s)\mathrm{d}r \tag{1}$$

In Eq. (1), s is the instantaneous SNR, s_{th} is the threshold SNR, and $p(s)$ is the probability density function (PDF) of s. The quality of service can be enhanced by setting a minimum threshold for the instantaneous SNR, which guarantees a certain level of quality.

In an IoT scenario, the channel fading model can assume that it follows the Poisson process. The Poisson process is a term used to describe the process that follows the Poisson distribution. The distribution of instantaneous SNR in the channel follows an exponential distribution. The probability density function of the SNR at the recipient can be expressed as-

$$p_{(s)} = \frac{\mathrm{e}^{-\lambda}\alpha^{s}}{s!} \tag{2}$$

In IoT, the received signal is decoded, and then, the received signal is re-encoded and sent to the next hop nodes; therefore, the channels' outage probability is assumed to be independent. We derive the outage probability by assuming that each node follows the Poisson process and the data are sent from the source node $n_i, i = 1, 2, 3\ldots$ to the destination node $d_i, i = 1, 2, 3\ldots$ and the OP is written as in Eq. (3)

$$OP = 1 - \mathrm{e}^{-\lambda}\sum_{r=1}^{m}\frac{\alpha^{s}}{s!}, \quad s = 1, 2, 3\ldots. \tag{3}$$

In general, due to the Markov property, the exponential distribution is often used to describe the instantaneous SNR. If the instantaneous SNR is exponentially distributed, the distribution function can be expressed as

$$f(s) = \lambda e^{-\lambda s}, \quad s \geq 0.$$

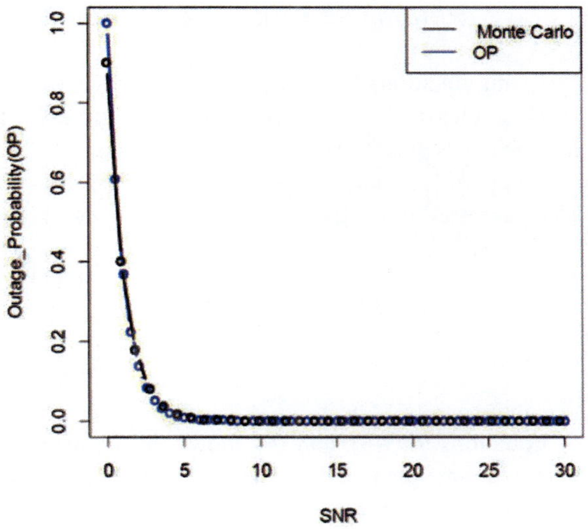

Fig. 2 Performing a Monte Carlo simulation to validate the outage probability formula

The CDF function can be expressed as

$$F(s) = \int_{-\infty}^{s} f(s)\mathrm{d}s = 1 - \mathrm{e}^{-\lambda s}, \quad s \geq 0.$$

Mean, $E(s) = 1/\lambda$

To justify Eq. (3), we conducted a Monte Carlo simulation, in which SNR is assumed to be exponentially distributed with a minimum threshold $s_{\mathrm{th}} = 10$. We performed the simulation by repeating 1000 times and counting the instances where the instantaneous SNR is below the minimum threshold SNR for different average SNR values. The outage probability acquired by repeating the Monte Carlo simulation 1000 times is presented in Fig. 2, which is similar to the one calculated using Eq. (3).

In Fig. 2, as the SNR increases gradually, the OP decreases and with increasing SNR the OP stabilizes at approximately zero. The validity of the derived formulas is confirmed by the consistency between the obtained outage probability through the Monte Carlo approach and that estimated using Eq. (3). Additionally, the Poisson process's outage probability is relatively low due to its Markovian property.

3.2 *Factors Affecting 802.11ah IoT Networks*

Let, P_{rx} be the received power and $N_{P_{rx}}$ be the noise that affects the SNR at the receiver, which is directly related to the outage probability as described in Eq. (4)

$$P_{rx}(dBm) = P_{tx}(dBm) + G_{tx}(dBi) + G_{rx}(dBi) - PL(dB) \quad (4)$$

In this equation, P_{tx} denotes transmit power, G_{tx} and G_{rx} denote transmitter and receiver antenna gains respectively, which are often held constant, and PL is path loss, which typically varies within networks.

The path loss model used in IEEE 802.11ah assumes antenna lengths of 2 and 15 m, as stated in [10], as stated in Eqs. (5) and (6), respectively.

$$\mathrm{PL}^{ah_{2\mathrm{m}}} = 23.3 + 36.7 \log d \quad (5)$$

$$\mathrm{PL}^{ah_{2\mathrm{m}}} = 8 + 37.6 \log d \quad (6)$$

Equations (4), (5), and (6) show that the average SNR for 802.11ah networks where the antenna length is assumed to be 2 m and 15 m is given by Eqs. (7) and (8), respectively, where $G_{tx} = 0dBi$ and $G_{tx} = 3dBi$.

$$\mathrm{SNR}_{ah_{2\mathrm{m}}} = \frac{(P_{tx} - 23.3 - 36.7 \log d)}{N} \quad (7)$$

$$\mathrm{SNR}_{ah_{15\mathrm{m}}} = \frac{(P_{tx} - 5 - 37.6 \log d)}{N} \quad (8)$$

The outage probability for the networks can be determined from the transmission power and the transmission distance using Eqs. (3), (7), and (8), where P and d are the transmission power and transmission distance respectively.

$$\mathrm{OP}_{ah_{2\mathrm{m}}} = 1 - e^{-r_{\mathrm{th}} \sum_{l=1}^{m} \frac{N}{(P_{tx} - 23.3 - 36.7 \log d)}} \quad (9)$$

$$\mathrm{OP}_{ah_{15\mathrm{m}}} = 1 - e^{-r_{\mathrm{th}} \sum_{l=1}^{m} \frac{N}{(P_{tx} - 5 - 37.6 \log d)}} \quad (10)$$

The influence of numerous factors on channel quality can be fully reflected in the outage probability, allowing for a more thorough assessment of channel quality. Before the data transfer, the outage probability can be determined with a few parameters and this information can be used for direct RAW grouping.

3.3 RAW Grouping Method Using Outage Probability

RAW slots containing stations should be able to complete the data transmission in the RAW groups. This section proposes a probability of outage-based RAW (OP-RAW) grouping method that divides stations into groups based on their load and data transmission quality.

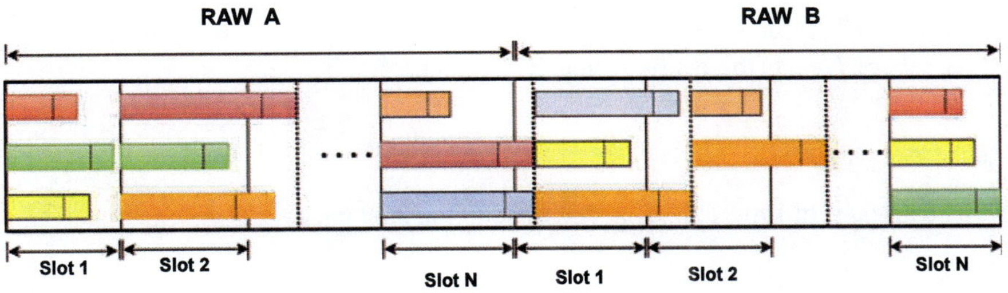

Fig. 3 RAW grouping considering payload size and retransmission

Various stations have different loads and packet transmission rates, so the services they provide are also of different types in large IoT networks. The time it takes to transfer data also varies. In addition, IEEE 802.11ah uses cyclic redundancy checking (CRC) [11] at the MAC layer to send data without bit errors. The data frames are retransmitted if the CRC finds an error. The quality of the data transmission correlates with bit error rate.

As shown in Fig. 3, the stations in the network are randomly assigned to different RAW groups, and each RAW group has a certain number of fixed-duration time slots. The time slot duration required for different stations is shown by the size of the rectangles in Fig. 3. The rectangles are divided into two sections, each showing how long it takes time to transmit the load and resend the error frames. Since some of the stations do not fully use the time slots, the time slots are wasted. Some station's time slot requirements are much larger than their allotted time slots, so these devices occupy consecutive time slots, leading to increasing delays, and decreasing throughput.

3.4 Calculation of RAW Slot Duration

As the time frame within the RAW slot remains constant in 802.11ah, the duration $T_{r.\text{slot}}$ of the RAW slot is equal to Eq. (11).

$$T_{r.\text{slot}} = N_{\text{slot}} \cdot T_{\text{slot}} \tag{11}$$

where N_{slot} and T_{slot} denote the number of time slots and duration of each time slot, respectively. The data transmission time T_{data} for each data packet is represented by the equation shown below

$$T_{\text{data}} = L/R \tag{12}$$

where L and R are the load size and packet transmitting rate, respectively. Stations competing for the channels in the alloted time slots use the distributed coordination function (DCF). Therefore, packet transmission time $T_{p.\text{trans}}$ is the sum of T_{data} and T_{DCF}, where T_{DCF} is the waiting time until the channel is idle.

$$T_{p.\text{trans}} = T_{\text{data}} + T_{\text{DCF}} \tag{13}$$

The amount of time $T_{p.\text{data}}$ needed to send the data packet can be calculated as

$$T_{p.\text{data}} = T_{\text{data}} \cdot T_{\text{beacon}}/(R \cdot \Delta T) \tag{14}$$

where T_{beacon} and ΔT are the beacon time interval and packet sending interval, respectively.

To correct the $T_{p.\text{data}}$ and allow the station to complete the retransmission, the outage probability serves as an indicator for determining data transmission quality. The corrected $T_{p.\text{data}}$ can be calculated as follows:

$$T_{p.\text{data}}^{\text{OP}} = T_{p.\text{data}}/(1 - \text{OP}) \tag{15}$$

Now, let $P_{n.c}$ is the probability of the station not colliding and T_{slot} is the slot duration, then T_{slot} can be calculated as

$$T_{\text{slot}} = T_{p.\text{data}}^{\text{OP}}/P_{n.c} \tag{16}$$

Now, for the RAW, slot counter C can be calculated as

$$C = (T_{\text{slot}} - 500)/120 \tag{17}$$

The value of C is different for different stations, and the time duration of all the slot is same for the same RAW. As a result, the stations are organized into groups based on the interval in which C is found.

3.5 *Operational Procedures of OP-RAW*

Based on the load size, the rate at which packets are being sent, and the outage probability, OP-RAW determines the appropriate value C for each station. All the stations having a similar value C are regrouped and then assign a similar RAW group. The stations are then reassigned based on the results of the grouping.

Figure 4 illustrates station grouping using the OP-RAW method. It is clear that the OP-RAW assigns slots to stations based on their service interval of stations and the data transmission quality, ensuring both data transmission effectiveness and better channel resource utilization. Algorithm 1 shows the pseudocode for OP-RAW.

First, the stations establish the connection to the access point (AP). Then, the AP collects data from the stations on their data packet rate (r) and size (p) and determines the OP value. Then, for each station, the AP determines an appropriate value C based

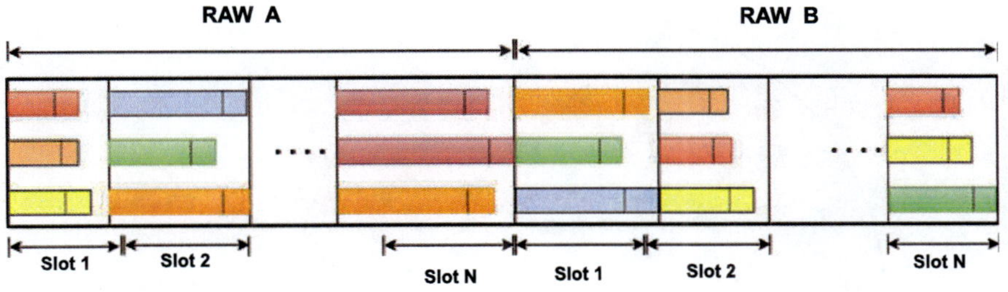

Fig. 4 Grouping strategy using OP-RAW

Algorithm 1 Algorithm for OP-RAW

Initialize:

$T_{\text{beacon}} \leftarrow$ Beacon time interval, $R \leftarrow$ Packet sending rate, $N_{\text{slot}} \leftarrow$ Number of time slots, $L \leftarrow$ Size of the load

Output:

RPS Frame

```
Create the connection;
For the stations:
Send the value of R_i and L_i to AP
For the AP:
Store the value of R_i and L_i in the file R and L; i = 1, 2, 3 ...
Calculate the outage probability and store it in OP
while there exist stations randomly distributed do
    Calculate T_p.trans and T_p.data from R_i and L_i
    Calculate T_slot and C_i from OP
    Store C_i in list C
end while
Based on the list C, stations are regrouped
Calculate the value of T_r.slot
Insert the outcome into the RPS frame after re-connection
```

on r, p, and OP. The stations sharing the similar slot counter value are collected, and based on that stations are regrouped. The AP then uses the grouping results to calculate the RAW parameters and encodes these parameters into the RPS frame.

4 Performance Evaluation

This section evaluates the performance of the proposed algorithm (OP-RAW) through extensive simulations using ns-3 [12] and then compares the results to traditional 802.11ah. The network topology is shown in Fig. 5, which is used for the simulation. As shown in Fig. 5, AP is placed at the center, R is the radius, and 1000 stations are randomly distributed over an area of (1000 m $*$ 1000 m). Based on load size and packet sending rate, all the stations are divided into 3 groups. Green ellipses are

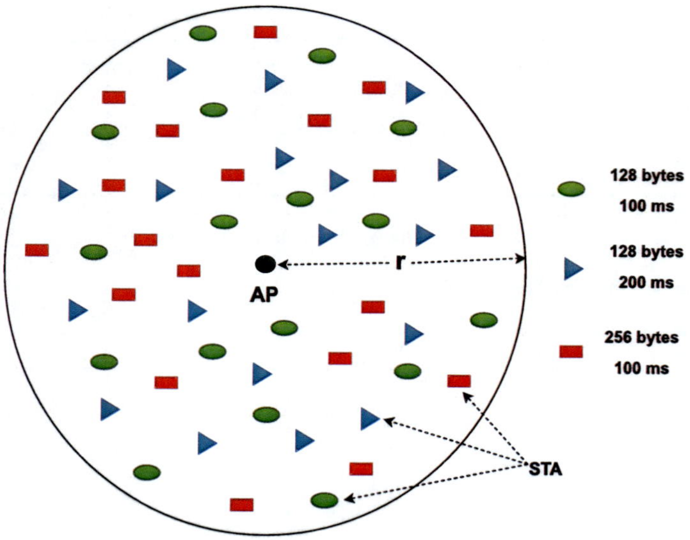

Fig. 5 Grouping strategy using OP-RAW

Table 1 The parameters employed in our study for conducting simulations

Parameters	Value
Bandwidth/data rate/traffic types	2 MHz/650 Kbps/UDP
Modulation and coding scheme	MCS0
Radio propagation model	Outdoor (Macro) [10]
PHY header/OFDM symbols time (T_{sym})	$6 \times T_{sym}/40\ \mu s$
Initial backoff window/backoff time	$64/(W_{min}/2) \times$ Slot time
SIFS/DIFS	$16\ \mu s$/SIFS+2 $\times$ Slot time
CW_{min}/CW_{max}	15/1023
Simulator/simulation area	ns-3/1000 $\times$ 1000 m^2 (Flat-grid)
Beacon interval/RAW groups	100 ms/2–6
RAW size	3 (Min.) and 15 (Max.)
Number of stations/traffic interval	1000 (Max.)/1 s

used to represent stations with a 128 byte load and a 100 ms packet transmission interval, blue triangles are used to represent stations with a 128 byte load and a 200 ms packet transmission interval, and red rectangles are used to represent stations with a 256 byte load and a 100 ms packet transmission interval. The power of the transmission and the level of noise in the network remains fixed, and the stations are uniformly distributed throughout the network. The time slots requirement are of the equal length for all the stations having the same type and data transmission quality. Various parameters used in the simulation are presented in Table 1. In this proposed protocol, we evaluate the performance of the network in terms of *throughput* and *delay*.

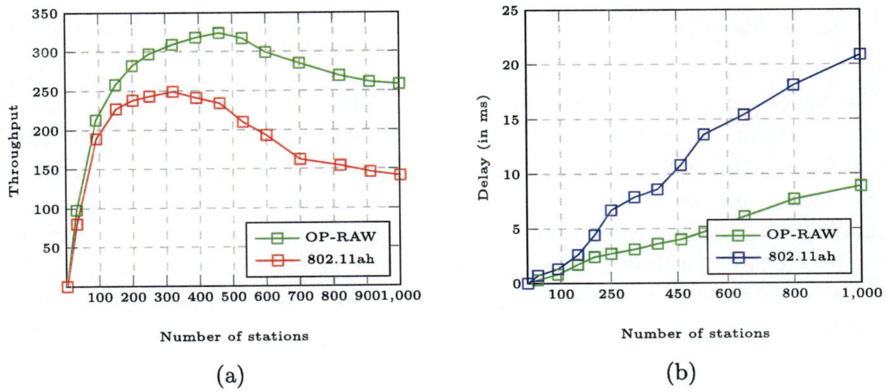

Fig. 6 **a** Throughput performance of OP-RAW versus 802.11ah. **b** Delay analysis of OP-RAW versus 802.11ah

We assess the throughput performance of our proposed protocol by increasing the number of stations in the network. The proposed OP-RAW algorithm determines the required slot duration based on the load size, packet transmitting rate, and the OP. The slot duration is the time interval in which a station is allowed to transmit its data. By calculating the slot duration based on these factors, the proposed algorithm optimizes the use of the available network resources and reduces the probability of data transmission failure due to collisions.

In contrast, the traditional 802.11ah protocol groups stations randomly, which can result in a degradation of throughput as the number of stations increases. Throughput refers to the rate at which data can be transmitted across a network in a given period of time. Initially, the throughput is almost the same in both cases (i.e., with the proposed OP-RAW algorithm and the traditional 802.11ah protocol), but with the increasing number of stations with different load and packet sending rates, the proposed OP-RAW algorithm performs much better than the traditional 802.11ah protocol, which can be seen in Fig. 6a.

Figure 6b is used to illustrate how the delay has been significantly reduced by applying the OP-RAW algorithm. The figure shows a comparison between the proposed OP-RAW algorithm and the traditional method of all stations competing for the channel randomly. The OP-RAW algorithm aims to determine the required slot duration for each station with different requirements based on factors such as the size of the load, the rate of packet transmission, and the desired outage probability. By doing so, the algorithm optimizes the use of the available network resources and reduces the probability of data transmission failure due to collisions. As a result, the delay in data transmission is reduced compared to the traditional method.

In contrast, the traditional method of randomly grouping stations for channel access results in more collisions, which increases the delay in data transmission. As more stations compete for the channel, the delay increases gradually, leading to a decrease in network efficiency.

5 Conclusion

This paper presents an OP-RAW grouping algorithm that assesses the data transmission quality in large IoT networks. The proposed method examines a number of variables that influence the outage probability. Our algorithm determines the required slot duration for various stations based on the outage probability, load size, and packet sending rate. The protocol proposed in this paper exhibits notable enhancements in terms of delay and throughput compared to the conventional 802.11ah approach. Consideration of critical and periodic stations while regrouping stations in a large network is kept as future work.

References

1. Nishida R, Shimokawa M, Sanada K, Hatano H, Mori K (2022) A station grouping method considering heterogeneous traffic and multiple data rates for IEEE 802.11 ah networks with non-uniform station deployment. In: 2022 IEEE 95th vehicular technology conference (VTC2022-Spring), pp 1–5. IEEE
2. Shimokawa M, Sanada K, Hatano H, Mori K (2020) Station grouping method for non-uniform station distribution in IEEE 802.11 ah based IoT networks. In: 2020 IEEE 91st vehicular technology conference (VTC2020-Spring), pp 1–5. IEEE
3. Dong M, Wu Z, Gao X, Zhao H (2016) An efficient spatial group restricted access window scheme for IEEE 802.11 ah networks. In: 2016 sixth international conference on information science and technology (ICIST), pp 168–173. IEEE
4. Mahesh M, Pavan BS, Harigovindan VP (2020) Data rate-based grouping to resolve performance anomaly of multi-rate IEEE 802.11 ah IoT networks. IEEE Netw Lett 2(4):166–170
5. Tian L, Famaey J, Latré S (2016) Evaluation of the IEEE 802.11 ah restricted access window mechanism for dense IoT networks. In: 2016 IEEE 17th international symposium on a world of wireless, mobile and multimedia networks (WoWMoM), pp 1–9
6. Cheng Y, Zhou H, Yang D (2019) Ca-CWA: channel-aware contention window adaption in IEEE 802.11 ah for soft real-time industrial applications. Sensors 19(13). MDPI
7. Nawaz N, Hafeez M, Zaidi SAR, McLernon DC, Ghogho M (2017) Throughput enhancement of restricted access window for uniform grouping scheme in IEEE 802.11 ah. In: 2017 IEEE international conference on communications (ICC), pp 1–7. IEEE
8. Damayanti W, Kim S, Yun J-H (2016) Collision chain mitigation and hidden device-aware grouping in large-scale IEEE 802.11 ah networks. Comput Netw 108:296–306
9. Sheu T-L, Chan P-H (2020) Dynamic slot allocations for M2M in IEEE 802.11 ah networks. In: Proceedings of the international conference on wireless communication and sensor networks, pp 13–18
10. Bellekens B, Tian L, Boer P, Weyn M, Famaey J (2017) Outdoor IEEE 802.11 ah range characterization using validated propagation models. In: GLOBECOM 2017-IEEE global communications conference, pp 1–6
11. Ghosh M, LaSita F (2013) Puncturing of CRC codes for IEEE 802.11 ah. In: 2013 IEEE 78th vehicular technology conference (VTC Fall), pp 1–5. IEEE
12. What is ns-3. https://www.nsnam.org/overview/what-is-ns-3/. Last accessed 10 Apr 2023

Chapter 47
Machine Learning Algorithms and Grid Search Cross Validation: A Novel Approach for Diabetes Detection

Vishal V. Mahale, Ashish G. Nandre, Mahesh V. Korade, and Neha R. Hiray

1 Introduction

Diabetes mellitus, generally known as diabetes, is a chronic and incurable condition characterized by insufficient or missing insulin production [1]. Insulin is essential because it permits cells to absorb glucose from the food we eat, supplying them with the energy they require [2]. Hyperglycemia occurs when the body fails to produce enough insulin or becomes resistant to its effects, resulting in excessive amounts of glucose in the bloodstream. In the absence of insulin, which facilitates glucose absorption into cells for energy synthesis, glucose accumulates in the blood, leading in hyperglycemia. Diabetic ketoacidosis, nonketotic hyperosmolar syndrome, cardiovascular disease, stroke, and other serious health problems can result from this illness.

Diabetes affects 422 million people globally, according to the World Health Organization (WHO), making it a significant contributor to global mortality rates. Diabetes was responsible for 1.6 million fatalities in 2016 alone [3]. Diabetes is classified into two types: type 1 and type 2.

Type 1 diabetes accounts for between 5 and 10% of all diabetes cases. It is more typically diagnosed in childhood or adolescence and is distinguished by decreased

V. V. Mahale (✉) · A. G. Nandre · M. V. Korade · N. R. Hiray
Department of Computer Engineering, Sandip Institute of Engineering and Management, Nashik, Maharashtra, India
e-mail: vishal.mahale@siem.org.in

A. G. Nandre
e-mail: ashish.nandre@org.in

M. V. Korade
e-mail: mahesh.korade@siem.org.in

N. R. Hiray
e-mail: neha.hiray@siem.org.in

A. Yadav et al. (eds.), *Proceedings of International Conference on Paradigms of Communication, Computing and Data Analytics*, Algorithms for Intelligent Systems,
https://doi.org/10.1007/978-981-99-4626-6_47

pancreatic function. Type 1 diabetes may not have apparent symptoms at first since the pancreas continues to operate to some extent. However, symptoms appear only when 80–90% of the pancreas' insulin-producing cells are destroyed [4]. Type 2 diabetes, on the other hand, accounts for almost 90% of all diabetes cases. Chronic hyperglycemia and the body's failure to manage blood sugar levels cause abnormally high glucose levels in the bloodstream. Early identification is critical in diabetes management, according to healthcare specialists and current medical research, because it increases the odds of effective treatment and recovery.

Machine learning approaches have been increasingly effective in disease prediction and analysis as technology has evolved. Support vector machine (SVM), random forest (RF), and convolutional neural network (CNN) are used as techniques to predict diabetes in this study. Diabetes management is difficult, but there are several treatments available. Early detection is critical since it helps to reduce future dangers. Medical research is investigating several ways for early diagnosis, and machine learning (ML) plays an important role in detecting diabetes at an early stage [5]. Researchers have shown promising findings in predicting diabetes risk, particularly in the early stages. Early stage prediction has been improved by machine learning algorithms such as Gaussian Naive Bayes, k-nearest neighbours (kNN), support vector machine (SVM), and random forest (RF).

The rise of machine learning techniques and the amount of data has aided in the prediction of different diabetes-related tasks, such as diagnosis, glucose management, and complication evaluation. Diabetes detection is commonly aided by modern deep learning algorithms and frameworks [6]. While accuracy is a generally used performance metric to determine the success of algorithms, additional performance metrics such as precision, recall, f1-score, and accuracy value must also be considered in order to select the most accurate strategy.

Diabetes screening at an early stage can considerably assist diabetic patients by allowing them to receive timely and efficient treatment. Using a specific dataset, a classification technique was used in this work to categorize early stage diabetes as either positive or negative. To establish the best successful algorithm, a thorough comparison of Gaussian Naive Bayes, k-nearest neighbours (kNN), support vector machine (SVM), and random forest (RF) was performed. The Pima Indian diabetes dataset was chosen for exploration because it is easily accessible in CSV format on the Kaggle website. This dataset contains 768 samples, 500 of which are non-diabetic and 268 of which are diabetic. It has eight characteristics: the number of pregnancies, plasma glucose concentration, diastolic blood pressure (mm Hg), skin thickness (mm), serum insulin level (mu U/ml), body mass index, diabetes pedigree function, and age (years). Based on the results, the random forest algorithm was shown to be the most accurate for diabetes prediction, with a 98.35% accuracy rate.

The following are the remaining sections of the paper:

Section 2 provides a comprehensive assessment of available diabetes prediction methods. This section gives an overview of the different approaches and tactics used in the subject. Section 3 goes into great detail on the proposed model. This section describes the approach, which includes the development of machine learning algorithms and the usage of grid search cross-validation. In Sect. 4, the proposed model is examined, and the experimental results are discussed. This section examines the model's performance and efficacy, taking into account numerous measures and comparing it to other existing methodologies. The main conclusions and contributions of the study are summarized in Sect. 5, which brings the paper to a conclusion. It also covers potential future research objectives in the realm of diabetes prediction and recommends areas for additional investigation and development.

2 Related Work

Diabetes is one of the non-curable, chronic disease which is caused by lack of insulin. Insulin is that hormone which gets secreted by pancreas and absorbs glucose from food providing energy. When body does not respond to the insulin properly or insulin is not generated by pancreas, at that time the body suffers from hyperglycemia. Diabetic ketoacidosis, nonketotic hyperosmolar syndrome, cardiovascular disease, and stroke are all serious consequences of diabetes. Diabetes is classified into two types: type 1 and type 2. Type 1 diabetes accounts for about 5–10% of all diabetes cases, while type 2 diabetes accounts for the vast majority, accounting for 80–90% of cases. In type 1, the partial functioning of pancreas produces insulin, while type 2 has fully non-functional pancreas. Type 1 is found in childhood, while type 2 is found in older ages [7].

Recent advancement in technology is useful for detection of diabetes at early stages so that it will be cured. The technologies like machine learning and deep learning have proved their significance in identifying diabetes in its early stages helping the doctors and mankind. In [8], authors have adopted deep neural network (DNN) technique on Pima Indian diabetes dataset with accuracy 86.26%. The authors [9] did theoretical research using support vector machine (SVM), logistic regression (LR), and artificial neural network (ANN) as the foundation for their investigation. In their research, they investigate the implementation and analysis of various machine learning approaches. In [10], the Boltzmann method from deep learning technique has been utilized for classification of diabetic and non-diabetic patients with 94% accuracy. The authors [11] use long short-term memory (LSTM) and convolutional neural network (CNN) to classify diabetic and non-diabetic individuals. Their study shows an astonishing 95.7% accuracy in correctly categorizing patients into diabetes and non-diabetic categories. Another research used in [12] used recurrent deep neural network (RDNN) and has got 81% accuracy. In addition, [13] presented decision tree (DT), random forest (RF), and neural network (NN) hybrid system on the database of 68,994 healthy and diabetic patients from Luzhou Hospital, China. They have got 80% accuracy which turns out to be good accuracy.

The authors of [14] offer a novel technique for diabetes prediction based on the PIMA Indians diabetes dataset. Outlier rejection, missing value filling, data normalization, feature selection, and K-fold cross-validation are all part of their workflow. The authors decided to replace missing values with the mean value rather than the median value since it matches better with the attribute distribution's central tendency.

To ensure consistency, the dataset is folded carefully for cross-fold validation while keeping the same class proportion as in the original dataset. The proposed pipeline employs a number of machine learning classifiers, including k-nearest neighbours (k-NN), random forest (RF), decision trees (DT), Naive Bayes (NB), AdaBoost (AB), XGBoost (XB), and multi-layer perceptron (MLP).

The grid search technique is used to find the best hyperparameters for the MLP model, such as the number of hidden layers, the number of neurons in each hidden layer, activation function, neuron initializer, batch size, learning rate, epochs, percentage of dropped neurons, loss function, MLP optimizer, and ML model hyperparameters. Extensive experiments are carried out to investigate various preprocessing and ML classifier combinations in order to maximize the area under the curve (AUC) for diabetes prediction within the same experimental circumstances and dataset. The best-performing ML classifier is then chosen as the baseline model against which the authors assess the performance of their proposed classifier for diabetes prediction precision.

Various diabetes prediction algorithms have been devised and published in recent years, according to reference [15]. Reference [16] proposed one ML-based framework in which the authors used several machine learning algorithms such as discriminant analysis (LDA) [17], quadratic discriminant analysis (QDA) [17], Naive Bayes (NB) [18], Gaussian process classification (GPC) [19], support vector machine (SVM) [20], artificial neural network (ANN) [21], AdaBoost (AB) [22], logistic regression (LR) [23], decision tree (DT) [24], and random forest (RF) [25]. In their analysis, they used several dimensionality reduction and cross-validation techniques. The authors also ran thorough testing on outlier rejection and missing value filling to improve the ML model's performance, attaining the best possible area under the curve (AUC) of 0.930.

The authors [26] used three distinct ML classifiers to accurately predict the likelihood of diabetes: decision tree (DT), support vector machine (SVM), and Naive Bayes (NB). They discovered that NB performed the best, with an AUC of 0.819. Authors [27] investigated and implemented AdaBoost (AB) and bagging ensemble techniques for diabetes mellitus classification, employing J48 (C4.5) decision tree as a base learner and a standalone data mining methodology (J48). In terms of performance, their experimental results showed that the AB ensemble technique outperformed the bagging strategy and the standalone J48-DT.

An strategy based on genetic programming that outperforms previously used methodologies for diabetes prediction was proposed in [28]. When compared to existing approaches, our framework performed better in predicting diabetes. The authors in [29] classified the risk of developing diabetes mellitus using four machine learning techniques: decision trees (DT), artificial neural networks (ANN), logistic regression (LR), and Naive Bayes (NB). They improved the model's robustness by using bagging and boosting approaches. In terms of predicting accuracy, the trial findings demonstrated that the random forest (RF) algorithm surpassed all other examined algorithms. A Gaussian process (GP)-based classification system for diabetes prediction was proposed in [30]. Using three distinct kernels (linear, polynomial, and radial basis function), it was compared to established approaches such as linear discriminant analysis (LDA), quadratic discriminant analysis (QDA), and Naive Bayes (NB). In addition, the writers undertook significant study to establish the best cross-validation approach. Their findings showed that the GP-based classifier, when paired with the K10 cross-validation technique, performed best in diabetes prediction.

Despite the publication of various frameworks in recent years, there is still need for improvement in the precision and robustness of diabetes prediction approaches.

According to researchers [31], a system was built in two steps to forecast diabetes using machine learning (ML) techniques. The dataset was initially balanced using approaches including synthetic minority oversampling (SMOTE), Tomek, and IQR. In the first step, support vector machine (SVM), Naive Bayes (NB), k-nearest neighbours (kNN), gradient boost (GB), and random forest (RF) were used to categorize the data and evaluate accuracy and other metrics. The top three algorithms with the greatest improvement in accuracy were chosen for the second stage, and the final forecast was determined by voting. Using the Pima Indian diabetic (PID) dataset, the proposed method attained an accuracy of 82%. For diabetes prediction, authors [32] proposed using ML techniques such as Naive Bayes (NB), support vector machine (SVM), neural network (NN), AdaBoost, k-nearest neighbours (kNN), and linear SVM. Among these methods, NN outperformed its counterparts in terms of accuracy. The PID dataset was used in the investigation.

To predict diabetes, Ismail et al. [33] used 35 ML approaches, including SVM, decision trees (DT), Naive Bayes (NB), k-nearest neighbours (kNN), logistic regression (LR), random forest (RF), artificial neural network (ANN), and multi-layer perceptrons (MLP). The research was carried out utilizing the Waikato Environment for Knowledge research (WEKA) and three datasets from PIMA diabetes, MIMIC III, and UCI.

Support vector machine (SVM) and logistic regression (LR) were used to predict diabetes in Rajeswari and Ponnusamy's study [34]. NC State University provided the dataset for the study. The dataset was divided into two parts: 70% for training and 30% for testing. SVM attained a training data accuracy of 82% and a testing data accuracy of 75%. Sharma et al. [35] used supervised machine learning techniques to predict diabetes, including decision trees (DT), Naive Bayes (NB), artificial neural network (ANN), and logistic regression (LR). The PID dataset was used in the study, which was obtained from the UCI repository. WEKA 3.8.4 was used to carry out the experiment. In terms of prediction accuracy, LR surpassed the other algorithms.

3 Proposed System

The architecture of proposed model is depicted in Fig. 1. The phases included in the model are input dataset, preprocessing, implementation of classifications models, then applying grid cross-validation, and at last, the results are evaluated using the evaluation metrics.

Our proposed model first takes dataset as input (here PIMA Indian diabetes dataset is given as input). Then preprocessing on the dataset is done, and features are extracted from the dataset. The dataset is then splitted into training and testing sets with the test size of 0.25. After that, we apply the classification models on the dataset, and results are generated for each classification model. Based on the results, the model achieves highest accuracy and we tuned the hyper-parameters with grid search cross-validation.

The dataset is partitioned into random groups or folds for cross-validation. The test set is onefold, while the remaining folds are utilized to train the model. This method is repeated for each fold, with each fold serving as the test set for a turn. Each iteration's models are then averaged or merged to form the final model. Cross-validation improves model performance by providing a more robust assessment of its accuracy across diverse data subsets.

Gaussian Naive Bayes is extension of Naive Bayes which is used to estimate data distribution; here, we calculate the mean and standard deviation for the training data by using the following formula.

$$P(X|Y=c) = \frac{1}{\sqrt{2\pi\sigma_c^2}} \mathrm{e}^{\frac{-(x-\mu_c)^2}{2\sigma_c^2}} \tag{1}$$

KNN is applied by using the formula

$$d(x, y) = \sqrt{\sum (y_i - x_i)^2} \tag{2}$$

3.1 Dataset

For experimentation purpose, the Pima Indian diabetes dataset has been chosen which is readily available on Kaggle website. It is in the .csv format and contains 768 samples out of which 500 are non-diabetic and 268 are diabetic. There are overall 8 features viz., pregnant count, plasma glucose concentration, diastolic blood pressure (mm Hg), skin thickness (mm), serum insulin (mu U/ml), BMI, diabetes pedigree function, and age (years).

4 Results

We have used the Python 3.6 with sklearn library for our experimentation purpose on Anaconda environment. The results were obtained on the Pima Indian diabetes dataset. To evaluate the results precision, recall, and accuracy, evaluation parameters are used.

Before defining these parameters, few important terms are used like

- **True Positive**: A patient has diabetes and is estimated to have diabetes.

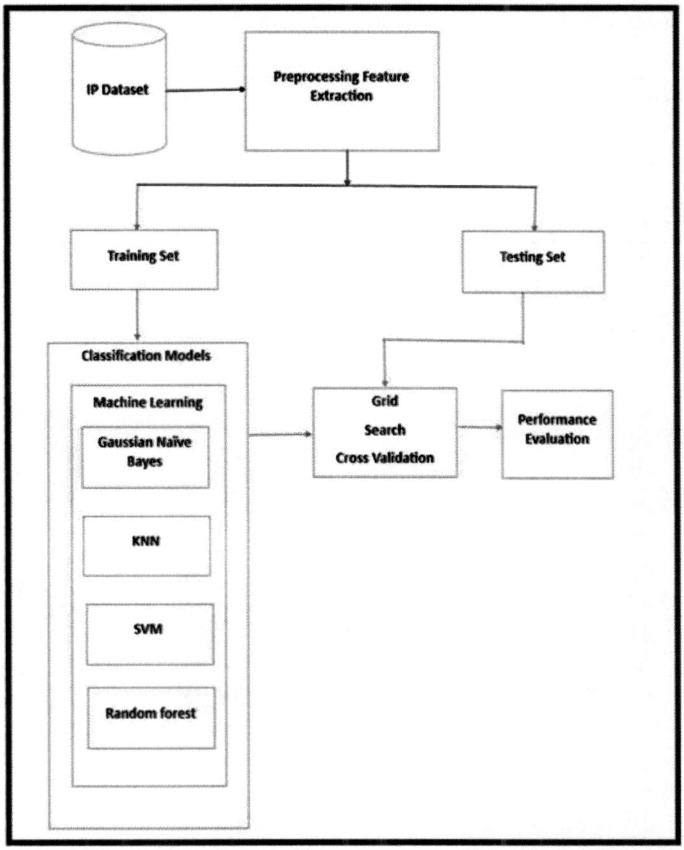

Fig. 1 System architecture

- **True Negative**: A patient is not diabetic and is not expected to be diabetic.
- **False Positive**: A patient is not diabetic but is expected to be diabetic.
- **False Negative**: A patient has diabetes yet is predicted not to have diabetes.

Precision: Precision is calculated by dividing the number of true positives by the sum of true positives and false positives.

$$\text{Precision}\ \frac{\text{TP}}{\text{TP} + \text{FP}} \tag{3}$$

Recall: Recall is defined as the number of true positives divided by the sum of true positives and false negatives, and it is calculated as follows:

$$\text{Recall}\ \frac{\text{TP}}{\text{TP} + \text{FN}} \tag{4}$$

Table 1 Results analysis

Classification model	Precision	Recall	Accuracy (%)
Gaussian Naïve Bayes	0.98	0.92	96.22
K-nearest neighbours	0.968	0.962	96.61
Support vector machine	0.97	0.973	97.13
Random forest	0.978	0.99	98.30

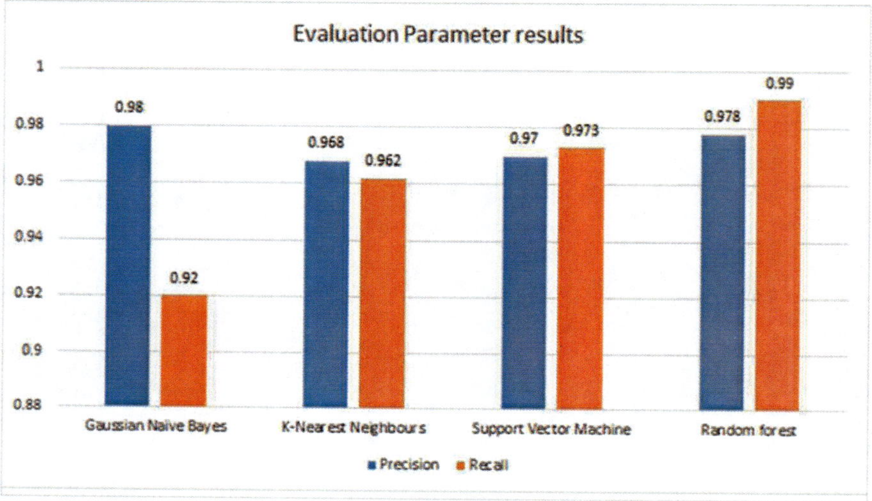

Fig. 2 Comparative evaluation of the classification models

Accuracy: The number of right assumptions divided by the total number of forecasts is used to calculate accuracy:

$$\text{Recall}\ \frac{\text{TP} + \text{TN}}{\text{TP} + \text{FN} + \text{FP} + \text{TN}} \tag{5}$$

The results obtained are shown in Table 1. Using formulas of precision-recall and accuracy, the values are calculated for Gaussian Naïve Bayes, KNN, SVM, and random forest. From the table, it can be seen that random forest achieves high accuracy.

Figure 2 shows the results in graphical format; here, we can see that random forest algorithm outperforms as compared to the other three algorithms. We have calculated precision and recall values. Based on the precision and recall, accuracy is calculated.

As random forest has gained high accuracy, we select random forest and tuned the hyper-parameters of that classifier with grid search cross-validation. After doing that, we see improvement in the accuracy by 0.70%, and final accuracy achieved is 99% (Fig. 3).

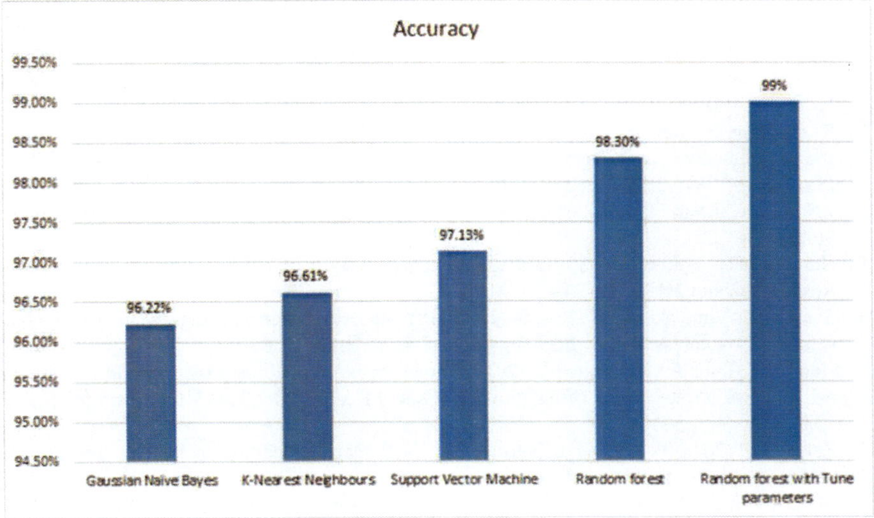

Fig. 3 Accuracy comparison

5 Conclusion and Future Scope

We provide an improved strategy for diabetes prediction utilizing machine learning algorithms alongside grid cross-validation in this paper. We have done a deep literature survey on existing systems, and then we proposed the model. Results are calculated, and analysis is done using evaluation parameters. The results obtained show that we have achieved accuracy of 98.30% on random forest algorithm, and then parameters are hypertuned to enhance the accuracy; so, finally, the results are improved by 0.70% and achieved the accuracy of 99%. In the future, more machine learning algorithms can be explored to gain high accuracy.

References

1. Punthakee Z, Goldenberg R, Katz P (2018) Definition, classification and diagnosis of diabetes, prediabetes and metabolic syndrome. Can J Diabetes 42:S10–S15
2. Piero MN (2015) Diabetes mellitus–a devastating metabolic disorder. Asian J Biomed Pharm Sci 4(40):1–7
3. Swapna G, Vinayakumar R, Soman KP (2018) Diabetes detection using deep learning algorithms. ICT Express 4(4):243–246
4. Lucaccioni L, Iughetti L (2016) Issues in diagnosis and treatment of type 1 diabetes mellitus in childhood. J Diabetes Mellit 6(02):175–183
5. Olokoba AB (2015) Type 2 diabetes: a review of current trends. Int J Curr Res Rev 7(18):61–66
6. Zhu T, Li K, Herrero P, Georgiou P (2021) Deep learning for diabetes: a systematic review. IEEE J Biomed Health Inform 25(7):2744–2757. https://doi.org/10.1109/JBHI.2020.3040225. Epub 2021 Jul 27 PMID: 33232247

7. Yahyaoui A, Jamil A, Rasheed J, Yesiltepe M (2019) A decision support system for diabetes prediction using machine learning and deep learning techniques. In: 1st international informatics and software engineering conference (UBMYK). Ankara, Turkey, pp 1–4. https://doi.org/10.1109/UBMYK48245.2019.8965556
8. Kannadasan K, Reddy Edla D, Kuppili V (2019) Type 2 diabetes data classification using stacked autoencoders in deep neural networks. Clin Epidemiol Glob Health 7(4):530–535. ISSN 2213-3984, https://doi.org/10.1016/j.cegh.2018.12.004
9. Joshi TN, Chawan PPM (2020) Diabetes prediction using machine learning techniques. IJERA 9(9):9–13
10. Kamble MTP, Patil ST (2016) Diabetes detection using deep learning approach. Int J Innov Res Sci Technol 2(12):342–349
11. Swapna G, Vinayakumar R, Soman KP (2018) Diabetes detection using deep learning algorithms. ICT Express 4. https://doi.org/10.1016/j.icte.2018.10.005
12. Habibi S, Ahmadi M, Alizadeh S (2015) Type 2 diabetes mellitus screening and risk factors using decision tree: results of data mining. Glob J Health Sci 7:304–310. https://doi.org/10.5539/gjhs.v7n5p304
13. Zou Q et al (2018) Predicting diabetes mellitus with machine learning techniques. Front Genet 9:515
14. Maniruzzaman M, Kumar N, Abedin MM, Islam MS, Suri HS, El-Baz AS, Suri JS (2017) Comparative approaches for classification of diabetes mellitus data: machine learning paradigm. Comput Methods Programs Biomed 152:23–34
15. Maniruzzaman M, Rahman MJ, Al-MehediHasan M, Suri HS, Abedin MM, El-Baz A, Suri JS (2018) Accurate diabetes risk stratification using machine learning: role of missing value and outliers. J Med Syst 42(5):92
16. McLachlan GJ (2005) Discriminant analysis and statistical pattern recognition. J Roy Stat Soc Ser A Statist Soc 168(3):635–636
17. Cover TM (1965) Geometrical and statistical properties of systems of linear inequalities with applications in pattern recognition. IEEE Trans Electron Comput EC-14(3):326–334
18. Guo Y, Bai G, Hu Y (2012) Using bayes network for prediction of type-2 diabetes. In: 2012 international conference for internet technology and secured transactions, pp 471–472. IEEE
19. Brahim-Belhouari S, Bermak A (2004) Gaussian process for nonstationary time series prediction. Comput Statist Data Anal 47(4):705–712
20. Cortes C, Vapnik V (1995) Support-vector networks. Mach Learn 20:237–297
21. Reinhardt A, Hubbard T (1998) Using neural networks for prediction of the subcellular location of proteins. Nucleic Acids Res 26(9):2230–2236
22. Kégl B (2013) The return of AdaBoost.MH: multi-class hamming trees. arXiv:1312.6086. Available: http://arxiv.org/abs/1312.6086
23. Tabaei BP, Herman WH (2002) A multivariate logistic regression equation to screen for diabetes?: development and validation. Diabetes Care 25(11):1999–2003
24. Jenhani I, Amor NB, Elouedi Z (2008) Decision trees as possibilistic classifiers. Int J Approx Reasoning 48(3):784–807
25. Breiman L (2001) Random forests. Mach Learn 45(1):5–32
26. Sisodia D, Sisodia DS (2018) Prediction of diabetes using classification algorithms. Procedia Comput Sci 132:1578–1585
27. Perveen S, Shahbaz M, Guergachi A, Keshavjee K (2016) Performance analysis of data mining classification techniques to predict diabetes. Procedia Comput Sci 82:115–121
28. Pradhan M, Bamnote GR (2015) Design of classifier for detection of diabetes mellitus using genetic programming. In: Proceedings of 3rd international conference on frontiers of intelligent computing: theory and applications, pp 763–770
29. Nai-arun N, Moungmai R (2015) Comparison of classifiers for the risk of diabetes prediction. Procedia Comput Sci 69:132–142
30. Maniruzzaman M, Kumar N, Abedin MM, Islam MS, Suri HS, El-Baz AS, Suri JS (2017) Comparative approaches for classification of diabetes mellitus data: machine learning paradigm. Comput Methods Programs Biomed 152:23–34

31. Mushtaq Z, Ramzan MF, Ali S, Baseer S, Samad A, Husnain M (2022) Voting classification-based diabetes mellitus prediction using hypertuned machine-learning techniques. Hindawi 2022. https://doi.org/10.1155/2022/6521532
32. Rawat V, Joshi S, Gupta S, Singh DP, Singh N (2022) Machine learning algorithms for early diagnosis of diabetes mellitus: a comparative study. Mater Today Proc 56(1):502–506. https://doi.org/10.1016/j.matpr.2022.02.172
33. Ismail L, Materwala H, Tayefi M, Ngo P, Karduck AP (2022) Type 2 diabetes with artificial intelligence machine learning: methods and evaluation. Arch Comput Methods Eng 29(1):313–333. https://doi.org/10.1007/s11831-021-09582-x
34. Rajeswari SVKR, Ponnusamy V (2021) Prediction of diabetes mellitus using machine learning. Ann Rom Soc Cell Biol 25(5):17–20
35. Sharma A, Guleria K, Goyal N (2021) Prediction of diabetes disease using machine learning model. Lect Notes Electr Eng 733:683–692. https://doi.org/10.1007/978-981-33-4909-4

Chapter 48
Environment Mapping Using Ultrasonic Sensor for Obstacle Detection and Navigation

Medha Wyawahare, Aditya Shirude, Akshara Amrutkar, Anurag Landge, and Ashfan Khan

1 Introduction

It is natural for humans to travel to new places frequently. Due to the brain's ability to fuse information on the fly, the intelligence of the natural senses makes this possible. Realizing how much humans take for granted in terms of our ability to perceive, orient, and process information is one of the most humbling aspects of bringing intelligence to inanimate objects. The advanced robots are able to build maps of their surroundings with the help of the data acquired through sensors. The generated map is nothing but the representational view of the human sight. These robots make use of the data from the depth cameras such as Infrared, LIDAR, LRF's, etc. The generated robot maps can either be 2D or 3D based on the requirements. 2D maps refers to terrain mapping and 3D refers to 3D mapping. The development in the technological fields have paved the way for humans to go beyond the boundaries of Earth's atmosphere and enter into space. A lot of exploration programs have been launched by various space agencies and to achieve this purpose, astronomers totally need to rely on autonomous robots in order to collect and share the relevant information from the surface of these external celestial bodies. With the help of image data with different resolutions and angles, captured with the help of rovers and landers, the researchers were able to design an autonomous system which can navigate itself. The system makes a connection between surface as well as orbital images with the help of SfM (Structure from Motion) algorithm which is then ultimately used for the terrain mapping for Mars exploration. Terrain mapping is a technique for capturing data across erratic or inclined surfaces while maintaining sample focus. Terrain maps are

M. Wyawahare · A. Shirude · A. Amrutkar · A. Landge · A. Khan (✉)
Department of Electronics and Telecommunication, Vishwakarma Institute of Technology, Pune, India
e-mail: ashfankhan313@gmail.com

A. Yadav et al. (eds.), *Proceedings of International Conference on Paradigms of Communication, Computing and Data Analytics*, Algorithms for Intelligent Systems,
https://doi.org/10.1007/978-981-99-4626-6_48

very unique and user-friendly. Data collected exclusively on the ground is used for these calculations. The most used technique for mapping terrain is contouring. Points of equal elevation are connected by contour edges. The perpendicular separation calculated from one contour line to the other is depicted using contour intervals. The Digital Elevation Model (DEM) is a common data source for topography mapping and analysis. Using stereo aerial photographs, satellite images, radar data, and other data sources, a DEM is made up of a regular array of elevation points. A DEM is initially transformed into an elevation raster for use in mapping and analyzing the terrain. An elevation raster's straight-forward data structure makes it relatively simple to carry out the calculations required to obtain slope and various different topographic characteristics.

Another visual technology, 3D mapping is another marvel of modern technology. The term "3D mapping" is also used to refer to "Projection Mapping" or "Video Mapping". The process of creating a 3D map involves creating a real-world representation of the object by rendering it in three dimensions. 3D profiling can be done in a variety of ways, including using a stereo camera. One way is to measure the depth of an object or object from a focal point. The main advantage of 3D mapping is the ability to visualize and store information using the latest technology. Having a 3D model available for the object/area under study facilitates knowledge visualization and scientific mapping. 3D maps give planners and local governments a realistic view of an area. 3D mapping is widely used in many industries, from science to entertainment to advertising.

2 Literature Survey

The author proposed the development of an instrument capable of long-distance travel and 3D modeling for use in remote sensing applications [1]. The machine can move in response to a command. There is also a reservation system for these plots, with options for later redemption. Since bots do not have much memory, they create a local system to keep this land. The authors present a new method for determining the surface elevation using the mapping method of localization drift [2]. The method uses legged robots to navigate the terrain. The map is only affected by the noise in the ultrasonic sensor and the uncertainty in the visible tilt angle. The authors proposed the use of 2D and 3D mapping for environmental modeling. The mapping is done using Bayes theorem, and the mapping data is stored in the OC-TREE data structure [3]. Localization mechanism is also designed to store data locally. The map is treated as a 2D mesh using the 2D SLAM algorithm. The authors present a new method for designing a robot that adapts to the orientation of the terrain in real time while still ensuring the stability of the robot [4]. The framework maps terrain for a vehicle for smooth movement. The mapping function uses a log-barrier function to calculate soil statistics. The system uses terrain mapping functions to improve vehicle stability and speed over challenging terrain. The authors proposed a 3D mapping system using an efficient network-based system. The algorithm uses

3 variables, namely slope, flatness, and reliability [5]. These are used to assess the navigability of the terrain. The proposed model uses a localization method to store the acquired data locally. The sensor model collecting data includes 2 laser scanner sensors (SICK LMS 291), and 1 (SICK LMS 200) mounted under the vehicle. The authors proposed the development of a scanning system for large-scale mapping of terrain mining operations [6]. The proposed model uses multiple scanners mounted on vehicles to produce close-up images of the terrain. All the vehicle robots are connected through a distributed system, which works in a mesh topology. There are several locally connected nodes in the system that form a subset map while the top node will store the map of the whole area [7]. The authors proposed a sensor system consisting of optical sensors and electromagnetic sensors for environmental mapping. This model uses RADAR- and LidaR-based systems to map the terrain [8]. The sensor architecture captures spatial characteristics of the soil which are then locally retrieved and stored for analysis. The data are analyzed using a special C library. A paper on incremental mapping and localization for mobile robots with a 2D laser range finder is presented. It also generates 3D maps of high circulation areas in real time [9]. A different approach is applied to the simultaneous localization and mapping (SLAM) problem of six degrees of freedom in this work [10]. Robot motion while moving over natural objects must account for yaw, pitch and roll angles, making pose estimation a six-dimensional mathematical problem. To develop soil models, a wide range of available spatial data including geology, soil topography, elevation, altitude, and soil color were combined with statistical analyzes of monitoring sites [11]. Furthermore, expert knowledge of geophysical relationships was used to optimize data and adjust spatial models. The paper proposes a solution to the problem using a depth-mapping based method [12]. This method is able to produce visual results with high accuracy and efficiency, and can be applied to landscape models, when materials, because the algorithm does not depend on geometry The research uses a methodology three-dimensional terrain modeling and classification in the park shows the system using laser rangefinders enables mobile robots to perform tasks [13]. Parts of the map have been classified as accessible, semi-accessible, or inaccessible by the authors using support vector devices. The authors developed a program that uses the geographical model as the main resource, and places images of real text on the map based on facts [14]. According to research, the program gives results that are true and superior, with efficiency, accuracy and precision. The system has demonstrated excellent performance and has potential for widespread use. The study addresses the first observations on the problem of elevation estimation in interferometric synthetic aperture radar using real-world data (InSAR) [15]. The proposed method uses maximum a posteriori (MAP) estimation and Markov random fields (MRF) image sampling methods, using multifrequency/baseline SAR raw data. The Maximum a Posteriori (MAP) estimate and Markov Random Fields (MRF) picture modeling techniques are also used in an approach, which uses multi-frequency/baseline SAR raw records. Take a look at a mapping gadget that consists of a mapping robot that could collect pertinent records and a mapping software which can use those records to map the surroundings [16]. Robotics also uses various techniques to characterize the separation of two regions in space, such as computer vision,

laser light, photovoltaic scanning, ultrasonic waves and uses a mobile robot equipped with two rotating laser planes and multiple photovoltaic receivers across industrial environments the environment is also used [17]. In an approach, the laser tracker is used in comparison tests to confirm the feasibility and accuracy at a location and along the orientation. Also, the authors present VPass, an algorithmic compass that predicts the movement of a mobile robot by missing space in confined spaces. VPass can have a loop closing effect even in open loop environments [18]. The authors in order to study the obstructions in residential and commercial surroundings, developed a new vision haptic sensor. It is made of a passive, flexible material that can be visually measured using a conventional, inexpensive camera. The camera records the immediate surroundings of the transition simultaneously to measure ego movement and environmental movement. This inexpensive sensor is available on a smartphone to collect environmental data [19]. In one approach, the UAV produces large-scale images of the terrain, which were then analyzed for angle and slope using the RMSE method. The inclination angle analysis was performed using an image captured by a UAV. It is an effective way to explore high and isolated terrain [20].

In a research paper, the authors reconstructed the images to the 3D images which will work on the texture mapping of the image [21]. The results show that instead of being monochromatic, the processed images also reflect the depth of the images. As the image quality is reconfigured, the overall size and accuracy increase. When Simultaneous Localization and Mapping Technique is used, the authors concluded, it is necessary to find the loop closure by canceling the errors in the process and the design [22]. Accelerometer sensors, magnetic sensors, gyro sensors, attitude and heading reference system, IMU and interior navigation systems were used to overcome the problem of unbalanced position estimation [23]. The authors in a study suggest a ground based interferometric synthetic aperture radar (SAR) method for mapping the landscape. It is based on the motion of a continuous wave step frequency (CW-SF) radar parallel to a linear horizontal rail. By comparing phases of microwave holographic images from different sources, height maps can be constructed [24].

Three-dimensional (3D) maps are an important tool for efficient and accurate visualization of geographic information. Due to the large number of data points in a 3D environment, data analysis using traditional methods is complicated and inefficient Consequently, 3D maps are used to efficiently analyze and study 3D spatial data, providing researchers and researchers be able to better understand the relationships and patterns between data sets focus on to explain the information easily to them It also enables filtering, which can be difficult to do with traditional methods Thus, 3D maps provide a valuable tool for applications such as urban planning, architecture, geography, environmental management, etc.

3 Technologies Used

In this proposed work, the following components were used to realize the sensor architecture.

3.1 HC-SR04 Ultrasonic Sensor

The HC-SR04 ultrasonic sensor is a commonly used distance sensor for measuring distances between 2 cm to 4 m. It emits ultrasonic waves that bounce off objects and then calculates the distance by measuring the time it takes for the waves to return. The sensor is widely used in robotics, automation, and other applications requiring distance measurement. It is relatively easy to use, low cost, and has a wide range of applications.

3.2 Servo Motor

A servo motor is an electrical device that can rotate to a precise position based on the input signal it receives. It is commonly used in control systems that require accurate and precise positioning, such as robotics, automation, and CNC machines. Servo motors have a built-in feedback mechanism that allows them to adjust their position based on the input signal they receive, ensuring that they reach and maintain the desired position. They come in various sizes and power ratings and can operate on different voltages and frequencies. Servo motors are widely used in industrial applications, as well as in hobbyist projects and educational settings.

3.3 Microcontroller

The Arduino Uno is a microcontroller board based on the ATmega328P microcontroller. It is one of the most popular boards in the Arduino family and is widely used in various projects due to its versatility and ease of use. The board has 14 digital input/output pins, six analog input pins, and a 16 MHz quartz crystal oscillator. It also has a USB port for connecting to a computer and a power jack for supplying power to the board. The Arduino Uno is a microcontroller board based on the ATmega328P microcontroller. It is one of the most popular boards in the Arduino family and is widely used in various projects due to its versatility and ease of use. The board has 14 digital input/output pins, 6 analog input pins, and a 16 MHz quartz crystal oscillator. It also has a USB port for connecting to a computer and a power jack for supplying power to the board. The Arduino Uno is a microcontroller board based on the ATmega328P microcontroller. It is one of the most popular boards in the Arduino family and is widely used in various projects due to its versatility and ease of use. The board has 14 digital input/output pins, 6 analog input pins, and a 16 MHz quartz crystal oscillator. It also has a USB port for connecting to a computer and a power jack for supplying power to the board.

3.4 Arduino IDE

The Arduino IDE is open-source software used to create and upload code to Arduino boards. It supports the programming languages C and C++. Here, IDE stands for Integrated Development Environment.

3.5 MATLAB

Using MATLAB, we will plot the points on the Cartesian plane. There are two items in the data that the Arduino sends. the servo's angle of rotation and the distance to an obstruction in that direction. This means that the data we currently have is in the Polar coordinate system. The Cartesian or *X*–*Y* coordinate system must be used in order for it to be viewed in a way that makes sense to human sight.

4 Methodology

In this study, we suggest that an Ultrasonic Sensor (SR04) be used to collect data, which will subsequently be sent to a host system, ideally a laptop. A Python script that computes directionality information and creates a graph for an *XY*-plane using the sensor data would provide a 2D feature for each plane. We will add an actuator mechanism to change the sensor assembly's height in order to improve measurement accuracy. Additionally, a servo motor will be used to position the sensor in a certain direction, enabling it to capture data at each step made and show a 360° view of an *XY* area. In this project, an ultrasonic distance sensor is being used. It generates audible sound waves to humans and utilizes them to estimate distance by timing how long it takes for the waves to pass an impediment and return. This is similar to how both cruise ships and bats work. As an additional component, a servo motor will be employed. It differs from a standard DC motor in that it can turn very precisely to a particular angular position and maintain that state. A servo motor response to pulses with a predefined time by angling to the corresponding position. After that, MATLAB examines the entire collection of sensor data that was sent to the PC through serial.

4.1 The Hardware Setup

The ultrasonic sensor and servo motor should first be connected to the microcontroller or computer that will be controlling them, as illustrated in Fig. 1. In order to be able to spin the ultrasonic sensor to various angles, it should be placed at the end of the

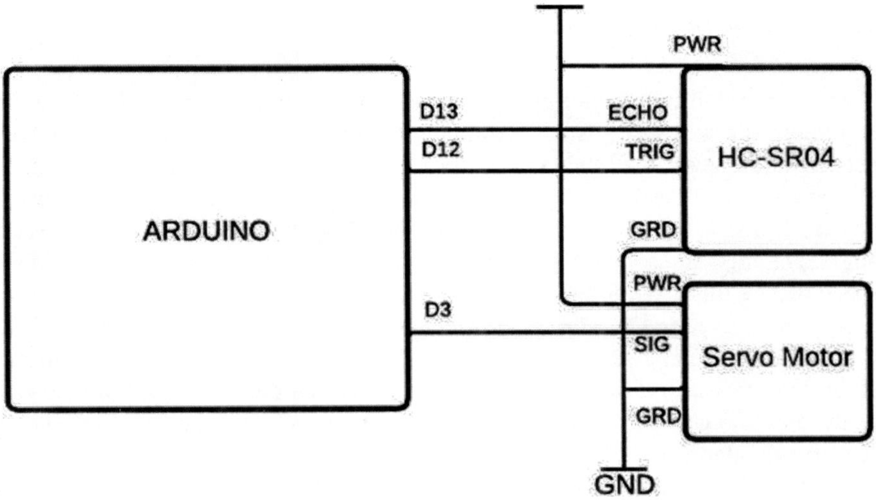

Fig. 1 Showing circuit diagram for the proposed system

servo motor arm. Verify that all connections are tight and that the components are receiving the necessary power.

4.2 *The Ultrasonic Sensor Setup*

The ultrasonic sensor has to be set up to produce sound waves with a certain frequency and duration. Additionally, the sensor should be configured to catch the sound waves that are reflected and gauge how long it takes for them to come back. The speed of sound in air may be used to translate a time measurement to a distance measurement. The microcontroller's signal may be used to trigger the ultrasonic sensor, which can be programmed to continually produce sound waves.

4.3 *The Servo-Motor Setup*

The servo motor should be set to rotate the ultrasonic sensor to particular angles. The range of angles and rotation speed should be chosen by the application's needs. The servo motor may be controlled using a microcontroller-generated pulse-width modulation (PWM) signal.

4.4 The Scanning Algorithm

The scanning algorithm begins by defining the scanning range, or the greatest distance that the sensor will scan. After determining the scanning range, the algorithm will determine the number of angles necessary to scan the whole range. The number of angles is calculated by dividing the scanning range by the sensor's minimum detectable distance. The program will decide how long the sensor will remain at each angle. The angular resolution of the sensor's stepper motor and the sensor's scanning speed dictate this. The difference between the sensor's rotational speed (scanning speed) and its smallest possible angle (angular resolution) is its rate of rotation. The sensor will rotate at a set of angles as it scans the environment, according to the algorithm. As was indicated previously, the angles can either be fixed or they can be dynamically changed depending on the surroundings being scanned. In the latter scenario, the algorithm will examine the surroundings to select the ideal range of scanning angles. This can be accomplished using a variety of methods, including examining the environment's geometry, examining its reflectivity, or utilizing machine learning algorithms to forecast the ideal combination of angles.

4.5 Data Acquisition and Processing

It is necessary to record and interpret the distance readings the ultrasonic sensor made at each angle in order to produce a map of the immediate area. The data can be analyzed immediately or saved for subsequent processing.

4.6 Mapping Visualization

Software technologies like 2D or 3D graphics can be used to visualize the environment's map. The map may be applied in a variety of ways, including navigation and obstacle avoidance.

4.7 Calibration

To guarantee precise distance measurements and rotation angles, the ultrasonic sensor and servo motor should be calibrated. One option for calibration is to use a known distance or a reference item.

4.8 Testing and Refinement

To find any problems or potential areas for development, the system should be tested in multiple settings and circumstances. To enhance performance, changes can be made to the hardware or software. Overall, hardware setup, sensor and motor programming, data collecting and processing, and mapping visualization go into environment mapping utilizing an ultrasonic sensor and servo motor. Depending on the particular needs of the application, the technique might be altered.

5 Results and Discussion

An ultrasonic sensor emits high-frequency sound waves that bounce off the surroundings and return to the sensor when the sound wave should return to the sensor is used to calculate the distance between the sensor and the object between by the sensor a servo motor robot can scan the surroundings on the environment and nearby maps Information can be used for avoiding navigation obstacles or other things Servo motor is used to rotate the sensor to different angles so that robot can explore a wide area Servo Motor rotation is controlled by microcontroller to determine the specific angle to be rotated.

As proven in Fig. 2. The Y-axis of the map diagram begins at a minimal distance, which is the nearest distance at which the sensor can discover items, and ends at a most distance, that's the farthest distance at which the sensor can discover items. The range of distances that the sensor can cover depends on the specific version and specifications of the ultrasonic sensor used. The x-axis of the map diagram represents the space at which objects are positioned relative to the sensor. Objects which can be in the direction of the sensor will have a smaller x-coordinate at the map diagram, even as gadgets that are farther away can have a bigger x-coordinate. The Ultrasonic sensor gathers facts of the environment and in step with that the graph is being plotted. While both strategies have their personal benefits and drawbacks, here are a few advantages of using an ultrasonic sensor and servo motor for surroundings mapping compared to the usage of a LiDAR sensor: Cost-effective: Ultrasonic sensors are usually much less high priced than LiDAR sensors, making them a greater price-effective option for some applications Low electricity intake: Ultrasonic sensors consume very little strength, making them suitable for battery powered robots or devices that want to perform for lengthy durations. Non line of sight detection: Ultrasonic sensors can hit upon items even though they are no longer in the line of sight, which may be beneficial in environments in which LiDAR sensors might not be effective. Simpler information processing: Ultrasonic sensors provide distance measurements immediately, at the same time as LiDAR sensors provide a 3D factor cloud that requires extra processing to extract useful statistics. Less tormented by ambient light: Ultrasonic sensors are much less affected by ambient light, making them greater reliable in out of doors or vibrant indoor environments in which LiDAR sensors can also war. That being stated,

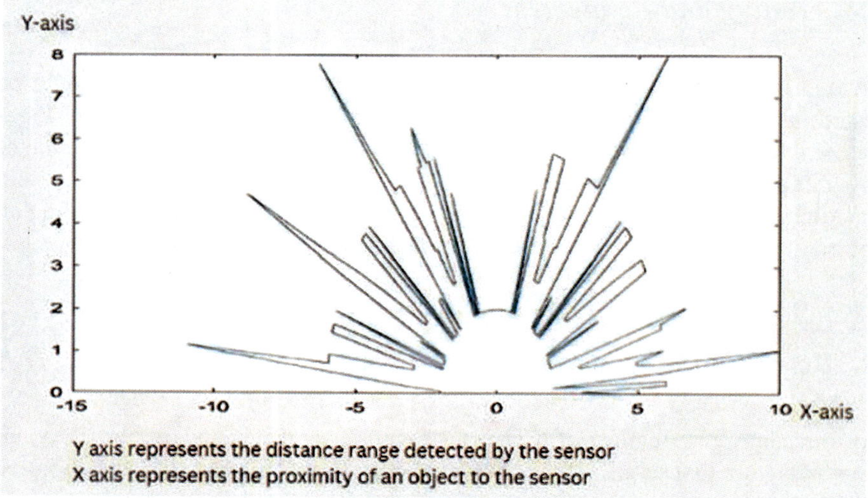

Fig. 2 Map diagram created using an ultrasonic sensor and MATLAB, showing the range of distances that the sensor can cover on the y-axis and the distances at which objects are located on the x-axis

LiDAR sensors have their personal unique advantages, such as higher accuracy and precision, longer range, and higher performance in sure environments. The desire of the sensor depends on the precise application and necessities.

6 Conclusion

The need for environmental simulation will increase in the future so there is a need for more construction that can do it in a more efficient and cost-effective manner. This proposed work uses 2 spatial dimensions to better simulate the vertical environment. This can be further enhanced by using fusion algorithms to capture complex spatial data such as depth. Further improvements can be made in data collection and mapping. Using LidaR sensors to generate texture maps to enhance the visualization of spatial maps.

References

1. Hata AY, Wolf DF (2009) Terrain mapping and classification using support vector machines. In: 2009 6th Latin American robotics symposium (LARS 2009), pp 1–6. IEEE
2. Park J, Kim JY, Kim B, Kim S (2018) Global map generation using LiDAR and stereo camera for initial positioning of mobile robot. In: 2018 international conference on information and communication technology robotics (ICT-ROBOT), pp 1–4. IEEE

3. Olson CF, Matthies LH, Wright JR, Li R, Di K (2007) Visual terrain mapping for Mars exploration. Comput Vis Image Underst 105(1):73–85
4. Kitayama D, Touma Y, Hagiwara H, Asami K, Komori M (2015) 3D map construction based on structure from motion using stereo vision. In: 2015 international conference on informatics, electronics & vision (ICIEV), pp 1–5. IEEE
5. Bi X, Li J (2008) The 3D terrain reconstruction algorithm based on texture mapping. In: 2008 34th annual conference of IEEE industrial electronics, pp 1942–1947. IEEE
6. Fankhauser P, Bloesch M, Hutter M (2018) Probabilistic terrain mapping for mobile robots with uncertain localization. IEEE Robot Autom Lett 3(4):3019–3026
7. Na HJ, Choe Y, Chung MJ (2014) Efficient 3D terrain mapping based on normal distribution transform grid. In: 2014 14th international conference on control, automation and systems (ICCAS 2014), pp 656–660. IEEE
8. Chou YS, Liu JS (2013) A robotic indoor 3D mapping system using a 2D laser range finder mounted on a rotating four-bar linkage of a mobile platform. Int J Adv Rob Syst 10(1):45
9. Steinbrücker F, Sturm J, Cremers D (2014) Volumetric 3D mapping in real-time on a CPU. In: 2014 IEEE international conference on robotics and automation (ICRA), pp 2021–2028. IEEE
10. Zhang Z (2012) Microsoft kinect sensor and its effect. IEEE Multimedia 19(2):4–10
11. Lee DH, Kweon IS, Cipolla R (1999) A biprism-stereo camera system. In: Proceedings. 1999 IEEE computer society conference on computer vision and pattern recognition (Cat. No PR00149), vol 1. IEEE
12. Ahmadabadian AH, Robson S, Boehm J, Shortis M, Wenzel K, Fritsch D (2013) A comparison of dense matching algorithms for scaled surface reconstruction using stereo camera rigs. ISPRS J Photogramm Remote Sens 78:157–167
13. Gohl P, Burri M, Omari S, Rehder J, Nikolic J, Achtelik M, Siegwart R (2014) Towards autonomous mine inspection. In: Proceedings of the 2014 3rd international conference on applied robotics for the power industry, pp 1–6. IEEE
14. Lefloch D, Nair R, Lenzen F, Schäfer H, Streeter L, Cree MJ et al (2013) Technical foundation and calibration methods for time-of-flight cameras. In: Time-of-flight and depth imaging. Sensors, algorithms, and applications. Springer, Berlin, pp 3–24
15. Gschwandtner M, Kwitt R, Uhl A, Pree W (2011) Infrared camera calibration for dense depth map construction. In: 2011 IEEE intelligent vehicles symposium (IV), pp 857–862. IEEE
16. Li Y, Ma L (2006) A fast and robust image stitching algorithm. In: 2006 6th world congress on intelligent control and automation. IEEE, vol 2, pp 9604–9608
17. Wyawahare MV, Patil PM, Abhyankar HK (2009) Image registration techniques: an overview. Int J Signal Process Image Process Pattern Recogn 2(3):11–28
18. Shapiro LG, Stockman GC (2001) Computer vision, vol 3. Prentice Hall, New Jersey
19. Snavely N, Seitz SM, Szeliski R (2008) Modeling the world from internet photo collections. Int J Comput Vision 80(2):189–210
20. Aulinas J, Petillot Y, Salvi J, Lladó X (2008) The SLAM problem: a survey. Artif Intel Res Dev 363–371
21. Taketomi T, Uchiyama H, Ikeda S (2017) Visual SLAM algorithms: a survey from 2010 to 2016. IPSJ Trans Comput Vis Appl 9(1):1–11
22. Zhu Z, Yang S, Dai H (2018) Enhanced visual loop closing for laser-based SLAM. In: 2018 IEEE 29th international conference on application-specific systems, architectures and processors (ASAP), pp 1–4. IEEE
23. Yuen DC, MacDonald BA (2003) An evaluation of the sequential Monte Carlo technique for simultaneous localisation and map-building. In: 2003 IEEE international conference on robotics and automation (Cat. No. 03CH37422), vol 2, pp 1564–1569. IEEE
24. Rusdinar A, Kim J, Kim S (2010) Error pose correction of mobile robot for SLAM problem using laser range finder based on particle filter. In: ICCAS 2010, pp. 52–55. IEEE

Chapter 49
Identification and Classification of Skin Diseases with Erythema Using YOLO Algorithm

C. Santhosh Kumar, K. Amritha Devangana, P. L. Abirami, M. Prasanna, and S. Hari Aravind

1 Introduction

Skin conditions are more prevalent than other illnesses. Viruses, germs, allergies, or fungi may bring on skin conditions. The texture or color of the skin may alter due to a skin illness [1]. Skin conditions are typically chronic, contagious, and occasionally can lead to skin cancer. Early diagnosis is crucial to halt the development and spread of skin problems. Skin conditions take longer to diagnose and cure, which can financially and physically cost the sufferer. Most everyday, individuals are ignorant of the type and stage of a skin ailment. Some skin problems don't show symptoms for several months, which causes them to deteriorate and advance.

This is a result of the public's lack of knowledge about medicine [2]. Diagnosing a skin ailment can occasionally be challenging for a dermatologist specializing in skin problems. Expensive laboratory tests may be required to determine the kind and stage of the condition. To recognize skin disorders, we advise employing picture processing [3]. Using a digital image or video frame of the diseased skin area, this method analyzes the image to identify the kind of disease. Our simple, rapid solution requires two expensive pieces of equipment: a camera and a computer.

To solve the issue, we are developing a model for the early identification and prevention of psoriasis and skin cancer [4]. In general, the diagnosis of skin diseases depends on various factors, including color, form, texture, and so on. Here, one may take skin-related pictures, which will be transmitted to trained models. The model analyzes the picture to determine whether the subject has a skin condition. Finally, the

C. Santhosh Kumar · K. Amritha Devangana (✉) · P. L. Abirami · M. Prasanna · S. Hari Aravind
Department of Information Technology, Sona College of Technology, Salem, India
e-mail: devangana.amritha01@gmail.com

C. Santhosh Kumar
e-mail: santhoshkumar.it@sonatech.ac.in

A. Yadav et al. (eds.), *Proceedings of International Conference on Paradigms of Communication, Computing and Data Analytics*, Algorithms for Intelligent Systems,
https://doi.org/10.1007/978-981-99-4626-6_49

model generates a result and determines whether the person has a skin disease. Image processing technologies significantly reduce the time spent on a specific activity by the customer. Hence, it is a time- and money-saving process.

The main objective of the model is to create one that forecasts skin conditions that can be avoided by examining the diseased area [5]. Here, skin color and tone are critical factors in identifying skin diseases. The person can take pictures of their skin, which will then be transmitted to a trained model for analysis.

The trained model determines whether the person has a skin ailment or not, and if so, provides a thorough explanation of the condition and possible treatments. It is quite accurate. Dermatologists (doctors who specialize in treating skin conditions) can utilize it when they are having trouble diagnosing a skin condition and may need to perform pricey laboratory tests to determine the kind and stage of the condition accurately.

2 Related Works

Tschandl et al. work [6] shows an innovative method to combine exact segmentation and classification models sequentially. We analyze a skin picture, derive high-level characteristics, and normalize the image. We first create a segmented picture map using a neural network-based segmentation model. Then, we combine the regions of aberrant skin and send this information to a categorization model. We divide each grouping into several common skin disorders using a different neural network model. Our segmentation model works better than other trials and receives a sensitivity score of almost perfect in difficult situations. We can categorize using our categorization algorithm. Our classification model is more precise than a baseline model trained without segmentation and can accurately identify multiple illnesses in a single picture.

Albawi et al. [7] propose a method using machine learning techniques to automate the analysis that could lead to a framework and system for the medical industry that would help identify diseases more easily. Creating a machine learning method that can categorize malignant and normal pigmented skin tumors is a move toward achieving this goal. The suggested study enables the earliest possible identification of malignant skin lesions by accurately classifying pigmented skin lesions in dermoscopic pictures using convolutional neural networks (CNN) and machine learning algorithms.

Codella et al. introduced a deep learning system (DLS) [8]. It offers a differential diagnosis of skin conditions based on 16,114 de-identified cases (photographs and clinical data) from a teledermatology practice serving 17 locations. According to the DLS, one of 26 frequently occurring skin diseases is present in 80% of patients seen in primary care, which offers a secondary prediction encompassing 419 skin illnesses. When a panel of three dermatologists sets the reference standard, the DLS did better than six primary care doctors (PCPs) and six nurse practitioners (NPs) in 963 validation instances and was comparable to six other dermatologists (accuracy:

0.66DLS, 0.63 dermatologists, 0.44 PCPs and 0.40 NPs). These results show how the DLS can aid other medical experts in identifying the skin condition.

Khan et al. [9] propose a method for producing "visual explanations" for the judgments made by the large set of CNN-based models. Gradient-weighted class activation mapping (Grad-CAM), which highlights the critical regions in the picture crucial for idea prediction, uses the gradients of any target technique. We create guided Grad-CAM, a high-resolution class-discriminative visualization, for visual question-answering (VQA) models, including ResNet-based architectures, by combining Grad-CAM with existing fine-grained visualizations. This visualization is suitable for image classification, image captioning, and VQA models. We create a technique to use Grad-CAM to find pertinent neurons and merge them with neuron names to provide textual justifications for model choices. Finally, we design and conduct human studies to determine whether Grad-CAM justifications give users the appropriate confidence in deep network predictions. Our results demonstrate that Grad-CAM, even in cases where both "stronger" and "weaker" deep networks generate the same results, aids new users in making this distinction.

In their comprehensive investigation of model scaling, Polat et al. [10] find that optimizing network depth, width, and resolution can improve performance. Based on this finding, we propose a novel scaling method that equally scales all depth, breadth, and resolution parameters using a simple yet very effective compound coefficient. By scaling up MobileNets and ResNet, we can demonstrate the viability of this approach. To go even further, we use neural architecture search to construct a fresh baseline network, enlarge it, and produce the EfficientNets family of models, which outperforms the prior ConvNets regarding accuracy and productivity.

3 The Proposed Methodology

3.1 Algorithm Used

The You Only Look Once version 3 (YOLOv3) technique, which recognizes objects in images by gridding the image and forecasting bounding boxes and class probabilities for each grid cell, finds objects in images.

The YOLOv3 algorithm follows a three-step process:

- Object detection: The algorithm divides an input image into a grid of cells. Bounding boxes and class probabilities are predicted for any objects that might be present in each grid cell. The enclosing boxes predict the locations of the object in the cell and the likelihood that it is present in the cell.
- Non-max suppression: The algorithm uses non-max suppression to remove duplicate detections of the same object. This involves comparing the confidence scores of all the overlapping bounding boxes and selecting the one with the highest confidence score.

- Classification: The algorithm assigns each remaining bounding box to a specific object class based on the highest class probability score. This is done using a softmax function that calculates the probability of each object class based on the scores of all the bounding boxes in the image.

YOLOv3 achieves high accuracy and speed using a deep convolutional neural network (CNN) to detect objects. The CNN consists of several convolutional layers and uses skip connections to incorporate information from earlier layers into later layers [11], improving the model's ability to detect objects at different scales and resolutions.

3.2 Pseudocode for YOLOv3 Algorithm

1. Load the pre-trained convolutional neural network.
2. Set the threshold for object detection confidence.
3. For each input image:

   a. Preprocess the image (resize, normalize pixel values, etc.).
   b. Forward the image through the network to obtain predicted bounding boxes.
   c. Apply non-maximum suppression to remove duplicate boxes and select the most confident boxes.
   d. Post-process the boxes (rescale, offset, etc.) to obtain final bounding box coordinates and class probabilities.
   e. Filter out boxes with confidence scores below the threshold.
   f. Return the final bounding boxes and class probabilities.

This pseudocode outlines the general steps, including loading the pre-trained neural network, preprocessing the image, detecting objects using the network, applying non-maximum suppression to remove duplicates, and post-processing the boxes to obtain the final coordinates and class probabilities. It also includes a step for filtering out boxes with low confidence scores, which helps to improve the accuracy of object detection.

3.3 Flowchart

Figure 1 depicts the high-level overview of the workflow for using YOLOv3 object detection algorithm with a UI for prediction:

1. Collect and prepare data: First, collect and prepare a dataset of images you want to detect objects. This involves labeling the objects in the images and creating a dataset in a format compatible with YOLOv3.

Fig. 1 Flowchart

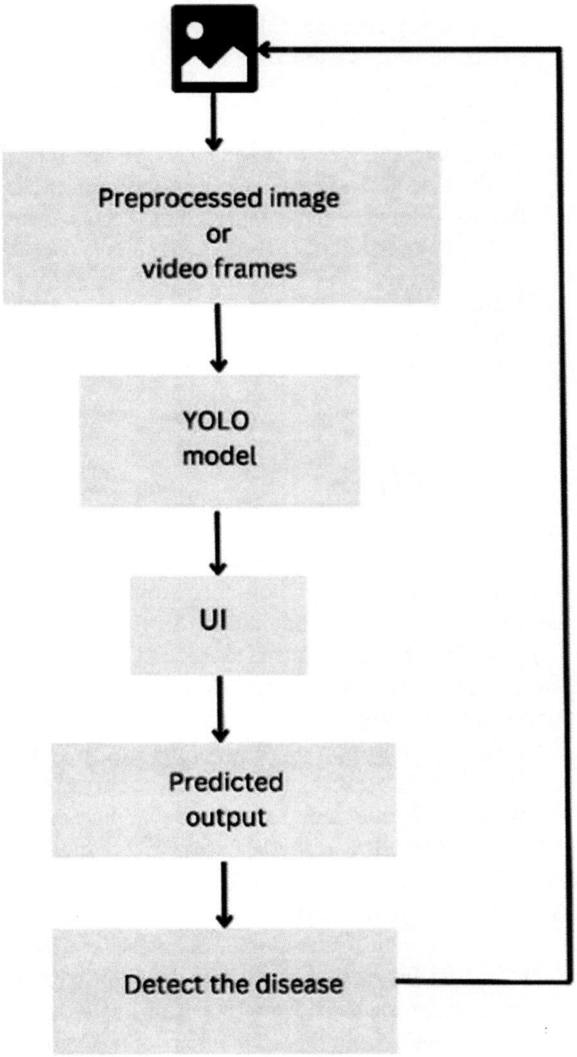

2. Train the YOLOv3 model: Next, train the YOLOv3 model using the prepared dataset. This involves configuring the model, setting hyperparameters, and training the model on a GPU using a deep learning framework such as TensorFlow.
3. Export the YOLOv3 model: Once the model is trained, export it in a format that can be used for prediction, such as a TensorFlow SavedModel.
4. Build a UI for prediction: Build a UI for prediction with the trained YOLOv3 model and exported files. This involves designing a UI where users can upload an image, select the YOLOv3 model and its configuration, and initiate the object detection process.

5. Perform object detection: When the user selects an image and initiates the object detection process, the UI will use the YOLOv3 model to detect objects in the image. The detected objects and their bounding boxes are then displayed in the UI for the user to view.
6. Post-processing and visualization: Finally, the UI can perform post-processing on the detected objects, such as filtering out false positives or grouping objects that are part of the same entity. The UI can also visualize the detected objects and their bounding boxes on the original image for better understanding and analysis.

Overall, the workflow for using YOLOv3 with a UI for prediction involves preparing data, training and exporting the model, building a UI for prediction, and performing object detection and post-processing on the detected objects.

3.4 Data Sources

Due to the lack of clinical datasets for EM pictures [12–15], a dataset was created from freely accessible photographs from several online sources. We have used a variety of search terms, including "Erythema migrans," "Lyme," "bullseye rash," "leg," "face," "hand," "normal skin," and "covid19 rash," to do Google searches on photographs of skin with normal skin (NO), EM, HZ, TC, IB, IB-T, CELL, and EMU. Afterward, we verified and eliminated any duplicate images, irrelevant photographs, and images with low probability for the EM, HZ, and TC images.

From Fig. 2, we can notice that normal skin has an even texture, color, and temperature. Normal skin can vary depending on age, ethnicity, and other factors, but it typically has a smooth texture, no redness or swelling, and is not itchy or painful. Erythema is a condition characterized by skin redness caused by dilation of the blood vessels in the affected area. Various factors, including infections, allergies, and autoimmune disorders, can cause it. Erythema can range from a mild rash to a severe, life-threatening condition, depending on the underlying cause. Erythema can be distinguished from normal skin by the presence of redness, warmth, and sometimes swelling in the affected area. The skin may also be tender to the touch and may itch or burn. In some cases, erythema may accompany other symptoms, such as fever, chills, or fatigue.

4 Results and Discussion

In this work, a large dataset of skin diseases with erythema is collected, including both normal and diseased images. These images will be annotated with labels indicating the presence or absence of skin diseases with erythema. The dataset will be split into training and testing sets to predict the region of skin affected using the deep learning algorithm YOLOv3. The dataset is inherently noisy because it has not

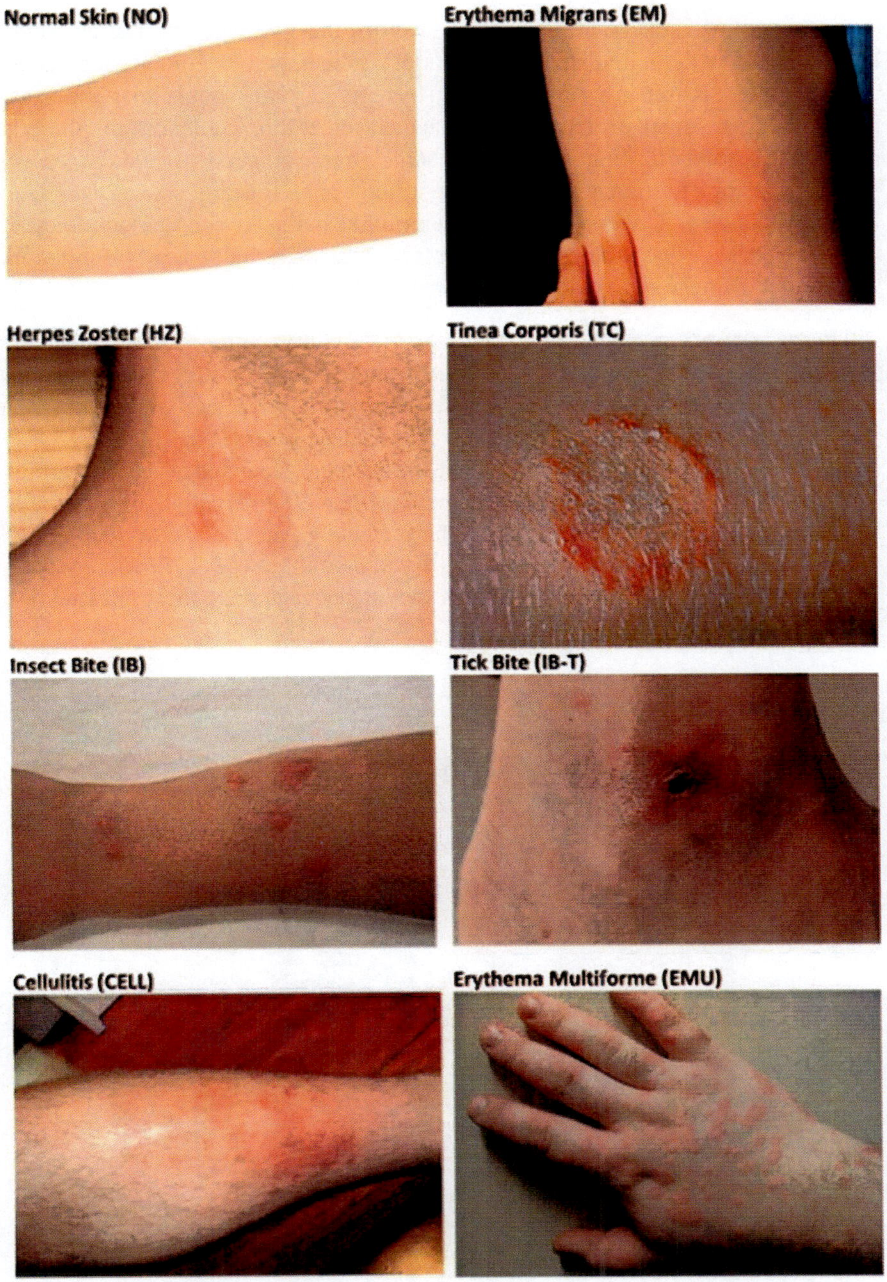

Fig. 2 Different types of skin images

been preprocessed, and we are currently converting that noise into the desired vector. Preprocessing is the initial step in the dataset's processing; it improves a picture's quality before it is used in a subsequent step by removing undesirable image information, which is frequently referred to as image noises. If this problem is not properly resolved, the categorization may contain some errors. Treatment is also necessary if there is a lack of difference between the lesion and the healthy skin around it, an irregular boundary, or abnormalities of the skin like hairs, lines, or dark outlines.

Figure 3 shows two different types of images with different types of information: A normal skin image and a gradient-weighted class activation mapping (Grad-CAM).

- Normal skin image: A normal skin image is a photograph or image of healthy skin tissue. It may show the skin's color, texture, and any distinguishing features or characteristics, such as pores, hair follicles, or moles. These images are often used in dermatology to identify skin conditions and assess skin health.
- Grad-CAM image: A Grad-CAM image is a heatmap generated by a deep learning model, such as a convolutional neural network (CNN), to highlight the regions of an input image that are most important for predicting a particular class or category. Grad-CAM images are used to visualize the neural network's decision-making process and understand which parts of the input image were most influential in the prediction.

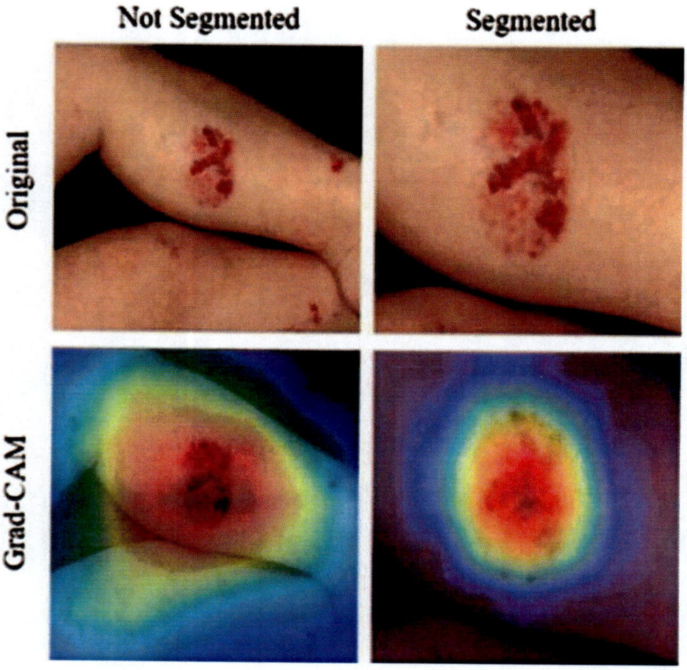

Fig. 3 Visual difference between original and output

While a normal skin image shows the physical characteristics of healthy skin, a Grad-CAM image highlights the regions of an input image that were most important in a deep learning model's decision-making process. They are two different types of images that serve different purposes and convey different types of information.

4.1 Experimentation and Results

The YOLOv3 object detection algorithm is the foundation of our proposed solution. With the ResNet-50 design, which greatly improves the neural network's performance with more layers, we used the YOLOv3 method to increase the model's precision. Figure 4 compares the performance of object detection models such as CNN, Fast CNN, and YOLOv3.

In Table 1, we have compared the 3 models with their accuracy, precision, recall, and F-measure. Here, the model using YOLOv3 shows the highest accuracy rate (90.09) compared to CNN and Fast CNN.

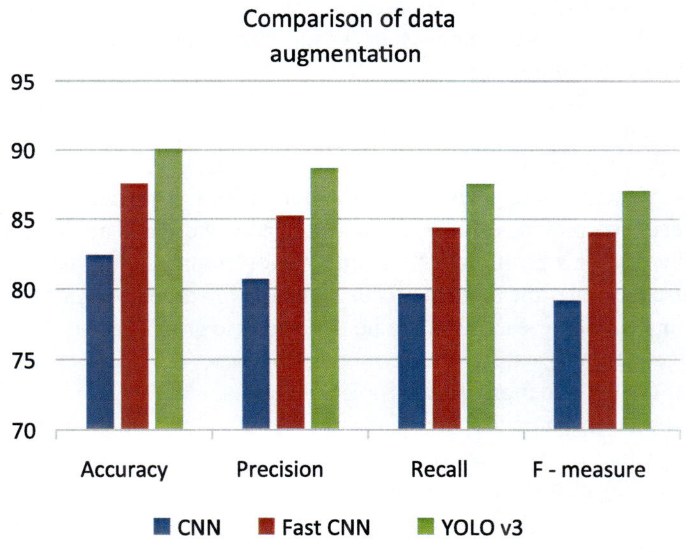

Fig. 4 Comparison graph

Table 1 Comparison table with YOLOv3 algorithm

	CNN	Fast CNN	YOLOv3
Accuracy	82.45	87.52	90.09
Precision	80.71	85.22	88.65
Recall	79.66	84.38	87.49
F-measure	79.23	84.07	87.02

Regarding accuracy, YOLOv3 tends to outperform both CNN and Fast CNN, especially on challenging object detection datasets. YOLOv3 also tends to have higher recall and precision, meaning it can detect more true positives while minimizing false positives and false negatives. However, the specific metrics can vary depending on the dataset and task.

CNN, Fast CNN, and YOLOv3 are different object detection algorithms that use convolutional neural networks as base architecture; however, they differ in terms of accuracy, recall, precision, and F-measure: In summary, while CNN and Fast CNN are general neural networks commonly used for image classification and object detection, YOLOv3 is a specific algorithm designed for real-time object detection. It achieves high accuracy, recall, precision, and F-measure.

5 Conclusion

In general, the diagnosis of skin diseases depends on many traits like color, form, texture, etc. Here, one can take pictures of diseased skin and email them to a trained model. The model examines the photograph sent and determines whether the subject has a skin condition. Dermatologists, or doctors specializing in skin conditions, can utilize it when determining the nature and stage of a skin condition proves challenging or when expensive laboratory testing is needed. Even without an extensive collection and high-quality images, it is possible to achieve adequate accuracy rates. Additionally, with proper data preprocessing, transfer learning, self-supervised learning, and specialty architectural methods, cutting-edge YOLO models can outperform models created by previous studies. In addition, accurate segmentation allows us to pinpoint the disease's position, which is useful for preprocessing the classification data because it enables the model to focus on the pertinent region. Last but not least, in opposition to other research, our method allows us to group various illnesses into a single image. Utilizing state-of-the-art models will make using CAD in dermatology with higher quality and more data feasible.

The proposed machine learning model may eventually connect to many sources that can offer real-time images for predicting skin diseases. The historical data on skin diseases may also assist in increasing the model's accuracy. To further improve performance, we may employ adaptive learning rates and train the model on data clusters rather than the entire dataset.

References

1. Son HM, Jeon W, Kim J et al (2021) AI-based localization and classification of skin disease with erythema. Sci Rep 11:5350
2. Shetty B, Fernandes R, Rodrigues AP et al (2022) Skin lesion classification of dermoscopic images using machine learning and convolutional neural network. Sci Rep 12:18134

3. Liu Y, Jain A, Eng C et al (2020) A deep learning system for differential diagnosis of skin diseases. Nat Med 26:900–908
4. Selvaraju RR, Cogswell M, Das A et al (2020) Grad-CAM: visual explanations from deep networks via gradient-based localization. Int J Comput Vis 128:336–359
5. Tan M, Le Q (2019) EfficientNet: rethinking model scaling for convolutional neural networks. In: ICML, pp 6105–6114
6. Tschandl P, Rosendahl C, Kittler HT (2018) HAM10000 dataset, a large collection of multi-source dermatoscopic images of common pigmented skin lesions. Sci Data 5:180161
7. Albawi S, Mohammed TA, Al-Zawi S (2017) Understanding of a convolutional neural network. In: 2017 international conference on engineering and technology (ICET), Antalya, pp 1–6
8. Codella N, Gutman D, Celebi ME, Helba B, Marchetti MA, Dusza S, Kalloo A, Liopyris K, Mishra N, Kittler H, Halpern A (2017) Skin lesion analysis toward melanoma detection: a challenge at the 2017 international symposium on biomedical imaging (ISBI), hosted by the international skin imaging collaboration (ISIC)
9. Khan MA et al (2021) Attributes based skin lesion detection and recognition: a mask RCNN and transfer learning-based deep learning framework. Pattern Recogn Lett 143:58–66
10. Polat K, Koc KO (2020) Detection of skin diseases from dermoscopy image using the combination of convolutional neural network and one-versus-all. J Artif Intell Syst 2(1):80–97
11. Santhosh Kumar C, Kumar KV (2023) Integrated privacy preserving healthcare system using posture-based classifier in cloud. Intell Autom Soft Comput35(3):2893–2907
12. Akilandeswari J, Dhayanithi J (2018) Interblend fusing of genetic algorithm-based attribute selection for clustering heterogeneous data set. Soft Comput 23(2):1–13
13. Saraswathi K, Mohanraj V, Suresh Y, Senthil Kumar J (2021) A hybrid multi feature semantic similarity based online social recommendation system using CNN. Int J Uncertainty Fuzziness Knowl Based Syst 29:333–352
14. Ramesh P, Jeba Emilyn J, Vijayakumar V (2021) Hybrid artificial neural networks using customer churn prediction. Wirel Pers Commun 124:1695–1709
15. Qu Y et al (2016) Product-based neural networks for user response prediction

Chapter 50
PSO-Based Controller for LFC of Deregulated Power System

Dharmendra Jain, M. K. Bhaskar, and Manish Parihar

1 Introduction

Load frequency management in a power network that is interconnected has two major goals: first, to keep the frequency of every region within a set range, and second, to keep the power exchanges between areas within the planned range as explained by Dharmendra Jain et al. [1]. Because of the large size and complicated structure of the electrical system, LFC has become more significant nowadays. It can be seen that the engineering mechanisms of making plans and procedures have been redeveloped in a deregulated electrical power system in the recent years even though vital thoughts stay unchanged. To enhance performance withinside the power system's functionality, few main adjustments in the shape of electrical power efficacies are added via liberalizing the electric power business and making it available for competition. The efficacies not anymore bundled as electrical power generation units, power transmission structure, and distribution of power to consumers; as a substitute, three separate entities are present, called as GENCO-Generation-Companies, TRANSCOs-Transmission Companies, and DISCOs-Distribution-Companies as explained in [2].

Since numerous DISCOs and GENCOs are present withinside the deregulated shape, a DISCO is free to enter into an arrangement for the sale of electricity with any GENCO. A DISCO and a GENCO may also have an agreement in a different control area. These dealings are termed as bilateral transactions. All the dealings ought to be supplied thru an independent object known as an Independent System Operator (ISO) as described by Donde et al. [2]. The ancillary services have to be controlled by the ISO. Load frequency control is also a very important ancillary service. The predominant purpose of the load frequency control is to maintain the frequency to its

D. Jain (✉) · M. K. Bhaskar · M. Parihar
Department of Electrical Engineering, M.B.M. University, Jodhpur, Rajasthan 342001, India
e-mail: dharmendrajainmbm@gmail.com

A. Yadav et al. (eds.), *Proceedings of International Conference on Paradigms of Communication, Computing and Data Analytics*, Algorithms for Intelligent Systems,
https://doi.org/10.1007/978-981-99-4626-6_50

particular value by maintaining the zero steady-state error for frequency deviation and minimizing the unprepared tie-line power flows among neighboring control areas.

In the present complicated deregulated structure, the classic AGC might not be practicable. To address this issue, intelligent control schemes, as well as methods from soft computing such as genetic algorithm, artificial neural network, bacterial foraging optimization algorithm, fuzzy logic, particle swarm optimization algorithm, firefly algorithm, and others, must be investigated.

Pal et al. [3] describe the construction of a multilayered perceptron neural network-MLPNN controller for AGC difficulties in a two-area restructured power system. S. Bhagya Shree et al. [4] described their hybrid neuro fuzzy technique AGC in a restructured system. Shekhar et al. [5] described how to manage the load frequency of a power system in a deregulated environment using the optimum firefly algorithm. Bhatt et al. [6] present the modeling of RGA, the real coded GA and BGA, a binary coded GA for getting the best gain parameters in a liberalized power system. Bhateshvar YK explains a GA-based PID controller for a multi-area deregulated power system [7]. Sood et al. [8] presented an optimum power flow (GA-OPF) based on GA for liberalized power systems. Jain et al. [9] explained and done the examination of LFC problem for two-area liberalized power system by means of genetic algorithm.

The basic idea of AGC of a linked power system beneath restructured situations has also been discussed by Kothari et al. [10]. Ravi et al. [11] studied the consequence of an energy storage device on load frequency regulation. The detailed study of thyristor-controlled series compensators for load following in a deregulated system has been done by Deepak and Abraham [12]. Abedinia et al. [13] used fuzzy PID using HBMO for LFC. The firefly algorithm is used by S. Load frequency controller design of a two-area network composed of a thermal generator and PV grid by Abd Elazim and Ali [14]. Babahajiani et al. [15] described intelligent demand response for LFC. Sahoo [16] explained the use of neural networks for line congestion study. Concordia and Kirchmayer [17] have described the tie-line power and frequency control of electric power systems. Reduced order observer method is used by Rakhshani et al. [18] for AGC of two-area systems. Adaptive decentralized LFC of multi-area power systems has been presented by Zribi et al. [19]. Pathak et al. [20] given the realistic model of centralized AGC. Reinforced learning neural network controller has been used for LFC by Saikia et al. [21]. Current operating problems associated with AGC have been described in [22]. Fosha and Elgerd [23, 24] used optimal control theory for the megawatt frequency control problem.

Mishra et al. [25] used PSO GWO optimized fractional-order PID-based hybrid shunt active power filter for power quality improvements. Suid and Ahmad [26] used optimal tuning of sigmoid PID controller using nonlinear sine cosine algorithm for the AVR system. Dhanasekaran et al. [27] used PSO-PID controller for a LFC of standalone multi-source power system. Bhatt et al. [28] proposed a craziness based particle swarm optimization-CRPSO for tuning the integral parameter of AGC loop and the control parameters of TCPS and SMES.

2 Design and Modeling of Deregulated Power System

The AGC work might or might not involve generation firms (GENCOs) in the deregulated power system, whereas distribution companies (DISCOs) are free to enter into contracts that involve any of the GENCOs in their respective or other sectors. This leads to various combinations of contract setups between GENCOs and DISCOs. This paper uses the idea of distribution participation matrix (DPM) to visualize all possible contracts in the two-area deregulated power system model as explained by Dharmendra Jain et al. [1, 28]. The number of rows in a DPM matches the quantity of GENCOs and the number of columns matches the quantity of DISCOs in the electrical system. Each DPM input is referred to as the CPF-contract participation factor. According to Donde et al. [2], CPF indicates the percentage of a DISCO contractual power demand satisfied by a GENCO. The '*ij*'th entry, cpf_{ij}, represents the percentage of overall electricity contracted by DISCO j from GENCO i. The addition of all items in a DPM column results in unity.

Equation 1 provides the DPM for a two-area system with two GENCOs and two DISCOs in each area.

$$\begin{aligned} \text{DPM } = & [cpf11cpf12cpf13cpf14cpf21cpf22cpf23cpf24 \\ & cpf31cpf32cpf33cpf34cpf41cpf42cpf43cpf44] \end{aligned} \tag{1}$$

Whenever, there is any alteration in power request of any of the DISCO, it can be seen as a local load in the same area to which the DISCO belongs to. This change in the load can be seen as local loads ΔPL_1 and ΔPL_2 for area-I and area-II, respectively. It must be linked to the block layout at the point of power system model input. Because there can be multiple GENCOs in every region, the ACE signal must be spread among them in accordance with their AGC share. ACE participation factors (*apfs*) are those that disseminate ACE to several GENCOs in the ratio specified by *apfs*.

Assume DISCO3 requires 0.1 pu MW of electricity, with GENCO-1 providing 0.025 pu MW, GENCO2 providing 0.03 pu MW, GENCO3 providing 0.035 pu MW, and GENCO4 providing 0.01 pu MW. Then, in (1), column 3 elements are readily defined as

$$\begin{aligned} cpf13 &= \frac{0.025}{0.1} = 0.25, cpf23 = \frac{0.03}{0.1} = 0.3, \\ cpf33 &= \frac{0.035}{0.1} = 0.35, cpf43 = \frac{0.01}{0.1} = 0.1 \end{aligned}$$

It is noted that $cpf_{13} + cpf_{23} + cpf_{33} + cpf_{43} = 1.0$, in general, $\sum cpf_{ij} = 1$. The scheduled steady-state tie-line power flow is given in Eqs. 2 and 3.

$$\Delta P_{\text{tie}1-2\text{ scheduled}} = (\text{demand of all the DISCOs in area } -2 \text{ from GENCOs in area} - 1) - (\text{demand of all DISCOs in area } -1 \text{ from GENCOs in area} - 2) \quad (2)$$

$$\Delta P_{\text{tie}1-2\text{scheduled}} = \sum_{i=1}^{i=2}\sum_{j=3}^{j=4} \text{CPF}_{ij}\Delta PL_j - \sum_{i=3}^{i=4}\sum_{j=1}^{j=2} \text{CPF}_{ij}\Delta PL_j \quad (3)$$

Equation 4 gives tie-line power flow error $\Delta P_{\text{tie}1-2,\text{error}}$ at any given time.

$$\Delta P_{\text{tie}1-2,\text{error}} = \Delta P_{\text{tie}1-2,\text{actual}} - \Delta P_{\text{tie}1-2,\text{scheduled}} \quad (4)$$

Equation 5 shows the contractual power delivered by the *i*-th GENCO for a two-area power system.

$$\Delta Pi = \sum_{j=1}^{n\text{disco}=4} \text{CPF}_{ij}\Delta PL_j \quad (5)$$

For $i = 1$, the contractual power supply by GENCO-1 is ΔP_1, which can be given as in Eq. 6.

$$\Delta P_1 = \text{CPF}_{11}\Delta P_{L1} + \text{CPF}_{12}\Delta P_{L2} + \text{CPF}_{13}\Delta P_{L3} + \text{CPF}_{14}\Delta P_{L4} \quad (6)$$

In the similar manner, ΔP_2, ΔP_3, and ΔP_3 can be calculated easily.

Figure 1 depicts a simulation schematic of a two-area deregulated electricity system for LFC. It is architecturally based on the notion of Donde et al. [2]. The local loads in first and second areas are denoted by $\Delta P_{1\text{LOC}}$ and $\Delta P_{2\text{LOC}}$, representing the local loads of area-1 and area-2, respectively. $\Delta P_{\text{uc}1}$ and $\Delta P_{\text{uc}2}$ represent uncontracted power (if any).

3 Controller

The PID controller is widely used in various applications in power systems. In this control scheme, there are three amending terms, proportional, integral, and derivative terms, whose sum give the manipulated variable (MV).

$$U(t) = \text{MV}(t) = Kpe(t) + Ki\int_0^t e(t)\text{d}t + Kd\frac{\text{d}}{\text{d}t}e(t) \quad (7)$$

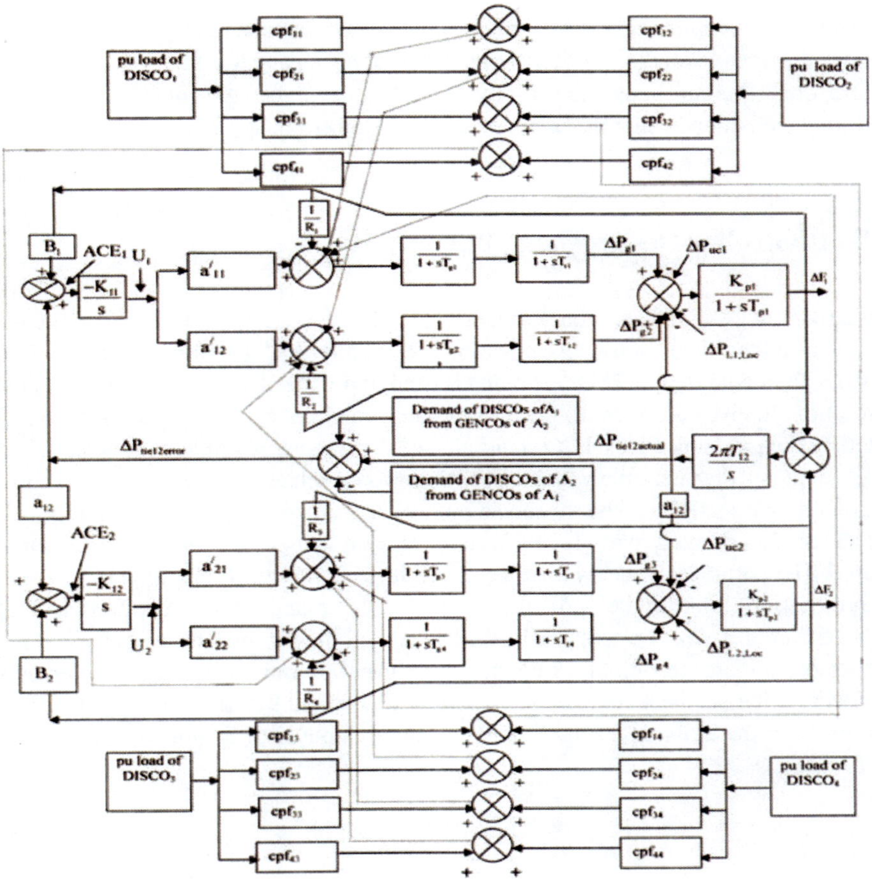

Fig. 1 Simulink diagram of two-area restructured power system

The PID controller's ultimate form is provided in Eq. (7), where $U(t)$ is the controller's output.

Adjusting the gains of the PID controller is called tuning of the PID controller. To produce the required response, the PID controller settings must be adjusted. This is called tuning of the PID controller. Tuning is necessary to get optimum value of the desired response as explained by Tan [29] and Arya [30]. Despite the fact that there are just three parameters that need to be tuned, PID tuning is a challenging operation. Designing and correcting a PID controller appears to be theoretically simple, but it may be tough in practice, especially if various and sometimes contradictory criteria, like, quick transient and good stability, are to be met. PID controllers often provide sufficient control when manually tuned, although performance can typically be improved by careful tweaking.

Various methods for fine-tuning the parameters of the PID controller are available. Z–N method and IMC methods are used by Jain et al. [31, 32]. Current research shows the use of soft computing methods in PID controller parameter tuning. These methods are very effective for discovering proper values of K_P K_I and K_D. In the work reported here, the PID controller settings for the LFC issue of a two-region deregulated power system were tuned using the particle swarm optimization approach.

4 Controller Design Using PSO

Eberhart and Kennedy are the developers of the widely used optimization approach known as "particle swarm optimization". Basically, the flocking of birds' social behavior encourages it. PSO technique is used in proposed work to explore possible solutions to given problems to find the optimum values of controller gains required to satisfy the LFC objectives. PSO is initialized with a group of random particles called solutions and then iteratively looks for optimum gain values by updating the solutions. Every single particle can be expressed by two vectors: position 'x_i' and velocity 'v_i'. The location of every particle at a certain moment is used to solve the issue at that time. The particles/elements then fly about the search region, changing their speed and location to locate the optimal place at every attempt. Each of the particles has values of fitness that are assessed by the function of fitness and have velocities which lead to the elements' flight. Following the current optimal elements, the elements fly across the problem space. The vectors in Eqs. 8 and 9 indicate the location and velocity of the individual i-th particle/element in a physical d-"dimensional" space of search.

$$X_i = [X_{i1}, X_{i2}, \ldots X_{id}] \tag{8}$$

$$V_i = [V_{i1}, V_{i2}, \ldots\ V_{id}] \tag{9}$$

The two "best" values for each particle are updated: pbest, the most effective response (fitness) it has so far acquired, and gbest, the most favorable value so far attained by any element in the entire population. The best location providing the greatest fitness value for the ith particle is denoted by pbest, while the best spot for the whole swarm population is denoted by gbesti. The best ith particle values are given by Eqs. 10 and 11.

$$\text{pbest} = [\text{pbest}_i^1, \text{pbest}_i^2 \ldots \text{pbest}_i^d] \tag{10}$$

$$\text{gbest} = \left[\text{gbest}_i^1, \text{gbest}_i^2, \ldots \text{gbest}_i^d\right] \tag{11}$$

The velocity updating is done as Eq. 12.

$$v_i^d(j+1) = w(j)v_i^d(j) + c_1 r_1\left[\text{pbest}_i^d(j) - x_i^d(j)\right]$$

$$+ c_2 r_2 \left[\text{gbest}_i^d(j) - x_i^d \right] \tag{12}$$

$v_i^d(j)$ denotes the velocity of i-th particle/element in d-th dimension and at jth iteration. After calculating the velocity for each element, the location of each element will be updated by adding the newly calculated velocity to the particle's previous position using Eq. 13.

$$x_i^d(j+1) = x_i^d(j) + v_i^d(j+1) \tag{13}$$

A performance index-based study is conducted to investigate and highlight the efficient use of PSO to optimize the proportionate integral gains for LFC in a reformed power system that operates under bilateral-based strategy. It should be emphasized that selecting the appropriate fitness function is critical in the synthesis phase, since various fitness functions stimulate distinct PSO behaviors, which yield fitness value, which provides an indicator of the performance of the issue under consideration.

ISE, IAE, ITAE, and ITSE are popular indices to depict the effectiveness of the PID controlled system. Here in this work, the performance index ITAE will be used as the objective function for fine-tuning gains of PSO-based PID controllers. The optimization issue is centered on minimizing the operating indicator or fitness function while ensuring that the PID parameters K_P, K_I, and K_D of the two controllers fall inside the smallest and largest constraints. Figure 2 depicts the main notion.

PSO algorithm steps that are continued until a halting condition is reached, which are given as follows.

Step 1: The first step is to initialize the particles.

Set the iteration number k to zero. Create n particles randomly, X_i, $i = 1, 2, \ldots, n$, with $x_i = [x_{i1}, x_{i2}, \ldots, x_{id}]$ and starting velocities $V_i = [V_{i1}, V_{i2}, \ldots, V_{id}]$.

Step 2: Second step is to update the iteration counting as $k = k + 1$.

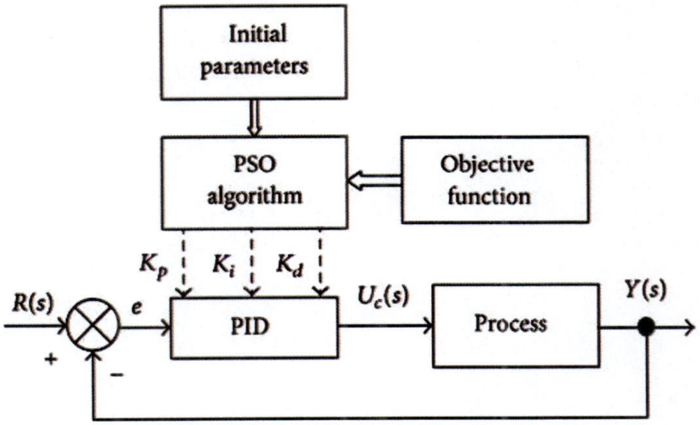

Fig. 2 Tuning of PID controller parameter by PSO algorithm

Step 3: Use velocity Eq. (12) to update velocity of the particle as a next step.

Step 4: In the next step, use position Eq. 13 to update the position of the particle.

Step 5: Now particle best can be updated,

$$\text{If eval}_i\left(x_i^k\right) > \text{eval}_i\left(\text{pb}_i^{k-1}\right) \text{ then } \text{pb}_i^k = x_i^k \text{ Else } \text{pb}_i^k = \text{pb}_i^{k-1}$$

Step 6: As a next step global best can be updated as:

$$\text{eval}\left(\text{gb}^k\right) = \max\left(\text{eval}_i\left(\text{pb}_i^{k-1}\right)\right)$$

$$\text{If eval}\left(gb^k\right) > \text{eval}\left(\text{gb}^{k-1}\right) \text{ then } \text{gb}^k = \text{gb}^k \text{ Else } \text{gb}^k = \text{gb}^{k-1}$$

Step 7: Last step is to reach decision to stop the algorithm:

If the overall value of iterations reaches the highest number of iterations or the total coverage is 100%, halt; otherwise, go to step 2.

4.1 Implementation of PSO Algorithm for Tuning of PID Controller Parameter

For optimum performance at nominal working conditions, using the PSO algorithm, PID controller settings may be tuned. System model shown in Fig. 2 is used to apply the PSO algorithm for tuning the PID parameters K_p, K_i, and K_d, with the performance index ITAE. The lower and upper bounds values are 0 and 10, respectively. The swarm population is the number of particles, here the population taken is 30. Every particle represents a candidate for PID controller parameters. PSO algorithms try to find a set of PID controller parameters which can provide a good system response and result in minimization of the performance index ITAE. Maximum number of iterations used is 100 in this case. Using the given algorithm, optimized parameters of the controllers are obtained.

5 Simulation and Result

5.1 Case-I

Total load demand of all the DISCOs is 0.005 pu MW. Comparative frequency response of both the areas, tie-line power response and GENCO responses of both the areas using GA-based controller and PSO-based controller are shown in Figs. 3, 4, 5, 6, and 7.

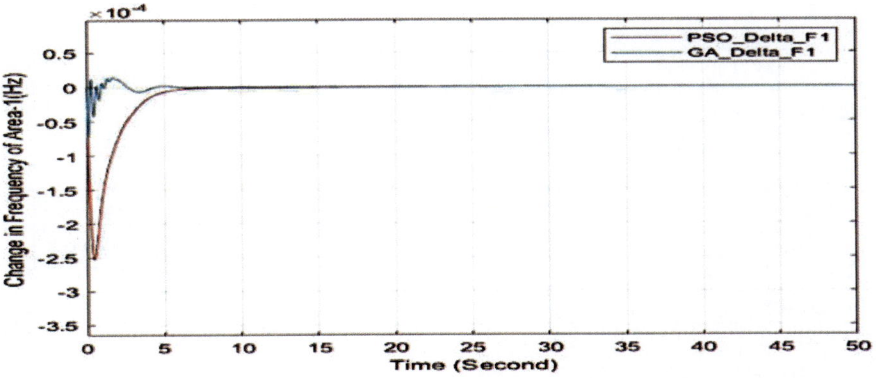

Fig. 3 Area-1 frequency change

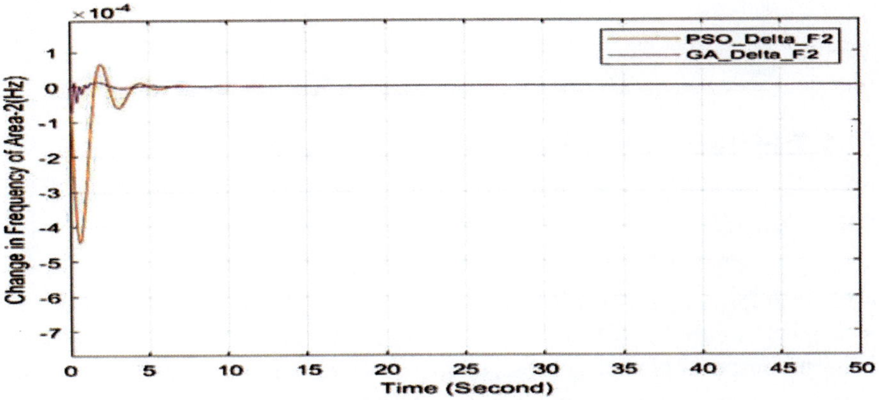

Fig. 4 Area-2 frequency change

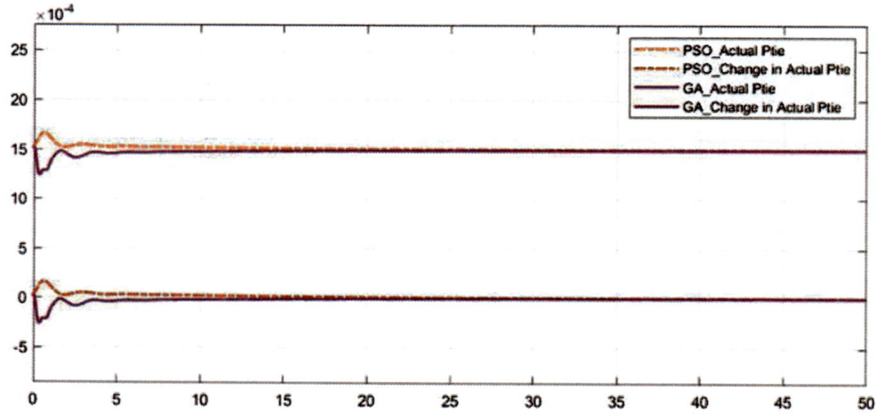

Fig. 5 Change in tie-line power and actual tie-line power

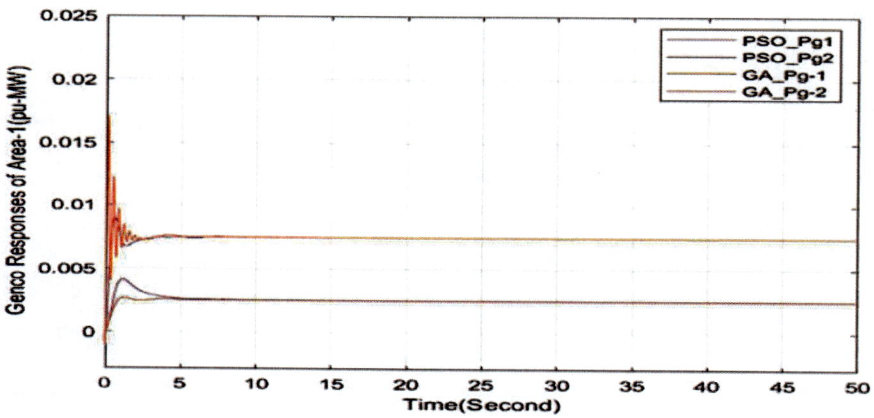

Fig. 6 GENCO responses of area-1

5.2 *Case-II*

Additional load request of 0.0025 pu MW is raised-up by area-1 at $t = 25$ s and it is supplied by only GENCO-1 of area-1. It is a contract violation case. Comparative frequency response of both the areas, tie-line power response and GENCO responses of both the areas using GA- and PSO-based controllers are shown in Figs. 8, 9, 10, 11, and 12.

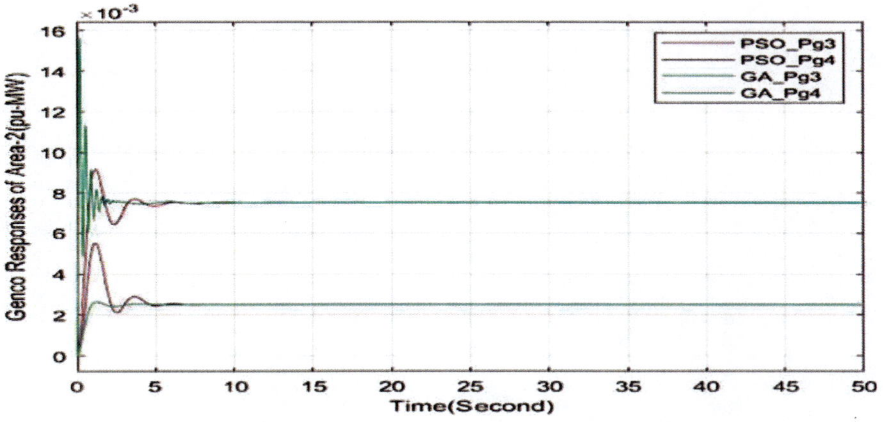

Fig. 7 GENCO responses of area-2

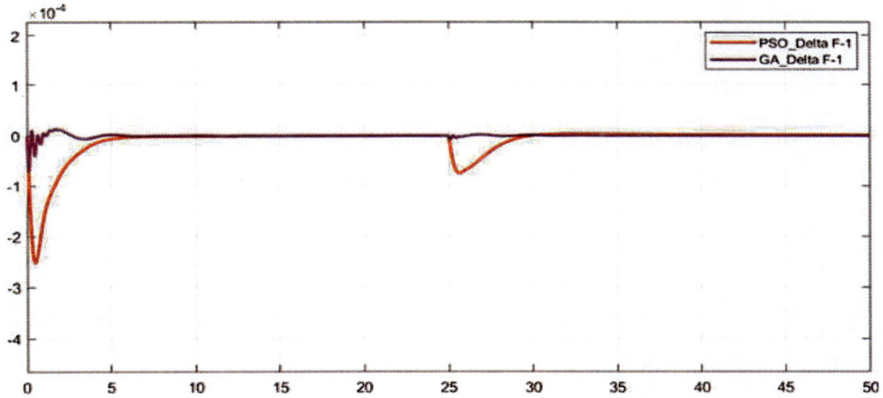

Fig. 8 Area-1: change in frequency

5.3 Parameter Variation

Here in this case, the parameter of the turbine has been changed by 50% in all the generating units to check the sturdiness of the controller when compared to parameter change. Responses are shown in Figs. 13, 14, 15, 16 and 17 for normal contractual conditions.

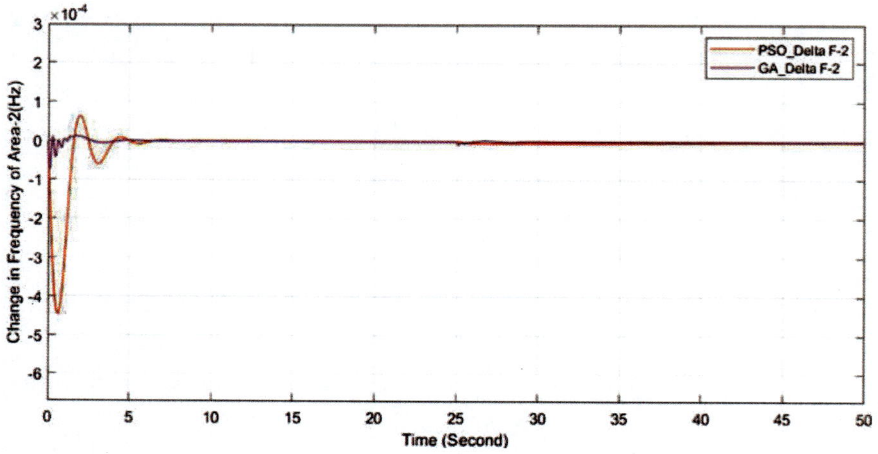

Fig. 9 Area-2: change in frequency

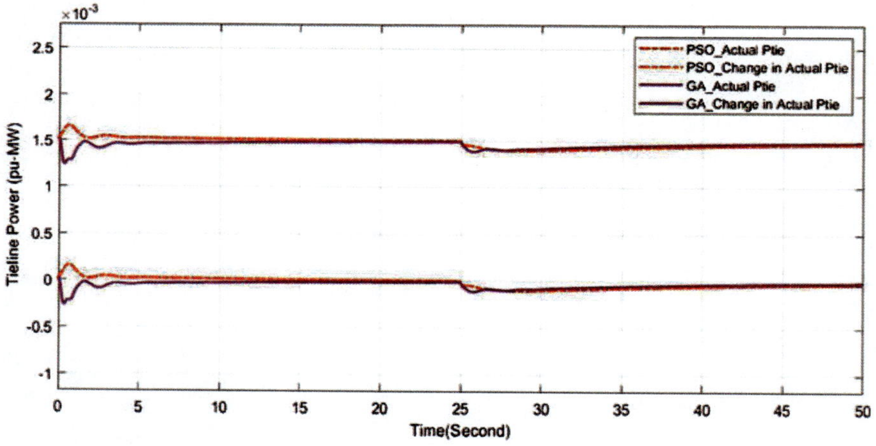

Fig. 10 Actual power flow in the tie-line and change in flow of tie-line power

5.4 Comparison with Respect to Time Response Specifications

The time response specifications of frequency responses $\Delta f1$ and $\Delta f2$ for case-I are given in Tables 1 and 2, respectively.

The time response specifications of frequency responses $\Delta f1$ and $\Delta f2$ for case-II are given in Tables 3 and 4, respectively.

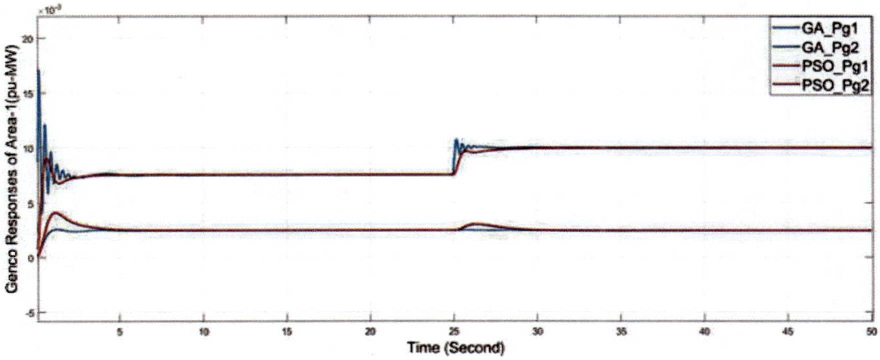

Fig. 11 GENCO responses of area-1

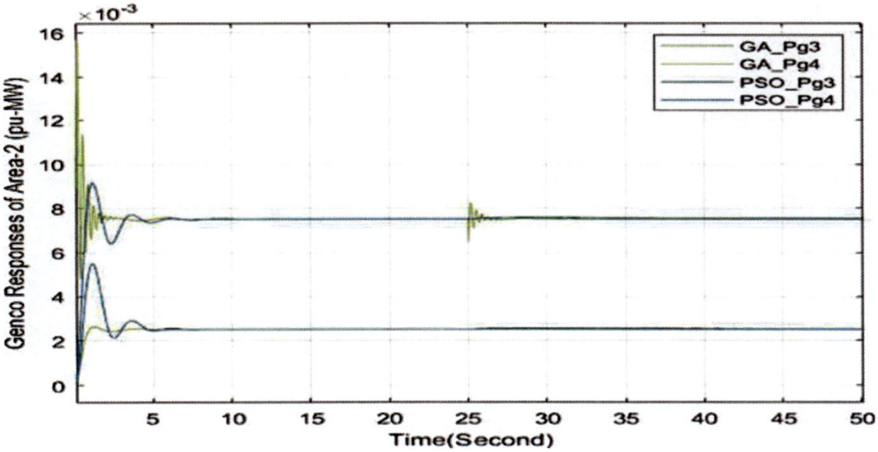

Fig. 12 GENCO responses of area-2

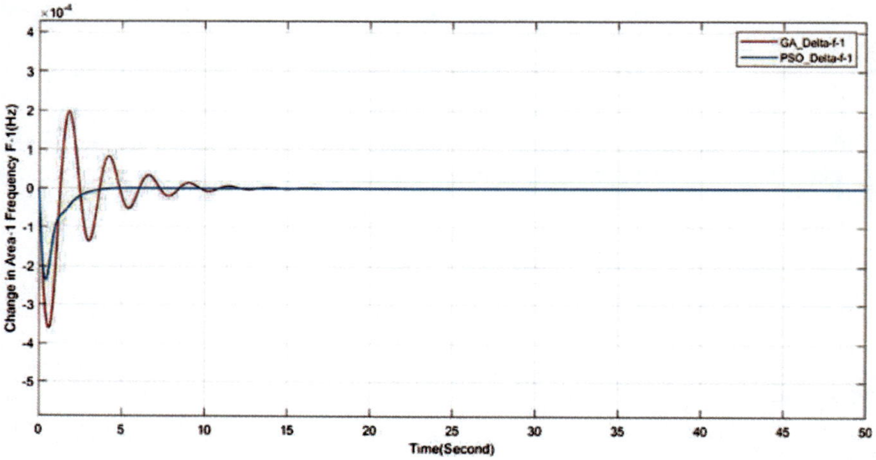

Fig. 13 Area-1: change in frequency

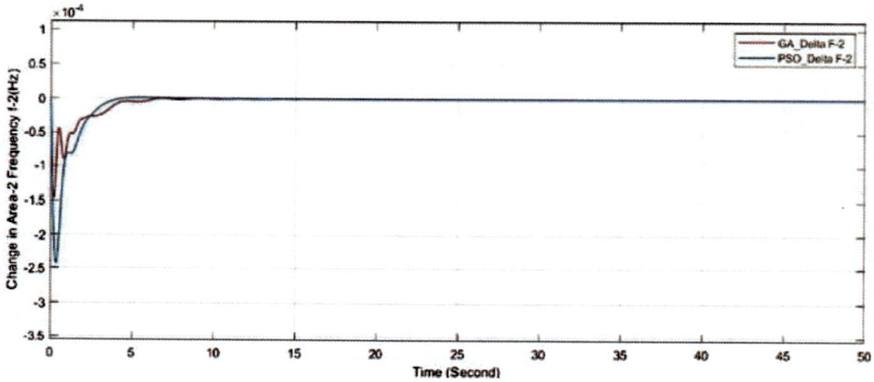

Fig. 14 Area-2: change in frequency

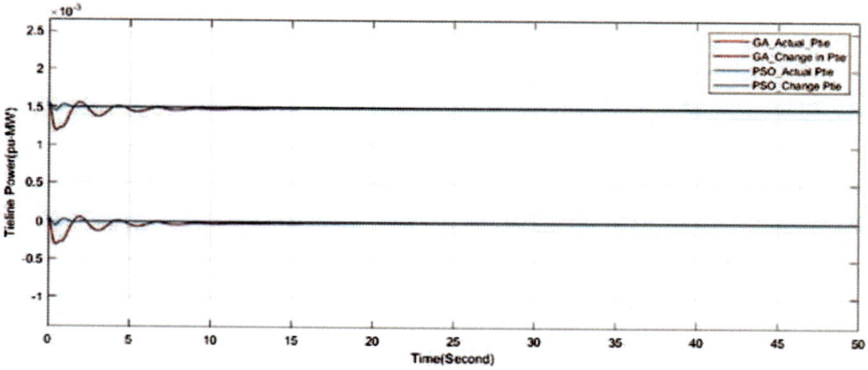

Fig. 15 Actual power flow in the tie-line and change in flow of tie-line power

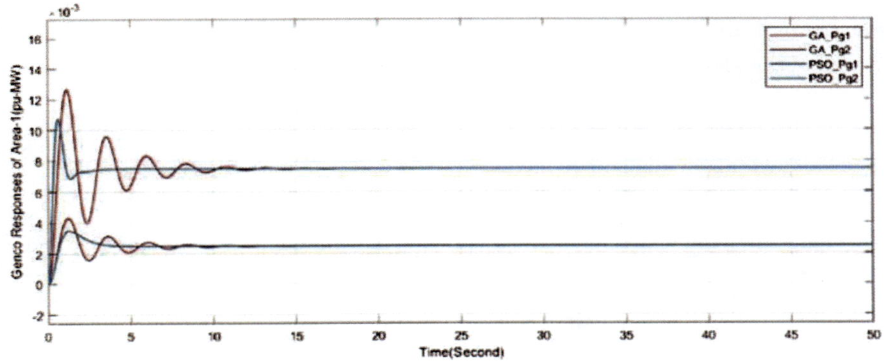

Fig. 16 GENCO responses of area-1

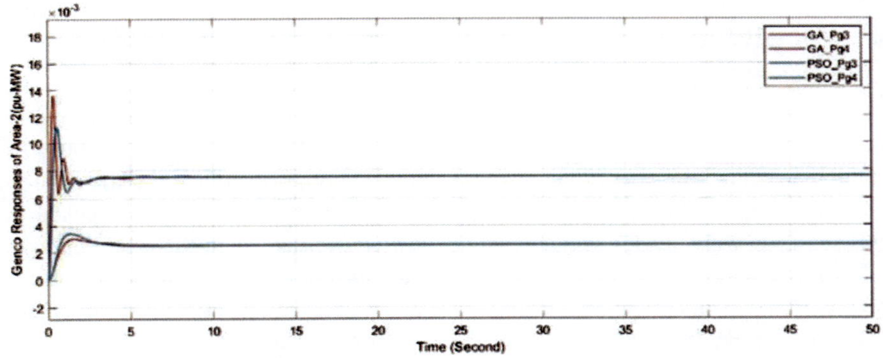

Fig. 17 GENCO responses of area-2

Table 1 Time response specifications for $\Delta f1$ (case-1)

S. No.	Controller type	Peak overshoot M_p	Peak time T_p (s)	Rise time T_r (s)	Settling time T_s (s)	Comment
1	PSO	$2.5 * 10^{-4}$	1.85	1.56	6.67	Stable
2	GA	$-0.72 * 10^{-4}$	0.27	0.22	7.82	Stable

Table 2 Time response specifications for $\Delta f2$ (case-1)

S. No.	Controller type	Peak overshoot M_p	Peak time T_p (s)	Rise time T_r (s)	Settling time T_s (s)	Comment
1	PSO	$-4.4 * 10^{-4}$	1.94	1.57	6.48	Stable
2	GA	$-0.725 * 10^{-4}$	0.28	0.23	6.95	Stable

Table 3 Time response specifications for $\Delta f1$ (case-II)

S. No.	Controller type	Peak overshoot Mp	Peak time T_p (s)	Rise time T_r (s)	Settling time T_s (s)	Comment
1	PSO	$-2.58 * 10^{-4}$	0.44	6.54	5.25	Stable
2	GA	$-0.72 * 10^{-4}$	0.27	0.23	7.5	Stable

Table 4 Time response specifications for $\Delta f2$ (case-II)

S. No	Controller type	Peak overshoot Mp	Peak time T_p (s)	Rise time T_r (s)	Settling time T_s (s)	Comment
1	PSO	$4.4 * 10^{-4}$	1.94	1.57	5.55	Stable
2	GA	$-0.725 * 10^{-4}$	0.27	0.23	6.95	Stable

6 Conclusion

The primary goal of load frequency management is to preserve the power system frequency and interarea tie-line power exchange as near to the planned values as feasible in a linked restructured power system. It is possible to achieve the desired response with the help of proper control methods. PSO technique is proposed in this research work to design an appropriate controller to catch the goals of LFC of two-area restructured power systems. Simulation prototype of a two-area linked power system in a deregulated environment has been developed so that the dynamic performance and the suggested controller's sturdiness can be checked. The genetic algorithm-based controller and PSO-based controllers are applied to the developed MATLAB/Simulink prototypical two-area liberalized power system and dynamic behaviors like frequency responses, tie-line exchange, and GENCO responses have been obtained for different contractual conditions. It has been seen that the PSO tuned PID controller has given the improved vibrant responses when compared with the controller built with GA. Parameter variation has also been considered in this work and responses have also been obtained for 50% parameter variation for the turbines of all the generating stations. Similarly, other parameters can also be changed and responses can be obtained. Comparison of PSO-based controller and GA-based controller has also been done with respect to time responses specifications. It has been discovered that PSO-based controllers provide superior responsiveness. With respect to genetic algorithm-based controllers in all respects, especially when the oscillatory nature of responses is compared.

References

1. Jain D, Dr MK, Bhaskar MP (2022) Comparative analysis of load frequency control problem of multi area deregulated power system using soft computing techniques. Math Statistician Eng Appl 71(4):10713–10729
2. Donde V, Pai MA, Hiskens IA (2001) Simulation and optimization in an AGC system after deregulation. IEEE Trans Power Syst 16(3):481–489
3. Pal AK, Bera P, Chakraborty K (2014) AGC in two-area deregulated power system using reinforced learning neural network controller. Proc IEEE 1:1–6
4. Bhagya Shree S, Kamaraj N (2016) Hybrid neuro fuzzy approach for automatic generation control in restructured power system. Electr Power Energy Syst 74:274–285
5. Sekhar GC, Sahu RK, Baliarsingh A, Panda S (2016) Load frequency control of power system under deregulated environment using optimal firefly algorithm. Int J Electr Power Energy Syst 74:195–211
6. Bhatt P, Roy R, Ghoshal SP (2010) Optimized multi area AGC simulation in restructured power systems. Electr Power Energy Syst 32:311–322
7. Cohn N (1957) Some aspects of tie-line bias control on interconnected power systems. Am Inst Elect Eng Trans 75:1415–1436
8. Sood YR (2007) Evolutionary programming based optimal power flow and its validation for deregulated power system analysis. Int J Electr Power Energy Syst 29(1):65–75
9. Jain D et al (2023) Analysis of load frequency control problem for two area deregulated power system using genetic algorithm. Int J Creative Res Thoughts (IJCRT) 11(2):c57–c64. ISSN:2320–2882
10. Kothari ML, Sinha N, Rafi M (1998) Automatic generation control of an interconnected power system under deregulated environment. Proc IEEE 6:95–102
11. Ravi S, Kalyan C, Ravi B (2016) Impact of energy storage system on load frequency control for diverse sources of interconnected power system in deregulated power environment. Int J Electr Power Energy Syst 79(1):11–26
12. Deepak M, Abraham RJ (2015) Load following in a deregulated power system with thyristor controlled series compensator. Int J Electr Power Energy Syst 65:136–145
13. Abedinia O, Naderi MS, Ghasemi A (2011) Robust LFC in deregulated environment: fuzzy PID using HBMO. Proc IEEE 1:1–4
14. Abd-Elazim S, Ali E (2018) Load frequency controller design of a two-area system composing of PV grid and thermal generator via firefly algorithm. Neural Comput Appl 30:607–616
15. Babahajiani P, Shafiee Q, Bevrani H (2018) Intelligent demand response contribution in frequency control of multiarea power systems. IEEE Trans Smart Grid 9:1282–1291
16. Sahoo PK (2018) Application of soft computing neural network tools to line congestion study of electrical power systems. Int J Inf Commun Technol 13(2)
17. Concordia C, Kirchmayer LK (1953) Tie line power and frequency control of electric power systems. Am Inst Elect Eng Trans Pt II 72:562–572
18. Rakhshani E, Sadeh J (2008) Simulation of two-area AGC system in a competitive environment using reduced-order observer method. Proc IEEE 1:1–6
19. Zribi M, Al-Rashed M, Alrifai M (2005) Adaptive decentralized load frequency control of multi-area power systems. Int J Electr Power Energy Syst 27(8):575–583
20. Pathak N, Nasiruddin I, Bhatti TS (2015) A more realistic model of centralized automatic generation control in real-time environment. Electr Power Compon Syst 43:2205–2213
21. Saikia LC, Mishra S, Sinha N, Nanda J (2011) Automatic generation control of a multi area hydrothermal system using reinforced learning neural network controller. Int J Electr Power Energy Syst 33(4):1101–1108
22. IEEE PES Committee Report (1979) Current operating problems associated with automatic generation control. IEEE Trans Power App Syst PAS-98
23. Elgerd OI, Fosha C (1970) Optimum megawatt frequency control of multiarea electric energy systems. IEEE Trans Power App Syst PAS-89(4):556–563

24. Fosha CE, Elgerd OI (1970) The megawatt-frequency control problem: a new approach via optimal control theory. IEEE Trans Power App Syst PAS-89(4):563–567
25. Mishra AK, Das SR, Ray PK, Mallick RK, Mohanty A, Mishra DK (2020) PSO-GWO optimized fractional order PID based hybrid shunt active power filter for power quality improvements. IEEE Access 8:74497–74512
26. Suid MH, Ahmad MA (2022) Optimal tuning of sigmoid PID controller using nonlinear sine cosine algorithm for the automatic voltage regulator system. ISA Trans 128:265–286
27. Dhanasekaran B, Kaliannan J, Baskaran A, Dey N, Tavares JMR (2023) Load frequency control assessment of a PSO-PID controller for a standalone multi-source power system. Technologies 11(1):22
28. Bhatt P, Ghoshal SP, Roy R (2012) Coordinated control of TCPS and SMES for frequency regulation of interconnected restructured power systems with dynamic participation from DFIG based wind farm. Renew Energy 40:40–50
29. Tan W, Zhang H, Yu M (2012) Decentralized load frequency control in deregulated environments. Int J Electr Power Energy Syst 41(1):16–26
30. Arya Y, Kumar N (2016) Fuzzy gain scheduling controllers for automatic generation control of two-area interconnected electrical power systems. Electr Power Compon Syst 44:737–751
31. Jain D et al (2014) Comparative analysis of different methods of tuning the PID controller parameters for load frequency control problem. IJAREEIE 3(11). https://doi.org/10.15662/ijareeie.2014.0311030
32. Jain D et al (2014) Analysis of load frequency control problem for interconnected power system using PID controller. IJETAE 4(11):2250–2459. ISO 9001: 2008 Certified J. https://ijetae.com/files/Volume4Isue11/IJETAE_1114_42.pdf

Chapter 51
Solar Maximum Power Point Tracking and Machine Learning-Based Forecasting

Akshay Pandya, Galav Bhatt, Jash Patadia, and Het Patel

1 Introduction

To derive utmost energy from solar system, we have implemented solar maximum power point technique. Solar cells composed of a forefront contact, p–n junction diode in midway section and bottommost contact. Solar modules are formed by attaching various solar cells [1]. Solar PV array is formed by joining solar modules which can be connected in series or parallel as per the power requirement. It is essential to study the characteristics of solar cell to understand working of solar ecosystem. Sunlight falling on Earth is essentially photon bundles, with photon in a packet comprises of definite proportion of energy. Extent of electrical energy depends on photon energy and band gap energy.

1.1 Working of Solar Cell and Its P–V and I–V Characteristics

The foremost part of semiconductor material soaks up photons of sunlight. Band gap present in solar cell soaks up photons, through which electron–hole pairs are formed. Here, electron symbolizes negative charge and hole symbolizes positive charge. When load is attached with solar system, it results into formation of electron–hole pair close by junction. At anode terminal, holes are accumulated, and at cathode terminal, electrons are accumulated. Electric potential gets formed at terminals, because of segregation of positive charge and negative charge.

A. Pandya · G. Bhatt (✉) · J. Patadia · H. Patel
Electrical Engineering Department, BVM Engineering College, Vallabh Vidyanagar, Anand, Gujarat 388120, India
e-mail: galav.bhatt2026@gmail.com

A. Yadav et al. (eds.), *Proceedings of International Conference on Paradigms of Communication, Computing and Data Analytics*, Algorithms for Intelligent Systems,
https://doi.org/10.1007/978-981-99-4626-6_51

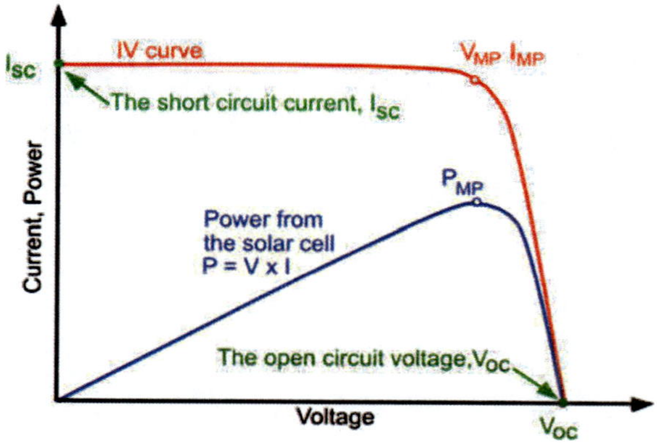

Fig. 1 P–V and I–V characteristics of solar cell [2]

In Fig. 1, I_{sc} depicts short-circuit current, V_{oc} depicts open-circuit voltage, initially as open-circuit voltage increases, short-circuit current remains constant, and then after saturation level, the current drops severely. The output power graph is represented by blue line we can see that a particular point of voltage and current we get the maximum power which is denoted by P_{MP}. The graph moves up and down depending upon the weather conditions like temperature and irradiance. As the temperature increases, voltage decreases, current increases up to 25 °C and above those current decreases. With the increase in irradiance, the output current increases and vice versa. To withdraw topmost power from solar panel, we need to operate system nearby voltage and current at maximum power point (mpp). The main purpose of solar MPPT is to help solar system operate around voltage and current which offers the maximum power point (mpp).

2 Solar MPPT System and Its Simulation

Before implementing the entire system on hardware, it become essential to analyze and simulate it on a software. Here, we have used MATLAB Simulink for software implementation of our system.

Description of block units present in Fig. 2.

Solar PV Panel: For our system, we have used 40 W solar panel with $V_{\text{maximum power}}$ 19.25 V, $I_{\text{maximum power}}$ 2.08 A, short-circuit current (Isc) 2.21 A and open-circuit voltage (V_{oc}) 22.5 V.

MPPT controller: We have implemented perturb and observe (P&O) algorithm for the adjustment of duty cycle, Arduino UNO is used as a MPPT controller, where P&O algorithm is implemented, and its input will be the solar panel generated voltage and current with respect to respective irradiance and temperature.

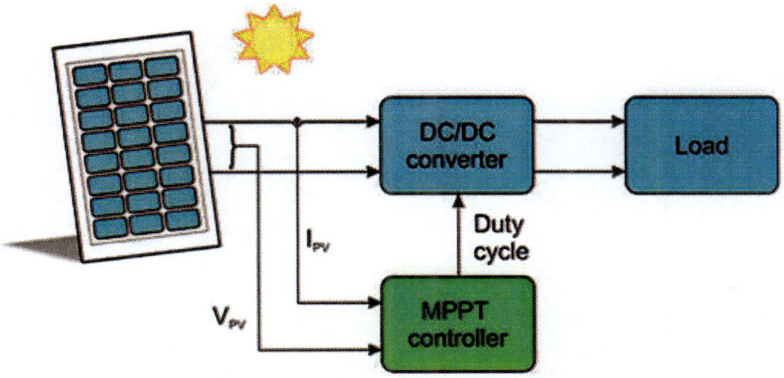

Fig. 2 Solar MPPT block diagram

DC/DC converter: We have used buck converter in our system, where the voltage steps down and current steps up in accordance with the duty cycle received from the MPPT controller.

Load: A DC bulb of 12 V and 35 W is used as our load.

2.1 Perturb and Observe Technique

The author from [3] finalized that P&O algorithm performs finer as compare to IC method and its reaction time is faster. Hence, we have implemented P&O algorithm on microcontroller and used DC/DC buck converter instead of boost converter as proposed by [3]. We found 180–200% increase in the solar PV panel output power with MPPT additionally we have trained a machine learning model which can predict the solar MPPT output power that depends on irradiance and temperature which is relatively new method than the given ones in [3]. As the magnitude of voltage enhances, output power magnitude inclines respectively. But at a certain point of voltage and current, we get the maximum power and our primary goal behind solar MPPT technique is to move around the maximum power point. At certain point as the saturation in the solar cell takes place, current moves down with the increase in voltage, as a result, we can see that power moves down after maximum power point. Figure 3 indicates when the operating point of graph is toward left of mpp, it depicts that voltage increases along with increase in power. According to Fig. 4, in 2 cases, we will enhance the duty cycle to reach at mpp, those are when rise in voltage depicts increase in power and fall in voltage depicts decrease in power. Thus, perturbation is on right side. Further, as per Fig. 3, at the time when operating point is toward right of mpp, it shows that voltage increases along with decrease in power. Therefore, according to Fig. 4, there are two situations where we will lessen the duty cycle to achieve mpp, when decline of voltage results into enhancement of power and incline of voltage results into decline of power. Therefore, here the perturbation

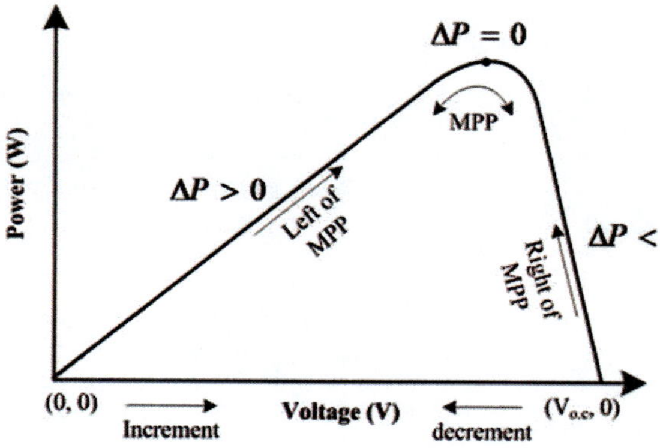

Fig. 3 Solar P–V graph and perturb and observe inference

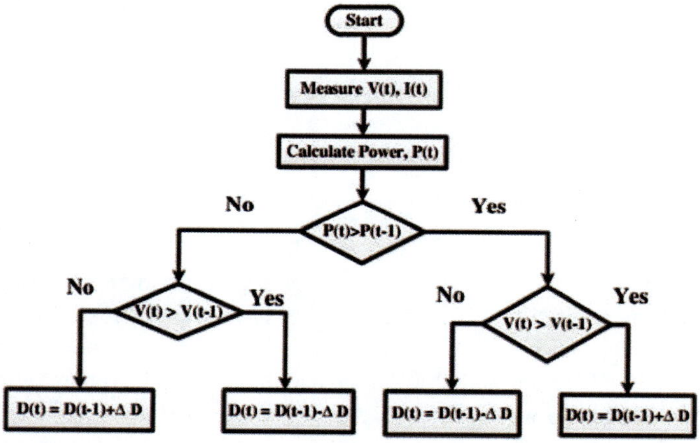

Fig. 4 Flowchart of P&O algorithm [4]

is on left side. The algorithm is loaded in Arduino UNO microcontroller and it is further utilized to obtain duty cycle which act as an input to buck converter switching part.

2.2 Solar MPPT System Simulation and Results

Figure 5 represents the simulation of solar MPPT system on MATLAB Simulink software. At standard test condition temperature (25 °C) and irradiance (1000 W/M^2), we found that without using MPPT, solar panel was generating maximum output power in the range of 24–26 W which is shown in Fig. 6. On the other hand, after using MPPT along with solar panel, we were able to achieve the maximum output power in the range of 38–40 W as shown in Fig. 7.

Figure 8 depicts the relation of solar MPPT output with temperature and irradiance. As temperature increases and irradiance decreases, the solar MPPT output decreases. If irradiance increases and temperature decreases, solar MPPT output increases.

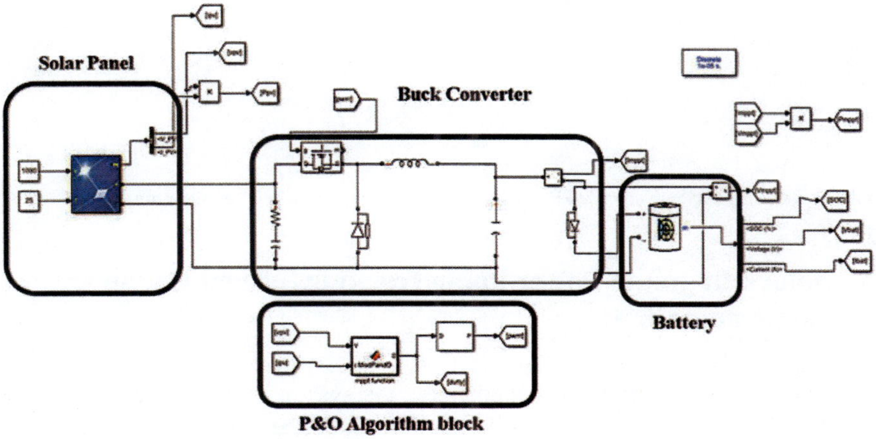

Fig. 5 Circuit/simulation of solar MPPT system

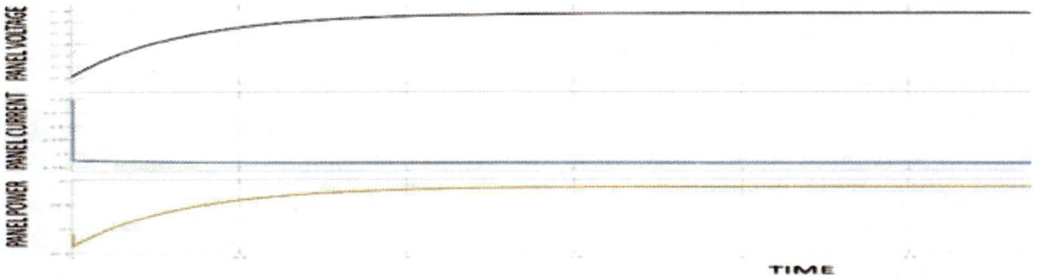

Fig. 6 Solar system power, current and voltage with STC temperature, irradiance for direct loading

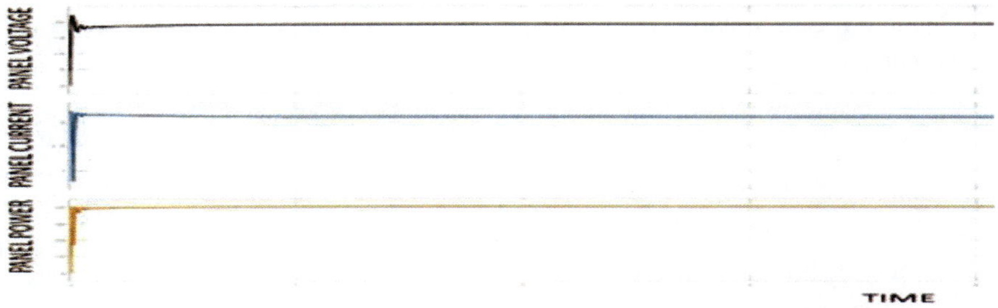

Fig. 7 Solar system power, current and voltage with STC temperature, irradiance for MPPT loading

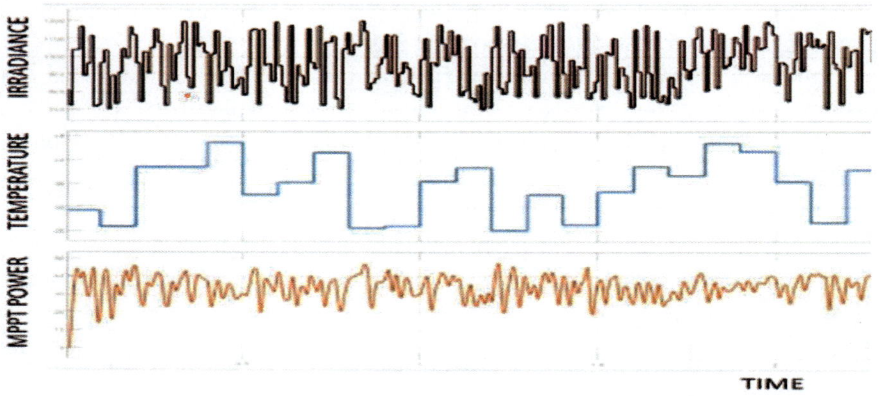

Fig. 8 Relation between solar MPPT output power versus irradiance, temperature

3 Solar Maximum Power Point Tracking System Hardware Implementation

As shown in Fig. 9, in order to implement the system on hardware, we initially interfaced 16 × 2 LCD display with Arduino UNO. Further, interfaced current sensor and sensing voltage through voltage divider circuit to calculate output power of solar panel and buck converter. Solar system output power is calculated to predict duty cycle through P&O algorithm. For voltage divider, we have used two resistors of 22 k ohm and 80 k ohm to step down 24–5 V which is the maximum limit of voltage that can be given to Arduino at analog pin. $V_{in} = V_{SOLAR} \times (R_2/R_1 + R_2)$. $(R_1 \to 80\ k\Omega$. $R_2 \to 22\ k\Omega)$.

The current sensor used is ACS712 hall effect sensor which converts current magnitude passing through it to equivalent voltage magnitude in the range of 0–5 V. Further, we designed the buck converter by fixing the output power associated with load (24 W). Then, fixed the input voltage (19.25 V) and output voltage (12 V), also find the load current by giving rated voltage to load. For the calculation of capacitor and inductor values we have assumed fixed input voltage of solar panel ($V_{maximum}$ 19.25 V), output voltage associated with load of DC bulb (12 V) and rated current (2 A) which had given us fixed ripple current and ripple voltage. If, we want less ripples in our output current and voltage, the values of inductor and capacitor will be higher.

$$C = \frac{\Delta I_l}{8 * F_s * \Delta V_{out}} \tag{1}$$

where

ΔI_l = approximate ripple current of inductor.
F_s = least switching frequency of converter.
ΔV_{out} = output voltage ripple.

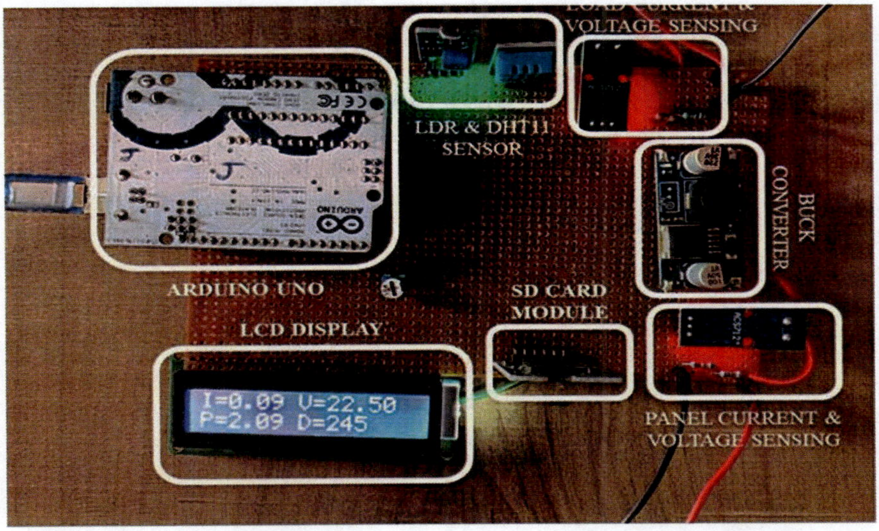

Fig. 9 Glimpse of solar MPPT system on brown board

$$L = \frac{V_{\text{out}} * (V_{\text{out}} - V_{\text{in}})}{V_{\text{in}} * F_s * \Delta I_l} \tag{2}$$

where

ΔI_l = approximate ripple current of inductor.
F_s = least switching frequency of converter.
ΔV_{out} = output voltage.
V_{in} = Typical input voltage.

For the calculation of inductor and capacitor values, the formula 1 and 2 were used, respectively. As described, the required ripple voltage and ripple current were fixed along with the switching frequency of MOSFET. Finally for the input duty cycle of input, we have implemented perturb and observe algorithm in Arduino UNO microcontroller. Additionally, we have integrated temperature and humidity sensor as well as SD card module for data collection in order to build robust machine learning model.

3.1 Hardware Results

Figure 10 shows the comparison between solar PV panel output power with and without MPPT device, and it is clearly visible that solar panel output power is increased by 180–200% with the help of MPPT device. Figure 11 represents the relation between solar panel output power with irradiance and temperature. As temperature increases beyond ambient temperature, solar PV panel power decreases. As irradiance increases, solar PV panel output power increases. Figure 12 represents the relation between solar PV panel power with duty cycle which will be obtained from P&O algorithm and it'll act as input for DC–DC buck converter.

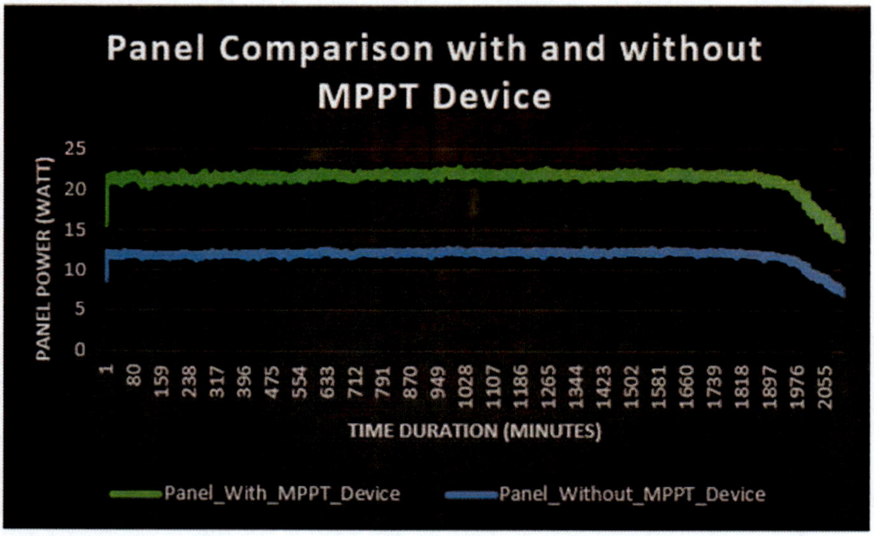

Fig. 10 Solar PV panel power with and without MPPT device

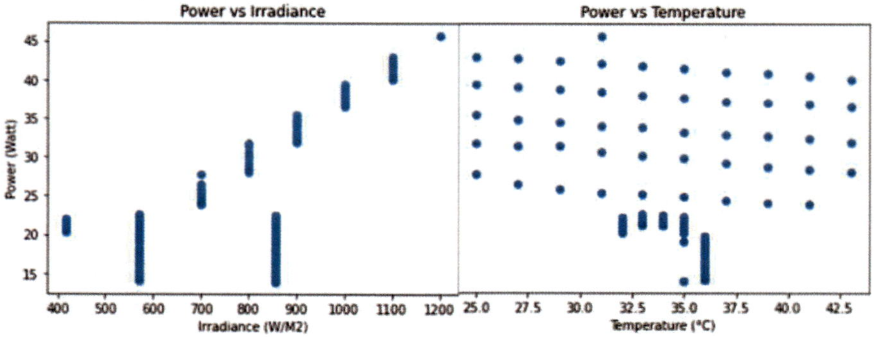

Fig. 11 Solar PV panel output power (W) versus irradiance (W/M^2), temperature (°C)

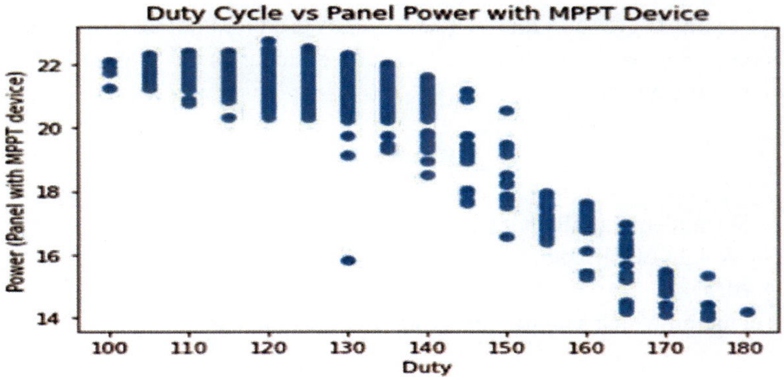

Fig. 12 Panel output power versus duty cycle

4 Machine Learning Implementation on the Collected Data

Machine learning is a concept initiated to leverage the power of data in order to forecast results based on data, and it is applicable in numerous fields. Here, we have input features in the form of temperature and irradiance sensor data and output as solar PV panel output power. Regression models will be used in order to build a robust prediction model which will predict solar PV panel output power. We have used relevant machine learning regression model as per our data distribution, and there are certain regression algorithms which we haven't used because of its irrelevance as per our dataset, i.e., linear regression depicts connection between self-supporting variable and target variable, logistic regression which is used when we have discrete values, ridge regression is used when there is significant amount of correlation among independent features, lasso regression is used when we want to select only specific features and avoid highly correlated features to avoid overfitting, polynomial regression is used to capture highly complex relationships. The regression algorithms which we have considered for out machine learning-based system are as follows:

Multiple regression model is a machine learning concept that leverages various self-supporting variables to forecast the target variable which is dependent.

$$Y = B_o + B_1X_1 + B_2X_2 + B_3X_3 + \cdots B_NX_N + E \tag{3}$$

$Y \rightarrow$ Dependent target attribute.
$X_1, X_2, X_3 \ldots X_N \rightarrow$ features/self-supporting attributes.
$B_0, B_1, B_2, B_3 \ldots B_N \rightarrow$ slope/weights of several features.
$E \rightarrow$ Error.

Support vector regression model. It is basically used for both the classification and regression problems. Here, there are basically boundaries based on which we will differentiate the errors.

$$Y = wx + b \tag{4}$$

$Y \rightarrow$ Dependent variable/target
$W \rightarrow$ Weight of variable
$B \rightarrow$ Bias.
The equation which satisfies the hyper plane is given as

$$-a < Y - wx + b < +a \tag{5}$$

Decision tree regression model Decision tree regression is a machine learning algorithm which distributes the data into tree-like shape and utilizes it to produce the output. The decision-making rules and its respective variables are decided by Gini impurity.

Random forest regression model is a machine learning technique which combines multiple decision tree in order to get low variance as it uses all the decision trees in parallel, and a particular decision tree is trained for a particular sample of data. Here, we execute bootstrapping (distribution of sample of data among multiple decision tree) and aggregation (combine decision tree results) for the final result.

In order to build a robust machine learning model, we have followed certain steps, i.e., import the necessary libraries (NumPy, Pandas, Matplotlib, ScikitLearn, etc.) then import the data into pandas' data frame and cleaned it by removing extreme outliers. Further, exploratory data analysis was performed to get insights of the data and understand correlation among machine learning model features. After analyzing data, we split the entire dataset into training and test dataset, feature sampling was also performed to limit the span of dataset discrete values distribution [5]. Training of dataset was done with various machine learning regression algorithms, and finally, evaluation of trained machine learning model was done using various evaluation techniques like mean absolute error, root mean squared error, r^2 score [6].

$$\text{Mean Squared Error (MSE)} = 1/n \sum_{i=1}^{n} (X_i - \overline{X})^2 \tag{6}$$

$$\text{Mean Absolute Error (MAE)} = \sum_{i=1}^{n} \left| X_i - \overline{X} \right| \tag{7}$$

$$\text{Root Mean Square Error (RMSE)} = \sqrt{1/n\sum_{i=1}^{n}(X_i - \overline{X})^2} \tag{8}$$

$$R\text{-square Error } \left(R^2\right) = 1 - \sum_{i=1}^{n}(X_i - \overline{X})^2 / \sum_{i=1}^{n}(X_i - \overline{X})^2 \tag{9}$$

n—total number of observations, X_i—Actual value, $\overline{X}$—Predicted value

As given in Table 1, random forest regression gives the least error for the sensor dataset we have collected. Hence, we have used the random forest algorithm in the backend of our user interface system. After training and developing the model, we have developed a user interface system using Streamlit open-source framework [7] as it easy to use, integrates better with machine learning model and has live web application preview facility which simplifies developer task. As shown in Fig. 13, in the user interface system, user or power operator gives input temperature and irradiance and, in the output, they get solar maximum power point tracking enabled system output power.

Table 1 Machine learning model performance

Regression algorithm	Mean absolute error	Mean squared error	R-squared (r^2 score)
Multiple regression model	2.63	18.69	0.02
Support vector regression	2.18	17.63	0.07
Decision tree regression	1.02	4.62	0.75
Random forest regression	0.75	2.14	0.88

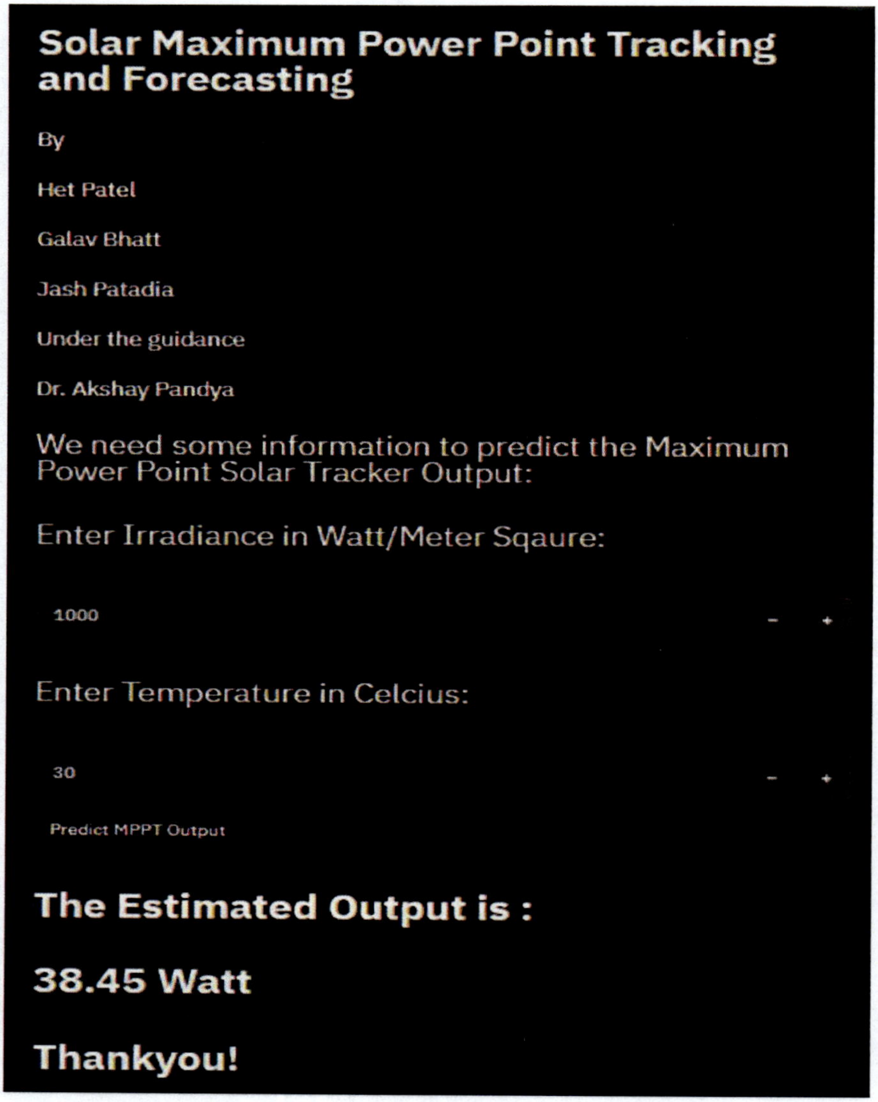

Fig. 13 Machine learning-based solar MPPT user interface system using Streamlit library

5 Conclusion

Solar energy is one of the solutions to pollution free energy, hence it is vital to enhance the efficiency of solar ecosystem. We found that by using MPPT as an intermediate stage, output energy from solar PV system gets increased by 180–200%. Hardware comprises DC-DC buck converter, Arduino UNO, current sensor, DHT11 sensor (temperature and humidity), SD card module, light intensity sensor and is tested for a DC bulb of 12 V and 35 W. According to our observation when the DC bulb is directly connected to the panel, it operates at 10 W, on the other hand, after introducing MPPT, the panel produces 20W power. Along with that we have created a machine learning model with 88% accuracy which will help user/solar panel operator to predict the generation of solar panel power based on irradiance and temperature. Additionally, a user interface was created using Streamlit open-source framework in which machine learning model runs in backend as per the user's input of temperature and irradiance for forecasting solar maximum power point tracking device enabled solar PV panel output power. Further, the system can also be used to charge the batteries in most efficient way to support standalone solar system.

References

1. Solar Photovoltaic Cell Basics. https://www.energy.gov/eere/solar/solar-photovoltaic-cell-basics
2. Solar IV Curve. https://www.pveducation.org/pvcdrom/solar-cell-operation/iv-curve
3. Villalva & Ruppert (2009) Analysis and simulation of the P&O MPPT algorithm using a linearized PV array mode 35th annual conference of IEEE industrial electronics, Porto, Portugal, pp 231–236. https://doi.org/10.1109/IECON.2009.5414780
4. Barbary & Alranini (2021) Review of maximum power point tracking algorithms of PV system. Front Eng Built Environ 1:68–80. https://doi.org/10.1108/FEBE-03-2021-0019
5. Predicting solar power output using machine learning techniques. https://towardsdatascience.com/predicting-solar-power-output-using-machine-learning-techniques-56e7959acb1f
6. Know the best evaluation metrics for your regression model. https://www.analyticsvidhya.com/blog/2021/05/know-the-best-evaluation-metrics-for-your-regression-model/
7. Streamlit open-source framework. https://streamlit.io/

Chapter 52
Comparative Performance Analysis of Various Controllers for Quadruple Tank System

C. Praveen Kumar and K. Ayyar

1 Introduction

In the process industries, the control of process liquid level in tanks is the basic problem as they have a large number of interacting control loops. It is a challenging task to control it due to inherent nonlinearity and existence of interactions among input and output variables. The voltage to the pumps and the water levels in the quadruple tank acts as its input and output. Due to the presence of nonlinearity in manipulated and controlled variable, it is not an easy task to control. Design methods of multivariable control include linear as well as nonlinear design methods.

The four tank quadruple process is a highly nonlinear system which has been used to test different MIMO controllers designed for this process. Some have used optimization methods like grasshopper algorithm [1] for the tuning and control of four tank system. However, they have restricted the number of search agents only up to 30. Many control strategies have been proposed [2] for the controlling of quadruple tank system. Control scheme like model predictive controller [3, 4] have also been used. Many other popular modelling methods [5–9] and controllers [8, 10–13] have also been used for the modelling and control of quadruple system. Here, a performance comparative analysis is evaluated for the quadruple system. Controllers like a conventional PID controller, sliding mode controller, a conventional PID controller with decoupler, a state feedback controller and MPC controller were proposed for the control of the quadruple tank process. The above different control schemes were designed using the linearized four tank model.

C. Praveen Kumar (✉) · K. Ayyar
SRM Valliammai Engineering College, Chennai, India
e-mail: chakrapraveen@gmail.com

A. Yadav et al. (eds.), *Proceedings of International Conference on Paradigms of Communication, Computing and Data Analytics*, Algorithms for Intelligent Systems,
https://doi.org/10.1007/978-981-99-4626-6_52

2 Process Description

A four tank process has been generally used in process control to demonstrate concepts like MIMO control with performance limitations. It consists of two pumps and four tanks which are interconnected with each other either directly or indirectly (Fig. 1). The inputs are the voltage to the two pumps and the outputs are the water levels in the lower two tanks. The input voltages to the pump and tank height are maintained at a range of 0–10 V and 20 cm, respectively. By interconnecting two interacting tank pairs, a quad tank system can be assembled. The primary goal lies in the lower two tanks level control. The process inputs and outputs are v_1, v_2 and y_1, y_2.

To define the flow rate, Bernoulli's principle and mass balance equation are implemented and the model is obtained using the below equations,

$$\frac{\mathrm{d}h_1}{\mathrm{d}t} = -\frac{a_1}{A_I}\sqrt{2gh_1} + \frac{a_3}{A_1}\sqrt{2gh_3} + \frac{{}_1k_1}{A_1}v_1$$
$$\frac{\mathrm{d}h_2}{\mathrm{d}t} = -\frac{a_2}{A_2}\sqrt{2gh_2} + \frac{a_4}{A_2}\sqrt{2gh_4} + \frac{{}_2k_2}{A_2}v_2$$
$$\frac{\mathrm{d}h_3}{\mathrm{d}t} = -\frac{a_3}{A_3}\sqrt{2gh_3} + \frac{(1-\gamma_1)k_1}{A_3}v_1$$

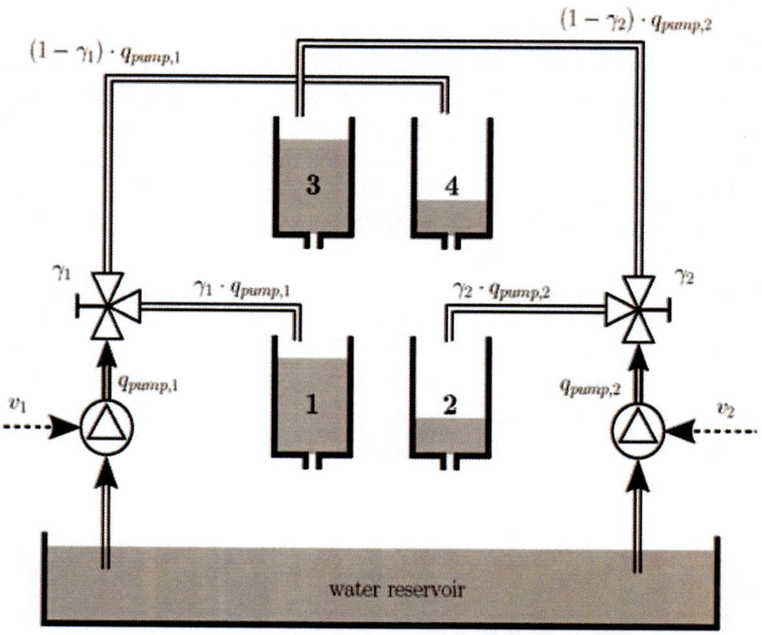

Fig. 1 Schematic diagram of quadruple tank system

Table 1 Process parameter values

Parameter	Values
A_I, A_3	28 cm^2
A_2, A_4	32 cm^2
a_1, a_3	0.71 cm^2
a_2, a_4	0.57 cm^2
k_c	0.50 v/m
g	981 cm/s^2

$$\frac{\mathrm{d}h_4}{\mathrm{d}t} = -\frac{a_4}{A_4}\sqrt{2gh_4} + \frac{(1-\gamma_2)k_2}{A_4}v_2$$

where

A_i cross section area of tank i
a_i cross section area of the outlet hole
h_i liquid level in tanks
g acceleration due to gravity

The flow through the pump is $k_i v_i$ and hence the flow through the tank 1 is $\gamma_1 k_1 v_1$ and the flow to tank 4 $is(1-\gamma_2)k_2 v_2$. It is similar for tank 2 and tank 3. Based on the valves position adjusted, the parameters $\gamma_1\gamma_2$ are determined. Parameters $k_c h_1$ and $k_c h_2$ are the measured levels. The process values are denoted (Table 1).

The corresponding linear transfer function matrix is:

$$G(S) = \begin{bmatrix} \frac{\gamma_1 c_1}{\tau_1 s+1} & \frac{(1-\gamma_2)c_2}{((\tau_1 s+1))((\tau_3 s+1))} \\ \frac{(1-\gamma_1)c_1}{(\tau_2 s+1)(\tau_4 s+1)} & \frac{\gamma_2 c_2}{(\tau_2 s+1)} \end{bmatrix} \tag{1}$$

For this system, time constants are:

$$\tau_i = \frac{A_i}{a_i}\sqrt{\frac{2h_{is}}{g}} \text{ and } c_j = \frac{\tau_j k_j}{A_j},\ j = 1, 2$$

On substituting the nominal parameter values (Table 1), the transfer function matrix is obtained (Eq. 2)

$$\begin{bmatrix} GP_{11}(S) & GP_{12}(S) \\ GP_{21}(S) & GP_{22}(S) \end{bmatrix} = \begin{bmatrix} \frac{2.6}{(62s+1)} & \frac{1.5}{(62s+1)(23s+)} \\ \frac{1.4}{(30s+1)(90s+1)} & \frac{2.8}{(90s+1)} \end{bmatrix} \tag{2}$$

By Taylor series approximation, the nonlinear model equations are linearized. The state space model is

$$\begin{bmatrix} \dot{h}_1 \\ \dot{h}_2 \\ \dot{h}_3 \\ \dot{h}_4 \end{bmatrix} = \begin{bmatrix} -\frac{1}{\tau_1} & 0 & \frac{A_3}{A_1\tau_3} & 0 \\ 0 & -\frac{1}{\tau_2} & 0 & \frac{A_4}{A_2\tau_4} \\ 0 & 0 & -\frac{1}{\tau_3} & 0 \\ 0 & 0 & 0 & -\frac{1}{\tau_4} \end{bmatrix}$$

$$\begin{bmatrix} h_1 \\ h_2 \\ h_3 \\ h_4 \end{bmatrix} + \begin{bmatrix} \frac{\Upsilon_1 K_1}{A_1} & 0 \\ 0 & \frac{\Upsilon_2 K_2}{A_2} \\ 0 & \frac{(1-\Upsilon_2)K_2}{A_3} \\ \frac{(1-\Upsilon_1)K_1}{A_4} & 0 \end{bmatrix} \begin{bmatrix} v_1 \\ v_2 \end{bmatrix}$$

$$y = \begin{bmatrix} 1\,0\,0\,0 \\ 0\,1\,0\,0 \end{bmatrix} \begin{bmatrix} h_1 \\ h_2 \\ h_3 \\ h_4 \end{bmatrix} \tag{3}$$

3 Controller Design

3.1 PID Controller Design

The proportional integral derivative (PID) control is the widely used industrial feedback control loop mechanism. The difference between a measured process variable and desired set point is calculated as error which is minimized by the controller by adjusting the process control outputs. A PID controller is designed for the quadruple tank process using Zeigler–Nichols closed loop method (Fig. 2) in MATLAB Simulink control tool.

PID represented mathematically in the following equation [4]

$$u = u_0 + k_p e + \frac{k_p}{T_i} \int_0^e e dt + k_p T_d \frac{d_e}{d_t} \tag{4}$$

where

k_p Proportional term
T_i Integral term
T_d Derivative term

The Simulink block diagram of PID controller for a MIMO process is shown below.

The PID controller tuning parameters are mentioned in Table 2.

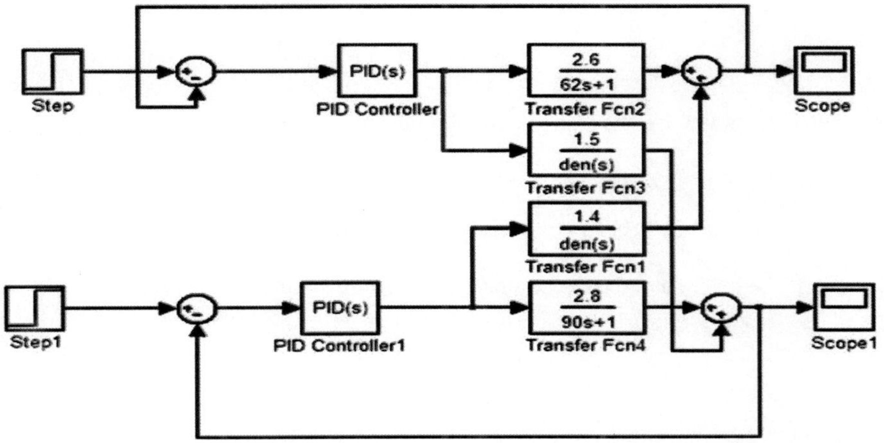

Fig. 2 Simulink diagram of PID controller

Table 2 Tuning parameters of PID controller

Controller	K_p	K_i	K_d
PID	1.5369	0.9487	2.7845

3.2 Decoupler Design

For a decoupler, additional cross controllers along with loop controllers are designed. The decoupler structure shown in Fig. 3 addresses the servo problems. The diagonal compensators can be recalculated if the loop controllers are tuned.

The plant of the process is taken in the form of equation given below.

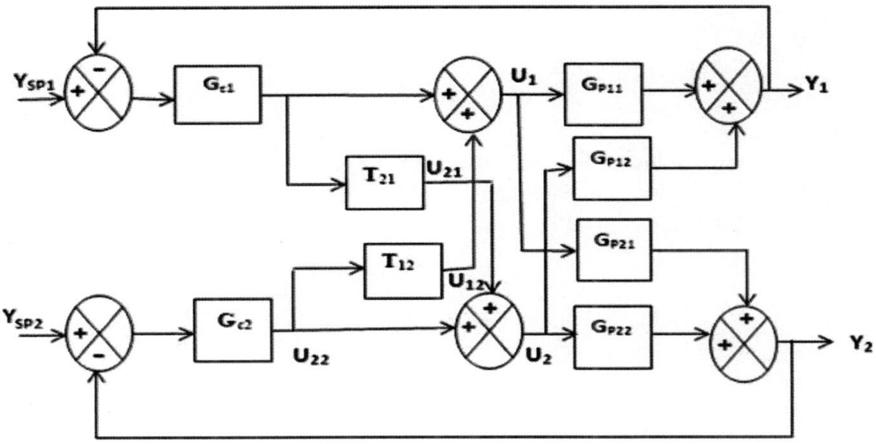

Fig. 3 Schematic diagram of PID decoupler

$$G(S) = \begin{bmatrix} GP_{11}(S)\ GP_{12}(S) \\ GP_{21}(S)\ GP_{22}(S) \end{bmatrix} \tag{5}$$

The control law for the static decoupler designed can be

$$\begin{bmatrix} U_1(S) \\ U_2(S) \end{bmatrix} = \begin{bmatrix} T_{11}\ T_{12} \\ T_{21}\ T_{22} \end{bmatrix} \begin{bmatrix} G_{C1}Y_{SP1}\ -G_{C1}Y_1 \\ G_{C2}Y_{SP2}\ -G_{C2}Y_2 \end{bmatrix} \tag{6}$$

Y_{SP} is the reference, the decoupler is denoted by a constant matrix

$$T = \begin{bmatrix} T_{11}\ T_{12} \\ T_{21}\ T_{22} \end{bmatrix} \tag{7}$$

Using cross decoupler, we get

$$T_{21} = -\frac{GP_{21}}{GP_{22}} = \frac{-0.025}{S + 0.04} \tag{8}$$

$$T_{12} = -\frac{GP_{12}}{GP_{11}} = \frac{-0.017}{S + 0.033} \tag{9}$$

3.3 State Feedback Controller

In full-state feedback control technique, all the state variables are feed back to the system's input. This method is known as using pole placement technique, where system is state controllable. The Simulink diagram of state feedback control for a four tank MIMO system is shown in Fig. 4

The control law is,

$$u = K[r(t) - x(t)]$$

By using the equation, Det ([$(SI - A) + (B)K$]) $= (s - p_1)(s - p_2)(s - p_3)(s - p_4) = 0$, the gain matrix can be obtained.

The state feedback controller gains are,

$$K = \begin{bmatrix} 3.17 - 0.45 - 0.56 - 0.54 \\ -1.944.8 - 3.852.45 \end{bmatrix}$$

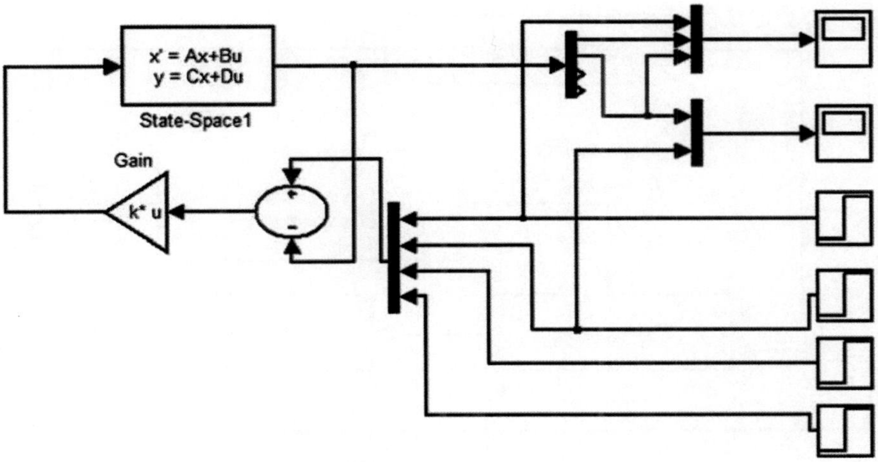

Fig. 4 Simulink diagram of state feedback controller for MIMO process

3.4 Sliding Mode Control Design

The system trajectory motion along the surface or plane of the state space is considered for sliding mode controller (Fig. 5).

General sliding surface equation

$$S = \left(\frac{\mathrm{d}}{\mathrm{d}t} +\right)^{n-1}$$

The sliding surfaces for this MIMO system can be written as,

$$S_1 = e_1 + \dot{e}_1$$
$$S_2 = e_2 + \dot{e}_2$$

where λ is a positive constant and $e = x - x_d$ is the tracking error.

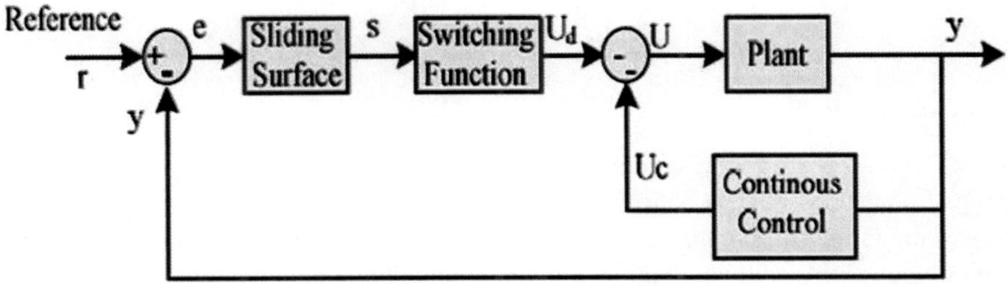

Fig. 5 Closed loop schematic diagram of SMC

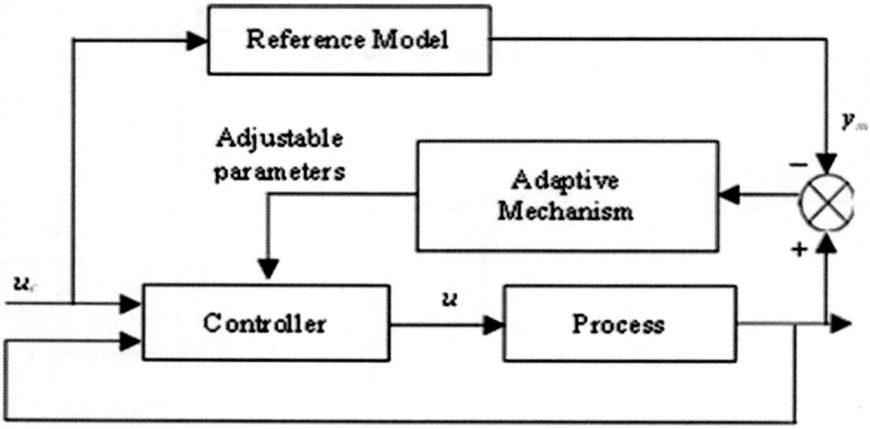

Fig. 6 Block diagram of model reference adaptive control

3.5 Model Reference Adaptive Control

Due to the nonlinear nature of level process in a quadruple system, model reference adaptive controller (MRAC) is designed and implemented (Fig. 6). The reference model block forms the major portion of control system, and its output acts as a set point which is used to adjust the parameters for controller tuning. This parameter adjustment can be obtained by either gradient method (MIT rule) or by Lyapunov Stability Theory.

On adjusting the parameters θ_1 and θ_2, the controller output u is calculated as below the command signal and the output y of the process as

$$u = \theta_1 u_c - \theta_2 y \tag{10}$$

where u_c is the command signal and y is the process output.

The parameters value with respect to time is

$$\frac{\mathrm{d}\theta_1}{\mathrm{d}t} = -\gamma_1 e \frac{\partial e}{\partial \theta_1} \tag{11}$$

$$\frac{\mathrm{d}\theta_2}{\mathrm{d}t} = -\gamma_2 e \frac{\partial e}{\partial \theta_2} \tag{12}$$

where γ_1 and γ_2 are the adaption gains for θ_1 and θ_2, respectively. The adaptation gain value determines the convergence speed. For a stable response, the gain should be low but it needs more output time to converge. If the gain is high, it will oscillates the output.

4 Simulation Results

The MATLAB Simulink is used to obtain the system's response. The response of PID controller with and without decoupler is shown in Figs. 7 and 8. The response of state feedback controller and sliding mode controller is shown in Figs. 9 and 10, respectively.

Table 3 gives the performance of various controllers for a quadruple system. Both the time domain and integral parameters are considered here for the performance evaluation.

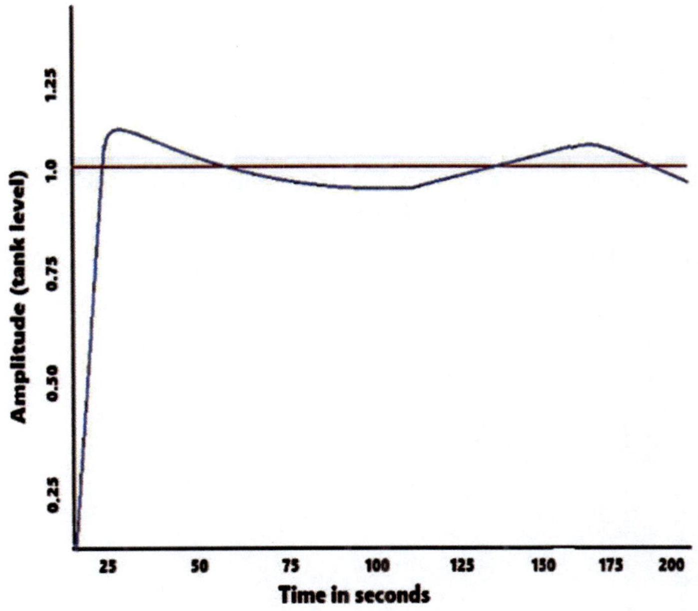

Fig. 7 PID controller response

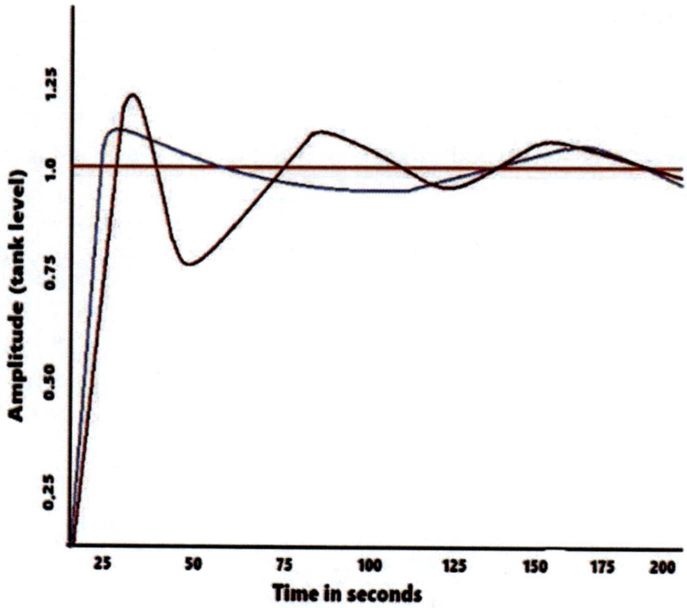

Fig. 8 Response of PID decoupler

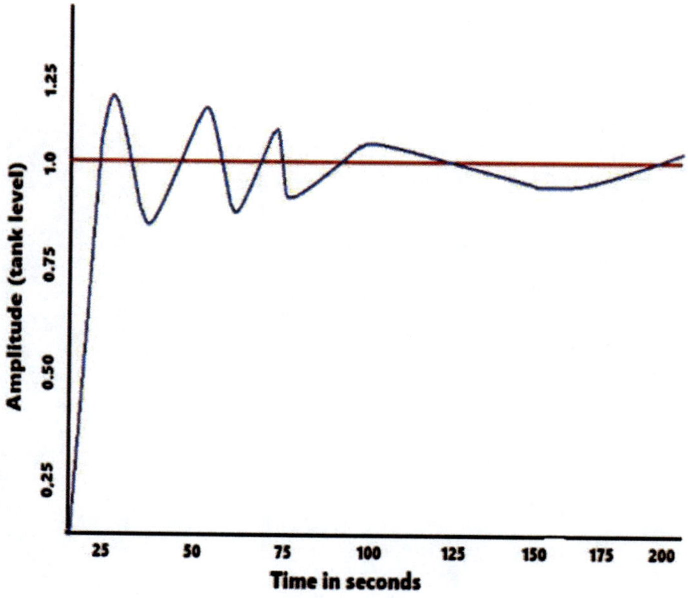

Fig. 9 State feedback controller response

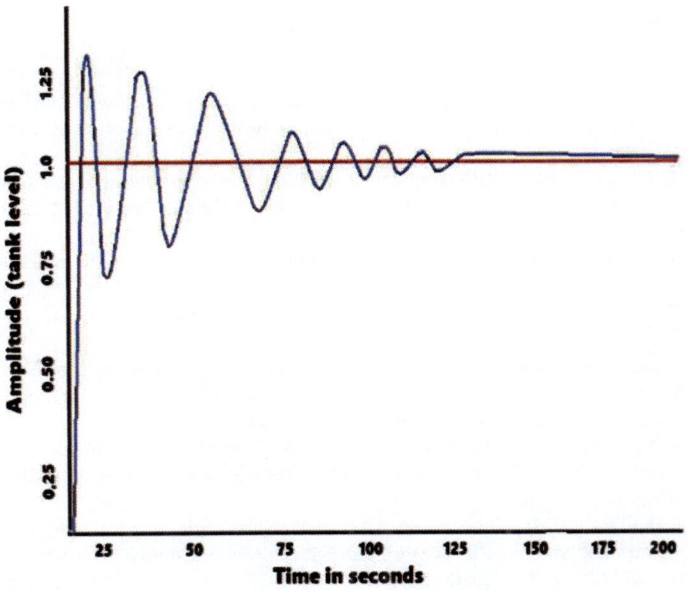

Fig. 10 Sliding mode controller response

Table 3 Performance analysis

Parameter	Conventional PID controller	De coupler	State feedback controller	Sliding mode controller	Model reference adaptive control
t_d	8.05 s	7.293 s	8.0 s	10 s	10 s
t_r	13.2 s	12.42 s	19.84 s	14.5	11.0
t_s	180 s	124 s	95 s	105 s	85 s
$\%M_p$	0.6	0.53	0.523	–	–
ISE	8.144	6.203	8.922	5.923	6.422

5 Conclusion

The behaviour and nature of the quadruple system was studied, and various controllers were designed and implemented to control its tank level.

From the comparative performance Table 3, it is observed that the conventional PID controller takes more time to get settle at the desired level and all its time domain parameters are high. In order to overcome these, a decoupler is implemented which reduces the time domain parameters to a value merely to the desired range. The state feedback controller offers better settling time than PID decouple design for MIMO process. Sliding mode controller for MIMO system is used to avoid the peak overshoot. The MRAC provides quick settling time when compared with other conventional controllers.

References

1. Nageswara Rao CV, Murty MSN, Potnuru D (2020) Control of four tank system using Grasshopper Algorithm. In: IEEE India council international subsections conference, pp 200–203
2. Shukla S, Chandra Pati U (2019) Implementation of different control strategies on a quadruple tank system. In: 6th international conference on signal processing and integrated networks, pp 579–583
3. Nirmala SA, Veena Abirami B, Manmalli D (2011) Design of model predictive controller for a four tank process using linear state space model and performance study for reference tracking under disturbances. In: IEEE xplore international conference on process automation, control and computing, 20–22 July 2011
4. Raff T, Huber S, Nagy ZK, Allgower F (2006) Nonlinear model predictive control of a four tank system: an experimental stability study. In: Proceedings of 2006 IEEE international conference on control applications, Munich, Germany, 4–6 Oct 2006, pp 237–242
5. Angeline Vijula D, Anu K, Honey Mol P, Priya P (2013) Mathematical modelling of quadruple tank system. Int J Emerg Technol Adv Eng 3(12). Website: www.ijetae.com. ISSN 2250-2459, ISO 9001:2008 Certified J
6. He Y, Wang QG (2006) An improved ILMI method for static output feedback control with application to multivariable PID control. IEEE Trans Autom Control 51(10):1678–1683
7. Johansson KH (2000) The Quadruple-tank process—a multivariable laboratory process with an adjustable zero. IEEE Trans Control Syst Technol 8(3)
8. Goyal N, Rai L (2015) Controller tuning of coupled tanks by astrom & hagglund using Matlab & simulation. Int J Res Manage Sci Technol 3(2). E-ISSN: 2321-3264
9. Shah P, Hanwate S (2020) Modelling and simulation of quadruple tank system using SBL-PI controller. In: International conference on Industry 4.0 Technology, 13–15 Feb 2020, pp 70–75
10. Parvat BJ, Jadhav VK, Lokhande NN. Design and implementation of sliding mode controller for level control. IOSR J Electr Commun Eng (IOSR-JECE) 51–54. ISSN: 2278-2834, ISBN: 2278-8735
11. Chekari T, Mansouri R and Bettayeb M (2021) Real time applications IMC-PID-FO multi loop controllers on the coupled tanks process. In: Proceedings of the Institution of Mechanical Engineers 2021, Part I: J Syst Control Eng 235(8):1542–1552
12. Ghazali MRB, Ahmad MAB, Raja Ismail RMTB (2022) Adaptive safe experimentation dynamics for data driven neuro endocrine-PID control of MIMO systems. IETE J Res 68(3):1611–1624
13. Davoodi M, Meskin N, Khorasani K (2018) Integrated fault detection and control design based on dynamic observer. Institution of Engineering and Technology (IET)

Chapter 53
Supervised Machine Learning Text Classification: A Review

Nisar Ahmad Kangoo and Apash Roy

1 Introduction

In this information era, electronic documents are getting stored and shared across various digital gadgets. These documents are produced via emails, blogs, news bulletins, and other web pages. Most of these documents contain unstructured text. Text mining is gaining more importance here as these text documents need to be classified so that users can extract information from these resources.

The main and important part of text mining is text classification [1]. It is very hard to manually classify the huge number of documents which are produced on a daily basis on the web and other electronic media. Thanks to machine learning which has made a drastic improvement in text classification. The automatic text classification classifies a document D1 from the document set $\{d1, d2, d3, \ldots dn\}$ to an already labeled class {C1, C2, C3}. The challenge for the classifier is to determine whether a particular document belongs to class C1 or C2, etc. For example, a new document is to be classified as a political or sports or cultural category. A text document can belong to a single label (single-label document classification), and in this paper, we are considering only those classifiers. But sometimes, a document can belong to more than one class or multi-label document classification.

We have divided this paper into the following sections: Sect. 1 Text cleaning and preprocessing which include document reading, tokenization, and stemming. Section 2 Enumeration of text classification algorithms. In Sect. 3, we will see the performance evaluation, and lastly, in Sect. 4, there will be a conclusion followed by references section.

In general, text classification can be illustrated below in diagram [1–3].

N. A. Kangoo (✉) · A. Roy
Department of Computer Science and Application, Lovely Professional University Punjab, Chaheru, India
e-mail: nisar.42000573@lpu.in; nisarphd2020@gmail.com

A. Yadav et al. (eds.), *Proceedings of International Conference on Paradigms of Communication, Computing and Data Analytics*, Algorithms for Intelligent Systems,
https://doi.org/10.1007/978-981-99-4626-6_53

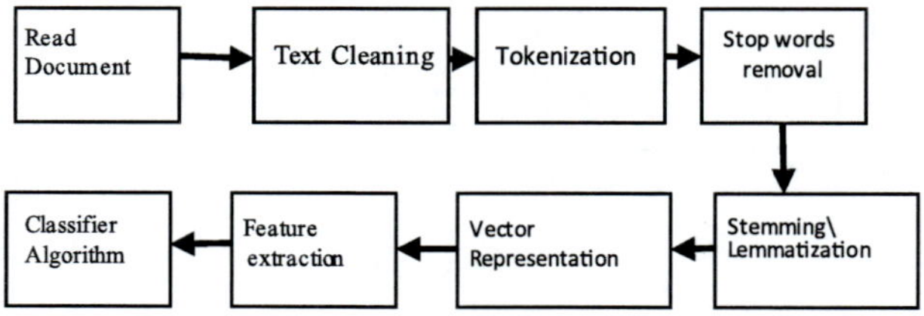

2 Text Cleaning and Preprocessing

The first and foremost step in text classification is document collection or datasets. After reviewing the literature, it has been found that in most of the text classifications, following datasets have been used: The TechTC-100 having around 150 documents [4–6], the 20 newsgroups dataset having 20 usenet newsgroups [7, 8], Stanford sentiment Treebank (SST) database having movie reviews which were parsed and labeled by Socher et al. (2013) [nlp.standford.edu/sentiment/], the YFCC100M dataset having almost 100M images caption, tags and titles [9], Reuters-21578 database a collection of news documents [10]. Other datasets include Library catalog records from National Technology Information Service, UPI newswire, and Reuters-22173 newswires [11], USENET and DigiTrad [12]. Web of Science datasets WOS-11967, WOS-46985, and WOS-5736 are also used in this domain. Amazon-12K [13], WIKI-30 [11–49], EUR-Lex [14], and Reuters-RCV1 [15] are some more datasets for text classification.

2.1 *Text Cleaning*

The removal of unnecessary, special characters, and punctuation marks is performed either before or after tokenization. This is done because special characters and punctuation marks seldom have any significance during feature extraction [16, 17].

2.2 *Tokenization*

The document is considered a string [1]. After reading the document, the sentences are then broken down into words or a list of tokens known as tokenization [2, 18, 19]. For example, “After finishing his examination, he started his coaching classes”. This sentence is broken into the following list of tokens: “After”, “finishing”, “his”, “examination”, “he”, “started”, “his”, “coaching”, and “classes”. In Python, this can be done by using the NLTK library.

2.3 Stemming

Since words in the documents are not always the root words, different stemming algorithms are applied to convert various words into basic forms or roots [1]. One such algorithm, Porter's stemmer, is mostly used for stemming the English text [20, 21]. In this suffixes of the words are stripped, e.g., "generalization" to "general" and "connection" to "connect". Sometimes, extra preprocessing is required for web pages to eliminate or modify HTML and script tags [22].

2.4 Lemmatization

Stemming may sometimes lead to the formation of words that do not hold any meaning like "ring" which will be stemmed to "r" and "several" to "sever" with no meaning. Hence, lemmatization is a good technique to get root words as it works more efficiently than stemmer. Lemmatization provides root words which are present in the dictionary, whereas stemming provides a root stem which might not be a dictionary word [textbook]. For example, after lemmatization, "been had languages cities mice" will be "been had language city mouse".

2.5 Stop Word Deletion or Removing Stop Words

Some words having little or no significance occur in the documents more frequently than the words with high significance. These words are known as stop words [16, 23]. During preprocessing, these stop words like a, the, me, myself so on are removed so that focus remains on the highly significant words during classification.

2.6 Vector Representation

A document is a sequence of words [24, 25], and every document typically consists of a list of words. Vocabulary, also referred to as a feature set, is the collection of all the words in a training set. A document can be represented by a binary vector, which assigns the value 1 if the feature word is present in the document and 0 otherwise. Putting a document in an R|V| space, where |v| represents the size of the vocabulary V can be expressed as this [2].

2.7 *Feature Selection*

To reduce overfitting and reduce the dimensionality of the dataset, the features which are considered irrelevant for classification are removed [26]. Reduced dataset size, lower processing needs for text classification algorithms, and a smaller search field are all benefits of this change.

Various feature subset selection and dimensionality reduction for text document classifications include term frequency-inverse document frequency (TF-IDF), information gain, chi-square, conditional mutual information, term frequency, term strength, probability ratio and length normalization [2, 26–34].

3 Text Classifiers or Text Classification Techniques

There are several text classification techniques or algorithms which include the following.

a. Naïve Bayes
b. Decision tree
c. Support vector machines
d. K-nearest = neighbors (KNN)
e. Random forest.

Due to the limitation in the number of pages, we have just enumerated these.

4 Performance Evaluation

To evaluate the performance of a text classification algorithm, there are several metrics for calculation. Among these precision, recall and accuracy are mostly used. Firstly, it is to be determined whether a document was classified as True Positive (TP), False Positive (FP), True Negative (TN), or False Negative (FN).

The number of specimens that were accurately identified to be positive and were positive is referred to as True Positive (TP). The amount of specimens that were accurately identified as negative and turned out to be negatives **is** True Negatives (TN). Additionally, False Positive (FP) denotes the number of specimens that were negative but were mispredicted and False Negative (FN) denotes the number of specimens that were positive but were mispredicted.

From the above features precision, recall and accuracy can be calculated as

$$\text{Precision} = \text{TP}/\text{TP} + \text{FP}$$

$$\text{Recall} = \text{TP}/\text{TP} + \text{FN}$$

$$\text{Accuracy} = \text{TP} + \text{TN}/\text{TP} + \text{TN} + \text{FP} + \text{FN}$$

In skewed datasets, accuracy is not a good metric for performance calculation and hence precision and recall are used in such cases. A better picture of performance can be achieved by combining precision and recall as

$$\text{F1} = 2 \times \text{Precision} \times \text{Recall}/\text{Precision} + \text{Recall}$$

5 Performance of Text Classification Algorithms

The text classification algorithms have shown different results on different datasets. Some algorithms are even better than others on a given dataset. Table 1 summarizes the performance of recent papers on various datasets. To minimize the table size, only the reference number of the paper has been given in the table and accuracy on the corresponding dataset is shown.

6 Notable Observations

After studying recent research papers regarding various text classification algorithms, it can be concluded that the accuracy of these algorithms lies in the range given against each algorithm in Table 2. The range just depicts the accuracy of algorithms from the papers mentioned in Table 1. It is observed that the random forest algorithm showed the highest accuracy of 98.7% on the CNAE9 dataset. The Naïve Bayes, SVM, decision trees, and KNN showed the highest accuracy of 97.6, 94.1, 82.0, and 95.4% on the BBC news dataset, Reuters-21578, Amazon Reviews, and popular news websites datasets, respectively. Figure 1 shows accuracy trends in the graphical representation of various text classification algorithms on different datasets.

7 Conclusion

Due to the increase in textual data via the number of blogs, websites, electronic storage, the text classification, an artificial intelligence research topic has gained much importance. After reviewing the literature, it can be determined that the structure of the related corpus affects how well text documents may be classified. Among various text classification algorithms, random forest provides the best performance in text classification concerning accuracy. The performance range of random forest lies between 88 and 99% as per the above literature review.

Table 1 Accuracy of various text classification algorithms on different datasets from recent research papers

S. No.	Paper references	Year of publication	Datasets	Algorithm/ model	Accuracy (%)
1	[36]	2019	News articles of 2018 (India)	Multinomial Naïve Bayes	73.40
				Bernoulli Naïve Bayes	69.15
2	[37]	2019	Movie review dataset V1.0	Naïve Bayes	70.9
			Movie review dataset V2.0	Naïve Bayes	79.7
3	[38]	2021	BBC news dataset	Complement Naive Bayes (MNB)	97.60
4	[39]	2018	Medical dataset	Gaussian event model	72.13
5	[41]	2021	OHSUMED-233445, Reuters-21578, TREC, and the WebACE project	Complement Naive Bayes (CNB)	83.80
6	[42]	2016	"la1s", "lla2s", and "new3s"	Complement Naive Bayes (CNB)	84.12
7	[43]	2019	News articles from an Indonesian news website	Multinomial Naïve Bayes (MNB)	86.0
8	[36]	2017	The dataset consists of approximately 4500 mobile phones and 4 lakh reviews on Amazon.com	Naïve Bayes	66.95
9	[44]	2019	Dataset V2.0	SVM	81.35
			Dataset V1.0	SVM	76.0
10	[46]	2018	Reuters-21578	SVM	83.33
			20ng dataset	SVM	43.79
			WebKB	SVM	53.07
11	[47]	2022	Reuters-21578	SVM classifier with a linear kernel	94.10
12	[48]	2021	Polarity Movie Dataset (PMD)	NS model	92.90
			Twitter-Sanders-Apple (TSA)	NS model	90.40

(continued)

Table 1 (continued)

S. No.	Paper references	Year of publication	Datasets	Algorithm/ model	Accuracy (%)
13	[49]	2019	Quora insincere question dataset	SVM	82.29
14	[50]	2019	Online social web blogs	SVM	83.36
15	[42]	2021	Database of JOBBKK Company recruitment agency in Thailand	SVM	92.73
16	[51]	2019	News articles from an Indonesian news website	SVM	93.0
17	[52]	2018	The dataset contains more than 16K news articles, collected from 'Daily Roshni', an Urdu newspaper from Srinagar, India	SVM	68.73
18	[36]	2017	The data set consists of approximately 4500 mobile phones and 4 lakh reviews on Amazon.com	SVM	81.77
19	[51]	2019	News articles from an Indonesian news website	Decision trees	80.0
20	[52]	2018	The dataset contains more than 16K news articles, collected from "Daily Roshni", an Urdu newspaper from Srinagar, India	Decision trees	62.37
21	[36]	2017	The dataset consists of approximately 4500 mobile phones and 4 lakh reviews on Amazon.com	Decision trees	74.75
22	[53]	2019	(Amazon Reviews: Unlocked Mobile Phones I Kaggle, 2016) consists of 400K reviews	Decision trees	82.0
23	[54]	2019	(Amazon Reviews: Unlocked Mobile Phones I Kaggle, 2016)consists of 400,000 reviews	Random forest	88.0

(continued)

Table 1 (continued)

S. No.	Paper references	Year of publication	Datasets	Algorithm/ model	Accuracy (%)
24	[33]	2018	Shenzhen regional medical information A total of 282K notes unfolding medical imaging diagnosis results	Random forest	91.0
25	[54]	2018	CNAE9 dataset	Random forest	98.7
26	[33]	2020	BBC news dataset	Random forest	93.0
27	[33]	2020	BBC news dataset	KNN	92.0
28	[55]	2019	Lao news text	KNN	71.4
29	[55]	2019	7 well-liked news websites (beinsports.com, tech-wd.com, skynewsarabic.com, Arabic.rt.com, cnbcarabia.com, arabic.cnn.com and youm7.com)	KNN	95.4

Table 2 Range of accuracy (percentage) shown by various text classification algorithms

Algorithm	Percentage range of accuracy (%)
Naïve Bayes	69.15–97.6
SVM	43.79–94.1
Random forest	88.0–98.7
Decision trees	62.37–82.0
KNN	71.4–95.4

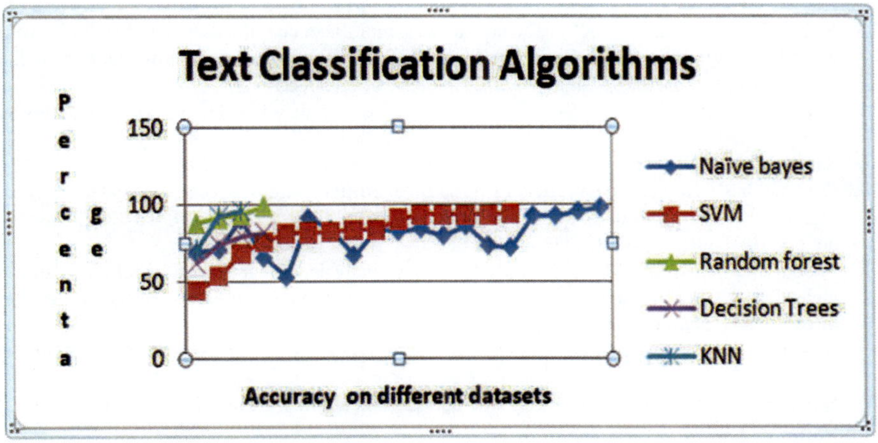

Fig. 1 Graph showing accuracy trends of different text classification algorithms

References

1. Korde V, Mahender CN (2012) Text classification and classifiers: a survey. Int J Artif Intell Appl 3(2):85
2. Ikonomakis M, Kotsiantis S, Tampakas V (2005) Text classification using machine learning techniques. WSEAS Trans Comput 4(8):966–974
3. Kumar J, Roy A (2021) DograNet—a comprehensive offline Dogra handwriting character dataset. In: International conference on robotics and artificial intelligence (RoAI)
4. Davidov D, Gabrilovich E, Markovitch S (2004) Parameterized generation of labelled datasets for text categorization based on a hierarchical directory. In: Proceedings of the 27th annual international ACM SIGIR conference on research and development in information retrieval, pp 250–257
5. Gabrilovich E, Markovitch S (2004) Text categorization with many redundant features: Using aggressive feature selection to make SVMs competitive with C4. 5. In: Proceedings of the twenty-first international conference on machine learning, p 41
6. Roy A, Ghosh D (2021) Pattern recognition-based tasks and achievements on handwritten Bengali character recognition. In: 2021 6th international conference on inventive computation technologies (ICICT). IEEE, pp 1260–1265
7. 20 Newsgroups Dataset. J. Rennie. http://people.csail.mit.edu/jrennie/20Newsgroups/. Accessed on 2023/03/15
8. 20 Newsgroups Dataset. UCI KDD Archive. http://kdd.ics.uci.edu/databases/20newsgroups/20newsgroups.html
9. Joulin A, Grave E, Bojanowski P, Mikolov T (2016) Bag of tricks for efficient text classification. arXiv preprint arXiv:1607.01759
10. Xu Z, Yu K, Tresp V, Xu X, Wang J (2003) Representative sampling for text classification using support vector machines. In: European conference on information retrieval. Springer, Berlin, Heidelberg, pp 393–407
11. Apté C, Damerau F, Weiss SM (1994) Automated learning of decision rules for text categorization. ACM Trans Inf Syst (TOIS) 12(3):233–251
12. Mansuy TN, Hilderman RJ (2006) Evaluating WordNet features in text classification models. In: FLAIRS conference, pp 568–573
13. McAuley J, Leskovec J (2013) Hidden factors and hidden topics: understanding rating dimensions with review text. In: Proceedings of the 7th ACM conference on Recommender systems, pp 165–172

14. Mencia EL, Fürnkranz J (2008) Efficient pairwise multilabel classification for large-scale problems in the legal domain. In: Joint European conference on machine learning and knowledge discovery in databases. Springer, Berlin, Heidelberg, pp 50–65
15. Lewis DD, Yang Y, Russell-Rose T, Li F (2004) Rcv1: a new benchmark collection for text categorization research. J Mach Learn Res 5(April):361–397
16. Sarkar D (2016) Text analytics with Python. Apress, New York, NY, USA
17. Deshmukh SV, Roy A, An empirical exploration of artificial intelligence in medical domain for prediction and analysis of diabetic retinopathy: review. J Phys: Conf Ser 1831:012012. https://doi.org/10.1088/1742-6596/1831/1/012012
18. Mishu SZ, Rafiuddin SM (2016) Performance analysis of supervised machine learning algorithms for text classification. In: 2016 19th international conference on computer and information technology (ICCIT). IEEE, pp 409–413
19. Roy A (2019) Handwritten Bengali character recognition a study of works during the current decade
20. Dalal MK, Zaveri MA (2011) Automatic text classification: a technical review. Int J Comput Appl 28(2):37–40
21. Porter MF (1980) An algorithm for suffix stripping. The Program 14(3):130–137
22. Changuel S, Labroche N, Bouchon-Meunier B (2009) Automatic web pages author extraction. LNAI 5822, Springer-Verlag, Berlin Heidelberg, pp 300–311
23. Roy A, Manna NR (2015) An Approach towards Segmentation of real-time handwritten text. Int J Adv Innov Res 4(5), (2278-7844)
24. Leopold E, Kindermann J (2002) Text categorization with support vector machines. How to represent texts in input space? Mach Learn 46(1):423–444
25. Roy A, Manna NR (2014) Handwritten character recognition with feedback neural network. Int J Comput Sci Eng Technol (IJCSET) 5(1)
26. Forman G (2003) An extensive empirical study of feature selection metrics for text classification. J Mach Learn Res 3(Mar):1289–1305
27. Brank J, Grobelnik M, Milic-Frayling N, Mladenic D (2002) Interaction of feature selection methods and linear classification models. In: Workshop on text learning held at ICML
28. Torkkola K (2002) Discriminative features for text document classification. In: Proceedings international conference on pattern recognition, Canada, 2002
29. Soucy P, Mineau GW (2003) Feature selection strategies for text categorization. In: Conference of the Canadian society for computational studies of intelligence. Springer, Berlin, Heidelberg, pp 505–509
30. Roy A, Manna NR (2013) Recognition of handwritten text: an artificial neural network approach. Int J Adv Innov Res 2(9), (2278–7844)
31. Ko S-J, Lee J-H (2001) Feature selection using association word mining for classification. In: Mayr HC et al (eds) DEXA 2001, LNCS 2113, pp 211–220
32. Dasgupta A (2007) Feature selection methods for text classification KDD'07, 12–15 Aug 2007
33. Roy A, Manna NR (2012) Handwritten character recognition using block wise segmentation technique (BST) in neural network. In: Proceedings of first international conference on intelligent infrastructure, held during, pp 1–2
34. Roy A, Manna NR (2012) Handwritten character recognition using mask vector input (MVI) in a neural network. Int J Adv Sci Technol 4(4)
35. Singh G, Kumar B, Gaur L, Tyagi A (2019) Comparison between multinomial and Bernoulli naïve Bayes for text classification. In: 2019 International conference on automation, computational and technology management (ICACTM). IEEE, pp 593–596
36. Kadhim AI (2019) Survey on supervised machine learning techniques for automatic text classification. Artif Intell Rev 52(1):273–292
37. Salman HA, Obaida TH (2021) BBC news data classification using Naïve Bayes based on bag of word. 湖南大学学报 (自然科学版) 48(9)
38. Ranjitha KV (2018) Classification and optimization scheme for text data using machine learning Naïve Bayes classifier. In: 2018 IEEE world symposium on communication engineering (WSCE). IEEE, pp 33–36

39. Roy A, Manna NR (2012) Handwritten character recognition using mask vector in a competitive neural network with multi-scale training. Int J Adv Innov Res 1(2)
40. Gan S, Shao S, Chen L, Yu L, Jiang L (2021) Adapting hidden Naive Bayes for text classification. Mathematics 9(19):2378
41. Zhang L, Jiang L, Li C (2016) A new feature selection approach to naive Bayes text classifiers. Int J Pattern Recognit Artif Intell 30(02):1650003
42. Londo GLY, Kartawijaya DH, Ivariyani HT, WP YSP, Rafi APM, Ariyandi D (2019) A study of text classification for Indonesian News article. In: 2019 International conference of artificial intelligence and information technology (ICAIIT). IEEE, pp 205–208
43. Singla Z, Randhawa S, Jain S (2017) Sentiment analysis of customer product reviews using machine learning. In: 2017 international conference on intelligent computing and control (I2C2). IEEE, pp 1–5
44. Roy A, Manna NR (2012) A competitive neural network as applied for character recognition. Int J Adv Res Comput Sci Softw Eng 2(3)
45. Goudjil M, Koudil M, Bedda M, Ghoggali N (2018) A novel active learning method using SVM for text classification. Int J Autom Comput 15(3):290–298
46. Al Hasan S, Hussain MG, Protim J, Rahman MM, Fahim N, Chowdhury MZ, Pritom AI, Classification of multi-labeled text articles with Reuters dataset using SVM
47. Asgarnezhad R, Monadjemi SA (2021) NB VS. SVM: a contrastive study for sentiment classification on two text domains. J Appl Intell Syst Inf Sci 2(1):1–12
48. Jain DK, Jain R, Upadhyay Y, Kathuria A, Lan X (2020) Deep refinement: capsule network with attention mechanism-based system for text classification. Neural Comput Appl 32(7):1839–1856
49. Asogwa DC, Anigbogu SO, Onyenwe IE, Sani FA (2021) Text classification using hybrid machine learning algorithms on Big Data. arXiv preprint arXiv:2103.16624
50. Panurug D, Rattanasiriwongwut M (2021) Text classification analysis by machine learning job segmentation algorithm. Int J Entrepreneurship 25:1–10
51. Rasheed I, Gupta V, Banka H, Kumar C (2018) Urdu text classification: a comparative study using machine learning techniques. In: 2018 Thirteenth international conference on digital information management (ICDIM). IEEE, pp 274–278
52. Guia M, Silva RR, Bernardino J (2019) Comparison of Naïve Bayes, support vector machine, decision trees and random forest on sentiment analysis. KDIR 1:525–531
53. Yang B, Dai G, Yang Y, Tang D, Li Q, Lin D, Cai Y (2018) Automatic text classification for label imputation of medical diagnosis notes based on random forest. In: International conference on health information science. Springer, Cham, pp 87–97
54. Shah K, Patel H, Sanghvi D, Shah M (2020) A comparative analysis of logistic regression, random forest and KNN models for the text classification. Augmented Hum Res 5(1):1–16
55. Al Qadi L, El Rifai H, Obaid S, Elnagar A (2019) Arabic text classification of news articles using classical supervised classifiers. In: 2019 2nd International conference on new trends in computing sciences (ICTCS). IEEE, pp 1–6

Chapter 54
Train Delay Prediction Using Machine Learning

Nilesh N. Dawale and Sunita Nandgave

1 Introduction

Train delay is a significant research topic in transportation organization and train dispatching management. Unexpected events can cause delays, and such delays have a tendency to spread throughout an area, affecting other trains' operations. Hence, predicting train delays is crucial in train dispatching as it can help minimize their impact on train operations. Accurate prediction of train delays can improve dispatching quality and minimize the potential impact of delayed trains on other trains [1, 2].

The main aim of train delay prediction is maintaining train operation without impact and delay dissemination. More precise prediction helps in risk analysis and early caution, enabling real-time adjustments of transportation schedules in emergencies. Train delay prediction can assist dispatchers in analyzing train operation status, estimating delay risk, and making informed traffic dispatching decisions [3]. Thus, studying the prediction model of train delay is essential for developing railway traffic command automation systems. Several research studies have been conducted to analyze train delay. These include the fuzzy Petri net model proposed by Milinkovi et al. [4] for simulating traffic processes and train operations in the railway system, the train delay analysis by Tikhonov et al. [5] that examined the relationship between arrival delay and various features of the railway system using SVM [1], and the train delay prediction model based on Bayesian networks by Corman and Kecman [6] and Lessan et al. [7]. Other models proposed include the high-precision ANN model by Yaghini et al. [8] for predicting the delay of Iranian railway passenger trains and the deep learning model established by Ping et al. [9] for predicting train delay time based on RNN [1].

N. N. Dawale (✉) · S. Nandgave
Department of Computer, G H Raisoni College of Engineering, Pune, India
e-mail: nilesh122n@gmail.com

A. Yadav et al. (eds.), *Proceedings of International Conference on Paradigms of Communication, Computing and Data Analytics*, Algorithms for Intelligent Systems,
https://doi.org/10.1007/978-981-99-4626-6_54

Goal of predicting the delay of individual trains does not consider various factors that contribute to train delay, such as route faults, train and network faults, extreme weather, and passenger flow [1]. Moreover, they rarely consider both the temporal and spatial [1] properties of trains and routes, which are essential for predicting the cumulative effect of train delay, particularly at junction stations where different routes can have varying effects.

The main goal of this paper is not only to predict the delay of individual trains since the transition decision is based on the train navigation department based on historical data like the delay of one train causes the delay of other trains but also to predict the number of delayed trains for each station in a certain period, which is more helpful for train dispatching [1]. The dispatcher on site determines the departure time of each delayed train based on the station's situation, such as commutator flow. Predicting the collective cumulative effect (i.e., the number of delayed trains) is more valuable as it helps the movement of train to make decision, especially in cases of worse weather condition and other factors that may cause train delays. Therefore, targeting the predicted delay of one train is less important than predicting the cumulative effect of train delay.

This paper explains the TSTGCN model to use to predict the total number of delayed trains in each railway station [1]. The main area of this research are as follows:

- Prediction of combined effect for train movement under delay scenario for the first time;
- Build a TSTGCN model (with factor temporal and spatial dependence) to predict arrival delays in one station during a certain period [1];
- Comparing our TSTGCN model with other models like ANN, SVR, RF, and LSTM baselines using metrics as MAE, RMSE, and MAPE to assess train delay prediction performance [1].

The local dispatcher determines when the delayed train will depart. For instance, there are four trains (numbered T1) in CSTM station reaching Mumbai, Chennai Central, and Howrah, respectively, in times 2, 3, and 4.

The departure details for these four trains are displayed in Table 1.

Table 1 Train operation record

Train number	Exp departure	Train status	Terminal	Current time	Departure	Actual departure time
T1	10.00	Delay	CSTM	10:32	Yes	10:40
T2	10.15	Delay	MAS	10:32	Yes	10:45
T3	10.20	Delay	HWH	10:32	No	10:50
T4	10.30	Delay	CSTM	10:32	Yes	10:40

The trains are delayed as a result of the severe weather. Trains T1 and T4 to CSTM may be given priority by the station dispatcher depending on station conditions (like passenger traffic).

2 History

There are many researchers who have done a lot regarding prediction of train delay. Conventional mathematical models commonly used to simulate trains. Zhaoxia and Zhongying [10] developed a graphical tracer for train delay propagation simulation system. Based on the discrete event dynamic system theory, Xin et al. [11] established a delay propagation model using a time-event network [1]. Kecman and Goverde [12] estimated the time of train operation. Based on the stochastic approximation method, Carey et al. [13] developed a train delay propagation simulation test system. All these discussed models are based on assumptions and are less effective in dealing with the complex data; using these models, we may not predict correct train delay [1, 14].

Yaghini et al. [8] developed an ANN model, which provides high accuracy results for train delay. Pu et al. [15] research on train delay based on SVM and design the "delay confusion matrix" to evaluate the model. Ping et al. [9] introduce the concept of time series based on RNN deep learning model. The study of transport has made extensive use of spatial and temporal data mining. Many studies have used graph neural network modeling in recent years to uncover intricate correlations in spatiotemporal data. Today, spatial dependency and spatiotemporal correlation are generally dealt with using graph neural networks for spatiotemporal data. The traditional models are the spatiotemporal convolution network (STGCN) [16] and the graph convolutional recurrent neural network [17], which integrate GCN and LSTM to establish spatial dependency and temporal correlation, respectively.

Machine learning models have the potential to provide better fitting results than statistical regression models. However, existing machine learning models for train delay prediction often require significant Feature Engineering, which involves the input of expert knowledge and does not take into account the spatiotemporal nature of train delay data [1].

The data on train delays are an example of spatiotemporal network data, which is commonly analyzed in transportation research and other fields. Recently, many studies have utilized graphs. To comprehend the complex interrelationships in spatiotemporal data, use neural network models. Analysis of spatial dependency and spatiotemporal correlation is the main application of graph neural networks for spatiotemporal data. The spatiotemporal convolution network (STGCN) [16] uses GCN and convolutional neural networks (CNNs) to build correlation, the graph convolutional recurrent neural network (GCRNN) [17], which combines graph convolutional networks (GCN) and long short-term memory (LSTM) to establish spatial dependence and temporal correlation, and the multi-component spatiotemporal graph convolutional network (MSTGCN), which models super long time.

The research methods discussed above suffer from one or more of the following issues:

1. They rely too heavily on expert knowledge to predict performance.
2. They excessively focus on predicting the specific delay time of individual trains, disregarding that dispatching strategies are usually determined by dispatchers.
3. There are certain restrictions on the suggested model inputs. Although having a relatively straightforward structure, the spatiotemporal graph neural network is unable to accurately represent the properties of high-speed rail networks.

This model is developed to anticipate delays at train stations. The original graph-based high-speed railway network data may be handled directly by the TSTGCN model, allowing for more precise analysis and prediction of spatiotemporal properties and dynamic spatiotemporal correlation.

3 Analysis

3.1 Train Delay Spatiotemporal Correlation

As data from nearby stations and time periods are constantly connected to one another, forecasting train delays is an example of a spatiotemporal data prediction issue. The geographical and temporal relationships that exist across various trains and routes must be considered while analyzing rail delays. Spatial dependency, temporal correlation, and spatiotemporal correlation are characteristics of train delay data.

1. **Spatial dependence**: The connection between a station and the stations next to it on the high-speed rail network leads to spatial dependency. A station often affects its first-order neighbors the most directly. We have included a visual representation of the effects of train delays from a spatial perspective in Fig. 1 to completely highlight this phenomenon. The degree of the interaction between the two stations is shown in this picture by the lines linking them, with darker lines denoting more intensity. As is evident, there is a link between the Howrah (HWR) and Mumbai (CSTM) stations, and delayed trains leaving HWR for CSTM may result in delays in the number of trains that arrive at CSTM. The amount of delays might directly affect various stations, such as Nagpur (NGP) station in Fig. 1, if the station is a junction and is close to other stations. Here, several trains traveling in various directions and routes halt, and if one train is delayed, it might affect several other trains traveling in different directions. The number of delayed trains is influenced by both the geographical qualities of stations and the proximity of stops [1].

2. **Temporal Correlation**: The impact of the trains varies depending on the direction they are traveling in and how they interact with one another. Due to dispatching, predicting a train's delay in a certain route is challenging. However,

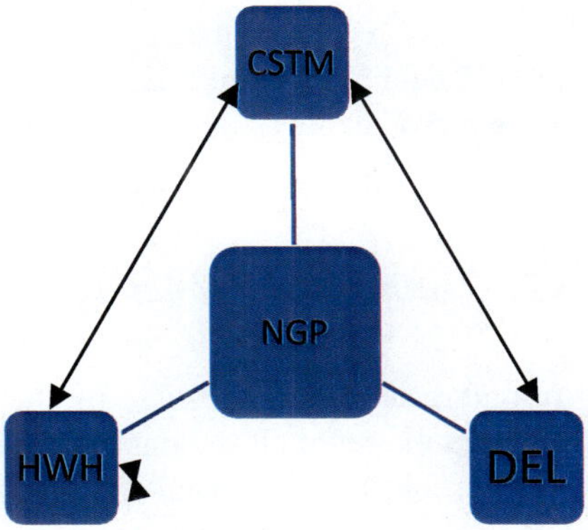

Fig. 1 Spatio-dependence diagram

the delay is unmistakably connected to the previous delays of one or more periods for each station on the high-speed railway network. This paper's emphasis is on the arrival delays at a single station. The delay also exhibits some periodicity, which indicates that a given period's delay will often follow the same pattern as recent days and weeks. In spatiotemporal network data, this closeness and regularity is referred to as temporal correlation. It is challenging to take rail delays into account [1].

3. **Spatiotemporal Correlation**: Even the same station might have varied effects on its neighbors depending on the date since the level of interaction between stations changes throughout time. The delay status of a station and its neighbors might change in a variety of ways depending on its previous data at various points in the future. As a result, both the geographical and temporal aspects of the train operation data show a high dynamic association. This implies that the complex nonlinear spatiotemporal network data must be taken into account in order to effectively estimate delays and that a single time series-based prediction model is insufficient. In order to forecast the cumulative effect for stations, the TSTGCN described in this study takes into consideration the spatiotemporal properties and dynamic correlation of train operation data [1].

4 Prediction Model

4.1 Collective Cumulative Effect

The network of high-speed rail lines may be visualized as an undirected graph, with each node denoting a station that connects to others based on the paths taken by the trains that pass through it. Each train has a set route that includes intermediate, departure, and destination stations. The estimated arrival and departure times of trains are listed on each station's timetable, and most trains arrive on time. The discrepancy between the anticipated and actual arrival or departure timings is used to quantify delays, which can also be caused by other variables like bad weather or passenger traffic. The dispatcher may develop better plans for train departure timings with the use of arrival delay analysis, assuring the smooth running of each train [1].

4.2 Train Delay Modeling

An undirected graph, denoted by the notation $G = (S, E, A, M)$, can be used to represent the set of all stations in a high-speed rail network, where S is the set of all stations, consisting of a total of N stations, E is the edges or routes between stations, A R is the connectivity between stations using an adjacency matrix, and M is the distance between stations using a distance weight matrix. Because the weight in the distance weight matrix is inversely proportional to the distance, longer distances have lower weights. For a certain time period, each station in G also includes unique statistical information, such as the volume of arrival delays and exit delays. Each station in G is time-dependent. X $i\tau \in R$ represents all eigenvalues of station i in the period τ, and $\chi = \left(X^1, X^2, \ldots, X^t\right)^T \in$ represents all eigenvalues of all stations in the period τ. We set $yi\tau \in R$ to represent the arrival delays of one station i in the future period τ. Using the past train data set, we generate the eigenvalue measures of all stations on the high-speed railway network in the past period τ. Given a fixed period τ and these eigenvalue measures, we predict the arrival delay sequence $Y = (y_1, y_2, \ldots, y_N)^T \in R^{N\times T_p}$ of stations in the future period of time T_p. Here, $y_i = \left(y_i^{\tau+1}, y_i^{\tau+2}, \ldots, y_i^{\tau+T_p}\right)$ represents the arrival delay sequence of station i in the future period T_p [1].

4.3 Attention Mechanism

The Time-Series Temporal Graph Convolutional Networks (TSTGCN) employed in this study include a general architecture that uses previous train operation data as training data to create a collective cumulative effect prediction model for train delays [18]. The structure of the three independent components of the prediction model, which was modeled after a prior paper's structure proposal (reference [19]), is the same. These elements each represent the dependency of historical train operation data for the last week, a day, and a week ago. Each part, which adheres to the same network structure, is made up of several spatial and temporal blocks, as well as completely connected layers. Spatial and temporal attention modules and convolution modules are included in each block. A residual learning framework is applied to each component to increase training effectiveness. In the next section, we will provide a detailed introduction of our TSTGCN model. TSTGCN takes in delay data from multiple stations in multiple time periods, which is a type of spatiotemporal network data. These data can be viewed as time series data made up of graph signals on the network. Each node on the network represents a station and has a time series of data, which exhibits complex correlations such as proximity and periodicity. Learning framework for efficiency. This paper is on analyzing time series data on a recent, daily, and weekly basis. Three time series segments with lengths of T_h, T_d, and T_w are extracted from the time axis to be used as inputs for the recent, daily, and weekly components, with the assumption that sampling occurs q times per day, the current time is $t0$, and the forecast window size is T_p. T_h, T_d, and T_w in this case are multiples of T_p. To represent the graph signal on the geographical network across the previous periods, we utilize the symbol X. The three time series components' precise specifications are explained below [1].

Graph time series Spatial and temporal network data are represented by delayed data from several stations collected across a range of time periods. This form of data may be thought of as time series data made up of networked graph signals. The network's nodes each have their own time series data with intricate relationships like proximity and periodicity. The time series data used in this study are current, daily, and weekly data. We take three time series segments from the time axis and use them as inputs for the recent, daily, and weekly components, assuming a sampling frequency of q per day and a forecast window size of T_p. These segments have lengths that are integral multiples of T_p, T_h, T_d, and T_w. We represent the graph signal on the spatial network in the past τ period as $X\tau$. The details of the three time series components are as follows [1].

Recent time series contemporary time series $Xh = (X_{t_0-T_h+1}, X_{t_0-T_h+2}, \ldots, X_{t_0}) \in R^{N\times F\times T_h}$. Specifically, the arrival delay of the next station may be somewhat impacted if a train traveling between fixed railway stations arrives late at one station for a variety of reasons, and this influence will be transmitted to numerous railway stations on the high-speed railway network through the connecting relations between stations. As a result, the arrival delays of several stations in the future will inevitably be impacted by the arrival delays of one or more stations in the past [1].

Daily time series Daily periodic time series $X_d = \left(X_{t_0-(T_d/T_p)\times q+1}, \ldots, X_{t_0-(T_d/T_p-1)\times q+1}, \ldots\right) \in R^{N\times F\times T_d}$ every day. Data from the same time period as the anticipated time period in the previous several days make up a daily periodic time series. Due to the regularity of people's daily travel plans, delays may happen within a comparatively constant time, such as between 14:00 and 15:00 every day in the afternoon. To replicate the day periodic of train arrival delay data, this component was built [1].

Weekly time series Weekly periodic time series $Xw = \left(X_{t_0-7\times(T_w/T_p)\times q+1}, \ldots, X_{t_0-7\times(T_w/T_p-1)\times q+1}, \ldots\right) \in R^{N\times F\times T_w}$. Fragments from the previous several weeks are combined to form a weekly periodic time series. These pieces share the same weekly characteristic and time span as the forecast period. In general, Monday's traffic pattern resembles Monday's past, but it could deviate significantly from Saturday and Sunday's. Many people will decide to use high-speed trains on Saturday and return on Sunday afternoon, which might cause quite heavy traffic at the terminals and cause train delays [1].

5 Experiments

To assess the predictive performance of the TSTGCN model, we train it on a dataset that we have constructed using real data. Additionally, we compare its performance against several baseline models.

5.1 *Data Processing*

The Indian Railway Passenger Ticket System, which is accessible at https://www.irctc.co.in, served as the data source for our analysis. During a period of time collect data from multiple stations with details attribute required for model. The original data are divided into 1 h time periods. For each station, the number of delayed trains is noted in each time slice (actual arrival time—expected arrival time > 0 and actual departure time—expected departure time > 0). The prediction aim for the experiment is the arrival delay, which is one of two types of train delay characteristics, the other being departure delays.

Our investigation used data from the Indian Railway Passenger Ticket System, which is available at https://www.irctc.co.in. Actual arrival time—expected arrival time > 0 and Actual departure time—expected departure time > 0 are used to identify the number of delayed trains for each station. The arrival delay, one of two types of train delay characteristics, serves as the experiment's prediction goal [1].

5.2 *Experimental Setup*

We used the MXNet framework to build the TSTGCN model. In our model, we employed 64 convolution kernels across all convolution layers and set the number of terms in the Chebyshev polynomials to 3. Additionally, 64 convolution cores were used for the temporal convolution layers. By adjusting the step size of temporal convolution, the time range of the data was altered. We determined the lengths of the three segments to be $T_h = 3$, $T_d = 1$, and $T_w = 1$. We wanted to forecast the stations' arrival delays for the next hour, therefore we set the prediction window's size to $T_p = 1$. During the training phase, the mean squared error (MSE) was employed as the loss function and was minimized using back propagation, a batch size of four and a learning rate of 0.000001 [1].

Algorithm 1 Splitting Time Serial Data

Steps:

Process input data and evaluate the length of it.
Determine how many windows can be made depending on the length of the input data, the size of the window, and the length of each step.
For each window, extract the corresponding sequence of data.
Append the extracted window to the input matrix x.
Extract the target data sequence for the predicted steps after the window.
Append the extracted target data to the output matrix y.
Repeat steps 3–6 for all possible windows.
Give the result in the form of input matrix x and the output matrix y [1].

Input: *data, window_si ze, step_length*
Output: *x, y*
1: $data\_length = length(data)$;
2: $window\_num = data\_length - window\_size - step\_length + 1$;
3: **for** $i = 1$ to $window\_num$ **do**
4: $window = data(i{:}\ i + window\_size - 1)$;
5: $x = [x; window]$;
6: $step = data(i + window\_size{:}\ i + window\_size + step\_length - 1)$
7: $y = [y; step]$
8: **end for**
9: **return** x, y; [1]

We created the ANN, SVR, RF, and LSTM models on a Windows 10 computer using the WEKA 3.8.5 platform. The ANN model specifically made use of a single hidden layer network structure with a learning rate of 0.01. The poly kernel function with a learning rate of 0.001 was utilized by the SVR model. The RF model has a batch size of 128 and a learning rate of 0.001 [1].

5.3 Evaluation Metrics

In this work, we used three widely used assessment metrics to evaluate the performance of TSTGCN, ANN, SVR, RF, and LSTM models in terms of prediction accuracy. The mean absolute deviation (MAE), root mean square error (RMSE), and mean absolute percentage error (MAPE) are some examples of these measurements. The equations used to calculate these metrics are as follows:

$$\text{MAE} = \frac{1}{n}\sum_{i=1}^{n}\left|x_i - \hat{x}_i\right| \quad (1)$$

$$\text{RMSE} = \sqrt{\frac{1}{n}\sum_{i=1}^{n}\left(x_i - \hat{x}_i\right)^2} \quad (2)$$

$$\text{MAPE} = \frac{1}{n}\sum_{i=1}^{n}\left|\frac{x_i - \hat{x}_i}{x_i}\right| \times 100\% \quad (3)$$

where x_i is the actual value, $\hat{x}_i$ is the predicted value, and n is the number of test samples [1].

5.4 Result Analysis

The processed station delay data set was used to compare TSTGCN to four baseline models. Table 2 [1] presents the outcomes of the performance of the train arrival delay forecast for the following hour. The highest scores among them were attained by our TSTGCN model. The prediction outcomes of conventional machine learning and deep learning techniques may be seen to be:

Traditional machine learning and deep learning approaches typically perform below par, demonstrating weak modeling skills for complicated and nonlinear train delay spatiotemporal network data. The best MAE, RMSE, and MAPE values among the four baseline models are 0.4447 (SVR), 0.8299 (SVR), and 53.6608 (ANN), respectively. With scores of 0.16, 0.45, and 34.36 for MAE, RMSE, and MAPE, respectively, TSTGCN performs better than the baselines and improves by around 64%, 46%, and 36% in contrast to SVR and ANN, the best baselines. Additionally, TSTGCN performs about 75, 50, and 46% better than the poorest scores of 0.6309, 0.9039, and 63.7141. In conclusion, when compared to earlier sophisticated models, our TSTGCN model performs better.

We use line charts to show the performance of the five techniques across various prediction periods in order to better evaluate the efficacy of our TSTGCN model in short- and long-term prediction. The five approaches' metric scores are used for estimating how many arrival delays will occur in the following one to twelve hours. It can be seen that when the forecast time lengthens, the performance of each approach shifts. The prediction complexity often rises with time, resulting in increasing mistakes. High amounts of errors are continuously seen in ANN, SVR,

Table 2 Comparison of Five Model

	TSTGCN	ANN	SVR	RF	LSTM
MAE	**0.1600**	0.6309	0.4447	0.6146	0.4960
RMSE	**0.4500**	0.8499	0.8299	0.9039	0.8507
MAPE	**34.3600**	53.6608	63.7141	54.9183	61.4930

RF, and LSTM. While LSTM performs better over time, RF's prediction skills drops abruptly as prediction time increases. Contrarily, our TSTGCN regularly exhibits higher performance throughout a range of durations. The inaccuracy persists even in long-term prediction [1].

6 Conclusion

This paper suggests a TSTGCN model with an attention mechanism for forecasting the cumulative impact of train arrival delays in railway dispatching based on the spatiotemporal features and dynamic spatiotemporal correlations evident in high-speed train operating data. The TSTGCN model successfully captures the spatiotemporal properties of train operation data through the integration of spatiotemporal attention mechanism and spatiotemporal convolution, leading to more precise predictions. Using MAE, RMSE, and MAPE as evaluation metrics, our TSTGCN model is contrasted with ANN, SVR, RF, and LSTM models at the experimental stage. The testing findings unmistakably show that our TSTGCN model works better than other models in estimating the overall impact of train arrival delays for railway dispatching.

References

1. Zhang D, Peng Y, Zhang Y, Wu D, Wang H, Zhang H, Train time delay prediction for high-speed train dispatching based on spatio-temporal graph convolutional network
2. Jing S (2019) Research on delay prediction of high speed railway train based on data analysis. Ph.D. dissertation, Southwest Jiaotong University, Chengdu, China
3. Xu RG (2015) Multiple traffic jams in full velocity difference model with reaction-time delay. Int J Simul Model 14(2):325–334
4. Milinković S, Marković M, Vesković S, Ivić M, Pavlović N (2013) A fuzzy Petri net model to estimate train delays. Simul Model Pract Theory 33:144–157
5. Marković N, Milinković S, Tikhonov KS, Schonfeld P (2015) Analyzing passenger train arrival delays with support vector regression. Transp Res C, Emerg Technol 56:251–262
6. Corman F, Kecman P (2018) Stochastic prediction of train delays in realtime using Bayesian networks. Transp Res C, Emerg Technol 95:599–615
7. Lessan J, Fu L, Wen C (2019) A hybrid Bayesian network model for predicting delays in train operations. Comput Ind Eng 127:1214–1222
8. Yaghini M, Khoshraftar MM, Seyedabadi M (2013) Railway passenger train delay prediction via neural network model. J Adv Transp 47(3):355–368
9. Ping H, Chao W, Zhongcan L, Yuxiang Y, Qiyuan P (2019) A neural network model for real-time prediction of high-speed railway delays. China Saf Sci J 29(S1):24–30
10. Zhaoxia Y, Zhongying D (1995) Simulation system of train delay propagation. J China Railway Soc 17(2):17–24
11. Xin W, Lei N, Wen-Jun L (2014) Study on robustness of high-speed train working diagram based on EMU utilization. Railway Transp Economy 36(11):50–55
12. Kecman P, Goverde RM (2015) Online data-driven adaptive prediction of train event times. IEEE Trans Intell Transp Syst 16(1):465–474
13. Carey M, Carville S (2000) Testing schedule performance and reliability for train stations. J Oper Res Soc 51(6):666

14. Chao W, Xiong Y, Ping H, Zhongcan L, Youhua T (2018) Review on conflict detection and resolution on railway train operation. China Saf Sci J 28(S2):70–77
15. Pu Z, Lingyun M, Baoxu L (2019) Prediction of high-speed railway train delay evolution based on machine learning. Electr Eng 20(z1):1–8
16. Yan S, Xiong Y, Lin D (2018) Spatial temporal graph convolutional networks for skeleton-based action recognition. Proc AAAI Conf Artif Intell 32(1):1–9
17. Seo Y, Defferrard M, Vandergheynst P, Bresson X (2018) Structured sequence modeling with graph convolutional recurrent networks. In: Proceedings of International Conference Neural Information Processing. Springer, Cham, Switzerland, pp 362–373
18. Guo S, Lin Y, Feng N, Song C, Wan H (2019) Attention based spatial-temporal graph convolutional networks for traffic flow forecasting. Proc AAAI Conf Artif Intell 33:922–929
19. Ning F, Sheng-Nan G, Chao S, Qi-Chao Z, Huai-Yu W (2019) Multicomponent spatial-temporal graph convolution networks for traffic flow forecasting. J Softw 30(3):759–769

Chapter 55
Logical Formalization for a HMDCS-UV

Salima Bella and Ghalem Belalem

1 Introduction

Maritime activity, particularly shipping and the transport of oil and other petroleum products, can contribute to water pollution in the form of oil spills. Oil spills are a result of oil leaks or spills from ships, offshore drilling platforms, or storage facilities. They can have severe environmental impacts, including harm to marine wildlife, damage to coastal habitats, and disruption of fishing and tourism industries. A concrete example is the 1989 Exxon Valdez oil spill in Alaska. The tanker spilled approximately 11 million gallons of crude oil into the Prince William Sound, causing widespread environmental damage to the region's coastline, wildlife and fishing industry.

In response to these risks, regulations have been implemented by many countries to reduce the likelihood of oil spills and mitigate their impact. These regulations can include requirements for ships to have spill response plans in place and for oil companies to invest in spill response technology. The planning of an oil spill response operation involves a coordinated effort among various organizations and agencies. It is documented that the cleanup costs rises at least ten times when the oil reaches a significant portion of the shore (Psaraftis and Tharakan [2]). This means that a well-planned operation should confront the oil when this is still in the sea and diminish its potential impact on the nearby coasts.

On the other side, Swarm robotics is a branch of robotics that involves the coordination of multiple robots to achieve a common goal. In the context of oil spill response, swarm robots can be used to cleanup contaminated areas more efficiently

S. Bella (✉)
Department of Computer Science, Université de Relizane, Relizane, Algeria
e-mail: salima@univ-relizane.dz

G. Belalem
Department of Computer Science, Université Oran 1, Oran, Algeria

A. Yadav et al. (eds.), *Proceedings of International Conference on Paradigms of Communication, Computing and Data Analytics*, Algorithms for Intelligent Systems,
https://doi.org/10.1007/978-981-99-4626-6_55

than traditional methods. Among the advantages of using swarm robots for oil spill response, increased efficiency by dividing the cleanup task among multiple robots, the cleanup can be completed faster and more efficiently than if done by a single robot or human.

Based on these concepts, this article presents and defines a logical formalization for the Hybrid Monitoring, Detection, and Cleaning System for unmanned vehicles (HMDCS-UV) (Bella et al. [1]). The proposed HMDCS-UV is based on a cooperative hierarchical architecture for heterogeneous semi-autonomous air-sea unmanned vehicles (UAV, USV swarm) to monitor, detect polluted/dirty zones (hydraulic spills), and clean them. Thus, the proposed logical formalization, inspired by the automated planning methods cited by the author Ghallab et al. [3, 4], allows simulating the behavior of these unmanned vehicles in different domains.

The paper is organized as follows: Sect. 2 presents some related work in the literature; Sect. 3 presents the proposed system for the different unmanned vehicles with the different steps in general; Sect. 4 illustrates the logical formalization proposed to simulate the HMDCS-UV proposal; finally, Sect. 5 concludes this work and cites some future work.

2 Related Works

In this section, various approaches for action and planning are proposed in the literature. For example, in Wang et al. [5] a real-time game-theoretic planning algorithm is proposed that supports a two-robot race scenario where robots have competing objectives and interact via shared collision avoidance constraints. The experimental results successfully validate the roach's applicability in a real-world scenario involving four quadrotor air robots and physical robotic hardware. Thus, an another approach is based on the work proposed in Patra et al. [6]. The authors propose using a single representation, the operational model, for both acting and planning by reasoning directly with the actor's operational models. The agent does not start a plan and then invoke the execution platform when needed.

A new action and planning system is proposed in Patra et al. [7] to integrate actions and planning using the participants' operating models. It uses a hierarchical task-oriented operational representation with an expressive, general-purpose language to provide a rich control structure for closed-loop online decision-making. The experiments show significant advantages in the efficiency of the acting and planning system. Then, a formalism is developed in Patra et al. [8] to synthesize hierarchically organized controllers to coordinate the agent's interactions. The formalism models agents as hierarchical input/output automata, and models a system of interacting agents as parallel compositions of automata.

A new system for integrating action and planning using a hierarchical refinement operation model is presented in Patra et al. [9]. The authors demonstrate that the proposed integration of action, planning, and learning can yield significant benefits reflected in improved efficacy or robustness. Thus, a new planning and learning algorithms are introduced in Patra et al. [10] to guide a acting procedure on what methods to use. The experimental results show that learning strategies and the acting procedure in different test domains using two metrics of efficiency and success rate.

In this sense, the present work aims to provide a logical formalization for the Hybrid Monitoring, Detection, and Cleaning System for unmanned vehicles (HMDCS-UV). The proposed HMDCS-UV allows for the cooperation and coordination of different unmanned vehicles to monitor and cleanup the polluted marine regions.

3 Presentation of HMDCS-UV System

This study presents a logical formalization of the to monitor, detect, and clean dirty marine regions system for unmanned vehicles (HMDCS-UV) [1]. The HMDCS system proposed in [1] is based on a process and a procedure for preventing and combating oil pollution in the port of Arzew (West Coast of Algeria). Thus, the improvements of the organization chart proposed for the Arzew Port Company (APC) are presented in Fig. 1 (further details in [1]).

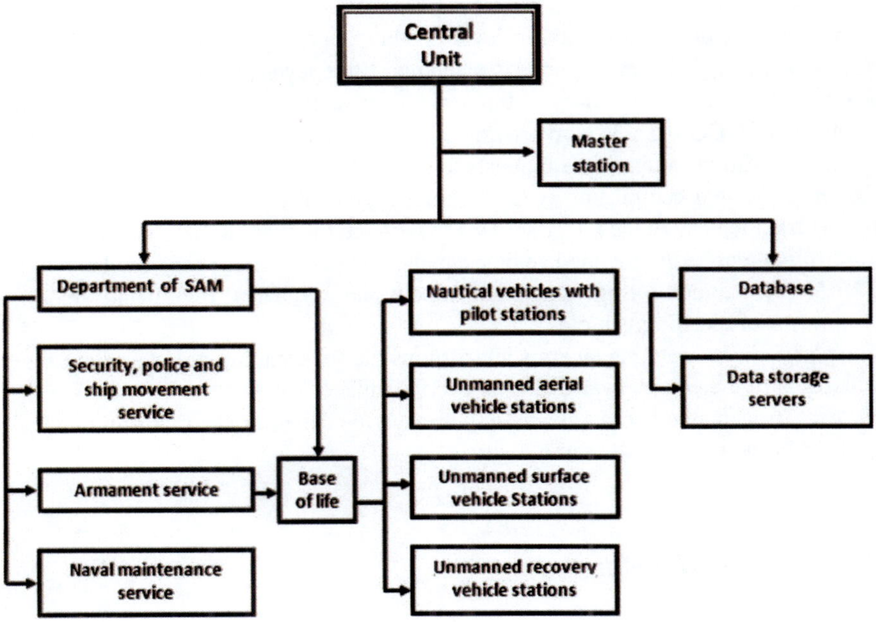

Fig. 1 Hybrid system central unit hierarchy [1]

3.1 System Architecture

The HMDCS-UV comprises a hierarchical hybrid architecture (Fig. 2) [1]. This architecture consists of a marine force base, a central unit, a monitoring vehicle and a swarm of cleaning vehicles. The marine force represents the base of the Territorial Grouping of Coast Guards (TGCG); it includes the coast guards who work in collaboration with the port to execute the various pollution control operations (surveillance, intervention, and response) in the port. The central unit comprises a SAM department (Security, Armament, and Maintenance Department), a master room, a database, and a base of life. The master room includes a general coordinator who consults and stores data in the database, and interacts with the TGCG base and the SAM department to execute the various tasks. The Very High Frequency (VHF) marine radio tool and other means of communication such as WiFi are used so that the entities of the system can exchange messages and information. For example, the SAM department interacts with its agents in the base of life by VHF, and with the master room by VHF and WiFi.

3.2 Hierarchical Decision-Making Level

The hierarchical decision-making level on the proposed hybrid architecture, presented by Fig. 2. This architecture is made up of five levels, which are:
First level: the central memory of the port is represented by the marine force which includes a TGCG base. It works with the port to monitor it, intervene in pollution control operations and initiate requests to the central unit (CU).
Second level: the central unit (CU) includes a general coordinator who represents the central memory of the CU. Its role is to perform the various tasks (launching and controlling manned/unmanned vehicles, using and storing data in the database, etc.).
Third level: the monitoring vehicle allows to monitor a marine region and supervise its swarms of cleaning vehicles.
Fourth level: the cleaning swarms are responsible for cleaning polluted/dirty zones according to the energy availability of each member.
Fifth level: each member of a cleaning swarm has local memory and can communicate with other members of the same swarm.

3.3 Applied Modeling

The modeling applied to the working environment is described in detail in the work [1]. The table of entity acronyms and the elements used in used in the HMDCS-UV are listed below (Table 1).

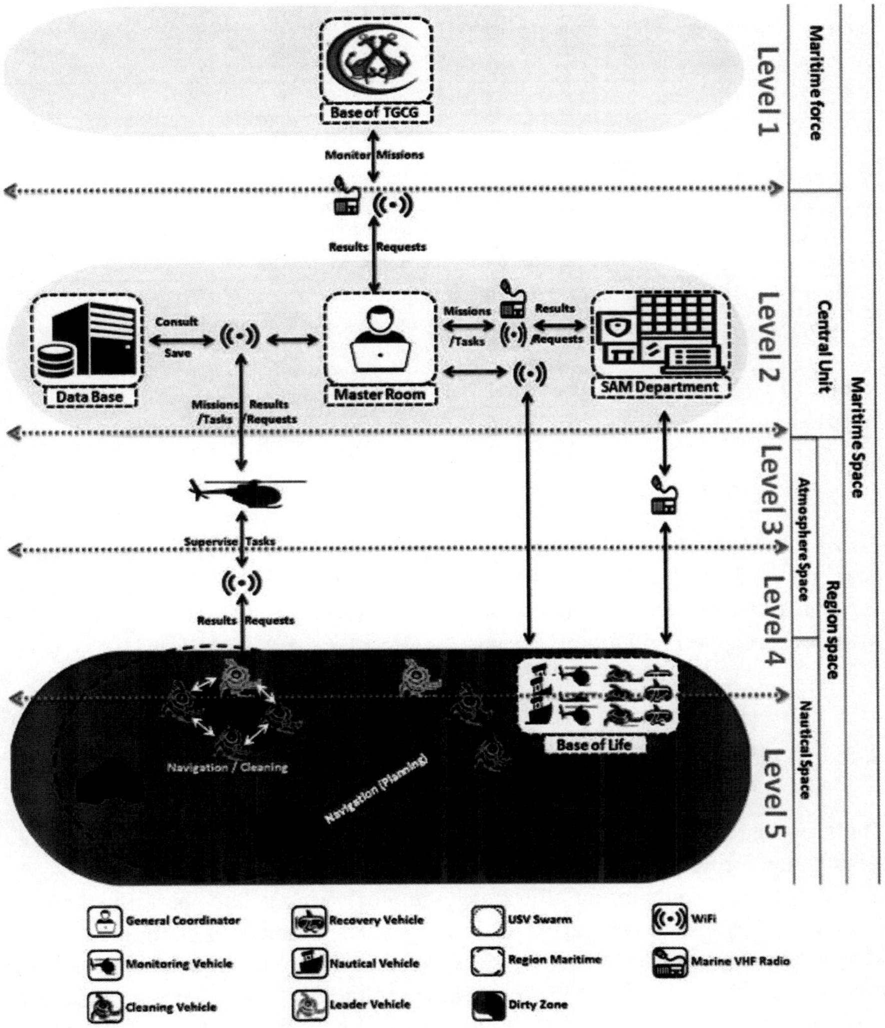

Fig. 2 Hierarchical hybrid architecture [1]

Set of tasks: allocation task, preparation task, monitoring task, cleaning task, cleaning supervision task, and launch task.
Set of vehicles: UAV_{mr}, USV_{cz}, $Leader_{cz}$, $Vehicle_{nt}$, and $Vehicle_{rec}$.
Set of agents: $General_{crd}$, Sup_{mr}, $Leader_{cz}$, and SAM_{agent}.
Set of regions: the marine region is within in the monitored marine space. It is composed of an atmospheric subspace and a nautical subspace.
Set of base of life: a base of life is an zone (a boat, a ship, etc.) to store a fixed number of manned/unmanned vehicles.

Table 1 Acronym description table [1]

Acronyme	Description
$TGCG$	Territorial Grouping of Coast Guards
$General_{crd}$	General coordinator (computer containing coordination software) of the marine space
UAV_{mr}	Unmanned Aerial Vehicle to monitor a region
USV_{cz}	Unmanned Surface Vehicle to clean a zone
Sup_{mr}	UAV_{mr} type supervisor to supervise UAVs/USVs
$Leader_{cz}$	Leader of USV_{cz} (intermediate between the swarm of USV_{cz} and Sup_{mr})
$Vehicle_{rec}$	Unmanned recovery vehicle (USV_{cz} type) to recover failed unmanned vehicles
SAM_{agent}	Security, Armament and Maintenance department agent
$Vehicle_{nt}$	Floating craft to board port officers, pilots, anti-pollution equipment, etc.

Set of database: the data of the marine space entities are stored in the database (servers).
Set of dirty zones: the set of dirty zones represents the part of water pollution. This work focuses on oil pollution.

3.4 Key Steps of the HMDCS-UV

The proposed HMDCS-UVs comprises different steps, but it is based on two key steps: "monitoring" and "cleaning", shown in Fig. 4 [1].

The first step "monitoring" illustrates the monitoring actions performed by the UAV_{mr}. These actions are carried out in two phases, Monitoring configuration, and Monitoring execution. The first phase is executed when the $General_{crd}$ assigns a monitoring vehicle UAV_{mr} for each marine region. The second phase consists of two sub-phases: Preparation/Navigation "base of life in the region" and Navigation in the region [1]. The first sub-phase presents a preparation method of a UAV_{mr} with a set of parameters by SAM_{agent} via the $General_{crd}$; a method for planning the trajectory of a UAV_{mr} to arrive at its region using its field of view, its sensory sensors (a camera and an ultrasonic sensor), an atmosphere space map, a nautical space map before arriving at the region, a rectilinear movement and a proposed algorithm based on Cartesian coordinates "Planning to the region" (Algorithm 1) [1], and a method of processing the captured data.

The second sub-phase is executed when the UAV_{mr} arrives at its region, it can plan its trajectory using its sensory sensors, the atmospheric map, and the "Boustrophedon modified [1]" algorithm proposed based on the "swipe" movement. Thus, UAV_{mr} can update its nautical map based on a proposed supervision and detection solution. This solution is based on an unsupervised classification "Kmeans Clustering" method

[1] to process the captured data using the "swipe" movement. In each sub-phase, the UAV_{mr} sweeps repeatedly and saves the energy consumed in its characteristics list until the end of the cleaning.

The second step "cleaning" describes the three-phase cleaning procedure; Cleaning configuration, Cleaning operation, and Cleaning termination [1]. The first phase presents a proposed cleaning process so that the agents of the HMDCS-UV can determine the cause of the pollution and limit its flow. In addition, a trajectory planning solution based on Cartesian coordinates (PCCA: Proposed Cartesian Coordinates Algorithm) [1] is proposed in this phase. This solution allows the swarms of USV_{cz} to move and plan their paths toward their dirty zones. The operation phase allows the swarm of USV_{cz} to move into the dirty zone and clean it by offering two solutions. The first solution is an improvement of the proposition (Algorithm 2) cited in Bella et al. [1, 11] by adding the notion of step between each USV_{cz}. The second solution is based on the method presented in Zahugi et al. [12] by proposing to clean the zone in two parts with two groups constituting the swarm. The last step consists of identifying the termination of the cleaning process [1].

4 Logical Formalization

4.1 A Conceptual Planning Model

A conceptual model is a simple theoretical device to describe the main elements of a problem [3]. Most of the planning approaches described in Ghallab et al. [3, 4] are based on model of state-transition systems. The technical terms of this formalization defined in this section are given in a general way (further details in [11]).

Example 1 presents the state-transition systems for the following applications: UAV-Monitoring (or UAV-M), Swarm (USV)-Cleaning (or SUSV-C) and USV (Substitute)-Cleaning (or USVS-C). Table 2 presents these three applications.

4.2 Application Execution: "UAV-M, SUSV-C and USVS-C"

The planning techniques are used in the three applications "UAV-Monitoring (UAV-M), Swarm (USV)-Cleaning (SUSV-C), and USV (Substitute)-Cleaning (USVS-C)", which complement each other. The sets of constant symbols is defined as representing an abstract version of these applications (Bella et al. [11, 13, 14]), namely a set of: regions (R_r), locations for each region (L_{lr}), base of life (B_{br}), dirty zones (Z_{zl}), monitoring vehicle (UAV_{mr}), cleaning vehicles swarm ($SUSV_{cz}$), cleaning vehicles in discharge state ($USVD_{cz}$), cleaning vehicles in free state ($USVF_{cz}$), cleaning vehicles in prepared state ($USVP_{cz}$) and a set of cranes (C_{nb}).

Table 2 Description of the applications: UAV-M, SUSV and USVS-C

Application	System representative	States	Actions	Action(State ⇄ State) Action
UAV-M (Fig. 15 [11])	A region involving: (2) locations, (1) dirty zone, (1) base of life, (1) UAV_{mr}	s0, s1, s2, s3	stayinbase, flapoutbase, move1∧start-monitor, move2∧end-monitor, discover, undiscover	*flapoutbase(s0 ⇄ s1) stayinbase, *move1∧start-monitor(s1 ⇄ s2) move2∧end-monitor, *discover(s2 ⇄ s3) undiscover
SUSV-C (Fig. 16 [11])	A region involves: (2) locations, (1) dirty zone, (1) base of life, (1) crane, (3) USV_{CZ}	s0, s1, s2, s3	take, put,start-clean,end-clean,move1, move2	*put(s0 ⇄ s1) take,*move1(s1 ⇄ s2) move2, *start-clean(s2 ⇄ s3) end-clean
USVS-C (Fig. 17 [11])	A region involves: (3) locations, (2) dirty zones, (1) base of life, (1) crane, (1) $USVD_{CZ}$	s0, s1, s2, s3, s4,s5	moveD1∧move2, moveD2∧move1, moveD1∧start-clean, moveD2∧end-clean, takeD∧move2, move1∧put,move1, move2, put, take	*moveD1∧move2(s0 ⇄ s1) moveD2∧move1, *moveD2∧end-clean(s1 ⇄ s2) moveD1∧start-clean,*takeD∧move2(s2 ⇄ s3) move1∧put,*move2(s3 ⇄ s4) move1,*take(s4 ⇄ s5) put

The instances of predicates are noted of applications, which represents the relationships that do not changes over time [11, 14]: *adjacent(E, E')*, *belong(C, L)* et *belong(B, R)*.

The instances of the following predicates for applications are shown, which represents the relationships that changes over time [11, 14]: *at(NAME, E)* where Vehicle *NAME* = *{NAME1*(UAV_{mr}, USV_{cz}), *NAME2*($USVF_{cz}$, $USVP_{cz}$, $USV\ D_{cz}$, $SUSV_{cz}$)} is currently at location E; *occupied(L)* where location L is already occupied by *NAME2* vehicles, $Z = \{Z_{zl}\}$ and *B*; *start-clean(NAME, Z)* where Vehicles *NAME1*.USV_{cz}, *NAME2*.$USVF_{cz}$ and *NAME2*.$USVP_{cz}$ is ready to clean a dirty zone *z*; *end-clean(NAME, Z)* where *NAME1*.USV_{cz} and *NAME2*.$USVF_{cz}$ is finished the cleaning of *Z*; *start-monitor(NAME, R)* where *NAME1*.UAV_{mr} is currently ready to start monitoring in *R*; *end-monitor(NAME, R)* where *NAME1*.UAV_{mr} has now completed monitoring of its *R*; *replace (Name, Name)* where *NAME2*.$USV\ F_{cz}$ replaces the *NAME2*.$USV\ D_{cz}$; *notreplace(NAME, NAME)* where *NAME2*.$USV\ F_{cz}$ does not replace *NAME2*: $USVD_{cz}$; discover (NAME, Z) where *Z* is discovered by the drone *NAME1*.UAV_{mr}; *undiscover(NAME, Z)* where *R* does not contain a *Z* or the *Z* is not discovered by the *NAME1*.UAV_{mr}; *supervise(NAME, NAME)* where *NAME1*.UAV_{mr} supervises the swarm *NAME2*. $SUSV_{cz}$; *notsupervise(NAME, NAME)* where *NAME1*.UAV_{mr} does not supervise the swarm *NAME2*. $SUSV_{cz}$; *holding(C, NAME)* where *C* is currently holding vehicle *NAME1*.USV_{cz} and $NAME2$; *stayinbase(NAME, B)* where *NAME1*. UAV_{mr} is currently in the base of life *B* and *flapoutbase(NAME, B)* where *NAME1*.UAV_{mr} is outside its base *B*. It flies in the air.

The possible actions in the three applications UAV-M, SUSV-C, and USVS-C, are : *flapoutbase(NAME, B)*; *discover(NAME, R, Z)*; *start-monitor(NAME, R)*; *start-*

clean(NAME, Z); *end-clean(NAME, Z)*; *end-monitor(NAME, R)*; *stayinbase(NAME, B)*; *undiscover(NAME, R, Z)*; *move(NAME, E1, E2)* where Vehicle *NAME* moves from a location $E1=\{L_{lr}, R_r\}$ to some adjacent and unoccupied location $E2=\{L'_{lr}, R'_r\}$; *moveD(NAME, E1, E2)* where $NAME2.USVD_{cz}$ moves from a location $E1 = \{L_{lr}, R_r\}$ to some adjacent and unoccupied location $E2= \{L'_{lr}, R'_r\}$; *take(NAME, C, R)* where empty *C* takes the $NAME1.USV_{cz}$, $NAME2.USVF_{cz}$ and $NAME2.USVP_{cz}$ found in the same *R*; *takeD(NAME, C, R)* where empty *C* takes a $NAME2.USVD_{cz}$ found in the same R.

4.3 Representations for Classical Planning

The ways to represent classical planning problems [3] are: Set-theoretic, Classical, and State-Variable representation. The present work focuses on the first and second representation (set-theoretic and classical). These representations are applied to our presentation [11, 13, 14].

Set-theoretic representation. The description of this representation is already defined in [3]. The Example 2 of a Set-Theoretic representation of the UAV- M application combined with the SUSV-C application and a representation of the SUSV-C application combined with the USVS-C application are illustrated in Example 1 (Bella et al. [11, 15]).

Example 2 Let $L = \{p_1,..., p_n\}$ be a finite set of proposition symbols for the:

UAV-M application combined with SUSV-C application: L = {*on_base, inair, on_watersurface, on_zone, holding, at_location1, at_location2*}, où: *inair* means that $NAME1.UAV_{mr}$ drone is in the air; *on_watersurface* means that $NAME1.USV_{cz}$ vehicle is on the water surface; *on_base* means that $NAME1.USV_{cz}$ / $NAME1.UAV_{mr}$ remains on the base of life; *on_zone* means that $NAME1.USV_{cz}$ is found in its dirty zone; *holding* means that crane *C* is holding $NAME1.USV_{cz}$; *at_location*1 means that $NAME1.USV_{cz}$ or $NAME1.UAV_{mr}$ is at location 1; *at_location*2 means that $NAME1.USV_{cz}$ or $NAME1.UAV_{mr}$ is at location 2.

USVS-C application combined with SUSV-C application: L = {*on_base, on_watersurface, on_zone, holding, at_location1, at_location2, at_location3*}, where: *on_watersurface* means that $NAME1.USV_{cz}$ or $NAME2$ vehicle is on the water surface; *on_base* means that $NAME1.USV_{cz}$ or $NAME2$ remains on the base of life; *on_zone* means that $NAME1.USV_{cz}$ or $NAME2$ is found in its dirty zone; *holding* means that crane *C* is holding $NAME1.USV_{cz}$ or $NAME2$; *at_location*1 means that $NAME1.USV_{cz}$ or $NAME2$ is at location 1; *at_location*2 means that $NAME1.USV_{cz}$ or $NAME2$ is at location 2; *at_location*3 means that $NAME1.USV_{cz}$ or $NAME2$ is at location 3.

Each state *S* is a subset of *L*, where for the:

UAV-M application combined with SUSV-C application: $S = \{s_0,..., s_6\}$, where: $s_0 = \{$*on_base, at_location1*$\}$; $s_1 = \{$*inair, at_location1*$\}$; $s_2 = \{$*inair, at_location2*$\}$; $s_3 = \{$*on_watersurface, at_location1*$\}$; $s_4 = \{$*on_watersurface, at _location2*$\}$; $s_5 = \{$*on_zone, at_location2*$\}$; $s_6 = \{$*holding, at_location1*$\}$.

USVS-C application combined with SUSV-C application: $S = \{s_0,..., s_5\}$, where: $s_0 = \{$*on_base, at_location1*$\}$; $s_1 = \{$*on_watersurface, at_location1*$\}$; $s_2 = \{$*on_ watersurface, at_location2*$\}$; $s_3 = \{$*on_watersurface, at_location3*$\}$; $s_4 = \{$*on_zone, at_location3*$\}$; $s_5 = \{$*holding, at_location1*$\}$.

The action $a \in A$ is a triple of subsets of L, which $a = (precond(a), effects^{-}(a), effects^{+}(a))$. The set $precond(a)$ is the preconditions of a, and the sets $effects^{+}(a)$ and $effects^{-}(a)$ are the effects of a. So, for the:

UAV-M application combined with SUSV-C application: $A = \{$*stayinbase, flapout-base, put, take, move1 , move2, discover, undiscover, supervise, notsupervise, start-monitor, end-monitor, start-clean, end-clean*$\}$, où: *stayintbase* = ({inair, at_location1}, {inair}, {on_base, at_location1}); *flapoutbase* = ({on_base}, {on_base}, {out-base}); *take* = ({on_watersurface, at_location1}, {on_watersurface}, {holding, at_ location1}), *put* = ({on_base, at_location1}, {holding}, {on_watersurface, at_loca-tion1}); *move1* ∧ *start-monitor* = [({at_location1}, {at_location1}, {at_loca tion2}) ∧ ({outbase, at_location1}, {at_location1}, {outbase, at_location2})]; *move2* ∧ *end-monitor* = [({at_location2}, {at_location2}, {at_location1}) ∧ ({outbase, at_ location2}, {outbase}, {on_base, at_location1})]; *discover* = ({outbase, at_loca tion2}, {outbase}, {at_location2}); *undiscover* = ({outbase, at_location2}, {at_loca tion2}, {outbase}); *supervise* = ({outbase}, {outbase}, {at_location2}); *notsuper-vise* = ({outbase, at_location2}, {outbase, at_location1}, {on_base}); *start-clean* = ({underwater, at_location1}, {underwater}, {on_zone, at_location2}), *end-clean* = ({on_zone, at_location2}, {on_zone}, {underwater, at_location2}); *move1* = ({at_location1}, {at_location1}, {at_location2}); *move2* = ({at_location2}, {at_ location2}, {at_location1}).

USVS-C application combined with SUSV-C application: A = *{put, putD, take, takeD, replace, notreplace, move1, move2, moveD1, moveD2, start-clean, end-clean}*, où: *take* = ({on_watersurface, at_location1}, {on_watersurface}, {hold-ing, at_location1}); *put* = ({on_base, at_location1}, {holding}, {on_watersurface, at_loca tion1}); *moveD1* ∧ *start-clean* = [({at_location1}, {at_location1}, {at_ location2}) ∧ ({underwater, at_location1}, {underwater}, {on_zone, at_location3})]; *moveD2* ∧ *end-clean* = [({at_location2}, {at_location2}, {at_location1}) ∧ ({on_zone, at_location3}, {on_zone}, {underwater, at_location3}); *replace* = ({underwater, at_location2}, {underwater}, {on_zone, at_location3}); *notreplace* = ({underwater, at_location2}, {at_location2}, {underwater}); *moveD1* ∧ *move2* = [({at_location2}, {at_location2}, {at_location3}) ∧ ({at_location3}, {at_location3}, {at_location2})]; *moveD2* ∧ *move1* = [({at_location3}, {at_location3}, {at_location2}) ∧ ({at_ location2}, {at_location2}, {at_location3})]; *takeD* ∧ *move2* = [({on_watersurface, at_location1}, {on_watersurface}, {holding, at_location1}) ∧ ({at_location3}, {at_ location3}, {at_location2})]; *move1* ∧ *putD* = [({at_location2}, {at_location2}, {at_location3}) ∧ ({holding, at_location1}, {holding}, {on_watersurface, at_

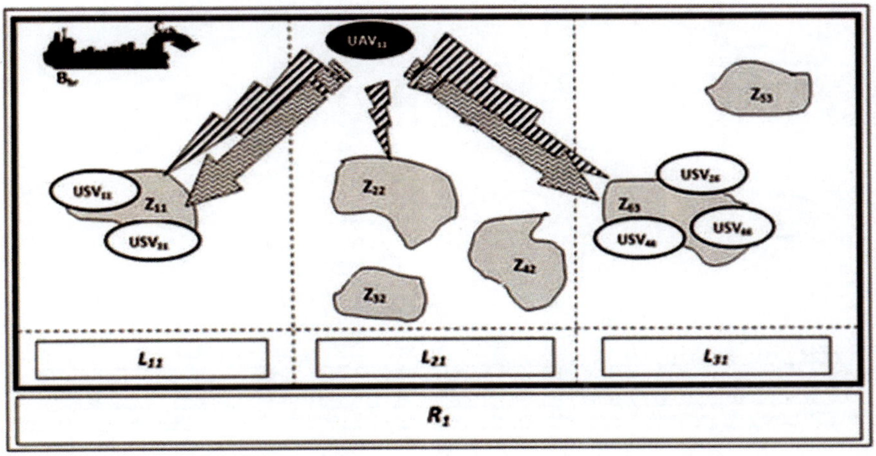

Fig. 3 UAV-M planning application

location1}); *move1* = ({at_location1}, {at_location1}, {at_location2}); *move2* = ({at_location2}, {at_location2}, {at_location1}).

Classical representation. this representation is already defined in [15]. Example 3 illustrates a classic representation of the UAV-M application and another for the SUSV-C application combined with the USVS-C application defined in Example 1 [13, 14].

Example 3 we suppose that for:
– *UAV-M planning application* is represented with one UAV_{mr} (UAV_{11}), five USV_{cz} (USV_{11}, USV_{26}, USV_{31}, USV_{46}, USV_{66}), one region (R_1), six dirty zones Z_{11}, Z_{22}, Z_{32}; Z_{42}, Z_{53}, Z_{63}), three locations (L_{11}, L_{21}, L_{31}) and one base (B_{11}). The set of constant symbols is a UAV_{mr}, a region, a set of locations, and set of USV_{cz}. Figure 3 illustrates a state of this application.
S3 = {belong(B_{11}, R_1), adjacent(L_{11}, L_{21}), adjacent(L_{21}, L_{11}), adjacent(L_{21}, L_{31}), adjacent(L_{31}, L_{21}), flapoutbase(UAV_{11}, B_{11}), start-monitor (UAV_{11}, R_1), discover(UAV_{11}, Z_{11}), discover(UAV_{11}, Z_{22}), discover(UAV_{11}, Z_{63}), undiscover (UAV_{11}, Z_{42}), undiscover(UAV_{11}, Z_{53}), undiscover(UAV_{11}, Z_{32}), supervise(UAV_{11}, E_{11}), supervise(UAV_{11}, E_{31}), occupied(L_{11}), occupied(L_{21}), occupied(L_{31})}.

– *USVS-C planning application* combined with SUSV-C planning application is illustrated with a region, three locations, eleven USV_{cz}, a crane (C_{11}), and five dirty zones (Z_{11}, Z_{22}, Z_{42}, Z_{63}, Z_{53}). The set of constant symbols is defined by {R_1, L_{11}, L_{21}, L_{31}, $USVP_{34}$, USV_{22}, USV_{52}, $USVF_{62}$, USV_{74}, $USVD_{44}$, USV_{25}, USV_{55}, USV_{85}, USV_{66}, USV_{76}}. The state is represented by the Fig. 4.
S3 = {belong(C_{11}, L_{11}), adjacent(L_{11}, L_{21}), adjacent(L_{21}, L_{11}), adjacent(L_{21}, L_{31}), adjacent(L_{31}, L_{21}), holding(C_{11}, $USVP_{34}$), end-clean(USV_{52}, USV_{62}, USV_{22}, Z_{22}),

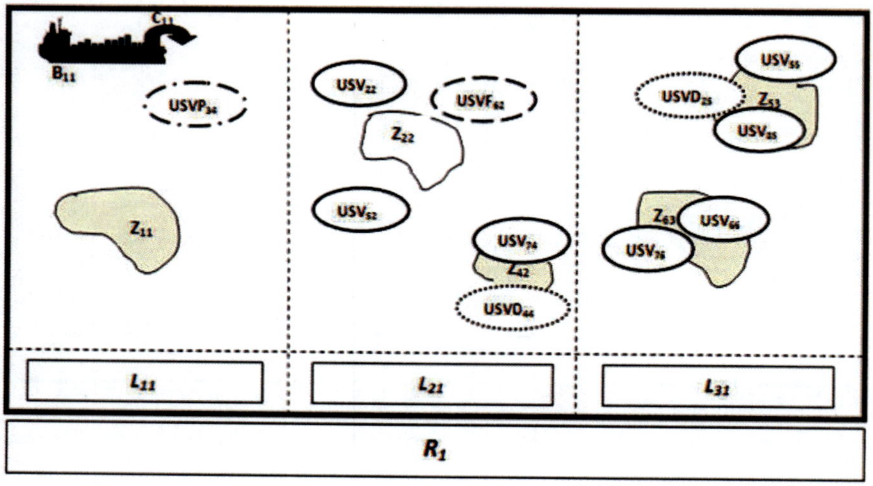

Fig. 4 USVS-C planning application combined with SUSV-C planning application

start-clean($USV_{74}, USVD_{44}, Z_{42}$), end-clean ($USV_{55}, USVD_{25}, USV_{85}, Z_{53}$), start-clean($USV_{66}$, USV_{76}, Z_{63}), at(USV_{22}, USV_{52}, $USVF_{62}$, USV_{74}, $USVD_{44}$ L_{21}), remplace($USVP_{34}, USVD_{44}$), at(USV_{34}, L_{11}), occupied(L_{11}), occupied(L_{21}), occupied(L_{31})}.

5 Conclusion and Future Work

This paper presented a logical formalization for the hybrid monitoring, detection, and cleaning system (HMDCS-UV) [1] of dirty marine regions, based on the cooperation and the coordination of heterogeneous semi-autonomous unmanned vehicles. This formalization, inspired by the automated planning methods cited by the author Ghallab et al. [3, 4], defines a conceptual model of classical planning for the various unmanned vehicles used in HMDCS-UV. Based on the work of Mallik Ghallab et al. , in future work, we intend to apply other classical planning methods by the different unmanned vehicles in performing monitoring and cleaning tasks in marine regions.

References

1. Bella S, Belbachir A, Belalem G, Benfriha H (2021) HMDCS-UV: a concept study of Hybrid monitoring, detection and cleaning system for unmanned vehicles. J Intell Robot Syst 102(44):1–35. https://doi.org/10.1007/s10846-021-01372-8
2. Psaraftis HN, Tharakan GG (1986) Optimal response to oil spills: the strategic decision case. Int J Oper Res 34(2):203–217. https://doi.org/10.1287/opre.34.2.203

3. Ghallab M, Nau D, Traverso P (2004) Automated planning: theory and practice. Elsevier, pp 1–635, ISBN 9780080490519
4. Ghallab M, Nau D, Traverso P (2014) The actor's view of automated planning and acting: a position paper. J Artif Intell 208(2014):1–17. https://doi.org/10.1016/j.artint.2013.11.002
5. Wang Z, Spica R, Schwager M (2019) Game theoretic motion planning for multi-robot racing. Int Symp Distrib Auton Robot Syst, pp 225–238. https://doi.org/10.1007/978-3-030-05816-6_16
6. Patra P, Ghallab M, Nau D, Traverso P (2019) Interleaving acting and planning using operational models. Association for the Advancement of Artificial Intelligence
7. Patra P, Ghallab M, Nau D, Traverso P (2019) Acting and planning using operational models. In: Proceedings of AAAI conference on artificial intelligence, 33(1):7691–7698. https://doi.org/10.1609/aaai.v33i01.33017691
8. Patra S, Traverso P, Ghallab M, Nau D (2018) Controller synthesis for hierarchical agent interactions. In: Annual conference on advances in cognitive systems (COGSYS). Cognitive Systems Foundation. United States, Stanford, pp 1–17
9. Patra S, Mason J, Ghallab M, Nau D, Traverso P (2021) Deliberative acting, online planning and learning with hierarchical operational models. J Artif Intell 299:1–68. https://doi.org/10.1016/j.artint.2021.103523
10. Patra S, Mason J, Kumar A, Ghallab M, Traverso P, Nau D (2020) Integrating acting, planning, and learning in hierarchical operational models. In: Proceedings of 30th International conference on automated planning and scheduling, vol 30, pp 478–487. https://doi.org/10.48550/arXiv.2003.03932
11. Bella S, Belbachir A, Belalem G (2020) HA-UVC: hybrid approach for unmanned vehicles cooperation. J Multiagent Grid Syst 16(1):1–45. https://doi.org/10.3233/MGS-200319
12. Zahugi EMH, Shanta MM, Prasad TV (2013) Oil spill cleaning up using swarm of robots. In: Meghanathan N, Nagamalai D, Chaki N (eds) Advances in computing and information technology. Advances in Intelligent Systems and Computing, vol 178, pp 215–224. Springer, Berlin, Heidelberg. https://doi.org/10.1007/978-3-642-31600-5_22
13. Bella S, Belbachir A, Belalem G (2019) A hybrid architecture for cooperative UAV and USV swarm vehicles. In: Renault É, Mühlethaler P, Boumerdassi S (eds) Machine learning for networking (MLN'2018), vol 11407, pp 341–363. Springer, Cham. https://doi.org/10.1007/978-3-030-19945-6_25
14. Bella S, Belbachir A, Belalem G (2020) A hybrid air-sea cooperative approach combined with a swarm trajectory planning method. Paladyn, J Behav Robot 11(1):118–139. https://doi.org/10.1515/pjbr-2020-0006
15. Bella S, Belbachir A, Belalem G (2018) A centralized autonomous system of cooperation for uavs-monitoring and usvs-cleaning. Int J Softw Innov (IJSI) 6(2):50–76. https://doi.org/10.4018/IJSI.2018040105

Chapter 56
SANKET—A Vision Beyond Gestures

Isha Gawde, Jisha Philip, Kanaiya Kanabar, Shilpa Tholar, and Shalu Chopra

1 Introduction

For the people who are deaf or hard of hearing, sign language plays a crucial form to communicate with the world. Worldwide, there are a lot of different sign languages in use, with American sign language (ASL) being the most used in the US. Despite the fact that sign language is used widely, communicating with deaf people and hearing people who are not familiar with sign language can be difficult. Systems for converting text into sign language have been created to close this communication gap. These programmes are designed to transform written information into a representation in sign language, either through animations or movies. Text analysis, lexicon mapping, gesture generation, and visualisation are some of the processes in the conversion process. There are approximately 466 million persons with hearing impairment or disabling hearing loss worldwide, according to the World Health Organization (WHO: World Health Organization 2018). Although not all of these people rely on sign languages as their primary form of communication, they are widely used. For example, there are an estimated 500,000 sign language users in the European Union and 151,000 British Sign Language users in the United Kingdom, according to the British people with hearing impairment association (BDA: British people with hearing impairment association, 2019). (EU: European Parliament 2018). People who have hearing loss or are hard of hearing use sign language, a non-verbal communication system that makes use of hand movements, facial expressions, and other body parts. Because sign languages lack a clearly defined structure or grammar, they are not or are only very marginally accepted outside of their own limited realm.

I. Gawde (✉) · J. Philip · K. Kanabar · S. Tholar · S. Chopra
Department of Information Technology, Vivekanand Education Society's Institute of Technology, Chembur, Mumbai, India
e-mail: 2019isha.gawde@ves.ac.in

A. Yadav et al. (eds.), *Proceedings of International Conference on Paradigms of Communication, Computing and Data Analytics*, Algorithms for Intelligent Systems,
https://doi.org/10.1007/978-981-99-4626-6_56

2 Objectives

2.1 Helping People with Hearing Impairment and People Who are Mute

A person's situation can be changed by taking a single little action of assistance. We can't imagine what these people with disabilities go through on a daily basis, but by developing this system, we hope to take a small step in their direction. Even a small amount of water can be used to start a lake.

2.2 Summarisation of News

News on the Websites is filled with a lot of content from their heads to their footers. Also, the body of those Websites contains such containers which suggest related ones that can be visited next. In this case, content that is extracted through Web scraping is not that useful. So, we'd like to introduce the feature which will summarise what has been extracted using Web scraping.

2.3 Easy-to-Use UI

The user interface (UI) is crucial to the efficient operation of any software. When done correctly, users are not even aware of it. Users cannot get over it to effectively use a product when it is not done in the right manner. Most designers adhere to interface design guidelines to ensure that the user interface is completed successfully. Software design is influenced by high-level notions known as interface design principles.

2.4 Converting Summarised News into Signs

Our main aim is to deliver signed text to the end user. Our Website creates a summary of the news article. This summary can then be converted into signed text. This will help people with hearing impairment or people who are mute to understand the news more clearly.

3 Literature Survey

Researchers' interest in sign language recognition has increased recently as a result of its potential to help people with hearing loss communicate with the general public. Alharbi et al. [1] offered a thorough analysis of deep learning-based sign language recognition. They explored the difficulties of dataset acquisition and sign language recognition while analysing several approaches and their drawbacks. Hoa et al. [2] used MobileNet and LSTM to develop a proposed sign language recognition system, which had a 90.67% accuracy rate. Onwukwe and Daramola [3] convolutional neural network (CNN) technology was used to create a real-time sign language recognition system that had an accuracy of 85.56%. Khan and Ngadi [4] achieved an accuracy of 97.16% using deep learning for hand gesture identification in sign language interpreting. Yang and Cheng [5] achieved an accuracy of 97.16% using deep learning for hand gesture identification in sign language interpreting.

Minocha et al. [6] developed a 90.62% accurate sign language recognition system using faster R-CNN and spatial-temporal LSTM. Mandal et al. [7] examined the effectiveness of several deep learning techniques, such as CNN and LSTM networks. With an accuracy of 95.6%, they claimed that CNN-LSTM performed better than previous methods. Chen et al. [8] suggested a spatial-temporal network-based and attention-based sign language recognition system. They achieved 98.62% accuracy.

Alghamdi and Alghamdi [9] created a hand gesture detection system for American sign language (ASL) utilising deep learning techniques. Duggan [10] used a CNN-based method to recognise ASL and attained a 94.6% accuracy rate. Dubey and Singh [11] built a real-time Indian sign language recognition system with 98.24% accuracy utilising CNN and LSTM networks. Wang et al. [12] proposed a multi-feature fusion and attention mechanism-based system for recognising signs that had a 98.5% accuracy rate.

Li et al. [13] suggested a multi-feature extraction and multi-task learning-based sign language recognition system. They had a 97.53% accuracy rate. Perera et al. [14] used deep learning to create a technique for Sinhala sign language recognition that has an accuracy of 88.2%. Mao et al. [15] suggested a deep learning model based on skeletal joint position for ASL identification that has a 98.4% accuracy rate. Azzaz and Ezzat [16] created a sign language recognition system with a 97.63% accuracy using a CNN trained by lifting wavelet transform. Nasir and Ahmad [17] presented a thorough analysis of deep learning's use in the recognition of sign language. They talk about the difficulties and various strategies used for sign language recognition systems. The dataset utilised for training and testing, the assessment criteria, and the shortcomings of the current systems are also included in the review. Wang et al. [18] proposed a deep learning and video-to-video translation-based sign language recognition system. They use deep learning to recognise spoken language after translating movies of sign language into spoken language. The accuracy of the suggested approach on the dataset for American sign language (ASL) was 87.5%. Alfaifi et al. [19] demonstrated a convolutional neural network (CNN)-based real-time sign language recognition system. On the Saudi sign language (SSL) dataset, they employed the Kinect camera to capture sign language motions and had a 91% accuracy rate. Tiwari and Nema [20] proposed a system for recognising sign language that makes use of transfer learning and a deep convolutional neural network for the purpose. On the Indian sign language (ISL) dataset, their method had a 92% accuracy rate. Peng and Yin [21] presented a hybrid deep learning network that extracts semantic features to recognise sign language. On the Chinese sign language (CSL) dataset, their system had an accuracy of 96.25%. Abdallah et al. [22] proposed a deep convolutional neural network-based attention mechanism for a sign language recognition system. On the Tunisian sign language (TSL) dataset, their method had a 95.5% accuracy rate. Panwar and Patel [23] offered a system for recognising sign language that combines convolutional neural networks with data augmentation methods. When applied to the Indian sign language (ISL) dataset, they attained an accuracy of 94.1%. Patel and Swain [24] conducted a comparison of various deep learning methods for recognising sign language. On the American sign language (ASL) dataset, they tested the performance of six alternative models and discovered that the CNN-based method had the best accuracy (96.8%). Ahmed and Albakri [25] presented a thorough analysis of deep learning-driven sign language recognition. The survey covers the most recent advancements in sign language recognition, including the sets of data utilised, the criteria employed for evaluation, and the shortcomings of the existing systems. Additionally, they offer a thorough evaluation of the effectiveness of the deep learning models utilised for sign language recognition.

4 Analysis of Design

4.1 Input

The user enters the url of the news article he wants to read and clicks on the submit button.

4.2 Data Preprocessing

Data extraction The url entered by the user is taken, and text is extracted from the news article by sraping the Web. For the purpose of extraction, we are using the Python library—Newspaper3k.The Newspaper3k package is a Python library used for Web scraping articles. It is built on top of requests and for parsing lxml.

Dataset We have used the 'News summary.csv' dataset for training the model. It is available at Kaggle. The dataset which we used consists of 4515 examples and contains Author name, Headlines, Url of Article, Short text, Complete Article. The information gathered is the summarised news from Inshorts and only scraped the news articles from Hindu, Indian times, and Guardian. Time period ranges from February to August 2017.

4.3 Algorithms

Spacy library A library that falls under NLP is called Spacy (natural language processing). The pre-processing of text, sentences, and information from the text utilising modules and functions is handled by this object-oriented library. After importing the library, we perform the following steps:

- The first step is the cleaning process in which we remove the stop words, punctuation marks and make the words in lowercase.
- The second step is word tokenization in which we tokenize each word from the sentences.
- The third step is counting the frequency of each word and then dividing the maximum frequency with each frequency to get the normalised word frequency count.
- The fourth step is to tokenize the sentences as per frequency of the sentence.
- The final step is to summarise the sentences and form a paragraph.

BERT model The BERT (bidirectional transformer) is a transformer that is used to get over the long-term dependencies that are a drawback of RNNs and other neural networks. It is a naturally bidirectional pre-trained model. This pre-trained model can be readily adjusted to carry out NLP tasks, in this example summarisation.

GPT-2 model The GPT-2 model, a sizable transformer-based language model, has 1.5 billion parameters. It has been trained to anticipate the next word. This model has been used to produce a text that has been condensed.

XLNet model The architecture of the XLNet model, an enhanced version of the BERT model, incorporates permutation language modelling. The following tokens are predicted by XLNet, a bidirectional transformer, in random order.

T5 model Also, we tried training the t5 model to produce brief summaries of news storeys. All of the downstream tasks are combined into a text-to-text format by the text-to-text transfer transformer. Investigating the limitations of transfer learning with a unified text-to-text transformer's studies was mostly coded in the t5 package.This generates a one line summary of the whole text. However for the final output that is for conversion to sign language, we need a summary which better describes the news article.

One of the examples from the dataset is shown below:
Predicted text = Malaika slams Instagram user for 'divorcing rich man'
Actual text =Malaika slams user who trolled her for 'divorcing rich man'
After testing the model with t5 model, the predicted output was nearly the same as the actual headlines, which is a great result. However, we haven't included this model with our actual project since the one line news summary might not illustrate the whole article, and hence, we just limited it for the testing purpose.

We tested these models on different news articles after scraping. We observed that the GPT-2 performs better than the other models. We observed that the extractive

Table 1 ROUGE metrics evaluation

Evaluation metric	BERT	GPT-2	XLNet
ROUGE-1 precision	0.5	0.475	0.5
ROUGE-1 recall	0.975609756097561	0.9743589743589743	0.975609756097561
ROUGE-1 F-Score	0.661157020312820	0.6686554577784055	0.6611570203128202
ROUGE-2 precision	0.4180327868852459	0.4262295081967213	0.4180327868852459
ROUGE-2 recall	0.9807692307692307	0.9629629629629629	0.9807692307692307
ROUGE-2 F-Score	0.586206892360946	0.5909090866554753	0.586206892360946
ROUGE-L precision	0.5	0.475	0.5
ROUGE-L recall	0.975609756097561	0.9743589743589743	0.975609756097561
ROUGE-L F-Score	0.6611570203128202	0.6686554577784055	0.6611570203128202

summary provided by the spacy model contrasts with the abstractive summary provided by the BERT, GPT-2, and XLNet models.Following the observation of such condensed results from various news articles, we discovered that GPT-2 model and XLNet model perform better.

4.4 Evaluation Metrics

For evaluating the algorithms, we have implemented the ROUGE metrics. Recall-Oriented Understudy for Gisting Evaluation, or ROUGE, is a metric that is used to determine how similar an automatically generated summary and a hand-written reference summary are to one another.

Table 1 shows that all the models' scores were quite identical. However, after examining the results for various inputs, we discovered that the GPT-2 model and XLNet provided the highest scores. It was discovered through manual comparison that the GPT-2 model produced superior results, and as a result, it was adopted for the project.

5 User Design

In accordance with the Fig. 1 work flow diagram above, the user must first enter the URL of the news article he wants to read. The Web scraping library Newspaper3K (a Python library) takes the user-provided url and scrapes the news article from it. The news is summarised using the summarisation model (GPT-2 model), which is retrieved from the news article's body. Eventually, this news summary has been translated word by word into sign.

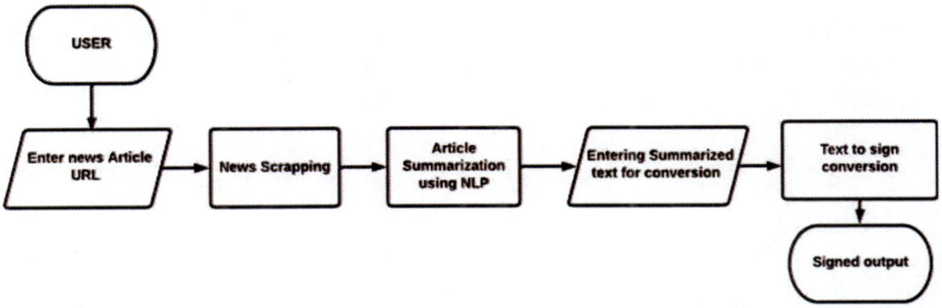

Fig. 1 Work flow diagram

Fig. 2 Results

The result for one of the news summaries is displayed in Fig. 2. Each word is translated into the corresponding sign language, as can be seen.The related gif is shown for every word whose gif is present in the dataset we created, and for all other words, each alphabet is transformed to sign using American sign language.

6 Conclusion

Although text-to-sign language conversion is a difficult topic, it has the potential to transform how deaf people and those who are not familiar with sign language communicate. While there are several methods for converting text-to-sign language, there are obstacles to their growth, such as a lack of standards and the scarcity of sign language data. To address these issues and enhance the effectiveness of text-to-sign language conversion systems, more study is required.

7 Impact

The main advantage of text-to-sign language translation is the improvement of communication between hearing people and non-sign language users. This may result in greater inclusivity and accessibility in a variety of contexts, including the workplace, healthcare, and education. The communication gap between hearing people and non-sign language users can be reduced with the aid of text-to-sign language conversion. This can promote deeper and more fruitful relationships by helping people better understand one another's needs and viewpoints. An effort to convert text-to-sign language can help raise awareness of the value of sign language and the difficulties experienced by the deaf. This may aid in removing obstacles and fostering an inclusive society. Additionally, text-to-sign language conversion can give deaf people access to a greater variety of information, such as books, news, and educational resources, which can enhance their quality of life. In conclusion, a text-to-sign language conversion project can have a big impact by enhancing communication, closing the gap in communication, raising awareness, and giving deaf people access to information.

8 Future Work

The system presented in this paper has the potential to revolutionise the way deaf individuals access and receive information. However, there is still a significant amount of work that can be done to further improve the system. In this section, we discuss several directions for future research and development.

- Integration with wearable devices: The system can be integrated with wearable devices such as smartwatches or augmented reality glasses to provide a hands-free experience for deaf individuals. This will allow them to receive information without having to stop what they are doing and look at a screen.
- Real-time news streaming: The system can be developed to provide real-time news streaming, which will allow deaf individuals to stay updated on current events as they unfold.
- Multi-language support: Currently, the system only supports English news articles. In the future, the system can be expanded to support multiple languages, which will make it accessible to a wider audience.

In conclusion, the system that summarises news articles and converts them into sign language has the potential to greatly improve the lives of deaf individuals. Further research and development in this field can lead to the creation of systems that are more efficient, effective, and accessible.

Acknowledgements We would like to express our sincere gratitude to everyone who supported us throughout this research project. Firstly, we extend our heartfelt thanks to our mentor, who provided us with invaluable guidance, feedback, and support throughout the entire research process. We are grateful for their encouragement, insightful comments, and constructive criticism, which helped us to improve our work significantly.

References

1. Alharbi R, Alhaisoni E, Alsolami F (2021) Sign language recognition using deep learning techniques: a review. J Ambient Intell Humanized Comput 12(5):4739–4760
2. Thanh Hoa P, Thang TQ, Anh PT, Vinh PT (2021) Sign language recognition system using MobileNet and LSTM. In: Proceedings of the 2021 5th international conference on innovation in artificial intelligence, pp 116–122
3. Onwukwe CE, Daramola JO (2021) Real-time sign language recognition using convolutional neural network. In: Proceedings of the 2021 IEEE 11th annual computing and communication workshop and conference, pp 001–006
4. Khan A, Ngadi MA (2021) Real-time hand gesture recognition for sign language interpretation using deep learning. In: Proceedings of the 2021 IEEE 17th international colloquium on signal processing & its applications, pp 105–109
5. Yang W, Cheng J (2021) Efficient sign language recognition using key frames selection and adaptive data augmentation. In: Proceedings of the 2021 IEEE international conference on multimedia and expo, pp 1–6
6. Minocha A, Bansal S, Sharma S, Mehra M (2021) Sign language recognition using faster R-CNN and spatial-temporal LSTM. In: Proceedings of the 2021 3rd international conference on emerging technologies in computer engineering: machine learning and internet of things, pp 53–58
7. Mandal P, Sankar S, Ravi S, Kumar P (2021) Deep learning based sign language recognition: a comparative study. In: Proceedings of the 2021 5th international conference on computing methodologies and communication, pp 111–119
8. Chen Z, Li W, Li W Wu L (2021) Sign language recognition based on spatial-temporal network and attention mechanism. In: Proceedings of the 2021 international conference on artificial intelligence and big data, pp 1–6
9. Alghamdi MM, Alghamdi M (n.d.) Hand gestures recognition using deep learning techniques for American Sign Language
10. Duggan D (2019) American Sign Language recognition using convolutional neural networks. J Electr Imaging 28:011012
11. Dubey A, Singh G (2021) Real-time Indian sign language recognition using CNN and LSTM networks. J Ambient Intell Human Comput 12:189–198
12. Wang Q, Yang Y, Li H (2021) A sign language recognition system based on multi-feature fusion and attention mechanism. IEEE Access 9:49633–49642
13. Li Y, Li Z, Hu J, Liu Y, Zhang D (2021) Sign language recognition system based on multi-features extraction and multi-task learning. In: 2021 IEEE international conference on artificial intelligence and computer applications (ICAICA), pp 326–332
14. Perera PD, Dissanayaka AT, Jayawardena N (2021) Sinhala sign language recognition using deep learning. In: 2021 7th international conference on information and communication technology for embedded systems (IC-ICTES), pp 1–6

15. Mao Y, Huang L, Qiao Y (2021) A deep learning model for American sign language recognition based on skeleton joint position. J Intell Syst 30(1):123–135
16. Azzaz HM, Ezzat MH (2021) Sign language recognition using a convolutional neural network trained by lifting wavelet transform. Wirel Personal Commun 118:2565–2583
17. Nasir SI, Ahmad N (2020) Sign language recognition using deep learning: a review. IEEE Access 8:191823–191843
18. Wang Z, Sun H, Liu M, Wu X (2021) Sign language recognition using video-to-video translation and deep learning. IEEE Access 9:64420–64429
19. Alfaifi M, Al-Nafjan A, Al-Dossari H (2021) Real-time sign language recognition using convolutional neural network. Sensors 21(4):1188
20. Tiwari A, Nema RK (2021) Sign language recognition using deep convolutional neural network with transfer learning approach. Wirel Personal Commun 118:3479–3494
21. Peng M, Yin C (2021) A hybrid deep learning network for sign language recognition based on semantic feature extraction. Int J Adv Rob Syst 18(1):1729881421989426
22. Abdallah IB, Amara Essoukri Ben N, Ben Mabrouk H (2021) Sign language recognition based on deep convolutional neural network with attention mechanism. Wirel Personal Commun 118:5243–5260
23. Panwar S, Patel S (2021) Sign language recognition using convolutional neural networks with data augmentation techniques. J King Saud Univ-Comput Inf Sci
24. Patel AK, Swain AK (2021) A comparative study on sign language recognition techniques using deep learning approaches. In: Advances in intelligent systems and computing, pp 267-274. Springer
25. Ahmed R, Albakri SH (2021) A comprehensive survey on sign language recognition using deep learning techniques. IEEE Access 9:25004–25028

Chapter 57
Assessing the Effectiveness of Different Mass Communication Approaches Used for Government Medical Programs in Rural Areas of Uttarakhand

Pradeep Joshi, Omdeep Gupta, Mayank Pant, Kartikeya Raina, and Bhanu Sharma

1 Introduction

Initiatives of information dissemination are ongoing efforts to advise, convince, or encourage behavioral change in a vast and well-defined public, generally for non-commercial advantages to people as well as society. Such initiatives are typically time-bound, entail organized communication efforts using mass media, and are frequently supplemented by social connectedness [1]. Because of their broad scope, attractiveness, and cost-effectiveness, mainstream media initiatives have been broadly used in promoting wellness and preventing illnesses. Mass media campaigns targets variables in each of the five fields of influence: individual, interpersonal, organizational, community, and policy. Mass media campaigns were described as organized efforts to disseminate communications that create sensitivity or behavior change in a broad audience through various channels, including traditional and newer technologies [2].

P. Joshi · O. Gupta (✉) · M. Pant
School of Management, Graphic Era Hill University, Dehradun, Uttarakhand, India
e-mail: ogupta@gehu.ac.in

K. Raina · B. Sharma
Department of Management Studies, Graphic Era Deemed to be University, Dehradun, Uttarakhand, India

A. Yadav et al. (eds.), *Proceedings of International Conference on Paradigms of Communication, Computing and Data Analytics*, Algorithms for Intelligent Systems,
https://doi.org/10.1007/978-981-99-4626-6_57

1.1 Components Influencing the Effectiveness of Public Health Mass Media Initiatives

The main goal of public health interaction works is to change the information environment to influence individual behavior and prevent chronic disease. Traditional campaigns focus on getting the right message to the right audience. Successful campaigns require control of the communication ecosystem, creative marketing and messaging, and creating a supportive environment. Additionally, campaigns should have a thorough consideration of the factors of health behavior and assess exposure to campaign messages to make mid-course corrections and explain final outcomes [2]. Individual citizens' motives and behaviors may be affected by four variables: perceived susceptibility to disease, perceptions toward the behavior, perceived normative beliefs affected by the group and social environments, and self-efficacy. To accomplish the objective of behavioral change, these variables and their predictors can be focused through advertising campaigns in the media [3].

2 Review of Literature

2.1 Factors Affecting Adoption of Government Schemes

Previous studies identified five types of operational obstacles and enablers: (1) the presence or absence of an empowering and positive environment; (2) the characteristics of community capacity; (3) healthcare system considerations; (4) characteristics of the interaction between the community and healthcare services; and (5) the programs' intercultural ability and responsiveness [4].

Enabling and Not-So-Enabling Environments

Collective responsibility social practices aided societies in addressing barriers to accessing quality care, with community awareness growing over time. Communities that shifted to valuing collective responsibility and action were more probable to preserve attempts to enhance health and processes such as transport systems.

Community Capacity

Studies show that community capacity development is crucial for successful implementation of maternal and child health programs. Factors that facilitate community capacity development include clear committee purpose and roles, strong and stable community leadership, partnerships among organizations, and regular meetings to monitor progress.

Health System Factors

Many studies highlighted limitations within health systems, such as incomplete and inconsistent data, resource constraints, and weak supervision systems for healthcare staff. Information on population health and medical facilities that is up to date is important for improving quality and planning within services, but incomplete data made it hard to prepare efficiently and evaluate the effects of changes made.

Intersection of Community and Healthcare Services

Poor communication, an absence of technical and financial support, and restricted availability of facilities make it difficult for communities and health services to collaborate to develop and execute programs. Combined evaluations between healthcare professionals and civic leaders can actually boost performance, and health professionals in the community, volunteers, and non-governmental organizations (NGOs) have significant responsibilities in connecting communities with healthcare systems.

Cross-Cultural Expertise and Responsiveness of the Programs

Culturally relevant items in native languages are required for community members to participate in health data analysis.

2.2 *Use of Social Media for Health Communication*

Health professionals may use social media to collect patient data and communicate with patients, but face limitations due to time and technological skills. Further work is needed to improve the "social presence" of online consultations and to develop mechanisms to evaluate tie strength in social media to improve its functionality for health communication. Training may be necessary for both patients and medical experts to completely utilize social media in healthcare [5].

2.3 *Measuring Effectiveness of Mass Media Campaigns*

Mass media campaigns have evolved into an important instrument for public health practitioners to promote healthy behaviors and discourage unhealthy ones. However, experience has shown that their effectiveness varies and is difficult to measure [6, 7].

2.4 *Different Methods Used for Communicating Mass Health Programs*

Previous studies have found that interpersonal communication has been the most trustworthy information source on child and maternal wellness, immunization, and neonatal care. For people who are illiterate about community-based intervention, folk media (traditional forms of mass communication) along with interpersonal means of communication may be effective in promoting healthcare awareness. The study envisions a need for improved access to healthcare services by offering particular attention to marginalized groups who provide more than 60% of the population. It also highlights informed knowledge as an essential factor in increasing demand for medical facilities/services among communities while improving their overall health-seeking behavior toward better outcomes [8]. Previous studies have suggested following channels of communication for health messages from government: mass media (print, radio, T.V), governmental bodies' and other international organizations website, tailored content provided by mobile telecommunications service providers, global and regional social networks (Facebook, Twitter, etc.), interactions in public spaces (like banners and holding), mobile health apps, and face-to-face communication (peers and authorities). Participants also highlighted networks with a local focus (municipalities, local health publications, and newsletters), as well as information sharing through businesses or groups [9].

The effects of individual components of communication such as sources of information, information channels, and texts on public information, perceived risks, behavioral and emotional reactions showed that the mediated (e.g., print newspapers, television, and the Internet) and interpersonal sources of information (e.g., friends and family) influence risk and general efficacy perceptions, leading to preventative behaviors [9–11]. Numerous research studies demonstrate that Web 2.0-based social networking platforms, particularly Twitter and Facebook, offer governments the greatest chance of connecting with their public [12–14]. The trustworthiness of social media influencers and the standard of knowledge supplied by the social media influencer are the most important elements influencing individuals' attitudes toward the information received. Furthermore, the way the person feels has an enormous effect on their desire to comply with the SMI's advice [15].

3 Research Methods

The current study recognizes, assesses, and evaluates various criteria and sub-criteria of public health mass media campaigns, as well as different mass communication alternatives. The research utilizes multi-criteria decision analysis, i.e., AHP methods, to rank five alternative communication methods. The AHP starts with defining decision problem followed by definition of the goal, identification of the expert's panel, identification if criteria and sub-criteria, construction of AHP model, pairwise

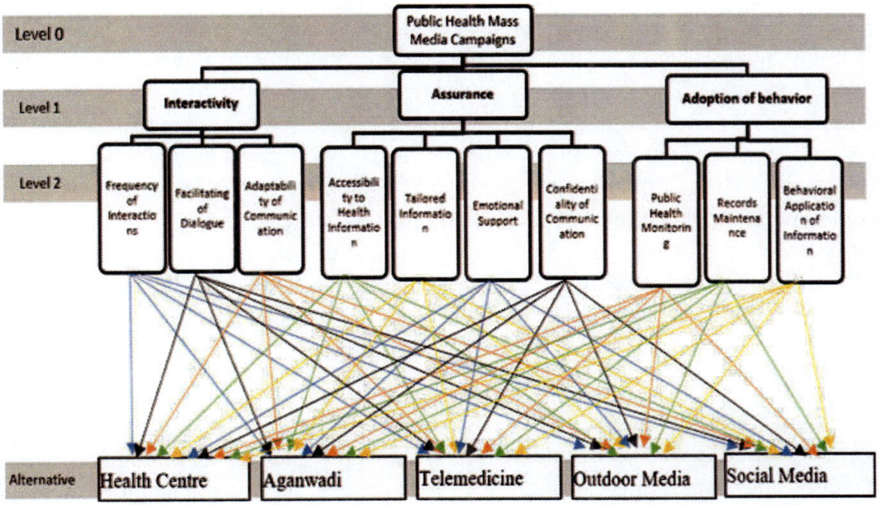

Fig. 1 AHP research design of present study

comparisons, common data agreed by experts, analyzing the consistency, developing priority index, ranking each alternative and finally selecting the best alternative. Once a hierarchy framework is constructed, users are requested to set up a pairwise comparison matrix at each hierarchy and a comparison is done by using a scale pairwise comparison as shown in Fig. 1. Finally, in the synthesis of the priority stage, each comparison matrix is then solved by an eigenvector method to determine the criteria importance and alternative performance.

3.1 Analytic Hierarchy Process (AHP)

The first stage in AHP is to formulate the issue in the context of a hierarchy framework, with the top level representing overall objectives, the middle levels representing criteria as well as sub-criteria, and the bottom level projecting decision alternatives. The perceived significance of characteristics is ascertained by pairwise comparisons conveyed in semantic judgment and transformed into quantitative data using Saaty's fundamental scale (Table 1).

Table 1 Saaty's fundamental scale

Importance	Definition	Importance	Definition
1	Equal importance	7	Demonstrated dominance
3	Moderate dominance	9	Extreme dominance
5	Strong dominance	2, 4, 6, 8	Intermediate values

The present research focuses on the effectiveness of communication on the basis of three dimensions. These three dimensions are interactivity, assurance, and adoption of behavior. The interactivity dimension focuses on the level of opportunity provided by a particular method to the people so that they are able to understand and respond to the communications as expected by the government programs. Assurance dimension focuses on the level of trust that people have in a particular source of communication. Adoption of behavior dimension focuses on how much a particular mode of communication will lead to the adoption of behavior by the people. Each of these dimensions were further assessed by certain sub-criteria. The sub-criteria in each dimension are discussed as follows.

Interactivity

The interactivity dimension of effectiveness consisted of three sub criteria. These were increased frequency of interactions with people, facilitating dialogue between patients and health professionals and adaptability of communication in local language/context.

Assurance

The assurance dimension of effectiveness consisted of four sub-criteria. These were increased accessibility to health information; more available, shared, and tailored information; peer/social/emotional support and confidentiality and privacy of communication.

Adoption of Behavior

The adoption of the behavior dimension of effectiveness consisted of three sub-criteria. These were public health surveillance; collecting data on patient experiences and opinions and how to apply information related to their personal health situation.

3.2 *Tools*

The suggested model treats the issue as a multi-criteria decision-making (MCDM) problem. The AHP technique was used to overcome this MCDM problem. AHP was applied to evaluate the weight of each criterion, and all options were ranked based on their criterion weights. For this study, three criteria were used which were interactivity, assurance, and adoption of behavior. Five options were considered as methods of mass communication for government health programs. These options (alternative) were health center (A1), Aganwadi's workers (A2), telemedicine (A3), outdoor media (A4), and social media (A5). The analytic hierarchy process (AHP) is a method for dealing with difficult decision-making situations. According to Satty [16], it is a method for organizing and analyzing multiple kinds of MCDM problems. This method relies on the eigenvalue and eigenvector of a matrix, which is built by performing pairwise comparisons of all factors used in the research.

The pairwise observations are built by comparing sets of components at every stage of the hierarchy to each and every element at the superior stage. These pairs are employed to determine the relative importance of every set of elements in each level of hierarchy. Comparing the pairs of elements is produced through the application of Saaty's nine-point measure to provide a comparative judgment of choices for each pair of components in every level. This method is employed to decide which of two elements in a set is more preferable compared to the rest. These similarities are arranged in a matrix of pairwise comparisons. The AHP's core idea is the development of preferences from a pairwise comparisons structure. The AHP enables decision-makers to deduce ratio scale priorities from pairwise comparison matrices. The priority vectors for each set in a level are predicted using the prioritization technique, which is the geometric mean technique in this research [17].

The AHP, unlike most other multi-criteria methods, allows for some discrepancy in expert judgments: the accuracy of pairwise comparison matrices is validated by calculating the consistency index (CI):

$$\mathrm{CI} = \lambda_{\max} - n/n - 1 \tag{1}$$

where $\lambda_{\max}$ represents the maximum eigenvalue, and n is the rank of the pairwise comparison matrix. With reciprocals pushed, the consistency index (CI) of a truly random reciprocal matrix is called the random consistency index (RI). A sample size of 100 was used to create an average RI for matrices of order 1–15. The final ratio to compute is the consistency ratio (CR). If the CR becomes less than 0.1, the judgments are largely consistent, and the obtained weights can be used. Otherwise, the decision-maker must start considering the pairwise comparison matrix entries. CI and CR are determined using the equations that follow to test the consistency of pairwise comparison judgment (Eq. 2):

$$\mathrm{CR} = \mathrm{CI/RI} \tag{2}$$

where RI is random consistency index, and its value can be obtained from Table 2.

In other words, the local priorities of criteria in a node are multiplied by the corresponding parent criterion's local priorities. Finally, sensitivity analyses are carried out in order to verify the solution and test for rank.

Table 2 Random consistency index

N	1	2	3	4	5	6	7	8	9
RI	0	0	0.58	0.9	1.12	1.24	1.3	1.41	1.45

3.3 Data Collection

The questionnaire was based on Saaty's fundamental scale and included pairwise questions about three criteria, ten sub-criteria, and five alternatives. It was distributed via Google Form to a group consisting of five subject matter experts, where three experts were from the medical profession and two experts were from reference groups.

4 Findings

The goal of the study, presented at level 0, is to evaluate mass media. The input of AHP shows that primarily there were three factors at level 1 and ten factors at level 2. The three factors at level 1 were interactivity, assurance, and adoption of behavior.

The first factor "Interactivity" has three factors: frequency of interaction, facilitation of dialogue, and adaptability of communication. The second factor "Assurance" has four factors: accessibility to health information, tailored information, emotional support, and confidentiality of communication. The third factor "Adoption of behavior" has three factors: public health monitoring, records maintenance and behavioral application of information, resulting in three factors at level 1 and ten factors at level 2.

On the basis of findings at level 1, the highest priority factor was "Adoption of behavior" (see Table 3). This finding is consistent with the literature which suggests that the campaign must be able to explain its outcomes A1 [2]. At level 2, the highest priority factor was behavioral application of information followed by confidentiality of information. This confidentiality of information may also influence the attitude of people toward the behavior which is important for the success of campaigns [3]. If we understand the priority at level 1, the group consensus was found to be low. On the basis of experts, it was found that medical experts gave higher priority to the adoption of behavior, while reference groups gave priority to the assurance and interactivity (Table 4).

Within the factor Interactivity, the medical experts gave priority to the factor adaptability of the communication, while the reference group gave priority to the facilitation of dialogue. Within the factor Assurance the group consensus was moderate, and both the groups gave priority to the emotional support and confidentiality of the communication. Within the factor Adoption of Behavior, the medical experts gave

Table 3 Priorities and ranking of Level 1 criteria's

Level 0	Level 1 criteria's		Rank
Public	Interactivity (6.8%)	(C1)	3
Health mass	Assurance (19.9%)	(C2)	2
Media	Adoption of behavior (73.3%)	(C3)	1

Table 4 Expert opinion for Level 2 criteria's

Group result (%)	Expert 1 (Reference leader) (%)	Expert 2 (Reference leader) (%)	Expert 3 (Medical expert) (%)	Expert 4 (Medical expert) (%)	Expert 5 (Medical expert) (%)
0.34	0.01	2.41	0.05	0.00	0.01
9.98	1.62	48.60	0.96	0.64	0.04
7.40	0.13	15.31	4.83	0.05	0.68
1.89	0.50	0.81	0.14	0.27	0.06
0.61	0.46	0.10	0.02	0.03	0.05
7.21	71.76	2.58	0.00	8.59	0.72
23.60	16.66	29.66	0.57	2.15	5.02
4.84	1.16	0.10	0.42	1.36	5.42
9.55	5.86	0.01	8.39	43.46	1.35
34.57	1.85	0.41	84.62	43.46	86.66

priority to the factor behavioral application of information, while the reference group gave priority to the records maintenance and behavioral application of information. (Table 5). On the basis of global importance, the group of medical experts gave priority to the behavioral application of information, while the reference group gave importance to the facilitation of information and emotional support (Table 6).

This study primarily focused on five alternatives of communication, which were health centers, Aganwadi, telemedicine, outdoor media and social media. The consolidated weights of alternatives show that health centers and Aganwadi were the most important alternatives with health centers having slightly more priority (Table 7). The importance for the rest of the alternatives was not even 5%. If we observe alternatives by participants, there was a high level of homogeneity with both the groups giving more importance to the health center and Aganwadi. For the factor frequency

Table 5 Priorities and local ranking of Level 2 criteria's

Level 1 criteria's	Level 2 criteria's (priorities %)		Local rank
Interactivity	Frequency of interaction (6.8%)	(C1.1)	3
	Facilitation of dialogue (19.9%)	(C1.2)	2
	Adaptability of communication (73.3%)	(C1.3)	1
Assurance	Accessibility to health information (7.4%)	(C2.1)	3
	Tailored information (5.5%)	(C2.2)	4
	Emotional support (25.1%)	(C2.3)	2
	Confidentiality of communication (62.0%)	(C2.4)	1
Adoption of behavior	Public health monitoring (18.8%)	(C3.1)	2
	Records maintenance (8.1%)	(C3.2)	3
	Behavioral application of information (73.1%)	(C3.3)	1

Table 6 Global priorities and global ranking of Level 2 criteria's

Level 2 criteria's	Global priorities (%)	Global rank
Frequency of Interaction	0.50	10
Facilitation of dialogue	1.30	8
Adaptability of communication	5	5
Accessibility to health information	1.50	7
Tailored information	1.10	9
Emotional support	5	5
Confidentiality of communication	12.30	3
Public health monitoring	13.80	2
Records maintenance	5.90	4
Behavioral application of information	53.60	1

Table 7 Priorities and ranking of alternatives

	Alternatives				
Level 2 Criteria's	A1	A2	A3	A4	A5
C1.1	0.36	0.344	0.122	0.133	0.04
C1.2	0.173	0.751	0.034	0.026	0.015
C1.3	0.58	0.276	0.058	0.062	0.023
C2.1	0.546	0.324	0.046	0.076	0.008
C2.2	0.191	0.726	0.027	0.038	0.018
C2.3	0.186	0.754	0.035	0.013	0.011
C2.4	0.289	0.617	0.048	0.026	0.019
C3.1	0.449	0.45	0.072	0.01	0.019
C3.2	0.635	0.298	0.035	0.019	0.013
C3.3	0.545	0.402	0.027	0.016	0.01
Alternatives priorities (%)	**48.00**	**44.60**	**3.90**	**2.10**	**1.40**
Alternatives rank	Rank 1	Rank 2	Rank 3	Rank 4	Rank 5

The higher value signifies higher priority of the alternative as comparied to others (as the percentage of the total) alternatives

of interaction, there was moderate level of consensus with both the groups giving more importance to the health center and Aganwadi. For the facilitation of dialogue, there was a moderate level of consensus. Aganwadi was the most preferred alternative followed by health center. Here, the preference of Aganwadi was quite ahead of the health center.

For the factor adaptability of communication, there was a moderate level of consensus. Health center was the most preferred alternative followed by Aganwadi. Here, the preference of health center was quite ahead of health center. For the accessibility of information, there was a moderate level of consensus.

The health center was the most preferred alternative followed by Aganwadi. Here the preference of health center was quite ahead of health center. For the factor tailored information, there was a moderate level of consensus. Aganwadi was the most preferred alternative followed by the health center. Here, the preference of Aganwadi was quite ahead of the health center. For the factor emotional support, there was a moderate level of consensus. Aganwadi was the most preferred alternative followed by the health center. Here the preference of Aganwadi was quite ahead of the health center. For the factor confidentiality of information, there was a high level of consensus. Aganwadi was the most preferred alternative followed by the health center. Here, the preference of Aganwadi was quite ahead of the health center.

For the factor of public health monitoring, there was a high level of consensus. Both Aganwadi and the health center were almost equally preferred alternatives. For the factor record maintenance, there was a high level of consensus. The health center was the most preferred alternative followed by Aganwadi. Here the preference of health center was quite ahead of health center. For the behavioral application of information, there was a high level of consensus. The health center was the most preferred alternative followed by Aganwadi. Here the preference of health center was quite ahead of health center.

The first three factors in the above table, i.e., frequency of interaction, facilitation of dialogue, and adaptability of communication, constitutes the interactivity of communication method. Frequency of interaction tells the number of times interaction takes place, facilitation of dialogue tells the ease of dialogue to the people, and adaptability of communication tells the variations in communications that can be brought keeping in view the target audience. The four factors—accessibility to health information, tailored information, emotional support, and confidentiality of communication, focus on assurance with respect to the method. Accessibility tells how easily the people can access the information, tailored information means the information according to the requirements of patients, emotional support tells the empathy of the source and confidentiality means the perceived privacy of information shared by the people. The three factors—public health monitoring, records maintenance and behavioral application of information, define the adoption of the behavior by the people. Public health monitoring tells that behavior of the people can be recorded. Records maintenance means that the record can be maintained. Behavioral application tells the chances of applying the behavior due to a given method.

5 Conclusion

With the continuous rise of new means of communications, developing an appropriate mix of all these means remains a big challenge for the organizations. Particularly, when these methods are to be used for mass communications and for rural people, the challenges are more because of low penetration of technology among such target audiences.

The above study aimed to assess the effectiveness of different approaches of mass communication used for health programs in rural areas of Uttarakhand. The study used the AHP method where certain methods and the criteria for the choice of these methods were determined on the basis of literature. The methods considered for health communications in this study were health center, Aganwadi workers, telemedicine, outdoor media, and social media. The criteria considered for the choice of these methods were interactivity, assurance, and adoption of behavior. The findings suggest that adoption of behavior is the most important criterion followed by confidentiality of information. It suggests that in such cases interpersonal sources of communication become more important [8]. In the same lines, the current research also concludes that health centers and Aganwadi's are the most preferred methods.

Further, it needs to be noted that despite the penetration of latest technology like social media in personal lives, its application in mass communication remains a challenge while it can be quite advantageous in such situations [12, 13]. Therefore, it is suggested that the government and other implementation agencies should focus on more utilization of social media in healthcare-related communications.

References

1. Rice RE, Atkin CK (2009) Public communication campaigns: theoretical principles and practical applications. In: Media effects. Routledge, pp 452–484
2. Randolph W, Viswanath K (2004) Lessons learned from public health mass media campaigns: marketing health in a crowded media world. Annu Rev Public Health 25:419–437
3. Fishbein M, Cappella J, Hornik R, Sayeed S, Yzer M, Ahern RK (2002) The role of theory in developing effective antidrug public service announcements. In: Crano WD, Burgoon M (eds) Mass media and drug prevention: classic and contemporary theories and research. Erlbaum, Mahwah, NJ, pp 89–117
4. Howard-Grabman L, Miltenburg AS, Marston C, Portela A (2017) Factors affecting effective community participation in maternal and newborn health programme planning, implementation and quality of care interventions. BMC Pregnancy Childbirth 17(1):1–18
5. Moorhead SA, Hazlett DE, Harrison L, Carroll JK, Irwin A, Hoving C (2013) A new dimension of health care: systematic review of the uses, benefits, and limitations of social media for health communication. J Med Internet Res 15(4):e1933
6. Hornik R (2002) Public health communication making sense of contradictory evidence. In: Hornik RC (ed) Public health communication: evidence for behavior change. Erlbaum, Mahwah, NJ, pp 1–22
7. Inst. Med., Comm. Commun. Beh. Change in the 21st century: Improv. Health Divers. Popul. (2002) Speaking of health: assessing health communication strategies for diverse populations. Natl. Acad. Press, Washington, DC
8. Kshatri JS, Palo SK, Panda M, Swain S, Sinha R, Mahapatra P, Pati S (2021) Reach, accessibility and acceptance of different communication channels for health promotion: a community-based analysis in Odisha, India. J Prev Med Hyg 62(2):E455
9. Tam LT, Ho HX, Nguyen DP, Elias A, Le ANH (2021) Receptivity of governmental communication and its effectiveness during COVID-19 pandemic emergency in Vietnam: a qualitative study. Glob J Flex Syst Manag 22(Suppl 1):45–64
10. Nazione S, Perrault E, Pace K (2021) Impact of information exposure on perceived risk, efficacy, and preventative behaviors at the beginning of the COVID-19 pandemic in the United States. Health Commun 36(1):23–31

11. Chen Q, Min C, Zhang W, Wang G, Ma X, Evans R (2020) Unpacking the black box: How to promote citizen engagement through government social media during the COVID-19 crisis. Comput Hum Behav 110:106380
12. Ma H, Miller CH (2021) The effects of agency assignment and reference point on responses to COVID-19 messages. Health Commun 36(1):59–73
13. Raamkumar AS, Tan SG, Wee HL (2020) Measuring the outreach efforts of public health authorities and the public response on Facebook during the COVID-19 pandemic in early 2020: cross-country comparison. J Med Internet Res 22(5):e19334
14. Zeemering ES (2020) Functional fragmentation in city hall and Twitter communication during the COVID-19 pandemic: evidence from Atlanta, San Francisco, and Washington DC. Gov Inf Q 38(1):101539
15. Gupta S, Dash SB, Mahajan R (2022) The role of social influencers for effective public health communication. Online Inf Rev 46(5):974–992
16. Satty TL (1980) The analytic hierarchy process. McGraw Hill, New York
17. Yadav A, Jayswal SC (2013) Using geometric mean method of analytical hierarchy process for decision making in functional layout. Int J Eng Res Technol (IJERT) 2(10):775–779

Chapter 58
Computer Vision-Based Virtual Mouse Cursor Using Hand Gesture

Tanmay Sonawane, Sarvesh Waghmare, Abhishek Dongare, Avadhut Joshi, and Anandkumar Birajdar

1 Introduction

Today, everyone needs technology, and technology is a necessary component of every day-to-day activity. Therefore, computers played a crucial role in meeting this need. The computer offers solutions to everyone, regardless of age or social class, in the industrialized world. Why are we using computers to interact? To simplify our lives and our job, obviously! Consequently, research on human–computer interaction (HCI) has become of utmost importance.

Computers have made huge progress in the last 2 decades. We utilize a mouse as an input device to interact with computers, yet its accuracy and capabilities have significant drawbacks. No matter how much the precision of the mouse has shown progress, since the mouse is an electronic hardware component and can occasionally experience problems, such as a malfunctioning click; in addition to being a hardware device with a limited lifespan, we must upgrade the mouse for better performance.

The entire world will become virtualized as a generation develops, including hand and speech recognition. In fact, we routinely use hand gestures in our daily conversations, which makes them a very popular and effective way of communicating with others. This type of interaction between people and computing devices can be carried out by leveraging the use of hand gestures for communications because they are strongly ingrained in our culture for communicating with one another.

How to employ computer systems to interpret hand motions is the most challenging issue. Data gloves are the answers. In this work, vision-based techniques for the hit upon hand motion detection and the execution of a few features, including left and right mouse actions, are demonstrated. Then, using artificial intelligence, I made a smart system that allows for right- and left-click control via hand gestures.

T. Sonawane · S. Waghmare · A. Dongare · A. Joshi (✉) · A. Birajdar
Pimpri Chinchwad College of Engineering, Pune, India
e-mail: joshiavadhut2001@gmail.com

A. Yadav et al. (eds.), *Proceedings of International Conference on Paradigms of Communication, Computing and Data Analytics*, Algorithms for Intelligent Systems,
https://doi.org/10.1007/978-981-99-4626-6_58

We will discuss the use of OpenCV to hand recognition in this paper. The Python OpenCV machine vision library is used to record webcam images. We can easily operate systems without involving humans thanks to two rising technologies: artificial intelligence and machine learning. As a result, after the technology is created, items like expert, more user-friendly software are created and built for this system's convenience. As a result, this system is a part of the technologies that can make it easier for us to work.

This system uses hand gestures to let users control the cursor, and it uses the system's webcam to track input for hand gestures. OpenCV is used in this situation because of its video capture module. Using OpenCV, we can collect data from real-time video in frame-by-frame manner.

2 Existing Method

According to past study, researchers use various methods, such as convex hull approaches, to designate the mouse pointer as the centroid of the palm and so forth. Certain cursor actions, such as right and left clicking, are reserved for specific fingers. We must raise our hands to the pointer's location (which is at the centroid of the palm) which elevates our finger-based commanding system over the screen. Our fingertips are outside the area where we can control the system, even though we can move about in the upper zone [1–4].

There are certain problems, such as the fact that some people fail to consider the worst-case situation of multiple target identification.

3 Proposed Solution

The index finger on my hand will serve as the cursor pointer, with the thumb serving as the left click and the middle finger serving as the right click, in order to address the problem. After that we will use our hand fingers according to their field of movements.

A more forgiving approach should be taken when detecting a specific color because several identifications may occur. As a result, we are utilizing the 21-segment identification system by MediaPipe to identify the fingers in this paper [5].

4 Requirements for Proposed System

We are using the Python package OpenCV for the project, which is primarily focused on real-time video capture and processing (computer vision). Python imports the OpenCV library using the import cv2 command. It also uses NumPy, a Python library designed primarily for advanced mathematical operations on arrays and matrices.

Table 1 Require specification table

Item	Specification
Microsoft Window	Support Architecture: 32 bit (×86) 64 bit (×86)
OpenCV	OpenCV 4.6.0
MediaPipe	MediaPipe 0.8.11
OS	Windows 7 and Linux OS
RAM	4 GB or greater
Camera	8 MP or higher
CPU	i3 generation or higher/AMD or higher
GPU	Not required

Table 1 includes the information on the prerequisites needed to complete this project. This environment of specifications is crucial for getting good results, avoiding some runtime issues and the shortcomings of earlier projects [4, 6–8].

5 Methodology

The methodology defines each component's technique level by level, and Fig. 1 can demonstrate how they operate.

5.1 Video Acquisition

This document describes the methods for acquiring video for video capture as well as several other video operations. Videos are captured with a frame rate of 40 and a resolution of 1920 by 1080 using the system's standard webcam by default (Standard setting) [9].

5.1.1 Capturing Real-Time Video

Real-time video is captured using a camera at consistent frame rates and resolutions. First, real-time video inputs are captured. Images are then extracted from the video frame by frame using a variety of methods, processed using the RGB shade layout of matrix ($m * n$), with each detail consisting of a ($1 * 3$) matrix of red, green, and blue channels. The most common and usually referred to as the "mother of all colors" are the RGB colors because all colors are created with them [10].

Fig. 1 Flow diagram

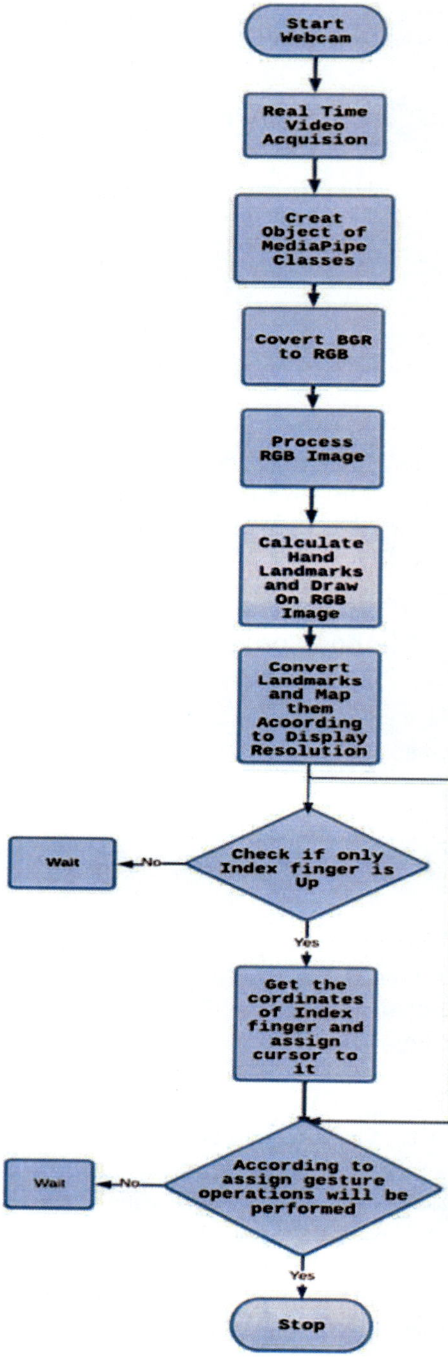

5.1.2 Flipping of Video

Moving our hand to the left in real-time video acquisition causes the picture of the hand to move to the right, and vice versa. A video is found to be horizontally inverted when it is previewed. As a result, the image must be horizontally flipped. It was done by utilizing OpenCV's flipping function [5].

5.2 Color Conversions

Colors are typically represented by red, green, and blue (RGB) numeric values in image processing software. Yet, there are other models besides RGB that can be used to quantitatively encode colors. These models are collectively referred to as "color spaces" since the majority of them can be converted into a 2D, 3D, or 4D coordinate system [11].

The many color spaces exist to provide a more intuitive way to identify colors or to present color data in a way that makes it easier to perform certain computations. For instance, the proportions of combined red, green, and blue colors are how the RGB color space characterizes a color.

For a better user experience, we convert the image file from BGR (blue, green, and red) to RGB when it is read by OpenCV.

5.3 Noise

Any computer vision system must clearly emphasize the essential and highly important role that noise plays in its operation. Small undesirable errors can emerge, especially when scanning images, due to dust or other issues [12].

5.3.1 Filtration

A few pixels are spread after the red and blue components from the image have been captured, creating noise that resembles salt and paper. OpenCV Median Filter is therefore used to remove those impacts [12].

5.3.2 Unwanted Objects from Image

The MediaPipe ML pipeline is made up of a number of connected models. A palm detection model makes use of the whole image and generates an oriented hand bounding box [12].

A palm detector model is created because it is considerably simpler to estimate the bounding boxes of stiff objects like fists and palms than it is to recognize hands with movable fingers. Due to the lower size of palms, the non-maximum suppression approach also works effectively in two-hand self-occlusion situations like handshakes.

Our subsequent hand landmark model uses the regression to precisely localize key locations for 21 3D hand coordinates after comprehensive palm identification across the whole image.

5.4 *Cursor Mapping*

Now that the image has been created utilizing the various above-mentioned techniques, cursor mapping is still not complete. To calculate the bounding boxes of solid objects like fists and palms, a palm detection model is used. Following that, the hand landmark model precisely localizes 21 3D hand coordinates inside the hand region that was detected [13, 14].

From these, we will get coordinates of all 21 landmarks. We assign the mouse cursor to the tip of the index finger. Due to continuously changing frame rate, we need to apply smoothing to get smooth movement of the cursor [10].

$\mathrm{curX} = 0, \mathrm{curY} = 0$
$\mathrm{prevX} = 0, \mathrm{prevY} = 0$
$\mathrm{smoothening} = 10.$

Smoothen value a bit

$$\mathrm{curX} = \mathrm{prevX} + (x3 - \mathrm{prevX})/\mathrm{smoothening} \tag{1}$$

$$\mathrm{curY} = \mathrm{prevY} + (y3 - \mathrm{prevY})/\mathrm{smoothening} \tag{2}$$

6 Perform Different Operations

We have implemented a function will return and array of size 5, saying whether the finger is up or down. Each finger of our hand is assigned 1 index. This will be very useful for further implementations of mapping gestures as shown in Fig. 2.

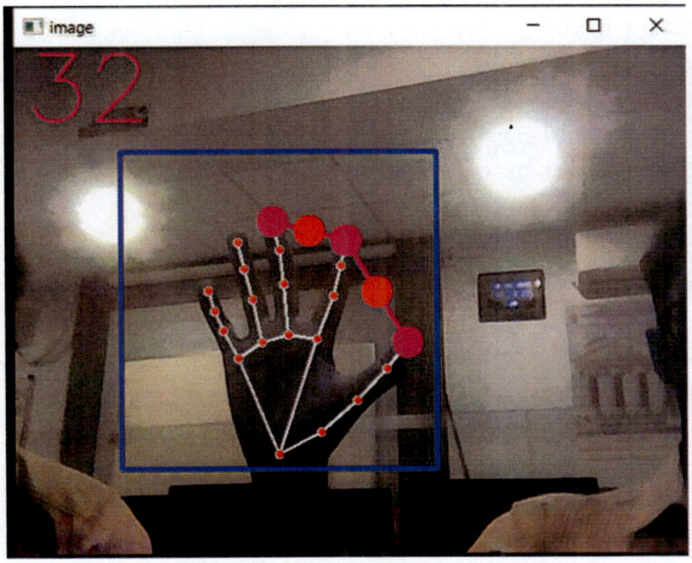

Fig. 2 Image detect

6.1 *Right Click*

On the basis of the image captured, we get the information about coordinates of index finger and middle finger. The distance between these two fingers is calculated and a click operation is performed if the distance is less than some threshold value.

$$\text{Condition: - Right Click} = \text{distance(index finger, middle finger)} <= \text{threshold} \quad (3)$$

6.2 *Left Click*

For the left click, we will be using the thumb and index fingers. Same as right action if distance between them is less than some threshold, left mouse click action is performed.

$$\text{Condition: - Left Click} = \text{distance(index finger, thumb)} <= \text{threshold} \quad (4)$$

6.3 Move Cursor

To move the mouse cursor based on the motion of our finger, we mapped the tip of the index finger to the mouse cursor. To move the cursor on screen, your index finger will be standing and all the other fingers are folded down.

6.4 Scroll Up

To perform the scroll operation, we use our pinky finger if our pinky finger is up the scroll-up operation get performed. It will move the page up by equivalent to 1 click at a time.

6.5 Scroll Down

To perform the scroll-down operation, we use our pinky finger and index finger. When both our pinky finger and index finger are up, the scroll-down operation is performed. Same as scroll up, it will move the page down by equivalent to 1 click at a time.

6.6 Copy Operation

We use index finger, middle finger, and ring finger to perform the copy operation. If all these 3 fingers are up, and the selected object gets copied to clipboard.

6.7 Paste Operation

To use the paste functionality of the mouse, we use index finger and middle finger. If these 2 fingers are up, then whatever is copied on clipboard get paste there. We must assure that after paste operation, clipboard will get empty.

7 Future Development

In order to control devices gaining knowledge of robots, we will leverage this technology to create graphical consumer interface software programs that we will pair with extremely good high-profile cameras. We can be capable of appointing a sophisticated graphical consumer interface once the videos have gone through additional optimization, at which factor this independent robot may be used as a home provider robot, a robot for dealing with complex activities, or a robot for protecting missions.

The domain of virtual painting in VR is where artists can create paintings of three-dimensional objects. These ideas could be used in gaming, augmented reality, and virtual reality. Snake games, adventure games, and many other types of games can be built with hand gestures.

8 Conclusion

Through the perception of simple colors from visual input, computer vision and machine learning methods offer promising ways to improve human–computer interaction. The successful and exact disjointing of primary colors is a crucial step in achieving this goal. By highlighting these situations, this work focuses on hand gesture cursor operation using image segmentation, machine learning, and vision-located color recognition.

In this paper, a technique for identifying objects and executing mouse actions, such as clicking, brightness control, volume control, dragging–dropping, and moving the pointer, some of keyboard shortcuts has been given. This device was created using NumPy, OpenCV, AutoPy, and MediaPipe. Due to its substantially lower cost than hardware devices, many users prefer using vision control devices. We have concluded from this study that this technology has a great potential in HCI-based systems. Numerous sectors, such as robotics, biomedical instruments, computer gaming, and others can make substantial use of it.

References

1. Shetty M, Daniel CA, Bhatkar MK, Lopes OP (2020) Virtual mouse using object tracking. 978-1-7281-5371-1
2. Varun KS, Puneeth I, Prem Jacob T (2019) Virtual mouse implementation using OpenCV. 978-1-5386-9439-8
3. Belgamwar S, Agrawal S (2018) An Arduino based gesture control system for human-computer interface, 978-1-5386-5257-2
4. Meena Prakash R, Deepa T, Gunasundari T, Kasthuri N (2017) Gesture recognition and fingertip detection for human computer interaction, 978-1-509-3294-5
5. https://google.github.io/mediapipe/solutions/hands.html. Last accessed 02 Jan 2023

6. Gajjar V, Mavani V, Gurnani A (2017) Hand gesture real time paint tool-box: machine learning approach, 978-1-5386-0814-2
7. Kour KP, Mathew L (2017) Literature survey on hand gesture techniques for sign language recognition by vision based hand gesture recognition, Paper Id: IJTRS-V2-I7-00
8. Kavarthapu DC, Mitra K (2017) Hand gesture sequence recognition using inertial motion units (IMUS). IEEE, Department of Electrical Engineering, Indian Institute of Technology Madras, Chennai, India, 2327-0985. https://doi.org/10.1109/Acpr.2017.159
9. Titlee R, Ur Rahman A, Zaman HU, Rahman HA (2017) A novel design of an intangible hand gesture controlled computer mouse using vision based image processing, 978-1-5386-2307-7
10. https://ai.googleblog.com/2019/08/on-device-real-time-hand-tracking-with.html. Last accessed 21 Jan 2023
11. Xue X, Zhong W, Ye L, Zhang Q (2015) The simulated mouse method based on dynamic hand gesture recognition (China), 978-1-4673-9098-9
12. https://www.pyimagesearch.com/2015/02/09/removing-contours-image-using-python-opencv/. Last accessed 11 Feb 2023
13. Mali Y, Malani H, Mahore N, Mali R (2022) Hand gesture controlled mouse. e-ISSN: 2395-0056
14. Gagnani L, Patel H, Chaturvedi S, Jaiswal R (2022) Gesture controlled mouse and voice assistant. ISSN: 2321-9653

Chapter 59
A Review of Machine Learning Models for Disease Prediction in Poultry Chickens

Divya Verma, Neelam Goel, and Vivek Kumar Garg

1 Introduction

Agriculture and related activities like animal husbandry and dairying have been essential to human existence since the dawn of civilization. By preserving ecological equilibrium, these operations have benefited not just the food supply but the animal power in draught equipment [1]. Humans today raise chickens largely as a food source (eating both their flesh and eggs). As per the OECD-FAO Agricultural Outlook 2030, poultry meat will account for 41% of all the protein from meat sources worldwide by the end of this century. According to the survey, consumers are drawn to poultry meat not just because of reduced pricing but also because of the consistency and adaptability of the product as well as the greater protein and lower fat balance [2] to official estimates, the agriculture sector accounts for 17% of India's GDP, of which animal husbandry accounts for 27%. The fact that the dairy, poultry, and aquaculture sectors together account for 4.4% of the country's GDP demonstrates the significance of these industries [2]. Over 16 million people nationwide rely on these industries for employment prospects, and hence, they are essential [3]. Globally, there are two basic types of chicken farms: egg-producing hens and broiler chickens. The former is used for egg production, while the latter is reared for meat.

In 2020, China produced more than 596 billion eggs than any other country [4]. Approximately 83 billion eggs were produced in the nation in 2015–2016, while 88 billion eggs were produced in 2016–2017, representing an increase of nearly 6%. In 2015–2016, there were 66 eggs available per person; up from 61 in 2013–2014, and

D. Verma · N. Goel (✉)
University Institute of Engineering and Technology, Panjab University, Chandigarh 160014, India
e-mail: erneelam@pu.ac.in

D. Verma · V. K. Garg
Chandigarh University, Gharuan, Punjab 140413, India

A. Yadav et al. (eds.), *Proceedings of International Conference on Paradigms of Communication, Computing and Data Analytics*, Algorithms for Intelligent Systems,
https://doi.org/10.1007/978-981-99-4626-6_59

by 2016–2017, it is 69. The nation's poultry meat production rose by about 6% to over 3.46 million tons in 2016–2017, up from 3.26 million tons in 2015–2016 [5].

On a large scale, poultry has lately eclipsed bacon as the most extensively produced sort of meat. The amount of meat produced in 2014–2015 was 6.69 million tons, and it is expected to increase to 8.80 million tons in 2020–2021. When compared to the supply of Major Livestock Products (MLPs) from 2011–2012 to 2020–2021, the production of meat has increased by 2.31% [1].

The poultry sector alone is estimated to have lost more than 22,000 crores as a result of the effects of Covid-19 and accompanying lockdowns. Due to the pandemic, the chicken business grew by just 2–3% in Fiscal Year 2020 (compared to an average annual growth rate of 7–8%) and declined by 4–5% in Fiscal Year 2021. The poultry industry is the most commercially supportive and developed, and broilers have become a significant food source for many households. The consumption of chicken is expanding at a rapid pace, which assists GDP recovery in some nations. Poultry screenings are carried out in compliance with the Poultry Product Inspection Act (PPIA). For the interrogation process, a team of FSIS inspectors is involved which detects the contagious status of poultry [6]. The spread of chicken disease has a negative impact on the farmer's health. As a result, it has become a severe concern to regulate the disease that is affecting chickens, which is explicitly or implicitly jeopardizing human health. Currently, disease diagnosis is done by observation. As manual assessment takes a long time, image processing technology is used through cameras and computers to have a digital image of chickens or poultry farms. Various studies have been conducted to detect how chicks behave, and the experiments were carried out by an expert and an image processing system. Kristensen and Cornou [7] conducted an experiment that automatically covers or detects broiler behaviors and observes them over 2 to 3 weeks, so that the chick's daily activity could be monitored. The frames were taken in order to identify the activities of chickens in terms of body structure and dynamics. To save the chickens from being infected, certain vaccines are injected at various stages. The broiler also underwent an antibiotic course for the first 5 days. Vaccines such as Livosin, Antrosin, and others are given to the chicken to prevent them from being infected. The chickens were infected with Newcastle disease, lameness, chronic respiratory disease (CRD), headstock, dehydration, drought, pneumonia, and many other diseases, and vaccines are given to safeguard them from all of these diseases [8]. According to the age of chickens, their temperature varies as mentioned in Table 1. Figure 1a and b show the healthy and lame chicken, respectively.

The objective of this paper is to review methods used to detect sick broilers in order to save the farm from going out of business because the poultry industry is one of the most promising sectors for the nation's GDP. On the basis of images, videos, and audio, machine learning algorithms are applied to the early disease classification of chickens. The following are the paper's main contributions: (i) the most frequently occurring diseases in poultry chickens are discussed. (ii) While the majority of the papers focus on a specific condition, the different diseases with their causes, symptoms, prevention, and therapy are described in this paper. (iii) Various models developed for chicken disease prediction are discussed with their strengths, limitations,

Table 1 Recommended temperature for chicks at various ages

Age	Temperature (°F)
Day 1–3	95
Day 4–7	92
Day 8–14	90
Day 15–21	84
Day 22–28	79
Day 29–35	74
Day 36 +	70

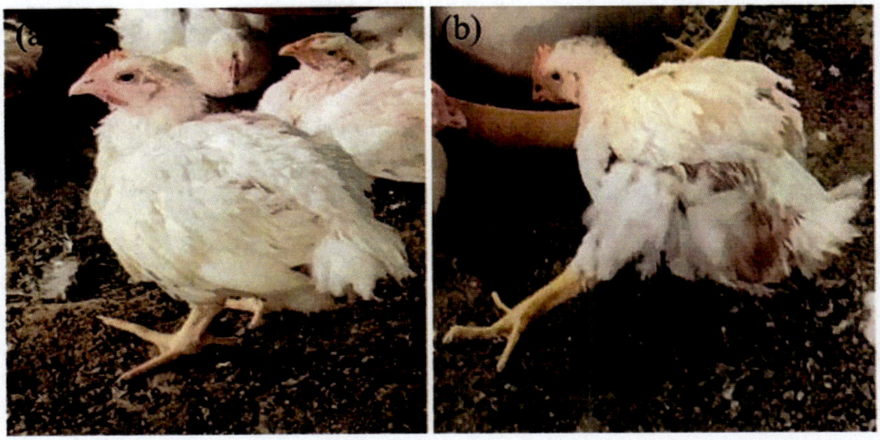

Fig. 1 **a** Healthy chickens, **b** diseased chickens

and future work. (iv) A summary table of different models along with their accuracy is discussed in this work.

The remaining portions of this paper are structured as follows: Sect. 2 summarizes common poultry diseases in chickens. Section 3 describes related work based on different technologies. The conclusion and future scope are covered in Sect. 4.

2 Common Poultry Disease in Chickens

Poultry bird illnesses are the most serious risks to poultry husbandry resulting in limiting productivity. Diseases in chicken farming cause significant economic losses to the farmers. The prevalence of poultry illnesses can endanger human health too. The common diseases found in chickens are described below.

2.1 Newcastle Disease (ND)

Newcastle disease (ND) is a prevalent chicken illness with serious consequences for poultry health and production. ND has a detrimental effect on the circulatory tract, changing the insulative impedance of bird speech patterns. Since its inception, ND has been a major health concern with massive financial damage because of significant mortality infection control expenses. Vaccination with attenuated, lentogenic, or mesogenic strains is used to manage ND [9]. Infected birds spread the virus in their feces, feces fluids (such as respiratory secretions from the mouth, and eyes), and eggs. When birds become infected with the Newcastle disease virus, they can shed a large amount of the virus. Avian paramyxovirus Type 1 strains with high virulence are the main causes of the disease. The Newcastle disease manifests itself in three different stages: lentogenic or mild, mesogenic or moderate, and exceedingly severe [10]. Although lentogenic strains are frequently found, only a few serious infections are induced by them. It is most commonly seen as respiratory sickness, although this can also be marked as depression, anxious symptoms, and diarrhea. It is an OIE-listed disease that needs to be addressed in its most virulent state. As ND can mimic the clinical presentation of avian influenza, laboratory testing is essential to establish the diagnosis. Aerosol vaccination is more effective and generates strong protection in a shorter amount of time than immunization through drinking water. Antibodies start to show up in the serum and mucosa 4 to 5 days after inoculation and attain their peak after the third or fourth week [11].

2.2 Avian Influenza

Avian influenza (AI) is an infectious avian respiratory illness characterized by the high or low pathogenic avian influenza virus. Highly pathogenic avian influenza viruses often induce moderate to severe infections with a high death rate, whereas LPAI viruses generally generate much milder, primarily respiratory sickness [12]. Influenza A virus (IAV) is a serious menace to poultry populations around the world. More than 10% of the domestic poultry population in South Korea had to be put to death as a result of an IAV outbreak in the years 2017 and 2018. In a similar vein, IAV outbreaks in the US in 2014–2015 resulted in economic losses of up to $3.3 billion. These included the expenses associated with the lost output, the adoption of suitable disease preventive measures, and the indirect costs associated with a decline in confidence in the industry [13]. IAVs are classified according to the antigenicity of two surface glycoproteins: hemagglutinin (HA) and neuraminidase (NA).

There are nine NA (N1–N9) and a total of 16 HA (H1–H16) subtypes in the wild waterfowl that serves as IAV's natural reservoirs. Numerous attempts have been made to employ genetic modification to guard gallinaceous fowl against IAV. According to reports, Lyall and colleagues launched the first such effort in 2011 [14]. AI viruses are transmitted primarily through the manual transmission of infectious fecal, in which viruses can be present in large quantities and live for extended periods of time. Water or food that has been shared may potentially become infected. Preventive sanitary procedures such as washing and cleaning are essential to prevent illness transmission on the farm [14].

2.3 *Chronic Respiratory Disease (CRD)*

CRD in poultry is essentially a respiratory disease. Mycoplasma gallisepticum, pathogenic bacteria, is the primary cause of CRD in hens. Mycoplasma gallisepticum, with or without subsequent problems, is the specific organism directly related to CRD. In straightforward outbreaks involving just pathogenic avian PPLC (Mycoplasma), the name "Avian Respiratory Mycoplasmosis" (ARM) should be used, while the word "CRD" should be used when PPLO infection is overlaid with another illness, according to the advice of the FAQ committee meeting in May 1969. Although the mortality from CRD is not significant, it is relevant because it makes the birds more susceptible to infection from other disease-causing infections [15]. CRD in poultry is a disease that affects broiler, layer, and breeder chickens, as well as other birds such as pigeons and ducks. Many additional diseases can emerge in birds if CRD is not properly treated. The total prevalence of CRD was 11.50%, summer had the highest occurrence, followed by winter and the wet season. CRD was most common in broiler-type birds, followed by indigenous and layer-type chickens. The estimate was 6.62% in chicks, 18.52% in growers, and 9.25% in adults, respectively. Eggs are the primary mode of transmission for M. gallisepticum, but droppings and nasal secretions are other ways for birds to spread bacteria. Furthermore, it can be transferred via the hands, feet, and clothing of visitors' companions [12]. To prevent birds and chickens from CRD illness, it is essential to always purchase chicks from reputable hatcheries, since the quality of the chicks has a direct impact on the health of the hens and the profitability of the poultry company. Respiratory herbs are a sure-fire and world-famous therapy for CRD sickness in hens. Feed 100 birds, in the morning, take 5 ml of respiratory herbs, and at night, take 15 ml of amino power. When combining respiratory herbs with water, keep in mind that the hens should consume/drink the mixture within an hour. The birds should be given respiratory herbs until the CRD is totally treated.

2.4 *Lameness*

Leg weakness is a comprehensive term that covers both infectious and non-infectious traits present in modern, quickly expanding broilers. Lameness is a broad term for a variety of problems affecting broiler chickens from both infectious and non-infectious sources. Lame broilers cannot walk freely and, as a result, cannot reach the feeder or drinker when hungry or thirsty. Their living quality is suffering because of their disability [16]. The presence of lameness is substantially associated with broiler weight and quick growth. However, broiler mobility issues might be unpleasant. It can reduce broiler activity and cause problems such as hock burns and chest dirtiness. Infectious, developmental, metabolic, and degenerative illnesses can cause leg weakness [17]. Lameness is associated with a higher frequency of morbidity than mortality. However, the handicapped bird suffers from discomfort and is unable to get food and water, and dies because of hunger. Infectious disorders, heredity, gender, weight and growth rate, maturity, nutrient utilization, diet, treatment, and mobility are aspects that relate to disability. Lameness can be minimized by improving one's surroundings and lifestyle [18]. There are 61 illnesses in all that have been classified. The ten most common diseases are given in Table 2 and are identified with the assistance of veterinary physicians, chicken farm supervisors, and owners.

3 Chicken Disease Prediction Approaches

Technological innovations assist poultry industrialists in establishing optimal bird well-being by setting up surveillance and monitoring techniques to observe the health of poultry chickens. The existence of flocks or clusters of poultry chickens allows for the examination and modification of the location of feeders and drinkers. Moving, perching, eating, and drinking are all activities that may aid in determining the health of the poultry chicken. The monitoring and tracking system is based on the early diagnosis of organisms in livestock production. Different methodologies are used to detect the illness in broilers. Figure 2 illustrates various techniques to identify chicken diseases.

3.1 Image Processing and Computer Vision

The practice of applying adjustments to an image in order to draw out pertinent information from it is known as image processing. It is a type of signal processing where an image serves as the input and either an image or image characteristics or features can serve as the output. By the use of processors, digital image processing techniques enable the modification of digital images. When adopting digital approaches, all types of data must go through three general processes: pre-processing, augmentation, and presentation, as well as information extraction [27]. Further, to discover turns and behaviors, the flock's activity can also be observed and examined in the future. Image processing makes it possible to autonomously investigate chicken behavior for little to no money, including the identification of health problems, forecasting weight, and continuous monitoring [12]. The detection of abnormal eating behavior within the flock could be taken as an ominous sign of poultry chicken health.

Table 2 Common poultry disease with causes, symptoms, prevention, and treatment

S. No.	Name	Cause	Symptoms	Prevention	Treatment
1	Newcastle disease [19] 	Para-myxo virus	Coughing, gasping, appetite loss, watery eyes, and vivid green diarrhea	Hygiene and immunization	No specific treatment, antibiotics can be prescribed
2	Avian influenza [20] 	Orthomyxoviridae family virus	Lack of energy, appetite, purple discoloration	Hygiene	No treatment
3	Chronic respiratory disease 	Mycoplasma gallisepticum bacteria	Sneezing, coughing, respiratory problems	Sanitation, proper vaccination	Antibiotics
4	Lameness 	Bird weight, bacterial illnesses, or litter condition	Trembling legs, Reluctance to relocate, Legs bent or twisted at the ankle joint	Antibiotics	Antibiotics or antiviral drugs, improving bird's diet
5	Fowl pox [21] 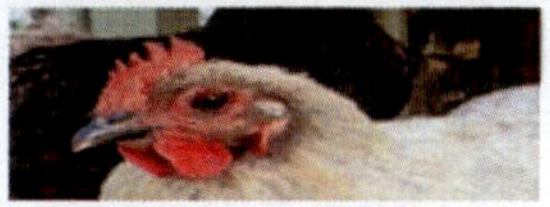	POxbridge virus and Avipox virus	Pimples or scabs on the skin, yellow lesions in the mouth	Vaccination by the wing web method	No treatment

(continued)

Table 2 (continued)

S. No.	Name	Cause	Symptoms	Prevention	Treatment
6	Cholera [22] 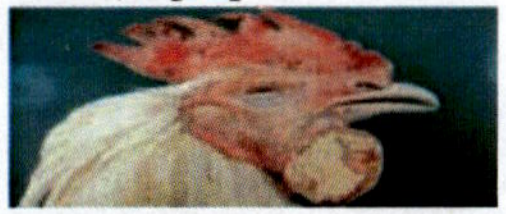	Pasteurella multocida bacteria	Ruffled feathers, mucoid discharge from mouth	Sanitation	Antibiotics, Vaccination, Sulfamethazine doses
7	Gumboro disease (IBD) [23] 	Infectious bursal disease virus	Sensatons, damaged feathers, appetite loss, dryness,	Vaccination	No treatment
8	Fowl typhoid [24] 	Gram-negative bacteria, Salmonella Gallinarum, and S.Pullorum	Anorexia, diarrhea, dehydration, weakness,	Sanitation, monitoring	Antibiotics
9	Coccidiosis [25] 	Apicomplexan parasite (Elimeria)	Diarrhea, depression	Sanitation and litter management	Sulfa-class antibiotic, Sulfa-di-methoxine
10	Infectious Coryzaa [26] 	Haemophilus paragallinarum bacteria	Nasal discharge, facial puffiness, anorexia	Biosecurity measures	Erythromycin and oxytetracy cline

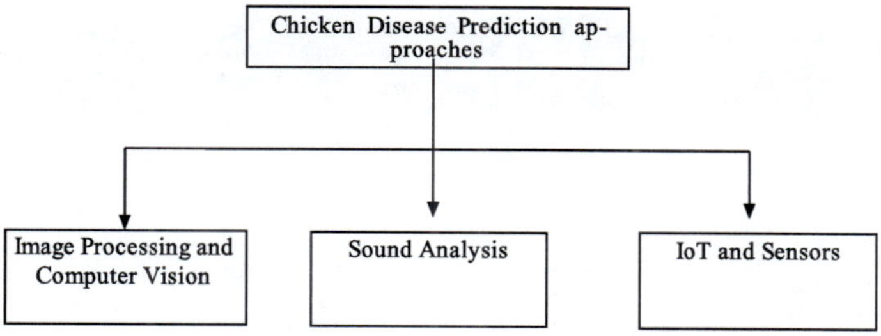

Fig. 2 Approaches used for chicken disease prediction

Fig. 3 Image processing process

The analysis of poultry bird excrement enables the monitoring of their activities as well as the early detection of sickness. The techniques of image processing were utilized to prepare the dataset for analysis. A 99.17 percent accurate convolutional neural network (CNN) implementation is used for a machine vision system for crowd surveillance of chicken chicks around feeders [12]. The illnesses identification and categorization are accomplished through the modeling of body position using an SVM classifier, which has 99.47% accuracy [8]. ResNet is a residual network-based automated approach for detecting sick chickens. The classic ResNet residual network served as the foundation for this system. By enhancing the overall architecture of ResNet, a more versatile ResNet-FPN sickness chicken classifier was created [4]. The various phases involved in image processing techniques are shown in Fig. 3. Image processing involves a series of steps that can be modified to the specific application and requirements. These steps can include preprocessing, segmentation, feature extraction, recognition, and many others, and they are critical in enabling machines to interpret and analyze visual data.

3.2 *The Sound Analysis*

The sound analysis of poultry hens can identify the social relationships of the broilers. The health of the chicken can be predicted using techniques such as energy distribution, amplitude, frequency, and frequency distributions. The pace of growth is significant in determining healthy and unhealthy growth patterns. The pecking habit of poultry hens is closely tied to their feeding activity. Furthermore, pecking noises successfully recognized 90% of the meal consumption. The peak frequencies emitted by the chickens were also used to assess the growth of the poultry birds. The peak

Fig. 4 Sound analysis process

frequency dropped as the poultry chicken's growth rate increased, allowing for the recognition of the poultry chicken's growth rate [13]. Sound analysis may also be used to make disease diagnoses. In poultry farms, healthy and ill hens make distinct noises depending on their health. A supervised learning neural network was used to analyze and classify healthy and ill hens with 100% accuracy. Using sound analysis, an experiment was carried out to determine whether or not the fowl chicken was contaminated with Newcastle, pneumonia, or bird flu. To identify avian influenza in poultry hens, sound (noise) analysis was used. RFID microchips and accelerometers were utilized in wearable sensing devices to control and manage spots. A message is sent from an RFID chip to an RFID reader. The RFID technology may be used to locate the poultry chickens when they are detected in a magnetic field, and the entire movement of the poultry chicken inside the group can be observed. The use of IoT devices as well as wearable sensor devices allows for real-time observation of poultry birds' mobility. An RFID reader was set in a poultry nest, and RFID tags were implanted in the poultry birds to track how often the hens enter and exit the nest [28]. The multiple steps in the process of sound analysis are shown in Fig. 4. In machine learning, audio signal processing involves capturing and preprocessing audio data, using a variety of signal processing techniques like pre-emphasis, framing, and the Hamming window, using the FFT to convert the signal to the frequency domain, applying a Mel filter bank to extract MFCCs, applying the DCT to decorrelate the features, and using the resulting MFCCs as input features for machine learning algorithms.

3.3 IoT and Sensors

The IoT-based decision and support systems rely on machine learning models due to the limited computational and processing resources within IoT peripheral systems. Real-world IoT devices also provide new data to machine learning models, allowing them to be trained for optimal performance. Transferring all of the data to the cloud as well as analyzing presents additional hurdles for researchers. It assists them in solving huge data transfer and computation challenges over the virtual machine. The IoT-fog-cloud ecosystem was described, which enables the creation of a ubiquitous cloud environment for data acquisition. The research also examined the issues of big data and diversity in order to develop autonomous fog resilience components [28]. A micro-service framework was designed to build IoT-based, context-aware decision systems with autonomous functionalities. The study presented the micro-service framework to handle the latency issue, as well as the difficulties of large data, device heterogeneity, and fog resiliency. The electronic skin idea, dubbed 'skin' was suggested by researchers, who projected the health of poultry hens using transistors attached to the skin of the poultry chickens and transmitting nonstop data [28]. Table 3 shows the methodologies used by researchers to predict chicken disease along with their precision, effectiveness, limitations, and future relevance.

4 Conclusion and Future Scope

Numerous methods are used to predict infections in chickens and to save them from becoming infected. The infectious birds can be recognized via images, videos, and sound analysis, and several procedures are used to ensure prediction accuracy. The applications of several machine learning algorithms help in the early disease identification in broilers and prevent farm economic losses. Early detection is essential to eliminate chickens with transmissible diseases. The paper describes various diseases, including their genesis, indications, prevention, and medication. The table analyzes the accuracy of the various disease prediction models based on machine learning algorithms. As compared to other models, CNN and SVM have the highest accuracy (99.7% and 99.469%, respectively), making them the best approach. In the future, researchers will be able to study different diseases and apply new technology to anticipate diseases in chickens. Most of the existing work is focused on images and videos, one can also work on sound analysis and various sensor devices, so that early prediction of chicks is possible and losses to poultry farmers are avoided, as this sector contributes significantly to the country's economy.

Table 3 Techniques used for disease prediction in chickens

Reference and year	Dataset	Algorithm used	Accuracy	Strength	Limitation	Future scope
[12] 2019	Broiler dataset (B1) (14,728 images) PASCAL VOC2012 (17,125 images)	Digital image processing, CNN	99.7%	Methods can be used to determine the health state of broilers based on their appearance	Cannot determine which illnesses are afflicting chicken	To conduct additional research on various types of poultry sickness
[4] 2020	5000 diseased chicks picture divided into training and verification set	ResNet residual network	95%	Improves broiler breeding survival rate, increases production efficiency	A camera	Camera device can be installed for real-time surveillance
[15] 2019	344 video surveillance 34,280 images(70% training set & testing set)	SVM, ANN	0.978	During course of birds' lives, proposed monitoring system can take measurements secretly and autonomously	This system must be evaluated on various chicken breeds and infections	System should incorporate new features bio-response surveillance systems
[8] 2018	200 images (training group) 565 samples (text group)	Digital image processing, SVM, K-means clustering	99.469%	Achieve the accuracy requirement, reduce false detections	When the hens look at the camera, either directly or indirectly	Extraction can be done on larger images
[29] 2017	700 images (training group) 200 images (test model)	Image processing technique	–	Determine the various diseases in poultry birds, and poultry breeders will take immediate action to cure or avoid diseases	The regional focus of study would be limited to state of Maharashtra. Service not currently available	The same study should be performed in multiple states based on the breeds and climate in each state

(continued)

Table 3 (continued)

Reference and year	Dataset	Algorithm used	Accuracy	Strength	Limitation	Future scope
[13] 2019	87.25%verification set, 88.12% and 87.12% (for S-kernel function	SVM	Between 84 and 90%	In large-scale chicken farms, early diagnosis of pandemic disease is done	–	–
[28] 2021	20-weeks dataset 10,000 entries in the fabricated database (originate through CTGAN)	GAN (Deep learning)	97%	A predictive service model based on industrial IoT that can categorize poultry birds more correctly in real time	Sound isolation, determining the specific broiler that created audio wave	Real-time predictive IoT service that uses Bayesian network model
[18] 2021	600 mixed-sex one-day-old chicks	Decision tree	91%	By monitoring the bird displacement velocity, an innovative methodology could be utilized	One attribute is considered	Allow automatic identification of flock lameness
[30] 2020	1590 images healthy 508 coccidiosis 516 salmonella 566	SVM & Decision tree, CNN	94%	Methods useful for reducing losses and increasing output	–	Obtain fecal photographs to enhance the collection, model will be applied to mobile devices for end-user engagement
[31] 2019	10,000 dropping images, 10 augmentation methods, Darknet53 (contains 53 convolution layers)	Image processing R-CNN YOLO-V3	99.1%	Technological support for the early diagnosis of digestive tract problem	–	–

References

1. Kara OAMAH (2014) Annual report of 2021–2022. Pap Knowl Towar Media Hist Doc 7(2):107–115 [Online]. Available: file:///E:/Annual Report 2021–2022.pdf
2. Indian poultry industry poised for growth—The Hindu BusinessLine (2021)
3. Bhosale J (2017) GDP: CLFMA of India calls for allied & integrated agriculture industry. The Economic Times. https://economictimes.indiatimes.com/news/economy/agriculture/clfma-of-india-calls-for-allied-integrated-agriculture-industry/articleshow/60701064.cms. Accessed 28 Jun 2022
4. Zhang H, Chen C (2020) Design of sick chicken automatic detection system based on improved residual network
5. Husbandry A (2022) National action plan for egg & poultry-2022 for doubling farmers' income by 2022. Department of Animal Husbandry, Dairying & Fisheries Ministry of Agriculture & Farmers Welfare Government of India
6. Yang CC, Chao K, Chen YR (2005) Development of multispectral image processing algorithms for identification of wholesome, septicemic, and inflammatory process chickens. J Food Eng 69(2):225–234. https://doi.org/10.1016/j.jfoodeng.2004.07.021
7. Kristensen HH, Cornou C (2011) Automatic detection of deviations in activity levels in groups of broiler chickens—a pilot study. Biosyst Eng 109(4):369–376. https://doi.org/10.1016/j.biosystemseng.2011.05.002
8. Zhuang X, Bi M, Guo J, Wu S, Zhang T (2018) Development of an early warning algorithm to detect sick broilers. Comput Electron Agric 144:102–113. https://doi.org/10.1016/j.compag.2017.11.032
9. Zhang J et al (2020) Transcriptome analysis reveals inhibitory effects of lentogenic newcastle disease virus on cell survival and immune function in spleen of commercial layer chicks. Genes (Basel) 11(9):1–15. https://doi.org/10.3390/genes11091003
10. Ball C, Forrester A, Herrmann A, Lemiere S, Ganapathy K (2019) Comparative protective immunity provided by live vaccines of Newcastle disease virus or avian meta pneumo virus when co-administered alongside classical and variant strains of infectious bronchitis virus in day-old broiler chicks. Vaccine 37(52):7566–7575. https://doi.org/10.1016/j.vaccine.2019.09.081
11. Cvetić Ž, Nedeljković G, Jergović M, Bendelja K, Mazija H, Gottstein Ž (2021) Immunogenicity of Newcastle disease virus strain ZG1999HDS applied oculonasally or by means of nebulization to day-old chicks. Poult Sci 100(4). https://doi.org/10.1016/j.psj.2021.01.024
12. Admassu B et al (2019) Detection of sick broilers by digital image processing and deep learning. Biosyst Eng 179:106–116. https://doi.org/10.1016/j.biosystemseng.2019.01.003
13. Huang J, Wang W, Zhang T (2019) Method for detecting avian influenza disease of chickens based on sound analysis. Biosyst Eng 180:16–24. https://doi.org/10.1016/j.biosystemseng.2019.01.015
14. Lyall J, Irvine RM, Sherman A, McKinley TJ, Núñez A, Purdie A, Outtrim L, Brown IH, Rolleston-Smith G, Sang H, Tiley L (2011) Suppression of avian influenza transmission in genetically modified chickens. Science 331:223–226
15. Okinda C et al (2019) A machine vision system for early detection and prediction of sick birds: a broiler chicken model. Biosyst Eng 188:229–242. https://doi.org/10.1016/j.biosystemseng.2019.09.015
16. Silvera AM, Knowles TG, Butterworth A, Berckmans D, Vranken E, Blokhuis HJ (2017) Lameness assessment with automatic monitoring of activity in commercial broiler flocks. Poult Sci 96(7):2013–2017. https://doi.org/10.3382/ps/pex023
17. Aydin A (2018) Leg weaknesses and lameness assessment methods in broiler chickens. Arch Anim Husb Dairy Sci 1(2):4–9. https://doi.org/10.33552/aahds.2018.01.000506
18. de Alencar Nääs I, da Silva Lima ND, Gonçalves RF, Antonio de Lima L, Ungaro H, Minoro Abe J (2021) Lameness prediction in broiler chicken using a machine learning technique. Inf Process Agric 8(3):409–418. https://doi.org/10.1016/j.inpa.2020.10.003

19. Lera R (2021) Newcastle [Online]. Available: data:image/jpeg;base64,/9j/4AAQSkZJRgABAQAAAQABAAD/2wCEAAoHCBYVFRgXFhYZGRgaGhodHBwYHBwcGhkYHCEaGRoaHhocIS4lHyErHxwaJjgmKy8xNTU1HCQ7QDs0Py40NTEBDAwMEA8QGhISHjQISExNDQ0MTQ0NDQxNDE0NDQ0NDQ0NDQ0NDQ0NDQ0NDQ0NDQ0NDQ0ND8/ NDQ/NDQxNDExMf/AABEIAMMBAwMBIgACEQEDEQH/
20. Aviaan Influenza (2021) [Online]. Available: https://cs-tf.com/wp-content/uploads/2021/09/how-to-prevent-bird-flu-in-chickens-scaled.jpg
21. Fowl-pox-in-chickens (2021) [Online]. Available: https://1.bp.blogspot.com/-aFQtOH-k0PA/YHzMUPj5ELI/AAAAAAAADEWk/WCp7sKbx39kgV9UUSL3ZQpiO5bdC82PfQCLcBGAsYHQ/w1200-h630-p-k-no-nu/fowl-pox-in-chickens.jpg
22. ByIVANDIVEN D (2005) Cholera [Online]. Available: https://cdn.globalagmedia.com/poultry/legacy/publications/images/image_Page_017_Image_0005.jpg
23. Gumboro Disease (2011) Vet Q [Online]. Available: https://encrypted-tbn0.gstatic.com/images?q=tbn:ANd9GcQdVKNm3Y8YA58I89FkgiLs_U4YRPQUC8v9AQ&usqp=CAU
24. Truche C (1923) Fowl typhoid. J Comp Pathol Therap 36:133–137. https://doi.org/10.1016/s0368-1742(23)80025-x
25. Lucyin (2021) Coccidiosis [Online]. Available: https://upload.wikimedia.org/wikipedia/commons/7/75/Coccidiôze_sblarixhaedje_houfès_plomes.jpg
26. LaceyHughett (2019) Infectious [Online]. Available: https://www.wattagnet.com/ext/resources/Images-by-monthyear/19_03/poultry/coryza-birds-white-discharge.jpg
27. Zhuang X, Bi M, Guo J, Wu S, Zhang T (2018) Development of an early warning algorithm to detect sick broilers. Comp Electr Agricult 144:102–113. https://doi.org/10.1016/j.compag.2017.11.032
28. Ahmed G, Malick RAS, Akhunzada A, Zahid S, Sagri MR, Gani A (2021) An approach towards IoT-based predictive service for early detection of diseases in poultry chickens. Sustainability 13(23). https://doi.org/10.3390/su132313396
29. Bhasvar D (2017) Diagnosis and Identification of visually example poultry bird diseases with the help of image processing techniques. IICMR Res J 11(2):32–38
30. Mbelwa H, Machuve D, Mbelwa J (2021) Deep convolutional neural network for chicken diseases detection. Int J Adv Comp Sci Appl (IJACSA) 12(2). http://dx.doi.org/10.14569/IJACSA.2021.0120295
31. Wang J, Shen M, Liu L, Xu Y, Okinda C (2019) Recognition and classification of broiler droppings based on deep convolutional neural network. Sensors 2019. https://doi.org/10.1155/2019/3823515

Chapter 60
Technological Approach Toward Smart Grid Security: A Review

Saish Kothawade, Akshat Dubey, Anush Shetty, Kartik Chaudhari, and Rachana Patil

1 Introduction

The term "grid" has historically been used to refer to an electrical system that supports all or more of the following four functions: power production, transmission, distribution, and control [1]. In a conventional transmission system, energy is produced in real time, taking into account seasonal and tidal variations in electrical consumption. When demand is at its highest, electricity production, transmission, and distribution systems must be able to keep up with it [2]. In order to make our twentieth-century energy system smarter and capable of making its own choices, it is urgently necessary to integrate IoT devices into the grid. However, since energy is created in real time, we need quicker and more effective data interchange in order to take use of the massive quantity of data arriving from IoT devices.

The use of information and communication technology in the smart grid has raised the risk of cyberattacks and unauthorized data access. As cloud computing offers a platform for data collecting and analysis, it can improve smart grid security by gathering data from all network areas, detecting threats, and weaknesses, and exchanging information with relevant users such as utilities, governmental bodies, and research institutions. Cloud-based monitoring technologies can increase smart grid visibility and detect potential faults. Smart grid security is essential as it is a critical infrastructure. Cloud computing can help secure the smart grid by offering a platform for data gathering, analysis, and network visibility.

The main goal of these initiatives is to create a sustainable society, but the centralized grid struggles with many connections. As a consequence, the smart grid topology

S. Kothawade (✉) · A. Dubey · A. Shetty · K. Chaudhari · R. Patil
Department of Computer Engineering, Pimpri Chinchwad College of Engineering, Pune, India
e-mail: saish.kothawade20@pccoepune.org

A. Yadav et al. (eds.), *Proceedings of International Conference on Paradigms of Communication, Computing and Data Analytics*, Algorithms for Intelligent Systems,
https://doi.org/10.1007/978-981-99-4626-6_60

is becoming decentralized. On the other hand, the blockchain with its excellent security traits is suitable for this, with its immutable transactions and history for auditing and resolving transactional disputes between producers and consumers.

So what is smart grid? It is basically a contemporary electrical power network that employs two-way digital communication to provide energy to its users [3]. To increase efficiency, lower energy usage and costs, and encourage the openness and dependability of the energy supply chain, this system provides monitoring, analytics, control, and communications across the distribution chain [3]. It integrates various components such as sensors, automation systems, energy storage devices, renewable energy sources, and intelligent devices that communicate with each other to optimize the grid's performance.

The remainder of the paper is organized as follows: Sect. 2 focuses on the literature review, subdivided into three sections, it addresses the previous work undertaken on securing the smart grid using cloud computing, IoT-enabled edge computing, and blockchain technology. The existing privacy and security concerns applicable to electricity grids are addressed in Sect. 3. The table in Sect. 4 provides a comprehensive comparative study of all the three studied technologies. Section 5 includes illustrations of the conclusion and future scope.

2 Related Work

In this section, we mainly describe the relevant work in three different areas of smart grid security using: cloud computing-based approach, edge computing-based approach, and blockchain-based approach.

2.1 Security of Smart Grid Network Using Cloud Computing

2.1.1 Role of Cloud Computing in Smart Grid

Cloud computing is a recent IT innovation that offers numerous advantages over traditional computer processing techniques. It provides scalable processing, storage, and network capabilities to any Internet-connected device, making it user-friendly and flexible. Cloud computing can help experimental smart grid initiatives overcome cost–benefit hurdles and improve data management. It consumes less power, has low operating costs, and is more adaptable and agile. It is expected to become the computing standard soon. Additionally, cloud computing improves smart grid data management by reacting to changes in data volume and effectively utilizing computational resources. It helps utilities speed up network performance and problem resolution by collecting and analyzing smart grid data to improve system management [4]. A landscape for integrating the cloud and smart grid is shown in Fig. 1.

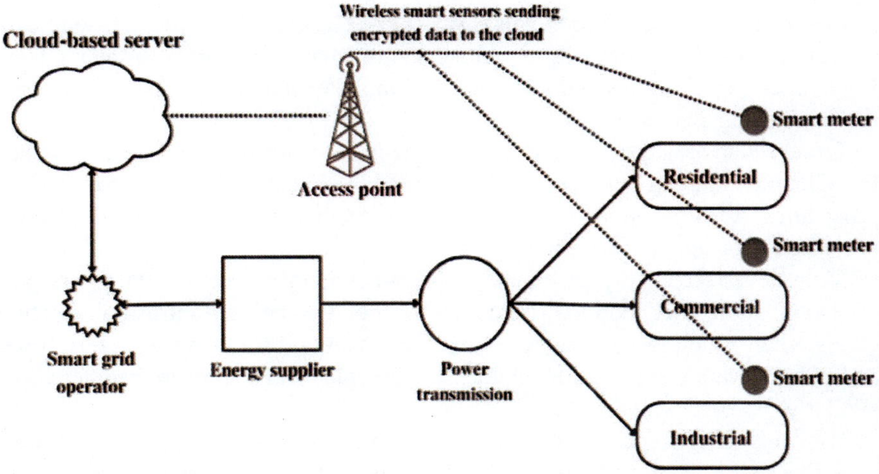

Fig. 1 Smart grid cloud integration landscape

Smart grids, with their interconnectedness, are vulnerable to cyberattacks and illegal power consumption. Manipulating data is also a security concern. A secure and reliable architecture is needed for complex grid systems.

Smart grids are more advanced than traditional ones due to intelligent sensor networks, wireless technologies, and smart meters, which complicate information security. With millions of smart meters in use, end-user protection is crucial as DoS attacks can disrupt smart grid apps. Utilities and other parties have access to users' personal data.

2.1.2 Various Smart Grid Security Technologies Using Cloud Computing

Electric power sector information protection solutions are insufficient for the continuously expanding complexity of security concerns [5]. In smart grid development, researchers have proposed numerous cloud-based application security techniques [5–10].

Yanliang et al. [5] present a cloud-based power transmission information safety and defense system. The authors split cloud security between server and client. Server data and outcomes drive client behavior. Using the cloud services platform to build distributed storage, the server makes smart decisions. Customers get results through their selected Web media.

Simmhan et al. [6] examined smart grid software designs deployed in different cloud environments for security and privacy vulnerabilities to address these issues. Private cloud systems manage enormous user data easily. Cloud computing can help electric utilities quickly and cheaply remove malware, improving network security and lowering maintenance costs.

Distributed verification protocols (DVPs) like Ugale et al. [7] protect cloud data storage. Cloud computing was recommended to smart grid clients to save money. Ugale et al. [7] recommended DVP protocol implementation for smart grid energy management and storage.

Power cloud services for smart grid development address several security issues [8]. Cloud computing increases confidence and participation in traditional security procedures. A trusted administrator's access to sensitive data and the cloud platform provider's data center are examples.

Maheshwari et al. [9] developed a cloud-based infrastructure for smart grid state estimation. PKI adoption overcomes failure tolerance and intrusion detection issues. Their state estimate technique maximizes cloud bandwidth for robust, safe, and failure-tolerant smart grid architecture. This platform's software handles cores similarly.

Wen et al. [10] propose a smart grid smart metering system that safeguards user data. The suggested method would transfer encrypted data from each smart meter to a cloud server. Thus, only authorized parties may access it. Data inquiries might be limited. The query is divided into two query tokens to collect the essential information while safeguarding users' identity. Table 1 displays a comparison of several smart grid cloud application security features. Cyber-physical, data security and privacy, and threat detection are some of the smart security aspects compared.

The authors of [5–10] offer a variety of security mechanisms for use in a cloud-based smart grid design; they are summarized and expanded upon in Table 2.

Yangling et al. [5] proposed smart grid server–client protection. Cloud software as a service can protect client data and drive smart grid adoption. Simmhan et al. [6] created a cloud-based security and privacy framework to resist harmful software attacks. This software platform uses public cloud computing and these services need strict privacy rules.

DVP is recommended for smart grid cloud data storage security [7] as it prevents accidental and intentional data leaks. Yang et al. [8] identified many power cloud computing security issues. Service quality improvement and technology security may

Table 1 Comparison of several smart grid cloud security features

Cloud security model name	Cybersecurity	Data security and privacy	Threat detection
Software platform for server and client security [5]	Yes	No	Yes
Software architecture for security and privacy [6]	Yes	Yes	Yes
Distributed verification protocol [7]	Yes	No	No
Power cloud computing [8]	No	Yes	Yes
State estimation method [9]	No	Yes	Yes
Privacy preservation [10]	No	Yes	Yes

Table 2 Various security mechanisms for use in a cloud-based smart grid design

Cloud security model name	Cloud computing applications	Future scope
Electric power grid data security system [5]	Server plays role of cloud and gathers information from its clientele	Potential for the development of cloud-based SaaS solutions to data privacy concerns raised
SG software's data privacy concerns [6]	Helps utilities rapidly and efficiently eliminate harmful programs	Smart meter data can be supported with proper security and privacy regulations
Distributed verification protocol [7]	Safety of data in cloud storage	Prevent data leakage with a distributed verification protocol
Security technologies for cloud computing [8]	Cloud services increase the radius of trust	Enhance QoS mechanisms and non-tech challenges via extended power cloud apps
Real-time state estimation for smart grid [9]	Cloud-hosted system's current state estimation method	Mission-oriented security solution assessment strategy for smart grid
Guaranteeing confidentiality range query [10]	Cloud-based information privacy preserving scheme	Incorporating a ranked range query, while guaranteeing users' confidentiality

be addressed together. Maheshwari et al. [9] recommend cloud applications for real-time state estimations. Wen et al. [10] present cloud range searches for data access. The recommended approach protects users' data in the cloud. Thus, the cloud-based solution helps support several enterprises on one platform while maintaining user privacy and data integrity.

2.2 *Security of Smart Grid Network Using Edge Computing*

With the advent of smart grids, the system's complexity is increased to a new level, allowing for a larger spread of inexpensive information technology (IT), two-way communication, energy flows, and interaction with other information systems [11]. We can employ Internet of Things (IoT) devices to establish a two-way communication channel. IoT devices produce enormous volumes of data, but have finite resources. IoT device application workloads should thus be transferred to distant cloud data centers. When all jobs are relocated to the cloud, the network is heavily used. Edge computing may be used to tackle this issue.

2.2.1 Edge Computing for IoT-Enabled Smart Grid Network

The Internet of Things (IoT) is rapidly expanding, with numerous smart devices being connected online. This results in bandwidth shortages, privacy concerns, security issues, and slow response times with traditional cloud computing. To address these problems, a new computing paradigm called "edge" has emerged [12]. Millions of interconnected smart devices in composite networks can provide crucial infrastructure and communication monitoring and control.

The processing and computation of client data closer to the data source rather than a centralized server are referred to as edge computing. The processed data can be then sent to the cloud for further computation as shown in Fig. 2. Essentially, edge brings computing resources, data storage, and business applications closer to the end-users, enabling them to access information more efficiently.

Edge computing offers various benefits to the smart grid, such as the following:

(1) Reduced Delay: When delivering power, the final few miles to end customers, power distribution and transmission networks are crucial. Edge computing can ensure low latency, enabling real-time grid frequency monitoring and proactive decision making to reduce power factor fines.
(2) Data Security: Smart grids are dealing with private and sensitive user data as smart houses and meters are becoming more common. Edge computing reduces data risk by processing it locally, deciding which data needs to go to the cloud and which can stay local. This is crucial as public cloud services are often located outside the region, limiting local control over data for residents and government.
(3) Lower bandwidth equals greater data collection: Renewable energy is popular because users can save or profit by selling excess electricity back to the grid. To do this, users need to estimate their energy generation or consumption. Edge computing can help create precise forecasts and models, taking into account factors like weather and location, and bypass the expense and resource drain of sending data to the cloud. This allows for more local data acquisition, filtering, and processing, which can enrich the models. As a result, the demand for renewable energy will increase, and people may prefer to purchase it rather than produce it themselves [13].

2.2.2 Smart Grid and 5G Enabled IoT

Future smart grid infrastructure incorporating IoT devices offers numerous benefits, such as enhanced supervisory control and data acquisition (SCADA) capabilities, advanced measurement infrastructure, and improved monitoring and management of operational assets. The digitization of power grids with 5G cellular networks provides low latency, ultra-high speed, and enhanced reliability, which can alleviate smart grid difficulties faced by energy firms. These difficulties include connecting a large number of sensors and providing universal coverage with high security and dependability. The 5G technology also enables an increase in the number of distributed generation sites that distribution networks can accommodate while fulfilling the required security criteria and providing solutions to discovered risks [14]. The edge computing-based computational framework is required for the real-time processing of a sizable volume of data produced by IoT devices in smart grids. The 5G specifications provide edge computing tools for data processing, application hosting and storage close to end devices [14].

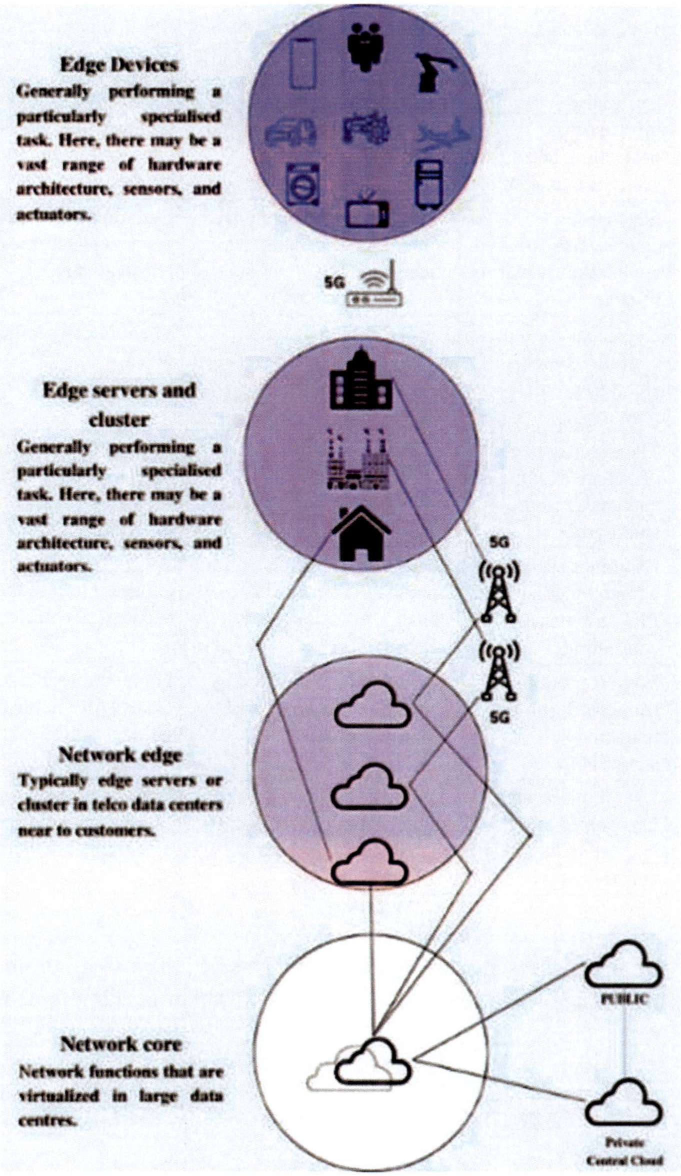

Fig. 2 Edge computing—processing data at the edge

Table 3 presents a summary of various research papers related to edge computing and 5G applications in smart grids. The advantages include improved efficiency, dependability, and fulfillment of security standards, while disadvantages include increased security risks and potential wireless security issues. Some papers do not

Table 3 Content, advantage, and disadvantage in research papers used in this survey

Reference	Content	Advantage	Disadvantage
[1]	Examines infrastructure, protection, and smart systems management	The paper introduces G2V/V2G and microgrid	A robust security system required to use such advanced technology
[15]	Explores characteristics-based application scenarios of edge	Thorough analysis of edge computing applications in smart grid	Geographic dispersal of edge resources increases risk of physical attack
[12]	Key needs for putting edge-IoT-based smart grid into practice	Resolves the real-time requirements, heterogeneous linkage of data & intelligence	There is no specific security given
[14]	The security benefits of 5G are examined and analyzed in smart grids	5G contributes to the fulfillment of security standards and the mitigation of recognized dangers	5G implementation may introduce new wireless security issues, not resolving all existing ones
[16]	Examines security technologies, such as PKI and trusted computing	Distributed intelligence and broadband capabilities enhance efficiency and dependability	By 2030, 3DES systems are expected to lose their security, according to NIST
[17]	Edge computing framework for real-time surveillance	The delay is minimized by 53–79% when compared to cloud surveillance	No mention of associated costs with establishing this entire system
[18]	User requests are maximized while respecting bandwidth constraints	More requests could be fulfilled by this algorithm than by earlier works	Issue of module deployment in edge computing with constrained resources

address security concerns and associated costs. Overall, they highlight the need for robust security systems and real-time capabilities in smart grid implementation.

2.3 *Security of Smart Grid Network–Blockchain Based Approach*

2.3.1 Employing Blockchain in Smart Grid

Blockchain, with its improved security features and functions, provides a secure solution in the intricate smart grid architecture. Prosumers and consumers are able to transact in a peer-to-peer environment without a centralized authority thanks to the inclusion of blockchain in smart grids.

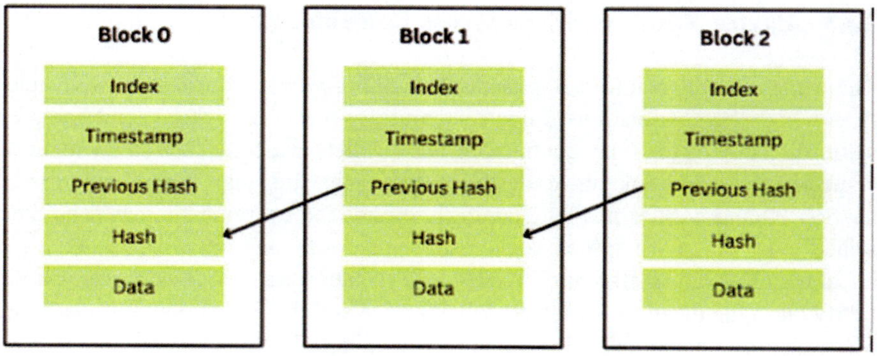

Fig. 3 Overview of blockchain technology

Blockchain is a blend of record-keeping concepts, digital certificates, asymmetric-key cryptography, and cryptographic hash functions. The data is handled automatically by the peer-to-peer architecture after being recorded in a public ledger, where any alteration to the ledger is duplicated in all versions throughout the network [19]. With the use of nodes, blockchain technology transmits data transactions within the grid network called a block. Each block carries the hash address information of the block preceding it, and these blocks are linked to one another as shown in Fig. 3. The network would not accept any changed blocks broadcast by any participant.

2.3.2 How Blockchain Technology Can Start a Microgrid Revolution?

Blockchain technology can simplify the process of monitoring and certifying renewable energy sources, which is necessary for energy grids. Currently, energy generators measure the energy they produce manually and enter the data into a spreadsheet, which is sent to a certifying organization, a process that can take several weeks. The government issues green energy certificates, which energy firms can resell on the open market after validating the data. However, connecting energy producers and consumers to trade certificates is difficult. Using blockchain can simplify the process, as electricity meters at power plants can directly and immediately record data to the blockchain. This can provide a dependable revenue flow for new energy enterprises, while homeowners with solar panels can benefit by selling their excess energy to their neighbors. Brooklyn already has a blockchain-based microgrid where neighbors can trade power, and blockchain can serve as the foundation for decentralized energy production. Blockchain technology has the potential to revolutionize the microgrid industry by streamlining certification, measurement, and sale of renewable energy [20].

2.3.3 Electric Vehicle Charging System Using Blockchain

The system comprises three components: the utility company or grid, electric vehicles (EVs), and charging stations. Initially, the utility registers each EV and charging station as a user on the smart contract and adds wallet balances. When an EV requests a specific price and amount of electricity to charge, the smart contract returns an auction ID and notifies nearby charging stations. The charging stations send their sealed bids, which are hashed versions of their prices, and get a bid id. After the auction closes, the smart contract selects the winner with the lowest price, and the utility takes the mean of the lowest seller's price and the EV's price. It updates the balances of the buyer and seller, and if the auction fails, the EV is notified to send another charging request [20].

2.3.4 Various Blockchain-Based Approaches for Smart Grid Security

Deep Coin Framework: The author suggests deep coin, a smart grid energy system that utilizes deep learning and blockchain through a recurrent neural network algorithm. The proposed model involves three stages: preprocessing, training, and testing of datasets. To sell excess energy to nearby users, deep coin users can use short signatures and hash functions, but this may compromise their privacy [21].

Decentralized NIST Conceptual Model: This paper outlines a theoretical framework by the National Institute of Standards and Technology for blockchain's three key characteristics: decentralization, incentive, and trust. It utilizes specific sub-domains from the NIST Smart Grid Conceptual Model, including customers, markets, generation, service providers, operations, transmission, and distributions. The paper identifies primary focus areas within each domain and specifies a NIST model version for each focus area that is decentralized and blockchain based [22].

Open Smart Grid Protocol (OSGP): This protocol produces security using encryption methods. But this comes with its fair share of limitations: Incorrect messages transmitted while using stream cipher encryption, one key is used to encrypt and one key is utilized for verification. Some other frameworks like open smart grid protocol (OSGP), WEP, ISO/IEC 14,908, and RC4 protocol are also proposed, but these too compromise on security [23].

DS^2 Approach: This paper gives a data-driven, smart, and safe solution for peer-to-peer trading in the local energy market. Smart contracts are used on a private Ethereum blockchain to optimize the trade procedures. DS2 protocol is implemented and evaluated for each household in the market. Smart contracts are used for transactions between prosumers and customers. The algorithms used are power usage prediction, demand prediction, and bipartite graph trading [24].

3 Privacy and Security Issues in Smart Grid Network

Implementation of smart grid technologies raise privacy and safety concerns as they are Internet-connected and susceptible to cyberattacks. These systems collect energy usage data, potentially revealing people's daily routines and behavior patterns, leading to questions about data access and usage [25]. This gives rise to questions such as who will have access to the data and how it will be utilized going forward.

In these systems, there is the potential for a wide variety of cyber and physical security problems to manifest. Among these problems are the following:

1. **Data Breaches**: Smart grid systems gather and store enormous quantities of data, which leaves the data open to the possibility of being breached. If these data are stolen or compromised in any way, it might have huge consequences, such as a delay in service or a power outage.
2. **System Vulnerability**: Because of the inherently complex structure of smart grid systems, there are a large number of possible failure sites. In the event that only one part of the system is hacked, the consequences may be severe enough to render the whole thing inoperable.
3. **Reliance on Technology**: Smart grid systems place a significant amount of reliance on technology, making them potentially vulnerable to cyber assaults. It is possible that a significant disruption to the system will have an effect on the electricity grid.
4. **Privacy Concerns**: Smart grid systems can potentially compromise personal privacy and lead to identity theft, as the information they collect may be exploited for malicious purposes.
5. **Security Breaches**: Smart grids may result in cyberattacks and other types of harm, including theft and property damage, as they can provide access to critical information and infrastructure.

4 Analysis of Related Work

Following our examination and analysis of the aforementioned research publications, Table 4 provides a detailed comparison of the works in question.

5 Conclusion and Future Scope

We studied different research papers on security of smart grid networks using three approaches: cloud computing based, edge computing based, and blockchain based. We mentioned their research findings in this paper. Despite cloud computing's limitations, it is expected to enhance smart grid pricing, computation, data management, power management, and security monitoring as it is superior to conventional technologies. Even though the use of edge computing in smart grids is at an early stage,

Table 4 Evaluation of cloud-based, edge-based, blockchain-based, and traditional surveillance techniques

Parameters	Traditional surveillance	Cloud-based approach	Edge-based approach	Blockchain-based approach
Manual or automated	Manual	Automated	Automated	Automated
Surveillance frequency	Long-term periodic inspection	Real time	Real time	Real time
Threat detection	Manual inspection	Approaches based on deep learning	Approaches based on deep learning	Threats not possible
Server hardware	No server required	Large storage	Medium storage	No server required
Server location	No server required	Placed in server farms by size	Put wireless access points close together	Decentralized system
Network latency	No network	Typically, between 50 and 500 ms	Few tens of milliseconds or less	A lot more than other two approaches
Network traffic	No traffic	Large traffic	Little network traffic	Little to no traffic
Data privacy	No worries about data leakage	Captured data may leak on the Internet	Safer data	Safest option for data protection
Reliability	Poor dependability	Low dependability	High dependability	Highest degree of reliableness
Price	Extremely high	Very high	Average price	High one-time costs

it is a promising one for tackling challenges, especially for energy management and improving sustainability. However, the blockchain approach with its distinct features like decentralized structure, traceability, and resilience is one of the most reliable techniques to overcome grid security related issues. Based on the related work, the privacy and security issues in smart grid networks are identified and the comparative study of all three approaches is done.

Going forward, a three-layer smart grid architecture can be designed consisting of cloud, edge, and blockchain for transmission of smart meter data and peer-to-peer energy trading while applying appropriate security protocols on it.

References

1. Fang X, Misra S, Xue G, Yang D (2012) Smart grid—the new and improved power grid: a survey. IEEE Commun Surv Tutor 14(4):944–980. https://doi.org/10.1109/SURV.2011.101911.00087
2. Alternative Fuels Data Center: electricity production and distribution. afdc.energy.gov/fuels/electricity_production.html
3. Techopedia (2017) Smart Grid. Techopedia.com, 26 Jan. 2017. www.techopedia.com/definition/692/smart-grid
4. Mohsenian-Rad A-H, Leon-Garcia A. Coordination of cloud computing and smart power grids
5. Yanliang W, Song D, Wei-Min L, Tao Z, Yong Y (2010) Research of electric power information security protection on cloud security. In: Proceedings of IEEE POWERCON, pp 1–6
6. Simmhan Y, Kumbhare A, Cao B, Prasanna V (2011) An analysis of security and privacy issues in smart grid software architectures on clouds. In: Proceedings of IEEE International conference on cloud, pp 582–589
7. Ugale B, Soni P, Pema T, Patil A (2011) Role of cloud computing for smart grid of India and its cyber security. In: Proceedings of IEEE NUiCONE, pp 1–5
8. Yang Y, Wu L, Hu W (2021) Security architecture and key technologies for power cloud computing. In: Proceedings of IEEE International conference on TMEE, pp 1717–1720
9. Maheshwari K, Lim M, Wang L, Birman K, van Renesse R (2013) Toward a reliable, secure and fault tolerant smart grid state estimation in the cloud. In: Proceedings of IEEE PES on ISGT, pp 1–6
10. Wen M, Lu R, Zhang K, Lei J, Liang X, Shen X (2013) PaRQ: a privacy-preserving range query scheme over encrypted metering data for smart grid. IEEE Trans Emerg Top Comput 1(1):178–191
11. Palensky P, Kupzog F (2013) Smart grids. Annu Rev Environ Resour 38:201–226. Available at SSRN: https://ssrn.com/abstract=2343668 or https://doi.org/10.1146/annurev-environ-031312-102947
12. Yasir Mehmood M, Oad A, Abrar M, Munir HM, Hasan SF, Abd ul Muqeet H, Golilarz NA (2021) Edge computing for IoT-enabled smart grid. Secur Commun Netw 2021:16. Article ID: 5524025. https://doi.org/10.1155/2021/5524025
13. STL Partners (2022) How can smart grids benefit from edge computing? STL Partners. stlpartners.com/articles/edge-computing/smart-grids-edge-computing
14. Borgaonkar R, Anne Tøndel I, Zenebe Degefa M, Gilje Jaatun M (2021) Improving smart grid security through 5G enabled IoT and edge computing. Concurrency Computat Pract Exper 33:e6466. https://doi.org/10.1002/cpe.6466
15. Feng C, Wang Y, Chen Q, Ding Y, Strbac G, Kang C (2021) Smart grid encounters edge computing: opportunities and applications. Adv Appl Energy 1:100006. ISSN:2666-7924,https://doi.org/10.1016/j.adapen.2020.100006
16. Metke AR, Ekl RL (2010) Security technology for smart grid networks. IEEE Trans Smart Grid 1(1):99–107. https://doi.org/10.1109/TSG.2010.2046347
17. Huang Y, Lu Y, Wang F, Fan X, Liu J, Leung VCM (2018) An edge computing framework for real-time monitoring in smart grid. In: 2018 IEEE International conference on industrial internet (ICII), pp 99–108. https://doi.org/10.1109/ICII.2018.00019
18. Sheu J-P, Pu Y-C, Jagadeesha RB, Chang Y-C (2018) An efficient module deployment algorithm in edge computing. In: 2018 IEEE wireless communications and networking conference workshops (WCNCW), pp 208–213. https://doi.org/10.1109/WCNCW.2018.8369032
19. Synopsys. (n.d.) What is blockchain and how does it work? Retrieved 24 Nov 2022, from https://www.synopsys.com/glossary/what-is-blockchain.html
20. Erturk E, Lopez D, Yu WY (2020) Benefits and Risks of using blockchain in smart energy: a literature review. Contemp Manag Res 15(3):205–225. https://doi.org/10.7903/cmr.19650
21. Ferrag MA, Maglaras L (2020) DeepCoin: a novel deep learning and blockchain-based energy exchange framework for smart grids. IEEE Trans Eng Manage 67(4):1285–1297. https://doi.org/10.1109/TEM.2019.2922936

22. Aderibole A et al (2020) Blockchain technology for smart grids: decentralized NIST conceptual model. IEEE Access 8:43177–43190. https://doi.org/10.1109/ACCESS.2020.2977149
23. Aldabbagh G, Bamasag O, Almasari L, Alsaidalani R, Redwan A, Alsaggaf A (2021) Blockchain for securing smart grids. Int J Distrib Sens Netw 21:255. https://doi.org/10.22937/IJCSNS.2021.21.4.31
24. Zeng Z, Dong M, Miao W, Zhang M, Tang H (2021) A data-driven approach for blockchain-based smart grid system. IEEE Access 9:70061–70070. https://doi.org/10.1109/ACCESS.2021.3076746
25. Boroojeni KG, Amini MH, Iyengar SS (2016) Overview of the security and privacy issues in smart grids. Smart grids: security and privacy issues, pp 1–16. https://doi.org/10.1007/978-3-319-45050-6_1

Chapter 61
Storage and Verification of Medical Records Using Blockchain, Decentralized Storage, and NFTs

Shubham Thakur and Vijay Kumar Chahar

1 Introduction

As we are heading toward digitization in every sector, there is a rapid surge in the digitization of the medical records too. But there are several factors that need to be taken care of, like privacy, security, ease of access, cost of operation, and long-term support. Previously, this digitization of records was being done using cloud, but the problem with this was that storage used was centralized; hence, this poses a security risk such as the central authority have the absolute power to change the stored records, without the patient ever knowing of the change and the centralized storage can at any time go out of service and therefore put the whole system into jeopardy. Blockchain seems to solve this problem quite fluently as it is a public ledger where what once is written and stored in a block can never be changed. This powerful aspect is what helps with the verification of records, and in this paper, we have compared and drawn a contrast between different techniques and technologies used to achieve this record verification and storage, doing so we encounter the situation multitude of times where the projects/approaches are using private blockchains. The justification posed by the authors for taking this approach is that the Ethereum uses a PoW consensus and thus uses high amounts of energy to maintain the consensus, and simultaneously, it has been quoted that the gas fees on Ethereum are high, which is the fees to send transactions on the network, thus making the overall operations costlier. Both cons of the Ethereum network have been solved recently because the network shifted from PoW to PoS which brought the energy consumption down by 99.9% from 93.98

S. Thakur (✉)
Computer Science and Engineering, NIT Hamirpur, Hamirpur, India
e-mail: st802156@gmail.com

V. K. Chahar
Computer Science and Engineering, NIT Jalandhar, Jalandhar, India
e-mail: vijaykumarchahar@nith.ac.in

A. Yadav et al. (eds.), *Proceedings of International Conference on Paradigms of Communication, Computing and Data Analytics*, Algorithms for Intelligent Systems,
https://doi.org/10.1007/978-981-99-4626-6_61

TWh to 0.01 TWh. Thus, given the decentralization and time-tested security of the Ethereum, it makes perfect sense to use that as the chain for our approach. There are a few approaches that have used Ethereum also, but they have rather focused on the storage and privacy of the data and not on the verification of it. In this paper, we have proposed a two-stage verification system which provides trustless security and verification between the two parties, being hospital and the patients. The approach makes sure that the patient cannot produce fraudulent reports or that some other patient is taking the place of the real patient, and additionally, the patient also is secured and hence cannot be conned by the hospitals. This two-stage verification is primarily supported by the fact that the blockchain is immutable and the records once stored cannot be altered ever again, and simultaneously, we have used IPFS as our decentralized storage, so that the storage service never goes down and no central entity can control the data. Additionally, the final piece to our approach and the major one is the use of the NFTs, which are unique non-fungible tokens, i.e., only one of its kind exists and thus makes the verification of the linked report easier and the patient is alerted whenever a new report is made on his/her account. The rest of the paper is laid out as follows. Section 2 contains some important concepts such as blockchain and its types, IPFS, the decentralized storage, consensus mechanisms, and previous approaches which are discussed and then compared against each other on some fundamental parameters. Section 3 presents our proposed approach step by step along with the flow diagrams which pictorially describe the various actors in the system and their interactions with each other; furthermore, Sect. 4 consists of the working of our approach in form of screenshots from all the various stages of the approach. Section 5 of the paper discusses how our approach is better than the closest existing approach in a tabular form stating the pros of our unique approach, and then, in the same section, the process to further improve our methodology is discussed.

2 Background Study

2.1 Components

1. Blockchain

Blockchain is a chain of blocks, and these blocks contain transactions; the number of transactions packed in a block are subjective to the implementation of the blockchain. Transactions are validated by the nodes and can only then be added to the block. This chain of blocks is important to be noted because if an attacker wants to corrupt some transaction in a block, then he/she needs to corrupt the blocks which are next in the chain to the compromised block, because the hash of the previous block is connected to the next block. This can be achieved if the attacker controls 51% of the computation power. Therefore, the more decentralized the chain is, the more secure it is. Now, blockchain can be subdivided into permissioned and permissionless categories.

Permissioned blockchains are those chains which are not open for everyone to join, rather there is some admin entity who can regulate the members and roles that they carry, while simultaneously maintaining the trustlessness among each other, e.g., Hyperledger by the Linux foundation. However, on the other hand, the permissionless blockchain are those chains which are open to all and hence anyone complying to their set protocols, e.g., having a wallet that can interact with the apps on the chain, and hence, in turn with the chains itself, can build and interact with the chain and their tools, e.g., Ethereum.

2. IPFS

Interplanetary file system is a decentralized file storage system. IPFS uses content addressing which implies that the files are not located by the address of the machine which they are on, like in the normal Web. But rather the files are located by the content they have. IPFS achieves this by hashing the entirety of files and when some user needs that file, then it asks all the nodes in the network if they possess any file with that hash. The files are transfer encrypted, and you can have the content encrypted with the protocol of your choice.

3. Consensus Mechanisms

Blockchains consist of transactions, and in the Bitcoin network, these transactions only consist of transfer of assets (BTC), while in the Ethereum network, this can consist of transfer of assets or some data storage only. To order these transactions so that the whole network agrees on a state is what is the job of the consensus mechanism. There are mainly two methods of achieving consensus, proof of work (PoW) where all the nodes/miners on the network try to solve a puzzle which is to find a number that when hashed with previous blocks hash and some other parameters give out a hash that starts with a certain number of zeroes and since this process can only be done through brute force thus the node that does this computation the fastest gets to add the block to the chain with certain defined number of transaction within the block, and simultaneously, all the other nodes have to agree to this change and add that block to their ledger too. The other method is called the proof of stake (PoS), here not all the nodes on the network try to solve the puzzle, rather an algorithm chooses a node to add the block based on several different parameters.

2.2 *Previous Papers and Approaches*

Paper [1] mainly deals with the storage of medical records in such a way that the patient is assured of his/her privacy while the subjective authorities are able to collect information for the greater good. Privacy is achieved by giving the control of the data to the patient itself. A healthcare data gateway (HDG) is discussed which consists of a storage layer which stores all the data in an encrypted manner on the private blockchain and data management layer works both as a medium to access the data and as a mediating interface which processes and serves the queries of external entities. Data usage layers consist of entities that use patient healthcare data. Thus, only storage of records and the distribution of the data in a secure manner is the focus of the approach, and no real, direct benefit to the patient is seen. In paper [2], a consortium blockchain termed HealthChain is proposed which consists of the following actors: hospitals, insurance providers, and governmental agencies. HealthChain implements the governance model via the combination of a membership service provider (MSP), chaincode, and consensus protocol. Peers can be both endorsing and committing peers. The approach draws its methodology and business logic from the Hyperledger and thus suffers from a major flaw, where the order, the one actor that orders the transaction into blocks using FCFS, is the single point of failure. The paper [3] has continued on the workings of the paper [2], i.e., they have used the same methodology of the consortium blockchain and the same set of peers, but here in this paper, they have developed a cryptographic layer that can be used on top of different EHR systems and can enhance their workings. This layer in question works as follows: before the patient data is stored onto the EHR, it is encrypted with a symmetric key which in turn is encrypted using an asymmetric key, yet again this suffers from the central point of failure because of the ordering mechanism. In paper [4] again, we can witness the use of the permissioned Hyperledger from the Linux foundation. The application proposed has patients, healthcare providers, and healthcare administrators. Healthcare admins register patients and doctors onto the platform after performing the necessary background checks. A patient's records are encrypted using the patient's key (symmetric key), and if a certain provider, say doctor, needs that patient's record, then he/she can send a request and the patient sends his/her key encrypted using the public key of the doctor. Only certain defining traits are stored on the chain to keep it light, and the rest of the data and reports are stored on the central database CouchDB. Thus, this approach is centralized in operation. Paper [5] uses Ethereum and smart contracts written in solidity. The doctors first need to register onto the website under the name of the hospital, and the patient also needs to create an account to have an identity and hence have some record on him. The doctors can view a patient's details only if the patient allows him/her, i.e., provides the private key to decrypt the reports. The validity of a report is being checked by checking if the hash of the key fields has occurred in any transaction which has already been confirmed, this is a unique property of the Merkle trees used by Ethereum. But there is no decentralized storage used for record keeping which is a problem. In paper [7], again a data storage and sharing mechanism is discussed based on a private implementation of a blockchain, with each hospital having its own private blockchain and then all the hospitals connected by one single chain. In [6], use of Ethereum, inter-planetary file system (IPFS), and cloud storage is seen here. The files/records are stored onto the IPFS, and the URI of IPFS and deterministic keywords are stored onto the cloud for later ease of access. A file presented by a patient can later be verified by calculating the IPFS hash of the file and comparing it against the one stored on the cloud. The blockchain again only stores certain key aspects of the report, which are hash of the patient ID and hash of the file stored onto IPFS, and hence can be used for verification of the report. Again, cloud is being used for a last step verification which can act as the central point of failure because of its centralized nature. However, [8, 9], and [10] provide the framework for the record keeping and verification in a more general sense and not just tied to the medical records. Additionally, [11] and [12] layout the logical and mathematical constructs of the Ethereum and the Smartcontracts which we are going to use in our proposed solution. Thus, we in our approach are mainly going to target and get rid of this flaw, we are going to use our NFT for this last step verification, and therefore, our solution is based on a public blockchain and hence has all the perks of it; we also use only IPFS as our storage, and therefore, the storage is completely decentralized and not dependent on any central source and thus does not suffer from any single, central point of failure. Since we are using an NFT issued by the hospital to the patient for each unique report as our last step verification, this makes our solution completely decentralized and unique.

2.3 Comparison Study

1. Parameters

The first three parameters to be discussed from what is known as the blockchain trilemma which states that any Byzantine fault tolerant system will have to choose two parameters among scalability, security, and decentralization; hence, it is stated that no chain can have all the three characteristics present, and therefore, there is a compromise to be made. This trilemma has evolved from the CAP theorem, because blockchain is a distributed system, which states that any distributed system can only provide two of the three guarantees simultaneously: consistency, availability, and partition tolerance (CAP).

1. **Decentralization**: This is the defining parameter that will distinguish these approaches from the centralized approaches. This takes the control from a central authority and provides it to the community, either in the form of direct decisions or in a way where the actions of the central authority are completely transparent and challenging.
2. **Security**: It is often seen that projects keep fewer number of nodes on the network and hence increase the speed of the transactions on the network because there is lower overall computational power of the network, and thus, a 51% attack is easy to cause; therefore, the network will behave as the attacker wishes to and the whole blockchain can be corrupted without any trace.
3. **Scalability**: This simply implies that the network should be able to handle the network loads that it will have to endure in the future with growing use cases and acceptability among people.
4. **Cost**: The cost to set up and maintain the Dapp or the chain should be low and then subsequently the cost to use should be also low, so that it stays accessible to laymen and gains wider acceptance.
5. **Energy**: This also needs to be low, so that the climate is least affected by the operations.

6. **TPS (Transactions per Second)**: The transaction settlement rate should also be high.
7. **Ease of Access (EoA)**: The application developed, be it a chain or a Dapp, should be easy to access; therefore, anyone who wishes to be part of the network can be on it.

With these parameters in mind, we can contrast and compare all the previous approaches in a qualitative and quantitative manner, and the results of that can be seen in Table 1 with the Reference No. indicating the reference number of the cited paper.

3 Proposed Approach

Our solution is a two-stage and a two-way security and verification approach with complete decentralization and no reliance on any centralized component, thus making our approach independent and sustainable in the long run. Two-stage: we, at stage one, store the records on the decentralized storage and then store the records tied up with the patient's unique ID (Aadhaar) on to the blockchain, which being an immutable ledger cannot be changed and thus the entries can be trusted and used for verification. Now, at the second stage, the NFT is generated and transferred to the patient, this further verifies the claims of the patient because the NFTs are unique (tokenID) and thus can only be tied to a single report. Two-way: the verifiers (hospital, clinics) can verify the record of the patient on chain, and simultaneously, and the patient also is relieved because a report cannot be forged on his/her identity because an NFT is transferred to his/her wallet if that happens. Additionally, we are using a completely decentralized and public chain for our operation which provides best in class decentralization and security due to its public nature and hence high number of network participants. The interacting actors of our approach are as follows: Patient, this is the person who goes in the hospital for diagnosis and the report gets generated on his/her account. Doctors and staff: These actors diagnose the patient and then generate the report in accordance with the diagnosis. Wallet operator, since only authorized hospitals are allowed to store records on the chain, there needs to be an operator in the hospital who punches in the data on the screen and then data is stored on the chain. Digression, these authorized hospitals can be chosen on the basis of a voting system among the hospitals, this is called a decentralized autonomous organization (DAO).

Stepwise approach:

1. Report is generated by the respective hospital in the name/identity of the patient, and then, the file is stored onto the IPFS which returns the hash of the file.
2. Certain defining parameters are taken from the report of the patient.
3. An NFT is generated which contains the IPFS hash of the report and these defining parameters.

Table 1 Contrast between the previous approaches based on the parameters defined above

Reference	Author	Cost of use	Energy consumed	Decentralization	Security	Scalability	TPS	EoA
[1]	Xiao Yue, Huiju Wang, Dawei Jin, Mingqiag Li, Wei Jiang	Low because private chain used	Low, because centralized nature	Poor	Poor, due to centralized nature	High, due to centralized nature	N/A	Poor, due to closed and centralized nature
[2]	Yonggang Xiao, Bin Xu, Wenhao Jiang, Yunjun Wu,	Low, because of centralized nature	Low, comparatively. Depends on the total peers	Poor	Poor, because centralized and hence easy to do 51% attacks	High, because centralized and hence is completely in control	N/A	Poor, due to centralized and closed nature
[3]	Dara Tith, Joong-Sun Lee, Hiroyuki Suzuki, Wijesundara, Naoko Taira, Takashi Obi, and Nagaaki Ohyama	Low	Low, total peers dependent	Poor	Poor, due to centralized nature hence a smaller number of peers	High	N/A	Poor, due to centralized nature
[4]	Muhammad Usmana, Usman Qamarb	Low, because private chain using Hyperledger	Low, because of centralized nature	Poor	Poor, due to centralized Hyperledger fabric	High	N/A	Poor
[5]	Kazi Tamzid Akhter Md Hasib et al.	Above the rest, because ETH is used	Low, because ETH rolled out PoS	High, because open to anyone	High, because anyone can create a node	High, layer 2 solutions can be used to lift load from ETH	12.23	Excellent
[6]	Norah Alrebdi, Abdulatif Alabdulatif	Average ETH transaction cost	Low	High	High	High	10	Excellent
[7]	Chaoran Li, Jusheng Liu et al	Low	Low	Poor	Poor	High	N/A	Poor

4. This NFT is then transferred to the patient's wallet by the hospital that generated it. Thus, the patient always gets to know whenever a report is issued on his/her name. And if the NFT is not transferred but the data on chain is present, then this would imply that the hospitals have misbehaved (Fig. 1).
5. The authorized hospital then hashes the IPFS hash of the report, defining parameters, and the Aadhaar of the patient and stores this hash in a key value pair against the wallet address of the patient. This way the report gets linked to the patient, and the report at any later stage if is altered, then the hash will not match, and the fraud is caught (Fig. 2).
6. Thus, we can see that this two-stage security helps in a trustless and confirmable transaction and setup between the hospitals and the patients.
7. Now, any hospital or clinic can verify the authenticity of the file. All they need to do is hash the IPFS hash, defining parameters and the Aadhaar together and check against the wallet of the patient; if values match, then the report is verified. And the patient can have this report be used in any location worldwide, and they will trust the report and proceed as such.

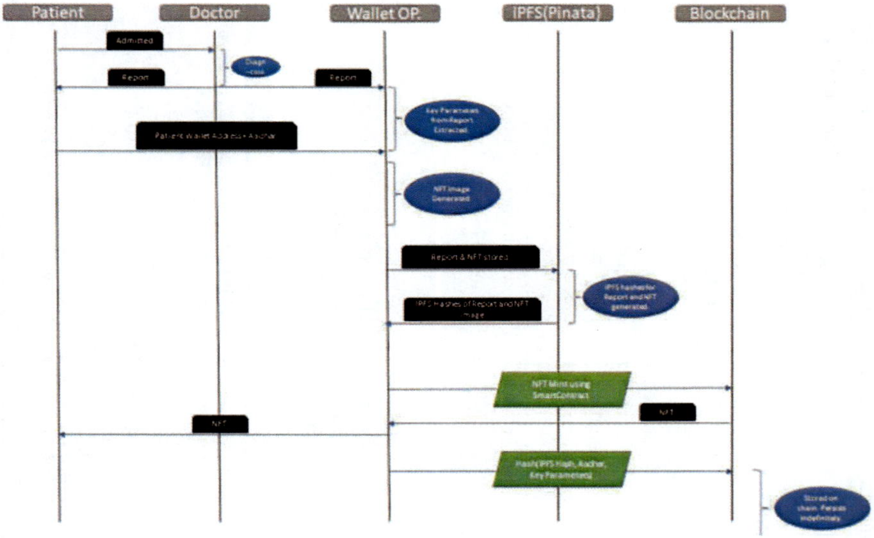

Fig. 1 This figure shows the pictorial representation of the flow of the steps described in this section. This figure specifically shows the different actors and their interactions during the phase where the report is generated and stored on a chain with the transfer of NFT. We can see that the patient is first admitted and then is diagnosed after the diagnosis is finished, and the report is given both to the patient and the wallet operator of the hospital; then, patient supplies his Ethereum (Metamask) wallet address and Aadhaar to the wallet OP who has already extracted the key parameters from the report and then stored the report and NFT on IPFS; moving forward, the IPFS hash of both the report and NFT is generated, and then, using the smartcontract, the NFT is minted and the report is stored on blockchain along with Aadhaar and key parameters. NFT is transferred to the patient's wallet

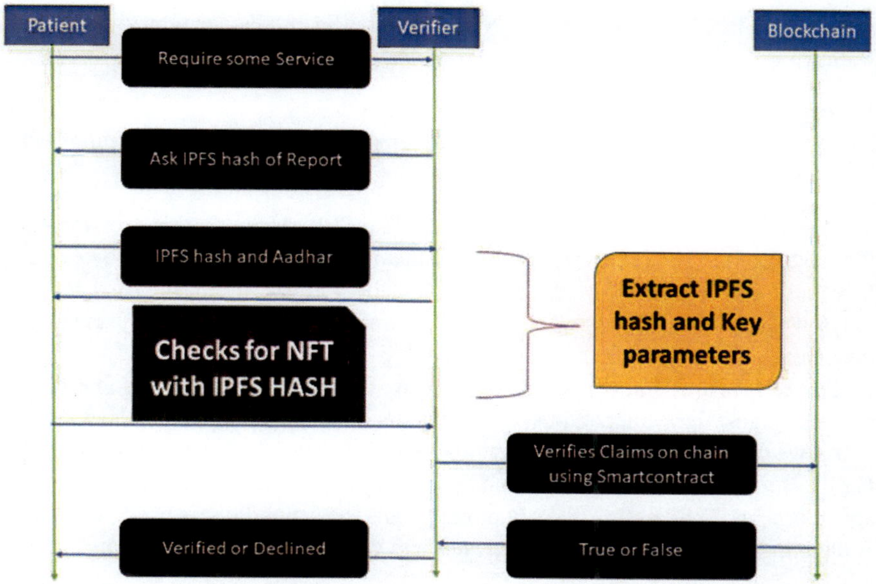

Fig. 2 This figure shows the steps and flow of the verification part of the methodology. A verifier could be a hospital or a clinic or some insurance company, i.e., anyone who wishes to verify the report. Patient asks for some service from the verifier, and to complete this, the verifier needs to verify the report of the patient, patient gives the hash of the report, or else, the verifier can also see that in the NFT of the patient. Verifier hashes the defining parameters, Aadhaar, and IPFS of file and checks on the blockchain. If the hash is found against the wallet of the patient, then the record is verified

Most approaches quoted above in the background study only claim the system, and there is no implementation given. We have given the complete proof of the implementation of the methodology, and hence, the approach is proved to be easier to implement and maintain.

4 Results

Now, let us look at the implementation of our approach. The smart contracts are written on, compiled, and deployed using Remix IDE in solidity which are shown in Figs. 5, 6, 7, 8, 9, 12 and 13, where we can see the above-quoted cases. Remix provides a front end after the deployment stage to interact with the smart contract, and hence, the Ethereum blockchain itself. We have used Pinata to upload our images, which can be seen in Figs. 3, 4, 10 and 11 below, to the IPFS which keeps record of all our uploads, and we can look up the hash anytime if needed. These files and images uploaded to the IPFS can be downloaded and verified using IPFS Desktop easily. Figure 14 is from Etherscan which shows the completion of the transaction which includes the transfer of the NFT from the hospital's account to the patient wallet.

1. *Uploading the Patient Report to the IPFS*
2. *Record on chain—Smart contract that stores the patient and report data on the blockchain.*

The **storeOnChain** function starting at line (10) stores the hash of the IPFS file hash, defining parameters, and the Aadhaar in a mapping, where the key is the Ethereum wallet address of the patient, and the value is the hash. Also note the onlyOwner specifier, this is what checks that only authorized hospitals can store on the chain.

The **readFromChain** function at line (18) verifies the records, if some hash is present against the wallet address of the patient, then it displays True else False.

As we can see in the figure below that the parameters last value was changed to HYPER from the original HYPO, and hence, the result is false (see left bottom on the above image). IPFS hash is the same that means the same file was used, and patient Aadhaar is also correct, but still the fraud was caught with such fine detail.

3. *NFT Minter—Smart Contract*

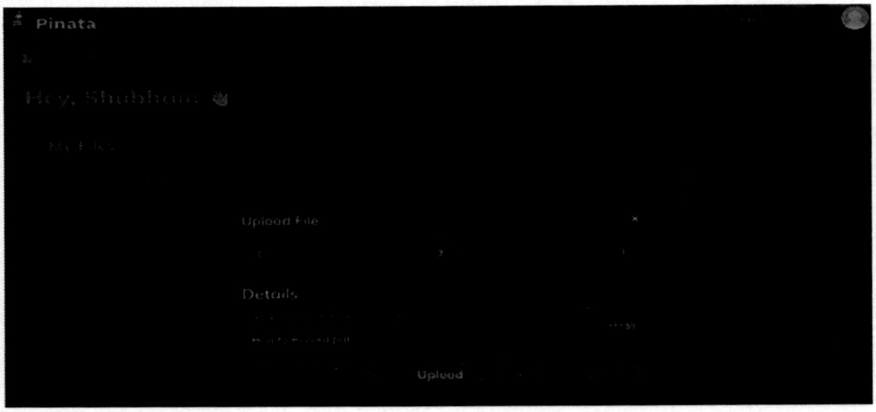

Fig. 3 File was selected from the local storage to be uploaded to IPFS using Pinata

Fig. 4 PDF file has been uploaded to the IPFS and the hash generated, which acts as the content address of the file

Fig. 5 Here, we describe the contract, and it takes the IPFS hash of the patient report, the defining parameters of the report, and the Aadhaar of the patient

Fig. 6 StoreOnChain function with its parameters populated

Fig. 7 Transaction has been sent successfully and the record (hash) has been stored in the mapping on the chain against the wallet address of the patient

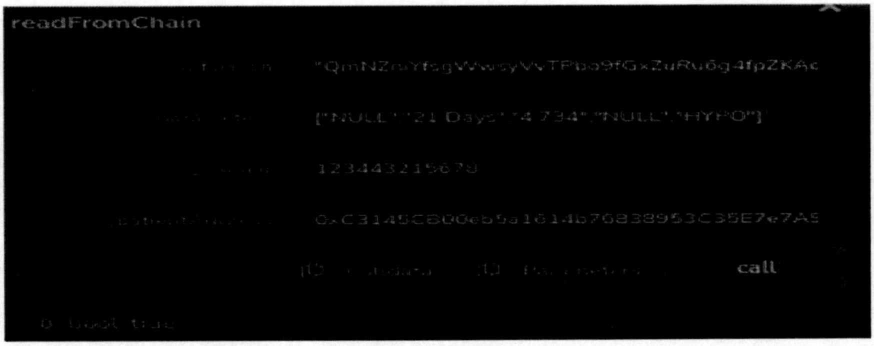

Fig. 8 This is the readFromChain function, and here, we verify that if these records are true for some given wallet address

Fig. 9 If there happens to be even a minor change in the data, then the result of calling the readFromChain function is false and the discrepancy is caught

Fig. 10 NFT image which will be transferred to the patient wallet

Here, we show the NFT minting Samar contract with the mint feature and the whole process of the transfer to the patient address by the hospitals where the patient gets the report from.

As we can see, it also keeps record of the token number of the NFT minted to the wallet of the patient; hence, the NFTs are all unique. It keeps a record of the NFT's IPFS hash and token ID against the wallet address of the patient. Thus, it provides even better security and protection from fraud.

Fig. 11 Patient NFT has been uploaded to the IPFS, and the hash is also shown for the upload

Fig. 12 Smart contract that mints the NFT token to the patient wallet directly

Fig. 13 Here, we pass the IPFS hash of the NFT to the minter and hit mint button

Fig. 14 NFT has been minted along with the address who minted the NFT

This will confirm that NFTs only from the authorized hospital entities are accepted. Here, we can also see the unique token ID of the NFT. Thus, whenever a report is published on account of a patient, he will always receive a NFT as confirmation.

Table 2 Describes the parameters that our approach performs better on than the closest existing approach

Parameter	Closest existing approach	Our approach
Implementation	Two-stage only. Because the patient has no record with himself about the report	Two-stage & two-way. Because the patient has unique NFT per unique report, hence, a record is in his wallet
Security	The flow of security is only in one direction, because the hospitals have the way to know if the patient has changed the report or the identity, but there is no security flowing toward the patient's side because the hospitals can produce reports on the behalf of the patient without them knowing	Security flows both ways. The hospitals are secured from fraud because they have the record on the chain and the patient is secured because he will receive the NFT if some report is produced on his account, thus in any case, he can refer and produce this NFT later from his wallet
Decentralization	The solution is not completely decentralized because it uses CouchDB during the second stage of the verification	Our solution is completely decentralized because it achieves the security at the second stage using decentralized IPFS and NFTs

5 Comparison and Conclusions

Let us now draw a contrast between our approach and the closest previous approach, so that we can point out distinctively how our approach is better by comparing on various parameters on which our approach betters the previous closest one using Table 2 shown below. Then, we will also discuss how our approach can be further enhanced which helps take our solution to the final form it can achieve.

Future aspect to our approach would be to have the decision of allowing new hospitals to be able to store on the chain be completely decentralized, i.e., there is no single entity or a group of entities which get to add some hospital as a new actor, who can store on chain, but rather all the participants of the network have a vote on this decision. This can be achieved by implementing a decentralized autonomous organization (DAO) into our solution, and this will then make the solution full proof, autonomous, and decentralized.

References

1. Yue X, Wang H, Jin D, Li M, Jiang W (2016) Healthcare data gateways: found healthcare intelligence on blockchain with novel privacy risk control
2. Xiao Y, Xu B, Jiang W, Wu Y (2021) The healthchain blockchain for electronic health records: development study
3. Dara Tith (2020) Application of blockchain to maintaining patient records in electronic health record for enhanced privacy, scalability, and availability

4. Usman M, Qamar U (2019) Secure electronic medical records storage and sharing using blockchain technology
5. Kazi Tamzid Akhter Md Hasib (2022) Electronic health record monitoring system and data security using blockchain technology
6. Alrebdi N, Abdulatif A, Iwendi C, Lian Z (2022) SVBE: searchable and verifiable blockchain-based electronic medical records system
7. Li C, Liu J, Qian G, Wang Z, Han J (2022) Double chain system for online and offline medical data sharing ***via*** private and consortium blockchain: a system design study
8. T. Rama Reddy (2021) Proposing a reliable method of securing and verifying the credentials of graduates through blockchain
9. Razatulshima Ghazali (2021)Blockchain for record-keeping and data verifying: proof of concept
10. Vijay Anant Athavale, Shakti Arora, Anagh Athavale (2022) Adoption of blockchain technology for storage and verification of educational documents
11. Wood G (2022) Ethereum: a secure decentralised generalised transaction ledger berlin version
12. Buterin V (2014) Ethereum: a next-generation smart contract and decentralized application platform

Chapter 62
A Study on Prediction of Temperature in Metropolitan Cities Using Machine Learning

Shweta S. Aladakatti, A. Bharath, V. T. Adarsha, B. J. Ajith, and H. R. Chaithra

1 Introduction

The most significant weather element is temperature, since it also serves a variety of energy, industrial, environmental, and agricultural needs. An urban or metropolitan area that, as a result of human activity, is much warmer than the nearby rural areas. Therefore, the dynamic nature of the atmosphere makes it challenging to anticipate weather makes factors with accuracy [1]. Predictions of atmospheric parameters including temperature, wind speed, rainfall, meteorological pollution, etc., are made using a variety of algorithms, including linear regression, long short-term memory (LSTM), ANN, DNN, and RNN, among others. For many applications, the prediction of atmospheric parameters is crucial [2]. Climate monitoring, drought detection, severe weather forecasting, agricultural and production, energy industry planning, aviation industry planning, communication, and pollution dispersal are a few of them.

Day-to-day business processes require detailed real-time weather information in the future. Operational decisions in many organizations are strongly affected by meteorological phenomena. The predicted temperature of this project will be reliable, easily understood, and thoroughly customized [3, 4]. By predicting the future temperature, it reduces the loss of life and the loss of crops associated with weather-related hazards. To achieve this project, first, we will pre-process the data of major metropolitan cities. Second, we will identify the dependent and independent variables. Third, we need study various parameters affecting temperature. Finally, we will predict the temperature. The main purpose of this project is to apply various

S. S. Aladakatti · A. Bharath (✉) · V. T. Adarsha · B. J. Ajith · H. R. Chaithra
Dayananda Sagar University, Bengaluru, India
e-mail: bharath11raaj@gmail.com

S. S. Aladakatti
e-mail: shweta-cse@dsu.edu.in

A. Yadav et al. (eds.), *Proceedings of International Conference on Paradigms of Communication, Computing and Data Analytics*, Algorithms for Intelligent Systems,
https://doi.org/10.1007/978-981-99-4626-6_62

machine learning models and compare the results among them to get the suitable model which provides less error [2, 5].

A time series is a variety of pieces of information that have been enumerated, graphed, or otherwise categorized chronologically. A time series is often a collection of images captured at a succession of moments in time that are evenly spaced apart. As a result, it is a collection of discrete-time data [6]. Techniques for obtaining practical statistics and other information from time series data through analysis are referred to as time series analysis. A model is used to predict future values based on values that have previously been observed. This technique is known as time series forecasting. Discrete symbolic data, discrete numeric data, and continuous real-valued data can all be subjected to time series analysis [7].

The most fundamental and popular predictive model for analysis is linear regression. The most common purpose of regression estimates is to describe the data and clarify the relationship between one or more independent and dependent variables. Visually, linear regression determines the best fit through the points. The regression line is the line through the points that fits the data the best.

Long short-term memory (LSTMs) is a kind of recurrent neural structure. In recurrent neural network, the yield from the previous advance is considered a contribution to the present advance. It addressed the problem of long-term RNN situations, where the RNN can provide more precise forecasts using the fresh input but cannot anticipate the value stored in the extended memory. RNN does not provide efficient performance as the hole length increases. The data can naturally be stored by LSTM for long periods of time. It is used for managing, predicting, and sorting-based on time series data.

2 Related Work

The literature review revealed that machine learning approaches are being applied to temperature prediction in several countries around the world. Comparing machine learning approaches to current prediction methods, the latter are proving to be considerably faster and more accurate.

In this field, a lot of work has been done worldwide. Which is evidenced by the work of Naïve Bayesian classification method and the linear regression technique have been introduced by Lakshmi et al. [1]. Machine learning makes use of weather predictions. It is a powerful method for making weather predictions that are more accurate. To anticipate the weather, the dataset of weather is gathered, examined, and algorithm are applied to it. They identify significant issues in synoptic climatology and meteorology, such as present patterns, air masses, weather kinds, and weather fronts.

Maddu et al. [2] have put forth the concepts of a long short-term memory, a recurrent neural network, an artificial neural network, BiLSTM, and iLSTM. These models' performances are compared using a variety of metrics, including NSE,

MSE, Norm, MAE, RSR, R2, and PBIAS. Overall, the study's integrated computation modelling structure showed that linking the LSTMs and BiLSTMs models was successful when forecasting surface temperature.

Huang et al. [3] have proposed the concept of multilayer perceptron (MLP) model and the long short-term memory (LSTMs) model. The intended system primarily forecasts temperature, pressure, humidity, UV radiation, and rainfall. It also gives a forecast of these variables for the following 24 h. Future temperatures can be predicted with the use of the LSTM approach.

Hoang [6] have provided examples of long short-term memory (LSTMs) and recurrent neural network (RNN). Numerous factors are included in the dataset, including UV, wind, humidity, pressure, chance, visibility, and many others. According to experimental, LSTM neural network outperforms other weather forecasting methods in terms of output quality and accuracy.

Park et al. [8] have discussed the theories behind deep neural network, recurrent neural network, and LSTMs. LSTMs were used as the foundation of the suggested model to suit the time series data. Specifically, an LSTM framework was used to correct the missing data, which is commonly occur in the gathered weather database. The missing data such as temperature, humidity, and direction of wind. The impact of the LSTM's hyperparameters on prediction accuracy was investigated.

Anjali et al. [7] have presented the concept of linear regression, artificial neural network, and support vector machine. By considering the error metrics and the correlation coefficient, we can say among the proposed approaches MLR is a more precise model than ANN and SVM.

3 Methodology

3.1 Dataset

Our project's first stage is to gather wind chill, heat index, UV index, wind speed, and temperature data. We have gathered and saved the data in a csv files. Information about the temperature is gathered from authoritative Websites [9]. In our project, each of the csv files is extracted into a distinct variable and put in a common folder. We collected the dataset from the year 2009 to 2022 of major metropolitan cities in India.

3.2 Time Series Model

The time series model is used to apply on the cleaned data. We determine the pattern in independent variables using time series analysis. As a result, this phase establishes the starting point for temperature forecast. The relevant data is saved in various

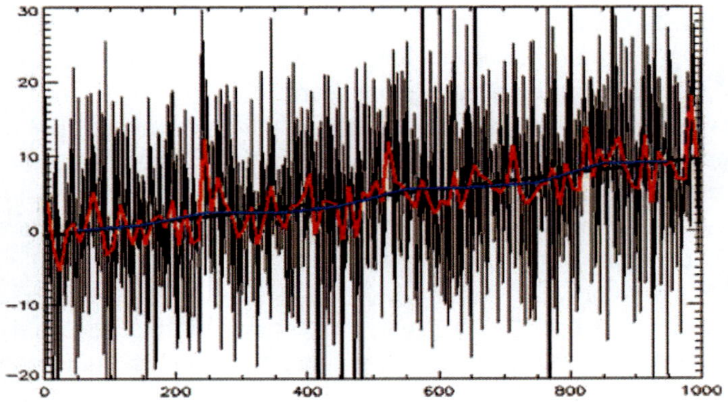

Fig. 1 Example of time series graph

different factors and transferred to the user interface (UI), where it is displayed in an approachable graphical format. We have implemented regression analysis and LSTM in addition to time series to predict the values based on the average temperature, wind chill, heat index, UV index, and wind speed as the independent factor variables. To implement time series, the series must be stationary [10].

Figure 1 shows the example of time series graph.

3.3 Linear Regression

Linear regression is a technique for simulating the connection between a scalar-dependent variable (*Y*) and one or more independent variables (*X*). In cases when there is only one explanatory variable, simple linear regression is employed. The procedure is known as multiple linear regression when there are several explanatory variables [5, 11].

From Fig. 2, the data from time a time series are subject to multiple linear regression analysis, with temperature serving as the independent variable (*X*) and average temperature, wind chill, heat index, UV index and wind speed as the dependent variables (*Y*). Upon application, the module stores the output in a variable along with information on the formula, the intercept value, and so forth. We get the expected temperature values after executing the predict function. The UI receives the result set and displays the results graphically. A method for modelling the connection between a scalar-dependent variable (*Y*) and one or more independent variables (*X*) is known as linear regression. In cases when there is only one explanatory variable, simple linear regression is performed. The process is referred to as multiple if there are several explanatory variables [4, 12].

Regression's basic mathematical formula is:

$$y = z\beta + \varepsilon$$

Fig. 2 Example of linear regression

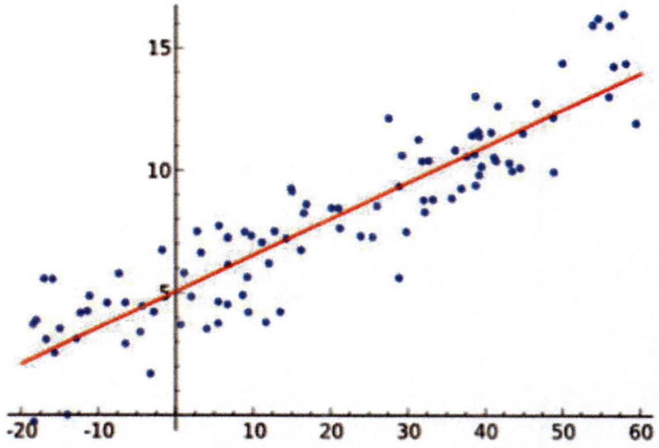

Assume we have an n sample. The design matrix X continues to have size $n * (r + 1)$. And now,

$$Y_{n\times m} = \begin{bmatrix} Y_{11} & Y_{12} & \cdots & Y_{1p} \\ Y_{21} & Y_{22} & \cdots & Y_{2p} \\ \vdots & \vdots & \vdots\vdots & \vdots \\ Y_{n1} & Y_{n2} & \cdots & Y_{np} \end{bmatrix} = \begin{bmatrix} Y_{(1)} & Y_{(2)} & \cdots & Y_{(p)} \end{bmatrix}, \tag{1}$$

And the ith variable's n measurements are represented by the vector $Y(i)$. Also,

$$\beta_{(r+1)\times m} = \begin{bmatrix} \beta_{01} & \beta_{02} & \cdots & \beta_{0m} \\ \beta_{11} & \beta_{12} & \cdots & \beta_{1m} \\ \vdots & \vdots & \vdots\vdots & \vdots \\ \beta_{r1} & \beta_{r2} & \cdots & \beta_{rm} \end{bmatrix} = \begin{bmatrix} \beta_{(1)} & \beta_{(2)} & \cdots & \beta_{(m)} \end{bmatrix}, \tag{2}$$

where the model's ith variable's $(r + 1)$ regression coefficients. Similarly, a $n * p$ matrix is also used to organize the p n-dimension vectors of defects $€(i)$, $i = 1,\ldots,p$.

$$\varepsilon = \begin{bmatrix} \varepsilon_{11} & \varepsilon_{12} & \cdots & \varepsilon_{1p} \\ \varepsilon_{21} & \varepsilon_{22} & \cdots & \varepsilon_{2p} \\ \vdots & \vdots & \vdots\vdots & \vdots \\ \varepsilon_{n1} & \varepsilon_{n2} & \cdots & \varepsilon_{np} \end{bmatrix} = \begin{bmatrix} \varepsilon_{(1)} & \varepsilon_{(2)} & \cdots & \varepsilon_{(p)} \end{bmatrix} = \begin{bmatrix} \varepsilon'_1 \\ \varepsilon'_2 \\ \vdots \\ \varepsilon'_n \end{bmatrix} \tag{3}$$

To get the Y value, or temperature, formula 3 incorporates the previous values.

Therefore, by using multiple linear regression, we can forecast the temperature value depending on several parameters, such as wind chill, average temperature, sun hour, wind speed, UV index.

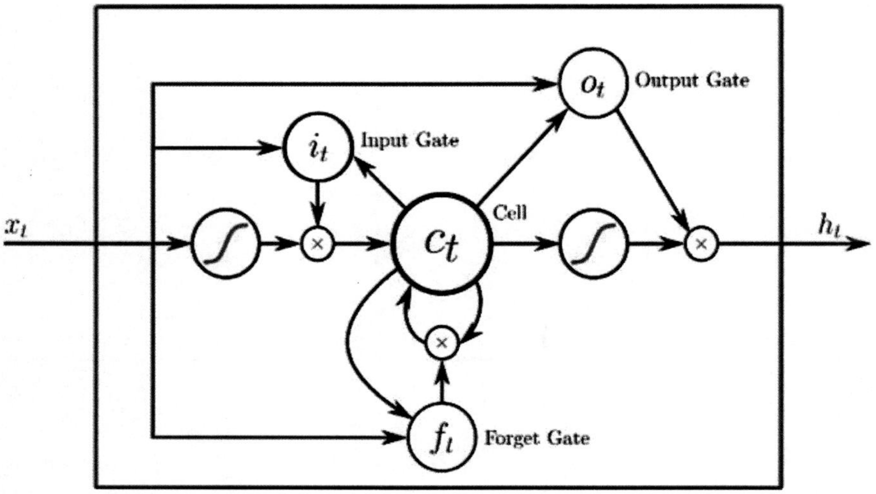

Fig. 3 LSTM gate

3.4 LSTM Model

Recurrent neural networks (RNN) are extended, and the memory is essentially extended, by long short-term memory networks (LSTM). As a result, it is appropriate for making conclusion from important events that occurred over extremely long stretches of time apart. Recurrent neural networks can recall inputs over a long period of time due to LSTM properties. This is because LSTMs contain a memory that saves data, much like an user's computer.

Data could be read, written to, and deleted from the LSTM's memory. This memory can be compared to a gated cell, which decides whether to store the information or delete the information based on the value it assigns to the information (i.e., whether it opens the gates or not). The system also learns parameters, which are used to quantify significance. It simply means that it progressively learns what information is important and what information is not.

In Fig. 3, the LSTM has three gates, those are input, forget, and output gates. These gates will determine whether to accept further input (input gate), delete the information because it is irrelevant (forget gate), or permit the information to influence the output at the current timestep (output gate) [13].

3.5 Proposed Method

Using a more advanced method, the suggested system is tested using data from the previous 11 years from 2009 to 2022. The outcomes are contrasted with those of earlier methodologies. The proposed improved method for forecasting the weather provides advantages over the established methods. In comparison with earlier methods, this model generates forecasts that are the most accurate. The meteorologist may easily and accurately anticipate the weather for the future with the use of this technology.

3.6 System Architecture

Python was used to create the system. To train our model, daily datasets spanning the previous 11 years (2009–2022) have been obtained. The system generates the output after receiving input from the datasets.

The steps that make up the system-building process are as follows:

1. Fetching the dataset
2. Cleaning the dataset
3. Selection of the features of dataset
4. Train model
5. Use the model to predict results.

In Fig. 4, we have discussed about the system architecture.

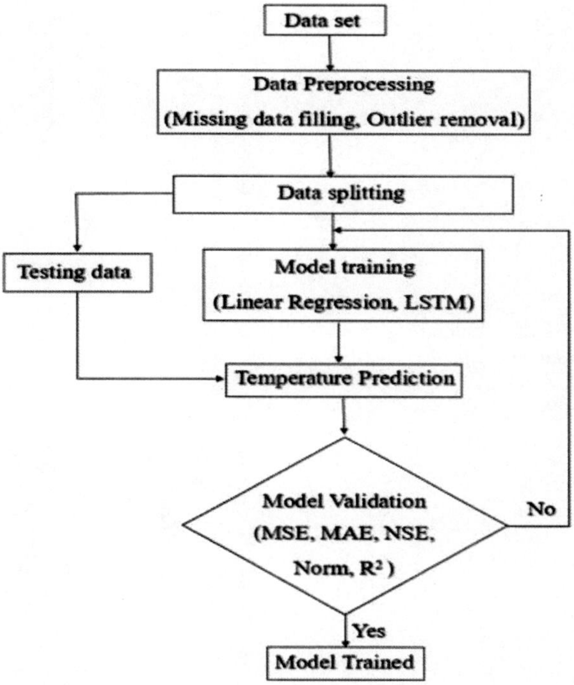

Fig. 4 System architecture

4 Experimental Results

For the experimentation, we have taken the Pune city dataset from the year 2009–2022. We have taken train dataset from the year 2009–2018 and test dataset from the year 2017–2022.

From the Pune city dataset, we will visualize the maximum and minimum temperature.

Figure 5 shows trend in maximum temperature, and Fig. 6 shows trend in minimum temperature.

4.1 Linear Regression Results

By using the above data, we determine the heatmap as shown in Fig. 7 [5].

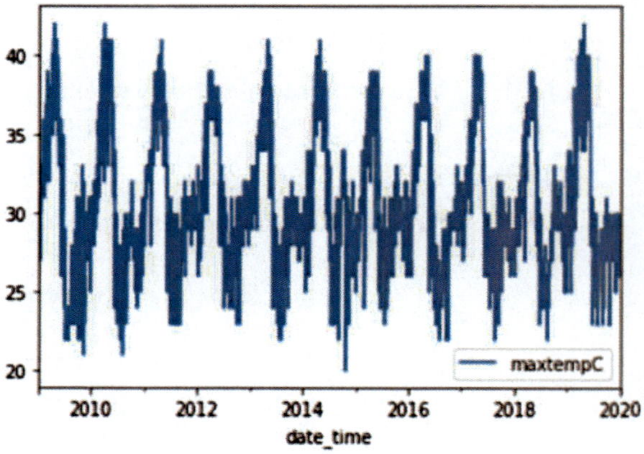

Fig. 5 Maximum temperature

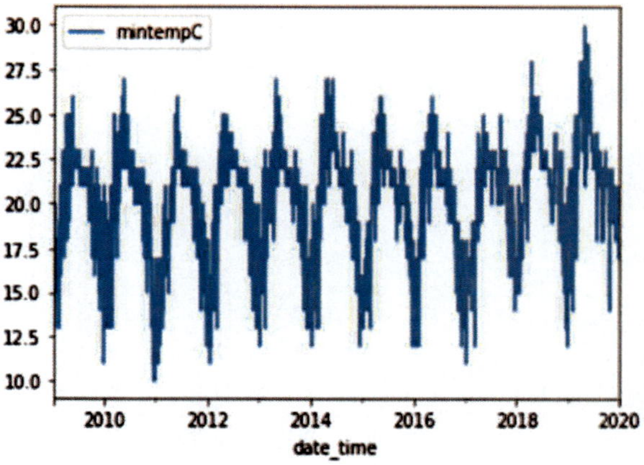

Fig. 6 Minimum temperature

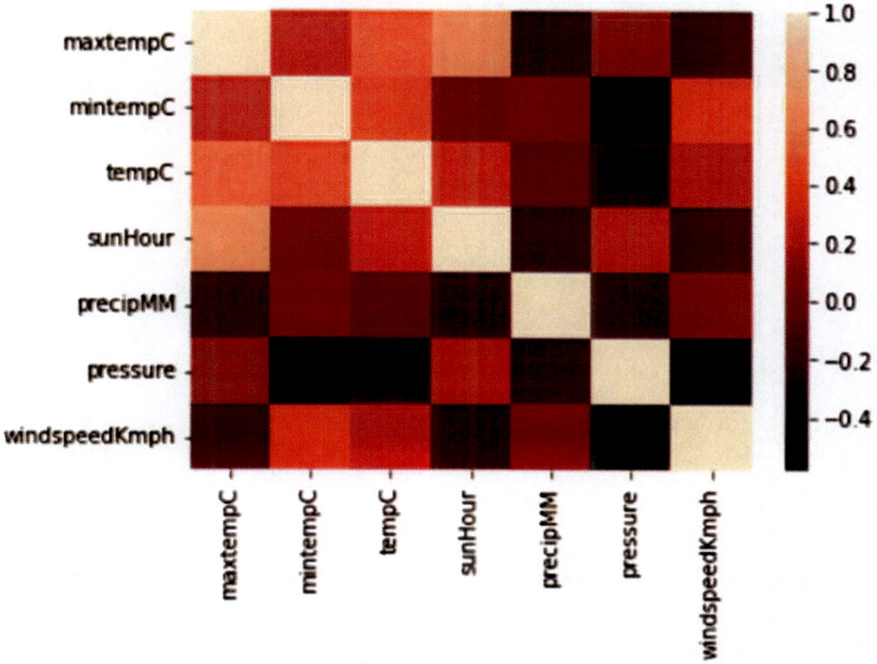

Fig. 7 Heatmap

Figure 7 is the heatmap that is used to show relationships between two variables, one plotted on each axis. Here, we will be using various factors such as max tempC, mintempC, tempC, sunHour, precipMM, pressure, and wind speed Kmph.

Table 1 shows the trend in actual value of temperature, prediction value of temperature, and the difference between actual and prediction values.

Table 1 Measured values

	ACT	PRD	DIFF
0	28	29.254407	−1.254407
1	31	28.893917	2.106083
2	28	27.078402	0.921598
3	24	26.993140	−2.993140
4	25	29.625130	−4.625130
...	...	...	...
28,925	27	27.730756	−0.730756
28,926	31	33.503686	−2.503686
28,927	37	35.832088	1.167912
28,928	26	24.688367	1.311633
28,929	29	27.918778	1.081222

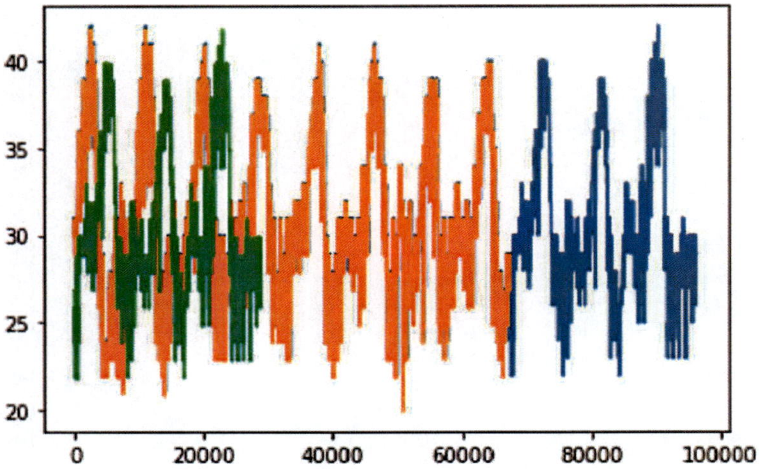

Fig. 8 Trend in temperature using LSTM

4.2 LSTM Model Results

The LSTM algorithm was applied to the dataset to forecast how the temperature would change over time, and the outcome is shown in Fig. 8 plot [5, 9].

The trained data is displayed in orange, the predicted value is displayed in green, and the actual trend is displayed in blue. These three lines proximity indicate how effective the LSTM-based model is. When a significant amount of time has passed, the prediction comes closer to the actual trend. The accuracy that is reached will increase when the system is taught more and larger dataset is used.

4.3 User Interface (UI) Snapshots

In this module, the UI components are defined, and the corresponding event handlers are defined. The UI consists of a sidebar panel where the user can provide input, and there is a main panel where the data is represented in the form of tables in various tabs. The data from 2009 to 2022 can be viewed on a monthly or year basis. The results of the time series, LSTM and regression models, are also represented in tabular. Finally, there is a comparison tab where the graph shows the accuracy of the prediction by comparing the measured versus the predicted values of the years from 2018 to 2022.

Figures 9 and 10 show the graph of minimum temperature and maximum temperature, respectively. The red colour will display the trend in maximum temperature, and blue colour will display the trend in minimum temperature.

Table 2 displays the max_Temp, min_Temp, Avg_Temp, wind chill, heat index, feels like, sun hours, UV index, and Wind speed parameters of the predicted temperature.

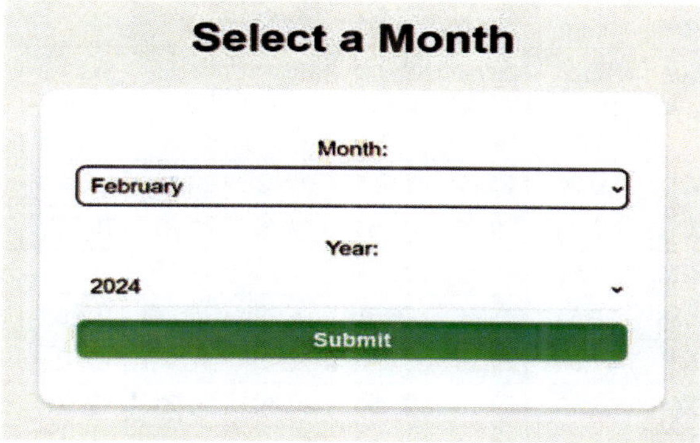

Fig. 9 Select a month

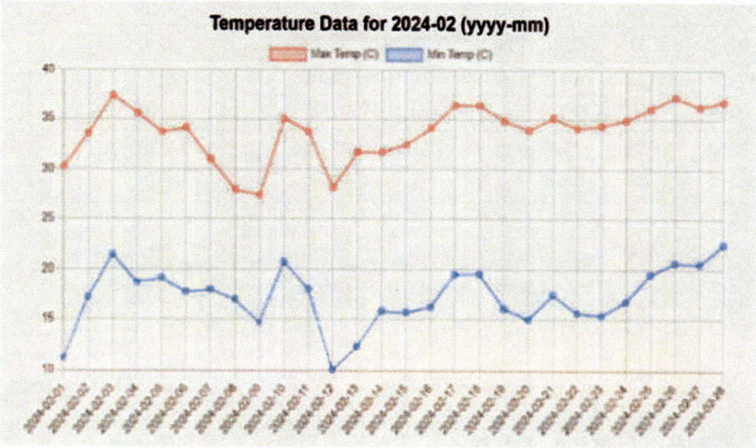

Fig. 10 Graph of temperature

5 Discussion of Results

By using linear regression method, the obtained accuracy is 91.13% from the r2 score validation as shown in Fig. 11.

By using LSTM method, the obtained test loss is 0.17% from the mean square error and Adam optimizer, and the obtained accuracy is 89.33% as shown in Fig. 12.

Table 2 Table of temperature

Date	Max Temp (C)	Temp (C)	Avg. Temp (C)	Wind Chill (C)	Heat Index (C)	Feels Like (C)	Sun Hours	UV Index	Wind Speed (km/h)
2024-09-01	29.31	10.26	17.07	22.85	22.83	22.55	10.58	6.37	9.75
2024-09-02	30.01	12.98	19.23	24.04	24.30	24.09	10.71	5.96	11.23
2024-09-03	38.15	23.91	30.98	31.78	31.63	31.70	11.38	6.84	5.77
2024-09-04	33.88	13.68	21.66	26.61	26.39	26.18	10.80	7.76	3.67
2024-09-05	37.81	21.86	29.18	31.15	31.45	31.40	10.97	7.92	2.68
2024-09-06	30.61	18.97	23.17	25.54	26.89	26.96	11.71	6.10	22.76
2024-09-07	28.44	13.37	18.48	23.50	24.33	24.17	10.03	5.24	16.24
2024-09-08	28.86	14.47	19.72	24.61	25.12	24.79	7.34	5.13	10.51
2024-09-09	29.70	7.25	15.13	22.55	22.60	22.23	11.86	6.17	4.01
2024-09-10	34.80	18.65	25.44	28.39	28.77	28.73	11.58	6.86	7.30
2024-09-11	33.41	20.39	26.07	28.24	28.74	28.59	8.80	6.71	8.27
2024-09-12	31.31	16.17	22.14	26.19	26.61	26.35	8.77	5.85	7.13
2024-09-13	29.19	7.10	14.73	22.19	22.31	21.94	11.69	6.14	5.47
2024-09-14	30.43	7.64	15.85	23.19	23.11	22.70	11.73	6.15	1.29
2024-09-15	30.91	10.85	18.26	24.21	24.47	24.12	11.52	6.11	3.41

```
In [14]: from sklearn.metrics import r2_score
         r2_score(X_test,x_pred)

Out[14]: 0.9113378379530118
```

Fig. 11 Accuracy of linear regression

```
In [18]: loss, accuracy = model.evaluate(X_test, y_test, batch_size=32)

         # Print the test loss and accuracy
         print('Test loss:', loss)
         print('Test accuracy:', accuracy)

603/603 [==============================] - 6s 8ms/step - loss: 0.0017 - accuracy: 0.8933
Test loss: 0.001703352201730013
Test accuracy: 0.8932648301124573
```

Fig. 12 Accuracy of LSTM model

6 Conclusion

The purpose of this study was to improve the reliability and accuracy of temperature predictions using machine learning methods. The key contribution of the researcher is the application of the novel LSTM model as a temperature forecast tool.

These strategies enhanced forecast accuracy and produced excellent results; however, the LSTM model turned out to be more efficient. The studies that demonstrate the improved effectiveness and accuracy of temperature prediction using machine learning approaches are quite positive.

By using a much larger dataset than the one being used right now, the accuracy of the temperature forecast system can be significantly increased in the future. Additionally, other machine learning models that are currently under development should be examined to determine how accurate they are.

References

1. Lakshmi NS, Ajimunnisa P, Prasanna VL, YugaSravani T, RaviTeja M (2021) Prediction of weather forecasting by using machine learning. Int J Innov Res Comput Sci Technol (IJIRCST)
2. Maddu R, Vanga AR, Sajja JK, Basha G, Shaik R (2021) Prediction of land surface temperature of major coastal cities of India using bidirectional LSTM neural networks. J Water Climate Change 12(8):3801–3819
3. Huang ZQ, Chen YC, Wen CY (2020) Real-time weather monitoring and prediction using city buses and machine learning. Sensors 20(18):5173
4. Aladakatti SS, Senthil Kumar S (2022) Raif-semantics: a robust automated interlinking framework for semantic web using mapreduce and multi-node data processing. J Interconnect Netw 22(Supp01):2141016
5. Aladakatti SS, Kumar SS (2023) PIRAP: a study on optimized multi-language classification and text categorization using supervised hybrid machine learning approaches. Int J Cooper Inf Syst
6. Hoang DT (2020) Weather prediction based on LSTM model implemented AWS machine learning platform. Int J Res Appl Sci Eng Technol 8(5):283–290
7. Anjali T, Chandini K, Anoop K, Lajish VL (2019) Temperature prediction using machine learning approaches. In: 2019 2nd International conference on intelligent computing, instrumentation and control technologies (ICICICT)
8. Park I, Kim HS, Lee J, Kim JH, Song CH, Kim HK (2019) Temperature prediction using the missing data refinement model based on a long short-term memory neural network. Atmosphere 10(11):718
9. Aladakatti SS, Senthil Kumar S (2022) Exploring natural language processing techniques to extract semantics from unstructured dataset which will aid in effective semantic interlinking. Int J Model Simul Sci Comput 2243004
10. Wikipedia Contributors (2023) Time series. In: Wikipedia, The Free Encyclopedia. Retrieved 10 Apr 2023, from https://en.wikipedia.org/w/index.php?title=Time_series&oldid=1146255868
11. Linear Regression (2022). Wikipedia. Retrieved 10 Apr 2023 from https://simple.wikipedia.org/w/index.php?title=Linear_regression&oldid=8356805

12. Kavitha S, Varuna S, Ramya R (2016) A comparative analysis on linear regression and support vector regression. In: 2016 Online International conference on green engineering and technologies (IC-GET), Coimbatore, pp 1–5. https://doi.org/10.1109/GET.2016.7916627
13. Wikipedia Contributors (2023) Long short-term memory. In: Wikipedia, The Free Encyclopedia. Retrieved 10 Apr 2023 from https://en.wikipedia.org/w/index.php?title=Long_short-term_memory&oldid=1148032239

Chapter 63
A Review of Secure Authentication Techniques in Fog Computing

Mahgul Afzali and Gagandeep

1 Introduction

Fog computing refers to a decentralized infrastructure situated in close proximity to IoT devices, facilitating management, storage, and communication for these devices. Being an extension of cloud computing, fog computing offers services similar to those provided by the cloud, such as data processing, storage, computation, and communication capabilities. The introduction of fog computing was driven by the need to address the limitations faced by cloud computing, particularly in terms of real-time response, complexity in distributed environments, mobility, latency, bandwidth, and location awareness of IoT applications. The data generated by IoT devices tend to be voluminous, requiring significant storage capacity to enable effective analysis processes. Additionally, numerous IoT applications necessitate swift or real-time analysis, such as gaming and big data analysis. However, the centralized nature of cloud data centers proves inefficient in addressing the processing and storage requirements of widely dispersed IoT devices. Consequently, network administrators face challenges in managing and processing this data while overseeing the network infrastructure. Fog computing, an extension of cloud computing, inherits certain security concerns from its predecessor [1]. The three layers of fog computing comprise the cloud layer, fog layer, and edge layer.

The main intention of this paper is to review various authentication techniques in fog computing context to identify the best solution for further process. We organized our paper as follows: In Sect. 2, the background of the authentication technique is presented. In Sects. 3 and 4, security aspects, types of fog-based authentication are discussed respectively. Then, a survey of literature on fog computing authentication

M. Afzali (✉) · Gagandeep
Department of Computer Science, Punjabi University, Patiala 147002, India
e-mail: mahgulafzali@gmail.com

A. Yadav et al. (eds.), *Proceedings of International Conference on Paradigms of Communication, Computing and Data Analytics*, Algorithms for Intelligent Systems,
https://doi.org/10.1007/978-981-99-4626-6_63

is given in Sect. 5. In the last section, a comparative result is given, and finally, the conclusion is followed by the references.

2 Background

Authentication poses a significant challenge in fog computing due to the sheer volume of data generated, making it difficult to manage effectively. Additionally, access control becomes a critical concern since a substantial portion of application processing occurs on user devices. In the realm of authentication, our primary objectives revolve around the authentication domain and addressing various attacks targeting the authentication process for users and devices. Authentication plays a vital role in establishing the primary relationship between end devices and fog devices, ensuring the authentication of user identities. This ensures that only legitimate end users who meet the necessary authentication requirements can access the fog nodes. Fog computing aims to address the shortcomings of cloud computing by enhancing communication latency, real-time processing capabilities, and privacy and security aspects. However, it is important to acknowledge that there are still unresolved issues in fog computing [2].

3 Security Aspects

Security issues in fog computing arise from the extensive connectivity of devices to fog nodes, which are distributed across different gateways. While authentication plays a crucial role in establishing the initial relationships between IoT devices and fog nodes, it is not sufficient on its own. Devices can still malfunction or become vulnerable to malicious attacks. The adoption of fog computing technology is growing rapidly in both industrial and social sectors. Consequently, developers face the challenge of creating applications that are resilient against system threats and attacks, ensuring the reliability of data storage and processing. With the increasing prevalence of cyber threats and malware, protecting data and the system within the fog environment becomes a formidable task.

4 Authentication in Fog Computing

Authentication is a fundamental process used to verify the identity of users who seek access to various systems, data, networks, servers, applications, Web sites, or devices. It involves establishing and confirming a user's identity. This is typically accomplished by presenting credentials, which are predetermined pieces of information shared between the user and the system. A common illustration of this process is accessing a user account on popular platforms like Facebook or Gmail. In order to gain entry to your account, you must provide the correct login credentials as a means of validating your ownership and identity.

4.1 Authentication Techniques

There are three fundamental types of authentication commonly used. The first type is knowledge-based authentication, which involves the use of passwords or PIN codes that only the authorized user should know. The second type is possession-based authentication, where the user possesses a physical item such as an access card, key, or authorized device that uniquely identifies them. The third type is biological-based authentication, which utilizes physical traits of the user, such as fingerprints or retinal patterns, for authentication purposes. These are some of the most frequently used authentication techniques.

i. **Password**: The most common form of the password which is used by the user is password. Users set a password for their account that only they know. When the password is interred by the user, the system checks if it matches the user's password in the database. If the password entered by the user matches the password in the system, the system grants the user access. A recent survey found that 31% of people write passwords in a notebook, and only 26% remember them without writing them down.
ii. **Biometric**: Organizations that need enhanced security might use behavioral biometric authentication solutions. Certain behavioral patterns are unique to individuals, such as how quickly and how hard they hit certain keys when typing, how fast or slowly they speak, and how big a stride they take when they walk. Behavior biometrics uses keystroke dynamics, voiceprints, and gait analysis to authenticate a user based on their unique behavioral patterns.
iii. **Multi-Factor Authentication**: Most companies are warming up to the fact that truly secure identity management requires multi-factor authentication (MFA). MFA requires two different authentication requirements, such as a static password and a text sent to a user's smartphone. MFA uses something the user knows and something they have like an email address or smartphone to protect the network from unauthorized users. Hand-in-hand with MFA is the use of time-sensitive, one-time passwords.

5 Review of Literature: Current State of the Art

A literature review serves as a comprehensive examination and analysis of existing authentication techniques within a specific area of study. Its purpose is to offer a concise summary of the research conducted and findings established regarding the mentioned topic. In our case, the primary focus is on fog node authentication. To gain a foundational understanding of fog computing, its applications, and authentication in fog computing, a literature survey has been conducted. This survey aids in obtaining essential knowledge and insights from previously published works.

In a recent development, a blockchain identity authentication scheme has been implemented in fog computing, incorporating 2-adic ring theory and related arithmetic algorithms. This novel scheme ensures that the generated signatures are immune to forgery attempts. However, a limitation of this scheme is its inability to generate blocks between master nodes [3].

Another proposal by Chen et al. introduces a secure authenticated and key exchange scheme for fog computing. This scheme utilizes operations such as elliptic curve cryptography point multiplication, bitwise exclusive OR, and one-way hash functions. To ensure its security, the automatic cryptographic tool ProVerif is employed for formal security analysis. The scheme demonstrates resilience against ephemeral secret leakage attacks. Through security analysis and performance evaluation, it has been determined that this scheme outperforms others in terms of both security and performance [4].

Weng et al. proposed a lightweight and anonymous authentication and secure communication scheme for fog computing services. This scheme employs one-way hash functions and bitwise XOR, along with BAN Logic, to ensure privacy preservation and key agreement. The aim is to address security issues present in existing authentication schemes [4].

Singh and Chaurasiya introduced an identity-based encryption scheme for fog computing environments. This scheme is lightweight, scalable, and cost-effective, utilizing the principles of identity-based encryption. It does not require bilinear pairing during encryption and avoids the storage of keys or IDs. The authors aim to incorporate blockchain technology and analyze the scheme's security, as it currently lacks resistance against various attacks [5].

Chen et al. proposed a secure verified and key trade plot for fog computing that offers multiple security services and withstands attacks [4]. However, Jia algorithm demonstrated that the plot, a confirmed key understanding scheme for fog-driven IoT healthcare systems, is vulnerable to an ephemeral secret leakage attack [6]. Thus, Chen et al. proposed a new verified key exchange scheme with optimal performance and improved security. However, the scheme is currently limited to IoT hardware with user input [7].

Ali et al. presented a congestion-resistant secure authentication scheme for fog computing services. This scheme employs symmetric key-based hash functions and elliptic curve cryptography to prevent insecure traceability and user impersonation attacks [8].

Hamada et al. proposed a lightweight anonymous mutual authentication scheme for securing fog computing environments. It relies on elliptic curve cryptography, hash functions, symmetric encryption/decryption, and XOR operations. The LAMAS scheme offers low computation cost and requires less storage capacity on the end-user side. It provides mutual authentication, user anonymity, user untraceability, perfect forward secrecy, and session key independence [9].

Stojmenovic et al. [10] identified authentication as a significant security challenge in fog computing. Conventional PKI-based authentication is not scalable enough to provide authentication to fog users at the edge of the network. Wazid et al. [11] proposed an identity verification scheme in the IoT environment that quickly obtains data. However, this scheme is vulnerable to various attacks and does not support user revocation. Banerjee et al. [12] introduced a lightweight three-factor authentication scheme for IoT environments. Existing solutions face security threats related to fog computing and are not secure against stolen identity verification programs and smart card attacks. Further improvement can be achieved by utilizing available computing resources from edge or smart devices [13]. As per their research on different domains. Research domains encompass the IoT, as well as the convergence of IoT with notable fields like cloud computing and mobile computing. Numerous emerging computing paradigms and technologies lie in those areas as it is emerging there is no standard definition for most of them [14].

The literature survey is organized into two subsections: The first focuses on fog computing environment-based approaches, while the second deals with application-based authentication. Table 1 provides a comprehensive study of proposed fog computing-based authentication techniques.

Table 1 Summary of authentication based on fog environment

References	Year	Algorithm used	Description
[9]	2022	The proposed scheme relies on elliptic curve cryptography, hash functions, symmetric encryption/decryption, and XOR operations to provide secure authentication in fog computing	The proposed scheme offers several advantageous features. It requires low computation cost, ensuring efficient processing on the end-user side. Additionally, it has minimal storage capacity requirements, reducing the burden on user devices
[8]	2021	The proposed scheme utilizes symmetric key-based hash functions and elliptic curve cryptography for authentication in fog computing	The clogging resistant secure authentication scheme is designed to overcome insecure traceability, attacks, and is also inefficient that the SAKA-FC scheme
[15]	2021	Used one-way hash function and bitwise XOR, BAN Logic for their work	The lightweight and secure authentication scheme aims to preserve privacy and establish key agreement in fog computing services. It addresses the security vulnerabilities present in existing schemes
[16]	2020	Algorithm used 2-adic ring theory and arithmetic	The proposed scheme has signatures that cannot be false, capable of transaction node anonymity, and forward security

5.1 Based on Fog Computing Environment

In this subsection of the paper, research findings are organised in terms of the authentication of fog nodes, algorithms implemented, and tools used for the authentication process in fog node computing various types of this scheme have been used by different researchers. Authentication is one of security issues in fog computing, which cannot be solved by using simple encryption algorithms as the data is generated by IoT devices in large scale which are closer to end users.

5.2 Related Work on Other Fog Application

In fog computing, network security is essential, and blockchain technology offers a trusted and decentralized solution. By using certificateless encryption and signature schemes, blockchain addresses the authentication needs and enhances security in the decentralized environment [3].

Tomar et al. [17] proposed a blockchain-assisted authentication and key agreement scheme for fog-based smart grid systems. The scheme utilizes the RoR model, elliptic curve cryptography (ECC), and the Hyperledger Fabric consortium open-source blockchain platform. It aims to enhance scalability and establish a trusted authority for secure communication between smart meters and fog nodes. The scheme ensures secure session key generation and also introduces a data offloading scheme for secure data transmission in fog-based smart grid environments.

Umoren et al. [18] proposed a decentralized user authentication approach for securing fog computing using the Ethereum blockchain and smart contracts. The paper presents a confirmation framework that leverages the features of blockchain and smart contracts to ensure secure user authentication. The implemented system utilizes user information such as email address, username, Ethereum address, password, and biometric data to register and verify users. Additionally, the paper suggests exploring alternative approaches, such as using NEO smart contracts, to further enhance the authentication process.

Singh and Chaurasiya proposed a mutual authentication framework for health care in fog computing. The framework utilizes elliptic curve cryptography, one-way hash functions, AVISPA for formal security analysis, and BAN Logic for informal security analysis. The proposed protocol is secure against unsolicited attacks and offers various security features such as scalability, authentication, forward secrecy, confidentiality, and integrity. The performance analysis shows that the protocol is more efficient in terms of communication costs, storage costs, and operating time compared to other existing schemes [19]. Table 2 provides a comprehensive study of fog computing-based fog applications.

Table 2 Summary of authentication based on other fog application

References	Year	Algorithm used	Description
[20]	2022	Used elliptic curve cryptography, one-way hash function, and AVISPA for formal security and BAN Logic for informal security	Singh and Chaurasiya et al. proposed a secure protocol for fog computing that ensures protection against unsolicited and intrigue attacks. The protocol supports scalability, authentication, forward secrecy, and confidentiality, making it efficient in terms of communication and storage costs
[17]	2022	Bilinear Maps, AggUn SignCrypt, KGC, PPK, UKGSignCrypt	The proposed PPAAS scheme for security caution framework in fog cloud-based VANTEs based on a unused mysterious lesson which fulfills information secrecy, unforgeability, sender namelessness, and key escrow freeness. It is more productive than the existing ones in terms of computation, communication, and capacity costs and accomplishes the security objective of productive conditional protection for the primary time
[21]	2022	The proposed approach employs SGX for secure storage of the fog node's private value and utilizes the ROR model, along with an informal analysis for security assessment	The proposed protocol demonstrates resilience against well-known attacks, offering enhanced security and lower computational costs
[20]	2022	Tomar et al. used the RoR model, ECC, and Hyperledger fabric blockchain platform for secure communication and scalability in fog-based smart grids	The proposed scheme aims to achieve scalability and reduce dependency on a single trusted authority for secure communication

6 Result

A number of research papers have focused on developing lightweight authentication schemes. In one such paper, Author [22] introduced a mutual authentication scheme that offers significant advantages in terms of storage and computation costs. This scheme allows fog users to seamlessly communicate with newly joined fog servers without incurring additional overhead. However, a notable limitation of this scheme is the lack of consideration for fog users' anonymity. Additionally, the mentioned scheme is susceptible to man-in-the-middle attacks and relies solely on a central registration authority. Consequently, if the central registration authority is compromised, the entire system becomes vulnerable to exploitation by attackers.

In their work, Rahman et al. introduced a lightweight mutual authentication security scheme that offers authentication anonymity and can effectively defend against man-in-the-middle attacks. The scheme achieves this using symmetric encryption and a hash-based message authentication code. However, one limitation of the proposed scheme is that it relies on the assumption that the secret key is pre-shared between the registration authority and fog users during system initialization. Consequently, the scheme becomes inconvenient when it comes to updating a new session key [17].

7 Conclusion

Authentication is an important part in improving fog security issues that could help to achieve confidentiality, integrity, and availability (CIA) of data in fog computing. Various authentication approaches have been employed to detect security attacks that evolved over fog computing and prevent unauthorized access to keep the users data secure. The paper emphasizes over distributing the existing methods based on different authentication techniques, key exchange and other approaches. Then, the entire survey is categorized based on authentication approaches such as fog environment-based and application domain so as to know what type of authentication scheme used and what are the issues faced by the researcher in a tabular manner. As per the survey conducted, various authentication-based scheme approaches are adopted by many researchers to prevent unknown attacks by monitoring network traffic and unauthorized users.

References

1. Roman R, Lopez J, Mambo M (2018) Mobile edge computing, fog et al.: a survey and analysis of security threats and challenges. Futur Gener Comput Syst 78. https://doi.org/10.1016/j.Future.2016.11.009
2. Patwary, Al-Noman A, Fu A, Naha RK, Battula SK, Garg S, Patwary MAK, Aghasian E (2020) Authentication, 0access control, privacy, threats and trust management towards securing fog computing environments: a review. arXiv Preprint arXiv:2003.00395
3. Wang H, Jiang Y (2020) A novel blockchain identity authentication scheme implemented in fog computing. Wirel Commun Mob Comput 2020:1–7
4. Chen C-M et al (2021) A secure authenticated and key exchange scheme for fog computing. Enterpr Inf Syst 15(9):1200–1215
5. Singh S, Chaurasiya VK (2021) Mutual authentication scheme of IoT devices in fog computing environment. Cluster Comput 24:1643–1657
6. Jia X, He D, Kumar N, Choo K-KR (2019) Authenticated key agreement scheme for fog-driven IoT healthcare system. Wireless Netw 25(8):4737–4750
7. Chen CM, Huang Y, Wang KH, Kumari S, Wu ME (2021) A secure authenticated and key exchange scheme for fog computing. Enterpr Inf Syst 15(9):1200–1215. https://doi.org/10.1080/17517575.2020.1712746

8. Ali Z, Chaudhry SA, Mahmood K, Garg S, Lv Z, Zikria YB (2021) A clogging resistant secure authentication scheme for fog computing services. Comput Netw 185:107–731
9. Hamada M, Salem SA, Salem FM (2022) LAMAS: lightweight anonymous mutual authentication scheme for securing fog computing environments. Ain Shams Eng J 13(6):101752
10. Stojmenovic I et al (2016) An overview of fog computing and its security issues. Concurr Comput Pract Exp 28(10):2991–3005
11. Wazid M, Das AK, Bhat V, Vasilakos AV (2020) Lam-ciot. Lightweight authentication mechanism in cloud-based IoT environment. J Netw Comput Appl 150:102496
12. Banerjee S, Odelu V, Das AK, Srinivas J, Kumar N, Chattopadhyay S, Choo K-KR (2019) A provably secure and lightweight anonymous user authenticated session key exchange scheme for internet of things deployment. IEEE Internet Things J 6(5):8739–8752
13. Umoren O, Singh R, Pervez Z, Dahal K (2022) Securing fog computing with a decentralised user authentication approach based on blockchain. Sensors 22(10):3956
14. Elazhary H (2019) Internet of Things (IoT), mobile cloud, cloudlet, mobile IoT, IoT cloud, fog, mobile edge, and edge emerging computing paradigms: disambiguation and research directions. J Netw Comput Appl 128:105–140
15. Weng C-Y, Li C-T, Chen C-L, Lee C-C, Deng Y-Y (2021) A lightweight anonymous authentication and secure communication scheme for fog computing services. IEEE Access 9:145522–145537
16. Wang H, Jiang Y (2020) A novel blockchain identity authentication scheme implemented in fog computing. Wireless Commun Mobile Comput 2020
17. Tomar A, Tripathi S (2022) Blockchain-assisted authentication and key agreement scheme for fog-based smart grid. Clust Comput 25(1):451–468
18. Singh S, Chaurasiya VK (2022) Mutual authentication framework using fog computing in healthcare. Multimedia Tools Appl 1–27
19. Singh S, Chaurasiya VK (2022) Mutual authentication framework using fog computing in healthcare. Multimedia Tools Appl 81(22):31977–32003. https://doi.org/10.1007/s11042-022-12131-8
20. Wu T-Y, Guo X, Chen Y-C, Kumari S, Chen C-M (2022) SGXAP: SGX-based authentication protocol in IoV-enabled fog computing. Symmetry 14(7):1393
21. Rahman G, Wen CC (2019) Mutual authentication security scheme in fog computing. Int J Adv Comput Sci Appl 10(11)
22. Ibrahim MH (2016) OCTOPUS: an edge-fog mutual authentication scheme. Int J Netw Secur 18(6):1089–101

Chapter 64
SafeMaps: Crime Index-Based Urban Route Prediction

Ria Singh, Shatakshi Mohan, Harsh Pooniwala, V. V. Gokul, and S. Shilpa

1 Introduction

Given the ever-increasing crimes in our society, a significant part of our daily routine, i.e. commute is being compromised every passing minute. Working through a solution for the same, we have devised SafeMaps. In our project, we have geo-visualized the crime statistics of a city on a Web application via tomtom API. We developed a risk model using a graph theory algorithm for urban street networks, allowing estimation of crime index on any road segment, while the route prediction was done via nested K-means clustering using the elbow method.

Our system can help people reach their destination via the safest and shortest path. It will be beneficial for tourists, college students, and working women who have to commute daily in busy and unfamiliar metropolitan cities. It is essentially a route prediction system. The map's display is set to a predetermined city. Parallelly a risk model has been designed to calculate the crime index of regions in order to save time consumption. The user inputs the source and destination address on its client side which is converted to floating-point coordinates and sent to the K-means model. The model computes the best possible routes with trade-offs between travel duration and safety and presents the alternatives to the user at the client side.

R. Singh (✉) · S. Mohan · H. Pooniwala · V. V. Gokul · S. Shilpa
PES University, Bangalore, India
e-mail: riasingh.rs12@gmail.com

S. Mohan
e-mail: mohan.shatakshi1234@gmail.com

H. Pooniwala
e-mail: pooniwalaharsh@gmail.com

S. Shilpa
e-mail: shilpas@pes.edu

A. Yadav et al. (eds.), *Proceedings of International Conference on Paradigms of Communication, Computing and Data Analytics*, Algorithms for Intelligent Systems,
https://doi.org/10.1007/978-981-99-4626-6_64

2 Background

In practically every corner of the world, criminal activity has reached record heights. Drastic times call for desperate measures. To ensure personal safety, particularly for travellers who frequent both known and unknown locations. Majority of the crimes happen while the victim is moving regardless of the transportation mode used such as walking, public transit, or personal vehicles. Thus, in such unsure times, there is a dire need for a trustworthy, and reliable method to navigate safely is prominent, and that is what we aim to address with our project.

Traditionally, it is the driver's responsibility to calculate the route by taking many factors into account. This parameter may also consider the road's infrastructure, safety record, distance, and traffic volume. This is where the idea to incorporate the safety factor comes to play. Our proposed solution is to provide not only the shortest but also the safest viable route to the user to reach their destination safely. Therefore, in such times, the need for a trustworthy and reliable method to navigate safely is prominent, and that is what we aim to address with our project.

We have considered the crime indices of each street that falls in a particular route to be able to deem it safe. Using the abovementioned A* algorithm, we have taken account of the distance between the start and end points, as well as the final index of the route.

3 Literature Survey

For the literature survey, we looked into projects similar to ours which have been developed. But each was employed using a different set of algorithms for the safest path prediction. Route application designed for Mexico City employs a Naive Bayes classifier to identify and geocode incidents based on social crime reports and tweets. Its major purpose is to find the safest path regardless of geographical distance, but its reliance on social media to get accurate data for prediction is a disadvantage. Similarly, *SocRoutes* processes geotagged tweets to recommend safe routes to users and to warn them to avoid dangerous areas. Another application, namely, *Be-safe* travel provides the safest path based on the shortest route and security level. However, it only counts the number of crime points along a path without taking the seriousness of the occurrence into consideration. *PreGo* algorithm suggests routes depending on the user's preferences, such as road conditions, scenery, and time. While this model may suggest routes based on danger indicators, it does not offer distances from risky edges, and therefore, there is minimal information on street safety.

Similarly, in [1], The SAFEPATHS issue is a bi-objective shortest path problem with the purpose of producing a set of pathways with trade-offs between length and danger. They investigated two forms of this basic problem that differ in their characterization of a path's risk. The major issue in constructing algorithms for these situations is that the solution space may be exponential. Using OpenStreetMap, they retrieved a city's real road network (OSM). The algorithms they demonstrated can report a representative collection of pathways in a very short span of time. They optimized these algorithms by adding early-stopping conditions, which resulted in a fourfold reduction in running time. Such technologies have the potential to considerably improve the liveability of urban areas, therefore contributing to the concept of smart cities. Their approach was designed to provide a quick and safe path, but instead, it generates routes with a significant compromise between both parameters. The proposed work can be made more user-friendly by adding more interactive components such as an emergency contact feature, and danger alert notification.

While in [2], since the shortest path may not always be the safest, the proposed system is built on the already existing technology of finding the shortest path using Dijkstra's algorithm with the addition of the safest route recommendation as an option for the user. The computed safe path is determined by the area's crime rate. The algorithm finds the route with the least number of registered crimes. The implementation is relatively simple and did not have to deal with the trade-off between the safest and shortest route. Users can choose to see the shortest or the most secure path based on their convenience. The proposed algorithm does not treat the problem as a bi-objective problem, i.e. the suggested route is either safe or short, not necessarily both. Thus, there is room for improvement.

Additionally, in [3], this study intends to offer a safe route visualized on maps to the user based on the geographical region's previous crime records, taking into account their age, gender, and time of travel, and assuring the supply of routes that are tailored to the needs of each user. The authors implemented the project in two stages, the first to realize user-specific functionalities using a decision network and the second to enable safe route generation using geospatial data analysis. Crime score calculation based on user-specific attributes gives the user a customized navigation experience based on age, gender, and time of the day. Similar to the papers we have discussed before, the issue lies in the safety and distance trade-off. The suggested route is not necessarily short and safe at the same time.

Furthermore, in [4], this research focuses on predicting the path based on latitude and longitude using hierarchical K-means clustering. For the first time, K-means clustering was used: After calculating the number of clusters to be generated using the elbow approach, this step is completed. For the second time, K-means clustering was used: Done in one of the already established clusters, centroids of freshly created clusters were discovered using K-means clustering. Finally, to determine the risk score associated with all the generated waypoints, K-nearest neighbour regressor has been used. Using layered K-means clustering yields better forecasts since it takes into account smaller crime regions as well, uses not only a crime dataset but also an accident dataset to determine the risk score of a route, and takes into consideration the severity of the crime as well. While the proposed method of route prediction seems fail-proof with two layers of algorithms being implemented, if the K-means clustering algorithm itself had assigned risk scores to areas, there would not have been a need to use KNN later. Hence, this complication could have been avoided.

Moreover, in [5], the final model has been decided after testing and comparing hierarchical clustering (agglomerate), spectral clustering, birch clustering, and K-means clustering. The results confirm that the K-means algorithm is highly suitable for the dataset utilized in this work. The suggested system provides users with a variety of paths by labelling them with different hazard indices. Multiple algorithms have been implemented out of which the best performing algorithm (K-means clustering) has been chosen as the final model which supports our decision to use K-means clustering for our project and displays the different routes with the associated time duration and danger index. It is up to the user to choose a route based on convenience as it does not suggest one single short and safe route. More factors for calculating the danger index can be added to make it fail-safe.

Further, in [6], the system has been designed using a K-means clustering algorithm to improve the accuracy and reduce the access time. The results show clustering of routes based on the following categories: best route, better route, average, less secure route, and least secure route. (No clear method of categorization of crimes specified) The algorithm is run over the dataset multiple times with different starting centroids to provide the best safe path, which is evaluated using the Kaufman evaluation model. The proposed solution allows the user to look at all possible safe routes from the source to the destination so the user can select the most suitable option. The work attempts to suggest the safe route along with the type of crime that happened on that route so that anyone can select the safest route from the source.

Given a source and destination, in [7], the application operates in three steps. When the application gets invoked, a request is sent to the data collection unit requesting up-gradation of the repository, their Internet is searched for new data. The data is then pre-processed with the objective of elimination of stop words.

Google API is invoked using the rest service to get alternatives from source to destination. Afterward, these routes are sent to the route selection unit, where the data get processed to extort the safest alternative. This step includes category identification and score calculation. In category identification, news related to the specific route is obtained. Category classification includes frequency of mishaps (Fm), time interval in which mishap occurs (Ti), severity of mishap (Sm), and last time event occurred.

Based on category classifiers, the data is segregated. Score calculation involves summation of different factors and category scores according to the formula (see Eq. 1):

$$\text{Route}_{\text{RS}} = \sum \text{Category}_{\text{RS}} \tag{1}$$

where $\text{Category}_{\text{RS}}$ is the risk score of each category, and Route_{RS} is the total score of a single route.

Lastly, in [8], to establish safety ratings for regions, which might be used to determine the route safety, the study introduces the vector-based diffusion and interpolation matrix (VDIM), a unique method to diffuse and interpolate information on spatial grid cells. According to the author, an ideal dataset was created using the diamond-square method [9] to test the algorithm. This dataset depicts a region where correct perceived safety reports are available at all feasible GPS values, subject to minor truncation because of the restricted GPS precision.

4 Proposed Solution

4.1 Dataset Used

We used the San Francisco crime dataset for our research. The dataset contained incidents derived from the SFPD Crime Incident Reporting system. Sourced from Kaggle, over 800,000 rows and 9 columns made for an abundant dataset, which gave us the freedom to be thorough with the data cleaning process.

Post-processing and analysis, column consisting of crime rating pertaining to each region is appended into the dataset.

4.2 Data Pre-Processing

We started off by dropping all 9985 rows that have other offences in the category column as we deemed it irrelevant to our research. Similarly, there were other category values that were unclear and therefore dropped-Trea with six entries and secondary code with 126,182 entries.

Next, we assigned crime scores to each crime type (see Table 1). The crimes were categorized for score assignment based on their intensity. For example, heinous crimes such as sexual offences were given a crime score of 10, whereas relatively harmless crimes such as gambling or family offences were given a score of 3.

Table 1 Crime scores assigned to each type of crime based on intensity

Crime	Score
Sexual Offences, Kidnapping, and Violent Crimes	10
Crimes against individuals, Assault, Missing Person, Robbery, Purse Snatching, and Weapon Law infringement	9
Arson, Murder, Vandalism, Vehicle Theft, and Carjacking	8
Driving Under Influence, Accidents, and Suicide	7
Bribery, Extortion, Drunkenness, and Liquor Law/Drugs	6
Disorderly Conduct, Loitering, Suspicious Occ., and Trespassing	5
Burglary, Fraud, Larceny, Theft (from the property)	4
Family Offence, Gambling, and Stolen Property	3
Bad check, Forgery, Counterfeit, and Runaway	2
Non-criminal/civil crimes	1

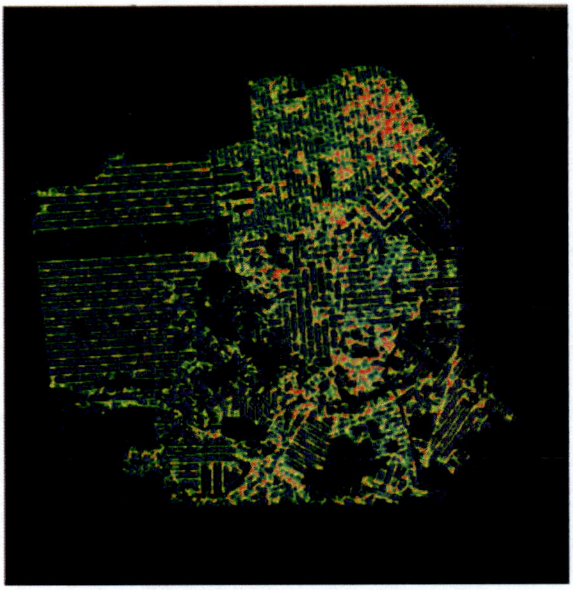

Fig. 1 Heatmap generated based on the San Francisco crime dataset

Data visualization was employed to conduct supplementary data analysis. Larceny emerged as the predominant form of crime within the city, closely followed by theft. Analysis of the heatmap unveiled that the north-eastern region of San Francisco exhibited a comparatively higher concentration of criminal incidents compared to other parts of the city, indicating heightened criminal activity in that area (refer to Fig. 1).

The dataset was further cleaned by removing outliers such as wrong dates and impractical longitudes or latitudes. Finally, after conducting a few other processes, we obtained a dataset with 566,976 rows and 10 columns.

4.3 Crime Rating Assignment

To calculate our path, we employed a heuristic-based methodology. K-means clustering was used to assign a crime rating to each neighbourhood. This crime rating was then utilized as a crucial heuristic. The A* algorithm, which we used for the final route prediction and computation and takes into consideration both distance and crime ratings, yielded the expected results.

We performed assignments of scores on three levels:

1. Crime scores were assigned to each crime type based on the intensity of those crimes (ranging from 1 to 10).
2. Crime ratings were assigned to each neighbourhood using the K-means algorithm (ranging from 1 to 5).

3. The crime index was calculated for each route predicted to make comparisons (ranging from 1 to 5).

As established before, crime scores were assigned to each crime type. Now, to assign crime ratings to each neighbourhood, we fed this data to the K-means clustering algorithm. A new column was created in the dataset to store the crime rating associated with each locality. Now, the task was to calculate the crime index of each route predicted so that the user could make comparisons to decide which route to take.

4.4 Crime Index Calculation

Along with the crime score, we have the frequency of crime occurred. To calculate crime index and for shortest/safest route calculation, we have come up with the following formula (as in Eq. 2):

$$\mathrm{CI} = \frac{\sum_i w_i f_i}{\sum_i f_i} \tag{2}$$

where CI = Crime index of the route; w_i = the weight of the occurred crime; f_i = frequency of the occurred crime, $i = 1, 2,.., n$.

4.5 Algorithms

K-means is a clustering algorithm under unsupervised machine learning. This kind of clustering divides the information or data into groups where the points inside each group have characteristics in common. K-means was chosen because it is straightforward, adaptable, and effective with larger datasets. In order for the K-means algorithm to function in relation to our project, the clusters had to be created using nested clustering.

Two sub-processes made up this process: First, we were able to segment the San Francisco map into smaller areas of risk by clustering based on the latitude and longitude of the locations where the crimes happened. To generate more accurate and meaningful regions of risk, clustering was applied in the already formed clusters in the map, this time with the crime scores of each type of crime fed to the K-means algorithm alongside the latitudes and longitudes.

The K value that we obtained using the elbow method was 5 for both layers of clustering, so the map was divided into five types of crime risk zones. Finally, we were able to obtain crime ratings associated with each locality to be used for final route prediction using the A* heuristics algorithm.

We began by populating an open list with the first node using the A* algorithm with the crime score heuristic. The approach computed the preliminary f-cost ($f = g + h$), where g was the distance from the start node, and h was the crime score heuristic. The path was deemed discovered if the node with the lowest f-cost was the objective. If not, the algorithm evaluated the neighbouring nodes and extended the current node with the lowest f-cost. The neighbours' f-costs were updated, and their parent nodes were recorded. The current node was put in a closed list, and the steps were repeated until either the goal node was achieved or the open list was exhausted (see Fig. 2).

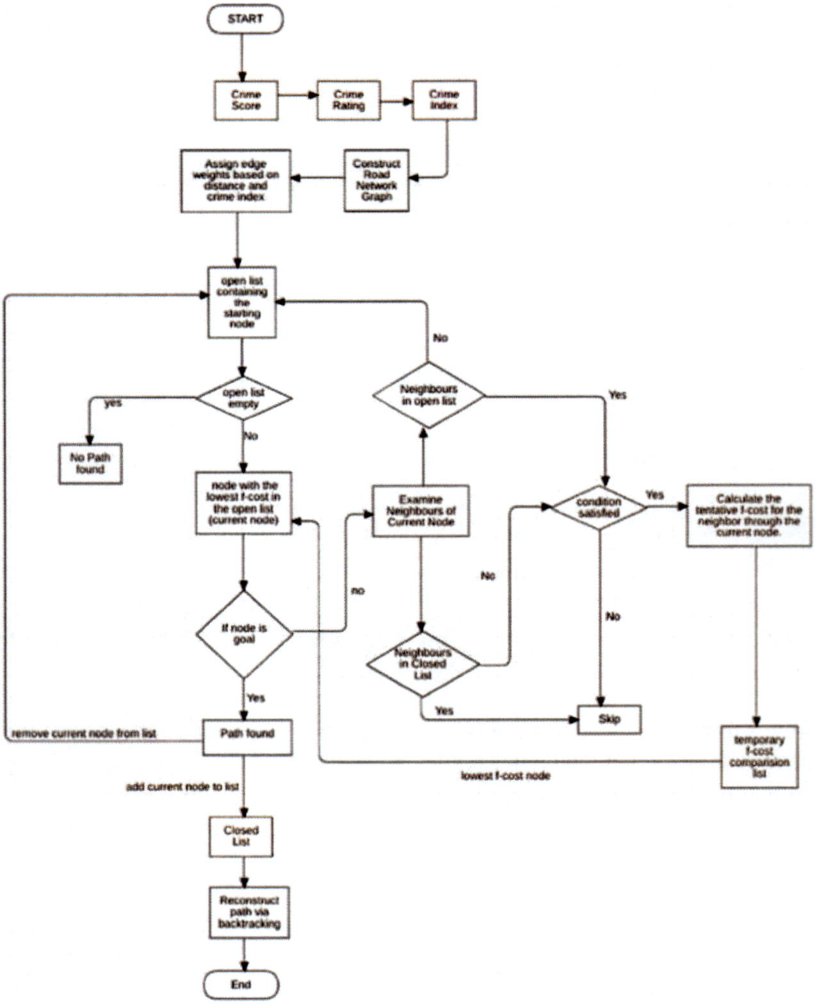

Fig. 2 Our proposed solution in standard algorithmic form

For predicting the final routes by taking the distance as well as risk factor into account, we decided to use the A* heuristic algorithm as it considers the weights associated with the segments in the route along with the heuristics of each segment. In our case, the heuristic was the crime rating of each locality. In this way, we were able to make a practical trade-off between safety and time.

5 Implementation and Results

By leveraging visualisation tools like histograms, scatterplots, and heatmaps for data analysis purposes, we were able to ascertain that Larceny/Theft was the primary crime committed throughout San Francisco. Further investigation through our heatmap revealed a higher density of criminal activity clustered together in one area: namely the north-eastern district of San Francisco.

As illustrated in the figures, we used K-means clustering on two levels (see Figs. 3 and 4).

Our system was tested on San Francisco streets and recommends multiple colour-coded routes between two points based on relative safety. Green being the safest path and red being the least recommended path.

The algorithm has been evaluated for a variety of user data use cases as well as distinct source and destination locations.

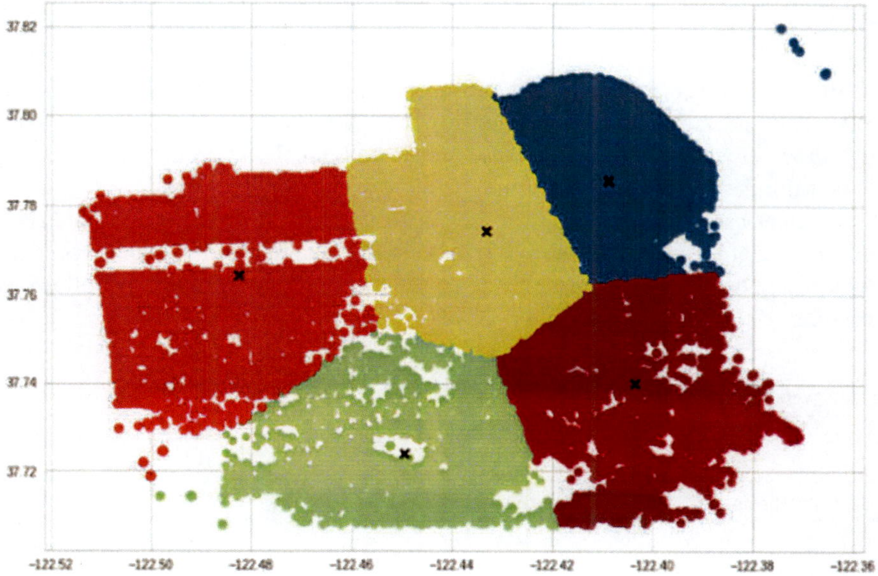

Fig. 3 First run of the K-means algorithm on coordinates alone

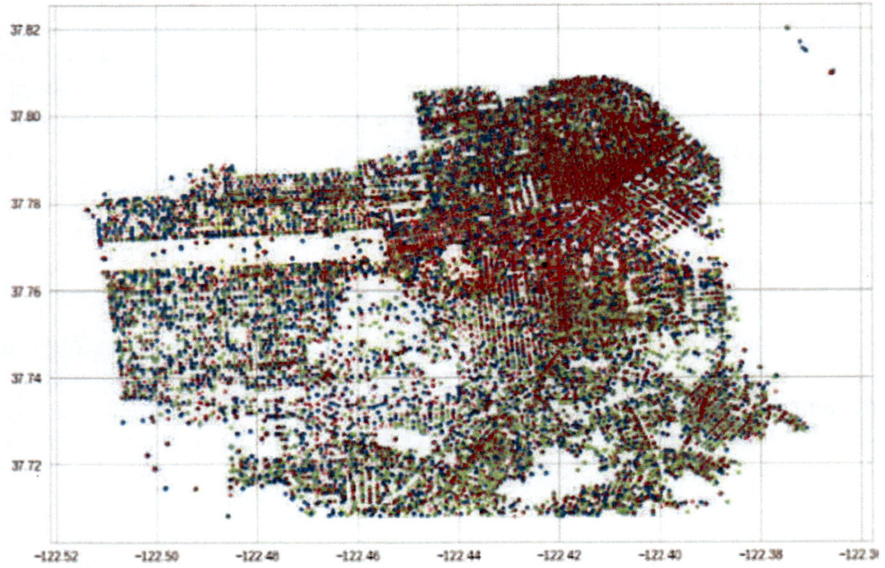

Fig. 4 Second run of the K-means algorithm on the coordinates as well as the *crime scores*

We recorded a comparative study of our application against traditional navigation system where we compared the durations of the best routes predicted by each system for the same source and destination (see Table 2). Here, the traditional system predicted the shortest routes while our application predicted the shortest and safest routes.

To compare our application's performance metrics, we have computed mean percentage error (MPE), which indicates the accuracy of a model's predictions regarding percentage differences between the traditional and predicted time durations. MPE is often used to assess the accuracy of demand forecasts.

We can observe from these measures that our application holds a lower MPE, indicating that its projected duration is closer to the real time in terms of percentage difference. To calculate the mean percentage error (MPE), we can use the following formula:

Table 2 Traditional system versus SafeMaps route duration

Source	Destination	Traditional system route duration (min)	SafeMaps route duration
Lyon street steps	Chrissy Fields Avenue	7	8 min 16 s
Inspiration point	West Bluff Picnic Area & Beach	11	10 min 43 s
Presido heights	USCF Medical Centre	22	24 min
Corona heights	San Francisco State University	16	21 min 48 s

Table 3 Calculation of MPE

Traditional route duration (s)	SafeMaps route duration (s)	lActual - Predictedl/Actual
420	496	0.180952381
660	643	0.025757576
1320	1440	0.090909091
960	1308	0.3635

$$\text{MPE} = \left(\sum (|\text{Actual - Predicted}|/\text{Annual})/n\right) \times 100\%$$

where:

Σ	Summation symbol
Actual	The actual value
Predicted	The predicted value
n	Number of observations

We converted the time durations of traditional system and SafeMaps to seconds and calculated (lActual - Predictedl/Actual) values to evaluate final MPE value (see Table 3).

Therefore, using the above-discussed MPE formula and the values obtained, we computed the mean percentage error to be 16.5%. This means that, on average, the SafeMaps route duration estimates differ from the traditional route duration by 16.5%.

6 Conclusion

Several societal factors, such as population growth, urbanization, and economic development, have led to an uptick in criminal activity. Crimes occurring on roads and against travellers demand urgent attention and mitigation. Our proposed solution aims to provide users with the safest and quickest route possible, tailored to their specific needs. This project combines the K-means algorithm and A* shortest path algorithm to deliver an optimized travel route in terms of safety, time, and distance. Our project achieved a remarkable MAPE score of 16.5%.

Future improvements to the algorithms could offer even safer and more efficient routes. Enhanced algorithms with superior time and space complexity could lead to more accurate calculations. Integration with cloud services would enable real-time updates, while incorporating user attributes, such as the time of journey, age, and gender, could yield more personalized route suggestions.

Acknowledgements This work was supported by Prof. Shilpa S who guided us during our research and implementation phase for our Capstone Project (UE19CS390SB) at PES University, EC Campus Bangalore.

Appendix

1. Risk Model—The model is responsible for the calculation and assigning of the crime indexes of the paths and areas which is a necessary aspect of route prediction.
2. Crime Score—Score associated with each crime based on the intensity of the crime.
3. Crime Rating—Value assigned to each street/locality by applying the K-means algorithm.
4. Crime Index—The mathematically obtained value indicating the safety level of a route based on which route prediction was performed.

References

1. Galbrun E, Pelechrinis K, Terzi E (2016) Urban navigation beyond shortest route. Inf Syst 57:160–171
2. Bura D, Singh M, Nandal P (2019) Predicting secure and safe route for women using Google Maps. In: 2019 International conference on machine learning, big data, cloud and parallel computing (COMITCon), pp 103–108
3. Asawa YS, Gupta SR, Jain NJ (2020) User specific safe route recommendation system. Int J Eng Res Technol (IJERT)
4. Soni S, Shankar VG, Sandeep C (2019) Route-the safe: a robust model for safest route prediction using crime and accidental data. Int J Adv Sci Technol 28:1415–1428
5. Puthige I, Bansal K, Bindra C, Kapur M, Singh D et al (2021) Safest route detection via danger index calculation and k-means clustering. Comput Mater Contin 69:2761–2777
6. Pavate A, Chaudhari A, Bansode R (2019) Envision of route safety direction using machine learning
7. Ruk S, Gul S, Mahoto N, Zia M (2018) Evaluating route security and predicting the safest alternative using risk factors. Indian J Sci Technol 11:1–6
8. Elsmore S, Subastian I, Salim F, Hamilton M. VDIM: vector-based diffusion and interpolation matrix for computing region-based crowdsourced ratings
9. Wang H, Chen W, Liu X, Dong B (2010) An improving algorithm for generating real sense terrain and parameter analysis based on fractal. In 2010 International conference on machine learning and cybernetics (ICMLC), vol 2, pp 686–691

Chapter 65
Controlling the Steering Wheel Using Deep Reinforcement Learning: A Survey

Narayana Darapaneni, Anwesh Reddy Paduri, B.G. Sudha, Vidyadhar Bendre, Midhun Chandran, M. Mohana Priya, and Varghese Jacob

1 Introduction

A machine learning method called reinforcement learning enables an agent to learn by doing by interacting with its surroundings. The agent receives feedback in the form of prizes for successful outcomes and penalties for unsuccessful ones. To maximise the cumulative reward over time, the agent's principal goal is to learn a policy that links environmental conditions to actions. A considerably larger range of issues are solved with RL, including those in robotics, control systems, and games [1, 2].

The origins of RL can be traced back to the 1950s and 1960s, with early work by researchers such as Widrow and Hoff, who developed the "ADALINE" and "PERCEPTRON" learning rules. However, it wasn't until the 1980s and 1990s that RL began to be studied extensively. In 1989, Sutton and Barto published the first edition of their book "Reinforcement Learning: An Introduction" which is still widely used as a textbook in the field today. This book introduced many of the fundamental concepts and algorithms of RL, such as Q-learning and SARSA, and laid the foundation for much of the subsequent research in the field.

Q-learning is a well-known RL algorithm that learns an action-value function, in a given state, it estimates the expected cumulative reward for taking an action and following the learned policy thereafter. SARSA is a similar algorithm, but instead

N. Darapaneni (✉) · B.G. Sudha (✉)
AIML, PES University, Hosur Rd, Konappana Agrahara, Banglore 560100, Karnataka, India
e-mail: darapaneni@gmail.com

A. R. Paduri (✉)
PES University, Hosur Rd, Konappana Agrahara, Banglore 560100, Karnataka, India
e-mail: anwesh@greatlearning.in

V. Bendre · M. Chandran · M. Mohana Priya · V. Jacob
AIML, PES University, Hosur Rd, Konappana Agrahara, Banglore 560100, Karnataka, India

A. Yadav et al. (eds.), *Proceedings of International Conference on Paradigms of Communication, Computing and Data Analytics*, Algorithms for Intelligent Systems,
https://doi.org/10.1007/978-981-99-4626-6_65

of using the learned policy to determine the next action, it uses the current policy. Both Q-learning and SARSA are off-policy algorithms, which means that they can learn about the value of actions that are not currently being taken. In the 2000s, RL saw increased interest and many new algorithms were developed. One important development was the actor-critic method, which separates the learning of the policy and the value function. The actor learns the policy, while the critic learns the value function. This allows for more efficient learning and better handling of continuous action spaces. Another important development was deep RL, which combines RL with deep neural networks. This allows RL to be applied to problems with large state and action spaces and has led to the development of many successful deep RL algorithms, such as DQN and A3C. Additionally, evolutionary methods were also incorporated in RL to improve the stability and sample efficiency of the algorithm [3, 4].

RL has advanced significantly in recent years, particularly in the field of deep RL. Deep RL algorithms like DQN, A3C and DDPG, PPO, TRPO, SAC, TD3, and D4PG are used for a variety of tasks, including operating robots, enhancing control systems, and playing Atari video games. For instance, a group of researchers at Google DeepMind trained an agent to play a variety of Atari games at a superhuman level in 2015 using a DQN algorithm version dubbed the asynchronous advantage actor-critic (A3C). This was a significant advance in the area since it demonstrated that RL could be used to resolve challenging, high-dimensional issues [3, 4].

In order to increase the sample efficiency and stability of RL, it has recently been combined with additional methodologies such as imitation learning, meta-learning, and inverse RL. In imitation learning (IL), a machine learns to mimic the actions of an expert presentation. By fast adjusting to new tasks, an agent uses the method of meta-learning to learn how to learn. A method called inverse RL seeks to deduce the reward function from the actions of an agent. These methods have demonstrated success in enhancing the RL algorithm's sample efficiency and stability [5, 6].

2 Literature Review

Reinforcement learning is a machine learning technique that enables an agent to learn by trial and error through interacting with its environment.

The origins of RL can be traced back to the 1950s and 1960s, with early work by researchers such as Widrow and Hoff, who developed the "ADALINE" and "PERCEPTRON" learning rules. However, it wasn't until the 1980s and 1990s that RL began to be studied extensively, with the development of Q-learning and SARSA, two popular RL algorithms.

In the 2000s, RL saw increased interest, and many new algorithms were developed, including actor-critic methods, deep RL, and evolutionary methods. RL is applied to solve a much wider range of problems, including robotics, control systems, recommender system, and gaming.

In recent years, RL has made significant progress, especially in the area of deep RL. Deep RL algorithms like DQN, A3C and DDPG, PPO, TRPO, SAC, TD3, D4PG have been successfully applied to different tasks, like playing Atari games, controlling robots, and optimising control systems [4]. One promising application is in the field of autonomous vehicles.

We first assess how well these algorithms perform in the well-known CartPole environment, demonstrating their capacity to learn intricate control strategies. On the basis of these findings, we talk about how the fundamental ideas and methods created for CartPole may be expanded upon and used to the trickier challenge of guiding a steering wheel in actual driving situations. This study attempts to bridge the gap between theoretical DRL research and practical implementations in autonomous cars, offering insightful information for academics and engineers working in this field. It does this by drawing comparisons between the reduced CartPole environment and steering control difficulties.

It is important to note that RL has faced some difficulties, such as sampling inefficiency, instability, and the difficulty of dealing with enormous state and action spaces. RL is still a vibrant area of study with numerous exciting applications in the world of autonomous cars, despite these difficulties.

In order to increase sampling efficiency and stability, RL has more recently been integrated with additional methodologies including imitation learning, meta-learning, and inverse RL.

3 Materials and Method

3.1 Dataset and Environment

The OpenAI gym settings and related data were used for this project. The TAXI-V3 environment is used to implement and test the Q-learning/Q-network algorithm. Utilising the Cartpole-v1 environment, the last three algorithms (DQN, DDQN, and duelling-DDQN) are built and evaluated.

The official documentation for the OpenAI gym contains information on the state and action space of the environment utilised in this work [7].

3.2 Core Reinforcement Learning Algorithms

- Q-Networks: Q-networks (also known as Q-learning) are a type of model-free, off-policy reinforcement learning algorithm that can be used to learn control policies for autonomous systems.
  In Q-learning, the agent learns a function known as the Q-function, which calculates the expected return (i.e. the total of future rewards) for each feasible action in each state of the environment. The Q-function is used to choose activities that are expected to produce high returns based on the agent's current awareness of the environment.

Using the Q-network, a policy is learned. The car may go through traffic safely and effectively thanks to the policy. A collection of driving situations is used to train the Q-network, and the agent is rewarded for choosing to drive safely and effectively in each case.

In general, learning control strategies for autonomous systems, such as steering control in autonomous driving, can be facilitated by Q-networks. To obtain good performance, it is necessary to design the reward function and the training procedure.

The mathematical formula for Q-learning:

$$Q(s, a) = Q(s, a) + \alpha[r + \gamma \times \max(Q(s', a')) - Q(s, a)] \quad (1)$$

$Q(s, a)$: The Q-function's most recent estimate for state s and action A is represented by this term. The state s indicates the vehicle's present position and orientation as well as a number of other characteristics, including its speed and acceleration, the status of the roads, and the presence of other cars or people. Action A denotes either the vehicle's steering angle or a more complicated action that incorporates the steering angle, acceleration, and braking as well as other elements.

$Q(s', a')$: This word denotes the highest possible anticipated benefit for all possible actions a' in the following state s'. The actions a' reflect the potential actions that might be taken from state s', and the state s' represents the condition of the vehicle after taking action a' in state s'. The performance of the vehicle in terms of safety, fuel efficiency, and other aspects would determine the anticipated future reward.

r: This phrase refers to the instant benefit experienced in state s after doing action a. Based on the distance driven, the fuel economy of the car, and any fines assessed for accidents or other risky activity, the instant reward is determined.

γ: This is the discount factor, which establishes the significance of potential benefits in comparison to existing rewards. If the vehicle's long-term performance is more crucial than the immediate reward, the discount factor is set to a higher value; otherwise, it is set to a lower value.

α: This is the learning rate, which controls how much the previous estimate of the Q-function influences the current estimate and how much the old estimate deviates from the desired value. If the agent is learning from scratch, the learning rate is set to a higher value; if the agent has already learnt a decent policy and simply needs to make minor tweaks, the learning rate is set to a lower value.

The steering wheel control task's Q-function estimates would be repeatedly updated by the Q-learning algorithm using the formula above, depending on the immediate rewards received and the anticipated future benefits for various actions in various states. The process would continue until the Q-function reached the ideal action-value function, at which time the agent would have discovered the best method for deciding which actions will result in secure and effective navigation.

- Deep Q-Network: A neural network is used by deep Q-network (DQN), an extension of the Q-learning technique, to approximate the Q-function. DQN may be utilised to address the issue of steering wheel control in autonomous vehicles, just like Q-learning. The Q-function is modelled in DQN as a neural network that receives the environment's current condition as input and predicts the expected return for each potential action. A collection of driving situations is used to train the neural network, and the agent is rewarded for choosing to drive safely and effectively in each case [8, 9]. A DQN is employed to learn a policy that permits the vehicle to go through traffic safely and effectively. A collection of driving situations is used to train the DQN, and the agent is rewarded for choosing to drive safely and effectively in each case [4, 6].
- Deep Double Q-Networks: Deep double Q-networks are a development of the Q-learning algorithm. For the purpose of solving Markov decision processes, DDQN is a model-free reinforcement learning technique. The best course of action to follow at each stage to maximise a reward signal is determined using DDQN, which is used to learn a policy.
  The action-value function is estimated via the DDQN method, and DDQN is a measure of the predicted future reward for performing a certain action at a specific state. The approach employs a neural network to approximate this function, and repeated training on a dataset of transitions encountered by the autonomous vehicle updates the network's weights.
  A "target" network and a "local" network are used by the "double" in the DDQN to estimate the action-value function. In order to stabilise the learning process, the target network, which generates target values for training the local network, is updated less often than the local network.
  When using DDQN to solve the issue of steering wheel control in an autonomous driving scenario, the states may be things like the location, speed, and heading of the automobile, and the actions could be things like rotating the steering wheel left or right or maintaining it straight. The smoothness of the car's motion or how close it is to other things in the area are two examples of the parameters that influence the reward signal. After then, the DDQN algorithm would figure out a strategy for deciding which actions to do at each stage to maximise the predicted return [10].
- Duelling Deep Q-Networks: The deep Q-network (DQN) algorithm is extended by the duelling deep Q-networks (duelling DQN). It is a method for solving Markov decision processes (MDPs) using reinforcement learning without the need of models. Duelling DQN uses a neural network, just as DQN, to estimate by roughly estimating the action-value function and develop a strategy for choosing actions at each stage. The manner in which the action-value function is calculated distinguishes DQN from duelling DQN in a significant way. Unlike DQN, which directly estimates the action-value function, duelling DQN divides the action-value function into two distinct functions.:

– A advantage function that determines the relative worth of each activity
– Value function, which calculates each state's total value.

Instead of simply estimating the action-value function, this decomposition enables duelling DQN to learn more effectively by describing the value of each action and state individually. The neural network architecture is changed to contain two distinct streams to achieve duelling DQN: one for the value function and one for the advantage function. A unique aggregation function is then used to combine the outputs of the two streams to produce the output.

It has been demonstrated that duelling DQN performs well in a variety of reinforcement learning tasks even when operating robots and playing Atari games. It has also been used to solve a variety of real-world issues. For instance, natural language processing, automated trading, and autonomous driving all use this technology.

4 Results

- Q-Networks: Q-learning starts with a trial-and-error method to reach the destination. From Fig. 1, it is evident that it requires 1000 iterations to converge. The learning between iterations 200–1000 is very minimal. The code to the Q-learning algorithm is published in GitHub [11].
- Deep Q-Network: From the Fig. 2, it is evident that deep Q-learning can learn from smaller amounts of data as it uses a neural network to arrive at the Q-function by approximation. It uses techniques like experience replay and fixed Q-targets to improve stability and reduce sensitivity to hyperparameters.
  The moving average is increasing even with minimal iterations as observed in Table 1 and Fig. 5. The code to deep Q-network algorithm is published in GitHub citech65projectCode.
- Double Deep Q-Networks (DDQN): From the Fig. 3, the presence of double neural network has aided in improving the moving average faster than in deep Q-network. The code to DDQN algorithm is published in GitHub [11].
- Duelling Double Deep Q-Networks: From the Fig. 4, it is evident that there is a steep increase in the score and the moving average due to the advantage and Q-value function. This method is performing better than prior methods. The code to the D3QN algorithm is published in GitHub [11].

5 Conclusions

While Q-learning, DQN, DDQN, and D3QN are all reinforcement learning algorithms that use value-based methods to solve Markov decision processes (MDPs), they differ in their approach to estimating the action-value function.

A table is used by Q-learning, a reinforcement learning system, to hold the action-value function. It is susceptible to instability and the overestimation issue frequently, and this can be seen clearly in the result as in Fig. 1.

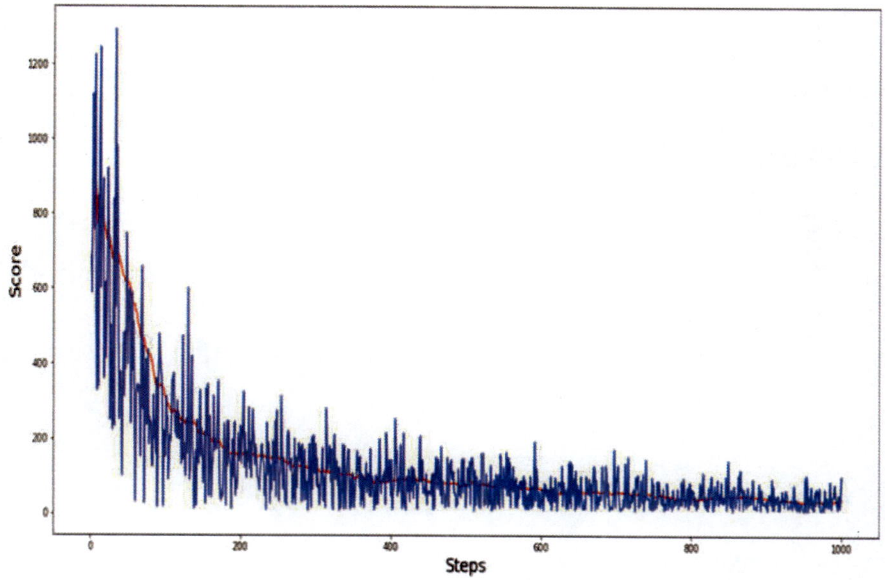

Fig. 1 Q-learning results

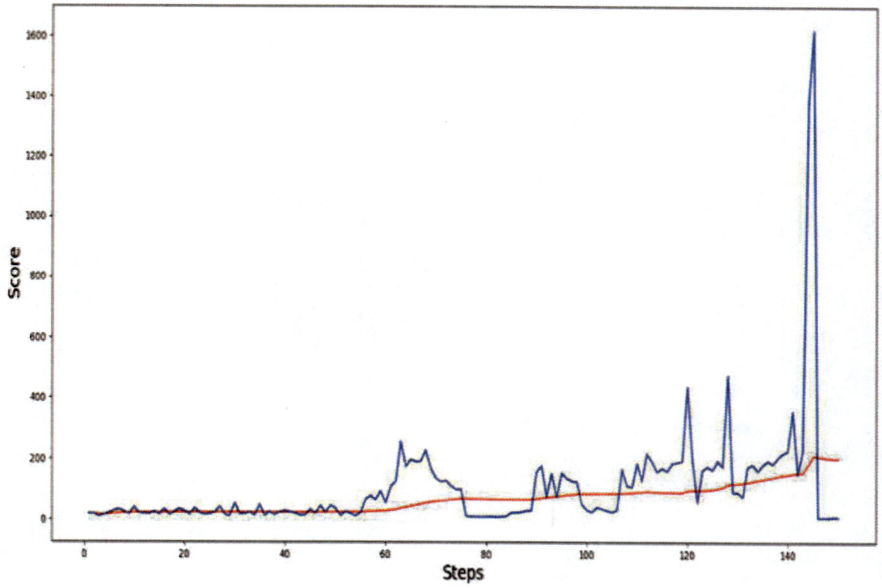

Fig. 2 Deep Q-network results

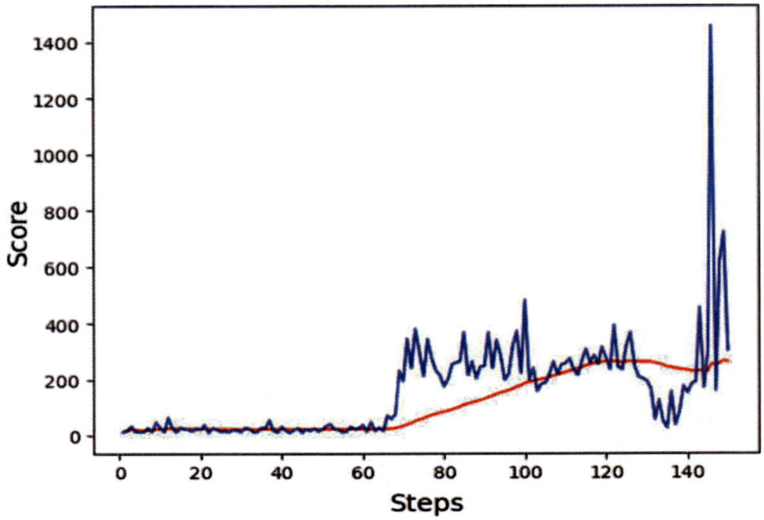

Fig. 3 Double deep Q-network results

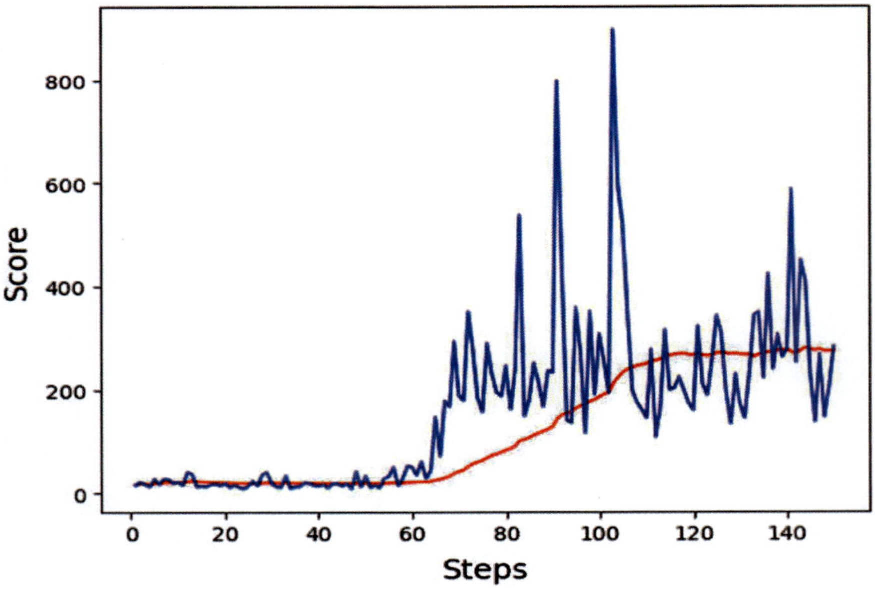

Fig. 4 Duelling deep Q-network results

Table 1 Moving average comparison table

Episodes	Average QN -ve scores	Average DQN	Average DDQN	Average D3QN
1	684.00	18.00	11.00	32.00
2	635.00	16.50	14.00	22.50
3	796.00	15.00	21.33	22.00
4	874.25	15.5	19.75	20.25
5	850.80	17.80	22.40	22.00
…	…	…	…	…
146	212.00	141.50	139.92	318.52
147	214.70	140.20	137.62	319.74
148	214.84	138.64	137.36	318.94
149	213.78	138.38	137.46	315.04
150	212.36	137.10	134.88	312.60

The deep reinforcement learning method deep Q-network (DQN) estimates the action-value function using a neural network. By utilising a fixed target network and experience replay, it tackles some of Q-learning's drawbacks. The improvement in the algorithm is shown in Fig. 2.

The deep reinforcement learning method deep Q-network (DQN) estimates the action-value function using a neural network. By utilising a fixed target network and experience replay, it tackles some of Q-learning's drawbacks. The improvement in the algorithm can be seen in Fig. 3.

The extension to DDQN is duelling double deep Q-network (D3QN). The estimation of the action-value function is made more effective by separating the estimation of the value function from the estimation of the advantage function. The improvement in the algorithm can be seen in Fig. 4.

Using neural networks and other methods, DQN, DDQN, and D3QN all offer superior stability and performance than the fundamental Q-learning algorithm. The most sophisticated of these algorithms, D3QN provides the best performance but also consumes most computing power. From Table 1 and Fig. 5 even with fewer episodes, it is clear that D3QN performs better than the other algorithms. Because Q-networks rely on trial-and-error techniques to populate the Q-table, they may need more training episodes than any other algorithm. D3QN is more stable in the presence of suboptimal convergence.

Future Scope: In this survey, we have done a comparison on only 4 algorithms

1. Q-networks
2. Deep Q-networks
3. Double deep Q-networks
4. Duelling double deep Q-networks

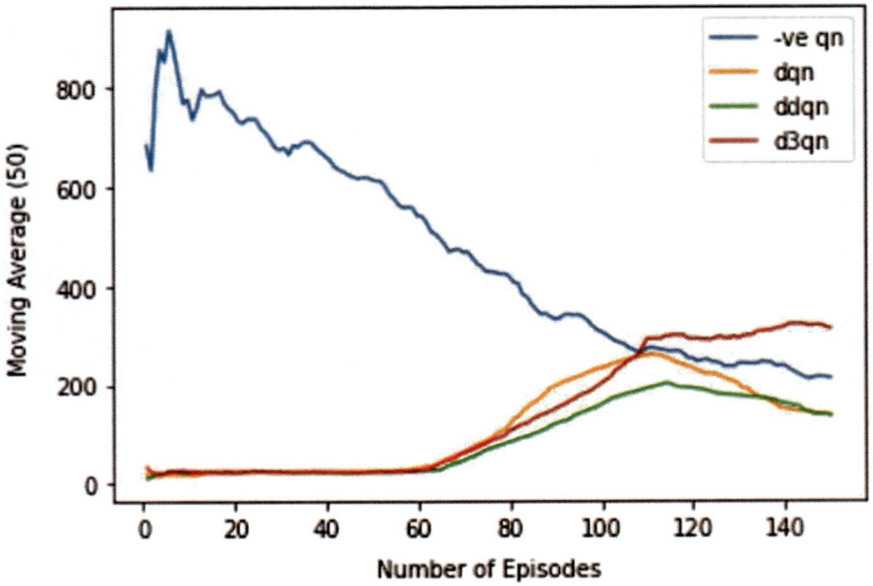

Fig. 5 Moving average comparison graph

We may expand the research to compare all of the steering wheel control algorithms that are described in this work. Because it requires a lot of processing, we have set the cap at 150 episodes. However, in the future research, a higher cap may be considered. We haven't thought about how algorithms interact when they work together to make up for each other's flaws. One may pick from a variety of algorithm combining strategies as an expansion to this study, such as

- Model Combination: It is possible to develop a new model that takes use of the advantages of both algorithms by combining the duelling DDQN model with other DRL models, such as soft actor-critic (SAC). The distinct representation of the Q-function in duelling DDQN and the entropy-regularised reinforcement learning in SAC can both be advantageous to the combined model [3].
- Technique Combination: Duelling DDQN's methods may be used with other methods, such as prioritised experience replay, to enhance performance. Duelling DDQN's sluggish convergence issue can be solved by prioritised experience replay, which samples transition from the replay buffer more frequently depending on their temporal difference faults.
- Output Combination: Duelling DDQN's outputs can be coupled with other outputs, such as the policy estimated by trust region policy optimisation (TRPO), to produce a new policy that takes use of the advantages of both methods. The duelling DDQN's stable and reliable Q-function estimations and TRPO's high-dimensional policy optimisation may both be used to benefit the new policy.

For those interested in reviewing the source code, it can be accessed via GitHub [11]. This will provide valuable insights into the implementation details and underlying algorithms discussed in this paper.

References

1. Michalík R, Janota A (2020) The pybullet module-based approach to control the collaborative yumi robot. In: 2020 ELEKTRO, pp 1–4. https://doi.org/10.1109/ELEKTRO49696.2020.9130233
2. Elallid BB, Benamar N, Hafid AS, Rachidi T, Mrani N (2022) A comprehensive survey on the application of deep and reinforcement learning approaches in autonomous driving. J King Saud Univ-Comput Inf Sci
3. Ultsch J, Mirwald J, Brembeck J, Castro R (2020) Reinforcement learning-based path following control for a vehicle with variable delay in the drivetrain. In: 2020 IEEE intelligent vehicles symposium (IV), pp 532–539. https://doi.org/10.1109/IV47402.2020.9304578
4. Emuna R, Borowsky A, Biess A (2020) Deep reinforcement learning for human-like driving policies in collision avoidance tasks of self-driving cars. Workingpaper
5. Osiński B, Jakubowski A, Zięcina P, Miłoś P, Galias C, Homoceanu S, Michalewski H (2020) Simulation-based reinforcement learning for real-world autonomous driving. In: 2020 IEEE international conference on robotics and automation (ICRA), pp 6411–6418. https://doi.org/10.1109/ICRA40945.2020.9196730
6. Zhang J, Chen H, Song S, Hu F (2020) Reinforcement learning-based motion planning for automatic parking system. IEEE Access 8:154485–154501. https://doi.org/10.1109/ACCESS.2020.3017770
7. Wu K, Abolfazli Esfahani M, Yuan S, Wang H (2018) Learn to steer through deep reinforcement learning. Sensors 18(11):3650
8. Pérez-Gil Ó, Barea R, López-Guillén E, Bergasa LM, Gómez-Huélamo C, Gutiérrez R, Díaz-Díaz A (2022) Deep reinforcement learning based control for autonomous vehicles in Carla. Multimedia Tools Appl 81(3):3553–3576
9. Maramotti P, Capasso AP, Bacchiani G, Broggi A (2022) Tackling real-world autonomous driving using deep reinforcement learning. In: 2022 IEEE intelligent vehicles symposium (IV). IEEE, pp 1274–1281
10. Kiran BR, Sobh I, Talpaert V, Mannion P, Al Sallab AA, Yogamani S, Pérez P (2021) Deep reinforcement learning for autonomous driving: a survey. IEEE Trans Intell Transp Syst
11. CodeBase by Vidyadhar Bendre, Mohana Priya M, Varghese Jacob, Midhun Chandran. https://github.com/midhuncnair/RL_SteeringWheelControlSurvey

Chapter 66
NL2SQL: Rule-Based Model for Natural Language to SQL

Kevin Tony, Kripa Susan Shaji, Nijo Noble, Ruben Joseph Devasia, and Neethu Chandrasekhar

1 Introduction

People create and interact with massive amounts of data every day. In this new Metaverse era, the vast majority of the population is getting online, resulting in an increase in data volume, necessitating the use of a data management and assessment tool. Any language with an appropriate database and corpus will be able to use this tool for natural language processing and querying. We have primarily used the English corpus made accessible by Oxford. The Oxford English Corpus is a text corpus of twenty-first-century English that is utilized by Oxford University Press' language research program and the developers of the Oxford English Dictionary. With approximately 2.1 billion words, it is the largest corpus of its sort. To address this issue, we developed a rule-based model for doing natural language processing on any language with an appropriate corpus for research. The rule-based model makes use of the analysis of the sentence structure and related vocabulary as provided by the relevant corpus. The existence, number, and sequence of the keywords recognized in the sentence submitted by the user have a significant impact on the query.

K. Tony · K. S. Shaji (✉) · N. Noble · R. J. Devasia · N. Chandrasekhar
Department of CSE, Amal Jyothi College of Engineering, Kanjirappally, Kerala, India
e-mail: kripasusanshaji2022@cs.ajce.in

K. Tony
e-mail: kevintony2022@cs.ajce.in

N. Noble
e-mail: nijonoble2022@cs.ajce.in

R. J. Devasia
e-mail: rubenjosephdevasia2022@cs.ajce.in

N. Chandrasekhar
e-mail: neethucsekhar@amaljyothi.ac.in

A. Yadav et al. (eds.), *Proceedings of International Conference on Paradigms of Communication, Computing and Data Analytics*, Algorithms for Intelligent Systems,
https://doi.org/10.1007/978-981-99-4626-6_66

2 Literature Review

Mahmud et al. [1] presents a context-free grammar (CFG) method for a rule-based paradigm. However, the generation of precise grammar is critical to the system's performance and is a time-consuming process in and of itself. The purpose of survey conducted in [2] is to present a detailed taxonomy of neural text-to-SQL systems that will enable a deeper study of all the parts of such a system. For the conversion of natural language query into Structured Query [3] utilized a lowercase conversion, removing escaped words, tokenization, PoS tagging, word similarity, Jaro-Winkler matching algorithm, and the method Naive Bayes. Kumar et al. [4] describes the D-HIRD system, which was created exclusively for the Hindi language. It is a system that is domain agnostic.

In this system, the input question is tokenized before being parsed by the system's natural language parsing module. In [5], they summarized 24 recently developed natural language interfaces for databases. Each of them was evaluated using ten sample questions to show their strengths and weaknesses. NADAQ [6] injects new dimensions of schema-aware bits associated with the input words into the encoder phase and adds new hidden memory neurons controlled by the finite state machine for grammatical state tracking into the decoder phase. Kumar et al. [7] presents a holistic overview of 24 recent neural network models studied in the last couple of years, including their architectures involving convolutional neural networks, recurrent neural networks, pointer networks, reinforcement learning, generative models, etc. [8] proposes a query processing system based on morphological analyzers. This model is based on a keyword and a set of words that may be used to extract the type of keywords using pattern matching. A survey study on natural language query interface techniques and challenges was performed in [9] and it introduces the recent research areas in that field.

NaLIX is an XML database-based general interactive natural language query interface. As query input, the system may accept any English sentence that includes aggregation, nesting, and value joins, among other things. This query is then transformed into an XQuery expression that can be tested against an XML database after reformulation. The translation is performed by mapping the grammatical proximity of tokens processed from natural language to the proximity of corresponding elements in the resulting XML. In [10], the authors define a set of rules based on the syntactic and semantic analysis of natural language queries. The proposed approach achieves high accuracy for a benchmark dataset and outperforms the existing methods in terms of efficiency. We discuss a basic approach in this work that can be used by those who have no prior expertise with structured querying or its syntax. This approach is created for the English language, but if the user uploads the necessary language's raw data - such as synonyms, thesaurus, and filter words - it can be used for cross-language querying.

The proposed approach utilizes natural language processing techniques such as part-of-speech tagging, and dependency parsing to parse the natural language statement and extract relevant keywords, entities, and relationships between them. The semantic representation captures the meaning of the statement and is used to generate the corresponding SQL query expression.The rules used in the proposed approach map natural language verbs to SQL operators, natural language nouns to SQL table and column names, and natural language phrases to SQL clauses. These rules can be customized based on the underlying data schema and query requirements.

3 SQL Queries

A database is a logically organized collection of structured data kept electronically in a computer system. Tables are used to hold data in a relational database. Tables are divided into columns, with each column storing a certain type of information (integer, real number, character strings, date). A row stores the data for a single "instance" of a table. One or more tables can be found in a database, which may or may not be linked to one another. A table is made up of columns that represent the same entity (object).

A schema, also known as a data model, is a visual or written representation of a database's data distribution and organization. It shows the features of each sort of data as well as their relationships. Structured Query Language (SQL) is an acronym for Structured Query Language. In a relational database management system, it is used to store and manage data (RDMS). It is a Relational DataBase Management System (RDBMS) standard language. It allows users to make relational databases and tables, as well as read, update, and remove them. The main purpose is to develop a querying application for a relational database. We solely deal with database interrogation, which is done with the SELECT command, which has the following syntax

SELECT column-list
FROM table-list
JOIN join-expression
$WHERE$ conditional$_e xpression$
GROUP BY group-by-column-list
$HAVING$ conditional$_e xpression$
ORDER BY order-by-column-list

To obtain data from a relational database, we usually utilise the "SELECT" keyword. The "FROM" keyword comes after the "SELECT" keyword, and it indicates the table we are referring to. The "WHERE" keyword is used to add constraints to the query in order to filter out only the relevant results. The rest of the lines, including the "WHERE" keyword, can be omitted.

So, when the user types in a query such as:

What is the average age of students with the first name Shawn?

A request like this must be fulfilled by the system:

SELECT age FROM student
WHERE FIRST NAME ='SHAUN';

4 Proposed System

To implement our model, we have built a rule-based system that can translate natural language to structured language queries (SQL) using Python, Treetagger, and Part-Of-Speech tags. We can apply our model on a client-line interface in its initial stages and provide appropriate SQLite data dump files which the user wants to query from. The model then iterates through the database structure to identify the various columns and table names. This information is then collectively used by the parser module to generate SQL queries for the provided input statement.

The parser has been formulated into various separate sub-parser modules, for example, separate parser modules for SELECT, FROM, WHERE, etc. All of these sub-parser modules function independently on the input statement and return specific results which is then collectively brought together by the main Parser module to form the respective SQL query. The result is stored in a JSON file, an instance of which is depicted in the later sections.

As shown in Fig. 1 flow chart, the user inputs a query which is processed by the model. The database selection takes place when the model decides which database the inputted query is referring to. It then refers to that database to fetch the required records. The query is then parsed by the model to extract the keywords and breaks down the request to form the conditions in the query if present. Information is extracted from the database schema simultaneously so that it can infer the words that are extracted and form meaningful queries over it. In order to bring about a certain level of uniformity, a Thesaurus is used to synchronize the words as per the database schema so that the querying gets simpler and enhanced, an SQL query corresponding to the words extracted is generated and executed.

4.1 Treetagger

The Treetagger is a program that adds information about parts of speech and lemma to text. TreeTagger uses a combination of rule-based and stochastic methods to tag words. It first applies a set of hand-crafted rules to identify the most probable POS [3, 5] tag for each word based on its context. The Treetagger tool then filters the filler words depending on where they appear in the phrase and then performs a stemming operation on the remaining words. Table 1 shows examples of Treetagger.

4.2 POS Tag

Part-of-speech (POS) [3, 5] tagging is a popular natural language processing process which refers to categorizing words in a text (corpus) in correspondence with a particular part of speech, depending on the definition of the word and its context. Part-of-speech tags describe the characteristic structure of lexical terms within a sentence or text, therefore, we can use them for making assumptions about semantics.

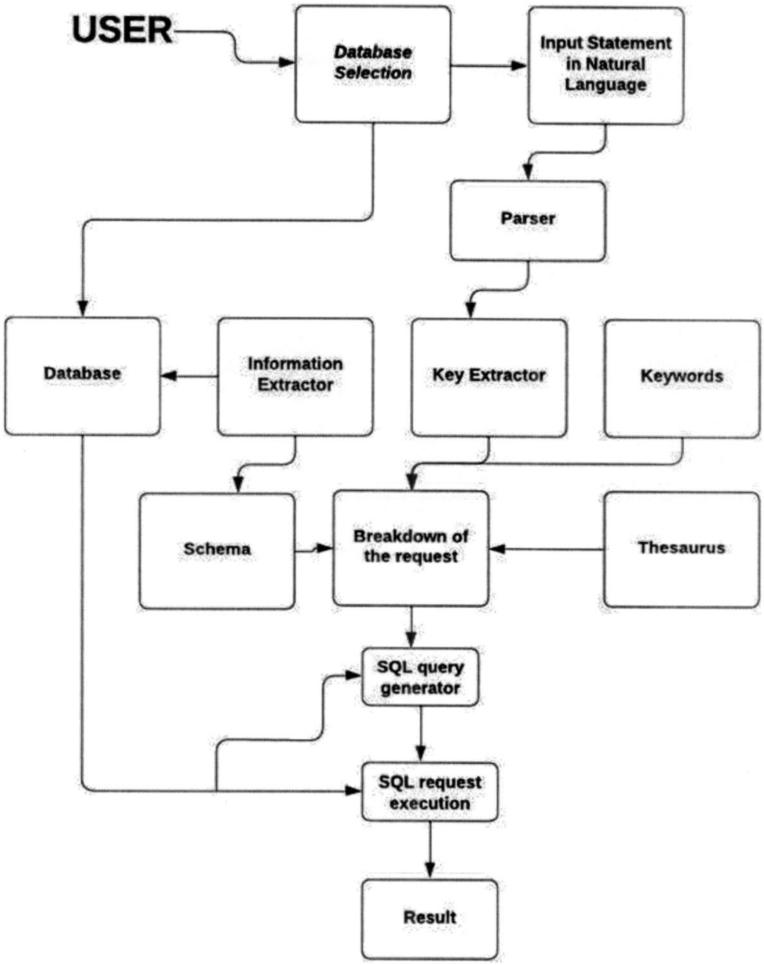

Fig. 1 Flowchart of the model

Table 1 Example of treetagger

Keywords	Operations
"what is the number", "how many are there"	Count
"do not[...], do not[...]"	Negation
"greater than", "greater than"	Superiority
"smaller than", "less than"	Inferiority
"what is the sum", "add"	Aggregate
"what is the average"	Average

Table 2 Example of POS tag

Word	POS tag
Why	Adverb
Not	Adverb
Tell	Verb
Someone	Noun
?	Punctuation mark, Sentence closer

Table 3 Example of thesaurus

Meaningful words	Substitution words
Age	Seniority, era, period, generation
Student	Apprentices, college students
First name	Baptismal name, nickname
John	–

PoS tagging is the process of annotating each token in a text with the corresponding PoS tag as given in Table 2.

4.3 *Thesaurus*

A thesaurus or synonym dictionary is a reference for finding synonyms. This will help the user freely type words that are similar to the words stored in the database dump file, hence making it more flexible for use. Table 3 shows a few examples of synonyms.

4.4 *Sample Database*

A database is used to test and feed input to generate a properly structured SQL query. For our project, we have 2 sample databases:-

1. School database
2. City database.

We have taken the sample databases from the internet as SQLite dump files, a few open-source libraries and repositories. We provide these sample databases in order to show the execution and output of our project. In order to make use of the tool for a different database, it is required that the user uploads an SQLite database dumb as input to the tool for further processing.

Table 4 Usage parameters

Parameters	Explanation
-h	Print this help message
-d ⟨path⟩	Path to SQL dump file
-l ⟨path⟩	Path to language configuration file
-i ⟨input-sentence⟩	Input sentence to parse
-j ⟨path⟩	Path to JSON output file
-t ⟨path⟩	Path to thesaurus file
-s ⟨path⟩	Path to stopwords file

4.5 Usage

Table 4 shows the related parameters that must be entered in order to obtain the desired result.

Here is the initialization format: python3 -m ln2sql.main -d ⟨path⟩ -l ⟨path⟩ -i ⟨input-sentence⟩ [-j ⟨path⟩][-t ⟨path⟩][-s ⟨path⟩]

5 Implementation

We only want to get the most important and meaningful words from the user's input. To do so, we keep a list of common nouns, proper nouns, adjectives, numbers, and other terms. The Treetagger is the tool that we're using for this. The Treetagger is a program that adds information about parts of speech and lemma to text. The Treetagger tool then filters the filler words depending on where they appear in the phrase and then performs a stemming operation on the remaining words.

Consider the following scenario.

What is the mark of students with the name Jake as their first name?

The result should include the items "mark, student, first name, Jake" after stemming. It's also worth noting that the sequence in which the words appear is retained because it's crucial for the accurate translation of the user's input sentence to a natural language question.

The next step in the process includes recovering the structure of the user-supplied database on which querying has to be performed. To get this result, two strategies have been used. The first approach entails querying the database using SQL statements like "SHOW TABLES, SHOW COLUMNS, DESCRIBE," etc. to gather the required data.

The second approach entails looking through the database's backup or construction file. This technique does not require a pre-processing connection to the database, but it does require a universal SQL schema (some commands differ in syntaxes under Oracle or MySQL for instance). Be aware that only SQL databases are compatible with NL2SQL.

If a user enters a phrase that is not spelt correctly or whose vocabulary is not exactly the same as that in the database, there will be no match between the user's phrase and items in the database, and no pertinent results will be produced. Therefore, it's critical to increase the number of words that a keyword in the input request will match with a relevant element in the database. A dictionary of synonyms is loaded to accomplish this in addition to the procedure described in the preceding sections. A concept table with all the words in the language that can be used in place of each word is thus available. The terms "pupils" and "students," for instance, refer to the same thing. An idea or meaning that is conveyed by a word or collection of words is referred to as a concept.

As a result, a notion is represented by a word with lexical meaning and any term that might possibly replace it in its table. This translator's goal is to make database interrogation accessible to someone who is unfamiliar with the structure and key terms (table and column names) and is therefore more likely to use a synonym of a word used in the database than the actual phrase. Therefore, it is wiser to represent a word by a concept rather than just by itself, such as a table of all the words that can be used in place of it (a table of its synonyms, which includes itself). The user can input the word student in this fashion to query the student table.

Thesaurus v.2.3 as of December 20, 2011, of LibreOffice v.3.4 was utilized as the dictionary in this study. On the Internet, this resource is freely accessible. An administration interface of the synonym dictionary has also been designed to support all naming nomenclatures of columns and tables of databases, enabling any user without specialized knowledge to add, delete, or alter the synonyms of each term as desired. The user can manually enter the word "student" as a synonym, and the equivalence will be automatically updated by adding all the synonyms of the word student contained in the synonym dictionary, if the table containing the students' information, for example, has the name studid, and no synonym has been automatically added to this table name.

Each keyword in the user-entered request is now extracted at this stage of the procedure. A list of synonyms for each of these keywords is also available to the program. In order to partition the request based on the found correspondences and determine the optimal structure to create, the aim is now to look for a correspondence between the keywords of the request (or their synonyms) and the database entities. All words are lower-cased and all diacritical marks (accents, cedillas, etc.) are normalized throughout the matching process. Each term is classified as belonging to a column, a table, or anything else that is currently unknown in the database being queried.

First, the input sentence is dissected in accordance with the labeled keywords. The first one outlines exactly what type of selection will be made and on which element(s), the second one outlines where to search and in which element(s), and the third one outlines exactly what type of selection will be made and on which element(s). A SELECT and FROM segment must both appear in the phrase; the first identifies the sort of selection to be made and specifically which element(s); the second indicates which table(s) to look in to find the element of the selection. The segments for TIER and WHERE are optional. The WHERE segment is used to describe, if there are any, the constraints on the selection, and the TIER segment is used for explicit joins. The splitting will not provide the same question structure in the output depending on the amount and placement of the keywords in the text. It should be noted that a request will be deemed incorrect and cause an error if it lacks any words that resemble tables.

Then, in each of the separated portions, we examine the previously unidentified keywords. When a constraint must be applied to a value, these words may be factors in the presence of a counting query, algebraic computations, a denial, etc. In this approach, the system recognizes the query to be generated as being a counting question, i.e., a COUNT(*) SELECT, if a word related to counting, such as "how many," is detected in the first segment of the phrase, the one corresponding to the SELECT. For many other types of operations, the application operates in the same way, using a keyword recognition mechanism in the SELECT and/or WHERE segments.

Only internal joins are covered in this paper (INNER JOIN). There are implicit and explicit types of internal joins. An implicit join occurs when a selection or constraint is placed on a column that is not a member of the FROM table, or when the table to which the targeted column belongs is not identified in the input. In this scenario, a join between the FROM table mentioned in the phrase and the table containing the target column is required. Which students have a teacher whose first name is John? is an example of an explicit join, in which the table to be linked to is stated explicitly in the question. In order to choose students while imposing a restriction on the first names of the teachers, we must here create a join between the tables of student and teacher. The query cannot be created if the column of the selection or constraint is neither in the FROM table nor in a table that can be accessed by a join from the FROM table. It enables the building of joins by the application since NL2SQL can implicitly determine the effective linkages between the tables by using the main and foreign keys of each table to determine whether a table may be linked to another and, if so, through which table (s). In fact, if necessary, NL2SQL can establish joins involving many tables.

Existing apps' permissiveness is caused by their matching module being overly "tolerant," since it looks for a limited set of data in a very vast space. Then, until the most logical query is presented, the space of potential questions is again reduced by the rules of their extremely rigid grammar. In a way, NL2SQL is the complete opposite. Since bidirectional matching performs an intersection between two weak data sets, it first lowers the space of potential queries. The output query is then produced using lax grammar, whose rules are only utilized to produce the question that was previously decided by matching and is not used to distinguish between or order the alternative inquiries. This method enhances output quality despite adding a sizable amount of processing time and pre-processing steps. Since the preceding steps will have mostly filtered out the causes of errors (false positive matches), the goal of the grammar is to simply provide an output query rather than to have the most discriminating rules possible.

We chose to index the rules using a hash table to enable their acquisition more quickly due to the huge number of rules, each reflecting the creation of a potential query (the search in an integer-indexed table is faster than the search for a key made up of a character string). To do this, each key element of the request is given an integer, such as 1 for table components, 2 for columns, 3 for values, 4 for a count word, etc. Thus, the input request generates a string of concatenated numbers to create an integer. The 0 indicates that the previous element in the chain may appear anywhere between 1 and N times in the rule. As a result, an array of rules with integer indexes is obtained. Simply look for the number matching the request's structure in the table, which also happens to be the key of the value representing the request's structure to be produced as output, to discover the rule comparable to an input request.

Keep in mind that NL2SQL does not support the issue of silent constraints. The application will not respond appropriately to inquiries like "Which students are named John?" or "Which students are 18 years old?" Since the column on which the requirement must be fulfilled in this case is implicit and not explicitly stated, the application cannot identify it. The system, as it is now built, has no method of determining the student's age in the first example or the object's age in the second.

Also keep in mind that if the same query is asked of multiple tables in the same request, like consider the following sentence, "Which students and teachers are over 25 years old?", "then the system is able to produce several output queries, here two, to answer the request."

Once we have a first candidate SQL query, we attempt to run it against the database. If the query fails, it is because it was poorly written (column name instead of a table, omission of a value, omission of a column, etc.). The system then makes an attempt to create it differently. If, after execution, it still produces an error, the system then sends back a message specifying the specific error type. It is possible to reduce output errors using this system, but it is especially possible to determine the type of returned faults.

6 Experimental Results

The result of the generated query is outputted to a JSON file, which adheres to the following syntax as depicted below:-

```
{"select": {
  "column": "age",
  "type": "AVG",
  },
  "from": {
    "table":"student",
    "join": {},
    "where": {
        "conditions" :  [
            {
                "column":"name",
                "operator":"=",
                "value" :"Doe"
```

```
            },
            {"operator":"OR"},
            {"column":"age","operator":">","value":"25"}
        ]
    },
    "group_by":{ },
    "order_by": { }
  }
}
```


The below examples show the result that we obtained after generating SQL queries from natural language through rule-based processing.

- EXAMPLE 1: Count how many city
  !python3 -m ln2sql.main -d database_store/city.sql -l lang_store/english.csv -j output.json -i "count how many city?"

  generates the following query as output:
  SELECT COUNT(*)
  FROM city;

- EXAMPLE 2 : Show city with cityname is USA
  !python3 -m ln2sql.main -d database_store/city.sql -l lang_store/english.csv -j output.json -i "count how many city?"

  generates the following query as output :
  SELECT *
  FROM city
  WHERE city.cityname='usa';

- EXAMPLE 3 : Show details of all student
  !python3 -m ln2sql.main -d database_store/city.sql -l lang_store/english.csv -j output.json -i "show details of all student?"

  generates the following query as output :
  SELECT *
  FROM student;

- EXAMPLE 4 : Find name of all student
  !python3 -m ln2sql.main -d database_store/city.sql -l lang_store/english.csv -j output.json -i "find name of all student?"

  generates the following query as output :
  SELECT student.name
  FROM student;

7 Conclusion

In this paper, we have successfully converted natural language to a structured language query using a rule-based system. A rule-based system is an easily implementable model which requires fewer parameters and CPU resources compared to other models. They can be more interpretable than other types of models, as the rules used to make predictions are often explicitly defined. The model makes use of techniques and tools such as Treetagger and the Part-Of-Speech tag for increased functionality. This model can be used by individuals, business owners, corporate companies etc., so as to facilitate the proper handling of data in today's society.

References

1. Mahmud T, Azharul Hasan KM, Ahmed M, Chak THC (2015) A rule based approach for NLP based query processing. In: 2015 2nd international conference on electrical information and communication technologies (EICT), pp 78–82. https://doi.org/10.1109/EICT.2015.7391926
2. Katsogiannis-Meimarakis G, Koutrika G (2023) A survey on deep learning approaches for text-to-SQL. The VLDB J. https://doi.org/10.1007/s00778-022-00776-8
3. Arefin, M, Hossen K, Uddin M (2021) Natural language query to SQL conversion using machine learning approach, pp 1–6. https://doi.org/10.1109/STI53101.2021.9732586
4. Kumar R, Dua M, Jindal S (2014) D-HIRD: domain-independent hindi language interface to relational database. In: 2014 international conference on computation of power, energy, information and communication (ICCPEIC), Chennai, India, pp 81–86. https://doi.org/10.1109/ICCPEIC.2014.6915344
5. Katrin A, Kurt S, Abraham B (2019) A comparative survey of recent natural language interfaces for databases. VLDB J 28. https://doi.org/10.1007/s00778-019-00567-8
6. Xu B, Cai R, Zhang Z, Yang X, Hao Z, Li Z, Liang Z (2019) NADAQ: natural language database querying based on deep learning. IEEE Access, pp 1–1. https://doi.org/10.1109/ACCESS.2019.2904720
7. Kumar A, Nagarkar P, Nalhe P, Vijayakumar S (2022) Deep learning driven natural languages text to SQL query conversion: a survey https://doi.org/10.48550/arXiv.2208.04415
8. Kumar R, Dua M (2014) Translating controlled natural language query into SQL query using pattern matching technique. In: International conference for convergence for technology, Pune, India, pp 1–5. https://doi.org/10.1109/I2CT.2014.7092161.
9. Mohey El-Din D (2019) A comparative study on natural language interfaces to database query between challenges and applications. https://doi.org/10.13140/RG.2.2.13316.68486
10. Mahmud T, Hasan KM, Ahmed M, Chak T (2015) A rule based approach for NLP based query processing, pp 78–82. https://doi.org/10.1109/EICT.2015.7391926

Chapter 67
VANET-Based Communication in Vehicles to Control Accidents Using an Efficient Routing Strategy

Humera Maahin, Deepthi Kondamuri, Sarvani Polisetty, and Sarada Devi Yaddanapudi

1 Introduction

The term "VANET" (Vehicular Ad-hoc Networks) describes a network set up on the spot where various moving vehicles and other connecting devices interact and share vital information. The vehicles and other equipment function as network nodes at the same time, forming a tiny network. It has been shown that vehicle-to-vehicle and vehicle-to-roadside communications architectures will coexist in VANETs to provide road safety, navigation, and other roadside services. VANETs are a crucial part of the intelligent transportation systems (ITS) infrastructure. Vehicle-to-vehicle (V2V) and vehicle-to- infrastructure (V2I) communication is used by the vehicular ad-hoc network to exchange data over a network. Numerous IT and automotive companies are pursuing the use of VANET in the market and the supply of user safety, reliable data interchange, route detection, security, etc.

NS2 software is used for simulation purposes. The network simulator (Version 2), also known as NS2, is a straightforward event-driven simulation tool that has been useful in studying the dynamic behavior of communication networks. The network functions and protocols of both wired and wireless networks can be simulated using

Humera Maahin (✉) · D. Kondamuri · S. Polisetty · S. D. Yaddanapudi
V. R. Siddhartha Engineering College, Kanuru, Vijayawada, India
e-mail: 198w1a0484@vrsec.ac.in

D. Kondamuri
e-mail: 198w1a0497@vrsec.ac.in

S. Polisetty
e-mail: 198w1a04b1@vrsec.ac.in

S. D. Yaddanapudi
e-mail: saradadevi@vrsiddhartha.ac.in

A. Yadav et al. (eds.), *Proceedings of International Conference on Paradigms of Communication, Computing and Data Analytics*, Algorithms for Intelligent Systems,
https://doi.org/10.1007/978-981-99-4626-6_67

NS2 (e.g., routing algorithms, TCP, and UDP). Network protocol configuration and general activity simulation are both possible with NS2. Modeling any network's behavior is done by this software suite. It is applied to evaluate novel network topologies, assess the effectiveness based on numerous metrics, and see the network and packet flow.

VANETs can be simulated using network simulation. It is an open-source event-driven simulator created especially for studies on computer communication networks. Both wired and wireless routing systems are simulated. Routing methods, router queue management, the transmission control protocol and UPD, and other protocols can all be simulated using NS2. Architecture of NS2 includes C++ and object-oriented tool command language make up the two main languages of NS2 (OTcl). While the internal workings of the simulation objects are defined by the C++ (i.e., the backend), the simulation is set up by the OTcl by gathering and setting the objects as well as scheduling separate events. By use of Tclcl, C++ and OTcl are linked.

2 Related Work

In 2021, a new hybrid boosted long short-term memory ensemble (BLSTME) and convolutional neural network (CNN) model were created to address the dynamic behavior of the vehicle and efficiently estimate traffic congestion on roadways [1]. Whereas CNN draws its features from traffic image data, the proposed BLSTME trains and improves classifiers to anticipate congestion. The TensorFlow Python libraries were used to develop the suggested model, which was then tested using SUMO and OMNeT++ in a simulation of actual traffic. Following extensive experimentation, the performance of the model is evaluated by likely prediction accuracy, precision, and recall performance measures.

In 2020, a prediction-based collision avoidance technique in VANET was put forth [2]. The proposed method effectively forecasts the chance of collision for cars beyond two hops by using a vector-based prediction model. The suggested method outperforms other direction-based algorithms in terms of accident rate since it is based on a vehicle's vector mobility, which has the advantage of forecasting vehicle movements in comparison with other protocols.

A multi-hop optimum forwarding method was proposed in 2018 [3] to present the accident management system. The optimal route planning algorithm (ORPA) is also suggested in this method in order to maximize the total spatial usage of a road network while minimizing the cost of vehicle operation. To evaluate ORPA, simulations are conducted, and the results are compared to other algorithms. The evaluation's results demonstrated that ORPA outperformed the competitors.

In 2016, an approach that combined the VANET and the Bat algorithm was proposed [4]. The best optimal option for locating the path is determined using the Bat algorithm. After obtaining the coordinates of the destination node, the proposed method divides the region first before determining which nodes are valid. Valid nodes are helpful for resource conservation because the provided technique defines the region such that Bat will only process them.

3 Algorithm

Step 1: The transfer of packets from source to destination begins.
Step 2: Setup the node to deliver the packet from source to destination.
Step 3: Configure each node to send a packet from source to destination.
Step 4: Utilizing a routing table in step four using advertise hop count, choose the shortest distance.
Step 5: Check each routing in the routing table and choose the shortest path for data transmission.
Step 6: Choose the following node in this step and utilizing the suggested route table.
Step 7: By using a routing table, measure the node's length.
Step 8: This node is chosen, and its information is entered into the routing table if the following node has a large bandwidth that is less than 50% of the current node. In the event that this node is the destination, packets will be transmitted there; otherwise, repeat step 6.
Step 9: Choose the adjacent node of the current node to provide a different path for data transmission if the next node's coverage area is more than 50% of the current node's coverage area.
Step 10: The process is completed.

4 Implementation

NS2 is programmed in Tcl language. Tcl allows for the storage of values in variables, which can then be utilized in commands. The proc function in NS2 can be used to develop new procedures. Lists and files are then defined. The command exec starts a subprocess and awaits its execution. Using exec is like giving a shell software a command line. A network topology must initially be constructed in order to conduct a simulation scenario. A group of nodes and links make up the topology in NS2. At the start of the script, a new simulator object needs to be created before the topology can be configured. The member functions of the simulator object make it possible to join agents, create nodes and links, and more. Four nodes are created and assigned to the handles by the simulator class "node" member function. UDP and TCP users are the two most widely used types of agents in NS2. Sources of traffic (FTP, CBR, etc.) must be configured, meaning that sources must be connected to the agents and the agents to the nodes, respectively. The UDP and TCP sources need to be connected to traffic sinks in order for the information transmission to stop without processing.

Both a TCP sink and a UDP sink are defined in the classes agent/TCP sink and sgent/null, respectively.

To finish the topology, links are needed. When building links in NS2, the user must also provide the queue-type because a node's output queue is structured as a link component. The data must be gathered in some way before it can be used to calculate the outcomes from the simulations. Traces and monitors are the two main monitoring features that NS2 offers. The traces make it possible to keep track of packets when a particular event, like a lost packet or an arrival, takes place in a queue or link. The monitors offer a way to gather data on quantities like the number of dropped packets or the number of arriving packets in the queue. Open and allocate a handle to the output file first. The events are then saved to the file that the handle specifies. Finally, the file must be closed, and the trace buffer flushed at the conclusion of the simulation.

In Fig. 1, the flowchart of main system functioning is explained. When the message transmission happens, a message request is sent to all the switches, and a reply is received, after which the throughput is updated. The data link's congestion level is decided, and if the link receives a congested reply, then the entire process is repeated till the link receives the non-congested link reply.

Fig. 1 System flowchart

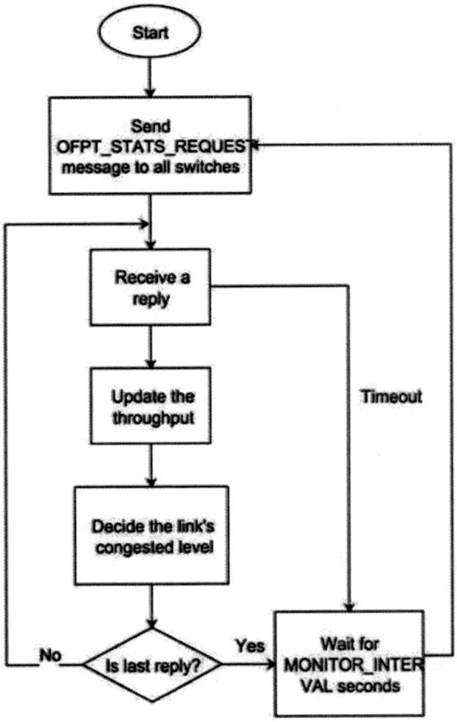

5 Protocol

5.1 Dynamic Source Routing Protocol (DSR)

The routing mechanism known as dynamic source routing (DSR), created specifically for multi-hop wireless ad-hoc networks is based on the source routing technique. DSR does not call for a network's periodic routing message, unlike other ad-hoc network protocols. Interior gateway protocols (IGP) and EGP (exterior routing protocols) are the two main classifications used to classify these dynamic routing protocols (exterior gateway protocols). An IGP is often utilized within a single business, but an EGP is typically used to link various organizations or major enterprises to the Internet. IGP examples include RIP and OSPF, although BGP and IS-IS are both capable of serving as IGPs and EGPs.

In Fig. 2, the dynamic source routing mechanism is explained. In it each source chooses the path that will be taken to send its packets to particular destinations. Route maintenance and route discovery are the two main parts. The best path for a transmission between a given source and destination is found using route discovery. Even if this means switching the route mid-transmission, route maintenance makes sure that the transmission path stays optimal and loop-free when network conditions change.

Route Mechanism

- The initiator node receives the message from the response package and stores the data in the packet in its own routing memory for future communication if this node is the target node for this route. If this node is the target node for this route, the node will send a "route response," which is collected by the reply bag from the response bag and contains the resulting routing record.
- The node will believe the request package has been received and destroyed if it discovers its address information in the request packet of the routing request package.

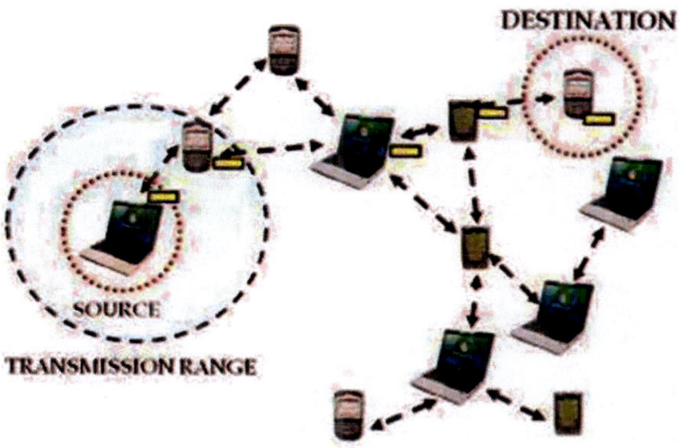

Fig. 2 Routing mechanism [5]

- If the routing request package does not contain this node's address information, the node will record that information there and then forward the request using the broadcasting method till it reaches the destination node.
- Advantages of DSR are scalability and adaptability which are two important benefits of dynamic routing versus static routing. A dynamically routed network has a faster rate of network expansion and is able to adjust to topology changes brought on by this expansion or by the failure of one or more network elements. As the topology of the network changes, it adjusts. It is appropriate for networks with large number routers.

6 Results

6.1 Simulation Results

Figure 3 explains the model of the accident control system. The simulation window has been set to use a standard Manhattan grid layout. The simulated output has three regions with 49 nodes. A base station is common for all the three regions. Internally five interfaces and a hospital node have been designed in each sub-region. The interfaces communicate with each other in the same region or different regions depending on the source and destination node's location. The interface nodes select the shortest path possible from source to destination using optimal route planning algorithm. The simulation we have developed on a dynamic path routing algorithm finds the optimal multi-path for avoiding the congestion and improving the QOS of the network.

6.2 Performance Analysis

In this section, four parameters, i.e., delay, energy consumption, packet delivery ratio, and packet loss ratio of the proposed routing strategy have been compared with the existing ones. In the above graphs, the red color line indicates the proposed strategy, and the green one indicates existing strategy.

Energy Consumption

Each node's energy consumption is used to compute the total energy used by the entire network. It is measured during packet transmission by each node. The energy consumption of the proposed strategy is less compared to the existing one. Less energy consumption indicates more efficiency and is shown in Fig. 4.

End-to-End Delay

The end-to-end delay indicates how long it takes for packets to travel from their source node to their destination. The comparison of E2E delay of existing and proposed strategy is shown in Fig. 5.

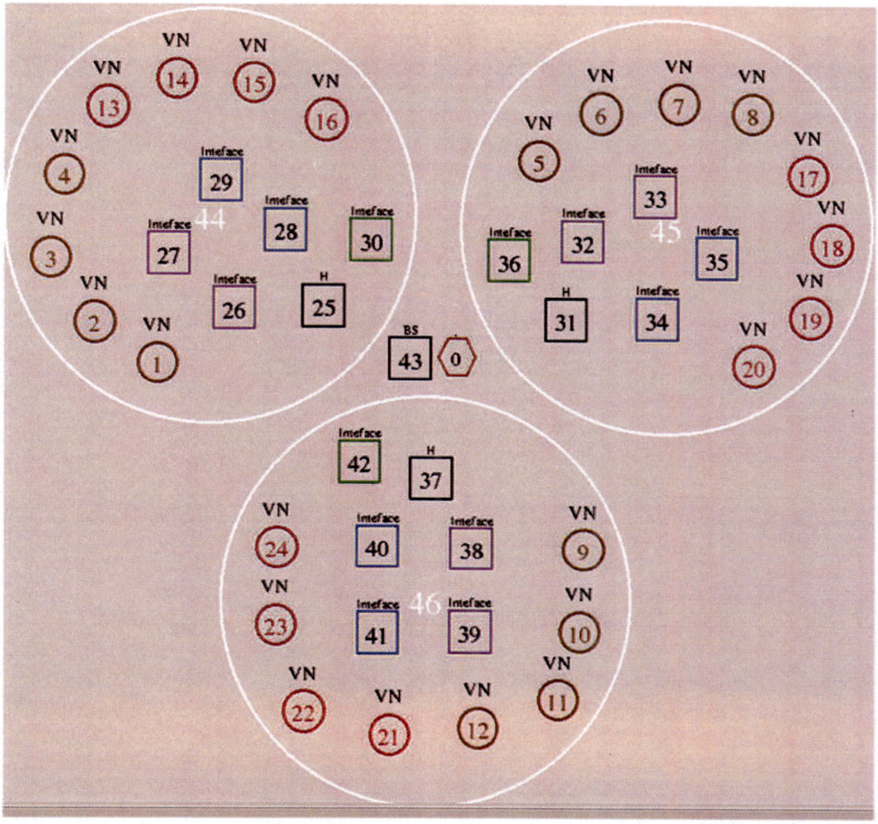

Fig. 3 Model of accident control system

Packet Delivery Ratio

The packet delivery ratio measures the proportion of packets delivered from the source to those received at the destination. The packet delivery ratio measures the proportion of packets delivered from the source to those received at the destination. Effective communication indicates good packet delivery ratio. The comparison of existing and proposed routing strategies is shown in Fig. 6.

Packet Loss Ratio

The number of lost packets to the total number of sent packets is represented by the packet loss ratio. When a network packet does not arrive at the destination it is supposed to, information is lost, and packet loss occurs. It is shown in Fig. 7.

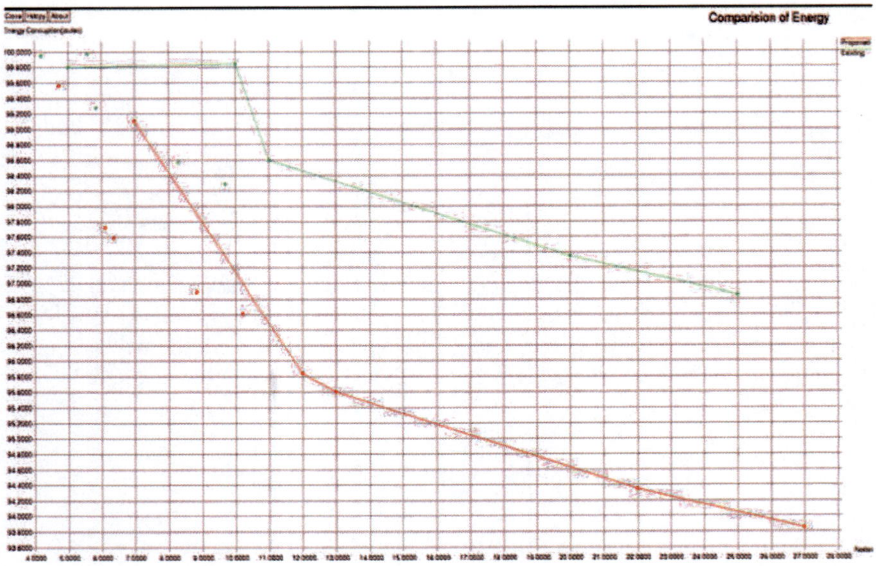

Fig. 4 Energy consumption comparison

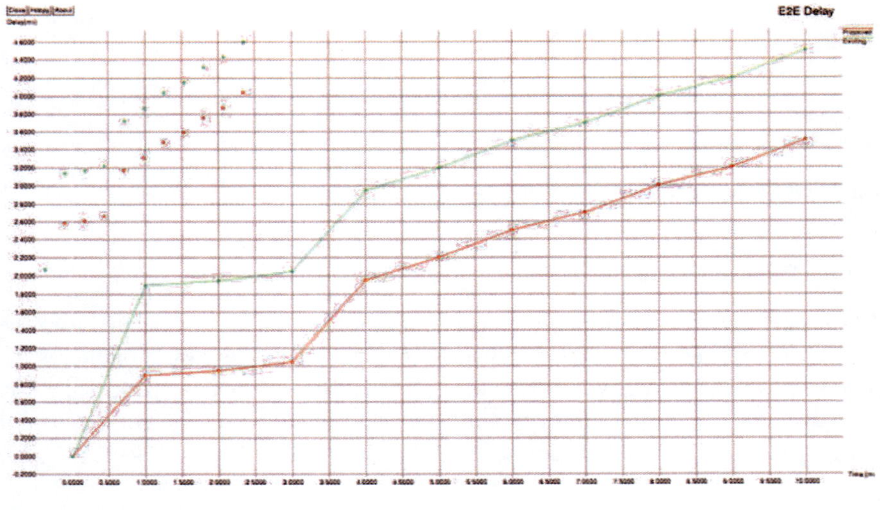

Fig. 5 End-to-end delay

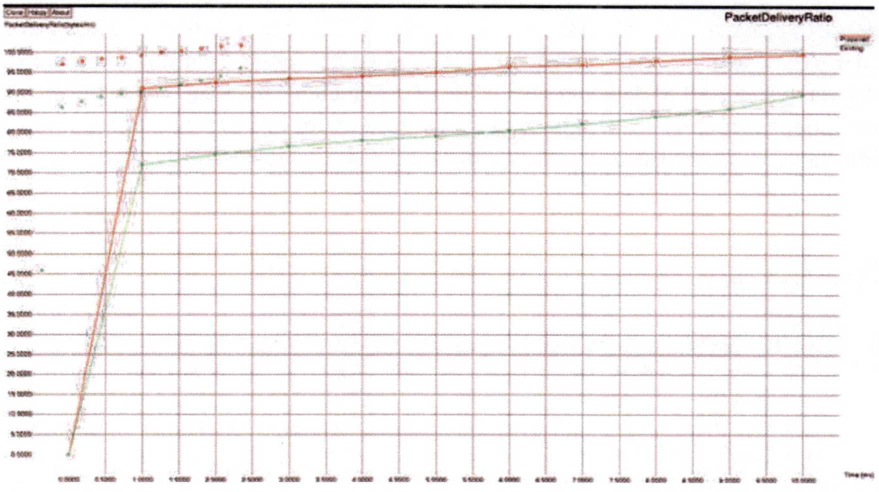

Fig. 6 Packet delivery ratio

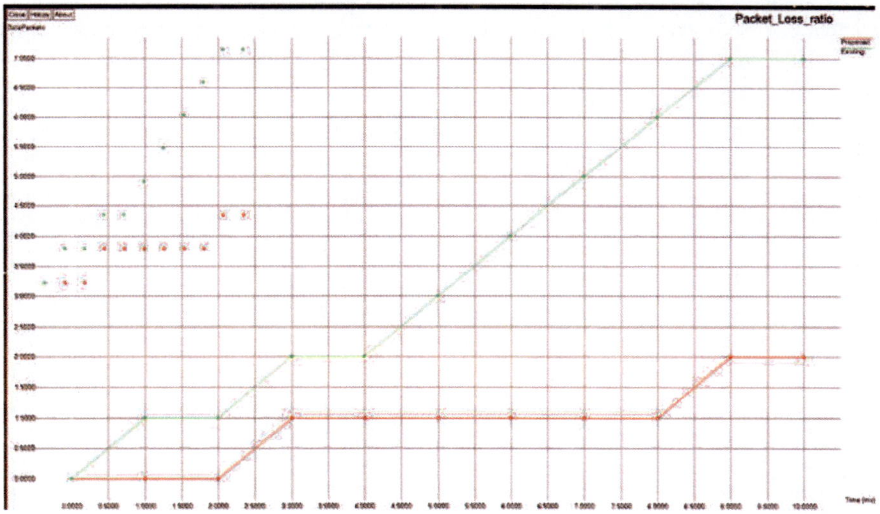

Fig. 7 Packet loss ratio

7 Conclusion

To start with, lack of proper communication between vehicles and inefficient route selection were the two main reasons for the delay in emergency services. To tackle these problems, for the lack of proper communication problems vehicular ad-hoc networks have been used to ensure that proper communication is established between vehicles. For the inefficient route selection, dynamic source routing strategy has been used. To prove that the proposed routing strategy is better than the existing strategies, four parameters have been considered. They are delay, energy consumption, packet delivery ratio, and packet loss ratio. All the parameters showcase better results for the proposed routing strategy. In conclusion, a better routing strategy is proposed for the emergency services to reach their destination at the time of an accident.

References

1. Kothai G, Poovammal E, Dhiman G, Ramana K, Sharma A, AlZain MA, Gaba GS, Masud M (2021) A new hybrid deep learning algorithm for prediction of wide traffic congestion in smart cities
2. Bang J-H, Lee J-R (2020) Collision avoidance method using vector-based mobility model in TDMA-based vehicular Ad Hoc networks
3. Cherkaouia B, Beni-Hssanea A, El Fissaouia M, Erritali M (2019) Road traffic congestion detection in VANET networks
4. Al-Mayouf YRB, Mahdi OA, Taha NA, Abdullah NF, Khan S, Alam M (2018) Accident management system based on vehicular network for an intelligent transportation system in urban environments
5. Tomar G, Shrivatava L, Bhadauria S (2014) Load balanced congestion adaptive routing for randomly distributed mobile adhoc networks. Wireless Pers Commun 77:2723–2733. https://doi.org/10.1007/s11277-014-1663-9

Chapter 68
ECG Image Classification for Arrhythmia Using Deep Learning

Shasmita Nair, Prerna Peswani, Jai Rohra, and M. Vijayalakshmi

1 Introduction

Electrocardiogram (ECG) signals are crucial tools for tracking the heart's electrical rhythm. The prevalence of any arrhythmias, which are irregular cardiac rhythms that can cause major health issues, is shown by the ECG signal. The identification and classification of arrhythmias are crucial for accurate diagnosis and effective treatment. Several types of arrhythmias can occur in the heart, including atrial fibrillation, ventricular tachycardia, and bradycardia. Atrial fibrillation is characterized by an irregular and rapid heartbeat, while ventricular tachycardia is a fast heart rhythm that originates in the ventricles. Bradycardia is defined by a slow heartbeat, often less than 60 beats per minute.

The traditional methods of ECG arrhythmia classification, such as time- and frequency-domain analysis and feature extraction techniques, have limitations in terms of accuracy and efficiency. In the research conducted by Lai [1], a comparison between traditional machine learning (SVM) and deep learning (CNN) for image recognition on MNIST handwritten digit data was performed. The results showed that deep learning achieved higher accuracy of 98.85% on the testing set and better feature extraction ability compared to traditional machine learning methods. Convolutional neural networks (CNNs) have demonstrated encouraging results in the categorization of ECG data and the identification of arrhythmia. A CNN model may be used to analyze ECG data by converting them into images and treating them as a 2D signal. The fully connected layers are utilized to forecast the occurrence of arrhythmias, and the CNN model learns to extract information from the ECG pictures. The CNN model is trained on a large dataset of ECG signals, and the accuracy of the models is evaluated on a separate test set. The experimental findings demonstrate that, in

S. Nair (✉) · P. Peswani · J. Rohra · M. Vijayalakshmi
Vivekanand Education Society's Institute of Technology, Mumbai, Maharashtra 400074, India
e-mail: shasmitanair77@gmail.com

A. Yadav et al. (eds.), *Proceedings of International Conference on Paradigms of Communication, Computing and Data Analytics*, Algorithms for Intelligent Systems,
https://doi.org/10.1007/978-981-99-4626-6_68

terms of accuracy and computing efficiency, CNN models outperform conventional machine learning techniques. In addition, the model can handle large amounts of data, making them suitable for real-time applications.

Due to their capacity to recognize intricate patterns in data, convolutional neural networks (CNNs) have recently been employed extensively in a variety of biological signal processing activities. The application created involves the use of CNN to learn and extract relevant information from ECG signals and categorize them into different types of arrhythmias. It allows a user to simply upload an image of the ECG signal and get the result instantly. The ability of CNN models to learn complex representations of ECG signals and their high accuracy make them a valuable contribution to the field of cardiology. The main purpose of this research is to demonstrate the feasibility and effectiveness of using CNNs for arrhythmia classification of ECG images. The findings of the tests demonstrate the proposed method's aptitude to be a credible tool for classifying ECG arrhythmias and emphasize the value of employing deep learning techniques for biomedical signal processing.

2 Literature Survey

Ye et al. [2] proposed a machine learning approach for the classification of ECG signals. The study's major aim is to create an accurate ECG classification system that can categorize ECG signals into different types of arrhythmias. To achieve this, the authors proposed a classification method that combines both morphological and dynamic features of the ECG signals. It involves four main stages: preprocessing, feature extraction, feature selection, and classification. In the preprocessing stage, the ECG signals are filtered and normalized to remove any noise and baseline drift. In the feature extraction stage, both morphological and dynamic features are extracted from the ECG signals.

The morphological features of the QRS complex include amplitude, width, and slope, while dynamic features include changes in the RR interval, heart rate variability, ST segment changes, and T-wave morphology. The authors used a combination of time- and frequency-domain techniques, such as wavelet transform and principal component analysis, to extract these features. In the feature selection stage, the authors used a genetic algorithm to select the most relevant features that contribute to the classification accuracy. Finally, in the classification stage, the selected features are fed into a support vector machine (SVM) classifier for arrhythmia classification. The proposed method was tested on publicly available ECG datasets, namely the MIT-BIH arrhythmia database, and reported an overall classification accuracy of 86.4% in the subject-oriented evaluation.

A classification method that combines the discrete wavelet transform (DWT) and the K-nearest neighbor (KNN) classifier has been developed by Toulni et al. [3]. It acts as an accurate and efficient ECG diagnosis system that can detect abnormalities in ECG signals. For the feature extraction stage, the DWT is used to decompose the ECG signals into different frequency bands, which are then used to extract the relevant features. The statistical metrics of the decomposed signal coefficients, such as mean, standard deviation, skewness, and kurtosis, are the features retrieved from the DWT. The author applied a genetic algorithm to choose the most relevant characteristics that improve classification accuracy during the feature selection stage. The selected features are fed into a KNN classifier for ECG signal diagnosis and reported an overall classification accuracy of 91.60%

Arrhythmia classification using pruned fuzzy k-nearest neighbor classifier [4] combines the concepts of fuzzy logic and k-nearest neighbor classification. PFKNN aims to improve the classification accuracy of FKNN by reducing the number of features used in the classification process, thereby avoiding overfitting and reducing computational complexity. In this paper, the classification time for the standard FKNN algorithm was 1116 s, while the classification time for the PFKNN algorithm with 11 features was 353 s. This indicates that the PFKNN algorithm is faster than the standard FKNN algorithm, which is an important consideration for real-time applications.

The number of features that are still preserved after pruning is indicated by the retained ratio (R). The PFKNN algorithm in this research retained 19% of the original characteristics. This shows that despite retaining a high degree of classification accuracy, the pruning procedure was successful in lowering the number of features utilized in the classification process. The overall accuracy of the PFKNN algorithm was 97.32%, which is slightly lower than the overall accuracy of the standard FKNN algorithm (97.63%). The results indicate that the PFKNN algorithm for arrhythmia beat classification achieves a high level of classification accuracy while reducing computational complexity and retaining a small number of features.

Gutiérrez-Gnecchi et al. [5] present a novel method for the automatic classification of arrhythmias in ECG signals using digital signal processing (DSP) techniques. The proposed method combines the wavelet transform with a probabilistic neural network (PNN) to accurately classify ECG signals into different types of arrhythmias. The wavelet transform is used to decompose the ECG signals into a set of wavelet coefficients that represent different frequency components of the signal. The authors used the Daubechies 4 wavelet as the mother wavelet in their study. These wavelet coefficients are then used as input features for the PNN. The PNN is a type of neural network that is particularly suited for classification tasks. It calculates the probability that a given input feature belongs to a particular class and then assigns the input feature to the class with the highest probability. The authors used a PNN with one hidden layer in their study.

The authors reported an overall accuracy of 92.746% for their method of classifying arrhythmias. The proposed method relies on the wavelet transform, which is a computationally intensive operation. This could make the real-time implementation of the method challenging, which may limit its practical applicability. The wavelet transform can be sensitive to noise in the ECG signals, which can affect the accuracy of the feature extraction process. Despite using a denoising procedure before using the wavelet transform, there could still be some residual noise that compromises the accuracy of the classification. As the PNN in use is a 'black box' model, it might be challenging to understand how decisions are made. This can make it more difficult for doctors to comprehend and interpret the classification's findings.

Kutlu et al. [6] propose a deep learning-based approach for automatic arrhythmia classification using raw ECG signals. The proposed model uses a deep belief network (DBN) with a greedy layer-wise training phase as a multistage classification system. The MIT-BIH arrhythmia database, a popular database for assessing ECG arrhythmia classification algorithms, was employed in the study. To eliminate baseline wander, the ECG data underwent preprocessing using a median filter. Then, using a window with a length of 501 data points and the R peak of the wave positioned in the window's center, the ECG waveforms from the long-term ECGs were segmented. With a high accuracy rate of 95.05%, a multistage arrhythmia classification system based on DBN was able to distinguish between five different types of heartbeats. The five types of heartbeats were defined by the ANSI/AAMI standards and include normal sinus rhythm (NSR), atrial premature contraction (APC), premature ventricular contraction (PVC), left bundle branch block (LBBB), and right bundle branch block (RBBB). It achieved high accuracy in arrhythmia classification, which demonstrates the potential of deep learning-based approaches in improving the accuracy and efficiency of ECG arrhythmia diagnosis. The paper highlights the importance of computer-assisted analysis of biomedical signals and presents a promising approach for automatic arrhythmia classification using deep learning-based models.

Table 1 provides a comprehensive comparison of various algorithms used for arrhythmia classification, showcasing their corresponding accuracy rates. The table demonstrates that several different features and algorithms, including wavelet transform, independent component analysis, fuzzy logic, and deep belief networks, have been employed. The highest accuracy achieved in arrhythmia classification was 97% using a binary classification approach based on fuzzy logic [3]. Other algorithms such as SVM, KNN, PNN, and DBN have also shown promising results, with accuracy rates ranging from 86.4 to 95.05%. Therefore, it is evident that different algorithms can accurately classify arrhythmia, and the choice of the algorithm may depend on the specific features and available data within a given study.

Table 1 Comparison of algorithms to perform arrhythmia classification

Authors	Features	Algorithm	Accuracy (%)
Ye et al. [2]	Wavelet transform and independent component analysis (ICA)	SVM	86.4
Toulni et al. [3]	Discrete wavelet transform (DWT)	KNN	91.60
Arif et al. [4]	Binary classification using fuzzy logic	PFKNN	97
Gutiérrez-Gnecchi et al. [5]	Wavelet transform, digital signal processing (DSP) platform	PNN	92.746
Kutlu et al. [6]	ECG Waveform	DBN	95.05

Xia et al. [7] presented an automatic cardiac arrhythmia classification system with wearable electrocardiogram (ECG) technology. The system utilized a stacked denoising autoencoder (SDAE) for feature representation learning and employed softmax regression for ECG beat classification. The study demonstrated the effectiveness of the proposed approach in classifying ECG signals, highlighting its potential for real-time monitoring and classification using wearable ECG devices.

3 Proposed Architecture

3.1 Architecture and Design of Deep Learning Model

The architecture of the model consists of a sequence of convolutional and pooling layers that extract features from the ECG images, a flatten layer that prepares the data for the dense layers, and two dense layers that make the final prediction of the class label.

Convolutional Layers (Conv2D): The Conv2D layers in the model are responsible for extracting features from the ECG images. The first Conv2D layer has 32 filters with a size of (3, 3), which means each filter covers a 3×3 area of the input image. The filters are used to detect different features in the ECG images, such as edges, curves, and patterns. The input shape parameter specifies the shape of the input images, which is (64, 64, 3) in this case, representing the height, width, and number of color channels of the image. The activation function used is rectified linear unit, which is a commonly used activation function in deep learning.

Max Pooling Layers (MaxPooling2D): The MaxPooling2D layers in the model are used for downsampling the input, reducing the size of the feature map while retaining important information. The pool size parameter specifies the size of the pooling area, which is (2, 2) in this case, meaning the input is reduced by a factor of 2 in both height and width. The MaxPooling2D layer helps to reduce the computational complexity of the model and prevent overfitting by aggregating information from neighboring pixels.

Flatten Layer: The multi-dimensional feature map is flattened using the flatten layer into a 1D vector, which is then sent into the fully connected layer as input. This layer is used to prepare the data for the dense layers.

Dense Layers (Dense): The dense layers in the model are fully connected neural network layers. The first dense layer has 32 neurons, and the second dense layer has 6 neurons, representing the number of classes in the classification task. The activation function used in the output layer is 'softmax', which is commonly used in multi-class classification problems. The softmax function maps the outputs of the dense layer to probabilities that sum up to 1, which can be interpreted as the confidence of the model in the prediction of each class.

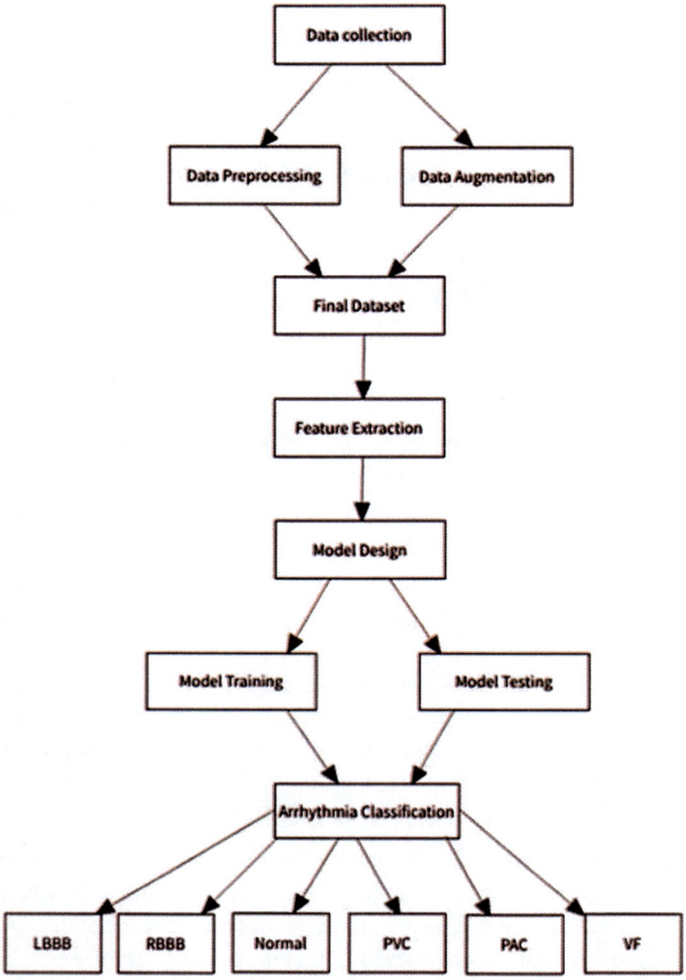

Fig. 1 Proposed architecture

As shown in Fig. 1, the model is designed to classify ECG images into six categories of arrhythmias: normal, ventricular fibrillation (VF), left bundle branch block (LBBB), right bundle branch block (RBBB), premature ventricular contraction (PVC), and premature atrial contraction (PAC).

3.2 Image Data Augmentation and Training

Data augmentation is used in the proposed methodology for ECG image analysis for arrhythmia to diversify the training dataset and enhance the classifier's performance.

The data augmentation was carried out using the image data generator class from the Keras package. For data augmentation, a distinct generator object with unique parameters was allocated to the train data and the test data, respectively.

The rescale parameter was set to 1/255 to rescale the pixel values from [0, 255] to [0, 1]. This is a common preprocessing step in deep learning tasks to normalize the pixel values. In addition to rescaling, random shearing, zooming, and flipping were introduced to the training data to increase the diversity of the images and make the model more robust to changes in the input data. The shear range, zoom range, and horizontal flip parameters were used to control the extent of these transformations.

The target size of the images was set to (64, 64), and the batch size for both training and testing was set to 32. The class mode was set to categorical to indicate that the task was a multi-class classification problem. The images were then loaded using the flow_from_directory method, which takes the path to the directory containing the images and loads them into the generator. The method automatically shuffles the images and generates batches of augmented images for training the classifier.

The training process involves using the augmented training data to update the model's weights using backpropagation and Adam. Adam is an optimization algorithm that demonstrates its effectiveness in stochastic optimization for deep neural networks. The results show that Adam outperforms other optimization methods, providing faster convergence and improved generalization, making it a highly recommended choice for optimizing deep learning models [8]. The model makes predictions on the training data, and the difference between the predicted class and the true class is used to compute the loss. To update the weights and lower the error, the loss is then back propagated across the network.

The training process is repeated for 10 epochs, to minimize the loss and improve the accuracy of the training and test data. The accuracy of the test data can be used as a measure of the model's generalization performance and its ability to make accurate predictions on unseen data. The training process involves using the augmented training data to train a deep learning model for ECG image analysis for arrhythmias.

3.3 Heartbeat Classification

Heartbeat classification is an important aspect of ECG analysis for diagnosing cardiovascular diseases and monitoring treatment efficacy. ECG signals are classified into different categories based on the morphological and frequency-domain features of the heartbeats. In the proposed model, heartbeats are classified into six categories: normal, ventricular fibrillation, left bundle branch block, right bundle branch block, premature ventricular contraction, and premature atrial contraction. These categories represent different types of cardiac arrhythmias and are critical for the diagnosis and treatment of cardiovascular diseases. The heartbeat classification system includes four crucial steps: preprocessing, QRS complex detection, feature extraction, and classification of heartbeats.

Preprocessing: The ECG signals are filtered and cleaned to remove noise and artifacts that can interfere with the accurate analysis of the heartbeats. This step also includes normalizing the ECG signals to eliminate the effects of different recording conditions.

QRS Complex Detection: The QRS complex is a distinct portion of the ECG signal that corresponds to the depolarization of the ventricles of the heart. Accurate detection of the QRS complex is crucial for the next steps in the heartbeat classification process.

Feature Extraction: After the QRS complex has been detected, features that characterize the heartbeats can be extracted. This can include morphological features, such as the duration, amplitude, and slope of the QRS complex, as well as frequency-domain features, such as the power spectral density of the ECG signal.

Classification of Heartbeats: Finally, the extracted features are used to classify the heartbeats into the six categories mentioned above. This can be done using machine learning algorithms, such as decision trees, support vector machines, or neural networks.

The heartbeat classification system is a robust solution for ECG analysis that can have a significant impact on the diagnosis and treatment of cardiovascular diseases. By classifying heartbeats, this system can provide detailed information about the type of arrhythmia and its severity, which can inform the clinical decision-making process.

4 Implementation

4.1 Dataset

The dataset used in this research is a collection of images belonging to 6 different classes. The classes are 'Left Bundle Branch Block', 'Normal', 'Premature Atrial Contraction', 'Premature Ventricular Contractions', 'Right Bundle Branch Block', and 'Ventricular Fibrillation'. The dataset was divided into a training set and a test set, with the training set containing 15,341 images and the test set containing 6825 images. Data augmentation was performed on both the training and test datasets using the image data generator class from the Keras library. The flow_from_directory method was used to generate batches of data from the given directories. The images were resized to a size of 64 × 64 using the target size argument. The batch size argument was set to 32, meaning that the data was divided into 32 samples per batch. The class mode argument was set to 'categorical' to indicate that the output should be a categorical variable, with each class represented by a one-hot encoded vector.

4.2 *Convolutional Neural Network Model*

The CNN model consists of multiple layers including Conv2D, MaxPooling2D, flatten, and dense layers. The input to the model is an image of size $64 \times 64 \times 3$, where the three channels represent red, green, and blue color channels. The first Conv2D layer has 32 filters of size 3×3 and uses the rectified linear unit (ReLU) activation function. The MaxPooling2D layer is used for downsampling the input image. The second Conv2D layer also has 32 filters of size 3×3 and a ReLU activation function. It is followed by another MaxPooling2D layer for further downsampling. The flatten layer is used to convert the 2D feature maps into a 1D array, which is then passed to two dense layers for a fully connected neural network. The final layer has 6 neurons and a softmax activation function, which is used for multi-class classification problems. Figure 2 illustrates the architectural visualization of the proposed CNN model. It showcases the various components and layers that constitute in the model, including convolutional layers for feature extraction, max-pooling layers, and fully connected layers for classification.

5 Results and Discussions

5.1 *Evaluation Parameters*

Accuracy: This evaluates how many of the model's predictions were accurate. It is derived by dividing the total number of predictions made by the number of predictions that were accurate.

Sensitivity: This evaluates the model's capacity to spot promising cases. It is derived by dividing the total of accurate positive and inaccurate negative predictions by the number of true positive predictions.

Precision: This measures the ability of the model to accurately identify positive cases. It is calculated as the number of true positive predictions divided by the sum of true positive and false positive predictions.

F1 Score: This is the harmonic mean of precision and recall and provides a single number that balances the trade-off between the two. It is calculated as 2 times the product of precision and recall divided by the sum of precision and recall.

Confusion Matrix: A classification model's performance is summarized in this table by the number of accurate positive, accurate negative, accurate false positive, and accurate true negative predictions. To determine the accuracy, recall, precision, and F1 score, utilize the confusion matrix.

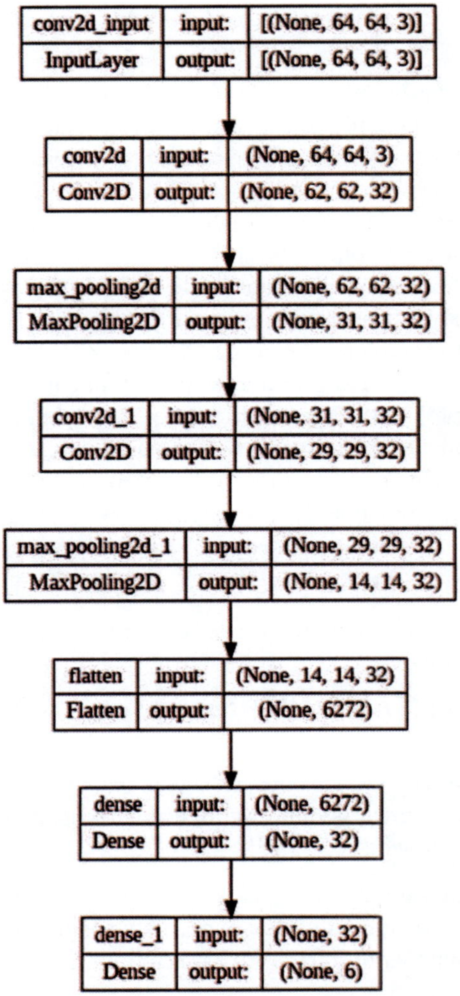

Fig. 2 Neural network architecture

5.2 *Performance Evaluation*

The metrics used for evaluating the performance of the model are accuracy and loss (cross-entropy loss). The model's accuracy for both the training data (steps per epoch) and the validation data (validation steps) is displayed in the output. As shown in Fig. 3, from epoch 1 through epoch 10, the training accuracy rises, reaching a high of 0.9778. On the other hand, the validation accuracy increases from 0.8889 in epoch 1 to 0.9111 in epoch 10, indicating that the model is generalizing well to new data. It is also important to observe the loss of values during the training process. The loss should decrease with an increasing number of epochs. In this case, the loss decreases from 0.1074 in epoch 1 to 0.0709 in epoch 10 for the training data, indicating that the

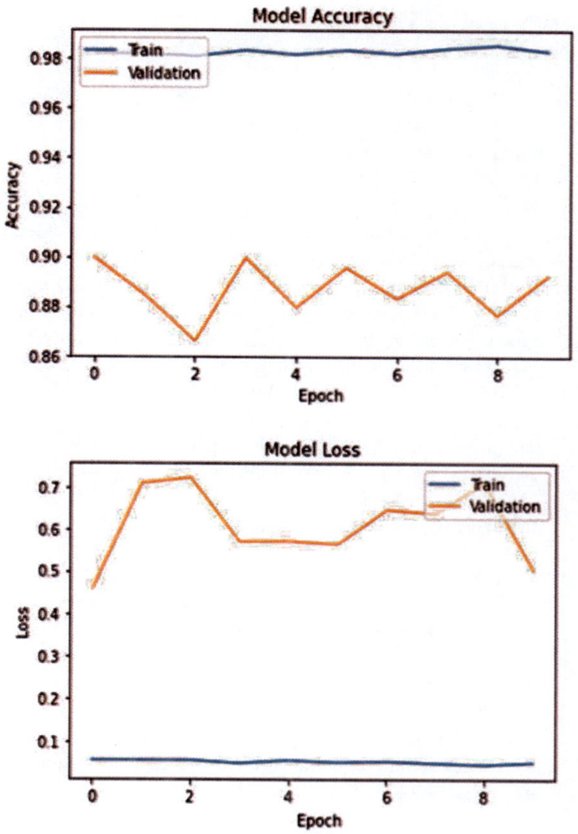

Fig. 3 Model accuracy and model loss

model is learning the patterns in the data. The validation loss decreases from 0.4744 in epoch 1 to 0.4170 in epoch 10, indicating that the model is not overfitting to the training data.

6 Conclusion

The proposed approach for ECG arrhythmia classification using CNNs has shown promising results in automating the diagnosis of heart conditions. The application's architecture and implementation have demonstrated the ability of CNNs to effectively learn and classify ECG signal images, outperforming traditional machine learning algorithms. As the deep learning model is viewed to have the potential to have a significant impact on the field of cardiovascular medicine, efforts are being conducted to further refine and improve it. With its improved accuracy and speed, the project has the potential to help healthcare professionals quickly and effectively diagnose heart conditions, leading to improved patient outcomes.

References

1. Lai Y (2019) A comparison of traditional machine learning and deep learning in image recognition. In: Proceedings of the International conference on electrical, mechanical and computer engineering (ICEMCE 2019). Association for Computing Machinery, pp 35–40. https://doi.org/10.1145/3345120.3345127
2. Ye C, Vijaya Kumar BVK, Coimbra MT (2012) Heartbeat classification using morphological and dynamic features of ECG signals. IEEE Trans Biomed Eng 59(10):2930–2941. https://doi.org/10.1109/TBME.2012.2213253
3. Toulni Y, Belhoussine Drissi T, Nsiri B (2021) ECG signal diagnosis using discrete wavelet transform and K-nearest neighbor classifier. In: Proceedings of the 4th International conference on networking, information systems & security (NISS2021). Association for Computing Machinery, pp 1–6. https://doi.org/10.1145/3454127.3457628
4. Arif M, Akram MU, Minhas FuAA (2009) Arrhythmia beat classification using pruned fuzzy K-nearest neighbor classifier. In: International conference on soft computing and pattern recognition, pp 37–42. https://doi.org/10.1109/SoCPaR.2009.20
5. Gutiérrez-Gnecchi JA, Morfin-Magaña R, Lorias-Espinoza D, Tellez-Anguiano AdC, Reyes-Archundia E, Méndez-Patiño A, Castañeda-Miranda R (2017) DSP-based arrhythmia classification using wavelet transform and probabilistic neural network. Biomed Signal Process Control 32:44–56. https://doi.org/10.1016/j.bspc.2016.10.005
6. Kutlu Y, Altan G, Allahverdi N (2016) Arrhythmia classification using waveform ECG signals. In: Proceedings of the 3rd International conference on advanced technology & sciences (ICAT'16). Konya, Turkey, pp 240–245
7. Xia Y, Zhang H, Xu L, Gao Z, Zhang H, Liu H, Li S (2018) An automatic cardiac arrhythmia classification system with wearable electrocardiogram. IEEE Access 6:16529–16538. https://doi.org/10.1109/ACCESS.2018.2807700
8. Kingma DP, Ba J (2014) Adam: a method for stochastic optimization. In: Proceedings of the 3rd International conference on learning representations (ICLR 2015). Retrieved from https://arxiv.org/abs/1412.6980